"*The Comprehensive Guide to Science and Faith* ... design, offering both scientific and philosophical discussions. ... well-known scientists provide scientific arguments on the big bang, the origin of life, fine-tuning, fossils, and more. A large part of the book is also devoted to philosophy: why making inferences to a nonmaterial cause is justifiable within science, the relationship between science and faith, overturning historical myths about the church's attitude toward science, and why a purely materialistic scientific position is self-defeating. Many subjects are covered by multiple authors, each with a different focus. The result is an overwhelming barrage aimed squarely at those who deny that intelligent design is science."

Ann Gauger, PhD (Zoology), Senior Fellow Discovery Institute

"This is a heroic encyclopedic work by some of the world's top thinkers on the topics of science, faith, and the cosmos, and the God who created them all. If anyone thinks that the study of science and faith should be disparate disciplines, here are more than 656 reasons why they're wrong. What is clear from this massive volume is that the alignment of science and faith is far closer to parallel than orthogonal, and it spans as far as from east to west across the tangents of science, life, and eternity."

James Tour, Professor of Chemistry,
Computer Science, Materials Science and NanoEngineering,
Rice University

"*The Comprehensive Guide to Science and Faith* is, quite simply, a *tour de force* whose publication is an event we must understand. The release of this book is a symbol of the decades-old growth, maturity, and rigor of the intelligent design (ID) movement. No longer can people employ dismissive slogans, handwaving, and ad-hominem attacks in response to ID claims. This book demands serious engagement. Filled with highly qualified scholars, detailed and first-rate argumentation, and solid engagement with virtually all the major issues, *The Comprehensive Guide to Science and Faith* throws down the gauntlet. If the reader does not agree with the points made within its pages, then let the reader bring forth his or her case. However, if someone ignores this book or simply brushes it aside, that person does so at the price of being intellectually irresponsible."

JP Moreland, Distinguished Professor of Philosophy, Biola University

"This is a fascinating book. Christians who read it—or even a few chapters—will find themselves rejoicing in the amazing consistency between God's words in the Bible and the most recent scientific discoveries. And any non-Christians who read it will likely be surprised to learn of the remarkable congruence between the Christian faith and numerous scientific findings. I am glad to recommend this book to Christians and non-Christians alike."

Wayne Grudem, PhD, Distinguished Research
Professor of Theology and Biblical Studies, Phoenix Seminary

"*The Comprehensive Guide to Science and Faith* is precisely the type of resource I recommend frequently to students, fellow educators, ministry leaders, and inquisitive laypersons. It has a fantastic breadth of scope, incorporating rigorous philosophical, theological, and scientific knowledge related to the major questions sparked by the intersection of science and faith. The depth of content offered by the contributing scholars is intellectually satisfying yet wonderfully accessible. This versatile book is a welcome addition to my academic and ministry tool kits."

Melissa Cain Travis, PhD, author of *Science and the Mind of the Maker*

"*The Comprehensive Guide to Science and Faith* is an essential reference for your personal library. The most valuable sections are those dealing with intelligent design, with contributions by the stellar specialists in this area—William A. Dembski and Stephen C. Meyer at the head of the list. The issues discussed cover a very wide range and the scholarship is impeccable."

John Warwick Montgomery, PhD (Chicago), D'Théol (Strasbourg, France), LLD (Cardiff, Wales, UK); Professor Emeritus of Law and Humanities, University of Bedfordshire, England; Director, International Academy of Apologetics, Evangelism and Human Rights (Strasbourg, France)

"This book is a wonderful guide to the range of important issues at the heart of the intersection of science and faith. From artificial intelligence to theistic evolution, the editors have put together a top-rate team who explore the biggest issues being discussed today. If you're looking for an introductory book that also has some depth, then *The Comprehensive Guide to Science and Faith* is for you."

Sean McDowell, PhD, professor of apologetics, at Biola University; author, coauthor, or editor of 20+ books, including *Understanding Intelligent Design*

"Anyone who desires to make sense of the world around them should read this book and follow the reasoning of these serious thinkers who have grappled with major questions about meaning and life. The existence of objective truth demands that there will be a coherence between the empirical and the theoretical, between facts and meaning. This compilation of essays provides an opportunity to view the world from a Judeo-Christian viewpoint that is both rational and empirically coherent."

Donald Ewert, who received his PhD in microbiology at the University of Georgia

"I have always felt that people of faith need to learn more science, not less. Sadly, the teaching of modern science often undermines rather than strengthens faith. *The Comprehensive Guide to Science and Faith* will help turn doubt into delight at the works of the Designer. Whether your questions are about the origin of life, or the fossil record, or the authority of science, or any of 44 other topics, you will find thoughtful, well-written articles that will answer your questions and strengthen your faith. I would highly recommend this volume for any pastor, or anyone with an ongoing interest in how understanding the works of God supports our understanding of the Word of God."

Ralph Seelke, Professor Emeritus of Biology at the University of Wisconsin-Superior

The
COMPREHENSIVE GUIDE
to SCIENCE
and FAITH

WILLIAM A. DEMBSKI
CASEY LUSKIN
JOSEPH M. HOLDEN
GENERAL EDITORS

HARVEST HOUSE PUBLISHERS
EUGENE, OREGON

Cover by Bryce Williamson

Cover photo © santoelia / Gettyimages

Interior design by KUHN Design Group

For bulk, special sales, or ministry purchases, please call 1-800-547-8979. Email: Customerservice@hhpbooks.com

ᴍ is a federally registered trademark of the Hawkins Children's LLC. Harvest House Publishers, Inc., is the exclusive licensee of the trademark.

The Comprehensive Guide to Science and Faith
Copyright © 2021 by William Dembski, Casey Luskin, and Joseph M. Holden
Published by Harvest House Publishers
Eugene, Oregon 97408
www.harvesthousepublishers.com

ISBN 978-0-7369-7714-2 (pbk.)
ISBN 978-0-7369-7715-9 (eBook)

Library of Congress Control Number: 2021931158

Printed in the United States of America

21 22 23 24 25 26 27 28 29 / KP / 10 9 8 7 6 5 4 3 2 1

This book is dedicated to the many scientists and academics who have sacrificed their careers because they wanted to investigate and declare the evidence for design in nature.

Acknowledgments

In a work like this one, which bears the marks of so many skilled individuals, it is difficult to adequately acknowledge everyone involved. We are deeply grateful to the team of distinguished contributors who caught the vision for this book and generously spent their time and energy to make the volume a truly collaborative effort by experts in the field of science and faith. Without their generous giving of labor and scholarly expertise, this project would have never made it to the press. Though our team members have their own thoughts on the various details presented in the book, we are united on the point that the cosmos provides a tremendous amount of information that points to an intelligent designer.

We also want to thank Bob Hawkins, president of Harvest House Publishers, and senior editor Steve Miller, for presenting us with the opportunity to share with our readers the wonders of the universe and believing in the project from start to finish. Their leadership and their continued attention to the needs of the reader made this project truly an enjoyable experience.

Contents

Part III: Science and Evolution

Part IV: Hard Questions on Science and Faith

General Editors

William A. Dembski, PhD, PhD
Board of Directors and Founding Senior Fellow of
Discovery Institute's Center for Science and Culture

Casey Luskin, PhD, JD
Associate Director of Discovery Institute's Center for Science and Culture,
cofounder of Intelligent Design and Evolution Awareness (IDEA) Center

Joseph M. Holden, PhD
Cofounder and President, Professor of Theology and Apologetics,
Veritas International University

Contributors

Douglas Axe, PhD
Maxwell Professor of Molecular Biology, Biola University, the founding director of Biologic Institute, and the founding editor of *BIO-Complexity*

Günter Bechly, PhD
Former curator for amber and fossil insects in the department of paleontology at the State Museum of Natural History in Stuttgart, Germany, and senior fellow of Discovery Institute's Center for Science and Culture

Michael Behe, PhD
Professor of biological sciences at Lehigh University in Pennsylvania, and senior fellow of Discovery Institute's Center for Science and Culture

John Bloom, PhD, PhD
Professor of physics and founding director of MA Science and Religion Program at Biola University, and Fellow of Discovery Institute

Walter L. Bradley, PhD
Fellow of Discovery Institute's Center for Science and Culture, and former professor of mechanical engineering at Texas A&M University, professor of metallurgical engineering at Colorado School of Mines, and distinguished professor, Baylor University

Frank Correa, DMin
Academic dean, professor of theology and apologetics, Veritas International University

David R.C. Deane, MA
Engineering contractor for the Royal Australian Air Force, author and speaker in Christian apologetics, Australia

Michael Egnor, MD
Professor of neurosurgery and pediatrics at State University of New York, Stony Brook, and senior fellow of The Walter Bradley Center for Natural and Artificial Intelligence at Discovery Institute

Miguel Angel Endara, PhD
Professor of philosophy and religion, Veritas International University

Terry Glaspey, MA
Author of *75 Masterpieces Every Christian Should Know* and *Not a Tame Lion: The Spiritual Legacy of C.S. Lewis*, DMin (in progress), professor at Northwind Seminary

Guillermo Gonzalez, PhD
Senior fellow at Discovery Institute's Center for Science and Culture and coauthor of *The Privileged Planet: How Our Place in the Cosmos Is Designed for Discovery*

Bruce L. Gordon, PhD
Associate professor of the history and philosophy of science, Houston Baptist University, and senior fellow at Discovery Institute's Center for Science and Culture

David Haines, PhD
Associate professor of philosophy and religion, Veritas International University

Christopher Travis Haun, MA (IP)
Senior editor of Bastion Books, and author, speaker, and graduate student at Veritas International University

H. Wayne House, JD, ThD
Distinguished research professor of theology, law and culture at Faith International University, and member of board of directors at Intelligent Design and Evolution Awareness (IDEA) Center

Richard G. Howe, PhD
Emeritus professor of philosophy and
apologetics, Southern Evangelical Seminary

Cornelius G. Hunter, PhD
Adjunct professor, Biola University and author
of *Darwin's God: Evolution and the Problem of
Evil*, *Darwin's Proof*, and *Science's Blind Spot*

Michael N. Keas, PhD
Adjunct professor of the history and philosophy
of science, Biola University, and senior fellow
of Discovery Institute's Center for Science and
Culture

Robert J. Marks II, PhD
Distinguished professor of electrical and
computer engineering, Baylor University, and
director of The Walter Bradley Center for
Natural and Artificial Intelligence at Discovery
Institute

Stephen C. Meyer, PhD
Director of Discovery Institute's Center for
Science and Culture and author of *Darwin's
Doubt: The Explosive Origin of Animal Life and
the Case for Intelligent Design*, *Signature in the
Cell: DNA and the Evidence for Intelligent Design*,
and *Return of the God Hypothesis: Three Scientific
Discoveries That Reveal the Mind Behind the
Universe*

Brian Miller, PhD
Research Coordinator for Discovery Institute's
Center for Science and Culture

Paul Nelson, PhD
Adjunct professor of science and religion, Biola
University, and senior fellow at Discovery Institute

Denyse O'Leary
Canadian journalist, blogger, science writer, and
coauthor of *The Spiritual Brain: A Neuroscientist's
Case for the Existence of the Soul*, *by Design or by
Chance*, and *What Are Newton's Laws of Motion?*

Fazale Rana, PhD
Vice president of research and apologetics,
Reasons to Believe

Jay W. Richards, PhD
Assistant research professor in the Busch School
of Business, The Catholic University of America,
executive editor of *The Stream*, and senior fellow
at Discovery Institute

Hugh Ross, PhD
Founder and president of Reasons to Believe,
and adjunct faculty member at A.W. Tozer
Seminary and Southern Evangelical Seminary

Henry "Fritz" Schaefer III, PhD
Graham Perdue Professor of Chemistry and
director of the Center for Computational
Chemistry, University of Georgia

Wesley J. Smith, JD
Chair and senior fellow at Discovery Institute's
Center on Human Exceptionalism

Richard Weikart, PhD
Professor of history and graduate program
director, California State University, Stanislaus,
and senior fellow of Discovery Institute's Center
for Science and Culture

Jonathan Wells, PhD, PhD
Senior fellow at Discovery Institute's Center
for Science and Culture and author of *Icons of
Evolution* and *Zombie Science*

John G. West, PhD
Former chair of Department of Political Science
at Seattle Pacific University, now vice president
and senior fellow at Discovery Institute, and
managing director of the Institute's Center for
Science and Culture

Foreword

Beginning in about 2006, a group of scientists and philosophers known as the new atheists ignited a worldwide publishing sensation. A series of bestselling books, led by Richard Dawkins's *The God Delusion*, argued that science, properly understood, undermines belief in God—and that science and theistic belief conflict. Other books—by Victor Stenger, Sam Harris, Christopher Hitchens, Daniel Dennett, Stephen Hawking, and Lawrence Krauss—followed suit.

The new atheists have explained the basis of their skepticism about the existence of God with admirable clarity. According to Dawkins and others, for a long time the evidence of design in life provided the best reason to believe in the existence of God because it appealed to publicly accessible scientific evidence. But since Darwin, Dawkins insists, scientists have known that there is no evidence of *actual* design, only the illusion or "appearance" of design in life. According to Dawkins and many other neo-Darwinian biologists, the evolutionary mechanism of mutation and natural selection has the power to *mimic* a designing intelligence without itself being designed or guided in any way. And since random mutation and natural selection—what Dawkins calls the "blind watchmaker" mechanism—can explain away all "appearances" of design in life, it follows that belief in a designing intelligence at work in the history of life is completely unnecessary.[1]

Although Dawkins allows that it is still possible that a deity *might* exist, he insists there is absolutely no evidence for the existence of such a being, thus rendering belief in God effectively "delusional." Popular TV figure, Bill Nye the "Science Guy," has echoed this perspective. In his book *Undeniable: Evolution and the Science of Creation*, he says, "Perhaps there is intelligence in charge of the universe, but Darwin's theory shows no sign of it, and has no need of it."[2] Consequently, Dawkins has concluded that "Darwin made it possible to be an intellectually fulfilled atheist"[3] and that a simpler and more parsimonious explanation of what we see in nature is that God does not exist.

Other new atheists, including Lawrence Krauss, say that physics also renders belief in God unnecessary. Krauss contends that the laws of quantum physics explain how the universe came into existence from literally nothing. Consequently, he argues, it is completely unnecessary, even irrational, to invoke a creator to explain the origin of the universe.[4] Stephen Hawking, formerly of the University of Cambridge and until his death in 2018 the world's best-known scientist, made a similar argument in his book *Brief Answers to the Big Questions*. There, he argued that "the universe was spontaneously created out of nothing, according to the laws of science."[5] Consequently, for him, that meant "the simplest explanation is that there is no God."[6]

This perception of a conflict between theistic belief and the clear implications of modern science has percolated into the popular consciousness. Recent polling data indicate that in North America and Europe, the perceived message of science has played an outsized role in the loss of belief in God. In one poll, more than two-thirds of self-described atheists and one-third of self-described agnostics affirm that "the findings of science make the existence of God less probable."[7]

Not all scientists accept the idea that the findings of science and faith in God conflict. Instead, some religious scientists believe that science and faith *cannot* conflict. They subscribe to an idea known as compartmentalism, or what the late Harvard paleontologist Stephen Jay Gould called *Non-overlapping magisteria*, or NOMA.[8] Compartmentalism holds that science and religion describe completely different realities. Proponents often support this view by quoting an aphorism used by Galileo that affirms that the Bible teaches "how one goes to heaven, not how the heavens go."[9] Others subscribed to a closely related idea called *complementarity*. Proponents of this view hold that science and religion may sometimes describe the same realities; however, they do so in complementary but ultimately incompatible or "noncommensurable" language.[10]

Proponents of both views deny that science contradicts religious belief, but they do

so by portraying science and religion as such totally distinct enterprises that their claims could not possibly intersect in any significant way.[11] In other words, both models assume the religious and metaphysical neutrality of all scientific knowledge. This assumption seems to insulate theistic belief from scientific refutation, but it also denies the possibility that scientific evidence could offer any support for theistic belief. Thus, until recently, few scientists have thought—as Newton, Boyle, Kepler and the other founders of early modern science did—that the testimony of nature actually supports important tenets of a theistic or Judeo-Christian worldview.

The authors of this volume deny that the evidence of the natural world, rightly understood, conflicts with theistic or Judeo-Christian belief. Many of them think that certain discoveries of modern science—in particular, discoveries concerning biological and cosmological origins or discoveries about the nature of the human mind—may actually support theistic or Judeo-Christian belief. Here, in this book, they have helped to develop another understanding of the relationship between theistic belief and science—one that I call "qualified agreement."

This idea maintains that, when correctly interpreted, scientific evidence and theistic belief can and do support each other. While accepting some disagreement about details as inevitable given the limits of human knowledge, advocates of this model affirm a broad agreement between the testimony of the natural world and the propositional content of Judeo-Christian theism—between science and religion so defined. Though advocates of qualified agreement acknowledge (with compartmentalism and complementarity advocates) that much scientific research and theorizing does address metaphysically and religiously neutral topics, we do not agree that *all* scientific theories have this characteristic.

Instead, the qualified-agreement model, like the conflict model, asserts that some scientific theories do have larger worldview or metaphysical implications. Nevertheless, unlike the conflict model, proponents of qualified agreement deny that the best or most truthful theories ultimately contradict a theistic or Judeo-Christian worldview. Instead, they view theological and scientific truth as issuing from the same transcendent and rational source—namely, God. Advocates of qualified agreement anticipate, therefore, that these two domains of knowledge, when rightly understood and interpreted, will come increasingly into agreement as advances in science and theology eliminate real points of conflict that sometimes have existed.

Many of the founders of early modern science held this view of the relationship between science and Judeo-Christian religion. Indeed, from the late-middle ages through the scientific revolution (roughly 1250–1750), scientists often affirmed the agreement between the book of nature and the book of Scripture, both of which were understood to be mutually reinforcing revelations of the same God. This present volume includes many authors who have helped to revive this perspective and includes entries from them that will help you to consider it as well.

In the Bible, in the book of Romans, St. Paul not only affirms that God created the world, but he also argues that the signs of God's handiwork are "clearly perceived" in "the things that have been made." The collection of essays that you have in your hands considers whether scientific evidence supports this biblical claim. The authors of this volume—who include a diverse group of scientists, scholars, and theologians—also consider many other relevant questions about the relationship between scientific knowledge and theistic belief (or even biblical teaching). I encourage you to read their entries and arguments. If you do, I think you will find yourself better equipped to dialogue with your friends, family, and colleagues about the relationship between science and faith and to explain how scientific evidence from the book of nature and insights from the book of Scripture can provide mutually reinforcing ways of knowing about our universe and its creator.

Stephen C. Meyer, PhD
Director of Discovery Institute's
Center for Science and Culture

Part I:

SCIENCE AND FAITH

1

What Are Science and Faith — and Are They Compatible?

William A. Dembski

If by science we mean modern science (not medieval or ancient science), and if by faith we mean classic, orthodox Christian faith (not, say, gnostic varieties of Christianity or other faiths entirely, such as Jainism or Islam), then science and faith are obviously compatible. After all, since the rise of modern science to the present day, top scientists have also been orthodox in their Christian theology, seeing no contradiction between their faith and their scientific work. Indeed, those most closely associated with the rise of modern science were overwhelmingly Christian.

Atheists might counter that if Christian scientists really understood the full implications of science, they would understand that their Christian faith is, in the end, unsustainable. But such accusations, on closer analysis, always ring hollow and display special pleading, trying to make obviously bright and reflective scientists of faith seem like idiot savants who happen to be really good at their science but really bad in thinking through its implications, especially for their faith.

As a case in point, while lecturing at the University of Toronto some years back, I encountered a biologist in the audience who claimed that evolution made it impossible to be a scientist of faith. Many scientists who are Christians believe in evolution. I'm not one of them, thinking the evidence for evolution to be sketchy at best. But it needs to be noted that many scientists hold to classic Christian orthodoxy (i.e., the incarnation of God in the person of Jesus, his bodily resurrection and ascension, etc.) and are also evolutionists—some even of a Darwinian stripe, seeing God as creating through an evolutionary process that gives no evidence of design. Of course, their interpretation of Scripture will be suspect to those who see Genesis as teaching that creation occurred a mere 6,000 years ago in six literal 24-hour days.

If you think that a literalist interpretation of Genesis is crucial to being a Christian, then you'll deny that evolutionary science and Christianity are compatible. But that seems to be asking too much for the compatibility of science and faith. The more things you require to be believed, whether on the side of science or on the side of the faith, the more incompatible you make the two. My own view, and the one I'm recommending in this chapter, is to take a minimalist approach to science and to faith. Don't make embracing science require holding on to too many controversial and suspect scientific views, and likewise, don't make faith major in minors, forgetting Christ, who purchased us with his blood.

In any case, sociology confirms that top scientists have also been orthodox Christians. Indeed, it's a matter of record that the society of scientists and the society of orthodox Christians intersect in a nonempty set. Certainly, we should be concerned about the compatibility of science and faith if no such people residing in both groups existed. As it is, I've personally engaged with not one but two Nobel laureates who were reasonably orthodox in their Christian faith (William Phillips at a conference in 1997, and Richard Smalley for lunch in Houston shortly before his death in 2005). So, I can attest from personal experience that such people exist.

Even so, it's always good to look deeper than sociology. What is modern science, and what is orthodox Christian faith, and are they the types of things that do well together, as with things that are genuinely compatible? Is the relationship between science and faith like a happy marriage, in which the spouses mutually support and reinforce each other and reside in wedded bliss? Or is it more like an uneasy marriage in which the spouses would just as soon be rid of each other? I personally think that modern science and orthodox Christian faith can reside in something like wedded bliss provided they are properly conceived.

Let's look at science and faith more closely, starting with science.

What Is Science?

Science is a sustained and systematic inquiry into nature. It tries to understand nature. It does so through observation, experiment, and theory construction. It is evidence- and reason-based. A scientific claim is considered true or compelling not because someone in authority asserts it to be so. Its strength in commanding our assent derives from the evidence for it and how it helps bring light to our understanding of nature. There are often aesthetic elements here, as in a scientific explanation being so pleasing or beautiful in how it ties things together and illuminates our understanding of nature that we think, *It must be so.*

Aesthetics, however, can also be misleading. A theory can be beautiful but untrue. Science is a fallible enterprise with many pitfalls. It has gotten some things astoundingly wrong (e.g., alchemy, phlogiston, phrenology, the luminiferous aether, etc.). Moreover, scientists have been implacably committed to wrong scientific theories, exuding a confidence in those theories that later was shown to be wholly unfounded. Indeed, the confidence with which scientists hold their theories is no gauge of their truth.

We need always to keep a critical eye on such overconfidence, especially in public policy discussions where scientists or politicians acting "in the name of science" attempt to steamroll people into some view or course of action because science says it must be so. Science has gotten many things wrong, especially when invoked in public policy recommendations. Yet provided that scientific discourse is allowed to be free and open, self-correction becomes a feature of science. Unfortunately, the self-correcting capacity of science is often more myth than reality, as when certain scientific theories attain a sacrosanct status and deny any place at the table to competitors.

What Is Faith?

The physicist Bob Bass was a friend of mine. A believing Roman Catholic, he died a few years back when in his eighties. He had been a Rhodes Scholar in the 1950s and had thereby gotten to meet Albert Einstein. As Bob shared with me, Frank Aydelotte, former president of Swarthmore College and head of the Institute for Advanced Study, was able to gather the 32 Rhodes Scholars at the time (one of them Bob) and have them meet with Einstein as the guest of honor. During their time together, Aydelotte posed this question: "Now, Einstein, can you give these young men any parting advice?" (Note that the Rhodes Scholarship did not start admitting women until 1977.) Einstein replied, "If I could give the young men any advice, it would be this: Don't believe anything is necessarily true just because you see it in the newspapers, or hear it on the radio, or everybody else believes it. Always think for yourself!"

Einstein was here giving crucially important and sound advice, and yet the spirit of this advice seems at odds with much of what typically goes by faith. Faith is often viewed as a matter of obedience, even slavish obedience, where you are expected to accept certain doctrines uncritically, and where any deviation marks you as a heretic and thus outside the body of true believers. Instead of giving pride of place to "think for yourself," faith is sometimes thought to underscore "trust and obey."

Now that I'm more than 40 years into my experience of the Christian faith (my conversion occurred in 1979), I don't see faith and science as conflicting in this way. The fact is, whatever we mean by an authentic or real Christian faith, it had better be a faith that you own, that you've thought through carefully for yourself, and where you're convinced

that it holds up and can sustain you through life's toils and snares. The flip side is going through the motions, pretending you believe something when you really don't, often to keep people in authority happy (your parents, perhaps, if you're younger; your deans and school presidents if, like me, you taught at a seminary).

If you are forced to pretend to believe something that you really don't think is true or holds up, you're asking for trouble. And this can hold as much in science as in faith. Science, as currently practiced, has its sacred cows and preferred dogmas, and if you step outside those boundaries, you can expect to be punished. Ditto for faith. The problem with pretending to be a believer is that you end up hating yourself for it. It makes for moral licensing—in other words, hypocritical behavior in which you allow yourself (i.e., give yourself the moral license) to do things contrary to what you say you believe because you don't really believe. Think climate change advocates flying in private jets to climate change conferences; think Christians advocating the sanctity of marriage but cheating on their spouses and getting divorced.

It's been said that God only has children, not grandchildren. You cannot inherit the Christian faith as you inherit a piece of land. Your faith is an intrinsic part of you. But it is truly your faith only if you have thought it through, worked it out, and embraced it deeply. Such a faith is not a matter of convenient assent, as in, "Gee, I better say I believe this and that because others will be unhappy with me if they suspect I have doubts about those things, to say nothing of my actively disbelieving them." So I would say Einstein's dictum about thinking for yourself holds as much for faith as for science. It must precede commitment and obedience.

✝ Science and Truth

The question of truth in science as well as in faith needs now to be addressed. For many, science seems less concerned about truth than faith, whose concern with truth is nonnegotiable. Orthodox Christian faith (I'm talking lowercase *o* orthodox as in the sense of the Christianity encapsulated in the early church creeds as opposed to that form of Christianity known as Eastern Orthodoxy) claims to know and articulate humanity's chief truth, namely, that God, by taking human form in Jesus, then dying on the cross, and then rising bodily from the dead, redeems fallen humanity. To this chief truth we can add the Apostles' and Nicene creeds. Lowercase *o* orthodox Christians believe these creeds to state the exact truth about their faith. This is not to say these creeds are exhaustive of Christian truth, but they do capture the basic core.

Science, on the other hand, often seems less about finding the exact truth about nature than about successfully understanding and explaining natural phenomena. Success here can refer to the ability of science to help us build stable bridges or predict certain measurements, but without any pretense for knowing the exact truth about nature. Some philosophers of science go further, contending that the success of science argues for its truth (how could science be so successful without being true?).

Some see science as approximating truth in an endless quest to approach the truth, yet never quite hitting it on the head. Others take a more empirical approach, seeing the value of science in its ability to faithfully describe and predict our observations of natural phenomena (sometimes called *empirical adequacy*). And others take a pragmatic approach, in which any ideas that help advance our understanding of nature are seen as legitimate. The philosopher Wittgenstein took this latter approach, arguing that neither Copernicus nor Darwin gave us the truth but rather a fruitful way of looking at nature.

My own sympathies lie more with the empirical and pragmatic approaches to science. Look where I might in science, and I see a fallible enterprise where yesterday's ways of viewing nature give way to today's more powerful ways of viewing nature. I therefore have no confidence that "we've arrived" and that science has come to the end of the line, with tomorrow's science identical to today's science. Science is fallible, as the history of science demonstrates. Some philosophers of science even write of a "pessimistic induction" in which the consistent failure of scientific theories to withstand the tests of time demonstrates that no scientific theory is to be taken too seriously. Even Newton's laws of motion, which seemed in their day to capture exactly how nature operated, have been displaced. Along came Einstein. Then came quantum mechanics. Newton's laws were shown to work for a certain range of phenomena, but not beyond. And even where Newton's laws work, newer theories suggested that they are approximations and not exact renditions of the natural phenomena in question.

Once one accepts that science is in a state of flux—that it is a fallible human enterprise and that it stands in constant need of correction—certain grandiose claims about science immediately fall by the wayside. Take, for instance, the claim that the only knowledge that deserves to be taken seriously is scientific knowledge. From such a claim, it obviously follows that orthodox Christian faith doesn't rank and needs to be discarded. But what is the nature of the claim that only scientific knowledge deserves credence? Is it a scientific claim? Obviously not. No experiment or observation or act of reason or theoretical insight justifies it. The view that only science constitutes legitimate knowledge is known as *scientism,* and it is self-referentially incoherent. In other words, it defeats itself, and because it defeats itself, scientism can be safely disregarded.

Methodological Naturalism

Another grandiose claim made in the name of science is that science is uncompromisingly committed to methodological naturalism (also known as methodological atheism or methodological materialism). The idea here is that science needs to treat nature as following unbroken natural laws without leaving any place for things like divine action or miracles. As the argument goes, if science were to allow for God to intervene in the world, then anything could happen, and science could offer no insight into what nature does and displays. This sounds good if one thinks, for instance, of a forensic scientist trying to understand money missing from a safe. Without methodological naturalism, would this forensic scientist need to take seriously the possibility that God (or perhaps Satan) removed the money by making it magically dematerialize from the safe?

But in fact, methodological naturalism is just an arbitrary rule for doing science. And who gets to set the rules? Our word *science* derives from the Latin word *scientia,* which was the generic Latin word for knowledge—all knowledge. In the nineteenth century, the term came to mean specialized knowledge of nature, but the fact is that what we call science today has a long history, and throughout much of that history, it was called *natural philosophy.* What's more, the rules of science or natural philosophy have changed over time. Scientists most responsible for the rise of modern science—such as Copernicus, Kepler, and Newton—rejected methodological naturalism outright, finding in nature clear, scientifically discernible marks of the divine intelligence.

At best, methodological naturalism should be regarded as a starting point for scientific inquiry, in which we try to understand phenomena in terms of undirected natural laws and processes. But if those do not suffice, then intelligently directed or even supernatural explanations may become reasonable. Jesus turning water into wine defies methodological naturalism. And even a forensic scientist trying to understand the disappearance of money from a safe might be convinced of a supernatural dematerialization of the money if sufficient safeguards were put on the safe to monitor the money during its disappearance.

Of course, in such circumstances, one can always appeal to unknown, and perhaps unknowable, natural laws that might have been at play. But such appeals to ignorance can be less insightful and less convincing than simply ascribing a divine miracle. We know, for instance, how nature brings about wine—namely, through a lengthy process of fermentation. We know of no natural process by which water is instantaneously transformed into wine. If it happened, it was a miracle. That's why metaphysical naturalists (i.e., those who think that nature is all there is) will not try to defend that some natural process did indeed turn water into wine at the wedding at Cana, but rather will argue that the Bible is in error when it attributes this miracle to Jesus.

Where methodological naturalism becomes especially problematic for science is with intelligent design. Intelligent design is the study of patterns in nature that are best explained as the result of intelligence. Yet from a naturalistic point of view, intelligence is a consequence of nature, not something that was present in nature from the start. Methodological naturalism is thus limited to seeing intelligence in nature as something that came about by an undirected or unintelligent natural process—in other words, by evolution.

Methodological naturalism requires that all intelligences in nature derive from processes that give no evidence of intelligence. Any intelligence studied by science is thus required to be an evolved intelligence. Intelligent design turns this view on its head, arguing that there can be good reasons and evidence to think that intelligence in nature can be scientifically studied and discovered even if the intelligence is unevolved. For the methodological naturalist, intelligent design is fine for archaeology, where evolved human intelligences create, say, burial mounds, but not for DNA, say, whose information-rich design would require an unevolved intelligence.

Turning Science into an Idol

Earlier I remarked that faith tends, more than science, to be about truth. I need to elaborate on this claim because there is also a sense in which both are essentially about truth. But as one first approaches faith and science, faith seems more about truth (especially the truth about the person and activity of God in history), and science seems more about explanation and understanding (especially gaining valuable insights into nature regardless of their truth status). Thus, orthodox Christian faith sees itself as "contending for the faith once and for all delivered to the saints" (Jude 3). It sees certain core doctrines as essential to faith and their repudiation as denials of truth.

Science, by contrast, as I noted earlier, tries to understand nature by means of theories that work within certain ranges of phenomena but not outside. Take superconductivity, for instance. The original theory of superconductivity was for extremely low temperatures (liquid helium temperatures, close to absolute zero). The newer theory of high-temperature superconductivity was for significantly higher temperatures (liquid nitrogen temperatures). The original theory could not explain high-temperature superconductivity. It was right for low-temperature superconductivity, but it could not be generalized to all cases of superconductivity. We therefore say that it was true for a certain range of phenomena, but not for others. That seems reasonable. Indeed, any scientific theories can only be tested within a certain range of phenomena and thus never verified to be exactly true across the board.

If the theories of science rarely seem exactly true, the implications of those theories and the claims made for them—and thus in the name of science—can be true or false. At the time of this writing, the COVID-19 pandemic continues to command worldwide attention. Models claiming to describe the expected rates of infection and death have been widely distributed and advertised. One such model made the dire prediction that, in the US alone, the coronavirus would quickly cause the death of over two million people. As a consequence of taking this model seriously, the US government made many public-policy decisions, not least to radically curtail the economy. Now it may be that most of those decisions should have been made even if the death toll was, as it now appears, only a tenth of that two million figure. But the point to note is that the model, inspired by science, was wrong. Rather than giving us true insight into the virulence of the coronavirus, the model vastly overstated it.

The lesson I want to draw here is that the invocation of science must not become a wand for magically giving credence to claims, especially for special interests who want certain things to be true and who use science to make it seem that they are true. Just because scientists have used science to draw certain conclusions does not mean those conclusions are justified or deserve credence. A kind of mystique has come to surround science, where, as soon as one says that a claim is scientific, its stock suddenly jumps in price. The first commandment of Moses is against idolatry—that is, against setting up anything in place of God. Science, in today's culture, often becomes an idol. ⅄

Science, it needs to be reiterated, is a fallible enterprise. Moreover, the marshaling of scientific evidence is itself more art than science. The same evidence in the hands of different scientists can point in radically different directions. Many of the scientific controversies of our age result from scientists looking at the same evidence and drawing different conclusions. Sometimes prior commitments make all the difference. In my own scientific work on intelligent design, I've consistently felt that the evidence and arguments are stronger on my side than that of the Darwinian naturalists, but precisely because they are committed to a naturalistic account of biological origins, no evidence can, for them, count in favor of intelligent design.

My point here is not to argue for intelligent design or for any method of advancing one scientific theory over a competitor, but rather to stress that science, as a fallible enterprise, needs to keep its options open and be willing to discuss alternatives and opposing views. Too often these days we see self-important scientists claim that there's only one legitimate scientific way to view a given phenomenon, and that anyone who dissents is too uninformed or dense to deserve a place at the table to discuss the matter scientifically. Thus, one hears about "consensus science" or "settled science" as though science, like a religious dogma, has permanently established the truth of some claim and that no questioning of it can henceforward be permitted.

The fact is that science advances by questioning settled opinion. Ptolemy, so it seemed, had settled the motion of the sun, moon, and planets. It took Copernicus to question the settled science of his day to unseat that old view and replace it with the true(r) view, in which the sun became the center around which the planets, including Earth, revolved. Science advances by revolutions, and revolutions overturn settled science. Thus, whenever I hear the words *consensus science* or *settled science*, I reach for my wallet because I know I'm about to be scammed. If some scientific claim is truly settled, you don't need to say so—there simply won't be any debate about it. It's only when a scientific claim is unsettled that advocates (ideologues) on one side invoke the term *settled science* to quash dissent from the other side.

There's a broader point here, and that's the need for freedom of thought and expression in science as well as in any other area of inquiry, and that includes faith. Are we ready to put all claims of science, faith, etc., on the table for discussion? Of course, we are entitled to be fully convinced of our understanding of science and faith and whatever. But are we willing to discuss our views and put them on the chopping block so that dissenters can subject them to scrutiny? Or do we have sacred cows that we are unwilling to subject to scrutiny because we are so certain of them or think it is somehow impious to question them? This is fundamentalism, and it's as much a problem for science as it is for faith.

Test Everything

In his autobiographical adventure *Surely You're Joking, Mr. Feynman!,* Richard Feynman, perhaps the greatest physicist of his generation, said that one's first task in explaining science is not to fool people, to which he immediately added that the easiest person to fool is yourself. When it comes to science or faith, our aim must be to tell the truth. It is not to fool people, and that means not fooling ourselves. It means being honest about what we truly believe, about the strength of the evidence for our beliefs, and where our beliefs may be firm and yet unjustified or even unjustifiable.

Too much of both science and faith is agenda-driven. We want to push through a certain public-policy proposal, so we misrepresent science to help us advance it. We want to usher someone into God's kingdom, so we misrepresent faith to elicit the commitment we are seeking. Both science and faith are easily abused. The challenge for scientists and Christians, and those who are both, is simply to be honest, neither overselling nor underselling their positions, and striving to convey the truth as they best understand it without insisting on any predetermined outcomes. The truth can take care of itself, in both science and faith, provided we get out of its way and let it have its say.

Instead, however, corrupt people can bend science and faith to their agendas. Christian faith takes the Bible as its point of departure, and yet the Bible can be misread in numerous ways. In the hands of a skillful skeptic, the Bible can be shown to be utterly unbelievable and discredited or to show things that no orthodox Christian has ever believed. It therefore takes someone to "rightly divide the word of truth" (see 2 Timothy 2:15) to refute such a skeptic. Likewise, science can be made to prove just about anything. Indeed, it's been widely discussed in the scientific literature just how many scientific studies are fraudulent, with data being falsified to establish claims that are known to be false.

We always need to take seriously Paul's dictum in 1 Thessalonians 5:21 to test all things and hold fast to what is good (and that obviously includes what is true). Paul places the burden to test, to check things out, on each of us individually. Verification is not something we can outsource. Each of us personally needs to test the things in life that confront us, discarding the bad, keeping the good. This is true even of scientific claims. We need to do our best to educate ourselves and figure out the truth of the claims being thrust at us. This is true even of nonscientists. As a nonscientist, you might not know the nuts and bolts of the science underlying, say, climate change. But you can understand the broad patterns of evidence and how they are being used or abused. Thus, whenever someone says to me "science has shown _____," I say, "Not so fast." If the claim is controversial, I check it out for myself and dig deeply enough to draw a conclusion: yes, it does show that; no, it doesn't show that; maybe, but the evidence is inconclusive.

Many scientific studies invoked these days to prove some point or other are bogus. Many use fraudulent data because it's easier to make stuff up rather than actually run experiments that may not confirm the result you're after. But even the absence of blatant fraud doesn't protect science. A widespread problem here is "the file drawer effect." Suppose I tell you that I just flipped a coin ten times and saw ten heads in a row. You might think that's an unusual result and question whether the coin is fair. But what if you learned that I had spent an hour flipping the coin until I witnessed ten heads in a row? In that case, getting ten heads in a row would be unexceptional—indeed, you'd expect to see such a run of heads if you had that long to flip the coin.

With many scientific studies that purport to establish some result, it's possible to keep running the study until you get the result you want. You then report the desired result to a scientific journal, but keep silent about all your failed attempts to achieve it (i.e., you put all the failures in "the file drawer," hoping that by leaving them there, they will be forgotten). In this way, many widely reported yet bogus scientific claims have crept into the literature. At the same time, it's hard to excise and correct them. Indeed, it's been found that many scientific results can't be replicated precisely because they bank on the file drawer effect, which can capitalize on one "success" in an experiment among what otherwise are all failures. The problem with getting these bogus results corrected is that it's much harder to get failures published than successes, and correction here requires public recognition of failure. "I wasn't able to replicate so-and-so's otherwise amazing study" typically won't get published, whereas so-and-so's amazing study, even if bogus, will.

I don't mean to cast a pall over science. Science is great. And faith is great. But as the preacher says in Ecclesiastes 7:29, humans have a tremendous facility for searching out and manufacturing shady schemes. Leaving aside such schemes, let's return to the compatibility of science and faith. If we think of science and faith each as a set of ideas, it's straightforward to make them compatible by simply not claiming too much for science or

for faith. The smaller each set, the easier it is for them to be compatible. Conversely, the more that is claimed in each set, the more you have to defend and the more opportunities there are for contradiction. Hence my advice about going minimalist with science and with faith—that is, going with the best-established science and the core doctrines of Christian faith. By conceiving of science and faith each as a set of ideas, a minimalist science and a minimalist faith can, it seems, be kept free of contradiction and thereby made compatible in the sense of maintaining logical consistency.

⋏ Christianity and Science

To change gears, the compatibility between science and faith can be considered less formally (less a matter of logic) and more organically. One point that my departed colleague Stanley Jaki (a Catholic priest who was a physicist and historian of science) made in many of his writings is how Christianity provided the matrix within which modern science could develop. "Why," he would ask, "did the ancient Greeks not create modern science?" Their math skills certainly were up to the task. But, as he noted, they viewed the universe not as a creation but as an eternal, sacred mystery.

The Judeo-Christian view of the universe as a creation meant that it was an invention, that it sprung from the mind of God, and that we, who are made in God's image, could likewise understand it. Moreover, because it was a creation, and thus not identical with God, it would not be impious to experiment on the universe. Christianity, according to Jaki, made modern science possible, overcoming all the past abortive attempts to get science off the ground by giving us permission to understand the universe, and thus the natural world, from the vantage of a creator God. Thus, when secular scientists suggest that Christian faith impedes scientific progress, the tables can legitimately be turned: It is reasonable to think that, without the Christian faith, science as we know it would never have developed in the first place. Accordingly, science and faith are compatible because faith gave rise to science.

But that was then, and this is now. What has faith done for science lately? Such questions prompt still another way of viewing the compatibility of science and faith: When confronted with a challenge, where do we go for help first—to faith and God, or to science and technology? I've purposely stated a false dichotomy here because the choice is never to go with science over faith or faith over science. If you're sick, pray and go to a doctor. The doctor, especially a good doctor, may be God's instrument for healing, and at the same time, as people of faith, we are called to lay all our concerns before God. Faith and science can thus work together. Faith and science constitute a both-and proposition, not an either-or opposition.

And yet, in our age, science and faith are often portrayed as being at odds, as though you're supposed to choose one over the other. On one side are those who elevate faith and denigrate science, such as faith healers who counsel throwing away one's meds and trusting divine miracles entirely for healing. On the other side are those who look only to science and have no thought of God or faith. I was particularly struck by the latter when I watched *The Martian*, in which the Matt Damon character is an astronaut stranded on Mars who must somehow survive. His approach is entirely secular. Without exhibiting any faith or offering any prayer, he comes to terms with his predicament by turning *science* into a verb and saying that he's going to have "to science"

his way out of his predicament. Many atheists and agnostics approve this attitude.

As a Christian who embraces both faith and science, I would, in such a circumstance, pray to God for deliverance and also make use of science as best I can to survive. The approach of Matt Damon's character, by contrast, is emblematic of our age—yet it need not be. Indeed, a healthy faith, and the gratitude it engenders, can maintain an optimism that helps us through life's challenges, even as we also creatively and energetically use science to meet those challenges.

This chapter has focused on the compatibility of science and faith, but in some sense it is as much about how Christians should think about their faith in relation to science. As I've shown, at issue here is much more than just trying to square certain scientific theories (such as Darwinian evolution) with certain understandings of faith (such as a literalistic reading of Genesis). Science holds tremendous prestige in our day. Whereas in centuries past it was the theologian and preacher who was the most respected figure in society, nowadays the scientist and technologist commands that position. Science has immeasurably increased our understanding of the world. It has helped cure our diseases. Thanks to science (and thanks to the God who gave us science!), we live on average much longer than in times past. Secular thinkers thus expect, and even demand, a hushed awe before science while at the same time being dismissive of faith. By contrast, I would urge that we give science and faith each their due and see them as mutually reinforcing each other.

Faith in a Person

In closing, I want to make a final point about science and faith. Faith is personal in a way that science is not. Sure, science is the result of human activity, and thus personal in that sense. But there's also a sense in which science is in the business of making propositional claims about the world and about how it works and about how to exploit its working through technology. Faith, too, can be viewed as making propositional claims, as with its core doctrines, but it is also radically personal, requiring a real-time direct relationship between the human person and the infinite personal God revealed in Christ. In John's Gospel, Jesus says that he is the way, the truth, and the life (14:6). The truth underlying Christian faith is the person of Jesus Christ, not some proposition. When we say we trust Jesus, we are not merely taking a cognitive stance toward some proposition (thinking of it as true versus false, plausible versus implausible, etc.). We are expressing confidence in the person of Jesus, believing that he will do right by us even when circumstances and evidence seem to the contrary.

Faith as a direct personal relationship with the living God must always take center stage in our understanding of faith. A lived faith is not one that depends on constantly reminding yourself of a set of propositions, even if those propositions come from the Bible or the Apostles' Creed. A lived faith springs from a living relationship with the living Savior, Jesus Christ. It is faith not in an idea or proposition, but in a living person.

2

How Do We Understand the Relationship Between Faith and Reason?

Joseph M. Holden and Christopher T. Haun

The modern age of scientific inquiry and technological advance has reinvigorated the perennial question of the relationship between faith and reason. If one's approach to reality is solely one of faith divorced from reason, it appears to lack any objective grounding to the real world. Faith of this sort is often discarded as an unjustified or unreasonable faith. If guided by reason alone, it would seem to destroy the confidence and presuppositions required to discover and inform truth. The tension between the two opposing camps is palpable, for with an emphasis on reason comes the charge that one fails to recognize and appreciate the essential role of immaterial realities necessary for scientific inquiry (e.g., logic, knowledge, truth, virtue, ideas, confidence, worldview, faith, hope, etc.). On the other hand, an overemphasis on faith seems to support the notion of one who is anti-intellectual and may be ignorantly oblivious to reality, holding a casual relationship to the facts. However, a closer look finds the need for faith and reason to work cooperatively together in the discovery of truth.

Faith and Reason Working Together

The normal relationship between reason and faith is simple: reason leads either toward or away from faith in some idea or object (R→F). This is true of all of life's decisions regardless of how mundane or profound, simple or complex. For any idea we believe, we can ask what reasons we have for believing that idea. The fact that there are always reasons underlying our beliefs indicates that *some kind of reasoning* helped establish those beliefs. Of course, some reasons are better than others. But faith without reason is called *unreasonable faith*.

⌐After reason leads us to a reasonable faith, reason should *not* be abandoned. Faith should proceed to inform reason (F→R). Reason should continue to understand, systematize, test, refine, and improve the mental models we believe correspond to reality (faith). So long as reason remains faithful and faith remains reasonable, both reason and faith should be held in high regard and encouraged to cooperate (R⇌F). While there are times when reason should take

33

priority over faith and other times when faith should be given priority over reason, we ultimately must reject both of the more extreme views that insist we should attempt to operate either on reason alone or on faith alone.

R	**Reason Alone** (extreme antagonism toward faith)
F	**Faith Alone** (extreme antagonism toward reason)
R>F	**Reason over Faith** (mild antagonism to faith)
F>R	**Faith over Reason** (mild antagonism to reason and/ or reason encouraged only after a foundation of faith is established)
R⇄F	**Reason and Faith** (reason leads to faith, faith informs reason, reason systematizes faith, reason continually tests and refines faith, etc.)

Table 1. The five basic positions in the faith and reason debate.[1]

Reasoning About Reason

Our species is called *Homo sapiens* ("men of wisdom") because our powers of reason vault us far above the beasts. We are the masters of the Earth not because of our physical superiority but because of our rational ability. We reason with one another, cooperate, and dominate all other living things. The next most intelligent animals—e.g., chimps, gorillas, pigs, porpoises, elephants, shepherd dogs, parrots, and ravens—can barely compete with human three-year-old children in cognitive tests involving symbolic logic.[2] All these animals are left far behind by the time human brains reach their adolescent level of maturity. When our brains reach their full maturity in our twenties, they contain around 86 billion neurons. There are more

connections between those neurons than there are stars in the observable universe!

Who or what caused the quantum leaps in our cognitive brainpower and wisdom? Some have reasoned that when the conditions are right, forces like random mutation and natural selection are freed to produce such leaps in complexity. Once our primitive ancestors learned how to cook food with fire and made a habit of doing so, all the energy formerly spent digesting raw meat could then be routed to building bigger and better brains. And by gathering shellfish on beaches during low tide, humans gained access to the fatty-acids and other raw materials that are needed for the building of those more complex brains.

But how can mindless forces—forces that are totally devoid of intention and intelligence—create minds with intention and intelligence?[3] Streams don't rise above their sources. And one cannot give what one doesn't possess. Today it takes a well-funded team of the most brilliant electronic engineers and computer scientists in the world to build massive AI supercomputer systems that can out-calculate us on mathematically rich games like chess, poker, or Go.[4] But our brains, which weigh a mere 2.8 pounds (1.27 kilograms), can do countless things that massive supercomputers cannot do. We can learn languages, for example, without even being given any instruction on how to do so.

Which is more reasonable: Mindless forces, cooking and eating shellfish, creating spectacular minds, or an even greater mind creating minds? A purposeful creator with an infinitely intelligent mind would surely be able to design and create finite beings with somewhat similar (but far less powerful) versions of the creator's mind. Some might argue that it is less reasonable to prematurely insert "the God of the gaps" into this particular gap than it is to wait for science to explain the gap in the future. But the hypothesis that explains humankind's greatness with shellfish consumption, cooking fires, random mutations, and natural selection over many generations in the last million or two years can only remain a hypothesis. It resists empirical verification and therefore can only be accepted as an article of faith today. Insofar as our experience offers data, we have all seen that minds create minds. But who has ever observed mindless matter creating mind? Until science can fill this gap with proven fact rather than unproven theory, it is more reasonable to hold to the adequate explanation that fits better with the observable facts.

Genesis indicates that being made in the image of God enables us to govern the Earth and rule over the animals. Ruling involves reasoning through complex problems, evaluating the possible solutions, and setting into motion the changes required to solve problems and optimize complex systems. If Moses is right, there is something very noble about reasoning (Genesis 1:26-27; Deuteronomy 8:11-16). The more we exercise our higher reasoning capabilities, the more we become like God intended us to be—and, in a limited way, the more like God we are. Conversely, the more we fail to exercise our reason as God intended, the more we become like mindless beasts (Daniel 4:16; 5:21).

The immediate challenge to this rather self-evident and simple R⇄F theory is that our reasoning often becomes messy. For example, even in those moments when we manage to reason very logically, we can still be led away from truth by incorrect premises, deceptive statistics, incomplete data, flawed methodology, and misinformation.

Three Types of Logic

Deductive and inductive reasoning. We are most obviously reasoning when we think slowly, carefully, and consciously about some question, problem, equation, or idea. We use the knowledge we have to try to understand it deeply, evaluate reasons for and against it, draw conclusions about it, explain it, and make predictions. We are performing *deductive reasoning* when we reason from general and known truths to a specific conclusion. For example, if A is greater than B and B is greater than C, then A must be greater than C. Mathematics and philosophical reasoning tend to be of this type. Assuming its premises are true and the structure of the argument is valid, deductive logic produces conclusions that are certain.

Reasoning from specific facts and data to general truths is called *inductive reasoning*. Scientific research involving experiments tends to be of the inductive variety—as is the inference from multiple observations that multiple chimpanzees were not able to reason on par with humans and, therefore, it is highly probable (and therefore reasonably believable) that the next chimp we test will also prove unable to reason at human levels. Inductive logic produces probable conclusions (not absolute certainty) and is useful for making reasonable predictions about the future.[5]

Abductive reasoning. Although it is a little more controversial and fuzzy, there is at least one more type of reasoning we routinely perform that has gained acceptance with many logicians. Abductive reasoning processes ideas in our subconscious minds and its hypotheses surface in our conscious minds as a flash of insight or intuition. The one having the insight may not even be able to articulate how he or she came to the idea, but it was arrived at logically.[6] This reasoning offers theories that fit observations. Some extend it to what happens when we have to make a best guess with limited or incomplete information—as with jurors who, after hearing arguments from the prosecuting and defense attorneys, realize they don't have all the information they would prefer to have and must make their decision based on gut instincts. They then make an educated guess even though evidence is incomplete, faith is imperfect, and certainty is impossible.

Some people's most brilliant insights may have been produced by abductive reasoning. This may be, for example, how Stephen Hawking realized that Roger Penrose's ideas about black holes might be applicable to the creation event of our universe. Some of Albert Einstein's most brilliant, creative, and enduring insights and theories seem to have come to him while his brain was in the barely conscious "theta-wave" state where the mind is free to use more creativity. Later, of course, he would scrutinize those insights while in the laser-focused "gamma-wave" state.[7] So, while abductive reasoning should warrant additional caution, it is not necessarily irrational or anti-scientific. (As Casey Luskin explains in chapter 16, the "inference to the best explanation" used by the theory of intelligent design and other historical scientific theories is based upon abductive reasoning.)

We're not championing one type of reason over another. To avoid having every problem look like a nail when the only tool in our toolbox is a hammer, we need to keep the right tools in out mental toolbox for different jobs. Keeping in mind their strengths and limitations, we respect and benefit from the understanding given by all types of reasoning. While scientific experimentation is associated with inductive reason, there are times, especially when the option for

experimentation is not available, when scientists use deductive reasoning. Archaeologists, for example, routinely analyze the material remains left behind by ancient peoples and are unable to duplicate the scenarios in experiments, yet they can still use deductive conclusions and abductive reasoning based on prior knowledge to make sense of the puzzle pieces. Such reasoning is neither strictly inductive nor necessarily antiscientific.

And where would science be without mathematical reasoning? When cosmologists and theoretical physicists attempt to reason about the time immediately before, during, and after the moment our universe began its expansion, they're switching from scientific thinking to philosophical reasoning and abductive guesswork.[8]

There are also less rational types of reasoning at work—emotional, social, and cultural reasoning, for example. When a member of a society that holds their cultural and religious traditions tightly hears intellectually appealing reasons to become a believer in Christ, he or she may reason toward doubt, saying, "If I receive Jesus in faith, my parents and family will be shamed and I will be ostracized. Shaming my family and being ostracized is not a good thing. Therefore, my receiving Jesus is not a good thing."

Despite the somewhat messy complexities of reasoning and faith, the simple $R \rightleftarrows F$ relationship remains defensible. While we should not blindly trust any of our attempts to reason, we can rely upon them and hold reason in high esteem.

The Controversy Is Not Between Faith and Reason

The "faith versus reason" tension ultimately centers on two crucial questions: Is there a supernatural God who created us and our world? And, if so, who or what best speaks for him today? Some have been convinced by their own set of reasons into a position of faith that there is no room for God (or anything supernatural) in our world. Others have become convinced by a different set of reasons that our world can best be explained by a supernatural being who designed and created it. Both camps have their reasons and believe they are being reasonable. Both positions have invested their faith in a conclusion. Even those devoted to scientism have faith in several foundational presuppositions that cannot be proved by science.[9] The controversy, then, is really between "natural-only" and "natural-plus-supernatural" positions.

Those who believe in a God tend to reason deductively (philosophically) from effects to their ultimate cause: from codes to Coder, from creation to Creator, from the dependent to the Independent, from design to Designer, from information to Informer, from law to Legislator, from meaning to Meaner, from mind to Mind.[10] While inductive (or scientific) logic is limited to drawing conclusions based on observations that our five senses (and our technologies) can give about this natural world, and not about anything supernatural, it can produce conclusions that can be used as premises for deductive logic to use that posit God as the logical conclusion.

Those who doubt the existence of a God also use deductive reason to arrive at and defend their faith. The existence of evil, suffering, and imperfections, for example, are effects that may seem, to some people, better explained by the lack of an all-powerful, all-good God than by his presence.[11] They may also use inductive reasoning. It seems generally true to them that in the past, scientific discovery has provided superior and natural explanations for many of the mysterious phenomena in nature that were formerly (and incorrectly) ascribed to supernatural beings. Assuming that is true, future scientific discoveries will explain all the mysteries of our universe as being caused by purely natural causes. Accordingly, we should invest our hope and faith in naturalistic science rather than supernatural philosophizing.[12]

Regarding the question of who or what best speaks for God today, many who believe in God reason along these lines:

1. It is reasonable to have faith in the message of messengers who are well authenticated by God by supernatural miracles, signs, and wonders, which no mere human can perform.

2. The main prophets and apostles who wrote the Bible were amply authenticated by God with miracles, signs, wonders, and fulfilled prophecies.

3. Therefore, it is reasonable to have faith in the message of the Bible as being a collection of messages from God.

4. Conversely, because none of the other prophets of any other known religious or philosophical tradition were authenticated by signs and wonders, it is not reasonable to invest any faith in them as God's messengers.[13]

Again, the controversy is not between reason and faith. It is between reasons for one position and reasons for an opposing position. It is between faith in one position or another. The real question, then, is which faith is the most reasonable.

Jesus's View of Reason and Faith

Miraculous signs as reasons for faith. It's one thing to claim to be a messenger of God. It's quite another to provide reasons for people to believe such a great claim. So, what convinced people to invest their faith into Jesus? He certainly didn't say the words they wanted to hear. He didn't raise an army to force others to accept his prophetic supremacy at the point of a sword. While he was on Earth, several thousand Israelites became intrigued by him. But only a few came to believe in him so deeply that they would leave everything behind to follow him and risk dying for his sake. Soon after his resurrection from the dead, 5,000 Israelites would come to believe in him. Was their faith based on evidence and reason?

Much like Moses, Jesus was authenticated as God's messenger by several supernatural signs, wonders, and miracles. Jesus also made predictions about the future that came true—predictions no mere man could have made. He also spoke to people with such authority, insight, and brilliance that some recognized that his words could only be the words of God. He constantly amazed people with his works and his words. He had been doing so since the age of 12, when he was found "in the temple, sitting among the teachers, listening to them and asking them questions. And all who heard him were amazed at his understanding and his answers" (Luke 2:46-47). He lived such a blameless life that many could see he wasn't an imperfect man like Moses had been. These are some of the most persuasive reasons that have convinced many people to believe in the messiahship and deity of Jesus.[14] The fact that Jesus performed several miraculous signs indicates that he assumed the normalcy of the need for reason to lead to faith.

As people pondered the significance of Jesus's signs, their reason led them in differing directions. Some reasoned to a strong faith. Most reasoned to a weak faith. Some reasoned their way to a position of no faith. Regardless, the miraculous signs Jesus performed convinced everyone that his claims needed to be taken seriously. His amazing words and works made him very popular with the common people of Israel, who were reasonable enough to accept him as a prophet on par with Elijah, John the Baptist, or Jeremiah.[15] But most did not accept him as a prophet on par with Moses, much less the promised Messiah—the ultimate prophet, priest, and king. Some of them reasoned with faulty premises when it came to their decision of no faith.[16] But "many of the people believed in him" because they reasoned, "When the Christ appears, will he do more signs than this man has done?" (John 7:31). The latter had good reasoning and, as a result, good faith.

The religious leaders of the day could not deny that Jesus was doing great miracles. They rejected his claim to be God after reasoning that he performed his works by the power of Satan rather than by the power of God's Holy Spirit.[17] Only one of them said in secret to Jesus, "Rabbi, we know that you are a teacher come from God, for no one can do these signs that you do unless God is with him" (John 3:2). The apostle who wrote the most about belief was also the keenest on recording the reasoning behind people's beliefs. Regarding Jesus's miracle of raising Lazarus from the dead, he wrote,

> Many of the Jews therefore, who had… seen what [Jesus] did, believed in him… So the chief priests and the Pharisees gathered the council and said, "What are we to do? For this man performs

many signs. If we let him go on like this, everyone will believe in him, and the Romans will come and take away both our place and our nation." But one of them, Caiaphas, who was high priest that year, [argued], "…it is better for you that one man should die for the people, not that the whole nation should perish."…Jesus would die for the nation…So from that day on they made plans to put him to death (John 11:45-53; see also John 18:12-14).

It's not that the leaders were being totally unreasonable. The sacrifice of one troublemaker for the many is, in a sense, quite reasonable. Crucifying the Messiah after he has been authenticated by many signs isn't, however. In the New Testament, the Jewish leaders provided several examples of partially reasonable reasoning leading ultimately to unreasonable doubts.

Jesus the master logician. There may be an even more convincing reason to believe that Jesus held reason (and its role in leading to faith) in very high regard. On several occasions, Jesus's enemies—the scribes, chief priests, Sanhedrin elders, Pharisees, Sadducees, and Herodians—attempted to use reason to discredit Jesus in the court of public opinion. They built logical traps for him. Some were meant to make him look foolish among the crowds who were fascinated by his words. Others were meant to get him arrested and, if possible, executed.

Jesus, who had done most of his ministry and miracles outside of Jerusalem, rode on a colt into the city, just as the prophet Zechariah had predicted he would do (Zechariah 9:9). Everyone was in Jerusalem for the Passover festival, and they knew Jesus claimed to be the Messiah. Many cheered him on. Jesus had entered the stronghold of enemy

territory as if he were the new king of Israel. He walked into the temple, turned over the tables of the money lenders, and sat down in the temple courts to teach the people. This was the last chance for the leaders to challenge him. Unable to challenge his miracles, they challenged his reasonableness.

The representatives of the Sanhedrin, the religious rulers of Israel, confronted him and demanded that he explain who authorized him to do what he was doing. But, in a move that sprung their trap, he would only give them a straight answer if they first gave him a straight answer about the authority of John, his forerunner. For fear of the people, they refused to answer; they ended up looking foolish and were put on the defensive. Jesus then told a brilliant parable that indirectly answered their question. Using a landowner as a symbol of God and his vineyard tenants as symbols of the rulers, he said, in effect, that they were guilty of killing the prophets whom God had sent, that they were going to be guilty of killing the landowner's son (who in that culture had the full authority of the landowner), and that the landowner would respond by destroying them. Jesus escaped their trap and was still able to affirm his deity and their condemnation. They had to look for another way to arrest him.[18]

On another occasion, as they again sought to "entangle him in his words," the Pharisees and Herodians gave Jesus one question: "Is it lawful to pay taxes to Caesar, or not?" They put Jesus between the two horns of a logical dilemma. If he said no, he could be arrested by the temple guard and handed over to the Roman justice system as an insurgent. If he said yes, the people, who hated the oppressive yoke that Rome had fastened around their national neck, would despise him as a Roman sympathizer and reject him as the Messiah who was supposed to send the Romans back to Rome on stretchers and restore Israel to its former greatness. It was a no-win situation. But Jesus, the master logician, gave an answer that sprang their trap, amazed them, and silenced them.[19]

It seemed like a no-win situation. But like a graceful matador, Jesus brilliantly escaped from both horns of the dilemma they had placed him in. His response, which concluded so famously with "Render therefore to Caesar the things that are Caesar's, and to God the things that are God's" not only sprang their trap but may have turned the ambush back on them, placing them between the same two horns they had intended to impale him on. His opponents were again silenced and the crowds were amazed. It was becoming increasingly clear who the real grandmaster of the logical chessboard was.

This masterful use of reasoning by Jesus had.[20] If his opponents had triumphed in their logical attacks, people would naturally have lost faith in Jesus. But when Jesus's opponents chose the ground to fight on and ambushed him, he didn't retreat. He took their logical arguments seriously, dismantled them, tackled faulty premises, exposed fallacies, ran circles around them, gave counterarguments, and invariably left them silenced. He himself exhibited the zeniths of both reason and faith.

The Apostles' View of Reason and Faith

Jesus chose a few of his disciples to be his apostles. The mission he sent them to accomplish was to convince people to invest their faith in the Lord Jesus Christ. Throughout the book of Acts, the eyewitness history book of the first-century church, we find that in all of their attempts to get people to believe

Jesus, the apostles reasoned with unbelievers using facts, logical argumentation, and evidence.

Peter reasoning toward faith. On the day of Pentecost, the miracle of the apostles speaking in foreign languages they hadn't learned had attracted the attention of several thousand Jews. Peter based a compelling argument upon a prophecy made by King David and the fact of Jesus's resurrection from the dead, which had happened 50 days earlier. The logical conclusion was, "Let all the house of Israel therefore *know for certain* that God has made him both Lord and Christ, this Jesus whom you crucified" (Acts 2:36). So persuasive was Peter's reasoning that 3,000 Israelites placed their faith in Jesus (Acts 2:41).

Later, Peter healed a paraplegic man in the name of Jesus. To the crowd that gathered over this "notable sign…evident to all" (Acts 4:16), Peter reasoned that he was an eyewitness of the fact that the same God who had healed the man had also raised Jesus from the dead and that Jesus's death fulfilled various prophecies. The conclusion was that the listeners should receive Jesus in faith as their Messiah. The number of people who had "heard the word [and] believed" increased to 5,000 (Acts 4:4). "And with great power the apostles were giving their *testimony* to the resurrection of the Lord Jesus" (Acts 4:33). The same apostle who challenges us to "*make a defense* to anyone who asks you for a *reason* for the hope that is in you" (1 Peter 3:15) gave reasons, evidences, and testimony to bring many to faith.

Luke records, "They did not cease *teaching and preaching* that the Christ is Jesus" (Acts 5:42). There seems to be a noteworthy difference between the two. Whereas teaching is presenting facts and helping others to understand them, preaching seems to attempt to convince people that the message is reasonable, true, and worthy of acting upon. It's not that Peter and the other apostles were brilliant logicians and persuasive speakers. They weren't. But Jesus had promised to send the Holy Spirit to speak through them.[21] And just as Jesus was able to use reason to silence his opponents, the Spirit enabled the apostles to silence their opponents in debate.

Paul reasoning toward faith among the Jews. According to Acts 9, Jesus wanted Saul of Tarsus, better known by his Roman name Paul, to move from a position of doubt to one of faith. He appeared to Saul with an overwhelming glory that Saul knew could only belong to God himself. Saul asked, "Who are you, Lord?" Jesus answered, "I am Jesus, whom you are persecuting" (verse 5). Jesus was not asking Saul to make an irrational leap of faith into the dark. He was nudging him to take a reasonable step into the light. Saul's reason realized: (1) Only the Lord himself appears in the shekinah glory. (2) This person who appeared to him in the shekinah glory was Jesus. (2) Therefore, Jesus is Lord. This is an example of R→F followed quickly by F→R. This one key realization that Jesus is Lord revolutionized his faith and reasoning.

A few days later, Saul *"proclaimed* Jesus in the synagogues, saying, 'He is the Son of God,'...Saul increased all the more in strength, and *confounded* the Jews who lived in Damascus by *proving* that Jesus was the Christ" (verses 20, 22). Proclaiming, confounding, and proving are all indicative of argumentative reasoning. As a rabbi trained in the Pharisaic interpretation of the law, Paul had almost certainly been trained in a sophisticated system of rabbinic logic. It is likely that he continued to use at least some of that training.[22] But the phrase "increased in strength" suggests that the Spirit himself

may have been the one strengthening Paul's power of reason. All of this was aimed at bringing many to faith in Jesus Christ.

In the synagogue of Pisidian Antioch, Paul spoke with logical arguments based on the eyewitness testimony of the resurrection of Jesus and on the prophecies fulfilled by Jesus, and "many...followed Paul and Barnabas" (Acts 13:43). In the synagogue at Iconium, Paul and Barnabas *"spoke in such a way* that a great number of both Jews and Greeks *believed*...speaking boldly for the Lord who bore *witness* to the word of his grace, granting *signs and wonders* to be done by their hands" (Acts 14:1, 3).

In Thessalonica, "Paul went [into the synagogue], as was his custom, and on three Sabbath days he *reasoned* with them from the Scriptures, *explaining and proving* that it was necessary for the Christ to suffer and to rise from the dead, and saying, 'This Jesus, whom I proclaim [preach] to you, is the Christ.' And some of them were *persuaded*..." (Acts 17:1-4). In Berea, "the word of God was *proclaimed* by Paul" (verse 13) and "they received the word with all eagerness, *examining* the Scriptures daily to see if these things were so. Many of them therefore believed" (verses 11-12).

In Corinth, Paul *"reasoned* in the synagogue every Sabbath, and *tried to persuade* Jews and Greeks...*persuading* people to worship God" and "many of the Corinthians hearing Paul believed" (Acts 18:4, 13, 8). The Lord even told Paul to "go on speaking, and do not be silent...for I have many in this city who are my people" (verse 10), indicating that bringing people to faith in God necessarily involves speaking persuasively to them. In Ephesus, Paul "went into the synagogue and *reasoned* with the Jews" (verse 19) and "did not shrink from declaring...teaching...*testifying* both to Jews and Greeks of repentance toward God and faith in our Lord Jesus Christ" (Acts 20:20-21). The ministry he received from the Lord was "to *testify* to the gospel of the grace of God" (Acts 20:24). Similarly, Apollos "spoke boldly in the synagogue…he powerfully *refuted* the Jews in public, *showing* by the Scriptures that the Christ was Jesus" (Acts 18:26, 28).

Apparently, Paul and his fellow preachers were unaware of the misguided maxim "You can't reason anyone into the kingdom."

Paul reasoning toward faith among pagan Greeks. While it is true that Paul preferred to reason the case for Christ in the synagogues among Jews and "God-fearing Greeks" who accepted the authority of the Hebrew Scriptures, he never hesitated to use reason with pagans who had no knowledge of the Scriptures. In Lystra, a sign convinced the pagan Greeks that there was something special about Paul, who reasoned with them about God. He gave them a logical argument from the effect to cause: "a living God, who made the heaven and the Earth and the sea and all that is in them…he did not leave himself without *a witness*, for he did good by giving you rains from heaven and fruitful seasons, satisfying your hearts with food and gladness" (Acts 14:15, 17). This argument was aimed toward establishing faith in an intelligent and benevolent God from purposeful designs in the world.[23] This also serves as proof that God has left "witnesses" (reminders that are difficult to ignore) in nature to point our reason, and subsequently our faith, to him.

In Athens, Paul "*reasoned* in the synagogue with the Jews…and in the marketplace every day…he was preaching Jesus and the resurrection" (Acts 17:17-18). Speaking to the Greeks, Paul reasoned in a way to move them from faith in their finite gods and idols to an infinite creator God. He argued that their gods were too small. He further stated

that the cause of the finite world we live in is an infinite God who doesn't fit in our world (verse 24), that living things need a living source but he does not need them (verse 25), and that rational beings like us need a cause that is also rational (verse 29). He reasoned that God has "fixed a day on which he will judge the world…by a man whom he has appointed [Jesus]; and of this he has *given assurance* to all by raising him from the dead" (verse 31). Some mocked him; others believed.

Reasoning among early Christian leaders. The apostles did not reserve reason only for nonbelievers. They held reason in high esteem as they tried to refine their own faith (R⇄F). The first church council in Jerusalem proves that thoughtful reasoning was also the preferred way of deciding which ideas were most worthy of investing faith in. The apostles, elders, and other members of the Jerusalem church met to deliberate over the first big problem that the churches faced. They met to "*consider* this matter" (Acts 15:6), and "after there had been much *debate*" (verse 7), Peter spoke very persuasively, offering, among other things, evidence from facts he and they had observed (verses 7-11). After Peter spoke, "all the assembly fell silent" (verse 12), suggesting that progress was made toward faith in a shared conclusion.

Next, Barnabas and Paul spoke and offered additional evidences they had observed (verse 12). Seeing that the argument was mostly settled, James then spoke, agreeing with Peter (verse 14), adding an argument of his own from a prophetic book they all agreed with (verses 16-17), and offering his final judgment, which was also based on the sojourner laws of Leviticus 17–20 (verses 19-21). This was an exercise of reasoning based on the logical interpretation of facts taken from multiple eyewitnesses as well as upon the logical interpretation of the Scriptures they had faith in. "Then it seemed good to the apostles and the elders, with the whole church," to act in accordance with the communal reasoning of Peter, Barnabas, Paul, and James (verse 22).

Weighing the Evidence: Faith and Reason Together

Reason naturally leads us toward a position of strong faith, weak faith, or no faith in a given idea. When our reason is faithful and when faith is reasonable, faith and reason are our best companions for life's journey. Exercising higher reason is what sets us apart from the animals, and, in a sense, allows us to be more like the God, who bequeathed us with the ability to reason.

The conflict is not between people of faith and people of reason. Everyone has faith.[24] And everyone uses reason. Some reasons are better than others. Some positions of faith are more reasonable than others. Some people are more reasonable than others. The contributors to this volume believe that after weighing the evidence, it is most reasonable to believe that we and our world were created by God, who has not been silent. He has communicated in an indirect way to us by placing us on a special planet in a part of a galaxy that allows us to see his creation. We are spectators in an amazing art gallery. In this time and place, we are left to reason in awe from artwork to Artist. The creation of our universe is the greatest miraculous sign for our reason to ponder. The creation of life on Earth is the second. The creation of the mysterious human mind, which can reason and believe, is arguably the third, though there are others. The sciences are bringing these miracles more and more into focus.

God also spoke intelligibly through a few messengers whom he validated through miraculous signs and fulfilled prophecies. This God—the God of the Bible—is not unreasonable. When he instructed Moses to confront the Pharaoh and lead the Hebrew slaves out to Canaan, he performed miraculous signs through him so that the Hebrews, the Egyptians, and Pharaoh could reason their way to faith. Moses then wrote down all that happened so it would serve as a universal witness (Joshua 4:23-24).

When Jesus of Nazareth presented himself to Israel as their Messiah (the ultimate king, priest, and prophet), he performed miraculous signs to back his claim. He too understood well that reasons and reasoning are necessary for people to change their minds from doubt to faith. Yet miracles were almost never enough. He almost always had to give other reasons for faith as well. He proved to be the master logician who took the winning of public battles of reason quite seriously. As a result, many came to faith. Some of the eyewitnesses who believed wrote accounts for the rest of us to read. And they wrote in a language that could be read across the Greco-Roman world. As a result, perhaps as many as 2.2 billion people on Earth (about 33 percent of the world's population) identify as Christian today.

Jesus's first messengers were witnesses who argued the case for him. In Acts, there are 14 clear instances of early Christians reasoning with non-Christians, urging them to come to faith in Jesus as the Christ. This appears to be the overwhelming pattern in Acts. While Acts does mention the work of the Spirit on the heart of the hearer (Acts 16:14), the focus is on the need to speak persuasively—preferably with words given by the Holy Spirit. The apostles and their coworkers preached in logically persuasive ways, and many people came to faith.

Certainly, Jesus recognized the *distinction* between faith and reason, but he did not radically *separate* them into two warring camps. The former occupies the realm of subjectivity while addressing the will (volition) of mankind, and the latter occupies the rational and objective domain, addressing the mind. Jesus never bypassed the mind to reach the heart; instead, he appealed to faith and reason as he justified his claims.

Has Science Refuted
Miracles and the Supernatural?

Richard G. Howe

Must a scientifically minded person reject miracles and the supernatural? The short and unequivocal answer is no. Yet there are some who insist that science has refuted miracles and the supernatural. Before we can show why they are wrong, it is essential to understand more about the nature of the question itself.

The Nature of the Question

It would be futile to try to referee a dispute between a mathematician and a historian as to whether Abraham Lincoln had a beard or referee a dispute between those two on the one hand and a political scientist on the other as to whether Lincoln was politically justified in suspending *habeas corpus*. Just as attempting to settle disputes such as these would be futile, we will see why a scientist as scientist is ill-equipped to make a judgment about the viability of any miracle or the reality of the supernatural.

The problem with the dispute between the mathematician and the historian is that the question of whether Abraham Lincoln had a beard is a historical question, not a mathematical one. Thus, the mathematician as mathematician is ill-equipped to weigh in on the issue. The same principle applies to the dispute over Lincoln and *habeas corpus*. This dispute has to do with political science. Therefore, both the mathematician and the historian are ill-equipped to address the question. Likewise, much of what goes into examining the case for the supernatural and miracles is philosophical (whatever that means). The scientist as scientist is likewise ill-equipped to adjudicate any debates over the supernatural or miracles.

It is hard to overestimate the significance of the role science plays in so much of the world. This rise to prominence can be tracked through history. Much has been written about the intellectual and cultural influences that have contributed to making the natural sciences so robust. While there is no doubt that fundamental elements of scientific thinking are traceable to the ancient Greeks, there also is no denying the explosive advances that the natural sciences began to make from the seventeenth century onward.[1]

If all we cared about were the benefits and amenities that science has brought into our lives, we likely would not need volumes such as this one to explore science's relationship to faith. But with the acceleration of scientific knowledge, there also began an increasingly hostile posture against the cherished beliefs of religion in general and Christianity in particular. Certain critics and skeptics of religion celebrate this trend, pointing to it as confirmation that the defining doctrines of traditional religion are not true. The more science tells us about reality (the skeptics' argument goes), the more one should realize that religion has been wrong all along.

For the skeptics, however, religion being false is not the worst of it. They claim its proliferation is extremely dangerous.[2] For our purposes, we will keep these two aspects separate and focus only on the question of the truthfulness of the supernatural and miracles. If it is true that the supernatural is real and that miracles have occurred, then it remains true no matter how dangerous an unbeliever may think those truths are. Surely no scientist will deny the truths of radiation discovered by Marie Curie despite the protests some may have about its dangerous application in nuclear weapons. This does not suggest that a discussion about the propriety of nuclear weapons is irrelevant. However, while a discussion regarding the propriety of nuclear weapons might be appropriate, it has nothing to do with the truth about radiation.

Certainly there are dangerous aspects to some of the world's religions. This says nothing in principle, however, about the truthfulness or falsity of the claims of those religions. There either is or is not a supernatural realm. It is either true or not true that miracles have occurred.

Historical Science Versus Experimental Science

An important distinction to make when it comes to understanding science is that between historical science and experimental science (or origin science and operational science, as others have called them).[3] The distinction seeks to identify, on the one hand, the principles and procedures that govern the study of the regularities (operation) of the natural world that are observable and repeatable, and thus susceptible to experimental/empirical confirmation and disconfirmation versus, on the other hand, the principles and procedures that govern singularities and how natural features originated and came into being historically (i.e., one-time events). Because such events are not, by definition, repeatable, their study in science falls more along the lines of extrapolating from current known principles and procedures determined by experimental/operational science.

For example, because events like the origin of the universe and the origin of life happened only in the past, these events are not directly observable and repeatable in a laboratory. Historical scientists investigate past events via the principle of uniformitarianism (or uniformity), which means they observe causes at work in the present day and then apply those observed causes to explain the historical record. Perhaps one can begin to see an implicit temporal element. Indeed, apologists Norman Geisler and Kerby Anderson flesh out the issue, resulting in four possible combinations: present repeatable, present singularities, past repeatable, and past singularities.[4]

What is at stake here is whether science is capable of, or allowed to weigh in on, subjects such as the origin of the universe and the origin of life. Because these events clearly are not repeatable in the laboratory, applying the distinction too strictly would seem to entail that such questions are out of bounds for the scientist.[5] But rather than come to that conclusion, scientists go more often in one of two opposing directions. Some claim these subjects are exclusively the purview of science in the form of historical science.[6] Others recognize that historical science can investigate these questions but that theology and philosophy are also crucial aspects of such historical investigations.[7] The latter position is the one espoused in this chapter.

Where does this take us with respect to the topic at hand? The historical scientist can take his understanding of how the physical world presently operates and seek to extrapolate back to suggest explanations of past events, such as the origin of the universe or the origin of life. In doing so, however, the scientist must make a crucial admission. The principles and procedures of experimental/operation science cannot account for the foundations of those very principles and procedures. Whatever it is (if anything) that brought about these principles (if, indeed, the principles were brought about), it cannot itself be explained by those very principles without committing the logical fallacy of circular reasoning. We will expand upon this point when we deal with how science compares to philosophy, especially with regard to the supernatural.

Some Definitions

Supernatural Versus Natural

Some use the term *supernatural* to refer to anything that has to do with unusual or mysterious objects or events, especially if that object or event is thought to relate to the spiritual realm. The term is even used to describe a genre of movies whose subject matter involves ghosts, demons, and the like.

Anything having to do with the occult is usually considered supernatural.[8]

For our present purposes, however, *supernatural* is typically defined as having to do with the existence and activities of angels, demons, or other nonphysical sentient beings. For the moment we will define supernatural in this spirit: as "things that exist outside the universe in a spiritual realm and are not subject to natural laws."

There is also a more technical sense of the term *supernatural*. As a philosopher and an apologist, I regret that the standard, popular use of the term has become so watered down that the importance of the technical sense has become obscured. Strictly speaking, *supernatural* means "beyond the natural." But even the term *natural* can be ambiguous. Sometimes the word refers to a location—i.e., existing within the physical universe. Other times we use *natural* to refer to what often or usually happens. It is *natural* for you to be tired after strenuous activity. Here, it is a synonym for *normal*. In this sense, it might not be normal or natural for one to be tired after mild activity.

Under these definitions, *natural* can be contrasted with either *artificial* or *supernatural*. In one sense, a stalactite is a natural object, whereas pottery is not. Here, *natural* is contrasted with *artificial*, where the difference has largely to do with an origin stemming from strictly blind laws of nature in contrast to an artificial origin—i.e., one resulting from intelligent causes and designers. In another sense, a piece of pottery is entirely natural and not supernatural—i.e., it exists in the physical universe and is subject to natural laws.

This distinction is especially relevant to apologetics. Here, for example, is where critics often misunderstand intelligent design arguments. When a scientist insists that the information content of a DNA molecule cannot be accounted for by natural causes, critics accuse the scientist of illicitly sneaking God into the argument and appealing to the supernatural. The critic then launches into the accusation that the scientist is no longer doing science but instead, is doing religion. The critic fails to appreciate the different meanings of *natural* and the distinction between *natural* versus *artificial*, or *natural* versus *supernatural*. As Table 1 explains, something can be intelligently designed and yet be natural in the sense of being within the universe and subject to natural laws—but that same thing might not necessarily relate to God or the supernatural.

Most aspects of the DNA molecule can be understood along the contours of the laws of nature. DNA itself is a natural structure in the sense that it exists within the universe and is subject to the laws of nature, but much of the information contained in DNA is not itself accounted for by those same laws—i.e., it does not originate from blind, natural mechanisms. As explained in Table 1, inferring intelligent design does not, strictly speaking, require appealing to the supernatural, although supernatural intelligent design is certainly a possibility. Similarly, one can understand all the workings of an automobile by the laws of nature. But these laws, by themselves, cannot explain the automobile itself. That can only be explained as the result of an automaker. The toner on a photocopy is subject to the laws of nature governing the mechanics and chemistry of the photocopy machine and the chemistry and physics of the paper and toner. But the information on the photocopy cannot be accounted for by those laws. That information is the result of an intellect.

How, then, should we understand that which is supernatural? As a philosopher and apologist, I recognize the popular or standard use of the term that we have used up to this point in this chapter. But it is also important to preserve a technical definition of the term *supernatural* to mean "beyond or transcendent to the created realm." In this respect, only God is supernatural. Under this definition, demons and angels, as created beings, are not supernatural inasmuch as they are no less created than mundane things like trees and rocks; thus, they are more accurately described as supernormal. The importance of sustaining this meaning will matter when we talk about miracles. Given, however, that so many use the term *supernatural* as a general reference to the spiritual realm, we will deal with both understandings of the term as we answer the question, "Has science refuted the supernatural?"

Natural vs. Supernatural	
Natural = within creation; i.e., not God; e.g., rocks, trees, animals, humans, angels, demons	Supernatural = beyond creation; i.e., God
Natural vs. Artificial	
Natural = according to the laws of nature; e.g., stalactites, erosion, crystals	Artificial = caused by intelligence; i.e., contrived, designed; e.g., arrowheads, DNA, writing

Table 1. Two uses of the term *natural*.

Miracle

We can appreciate what people mean when they talk about the miracle of falling in love or of childbirth or of a beautiful sunset. Things or events that evoke awe or wonder are often called *miracles*. But in the context of apologetics, we should insist upon a precise definition. Simply put, a miracle is an act of God whereby he suspends the normal laws of nature for the purpose of vindicating a messenger.[9] Norman Geisler said, "In brief, a miracle is a divine intervention into the natural world. It is a supernatural exception to the regular course of the world that would not have occurred otherwise."[10] Richard Purtill adds an important extra element in his definition: "an event in which God temporarily makes an exception to the natural order of things, *to show that God is acting.*"[11] This "showing" has everything to do with God vindicating his messenger.[12]

The key element to note here is that because a miracle is, by definition, an act of God, then, at least as a matter of principle, the question of God's existence must be settled before the possibility of miracles can be addressed.[13] If an atheist is consistent with his atheism, then he could never acknowledge an event as a miracle no matter how otherwise inexplicable it is.[14] What is more, in defending the truth of Christianity, the proper definition of *miracle* is how we are to answer the common objection that because so many other religions appeal to miracles, Christianity is not unique. Such an appeal to miracles, the critic argues, cannot distinguish Christianity from the world's religions and, thus, does not confirm Christianity's truth.[15]

Science Versus Philosophy

What are we to make of the suggestion that there are aspects of reality that lie beyond the purview of the natural sciences? Such a suggestion likely will worry only certain scientists. But these would not be average scientists. Stephen Hawking and Leonard Mlodinow are quite direct on this point:

How can we understand the world in which we find ourselves? How does the universe behave? What is the nature of reality? Where did all this come from? Did the universe need a creator?…Traditionally these are questions for philosophy, but *philosophy is dead*. Philosophy has not kept up with modern developments in science, particularly physics.[16]

While it might be understandable that a scientist would balk at the idea that there are aspects of reality that lie beyond the purview of science, I submit that any scientist who denies this is tragically failing to "see" what is staring him right in the face.

During a debate on the existence of God at a state university, I stressed that the question of God's existence was decidedly a philosophical one. In the question-and-answer segment, I was taken aback when asked who I was to make such a pronouncement. This was like asking me "Who are you to say that a question about a plant is a botanical question?" The fact that question is about plants makes it a botanical question. The fact that the question of God's existence involves, among other things, metaphysics, makes it a philosophical question. The only way to deny botanical questions is to deny the reality of plants. The only way to deny philosophical questions would be to deny the reality of areas of study like metaphysics (which asks, "What is real?"); epistemology (which asks, "How do we know what is real?"); ethics (which asks, "What is right action?"); logic (which asks, "What is right reason?"); and more.

We certainly can challenge each other's specific answers to these questions. But it is absurd to think that the answers are something that only science can provide. Indeed, the very dispute as to whether science can answer such questions is itself a philosophical question.[17] It remains, therefore, to see how we are to respond to the challenges that science thinks it poses for the reality of miracles and the supernatural.

Specific Challenges from Science

Using the technical sense of the term *supernatural*, angels and demons are not supernatural because they are part of God's creation. Nothing except God is supernatural. But this is not how many people use the term. Perhaps one argument for the popular definition is that surely demonic and angelic activity are not natural.[18] Waters that are troubled by the wind are certainly different than waters that are troubled by an angel, as in John 5:4.[19] Regardless of what term one uses, the question remains: "Has science refuted the supernatural?" To address this question, we must consider both senses of the term *supernatural*. We will begin with the technical sense because this answer will have a direct bearing upon several other answers elsewhere.

Has Science Refuted the Supernatural in the Technical Sense?

The question about whether there is a supernatural is the same as the question about whether God exists. Except for antecedent foundational issues that are necessary for intellectual discourse in the first place, the question of God's existence precedes virtually every other apologetic question.[20] This will especially be true for our questions about miracles. Given the definition of miracles, it is manifest that if there is no God who is transcendent over creation, then miracles are impossible. What is needed, then, are demonstrations that God exists.[21]

Piles of books have been written about God's existence. Here we touch on only the barest elements of the arguments. The arguments or demonstrations for God's existence appeal to both philosophy and science. Those based primarily on science are more prominent in contemporary apologetics.[22] They generally take the form of "argument to the best hypothesis." These types of arguments appeal to features or sets of data about the physical universe—usually that are themselves not in dispute—and try to show that the "hypothesis" of God, or at least intelligent agency, is the best explanation for those features or sets of data.

Some of these arguments marshal the contemporary scientific evidence that the universe began to exist a finite time ago.[23] Because the universe came into existence, the argument goes, there must have been a cause that is itself not part of the universe. Other arguments seek to show that God is the most reasonable explanation for various features of the universe, including the initial physical conditions that made biological life possible,[24] the origin of biological life,[25] and the information content of DNA.[26]

Several challenges present themselves in making the scientifically based arguments. While the arguments try to focus on features or sets of data that are not themselves in dispute, this is not always possible. In those cases, one would need to be trained in the relevant science in order to carry out the argument. In addition, some critics will counter that the conclusion of these arguments is not necessarily God in the traditional sense of the term.[27] Indeed, proponents of intelligent design readily acknowledge that as a science, ID theory only infers an intelligent designing agent and does not specify the identity of the designer.[28] Last, it might not be obvious to the critic why the supposed cause of these various features or sets of data is immune to needing an explanation. This is the "Who designed the designer?" objection.[29] While many of the scientifically based arguments have an initial strength inasmuch as they resonate with many who already have a familiarity with certain basic contemporary scientific views (e.g., the universe began; biological life exhibits amazing complexity), additional arguments beyond intelligent design are needed to show that the designer is God, and many second-order objections may need to be addressed as well. Such objections can be answered, but doing this requires some understanding of the basics of intelligent design arguments—a problem this volume aims to help rectify.

Has Science Refuted the Supernatural in the Popular Sense?

The arguments for the supernatural in the technical sense (God) are relatively direct (even if not always easy). They work from effect to cause, which is to say they take some fact about creation and argue that God is the cause, reason, explanation, or best hypothesis for that fact.

Arguments for the existence of the supernatural in the popular sense are more indirect. Here, one needs arguments or reasons to believe in the realm of angels and demons. Those arguments and reasons show why it is that science has not refuted the supernatural in the popular sense.

Here is a relatively straightforward approach: We can know that angels and demons exist because the Bible teaches us they exist.[30] Once we know God exists, we can make our case that this God—the only true God—has revealed himself to humans in history. God revealed himself to Abraham, Moses, and the ancient Hebrew prophets.[31] Through these prophets, God revealed truths about himself and his will that could not be known merely from his creation.

Eventually, God took on the form of a man in Jesus Christ. Jesus established his church and taught his church further truths through his chosen apostles. All these together constitute what Christians now know as the Bible. These points would themselves have to be demonstrated, but once the case is made, it is manifest that angels and demons are real.

Further, because of the narrow limits of the methods, tools, and protocols of science, it is incapable of weighing in on the subject. By analogy, the fact a metal detector cannot notice a seashell (composed as they are of calcium carbonate) is not an argument against the existence of seashells. Precisely because science as science has aligned itself with certain assumptions about what the nature of things must be, science cannot treat things that fall outside its methods. This point harkens back to the initial illustration about a mathematician's ability to render a relevant opinion about whether Abraham Lincoln had a beard. Such a question is outside the abilities of mathematics to answer. The same is true about science and the spiritual realm. Science is ill-equipped to make any judgments about the truths of the spiritual realm.

The Supernatural Is, by Its Very Nature, Unknowable

If the physical world and its properties are all that is, then it would certainly be true that the supernatural is unknowable, for it would not be real. Even if it was real, constituting some aspect of reality beyond the physical (natural) world, it would still be unknowable within the limits of science. Still, Christians can celebrate that some scientists repudiate the increasing relativism of today. More often than not, scientists—atheist or Christian—are committed to the idea that reality is, at least to an important extent, objectively knowable.

Richard Dawkins is no friend of postmodernism's relativism. He once observed, "Unlike some of his theological colleagues, Bishop Montefiore is not afraid to state that the question of whether God exists is a definite question of fact."[32] We would certainly agree with both Bishop Montefiore and Richard Dawkins. To say that God's existence is a question of fact means that it is either true that God exists or it is false that God exists. It is not possible to say that God objectively exists for one person and God objectively does not exist for another person. People may disagree about tastes—for example, one person may believe asparagus tastes good and another may not. The claim that asparagus tastes good can be true for one person and not true for another. But this cannot be the case regarding objective facts of reality. Either atheism is true or it is not. Either God exists, or he does not. So far, so good.

But compare this to how Dawkins makes the same claim in a later writing: "The presence or absence of a creative super-intelligence [God] is unequivocally a scientific question, even if it is not in practice—or not yet—a decided one."[33] It is important to note that Dawkins did not change his position. He means the same thing in both statements. But this latter statement reveals the underlying problem with Dawkins's view of how we can know objective facts about reality. The question of God

is a definite question of fact. But whether that fact is a *scientific fact* is a matter of dispute. I would argue that it is not at all a scientific fact. Without any argument or justification, Dawkins imperializes over the nature of the question itself by insisting that the principles, methods, tools, and protocols of science are sufficient to discover the answer to the question of God's existence.[34] They most certainly are not. While there may be some overlap, science is woefully ill-equipped to address, much less answer, philosophical questions.[35]

Miracles Violate the Laws of Thermodynamics

Many of us perhaps have grown up occasionally citing one or more of the fundamental laws describing thermodynamics, even if not fully appreciating their meaning. These laws describe the relationships within a system of its energy, temperature, entropy, and more. The pioneer of thermodynamics, Rudolf Clausius (1822–1888), defined thermodynamics this way: "We can express the fundamental laws of the universe which correspond to the two fundamental laws of the mechanical theory of heat in the following simple form: 1. The energy of the universe is constant. 2. The entropy of the universe tends toward a maximum."[36] Stated informally, the first law maintains that energy can be neither created nor destroyed.[37]

The informal way of stating the first law unfortunately uses the term *created*. The mistake some people make in thinking that this fundamental law of science precludes the possibility of miracles arises from their failure to understand that the law is a statement about the physical universe. As such, it not only has nothing to do with the Christian doctrine of miracles but is entirely consistent with that doctrine. Christians certainly agree that there is nothing within the physical universe that has the power or ability to create energy *ex nihilo*. But God is not a member of the physical universe. Rather, God is transcendent to the universe. The universe has existence. God *is* existence itself—*ipsum esse subsistence*: substantial existence itself. God is not merely one more thing in the universe whose existence we can debate. For the atheist to say God does not exist and miracles are not possible is as absurd as saying there is no such thing as existence at all.

Miracles Stand Unrefuted by Science

Certain facts about the universe (about which science can tell us much) are facts that point to the existence of the Creator. Further, certain facts of metaphysics demonstrate the existence and attributes of the one true God. This God has revealed himself not only through "the things that have been made" (Romans 1:20), but also through his prophets, apostles, and the Lord Jesus Christ. We now have his Bible. This Bible tells us the truth about the spirit realm of angels and demons.

A proper understanding of miracles requires the existence of God. The significance of this is key to understanding why scientific objections against miracles completely miss the mark. The existence of God—that is, the reality of the supernatural—has been demonstrated. God is the creator and sustainer of all else that exists, including the very laws of science. Far from science being able to refute miracles and the supernatural, the laws of science can be part of the demonstration of the God whose existence makes the possibility of miracles undeniable. All that remains would be to show, in a full defense of miracles, that miracles have occurred.[38] Miracles and the supernatural stand unrefuted by science.

Is Christianity at War with Science?

Michael N. Keas

The illusion that Christianity has been at war with science has been created and sustained by many specific myths about the history of science. In this chapter, I will refute the following myths:

1. *The Dark Myth:* Christianity produced 1,000 years of antiscience "Dark Ages."

2. *The Flat Myth:* Church-induced ignorance caused European intellectuals to believe in a flat Earth.

3. *The Big Myth:* A big universe is a problem for Christianity.

4. *The Demotion Myth:* Copernicus demoted us from the cosmic center, and thereby destroyed confidence in a divine plan for humanity.

5. *The Skeptic Myth:* The main heroes of early modern science were skeptics, not believers in God.

The Dark Myth: Christianity Produced 1,000 Years of Antiscience "Dark Ages"

Atheist biologist Jerry Coyne wrote, "Had there been no Christianity, if after the fall of Rome atheism had pervaded the Western world, science would have developed earlier and be far more advanced than it is now."[1] Did Christianity really drag the West into an antiscientific Dark Ages, a period said to stretch from the fall of Rome to AD 1450? This and other anti-Christian myths crash and burn against the facts of history.

Early Medieval Light: 400–1100

The great Church Father Saint Augustine (354–430) laid some of the foundations for science. He contributed to Aristotelian physics in his *Literal Commentary on Genesis*.[2] More broadly, Augustine expressed confidence in our ability to read the "book of nature" because it is the "production of the Creator."[3] He insisted we should proceed "by most certain reasoning or experience" to discern the most likely way God established "the natures of things," a phrase that became a popular medieval book title for works emulating Augustine's investigative approach.[4]

The English monk Bede (673–735) studied and wrote about astronomy in the tradition of Augustine and Ptolemy. Historian Bruce Eastwood called Bede's book *The Nature of Things* (ca. 701) "a model for a purely physical description of the results of divine creation, devoid of allegorical interpretation, and using the accumulated teachings of the past, both Christian and pagan."[5]

Note how Bede's Christian worldview was compatible with analysis of the natural world as a coherent system of natural causes and effects.

The Light of the High Middle Ages: 1100–1450

Around 1100, European intellectuals graduated from limited translations and commentaries on Aristotle to a more extensive recovery and further development of Aristotelian logic. As refined within a Christian worldview, this advance included a reasoning method well suited to natural science.

Scholars called this form of argument *ratio* (reason), contrasting it with mathematical demonstration. Mathematics begins with first principles thought to be certain and deduces conclusions that carry the same certainty. Ratio, in contrast, uses premises inferred as likely true from sensory experience, and then reasons from there to probable conclusions.[6]

Ratio, a logic appropriate to observational science, enriched the study of motion and change in the natural world. Historian Walter Laird writes, "The study of motion in the Middle Ages, then, was not a slavish and sterile commentary on the words of Aristotle...Part of the measure of their success...is that some of these insights and results had to be rediscovered later by Galileo and others in the course of the Scientific Revolution."[7]

The institution in which most scholars investigated natural motion is also noteworthy—the university. This Christian invention began with the University of Bologna in 1088, followed by Paris and Oxford before 1200 and more than 50 others by 1450. The

papacy supported this unprecedented intellectual explosion.

Universities provided additional stimulus to the medieval translation movement already underway, in which Greek and Arabic texts were rendered in the common European intellectual tongue of Latin. This movement greatly outperformed the comparative trickle of imperial Roman translations. If European Christians had been closed-minded to the earlier work of pagans, as the Dark Ages myth alleges, then what explains this ferocious appetite for translations?

The Franciscan cleric and university scholar Roger Bacon (ca. 1220–1292) read much of the newly translated work of earlier Greek and Islamic investigators, including Euclid, Ptolemy, and Ibn al-Haytham, or Alhazen (ca. 965–1040). By evaluating them and introducing some controlled observations—what we now call experiments—Bacon substantially advanced the science of light.[8]

Subsequent authors summarized and reevaluated Bacon's work, transmitting it through books used in university instruction. That is how it came to the attention of Johannes Kepler (1571–1630), whose account "helped spur the shift in analytic focus that eventually led to modern optics," in the words of historian A. Mark Smith.[9]

By one estimate, 30 percent of the medieval university liberal arts curriculum addressed roughly what we call science (including mathematics).[10] Between 1200 and 1450, hundreds of thousands of university students studied Greco-Arabic-Latin science, medicine, and mathematics—as progressively digested and improved by generations of European university faculty.

Contrary to the Dark Ages myth, medieval European Christians cultivated the idea of "laws of nature," a logic friendly to science, including the science of motion, human dissection, vision-light theories, mathematical analysis of nature, and the superiority of reason and observational experience (sometimes even experiment) over authority in the task of explaining nature.

Medieval trailblazers also invented self-governing universities, eyeglasses, towering cathedrals with stained glass, and much more. Although labeling *any* age with a single descriptor is problematic, the so-called Dark Ages would be far better labeled an Age of Illumination or even an Age of Reason.

The Flat Myth: Church-Induced Ignorance Caused European Intellectuals to Believe in a Flat Earth

In November 2018, I spoke in Denver on myths about science and religion. I did not realize that a few weeks prior, the same city had hosted the Flat Earth International Conference. Many of the 650 people who attended the wacky flat-Earth event told a reporter something notable. They said they had "been kicked out of churches, or lost jobs with churches, or suffered broken relationships with family members" because of their belief in a flat Earth.[11] Ignorance has consequences, apparently.

Celebrity astrophysicist Neil deGrasse Tyson is a smart man. However, he is evidently ignorant about at least one aspect of the early history of his own scientific discipline. Back in 2016, he responded to rapper B.o.B, a flat-Earth promoter, with a tweet. Tyson wrote, "Duude—to be clear: Being five centuries regressed in your reasoning doesn't mean we all can't still like your music." Tyson follower Andy Teal responded, "Five centuries? I believe the knowledge of Earth's shape goes back a bit farther than that."

Tyson tweeted back: "Yes. Ancient Greece inferred from Earth's shadow during Lunar Eclipses. But it was lost to the Dark Ages."[12]

Church-Induced Ignorance?

People stopped believing in a spherical Earth during the Middle Ages? Not really. Medieval intellectuals had many reasons for grasping that the Earth is round. Those reasons included the curved shadow of the Earth projected on the moon during a lunar eclipse. To deny medieval belief in a round Earth is to be guilty of what I call the *flat myth*. This is the most enduring component of the larger myth of the Dark Ages. The allegedly anti-science Christian Dark Ages never happened, as I demonstrated above.

Tyson is correct about how ridiculous contemporary flat-Earth belief is. Some "believers" such as Shaquille O'Neal and Kyrie Irving of NBA fame have said they were only joking. And who can tell what the small number of people behind today's flat-Earth societies *actually* think? If many of them are joking, it would come as little surprise.

But the fact is that Tyson, probably one of the world's most influential voices for science, spread misinformation about medieval views.

Tyson's misperception has a history. They trace back to writers in the 1800s. For example, the nineteenth-century chemist-historian John William Draper claimed that medieval Christians believed "the Scriptures contain the sum of all knowledge." They therefore "discouraged any investigation of Nature," including the study of the Earth's shape.[13] Supposedly, this ignorance continued until the time of Columbus.

Consider the 1,200 American college students I have taught astronomy to over the past quarter century. The vast majority learned something false from their precollege teachers. They were told that in the Middle Ages, Europeans were ignorant of the Earth's roundness until Christopher Columbus proved it in 1492. Only some of my students had previously detected the typical anti-Christian slant to the story. After I did my job, they all understood how this fake history perpetuates the myth of warfare between science and Christianity.

Medieval Round-Earth Arguments

Imagine the year is 1300. You are a student at the University of Salamanca, Spain's oldest university. In class, you have studied Aristotle's argument for a spherical Earth based on the changing positions of the stars as one travels north or south. This was standard in the medieval curriculum. You wish to demonstrate it for yourself. How will you do this?

First you note that the apparently motionless North Star is located about 40 degrees above your horizon in Salamanca. Then you travel to the southernmost point of Europe. There, you find that this star appears only about 35 degrees above the horizon. Why the change of angle?

Almost every medieval university student learned a simple explanation: the Earth is round. This and other reasonable arguments were combined to present a very strong case.

Back Around to Today

Here's a surprise: My students have typically been *less* able to defend the Earth's roundness by such scientific arguments than the average medieval student. Upon completing my astronomy course, they finally caught up to the Middle Ages!

Why couldn't my students articulate these arguments? For one, the roundness of the Earth has been well-known for centuries—since the Middle Ages and even earlier—and

as such, most modern science curricula see no reason to inform students about the arguments of this long-settled debate. Thus, today, most students accept the roundness of Earth as a mere fact without knowing the reasons. Unfortunately, they also frequently lack the ability to reason from observations to this conclusion. This is because science today is typically taught this way: as something to be accepted (for example, contemporary Darwinism), not understood through arguments for and against particular theories. That's a loss for students. Things were, in this respect, brighter in the so-called Dark Ages.

We have debunked the anti-Christian myth about Earth's *shape*. Let's examine a popular myth about its *size*—its utter smallness in relation to a really big universe.

The Big Myth: A Big Universe Is a Problem for Christianity

Self-appointed spokesmen for science often use the enormous size of the cosmos, with its billions of galaxies, as a club to beat up Christianity. They say people in the Western tradition had to wait till modern science to grasp that the universe was huge, and had to shed their historic Judeo-Christian views to do so. Not true.

Bill Nye, the Scientism Guy

Prominent scientists from centuries past, including Nicolaus Copernicus (1473–1543) and Blaise Pascal (1623–1662), recognized that the universe is vast. They saw in this no contradiction with their Christian beliefs. Yet celebrity TV science educator Bill Nye, the "Science Guy," is among those who suggest that the sheer scale of the cosmos means humans are insignificant. In the last minutes of his 2010 Humanist of the Year acceptance speech, Nye—speaking for science and all humanity—delighted the American Humanist Association with this:

> I'm insignificant…I am just another speck of sand. And the Earth really in the cosmic scheme of things is another speck. And the sun an unremarkable star…And the galaxy is a speck. I'm a speck on a speck orbiting a speck among other specks among still other specks in the middle of specklessness. I suck.[14]

Nye's audience laughed approvingly, no doubt because they believed that "I suck" really means religion (which teaches that we don't suck) sucks.

But Bill Nye isn't so much *the science guy* as he is *the scientism guy*. Scientism is atheistic dogma masquerading as objective science.

C.S. Lewis on Dogma and the Universe

C.S. Lewis, in his 1943 essay "Dogma and the Universe," demolished Nye's way of thinking. Lewis begins with an analogy. Consider how a doctor determines that someone has been poisoned to death. The doctor can conclude this reasonably if "he has a clear idea of that opposite state in which the organs would have been found if no poison were present." Similarly, if we try to disprove God by pointing out how small we are in a huge cosmos, we should clearly identify the kind of universe that is expected if God did exist.

But Lewis argues that such a project fails. "Whatever space may be in itself…we certainly perceive it as three-dimensional, and to three-dimensional space we can conceive no boundaries," he writes. So we naturally feel that the cosmos is huge. What if we discovered nothing but our own sun and moon in such seemingly infinite space? "This vast emptiness would certainly be used as a strong argument against the existence of God," Lewis notes. In that case, atheists would argue that no God would create such vast amounts of wasted empty space.

Lewis runs through the other options: "If we discover other bodies, they must be habitable or uninhabitable: and the odd thing is that both these hypotheses are used as grounds for rejecting Christianity." If there are billions of habitable planets, then the skeptic would likely say that this means humans are not special. We would be lost in a crowd of aliens, or so the story goes.

Lewis continues: "If, on the other hand, the Earth is really unique, then that proves that life is only an accidental by-product in the universe, and so again disproves our religion."[15] In that case, atheists might further complain that no God would create trillions of sterile planets—what a lousy design.

Do you see the problem? No matter how God might have made the universe and life, skeptics would surely complain about something to the point of disbelief. What we have here isn't truth-seeking, but rather, game rigging.

Spinning the Universe

Atheists would find ways to spin a story that ridicules belief in God no matter what the size or contents of the cosmos. Bill Nye's God-bashing cosmic storytelling fails the credibility test. Keep all this in mind the next time you hear this popular myth invoked to mock religious believers.

For both Jews and Christians, here is the situation: We believe in an omnipotent, infinite God, and modern astronomical discoveries have confirmed that we inhabit a majestic universe befitting just such a creator. The psalmist got it right 3,000 years ago: "The heavens declare the glory of God" (Psalm 19:1).

But what about our position within this enormous universe? Has modern science shown it to be mediocre? Our next myth gets the wrong answer.

The Demotion Myth: Copernicus Demoted Humans from the Cosmic Center, and Thereby Destroyed Confidence in a Divine Plan for Humanity

In episode 8 of the 2020 season of *Cosmos*, "Possible Worlds," host Neil deGrasse Tyson spreads fake history.[16] I'm talking about the claim that astronomer Nicolaus Copernicus (1473–1543) demoted humans from the center of the universe. This supposedly challenged the religious idea of human importance. "Demoting the earth from the center of the universe was a severe blow to human self-esteem," Tyson claims on this *Cosmos* episode.

Tyson is among the many promoters of this falsehood. Atheist Christopher Hitchens, for instance, called Christianity's Earth-centered view its "greatest failure."[17] But in this case the atheists failed to do their homework.

The Real Copernicus

Copernicus argued against Earth-centered astronomy, but he didn't think this challenged Christianity. He even once said that God had "framed" the cosmos "for our sake."[18] Copernicus was not alone in this opinion. Most other early supporters of sun-centered astronomy thought the Bible and science are in harmony.

The myth that Copernicus demoted humans makes a false assumption. It assumes that earlier Earth-centered astronomy exalted humans. But according to the Earth-centered astronomy of the ancient Greeks, Earth was at the bottom of the universe. "Up" pointed to the perfect cosmic heaven. Earth was in the "dead center" of corruption, they thought.

This makes sense of what Galileo wrote in the century after Copernicus. He said, "I will prove that the Earth does have motion... and that it is not the sump where the universe's" filthy things "collect."[19]

Distorting Copernicus

The idea that Copernicus demoted humans was invented in the mid-1600s to bash Christianity. By the mid-1800s, the myth had entered astronomy textbooks, and by the 1960s, it had become the majority view.

The latest version of this story in a college textbook might surprise you. It is built on the view that exotic dark matter is more common than ordinary matter. "It is interesting to consider how far we have moved from our Earth-centered view," write Stephen Schneider and Thomas Arny. They claim that our cosmic location is not special, despite recent evidence otherwise that they ignore.[20] They continue, "And now we are realizing that the kind of matter that makes up everything we know is just a minor kind of matter in the universe. This is the Copernican revolution taken to extremes!"[21]

This is subjective storytelling, not science. One could just as easily have declared humans unimportant because our bodies are mostly composed of common, ordinary hydrogen, oxygen, and carbon. "See, there's nothing special about the material me!" Thus, one could make use of either the commonness or rarity of our material parts as grounds for our unimportance. Heads or tails, humans are losers either way. It's a rigged game, not a serious argument for a godless view of life and the cosmos.

Spiritual Atheism

Some atheists attempt to salvage some meaning and specialness from the Copernican "demotion." Consider a recent book by Neil Tyson. He argues that despite a series of scientific discoveries humiliating humanity, there's hope without God. We can still find meaning and purpose. How? Because "the cosmic perspective is spiritual—even redemptive—but not religious."[22] Science is our savior!

Tyson insists the Copernican demotion story redeems us from religious ignorance. He writes, "The cosmic perspective opens our eyes to the universe, not as a benevolent cradle designed to nurture life but as a cold, lonely, hazardous place." If Tyson had discovered that our solar system was jam-packed with dozens of smart extraterrestrial species (and billions of other kinds of intelligent life farther "out there"), he surely would have used this as an excuse to remove humanity from any special place within a divine plan. Because of never-ending godless stunts like this (regardless of the evidence), C.S. Lewis (as cited earlier) identifies such storytelling as hopelessly subjective.

Tyson's *Cosmos* series broadcasts this spiritual atheism to middle school and high school students. How so? In it, Darwin is said to be the "greatest spiritual teacher of the last 1000 years." He "worshiped nature," Tyson proclaims approvingly. "Life is an emergent property of chemistry. Science is an emergent property of life. Life can begin to know itself." In *Cosmos* episode 7, Tyson shares his strange belief that bees and trees can think much like us. But, of course, they can't do science. Apparently we are special *after all* in the spirituality of the new cosmic atheism.

We figured out we're insignificant—if we confuse significance with size and being in the exact center. And this somehow makes us significant?

This is not science. This is confused atheism dressed up as science. Such thinking shouldn't shock us. Anthropologist Gregory Schrempp explains how popular science writing often works.[23] It creates new myths to replace ancient religious ideas. This is a dead-end path, and once we've done our homework, an unbelievable one.

Atheists and agnostics, including the makers of the *Cosmos* TV series, do not inspire confidence today. But what about skeptics at the time of the scientific revolution? Were they primarily responsible for forging modern science?

The Skeptic Myth: The Main Heroes of Early Modern Science Were Skeptics, Not Believers in God

Episode 1 of the 2020 *Cosmos* TV series gives the impression that the main heroes of early modern science were skeptics, not believers in God. The series designates Baruch Spinoza (1632–1677) as the next-greatest persecuted hero of science after Giordano Bruno (as depicted in the *Cosmos* 2014 season). Although Bruno was burned to death in 1600 for his religious (not scientific) views, the attempted murder of Spinoza, if it occurred, was likely due to a disputed business transaction, not having anything to do with science or religion.[24] For *Cosmos* to suggest that Spinoza's life was threatened because of his scientific views is just the beginning of an enormous misrepresentation.

Like the heretical Catholic philosopher Bruno, Spinoza traded belief in the biblical God for a necessitarian philosophical creed. Both believed "God" had no choice in creation, and an infinite cosmos resulted.

Consequently, both had philosophical reasons for believing in an infinite number of inhabited worlds. There was, and still is, no scientific support for the idea of an infinite cosmos. Science is not well-equipped even to address this kind of question.

In *Cosmos*, Tyson equates traditional religion with ignorance, especially the biblical religions of Judaism and Christianity. Spinoza was a wayward Jew whom Albert Einstein, a secular Jew, later celebrated as likeminded. *Cosmos* depicts this connection with film footage of Einstein visiting the Spinoza museum.[25] Indeed, Einstein publicly confessed his faith in "Spinoza's God."[26]

Spinoza's God

Spinoza's God was nature, or some aspect of it. (Scholars debate how to interpret his ambiguous views.[27]) "From the necessity of the divine nature there must follow infinitely many things in infinitely many modes," Spinoza wrote.[28] Nature could not have been other than what it is.

This necessitarian vision, which traces back to the ancient Greeks, is precisely the view that the Judeo-Christian tradition overcame, leading to the birth of modern science. This transformation was one of the key ingredients for a cultural context conducive to modern science (as I explain in my video "Three Big Ways Christianity Supported the Rise of Modern Science").[29] Consequently *Cosmos 3.0* celebrates as a science hero a philosopher who opposed the very Judeo-Christian cultural context that helped make modern science possible. Oops!

Spinoza's God Versus Science

The Christian belief in divine freedom undercut the view, established by Plato and Aristotle, that the structure of the cosmos is a necessary one. Christians insisted that God could have created a universe quite different from ours, and so testing multiple hypotheses by experiment was an effective way to determine which set of natural laws God actually created to govern our cosmos. In his departure from theism, Spinoza undercut some of this science-fostering culture.

Let's go deeper as to why Spinoza was no science hero. In the last decade, scholarship on Spinoza has increasingly recognized that he opposed the observational (empirical) and mathematical analyses of nature advanced by the likes of Johannes Kepler and Galileo Galilei. "Skepticism about the very *possibility* of empirical knowledge of nature runs through Spinoza's books," notes Eric Schliesser in *The Oxford Handbook of Spinoza*.[30] More specifically, "Spinoza was very critical of applying mathematics and measurement in understanding nature." That's even more damaging! Similarly, Alison Peterman writes, "Spinoza took a dim view of the extent to which the application of mathematics to physics and the empirical investigation of the physical can give us knowledge of nature."[31]

Here is one memorable expression of Spinoza's criticism of the application of mathematics to science: "There are men lunatic enough to believe, that even God himself takes pleasure in harmony; indeed there are philosophers who have persuaded themselves that the motions of the heavens produce a harmony."[32] Spinoza attacked the view of Johannes Kepler and Christiaan Huygens (the leading Dutch scientist and a Spinoza acquaintance, also highlighted in the 2020 version of *Cosmos*) that God infused mathematical harmonies into the fabric of the cosmos. This is a projection of mathematical harmony and beauty in nature where none exists, Spinoza insisted.[33] Fortunately, astronomy textbook authors over the past four centuries ignored Spinoza's attack on Kepler and instead affirmed Kepler's third mathematical law of planetary motion, also called *the harmonic law*.

The Book of Nature

Christianity has a long and remarkable track record of contributing to the foundations of science. As mentioned earlier, Augustine expressed confidence in our ability to discover and read the "book of nature" because it is the "production of the Creator."

Galileo and many other early modern scientists used this traditional Christian metaphor of the "book of nature." They sought to convey the idea that God wrote two books that are consistent with one another: nature and the Bible. Nature is largely written in the language of mathematics, many of these scientists argued, and so it can be read only by those who know this language. Galileo argued as much in his book *The Assayer* (1623). He wrote, "Philosophy [natural science] is written in this all-encompassing book that is constantly open before our eyes, that is the universe; but it cannot be understood unless one first learns to understand the language and knows the characters in which it is written. It is written in mathematical language."[34]

Consider also these three remarkable utterances by Kepler that similarly affirm the theological foundations of science. On October 3, 1595, Kepler wrote in a letter to Michael Maestlin,

> I am eager to publish (my observations) soon, not in my interest, dear teacher…I strive to publish them in God's honor who wishes to be recognized from the book of nature…I had the intention of becoming a theologian. For a long time I was restless: but

now see how God is, by my endeavors, also glorified in astronomy.

Kepler further wrote in a letter to Herwart von Hohenburg, dated April 9-10, 1599:

> God wanted us to recognize them [i.e., mathematical natural laws] by creating us after his own image so that we could share in his own thoughts.

And in a letter to Galileo, in 1610, Kepler explained:

> Geometry is unique and eternal, and it shines in the mind of God. The share of it which has been granted to man is one of the reasons why he is in the image of God.

Here is how I explain the significance of these Keplerian sayings in my book *Unbelievable: 7 Myths About the History and Future of Science and Religion*, which tells the true story of science and God—the story Tyson tries to suppress with atheistic mythology:

> Kepler was a devout Christian who believed that the Bible and the "book of nature" were fully compatible and mutually supportive. He recognized them both as God's revelation. He studied both intensely. In fact, he almost finished a doctoral degree in theology before he turned to a career in mathematics and astronomy. Kepler believed that mathematical ideas exist eternally in the divine mind and that God freely selected some of these principles to govern his creation. Because God created humans in his image, we have the intelligence needed to discover those natural laws, and in so

doing, Kepler announced, we "share in his own thoughts." The human mind emulates God's thoughts in ways that reveal the deep structure of the cosmos. Thus God is "glorified in astronomy," Kepler concluded.[35]

The Bible and Aliens

Kepler also considered the existence of intelligent extraterrestrial life to be consistent with the Bible, even though Scripture does not specifically address this issue. This leads me to identify a final related error about science and religion in the 2020 *Cosmos* documentary. Tyson suggests that there was a contradiction between biblical faith and science given that the Bible does not mention extraterrestrial life (which Tyson thinks is established by science). There are countless aspects of the universe that the Bible does not address (e.g., quantum mechanics, thermodynamics, and electromagnetic radiation), but that does not mean scientific theories about such topics are in contradiction with the Bible. It simply was not within the communicative intent of the Bible to address whether there are other inhabited worlds—and many other interesting topics. Leading scientists such as Galileo, Kepler, Huygens, Descartes, and Newton understood this about the Bible and science.

For example, after affirming that Christ's sacrifice on the cross had redeemed many humans, René Descartes (whom Spinoza carefully studied) remarked,

I do not see at all that the mystery of the Incarnation, and all the other advantages that God has brought forth for man obstruct him from having brought forth an infinity of other great advantages for an infinity of other creatures. And although I do not at all infer from this that there would be intelligent creatures in the stars or elsewhere, I do not see that there would be any reason by which to prove that there were not; but I always leave undecided questions of this kind rather than denying or affirming anything.[36]

Tyson's alleged contradiction between Christianity and extraterrestrial life ignores centuries of often-friendly dialogue among theologians and scientists about this topic (which is a major theme of my book *Unbelievable*). It is simply not true, as Tyson asserts, that there was "only one man," Spinoza, who "dared to address" such questions during the seventeenth century. Such simplistic hero worship is laughable to professional historians of science.

Furthermore, Tyson attributes to Spinoza a view of nature that he makes sound daringly novel: "His sacred text," he says of Spinoza, was "the book of nature." But for most early modern scientists, there were *two* sacred texts: the Bible and nature. By turning his back on the former, Spinoza undermined some of the theological foundations for the scientific study of the latter. That's tragic, not heroic.

Of course it is possible that humans, on many occasions, have misinterpreted either the book of nature or the book of Holy Scripture—or both. In such cases, there might *appear* to be a conflict between science and religion. It is also reasonable to conclude that many "holy books" are not actually inspired by God—particularly because they make conflicting claims about reality. Such books might *actually* conflict in many respects with the way God made the natural world. My colleagues in the history of science community typically do not address such issues, but surely a comprehensive search for truth would not allow us to ignore them.

No War with Science

Has Christianity been at war with science? No.

If you think Bruno's and Galileo's encounters with the Catholic Church are major counterexamples to this conclusion, or if you think "alien contact" would be the ultimate Christianity killer, then read my book *Unbelievable* to discover otherwise. There, I identify an "extraterrestrial (ET) enlightenment myth" that is rooted in both atheist historical fiction (especially a "serial Copernican demotion" myth) and imaginary atheist futuristic storytelling. The fake history alleged by atheists distorts our vision of the true relationship between science and religion.

Does Science Conflict with Biblical Faith?

David Haines

In his article "Science Is at Odds with Christianity," philosopher Julian Baggini argues that when Christianity makes factual statements about the sensible universe, it enters into conflict with the truth claims of the natural sciences.[1] For Baggini, the only way that Christianity can avoid such conflict is by adopting the position that none of its claims are truth claims about the sensible world. This would imply that Christianity is essentially a myth designed to help us live rightly and cope with reality.[2]

Baggini represents one of the three common approaches within the philosophy of science to the relationship between science and religion: (1) the Conflict view, (2) the Complementary but differing spheres of discourse view, and (3) the Interaction and Collaboration view.[3]

The Conflict view, concisely stated, holds that there is a fundamental conflict between the natural sciences and Christian truth claims. This view may draw examples from specific claims, or cite historical conflicts or skirmishes between natural scientists and Christianity; however, such claims are, in fact, of relatively little use in supporting the

basic claim that there is an intrinsic conflict. To truly uphold the Conflict view, one must be able to show that the natural sciences, by their very "nature," are inherently opposed to biblical Christianity; here, we will argue that such an approach is in fact not viable.

The Complementary view states that the natural sciences and Christian theology complement each other, but neither interact nor contradict each other. John Hedley Brooke articulates this view as follows: "Scientific and theological language have to be related to different spheres of practice. Discourse about God, which is inappropriate in the context of laboratory practice, may be appropriate in the context of worship or of self-examination."[4] This view has been taken by Christian theologians such as Rudolf Bultmann. On this view, Christian theology is not an empirically verifiable religion, at least in regard to its central claims (i.e., creation, resurrection, immortality, miracles, etc.). If the Complementary view were true, then there is clearly no conflict between science and Christianity. However, Christianity would become nothing more than a phantom of its former self, bearing no resemblance to historic Christianity, in which the apostles, the early church fathers, and the medieval and reformation theologians all believed that Christianity was both historically verifiable and rationally coherent. The Complementary view emasculates biblical Christianity. To remain faithful to historic Christian belief, therefore, the Complementary view must be abandoned.

The third view, the Collaborative and Interactionist view, states that both the natural sciences and theology have something to say about the universe. There is no intrinsic contradiction between them, insomuch as they are properly articulated. This view is well-stated by John C. Polkinghorne, who says, "Science and theology seem to me to have in common that they are both exploring aspects of reality. They are capable of mutual interaction which, though at times it is puzzling, can also be fruitful."[5] This is the view which, though articulated in different ways throughout the centuries, recognizes that historic Christianity does not conflict with the natural sciences. This is the position that we will defend in this article. In what follows, we will begin by considering the reasons given for upholding a Conflict view, pointing out the lacunas of each reason. We will then discuss some of the important foundations for the Collaborative view.

The Appearance of Conflict

The two most common proofs that are provided by proponents of the Conflict view are (1) historical conflicts between Christianity and the practitioners of the natural sciences (who were called *natural philosophers* until the mid-nineteenth century, when the term *scientists* was coined), and (2) doctrinal conflicts, where the natural sciences make claims that appear to be contradicted by

claims made in the Christian Scriptures, or, at least, by particular interpretations of the passages in question. Neither of these claims, however, proves that there is any intrinsic conflict between the natural sciences and Christian theology. Furthermore, many of the test cases cited to prove conflict, upon closer examination, end up not supporting the Conflict theory. Examples abound, but it is sufficient to consider a few important ones.

Historical Conflict Between Christianity and the Natural Sciences

Probably the most popular cases of conflict between the Christian church and natural philosophers are those of Giordano Bruno (1548–1600), Michael Servetus (1511–1553), and Galileo Galilei (1546–1642). A certain tradition (informed by the Conflict view and popularized primarily through two books published in the nineteenth century: *A History of the Warfare of Science with Theology in Christendom* by Andrew Dickson White, and *History of the Conflict Between Religion and Science* by John William Draper) tells us that these great men of science were persecuted by the Christian church because their "science" threatened the traditional views about the world that were apparently anchored in the Christian Scriptures. The church, therefore, squelched, exterminated, and otherwise silenced these scientists who were opposing infallible church doctrines. Bruno was burnt at the stake, Servetus was executed by John Calvin at Geneva, and Galileo was persecuted by the Pope.

Of course, as recent historical scholarship has shown, the stories of these unfortunate events has been somewhat exaggerated and are not nearly as simple as Conflict theorists would have us believe. There are two reasons for this: (1) In some cases, although these authors are elevated to the status of scientists in order to fit the Conflict narrative, these men were actually engaged in practices that are closer to witchcraft and sorcery than to science (even according to medieval standards). (2) It is far from clear that they were persecuted for their science. Rather, it appears that they were punished less for their scientific research than for their theological claims.

Giordano Bruno, though he is often touted as an important astronomer (engaging in speculative astronomy and discussing questions such as the eternity of the cosmos and a theory of many real worlds[6]), was not declared a heretic and burned at the stake for his astronomical theories. Rather, it seems more likely that he was persecuted for his heretical theological claims concerning Christology.[7] Brooke notes this about Bruno: "A renegade monk, he made no secret of his unorthodox Christology. It was rumoured that he had declared Christ a rogue, all monks asses, and Catholic doctrine asinine."[8] Beyond heretical theology, Bruno, far from being a practitioner of the natural sciences, was actually a proponent of magic and sun worship. Hannam notes that "Bruno went further than many of the magicians of his time by trying to add an entirely new religion of his own creation to the existing magical doctrines."[9]

Michael Servetus was a Spanish medical practitioner who was forced to flee the Spanish Inquisition because of his theological beliefs about Christ and infant baptism. He had done work in physiology, on blood circulation, and contributed to rejecting some of the views proposed by Galen (an ancient Greek medical practitioner whose works were in vogue during the medieval and modern periods). It was not, however, because of his medical practices, nor because of his rejection of Galen, that Servetus was eventually burnt at the stake, but for his denial of the divinity of Christ and of the importance of infant baptism.

Galileo Galilei is one of the most popular examples of ecclesiastical persecution of well-meaning scientists. Yet once again, the facts of the story, properly nuanced, present a far more complicated narrative. First, although Galileo made many important discoveries, it is not for his discoveries that he was "persecuted" by the church. Second, it is vital to remember that Galileo, a personal friend of Pope Urban VIII,[10] had successfully presented his theories to the Pope (who was an admirer of Galileo's work[11]), remained a convinced and devout Catholic to the end of his life,[12] and was, in fact, generally well received in Rome, in part because of his work in refuting Ptolemy.[13]

What seems to have gotten Galileo in trouble with the ecclesiastical powers was actually more complicated than just his scientific theories. There are a number of elements that contributed to his censure and house arrest: (1) He made poor choices of patronage for his scientific endeavours—the Medici family—which took him from Venice to Florence, where he would no longer be protected from the Inquisition.[14] (2) In the *Dialogue*, more a polemical than scientific work, Galileo not only openly ridiculed the Pope's view (making him look like an idiot and an opponent of the sciences[15]), but also the Aristotelian natural philosophers of his time (who had a great deal of power in the church). In other words, Galileo was better at making enemies than friends in high places.[16] (3) At the time, the primary church authority in charge of ensuring that heretical claims were not published was Robert Bellarmine, who held an unusually rigid view of the relationship between biblical interpretation and scientific investigation.[17] (4) His book the *Dialogue* used questionable and unproven arguments to advance a theory that was rejected by most scientists of the time,[18] not the church per se. In Galileo's day, his theory might have been viewed as the modern equivalent of trying to prove that the Earth is flat by using pure conjecture. As Brooke notes, "Galileo seems to have felt that his difficulties with the Catholic church had their origin in the resentment of an academic philosopher who had put pressure on ecclesiastical authorities to denounce him."[19]

For his faults, Galileo was put under house arrest, but he was allowed to continue his research and writing, to see his family and friends, and essentially live out his days in relative luxury.[20] It was during this period that Galileo wrote his most important work, *The Two New Sciences*, which was published and widely accepted in Protestant countries.[21] Whitehead's observation is worth quoting: "In a generation which saw the Thirty Years' War and remembered Alva in the Netherlands, the worst that happened to men of science was that Galileo suffered an honourable detention and a mild reproof, before dying peacefully in his bed."[22]

In considering the three primary examples of alleged historical conflict between the church and science, we discover that whatever conflict there was, in fact, wasn't between the church and the natural sciences per se, but between certain ecclesiastical powers and intellectual and ecclesiastical leaders (who also happened to be natural philosophers) who were frequently promoting heretical doctrines in their theological work. In other words, that these men were natural philosophers had little, if anything, to do with their poor treatment—it had to do with their poor theology.

Now, this does not excuse the way these men were treated, but it does extinguish the fires of those who would use these unfortunate events to posit widespread conflict between the church and the natural sciences. Where there was conflict in relation to the scientific claims of these men, it was no more intense than what we see in contemporary academia when a scholar advances an unfounded theory or one that goes against the majority opinion. (This brings to mind how some neo-atheists accused the celebrated atheist, Antony Flew, of going senile when Flew published a book in which he affirmed the existence of a creator God; or, of how proponents of intelligent design face discrimination and persecution at the hands of mainline evolutionists and science journals.) Commenting on the church's reaction to Copernican theory, Brooke notes that it was not a case of religious opposition to science, but of "science of the previous generation" rejecting "new science."[23] This section is nicely summed up by Brooke's comment that "the dependence of the conflict thesis on legends that, on closer examination, prove misleading is a more general defect than isolated examples might suggest."[24] Conflict theorists are hard-pressed to find any examples of true historical conflict between the church and natural philosophers.

But what about doctrinal conflict?

Doctrinal Conflict Between Christianity and Science

A second argument used to support the Conflict model is the apparent existence of important doctrinal differences: truth claims made by the Christian Scriptures that are apparently refuted by the discoveries of the natural sciences. Some of the most-often-cited doctrinal conflicts are (1) creation versus evolution, (2) the existence and immortality of the soul, (3) the virgin birth and resurrection of Jesus, and (4) miracles in general. Other so-called conflicts have been cited in the past, such as the shape of the Earth, the movements of the heavenly spheres, and so on. However, it has been sufficiently shown, through in-depth historical research, that these views were either not held by most Christian theologians (as in the shape of the Earth, which all important Christian theologians agreed was spherical, not flat), or they were held not for theological reasons but for philosophical or scientific reasons. Regarding the latter, the general scientific consensus throughout the Middle Ages and into the early modern period accepted the Ptolemaic model of the universe until enough empirical evidence was discovered to cause it to be doubted, at which point the church theologians and natural philosophers moved toward either the Copernican model or the model of Tycho Brahe.[25]

In fact, the ways in which the early modern church interacted with the astronomical models of Ptolemy, Copernicus, Tycho Brahe, Galileo, and Kepler, provide interesting test cases for showing how the church has traditionally interacted with the discoveries of the natural sciences. The church has, throughout the centuries, promoted and financially supported scientific research, providing a great deal of academic freedom—so long as the natural scientists don't touch the core theological doctrines of the church (such as, for example, those found in the Apostles' Creed). A fair and informed view of Western history leads us to the conclusion that, as David Lindberg notes, the medieval universities (founded by Christian scholars, and originating in cathedral schools) provided scholars with a great deal of liberty:

Broad theological limits did exist, of course, but within those limits the medieval master had remarkable freedom of thought and expression; there was almost no doctrine, philosophical or theological, that was not submitted to minute scrutiny and criticism by scholars in the medieval university. Certainly the master who specialized in the natural sciences would not have considered himself restricted or oppressed by either ancient or religious authority.[26]

In fact, many of the natural philosophers working on science in the High Middle Ages were also trained theologians and churchmen. Understanding that the church has typically not only supported and encouraged scientific research but was also willing to "reinterpret" certain ways of understanding the Scriptures in light of scientific discoveries, we can now consider some of the main areas of doctrinal conflict.

Creation and Evolution

Probably the biggest area of today's contention between the natural sciences and Christian theology is the question concerning creation and evolution. Christian doctrine teaches, according to at least one interpretation of the Scriptures, that God created the entire physical universe, including all the life on Earth, in the space of six 24-hour days. Evolutionary viewpoints claim that the empirical data show that all life on Earth evolved over a long period of many millions of years,[27] through strictly natural undirected processes. Sometimes the matter of the age of the Earth is unceremoniously dropped into the discussion: young-Earth theorists propose that the cosmos is probably 6,000 years old, but certainly no more than 10,000 years old (give or take), whereas mainstream scientists propose many billions of years.[28] Because the question of the age of the cosmos is not necessarily related to the creation versus evolution debate, we will leave it to one side for now.

It is said that biological evolution accounts for the emergence of all life, and a form of cosmological evolution from a big bang can account for the appearance of the cosmos in its present form.[29] Consequently, there is no need for an account of the origins of the cosmos and biological life that calls upon a divine intellect. It is also suggested that if the Bible is true, then evolutionary theory is both wrong and unnecessary. Is this not a prime example of a serious doctrinal conflict between science and biblical faith?

To claim that there is a necessary conflict here is to hold a very simplistic understanding of Christian theology and a naïve view of science. As most historians and philosophers of science recognize, science is a tumultuous enterprise. Well-researched and commonly believed explanations of observed phenomena, which may have impressive predictive power, are often supplanted by later theories that explain the phenomena better—e.g., Ptolemy's model of the universe, which was supplanted by the Copernican model as refined by Kepler, or Newtonian physics, which required major revisions in light of Einstein's work. More powerful observational equipment that allows for greater precision has provided us with new details about the cosmos that were previously unknown, which, in turn, has entailed important revisions or wholesale replacement of theories (e.g., better telescopes have allowed astronomers to better explain the movements of the heavenly spheres and confirm theories, and better microscopes and other analytical techniques have allowed biologists to peer deeper into the cell).

In light of the tumultuous progression of the natural sciences, it is unwise to put too much confidence in their so-called established theories—especially broad, overarching theories such as the theories of cosmological and biological evolution. Future discoveries may require (and have required, since Darwin first proposed his theory of biological evolution) important qualifications, nuances, or large-scale rejection of these theories.

In relation to the Scriptures, the church has historically held to many different views on the proper interpretation of the creation narrative, and though not all of these views agree with or allow room for some variation of evolutionary theory, some do. Now, on the one hand, the fact there are some Christian interpretations of the creation narrative that not only "allow" for evolutionary theory as a possible explanation of the emergence of life but also seek to show how the Scriptures can be interpreted rightly while accepting some variant of evolutionary theory demonstrates that there is no necessary contradiction between the Bible and science. (It should be noted that many Darwinian biologists would likely be unsatisfied by some versions of "guided evolution" that are promulgated in these reconciliation attempts.) However, if there is no necessary contradiction, then this area of research cannot be used as an example of doctrinal conflict.

On the other hand, there are many areas of our experience that the natural sciences have not been able, and may not be able, to explain away—e.g., the origin of the cosmos, the intentionality and rationality of the human spirit, the experience of and desire for beauty, and so on. Though Christian theology (and many nonmaterialistic approaches to philosophy, such as those rooted in

Platonism, Neoplatonism, or Aristotelianism) provides clear explanations of these phenomena that are coherent and helpful, many scientists withhold belief in the hopes that scientific research may someday explain these phenomena by calling entirely upon material causes.

In summary, there is no necessary contradiction between a biblical faith and the natural sciences on the question of creation and evolution. However, even allowing that there is a conflict between current science and certain interpretations of the creation narrative, there is no reason to think that the biblical narrative cannot be vindicated by future research. Due to the nature of scientific research and biblical exegesis, a current conflict may be found to be superficial, based on a lack of knowledge in either the natural sciences or biblical exegesis. Finally, it should be noted that to maintain Christian orthodoxy on creation, Christians are committed only to the claim that God created the world as well as two humans who served as the original representatives of the human race. These humans then disobeyed God and plunged humanity into sin. Christians are not, according to the historical creeds of the church, committed to any claims about *how* God created, or how long it took for him to do the work of creation.[30]

The Resurrection of Jesus and Other Miracles

In relation to the resurrection of Jesus Christ, and miracles in general, many scientists suggest that the natural sciences demonstrate that miracles simply cannot happen and that dead men do not come back to life. The circularity of these claims is so apparent that it is not necessary to go into much detail in responding to them. After all, if miracles cannot happen, then they do not happen. As C.S. Lewis notes, "Now of course we must agree with Hume that if there is absolutely 'uniform experience' against miracles, if in other words they have never happened, why then they never have."[31]

However, the claim that miracles cannot happen is not self-evident. There are two ways of responding to this claim. The first is to note that this is an instance of circular reasoning. Lewis writes, "Unfortunately we know the experience against them to be uniform only if we know that all the reports of them are false. And we can know all the reports of them to be false only if we know already that miracles have never occurred."[32] Second, one would have to prove that we live in a universe that contains no immaterial and voluntary agents in order to prove that the miraculous cannot happen. However, if God (or other immaterial voluntary agents, such as angels or demons) exists, then we should expect the occasional event that is unexplainable by the natural sciences. Indeed, as soon as a voluntary, rational agent acts, the so-called laws of nature are incapable of explaining what happened.[33] Thus, the natural sciences cannot rule out the possibility of the miraculous.

Now, if only that which can be studied by the natural sciences exists, then, of course, God and other immaterial agents (including the human soul) simply cannot exist. The antecedent, however—the naturalistic claim that only the material world exists—is a statement that must be demonstrated to be true, not assumed; and it cannot be demonstrated by the natural sciences. There are, it seems, some truth claims that cannot be established by the natural sciences, yet which materialistic naturalist scientists take for granted. In order to avoid a vicious circularity, therefore, we need to either prove conclusively that there are no immaterial causal agents (a philosophical question), or accept the fact that miracles are possible. Until it has been proven that no immaterial causal agents exist, there can be no conflict between the natural sciences and Christian theology on the question of miracles. The thing is, there is a wealth of evidence pointing toward the existence of immaterial causal agents (including the existence of the human soul). Some philosophers, such as those in the Thomistic-Aristotelian school of philosophy, claim that we can demonstrate conclusively that God exists. If this is the case, then miracles are to be expected, and the only "conflict" is between those natural scientists who erroneously adhere to the view that science has proven that miracles cannot happen.

Yet some may repeat the old adage—often attributed to the atheistic astronomer Carl Sagan—that "extraordinary claims require extraordinary evidence." David Hume, for example, stated that the reasonable person "proportions his belief to the evidence."[34] Hume went on to propose that the universal experience of the human race provides a high probability that miracles do not happen.[35] As such, the wise man, confronted with the claim that a miracle has happened, will, at best, withhold belief. Based upon such reasoning, Hume suggested that

it is a miracle, that a dead man should come to life; because that has never been observed in any age or country. There must, therefore, be a uniform experience against every miraculous event, otherwise the event would not merit that appellation. And as a uniform experience amounts to a proof, there is here a direct and full *proof*, from the nature of the fact, against the existence of any miracle; nor can such

a proof be destroyed, or the miracle rendered credible, but by an opposite proof, which is superior.[36]

Again, there are multiple responses we can provide here—both factual and logical.

First, Hume's argument assumes a factual premise—the resurrection of the dead "has never been observed in any age or country"—that all evangelicals who believe that the early apostles observed the death and resurrection of Christ would roundly reject.

Beyond this, on a logical level, Hume is illicitly moving from highly improbable by definition to a *proof* of impossibility. If it was the case, in fact, that the high improbability of some event coming about necessarily implied that it was in fact impossible, then it follows that we should also reject as impossible a number of other highly improbable events, such as (1) the existence of the cosmos, (2) the existence of life on Earth, (3) all the events in the lives of Napoleon Bonaparte or Hitler,[37] (4) Hannibal's crossing of the Alps, and, in fact (5) most of the incredible and notable events of human history. The chances that any single event in our past would come about as it did are so infinitesimal as to be absolutely improbable, and if so improbable, according to Hume, then impossible. The miraculous cannot be ruled out on the claim that miracles are improbable; they *are* improbable, but so are most of the other important events of history. Hume has attempted to prove more than he can.

Third, it should be noted that Hume holds miracles up to an unreasonably high standard of proof. As historical events that are no more improbable (if God exists) than many other incredible and notable events of our past, miracles should be verifiable in the same way that we verify other events of history: by eyewitness testimony. The evidence required for miraculous events need be no more extraordinary than that which we require for belief in the astounding events of human history mentioned above.

In relation to miracles (and, in fact, in relation to the existence and immortality of the human soul), it turns out that the only conflict between the natural sciences and Christian theology is between one approach to the natural sciences (that of a dogmatic materialistic naturalism) and Christian theology. The history of philosophy reveals that not only Christian theologians, but some non-Christian philosophers and theologians as well, have argued that immaterial causally efficient agents exist, that humans have souls, and that these souls are immortal.[38] The supposed conflict, then, is not between Christian theology and the natural sciences, but between materialistic naturalism and everybody else.

In summary, then, not only are the four main examples of doctrinal conflict not truly contentious, but the so-called conflict is not actually between Christianity and the natural sciences, but, in some cases, between one dogmatic understanding of science and everyone else, or, in other cases, between one dogmatically held understanding of the Scriptures and everyone else. In both cases, it seems best to advance with caution and humility, recognizing the complexity of the issues and hoping that future discoveries will help to clear up present incertitude. An admirable attitude to have in relation to doctrinal conflict can be found in one of the forefathers of modern natural science, Roger Bacon. Lindberg notes,

As for the points of alleged conflict with Christian belief, Bacon dismissed them as problems arising from faulty translation or ignorant interpretation;

if philosophy [which, for Bacon, includes the natural sciences] is truly God-given, there can be no genuine conflict between it and the articles of faith. To reinforce this point, Bacon marshaled the authority of Augustine and other patristic writers who urged Christians to reclaim philosophy from its pagan possessors.[39]

Conflict Based on the Nature of Scientific Inquiry and Theological Explanation

Julian Baggini, in his article "Science Is at Odds with Christianity," presents two arguments concerning the nature of the scientific enterprise—arguments that he believes show that the natural sciences and religion are in conflict. The first is based on the nature of the questions that are asked by the natural sciences and theology, and the second is a consideration of the problems created by including God in scientific explanations. We will now look primarily at the first problem, as the second problem is essentially rendered unimportant in the light of our answer to his first argument.

In his first argument, Baggini begins by proposing that distinguishing between *why* questions (which are supposedly asked primarily by theologians and philosophers) and *how* questions (which are asked by scientists) is unhelpful because most *how* questions can be formulated as *why* questions.[40] Having pointed out the ambiguity between *why* and *how* questions, Baggini goes on to note that theology is primarily concerned with two subcategories of *why* questions: (1) "agency why" questions,[41] and (2) "the search for meaning why" questions (as in, *Why did this evil event happen to me?*).[42]

These two types of *why* questions, though they at first appear entirely different from *how* questions, turn out, for Baggini, to be less mysterious than we may at first think. In fact, suggests Baggini, on the assumption of materialistic naturalism, the question of meaning is in fact meaningless: "An atheist can conclude that suffering serves no teleological purpose. Pain and suffering are natural phenomena that evolved because they added survival value to the conscious creatures who felt them, alerting them to damage in the body. Nevertheless, the atheist can resolve to find meaning in suffering, or to give it meaning."[43] In this way, thinks Baggini, we realize that the "why-meaning" question is a chimera—there is no higher purpose for your suffering other than that which you give it.

Concerning the "agency why" question, materialistic naturalism must assume that the natural sciences are able to provide an explanation for how agency happens—that is, there must be a natural explanation as to why people do certain things instead of others. Though we often call upon the will to explain why a person acts in a certain way, says Baggini, "in principle, however, we have every reason to think that there is a level of description that would account for any human choice or behaviour in purely scientific terms. As physical beings in a physical universe, we have to assume that a complete physical science would be able to fully account for everything we do."[44] He goes on to make this illuminating statement:

It may not be practically possible to ever get to this complete description, and I suspect it will never be the most useful way of thinking about ourselves and others if we could, but in principle, such a level of description must exist. *If it does not, then by definition, the natural does not explain everything and there*

must be some supernatural forces at work in the universe too.[45]

In other words, on the assumption of materialistic naturalism, "agency why" questions are, in principle, capable of being answered as *how* questions—even though we have no idea how, we must assume such an explanation exists, because, well, we are assuming materialistic naturalism. Of course, if materialistic naturalism is true, then one would have to accept an absolute physical determinism. It is worth noting, in passing: (1) that materialistic naturalism is a philosophical position that is not necessary for scientific research, (2) that has never been proven, and (3) that cannot be proven by scientific research. There is, therefore, no reason that we should assume it in scientific research, and there are many very good reasons for rejecting it (e.g., the intentionality of the human intellect, the human ability to reason, etc.). If materialistic naturalism is rejected, then there is no reason to think that "agency why" questions can be answered entirely by the natural sciences. Thus, Baggini is asking us to have faith in the natural sciences based upon a highly debatable philosophical theory—faith that we will eventually find a purely natural—that is, material—explanation for "agency why" questions.

But also, in the same breath, we are told that because theological explanations call upon immaterial causes (which, in principle, cannot be studied by the natural sciences) to explain certain "agency why" questions, we must reject theological explanations. If the existence of immaterial rational and voluntary agents can be proven, then materialistic naturalism is false. Interestingly, historically speaking, the most prominent schools of philosophy (Platonism, Neoplatonism, and Aristotelianism) have thought that such demonstrations were possible. The debate between strict materialism and every other approach to human nature is still alive and well.[46] But, says Baggini, we must reject these arguments and put faith in the natural sciences. Leave your reason at the door of the laboratory and believe.

Turning to the question of divine agency, Baggini argues that on Christian theism, *how* questions are just "agency why" questions. This, thinks Baggini, implies a necessary conflict between the natural sciences and theology. The natural sciences are looking to answer these questions by providing natural mechanisms, and theologians want to say, "God did it." Baggini says,

> The Christian God is both creator and ruler of the universe. So if things happen in that universe that have meaning and purpose, then clearly it is because this God has set things up in some way, or intervened in some way, to make sure that purpose is achieved or meaning realized. The neat division between how and why questions therefore turns out to be unsustainable.[47]

Baggini's conclusion is both right and demonstrates a lack of historical knowledge. Modern science, as has been amply demonstrated by historians of science, was born from and raised upon the theological and scientific theories of the Middle Ages. The great thinkers of the twelfth and thirteenth centuries answered Baggini's worries in such a way as to preserve both the ultimacy of divine agency and the human ability to engage in the natural sciences by searching for natural mechanisms and causes. As Lindberg points out, the naturalistic approach to explaining observed effects was "one of the most salient features of twelfth-century

natural philosophy."[48] Though the different scholars did not all agree on the distinctive features of their explanations, they agreed that natural philosophy was looking for secondary causes, and that the discovery of secondary causes did not eliminate the need for a primary cause. In fact, "Searching for secondary causes is not a denial, but an affirmation, of the existence and majesty of the first cause."[49] Ronald L. Numbers notes this about the late medieval natural philosophers: "although characteristically leaving the door open for the possibility of direct divine intervention, they frequently expressed contempt for soft-minded contemporaries who invoked miracles rather than searching for natural explanations."[50] This approach to the natural sciences continued well into the seventeenth century.

Appealing to secondary causality in the natural sciences implies that for every question that can be answered by the natural sciences (and clearly, there are some questions that the natural sciences cannot answer) by appealing to secondary causality, the Christian theologian is also able to argue that God is the first cause who not only gives existence to secondary causes but ensures that they act according to their nature. This theological statement cannot be disproved by appeal to the natural sciences, though it can be discussed in philosophy or theology. Making an appeal to secondary causality answers Baggini's concerns and implies that there are no conflicts between theological and scientific explanations. Baggini's second argument, which states that theological explanations have no place in the natural sciences,[51] is seen to be of little effect, in light of our response to his first argument. (Indeed, the modern theory of intelligent design described in this volume has developed methods to reliably distinguish between material/secondary causes versus intelligent agency—the latter which we might observe during direct divine intervention or miracles. This too remediates Baggini's concern that inferring divine agency must interfere with our ability to scientifically study nature and discover material causes.) There is, therefore, no conflict between the natural sciences and Christian theology, even at the foundational level of how they explain natural phenomena.

Foundations of a Collaborative View

We have shown that the Conflict model fails to support the weight of its claims by pointing out the weaknesses of its three main arguments. Next we come to support for the Collaborative view. We have already mentioned a number of important features of the Collaborative approach: (1) the notions of primary and secondary causality; (2) that it is not necessary to assume materialistic naturalism in order to engage in the natural sciences; and (3) that the possibility of the miraculous does not preclude scientific research. In this section, we will restrain our comments to two foundational themes that provide support for the Collaborative view: (a) a single cause of both nature and Christian scriptures, and (b) a view of science that puts both the natural sciences and theology in their appropriate places.

Two Books, One Author

Christian theologians have historically taught, based upon what the Christian scriptures teach about God's revelation of himself in both creation and in Scripture,[52] that there are two books (nature and Scripture) that were "written" by one author—God. What this means is that because there is one inerrant and omniscient author of both nature

and Scripture, there can be no true contradiction between these sources of knowledge. Apparent contradictions are not due to contradictions in the sources themselves, but to interpretative errors on the part of those reading the books. In other words, apparent contradictions between theology and the natural sciences are not due to actual contradictions between scriptures and the natural world, but to how we understand these sources.

Thomas Aquinas, in his commentary on Boethius's *De Trinitate*, says of the relationship between philosophy and theology that "the light of faith, which is imparted to us as a gift, does not do away with the light of natural reason given to us by God. And even though the natural light of the human mind is inadequate to make known what is revealed by faith, nevertheless what is divinely taught to us by faith cannot be contrary to what we are endowed with by nature."[53]

The Christian description of nature and Scripture as two books written by one author has inspired Christian theologians to engage in natural philosophy, for they believed that if the cosmos was created by an intelligent being, then the cosmos was highly intelligible and would render many truths to those who studied it. For example, it was Kepler's conviction that God doesn't make mistakes that inspired him to rework the Copernican theory. The conviction that there was a designer inspired Christian scientists to look for the "final causes" of the things and events of the natural world.

Two Sciences: Theology and Natural Philosophy

A second important concept that undergirds the Collaborative approach to the relation between Christian theology and the natural sciences is a clear notion of what is meant by the term *science*, how different areas of research are said to be science, and how they are related to each other. In contemporary discussions, the term *science* is used essentially as synonymous with the term *natural sciences*. A scientist is an individual who studies the natural sciences. This is not, however, the way earlier scholars understood the term *science*.

Returning to Aristotle, we find science described as a systematically organized body of truth claims that explain the causes or reasons as to why something is, always or for the most part, as it is. Note that such a definition determines neither the method nor the object that can be studied. The reason is because there are many different methods for discovering the causes or reasons as to why something is the way it is, and these methods depend upon the nature of that which is being studied. This implies that, properly speaking, more than just the natural sciences should be considered sciences.

In his *Summa Theologiae*, Aquinas asks whether theology is a science, and then goes on to make some distinctions that help us to understand the respective relation of theology and the natural sciences. First, he explains,

> Sacred doctrine is a science. We must bear in mind that there are two kinds of sciences. There are some which proceed from a principle known by the natural light of the intelligence, such as arithmetic and geometry and the like. There are some which proceed from principles known by the light of a higher science: thus the science of perspective proceeds from principles established by geometry, and music from principles established by arithmetic. So it is that sacred doctrine is

a science, because it proceeds from principles established by the light of a higher science, namely, the science of God and the blessed.[54]

In these comments, Aquinas explains that there are certain sciences that are fundamental for, or higher than, other sciences, which draw their principles from the higher sciences (e.g., geometry provides the first principles of optics). The various natural sciences can be analyzed into those that are higher and lower. Now, based upon the definition above, theology is clearly a science (a systematized body of knowledge), and, for Aquinas, it is a lower or subalternate science because it receives its principles from a higher science: divine knowledge itself.

Having seen that theology is as much a science as the natural sciences (differing, of course, in method and in object), Aquinas goes on to ask about the relationship between the natural sciences and theology. He argues that the sciences can be arranged hierarchically based upon the amount of certitude they provide concerning their conclusions and the "nobility" of their objects. The question of certitude depends upon the intellect containing the knowledge in question. Aquinas argues,

> In both these respects this science [theology] surpasses other speculative sciences; in point of greater certitude, because other sciences derive their certitude from the natural light of human reason, which can err; whereas this [theology] derives its certitude from the light of the divine knowledge, which cannot be misled; in point of the higher worth of its subject-matter, because this science treats chiefly of those things which by their sublimity

transcend human reason; while other sciences consider only those things which are within reason's grasp.[55]

From there, Aquinas goes on to discuss how theology, even in its practical elements, is superior to the other sciences.

For Aquinas, then, the natural sciences are inferior to theology, philosophy, mathematics, and even the intermediate sciences (e.g., theoretical physics)—both because the natural sciences provide less certitude and because their object is less noble than that which is divine. Of course, we may have less knowledge of the divine than of the cosmos, but "the slenderest knowledge that may be obtained of the highest things is more desirable than the most certain knowledge obtained of lesser things."[56] It is in part due to these reasons that all sciences other than theology, grouped under the title *philosophy*, have been called the handmaiden of theology, which is said to be the queen of the sciences.

In this approach to the sciences, each science has its place and purpose in human knowledge. Each has its own object and method. Each one, studied for itself, contributes to the human knowledge of the cosmos and its creator. There can, in such a conception of the sciences, be no conflict between

them, for each of the sciences is studying something of the divine: whether it is the divine nature as it is revealed in Scriptures (the artist), or it is the divinely created cosmos (the work of art).

A Collaborative Approach Is Superior

The Conflict approach to the relationship between the natural sciences and Christian theology is convenient for those who are against religion or God, and for those religious persons who are scared of having their heartfelt convictions shaken by the discoveries of the natural sciences. This approach, however, lacks both historical and doctrinal nuance, and a Collaborative approach is superior. A closer look at the historical development of the natural sciences (in their relationship with theology), at the apparent doctrinal conflicts, and at the very nature of both scientific and theological explanation reveals that the most appropriate way of relating them is to say that they interact and collaborate, maintaining a nondogmatic dialogue in which both sciences influence each other by helping to provide nuances, refute errors, support truth, and provide humans with a greater appreciation of and a united vision for the cosmos and its creator.[57]

Did Christianity Help Give Rise to Science?

John A. Bloom

Given today's politically correct mentality that all views are equally true and the contemporary cultural stereotypes that exist against Christianity, it may seem odd to answer the question in this chapter's title with a *yes*. But if we go beyond the stories told by those holding the microphone in today's public square and look to the actual people and events in the past, historians find clear indicators that Christianity provided the fertile soil for modern science to flourish, and that the two were partners in the pursuit of truth far more often than they were adversaries.[1]

First, it is important to recognize that what we call science today has not always been with us. In Greek and Roman times, the few people with the means and time to be more than subsistence farmers, merchants, or craftsmen focused their mental energies on politics and philosophy. Aristotle, standing on the shoulders of Plato, developed observational approaches for categorizing nature, especially in zoology, and his systematic logic would later prove helpful in the organizing of the academic curricula of universities. But for Greek and Roman thinkers, getting one's hands dirty in order to understand better the physical world was generally not seen as beneficial or valuable. Math was useful for accounting in business, geometry for surveying property, and medicine for simple remedies, but the physical world was seen as capricious and messy, influenced by the whims of the gods. So there was not much out there to be studied and then applied to make our lives better. Sure, there was the occasional science/math genius like Archimedes or Euclid, but aside from some innovations by craftsmen that improved ancient technologies, nothing like our modern sciences existed.

So where did science come from? It started with determining the right way to look at nature: the world was not a casino filled with capricious spirits in various physical manifestations; rather, it was created by a single God who established it to work via fixed regular laws. One didn't have to pray that the wood and water gods would be in a good mood today; the things around us weren't gods at all but were created by God for his glory and our benefit. This is the clear inference from Genesis 1:16, which says God made "the greater light to rule the day" and "the lesser light to rule the night." Had the sun and moon been named in the text, the ancient reader would have assumed God was making these lesser deities. The fact they are described and not named implies they are things, not beings. Many other Bible passages make the same point, such as Jeremiah 10:2-3: "Learn not the way of the nations, nor be dismayed at the signs of the heavens because the nations are dismayed at them, for the customs of the peoples are vanity." The heavens are filled with things, not populated with deities or personal beings.

Moreover, these created things follow regular patterns: When he creates the sun and moon, God declares they are to "be for signs and for seasons, and for days and years" (Genesis 1:14). God tells Noah in Genesis 8:22, "While the earth remains, seedtime and harvest, cold and heat, summer and winter, day and night, shall not cease." Speaking through Jeremiah, God says that if we can change the covenant for day and night, he will break his covenant with David (Jeremiah 33:20-21). These passages imply that God created the world to follow fixed laws (covenants).

Because the first goal of science is to study the physical world to find regular patterns, one must presume there are patterns out there to be found in the first place. Incidentally, this is not an obvious fact, as most of these patterns are subtle and even counterintuitive. Few cultures have held to this view of nature; animism or polytheism, with the attendant expectation of the gods' personal vagarity, have tended to dominate the religious scene, making science's first premise and goal difficult to conceive, let alone assume.

In addition to possessing correct views about the impersonal and regular nature of the physical world, one also needs to have a motivation to study it. Again, many religions focus on the spiritual realm and see the physical world as dirty, bad, and evil—hence, to be avoided. But Christianity sees the world as good (although corrupted in some way due to the fall) and therefore worth studying to gain wisdom and to glorify God. Psalm 19:1 famously proclaims, "The heavens declare the glory of God, and the sky above proclaims his handiwork." These ideas gave rise to the "two books" view of God's revelation: God has spoken through his prophets and apostles, and we study their writings; and we also see his handiwork in the book of nature, the things he has created. Thus, we can study creation itself to learn more about the power and wisdom of God and to worship him more. Galileo speaks of discovering the language of God (mathematics), Kepler of "thinking God's thoughts after Him," and Copernicus writes,

> To know the mighty works of God, to comprehend His wisdom and majesty and power; to appreciate, in degree, the wonderful workings of His laws, surely all this must be a pleasing and acceptable mode of worship to the Most High, to whom ignorance cannot be more grateful than knowledge.[2]

Because of this two books thinking, science was able to grow and flourish alongside theology in the early church-supported universities in Europe: studying the Scriptures and God directly were understandably the university's highest academic priority, but studying the creation was also valuable because it pointed people to God.

Most importantly, Christianity's motivation for doing science is not just ethereal: with our knowledge, we can make the world a better place for others, and thus love our neighbor by developing medicines and technologies that improve everyone's quality of life. Although he thought in terms of regaining the knowledge and dominion over nature that we lost at the fall, Francis Bacon stressed this practical value to studying God's world.[3] For Bacon, unlocking nature's secrets and applying that knowledge to heal diseases and live more comfortably was paramount.

To give a modern example, total internal reflection, the observation that light traveling inside of glass (a high-refractive-index material) cannot escape out into the surrounding air (which has a low-refractive index) if it strikes the interior surface at too high of an angle and thus remains trapped inside the glass, has been a standard textbook optics exercise and classroom demonstration for centuries. However, it wasn't until 1970 that researchers at Corning Inc. succeeded in developing an ultrapure glass that reduced transmission losses to the point where fiber optics became practical over long distances, ultimately fueling the Internet and modern data communications revolution.[4] Science's and engineering's incredible success in turning knowledge into practical applications is certainly one of their greatest triumphs: You (or I) probably would not be alive today and able to read (or write) these words but for the medical, technological, agricultural, educational, and communicational advances that science has given us.

But back to Bacon and the Christian motivation for science: Why do we even think we can unlock nature's secrets, which is the first step of science? God's ways are not our ways, so why should we think that we can understand his creation at all? Here is where early scientists appealed to the fact that we are made in his image—this implies that he has given us the mental abilities to

comprehend at least some of his handiwork. Thus the unreasonable effectiveness of mathematics[5] at describing the world and our conceptual ability to do math follow from God creating us with a tiny version of some of his abilities. And here is where the Greek philosophers erred: We shouldn't think that we can know the mind of God well enough to derive how the world works solely from logical axioms while sitting on the couch. God can create whatever he wants however he wants, so it is our task to examine and discover what God has actually done. "It is the glory of God to conceal things, but the glory of kings [those with the time and funding] is to search things out" (Proverbs 25:2). Note that there's no promise that this will be easy, so science draws upon the virtues of patience and persistence.

Another key ingredient provided by Christianity for the success of science is the concept of the linear motion, or directionality, of time.[6] Most non-Christian cultures view (or viewed) time as an endless cycle of progress and decay, like the seasons, and see no long-term goals for humanity other than caring for and feeding their gods. The rest take (or took) a fatalistic view of life and don't see progress as humanly possible. In contrast, Christianity has a beginning, observes that knowledge can accumulate over time (progressive revelation), and sees a glorious light at the end of the tunnel. History is going somewhere, implying that we can learn, grow, pass things along to future generations, and improve both our lives and theirs. Thus the difficult struggle to discover new scientific truths has a long-term payoff, making it worth the effort.

Christianity's perspective on human nature is also a key factor in the rise of modern science and its fruitful application: Because people don't seem naturally inclined to work hard and selflessly to help others, Christian societies encouraged financial motivations to innovators who worked to improve the lives of others. While altruism is great, most progress has come about by allowing the inventor to get rich off his idea. In particular, this involves protecting the inventor's intellectual property rights so that another person, company, or even the government cannot steal the idea. This hope of reward for one's efforts (the fruitful harnessing of selfishness) has helped technology and medicine to flourish.

A similar motivator is what's known as the Protestant work ethic: the broad concept that God has given each of us skills that we can use to glorify him and to help others. Theologians note that God gave Adam and Eve work to do *before the fall*, so work must be a good and satisfying thing that we were created to do, even though it was made frustrating and difficult to do after the fall. Viewing one's scientific research as a gift and calling from God is strong incentive to do one's work well.

Although textbooks might lead us to think that science progresses through the efforts of a few great geniuses (Einstein, Maxwell, Newton, Galileo, etc.), the real history is more complex and shows the need for a community of scientists collaborating together with high standards of honesty and transparency. Science builds on the work of the community, as no individual can do every experiment and test every hypothesis. Trust in each other's work is critical, as errors or deception can lead to years of wasted effort and resources. Thus the Christian virtues are foundational to good science practice.

We need to be careful not to let the tremendous blessings from science lull us to ignore its dark side (pollution, nuclear weapons, reductionism, dehumanization) or to assume that science alone gives us truth and religion is no longer helpful. Science does not provide itself with a moral compass or offer guidance as to which applications are best for us. For example, Fritz Haber, who received the Nobel Prize for inventing the chemical process to synthesize ammonia from nitrogen in the air, thus making cheap fertilizer available to help feed billions of people, also developed explosives and poison-gas agents for the Germans to use during World War I.[7] Scientific knowledge can be used for good or evil.

In summary, Christianity fostered early science by providing the right lens to view nature (it's impersonal and follows regular laws, good things can come from its study, and progress is possible), the best cultural incentives (uplifting and praising selfless work, making personal gain possible, fulfilling one's calling, and requiring trust and teamwork), and the highest goals (alleviating famine, suffering, and death; and promoting human flourishing). All these factors have served as critical nutrients for science to grow and flourish. In non-Christian cultures where one or more of these key features was missing, we may see a flash or two of insight (Arabic numerals, algebra), but not sustained growth.

Given this nurturing past, why does the relationship between Christianity and science seem so strained today, and why is their past collaboration rarely recognized? Largely because the success of mathematics at predicting behavior and because the power of looking at the world as a purely physical chain of cause and effect appear to make the Master Architect unnecessary, creating a reaction somewhat like what one observes when a teenager is embarrassed to be seen with their parents. But this purely materialistic approach to science loses its foundational reasons for why the universe is intelligible or follows regular laws and can end up directing science's goals toward only short-term profit and political power. Historians and philosophers have shown that this modern tension is rarely between science and Christianity per se; rather, it is between two competing religious views: materialism and Christianity. Placing science and Christianity in their historical and philosophical contexts allows us to appreciate their deep partnership, recognize the main source of their tensions, and allow them to guide us as we move forward.

Can a Christian Be a Scientist (and Vice Versa)?

David Haines and Frank Correa

To some extent, the question "Can a Christian be a scientist?," measured against the testimony of history, appears somewhat pointless. Of course a Christian can be a scientist, and many well-known scientists have been, and are, devout Christians. However, in light of some recent neo-atheistic polemics, this question makes more sense.

Richard Carrier, for example, in a particularly polemical article, allows that science can develop in a Christian atmosphere and be engaged in by Christians; however, he immediately qualifies this with the contentious claim that though science was practiced by Christians, "this was a development *in spite* of Christianity's original values and ideals."[1] So, according to Carrier, Christians can do science, but only by putting their most fundamental Christian beliefs to one side. Carrier proposes, in a somewhat simplistic reading of history, that it was Christianity that slowed down the steady advance of the natural sciences by about 1,000 years: "Had Christianity not interrupted the intellectual advance of mankind and put the progress of science on hold for a thousand years, the Scientific Revolution might have occurred a thousand years ago, and our science and technology today would be a thousand years more advanced."[2]

Many examples of this attitude toward Christian involvement in science or philosophy can be furnished. The well-known German philosopher Martin Heidegger, in his *Introduction to Metaphysics*, famously said, "A 'Christian philosophy' is a round square and a misunderstanding."[3] Though Heidegger is not speaking directly to the natural sciences, the context of this statement (asking the question "Why is there something rather than nothing?") leads one to wonder if it is possible to pursue knowledge as a Christian at all, philosophical or scientific, or if it is necessary to put aside one's Christian convictions in order to engage in any truly scientific or philosophical enterprise.

In this chapter, we will first consider some of the philosophical questions related to being a Christian scientist. We say philosophical because the so-called theoretical problems are not scientific per se, but are related to philosophy of science and religion. We will then look at this question from a historical perspective, providing a broad survey of the role played by Christians in the development of modern science.

The Christian Scientist: A Misunderstanding?

Faith Versus an Inquisitive Spirit

In *The God Delusion*, Richard Dawkins explains that one of the reasons he is so hostile to religion in general is that fundamentalist versions of the various religions are destroying the scientific curiosity of believers, and the more liberal versions of these religions are empowering the fundamentalists.[4] How are the fundamentalists destroying the scientific spirit? By insisting on a blind adherence to the Bible, which requires the believer to reject "irrefutable" scientific evidence in the name of faith.[5] Dawkins says he is "hostile to fundamentalist religion because it actively debauches the scientific enterprise. It teaches us not to change our minds, and not to want to know exciting things that are available to be known."[6]

What it all comes down to for Dawkins is that whereas scientists are required to "believe" theories because of the evidence, "Fundamentalists know they are right because they have read the truth in a holy book and they know, in advance, that nothing will budge them from their belief."[7] This critique is a feeble echo of Martin Heidegger's claim that a Bible-believing Christian cannot authentically engage in a philosophical questioning of being because the believer already "knows" the answer to the question. Heidegger says,

Anyone for whom the Bible is divine revelation and truth already has the answer to the question "Why are there beings at all instead of nothing?" before it is even asked: beings, with the

exception of God Himself, are created by Him...One who holds on to such faith as a basis can, perhaps, emulate and participate in the asking of our question in a certain way, but he cannot authentically question without giving himself up as a believer, with all the consequences of this step.[8]

According to Dawkins and Heidegger, religious faith precludes a person from being able to engage in scientific or philosophical inquiry. For Dawkins, this is because faith keeps such a person from objectively considering the evidence; for Heidegger, this is because to authentically ask questions, one must not begin with the answer. There are a number of elements that must be considered as we respond to these critiques.

First, as Augustine famously stated in his *De Trinitate*, "before we can understand, we have to believe."[9] This statement was echoed by many important medieval scholars, including Anselm in his *Proslogion* and Thomas Aquinas in his *Commentary on Boethius's De Trinitate*. Faith, or belief, according to these great medieval theologians, is voluntary assent to the truth of a statement that is proposed for belief by a trustworthy authority.[10] So defined, it becomes immediately obvious that faith is as important to the natural sciences as to any other domain of research. This point can be seen through a variety of different examples. Consider how one gains sufficient knowledge in an area of study in order to be able to begin practicing in that domain. In order to learn enough to be able to engage in scientific inquiry and profit from the work of other scientists, one must first believe their teachers, who provide them with the intellectual tools necessary for the scientific enterprise (e.g., mathematics, language skills, observational skills, etc.).

This belief leads to knowledge. This knowledge, if pursued, leads to more knowledge, and may even bring one to be able to refute one's teachers on certain subjects.

A second example can be found in the verification of experimental or observational data. When we read about the discoveries of a scientist in a domain that is not our own area of expertise, we tend to believe them unless we are aware of some reason to distrust their findings. We may ask for the opinion of a colleague or friend who specializes in that area, and when we accept our colleague's or friend's opinion as true, we are "believing" them. This is an act of faith. If a researcher points out an interesting phenomenon and we attempt to verify their findings on our own, we are acting out of faith. This is analogous to the birdwatcher who, upon seeing a gorgeous specimen, invites his companion to look and be amazed. The companion who looks in the direction to which he has been pointed acts out of faith in the truthfulness of what his companion has said.

These acts of faith produce understanding. This is how science has continually advanced throughout the centuries. As such, faith, properly defined, does not preclude scientific inquiry (nor destroy it), but makes it possible. Faith as a voluntary act of the rational human being can, of course, be misplaced or deceived (as when we trust a person who is not an authority or who intends to deceive). Reason can also be misled or arrive at erroneous conclusions (as when an error is introduced into our reasoning, or when we reason poorly and arrive at false conclusions, or when we accept a falsehood as truth). As Thomas Aquinas says in the prologue to his *On Being and Essence*, quoting Aristotle, "A slight initial error eventually grows to vast proportions."[11]

It follows that faith in a trustworthy authority, far from hindering the scientific enterprise, actually helps it. In fact, historically, the fathers of contemporary science were motivated to study the universe because of their belief that it was created by God, and, if created, then endowed with a discoverable and intelligible order. For example, Johannes Kepler, a devout Protestant astronomer who was unable to make physical observations himself due to poor eyesight,[12] proposed theories concerning the solar system that proved more accurate than those of any scientist before him. His theories were based upon (1) the belief that the cosmos was created by God, and (2) his faith in the work of, and data gathered by, the Protestant astronomer Tycho Brahe. Kepler could not verify the data provided by Tycho, but voluntarily assented to its truthfulness, and, through his great mathematical ability, proposed the most accurate model of the solar system up to his time.

Hannam notes that "for Kepler, the most important fact about the world was that God had created it. Like Copernicus, he was convinced that the structure of the heavens had to reflect the perfection of its creator."[13] It is this conviction that led him to study the data supplied by the stargazers (data he could not verify, implying that he had to voluntarily accept the truthfulness of that data) and to work out a refined model in which the planets moved around the sun in elliptical orbits. Hannam notes, "His ideas about God provided his hypothesis, he had the mathematical ability to turn his ideas into a system and, at last, Tycho's data meant he could check to see if his system was actually true."[14] This is one example among many, but the point is clear. Far from hindering his scientific inquiries and theories, Kepler's Christian faith, and the faith he placed in his Christian colleagues, provided both the theoretical stimulus and the hard data that was needed to make his research and theorizing possible. Faith, then, is not necessarily a hindrance to scientific research, but can be a powerful ally.

A second element that must be considered in relation to the claims made by Dawkins, Heidegger, and Carrier is that knowing *that* something is true does not keep one from engaging in research in order to discover *why* it is the case, and *how* it came about.[15] For example, receiving the correct answer for a problem from the professor may help the student discover why it is the correct answer. In fact, when we consider the nature of scientific research, it turns out that this is, in many cases, how the scientific enterprise begins and is driven. The scientist discovers some phenomenon, is amazed, and engages in observation and experimentation to discover why and how that phenomenon came about. The scientist begins with the *that* and works to discover the *why* and the *how*. Aristotle famously said that all philosophy begins in wonder. It is the discovery that something is the case, and we don't understand why or how, that drives our desire for knowledge. In fact, even when we know why, we may continue our research to discover how, and vice versa.

The cosmos and the incredible inhabitants of this universe confront us constantly, pushing us to ask the primordial philosophical question: Why is there something rather than nothing at all? The Christian believes the biblical answer to this question is like the student who accepts the answer given by the professor, then attempts to work out why it is the right answer and how to get that answer for himself—*Credo ut intelligam* (believe in order to understand). Faith does not preclude knowledge or scientific inquiry, but rather, makes it possible.

Naturalism, Theism, and the Sciences

The term *naturalism* is used in a number of ways in different domains of study and is constantly being subjected to a number of important qualifications (e.g., scientific naturalism, methodical naturalism, materialistic naturalism, and so on). In the natural sciences, scientific naturalism is often defined as "the absolute ontological and epistemological primacy of the natural sciences *as a whole*, whether the other natural sciences are reducible to physics or not."[16] In other words, scientific naturalism states that the natural sciences are the primary source and standard for all knowledge. This definition of scientific naturalism roughly corresponds to J.P. Moreland's definition of *scientism*: "the view that the hard sciences alone have the intellectual authority to give us knowledge of reality."[17] Such a perspective is, as Moreland rightly argues, self-contradictory, because this statement, being a statement about reality, is not a conclusion that can be arrived at through the hard sciences, such as physics, biology, or chemistry. Rather, it is a philosophical statement about the relationship between reality and our knowledge of it.

For our purposes, more interesting is the purported claim of modern methodical naturalism that "science must proceed as if nature is all there is."[18] Though this is a widespread contemporary understanding of the term *naturalism*, and clearly no religious believer could adhere to such a definition, it is not the only way in which this term can be understood. Nor must one adhere to such a claim to participate in the scientific enterprise.

As the study of medieval and modern science clearly demonstrates, it is possible to be both a devout Christian (thus proceeding as if nature is not all there is) and a scientist who seeks for natural causes of observed effects. During the Middle Ages, for example, a number of important Christian theologians sought natural explanations for all kinds of observed phenomena (and even for the events described in the Bible). "This naturalism is one of the most salient features of twelfth-century natural philosophy," says David C. Lindberg.[19] He lists a number of important twelfth-century theologians, devout Christians convinced of the truth of the Christian Scriptures, who engaged in this naturalistic approach to the natural sciences: "William of Conches, Abelard of Bath, Honorius of Autun, Bernard Sylvester, and Clarenbald of Arras."[20] Lindberg goes on to note that though they differed in the details of their theories, "they shared a new conception of nature as an autonomous, rational entity, which proceeds without interference according to its own principles."[21] Concerning William of Conches's view, Lindberg notes that "William's purpose, as he made clear elsewhere, was not to deny divine agency, but to declare that God customarily works through natural powers and that the philosopher's task is to push these powers to their explanatory limit."[22]

What William describes as natural powers has, elsewhere, been described as "secondary causation." Without denying the primacy and everywhere-present reality of divine agency, medieval and modern natural philosophers (what we today call scientists) proposed that God worked through secondary—natural—causes, and that the purpose of the natural sciences was to explore and understand secondary causation. As Ronald L. Numbers has noted, "Long before the birth of modern science and the appearance of the 'scientists' in the nineteenth century, the study of nature in the West was carried out primarily by Christian scholars known as natural philosophers, who typically expressed a preference for natural explanations over divine mysteries."[23] By the late medieval ages, Christian natural philosophers scorned appeals to the miraculous when a natural explanation—by appeal to natural, secondary causation—was possible.[24]

Later, modern natural philosophers maintained this naturalistic stance in the study of the cosmos, which they saw as entirely compatible with devout religious belief. Robert Boyle (1627–1691), for example, a devout Protestant chemist and physicist (known not only for the discovery of Boyle's law, but also for discovering the elastic properties of air, that fire needs oxygen, and that sound does not pass through a vacuum), thought that science—the study of natural laws and causations—was an act of religious worship. According to Numbers, Boyle further believed that "God's customary reliance on natural laws (or secondary causes) did not…rule out the possibility of occasional supernatural interventions, when God (the primary cause) might act 'in special ways to achieve particular ends.'"[25] The study of the cosmos—of the natural laws that describe its basic workings, and of the natural causes that typically or, for the most part, bring about the observed effects—did not rule out the possibility of the supernatural, or of divine special causation. There could, of course, be no "science" of miracles, as they do not always, or for the most part, happen (the basic requirement for scientific study, according to Aristotle), but the natural sciences need not rule them out.

Historically, therefore, there is no conflict between a naturalistic approach to the natural sciences and devout Christian belief, so long as the naturalistic approach is not defined in such a way as to exclude (by definition) the existence of God and divine causality. An antitheistic approach to the natural sciences is not only unnecessary, but is also not necessarily more helpful to the scientific endeavor than a theistic approach to the natural sciences. The history of the natural sciences demonstrates, quite clearly, that belief in a divine creator cannot only go hand in hand with rigorous scientific research, but that such a belief can and has stimulated and motivated rigorous scientific research.

We have seen the examples of Kepler, Boyle, and the twelfth-century natural philosophers. In the next section, we will use the history of the natural sciences to further illustrate our point by noting some prominent Christian scientists who were both devout believers and famous scientists motivated by their religious beliefs. Note that we are not saying Christianity saved the natural sciences from stagnation and almost certain demise, nor that one must be a Christian in order to engage in scientific inquiry. We are making the less-contentious affirmation that it is possible to be both a devout Christian (who believes that the Bible is God's inerrant Word, that Jesus is God, that the cosmos was created by God, and the basic tenets of the historic Christian faith) and a rigorous natural scientist who does not put his or her religious beliefs to one side in order to engage in scientific (or even philosophical) research.

The Testimony of History

In this final section, we will consider a representative sampling of notable natural philosophers who were instrumental in the development of what is now known as the natural sciences.

Robert Grosseteste (1168–1253) studied law, medicine, and theology at Oxford and became the first lecturer in theology for the Franciscans at Oxford. He was also the archbishop of Leicester from 1235 to the end of his life. Grosseteste contributed to the natural sciences in many ways, not least by translating and commenting on the works of Aristotle, and the publication of his own treatises on color, light, rainbows, comets, mathematics, and the movements of the heavenly spheres. He taught Roger Bacon and was one of the major medieval scholars who had an influential role in stimulating research in the natural sciences.

Albert the Great (1205–1280) joined the Dominicans in 1223 and is now known primarily as the mentor of Thomas Aquinas. In his own time, he was known as the "universal doctor" because of the incredible extent of his knowledge. Albert commented on all of the available works of Aristotle, wrote two *Summas* of theology, commented on the *Sententiae* of Peter Lombard, wrote commentaries on the works of Pseudo-Dionysius, wrote numerous commentaries on the Christian Scriptures, and wrote a number of important treatises on natural philosophy (such as *On Animals*, *On Plants and Vegetables*, *On Minerals*, *On the Causes of the Properties of the Elements*, and *On the Nature of Location*).

Not only is Albert rightly identified as the grandfather of modern botany, biology, and zoology, he is also one of the first medieval writers to clearly articulate the importance of firsthand observation and experience for the study of nature. We find this, for example, in his introduction to his treatise *On the Causes of the Properties of the Elements*, written c. 1251–1254, where he says, "And we will confirm whatever we will say about the elements by reference to visible and sensible evidence. For in the natural sciences [*in physics*], those things that are in accord with sense knowledge are the most certain."[26]

Roger Bacon (1214–1292), a Franciscan theologian, following in the steps of his mentor Robert Grosseteste, is known for his defense of the need for experience in the natural sciences and not just theory. He contributed to the natural sciences primarily in optics, mathematics, and physics.

Nicolas Copernicus (1473–1543) was a devout Catholic, serving as a canon of the church for a number of years, as well as a respected mathematician, astronomer, and practicing medical doctor. He is known primarily for his revolutionary, at the time, astronomical model called *heliocentrism* (which has the sun at the center of our solar system and the planets rotating around the sun).

Galileo Galilei (1546–1642) and his so-called conflict with the church at Rome has been the source of much theorizing about the negative influence of religion on the natural sciences. However, those who see in Galileo a staunch defender of the natural sciences against oppressive religion need to be reminded that not only was Galileo a devout Catholic and a personal friend of the Pope (who admired Galileo's work),[27] but he also acknowledged that God reveals himself both in the Scriptures and the divinely created cosmos. As far as Galileo was concerned, his conflict was not so much with the church as it was with certain natural philosophers who held positions of power in the church and who disagreed with Galileo on a philosophical level (and used their ecclesiastical influence to silence their philosophical and scientific opponent). In other words, the conflict was primarily between scientific and philosophical theories, and not between religion and science. John Hedley Brooke notes that "Galileo seems to have felt that his difficulties with the Catholic Church had their origin in the resentment of academic philosophers who had put pressure on ecclesiastical authorities to denounce him."[28]

We have already mentioned Johannes Kepler and Robert Boyle, both of whom were devout Protestants, and we could name many more, such as William Harvey (1578–1657), the devout Protestant credited with the discovery of the blood circulation system; René Descartes (1596–1650), the devout Catholic philosopher whose influence is still felt in contemporary philosophy, theology, and mathematics (he essentially invented analytical geometry); Blaise Pascal (1623–1662), the Catholic philosopher and theologian whose influence is as extensive as Descartes's and who is known for his important contributions to geometry, probability theory, physics, and mathematics; John Ray (1628–1705), the Anglican minister who is known as the father of natural history and made important contributions to embryology and the physiology of plants; Willem Jacob's Gravesande (1688–1742), the Dutch mathematician and natural philosopher who was instrumental in increasing the influence of Newton's physics outside of England and also authored an important textbook, *Mathematical Elements of Natural Philosophy*, in which he argued that an in-depth examination of our sensible world brings us to recognize the axiom of natural philosophy that states that God governs the entire natural universe via the laws of nature;[29] Joseph Priestley (1733–1804), the separatist theologian, clergyman, and celebrated natural philosopher who not only made revolutionary discoveries in relation to electricity and botany (discovering photosynthesis), but was also one of the fathers of modern chemistry (being one of the first scientists to discover oxygen, and the first to discover how to make a fizzy drink by dissolving carbon dioxide in water); Georges Lemaître (1894–1966), the Belgian Catholic priest, astronomer, and professor of physics who first proposed what came to be called the *big bang model*, which, though controversial at the time (1930s), was eventually applauded by Einstein and has come to be something of a consensus view; Francis Collins (1950–present), the Protestant geneticist whose leadership of the Human Genome Project led to the sequencing of the complete human genome, discovered the genes associated with numerous diseases and is the director of the National Institute of Health; and so on.[30] Of course this list is incomplete and says nothing of the numerous less-famous scientists who themselves were or are devout Christians. A list of additional Christians and other devoutly religious persons who were prominent scientists appears in the appendix of this book.

Can Atheists Be Scientists?

In this chapter, we have answered the question "Can a Christian be a scientist?" We considered and responded to some important theoretical difficulties, showing that faith is not antithetical to science, and that it is possible to be both a devout believer and an influential and important natural scientist. We then showed, through a brief overview of some important natural scientists, that historically some of the greatest natural scientists were devout Christians, and were, in fact, motivated by their religious beliefs to engage in the study of the cosmos.

One of the primary beliefs of the majority of late-medieval and early modern natural philosophers was that the cosmos and the Bible had the same source—the triune God of Christian Scriptures. They believed that because God was the creator of the cosmos, and had divinely inspired the Christian Scriptures, the two could not contradict each other. Our understanding of nature or interpretation of the Scriptures could be faulty or erroneous (and, therefore, our statements about nature or the Scriptures can contradict each other and be in need of correction), but nature itself and the Scriptures, as they find their source in the same author, cannot be in opposition. With these firmly held beliefs, these scientists remained devout believers and laid the foundations for modern science. These great individuals of the past should serve as examples and models for Christians today and as a warning to unbelieving scientists that they should not so easily reject the work of believing scientists on account of their belief.

On the other hand, although those with an antitheistic worldview see theists as handicapped by their alleged enslavement to superstitious beliefs, is it possible that the opposite is true? It may be, in fact, that the atheist or naturalist is the one who is handicapped by his own *a priori* conclusion that only naturalistic causes can be accepted or believed. Therein lies the problem. The conclusions regarding causes of things in our material world are predetermined by their atheistic philosophy, which requires them to exclude any supernatural causes.

Science can tell us the *how*, but not the more important fundamental question of *why*. John Lennox, in his book *Can Science Explain Everything?*, addressed this very question using an analogy. He tells the story of his Aunt Matilda baking a cake.[31] Scientists can conduct all kinds of sophisticated tests to determine an enormous amount of information about the cake, such as the nature and origin of all the ingredients. And yet even with all this scientifically derived data in hand, a fundamental question is still left unanswered: Why did Matilda bake the cake in the first place? The universe and life itself exist, but why?

The more important part of the answer lies in acknowledging the limits of science itself. Science is trying to provide answers to the fundamental questions about our very existence. Attributing the existence of the universe to pure chance, besides being unrealistic, fails to answer the questions most important to human beings. Random chance doesn't provide the answers to our origin, identity, meaning, morality, and destiny.

The goal of science is to discover important truths about the cosmos. The strict exclusion of explanations that appeal to unnatural (or miraculous) causes is like telling a homicide detective, before he even arrives at the crime scene, that the cause must be ruled a suicide. But we know there are other possible explanations. And just like the detective, the scientist is restricted in his possible conclusions because of an underlying motive. The fear of the existence of a God who is both creator and judge is a powerful motivator to exclude that possibility being discussed. Stephen Hawking famously stated that heaven and an afterlife were a "fairy story for people afraid of the dark,"[32] to which John Lennox replied, "Atheism is a fairy story for people afraid of the light."[33] Perhaps the better question should be "Can atheists be scientists?"[34]

What Is the Biblical and Scientific
Case for a Historical Adam and Eve?

Fazale Rana

D id Adam and Eve really exist? Did all humanity come from a single couple? Is there any scientific evidence that supports the biblical narratives for human origins? Can a scientific case be made for the biblical concept of human nature? Does all this really matter?

It does matter. Answers to these questions are of central importance to the Christian faith. The existence of a historical Adam and Eve as sole progenitors of all humanity bears directly on a number of key doctrines, including[1]

- the inspiration and authority of Scripture

- the nature and identity of human beings

- original innocence

- the fall

- origin of sin and transmission of original sin

- the atonement

- biblical ethics

- family and marriage

- human sexuality

All of these doctrines find their basis in Genesis 1–5, in the teachings of Jesus (Matthew 19:4-6), and in the apostle Paul's interpretation of Christ's life, death, and resurrection (Acts 17:26; Romans 5:12-19; and 1 Corinthians 15:21-22, 45-49). Significantly, these doctrines rest on the historicity of Adam (and Eve). They seamlessly follow from the biblical text if, and only if, Adam and Eve were the first humans—created uniquely to bear God's image—who gave rise to the entirety of humanity as their sole progenitors.

A Biblical Case for the Traditional Theological View of Adam and Eve

It's not hyperbole to say that over the centuries, most theologically conservative Christians have regarded Adam and Eve to be real, historical individuals based on a traditional understanding of specific Old and New Testament passages.

As a case in point, the Hebrew word used in Genesis 2:7 to describe Adam's creation from the dust of the Earth is *yatsar*. This would be the verb used to describe a potter shaping and molding a piece of clay.[2]

Later, in Genesis 2:22, the verb *banah* is used to describe Eve's creation from Adam's side. This verb (which means "to build, to repair") implies Eve was created by reworking a piece of Adam.[3] In both instances, the text describes God's direct and intimate involvement in Adam and Eve's creation. This passage leaves little room, if any, for an evolutionary origin of Adam and Eve from preexisting hominins. Adam is made from inanimate material that is fashioned and shaped by God and then animated by the divine breath. This passage largely precludes the idea of the Creator selecting Adam and Eve from an already-existing group of humans. It clearly describes Adam and Eve as the *first* humans. They had no parents or evolutionary ancestors of any sort.

In many respects, Genesis 2 echoes Genesis 1. Genesis 1 uses the verbs *asah*[4] (in verse 26; to build and construct with an emphasis on fashioning) and *bará*[5] (in verse 27; to initiate something new instead of manipulating preexisting material) to describe the creation of male and female in God's image and likeness. As with *yatsar* and *banah*, these two verbs connote God's direct and intimate role in the creation of the first humans.

On a related note, Genesis 2:19 teaches that the beasts of the field and the birds of the air were also formed (*yatsar*) from the dust of the Earth. However, these creatures do *not* receive the divine breath. The passage teaches that Adam and the animals are made from the same "stuff" and implies that humans and animals share biological features. This trait anticipates modern scientific discoveries that reveal anatomical, physiological, cellular, biochemical, and genetic similarities among humans and other animals. Within the evolutionary framework, these similarities are viewed as *prima facie* evidence for common descent and a shared evolutionary history.

But from a biblical perspective, these shared similarities connote creation from the same materials—that is, common design.

These two passages also highlight a key distinction between humans and the animals. Only Adam receives the divine breath. And only male and female humans are made in God's image. So, even though humans share biological features with other animals, they stand apart from all other creatures as the ones who uniquely bear God's image. Again, this idea leaves little room to view humans as having evolved from intermediate forms, differing only in degree (not in kind) from these predecessors, and lacking the cognitive and behavioral features that make humans exceptional.

The genealogies in Genesis 5 and Luke 3:23-38 include Adam, strongly implying that Adam was a historical person, the first of his kind created by God to bear his image. The Genesis 5 genealogy indicates that through Adam, both God's image and the effects of sin were transmitted to Adam's descendants. In Romans 5:12, Paul echoes this point, teaching that Adam's fall brought sin and death to the entire world, indicating that he understood Adam's descendants to encompass all of humanity.

In other words, it is evident from the Pauline epistles to the Roman (5:12-21) and Corinthian (1 Corinthians 15:22) churches that Paul regarded Adam as a historical figure, the first human being, and the sole progenitor of all people. For these reasons, Adam's actions in the Garden of Eden put all of humanity at odds with God. Paul reiterates his view of Adam's historicity and progenitorship in his sermon on Mars Hill at the Areopagus, proclaiming that God "made from one *man* every nation of mankind to live on all the face of the earth" (Acts 17:26). And lastly, Genesis 3:20 functions as a counterpart to Acts 17:26 by describing Eve as "the mother of all living."

Two Challenges from Mainstream Science

Despite this widely held view, a growing number of evangelical and conservative Christians question the traditional treatment of the biblical account of human origins and the historicity and sole progenitorship of Adam and Eve. The primary motivation behind their uncertainty about the Bible's human origins story is the seemingly unanimous consensus of the scientific community that humanity began from a minimum population size of several thousand individuals, not a single pair.[6]

In part, this view flows out of a deep-seated commitment to an evolutionary interpretation of humanity's origin. Evolution is a population-level phenomenon; thus, if humans evolved with an evolutionary ancestry that is shared with the great apes, then by definition humans must have started with a population. To think otherwise would be, from an evolutionary vantage point, an absurdity.

A second reason to question the traditional theological view of Adam and Eve stems from the genetic diversity observed among people spread all over the world. Many scientists claim humanity's genetic diversity is far too expansive to have come from an original couple.

Are these reasons valid? Not necessarily. The scientific case for the biblical account of human origins will be discussed shortly. Still, for those Christians who accept the consensus of the mainstream scientific community, they have no choice other than to move on from the traditional theological view of Adam and Eve while, at the same time, hoping to preserve key biblical doctrines.

Alternative Proposals to the Traditional Theological View of Adam and Eve

So, how should the Genesis 1 and 2 creation accounts of humanity be understood if humanity does, indeed, share an evolutionary ancestry with the great apes, arising out of an initial population size of several thousand individuals? Evangelical scholars have proposed several possible interpretations.[7] Four of the most prominent theological models are described below.

Adam and Eve Were Ancient Representatives of Humanity

One common version of this model considers Adam and Eve to have been part of the emerging population of the first modern humans (around 200,000 years ago). In some versions of the model, modern humans developed the capacity to relate to God (the image of God) through the evolutionary process. Other accounts maintain that God intervened in a miraculous way to endow humans with the image of God and the capacity to relate to him. Either way, God chose Adam and Eve from among a preexisting population of hominins to be representatives for all humanity. In a sense, they were ancestors of humanity because they were part of the initial population of modern humans. God placed them in the Garden of Eden, where they received moral instruction. Yet they chose to disobey him and they fell. Original sin was then spread to all of humanity as Adam and Eve had offspring, who, in turn, had children with those outside the Garden. In this way, the effects of original sin disseminated to the rest of humanity through genetic and cultural transmission.

Adam and Eve Were Recent Representatives of Humanity

The second model also has a number of variants. One common adaptation places Adam and Eve in the time of the Neolithic period (around 15,000 years ago). Accordingly, God selected them to serve as humanity's representatives, but not ancestors (either genealogical or genetic). God placed them in the Garden of Eden, where they received special revelation and, ultimately, sinned. Because they served as humankind's representatives, guilt was imputed to all humanity (under this view, "all humanity" is a population that descended through evolution from a common ancestral population we shared with apes). Some versions of this model regard the creation account in Genesis 2 to be situated in the Neolithic period and Genesis 1 (verses 26-31) in the Paleolithic period around 200,000 years ago.

Adam and Eve Were Genealogical (Not Genetic) Ancestors of Humanity

A third model—most recently advanced by computational biologist and physician Joshua Swamidass—shares some features with the two representative models. However, it differs by viewing Adam and Eve as the genealogical, not genetic, ancestors of all humanity.[8]

At first glance, the difference between genetic ancestry and genealogical ancestry can appear unclear. Genetic ancestry refers to specific genes an individual receives from his or her ancestors. Genealogical ancestry refers to a person's family tree. The specific genetic contribution a person's ancestors make to his or her genome becomes increasingly imperceptible as ancestors become more distant in time. Not so for genealogical ancestors who are part of an individual's family tree.

When tracing a person's genealogical ancestry back through time, the number of his or her genealogical ancestors grows exponentially. At the same time, the further one goes back in human history, the smaller the human population size becomes. Due to these two effects, there is a point in human history at which every person's genealogical ancestry collapses into a single genealogy. As a consequence of this collapse, every person who existed during that window of time (who also has descendants living today) will, in effect, be a genealogical ancestor of everyone alive today. However, this coalescence does not require that only two people existed at that time. Rather, a population of individuals existed, and all of them will be the genealogical ancestors of everyone alive today (provided they have living descendants). Of this group of people, it is quite possible that a couple existed—an Adam and Eve—that, if they have living descendants today, would be ancestors to us all.

Surprisingly, the time window for the collapse of the genealogies into a single genealogy can be as recent as 3,000 years ago for people alive today. And for people alive at AD 1, when Jesus was on Earth, the time window for the universal genealogical ancestor can be as recent as 5,000 years ago.

Again, it is important to emphasize that in this model, Adam and Eve are *not* the sole genealogical ancestors of humanity. There would have been other people, contemporaries, who would also be the genealogical ancestors of everyone alive today. And there would have been people living *before* Adam and Eve lived. Also, at various earlier periods of human history, Adam and Eve would *not* have been the genealogical ancestors of *all* humans alive at that time. Thus, the genealogical Adam and Eve model shares features with the representative models. But it is also distinct in that the genealogical model regards Adam and Eve as the ancestors of everyone alive today and at the time of Christ (though not necessarily ancestral to everyone alive at early periods of human history). As with the representative models, Adam and Eve's immediate offspring would have culturally and genetically intermingled with their contemporaries, spreading the image of God and the effects of original sin to other humans.

The genealogical approach allows for Adam and Eve's contemporaries to have evolved while they themselves were created *de novo* (i.e., specially and miraculously created, without any physical ancestors), as Scripture teaches.

Adam and Eve Were Literary Figures

This fourth model rejects the historicity of Adam and Eve altogether. Instead, this approach considers the events of Genesis 1–11 to be mythic, with Adam and Eve functioning as a literary typology for humanity—a theological "everyman and everywoman." Some versions of this model claim there is no scientific or historical content in the Genesis creation accounts. Other versions might grant Adam and Eve (and the people mentioned in Genesis 1–11) some type of historical status, with an important caveat. The literary account in the first several chapters of Genesis alludes to people and events that may well have existed and happened. However, under this view, these events have been "fictionalized" and compressed together in a narrative designed to communicate theological truths.

A Critique of the Alternative Theological Views of Adam and Eve

These four alternative theological approaches attempt to reconcile human evolution with the biblical account of human origins. At the same time, they seek to retain key doctrines of the Christian faith. Are these models viable? One important criterion these views must meet is compatibility with Old and New Testament passages that refer to Adam (and Eve). In light of this standard, how do these alternative accounts of humanity's origin fare? Let's begin by first considering the fourth approach.

Were Adam and Eve Literary Figures?

Properly identifying the genre of the Genesis creation accounts is key to evaluating the literary hypothesis. Are they merely literary constructs? Two reasons suggest it is best to view these passages as historical narratives.

First, while it is true that scholars have debated the literary genre of Genesis 1–3, there are good reasons to think that these passages are historical narratives. It is quite possible that they reveal a compressed prehistory. If so, they provide a snapshot of historical events that were separated in time, juxtaposed in a contiguous chronological order. Admittedly, these passages also have a poetic structure to them, with such devices most evident in the Genesis 1 creation account. But, given the way poetry functions in biblical Hebrew, the use of poetic devices in these passages isn't unexpected and doesn't automatically render them literary constructs exclusively. The same could be said for the allusions these passages make to ancient Near Eastern creation myths. As Old Testament scholar John Currid points out, "Genesis 1–3 is at its very core anti-mythological, and this can be seen in its polemic quality and disposition."[9] Currid argues that Genesis 1–3 must be rightly viewed as a historical narrative because of the grammatical devices used in these passages, which are common to all historical narratives in the Old Testament.

Second, the witness of the biblical authors throughout the Old and New

Testaments indicates they understood Adam and Eve to be historical—not literary—figures. The appearance of Adam in the genealogies of Genesis 5, 1 Chronicles 1, and Luke 3 seems out of place if Adam were mythical, given that all the other names in the genealogies are historical figures. Jesus's allusion to Adam and Eve in Matthew 19:4-6 indicates that he regarded Adam and Eve as real, historical figures who lived at the source of humanity. Likewise, Paul's reference to Adam in his speech on Mars Hill (Acts 17:26) and his direct mention of Adam in Romans 5 and 1 Corinthians indicates that he, too, regarded Adam (and Eve) as actual humans who lived at humanity's onset. In fact, a historical Adam seems to be essential to Paul's interpretation of Christ's life, death, and resurrection. As theologian Kenneth Samples points out, "Paul's theology seems to require a historical Adam as integral to the gospel's storyline. The mythical Adam view distorts the gospel to such an extent that historic Christianity is fundamentally changed."[10]

Were Adam and Eve Representatives or Our Genealogical Ancestors?

The other three models are not without theological merit. They recognize Adam and Eve as historical figures. But despite this positive feature, all three models still fall short for both biblical and theological reasons. One biblical reason concerns the previously mentioned Hebrew words used to describe Adam and Eve's appearance in Genesis 2. They preclude viewing them as two humans who were *selected* by God out of a preexisting population, or who had parents, human or otherwise. These verbs used to describe their origin unequivocally indicate that both were *created*.

Second, the Genesis 2 account also strongly implies that, initially, Adam was truly alone. Through the process of naming

the animals, Adam had a moment of self-discovery. There was no other creature like him. Only he uniquely possessed God's image. And his response to Eve's creation—characterized by delight—reinforces the idea that without her, Adam lacked a suitable helpmate. This account of Adam and Eve's creation makes little sense if Adam and Eve were part of a population who were then chosen as humanity's representatives to take up residency in Eden.

A third biblical reason rests on the observation that there is no direct mention of any other humans who coexisted with Adam and Eve in Genesis 1–3. There is also no scriptural evidence that supports the idea that Adam and Eve and their offspring intermixed with humans outside the Garden of Eden. At best, there is a weak inference for the existence of other humans outside the Garden based on the account of Cain and Abel in Genesis 4. According to this passage, Cain was worried that he would be attacked and killed by other people after God punished him for killing his brother by forcing Cain to wander the land. In the process, Cain found a wife, had children, and built the city of Enoch.

Some scholars have argued that this passage indicates there were other humans who coexisted with Adam and Eve and Cain and Abel. If there weren't, then who was it that Cain feared? Where did he get his wife? Who would have helped him build a city? Ready answers to these questions are found in Genesis 5:4, which notes that Adam and Eve had *many* sons and daughters. It makes sense that their many children were the people whom Cain feared.

Finally, Paul's theology of the atonement appears to require an Adam who is the *sole* progenitor of all humanity. As theologian Kenneth Samples puts it:

With the historical Adam who is also humanity's physical progenitor there is indeed a common humanity in terms of origin, unity, and sinfulness…To propose that Adam was a representative of humanity but not the physical progenitor damages the solidarity of the human race as reflected in Scripture…An Adam who is merely a historical representative but not a physical ancestor of human beings may also compromise the logic of Paul's comparison between Adam and Christ in the book of Romans.[11]

This progenitor position is also affirmed by Genesis 3:20, where Eve is designated the "mother of all living." This seemingly stands in direct contrast to the genealogical ancestor model, where Eve could not have been the "mother of all"—or for that matter, any—living humans at the time she was created and given that title. Indeed, a major theological difficulty of this model is that neither Adam nor Eve would have been genealogical ancestors of *all* living humans—i.e., not only *not* the sole progenitors of all humans, but not even among any group of progenitors of all humans—until a long period of time elapsed after their creation.

What's the Relationship Between the Creation Accounts in Genesis 1 and 2?

A related question that arises with these three models has to do with how to place the two creation accounts in history. Some versions of the representative and genealogical models require the Genesis 1 and 2 creation accounts to be either sequential or separate. But neither view is borne out by the witness of other biblical authors. For example, when Jesus discusses the origin of marriage (Matthew 19:4-5), he overlays the

creation accounts in Genesis 1 and 2, clearly indicating that he understood them to be complementary accounts, not sequential or distinct descriptions of humanity's origin. The genealogy of Genesis 5 and Psalm 8 also intertwine the Genesis 1 and 2 accounts, again, indicating that Genesis 1 and Genesis 2 are complementary.

In other words, any model that depends on Genesis 1 and 2 being either unique or sequential accounts of creation simply cannot be valid.

When Were Adam and Eve Created?

The timing of Adam and Eve's creation bears significantly on the question of their historicity. However, as important as this question may be, Scripture gives no clear guidance for gauging when the first humans appeared. In fact, the best scholarship on this matter suggests that Scripture is silent regarding the date for Adam and Eve's creation.

Nevertheless, some Bible interpreters treat the genealogies in Genesis 5 (Adam to Noah) and Genesis 11 (Noah to Abraham) as exhaustively complete chronologies and have attempted to determine the date for Adam and Eve's creation from them. This approach, however, is questionable for several reasons.

First, the Genesis 5 and 11 genealogies were not intended as timekeeping constructs. Rather, like all genealogies found in Scripture, these accounts were meant to communicate theological truths.[12] At the onset of the Genesis 5 genealogy, the author makes reference to Genesis 1:26-27 with the clear implication that the image of God imparted to Adam was passed on to all his descendants. In like manner, the expression "and then he died" (which is not used in any other genealogy found in Scripture) teaches that the death that Adam incurred when he sinned against God would be passed along to all of his descendants, for all are "in Adam" (1 Corinthians 15:22).

Second, this expression also reinforces the idea that God keeps his promises. God informed Adam that if he ate from the tree of the knowledge of good and evil, he would surely die. And yet, he didn't die that day. He lived 930 years. Still, he died. And so have (and will) all of those who are in him.[13] God's faithfulness in fulfilling his promises is also part of the message of Genesis 11. In Genesis 6:3, God declares that he intends to limit human life expectancy to no more than 120 years. And, it is clear from the Genesis 11 genealogy that God kept his promise as the ages of the patriarchs successively declined from Noah (950) to Abraham (175).[14]

Third, it is commonplace for biblical authors to telescope genealogies, meaning that they intentionally omit names for a variety of reasons.[15] One reason for telescoping is to ensure that the genealogy conforms to a pattern. The Genesis 5 and 11 genealogies both display the same pattern: a tenfold linear genealogy (identifying ten patriarchs), with the eleventh entry segmented into three individual names.

Finally, the ambiguity surrounding biblical Hebrew also makes it clear that the genealogies shouldn't be used to mark time. The range of meaning for the Hebrew words translated as "father" (*ab*) and "son" (*ben*) can include "ancestor" and "descendant," respectively.[16] Similarly, the Hebrew word translated as "begot" or "become the father of" can mean to father an individual or to bring forth a lineage.[17] In Hebrew thought, a father is not only the parent of his child but also of all his child's descendants. In other words, the word usages associated with the genealogies are perfectly consistent with telescoping. According to biblical scholar

K.A. Kitchen, the genealogies of Genesis 5 and 11 should be read as, "*A* fathered [*P*, who fathered *Q*, who fathered *R*, who fathered *S*, who fathered *T*, who fathered…] *B*." This means that Genesis 5 and 11 should then be read as "*A* fathered the lineage culminating in *B*, and after fathering the line, lived X years."[18]

Bottom line: A careful reading of the Genesis 5 and 11 genealogies makes it evident that Scripture places no constraints on the timing of Adam and Eve's creation. This relaxation means that the scientific evidence should be a significant factor for determining the timing of humanity's first appearance on Earth.

But, before we turn to the scientific evidence to estimate the timing of humankind's creation, it is imperative to know precisely what is intended by the term *human*.

The Science of Human Origins: Who Was Adam?

One of the challenges in reconciling the science of human origins with the biblical accounts relates to the ambiguity of the word *human*. Sometimes *human* refers to all of the ostensibly bipedal primates (hominins), including members of the genera *Sahelanthropus*, *Orrorin*, *Ardipithecus*, *Australopithecus*, *Paranthropus*, and *Homo*. Sometimes paleoanthropologists use *human* as a designation for all members of the *Homo* ("*H.*") genus. Sometimes the term applies only to the species *Homo sapiens*.

The concept of *H. sapiens* can be confusing as well. In some instances, paleoanthropologists will use the term *H. sapiens* to classify hominins that lived between 500,000 years ago up to the appearance of modern humans. This list can include some specimens of *H. erectus*, *H. antecessor*, *H. heidelbergensis*, and *H. neanderthalensis*. Sometimes paleoanthropologists use the phrase "archaic *H. sapiens*" in reference to these hominids, in contradistinction to modern humans. When referring to human beings (as popularly understood), paleoanthropologists use terms like *modern human, anatomically modern human*, or *anatomically and behaviorally modern human* interchangeably. Scientifically speaking, modern humans are classified as *H. sapiens sapiens*.

So, which of these concepts of human corresponds to Adam and Eve and their descendants? Scholars committed to the traditional theological view of Adam and Eve hold varying perspectives. Some take Adam and Eve to correspond to both archaic *H. sapiens* and *H. sapiens sapiens*. Others take Adam and Eve to correspond solely to *H. sapiens sapiens*. The latter view seems to make more sense. People alive today are all anatomically and behaviorally modern humans. If we are descendants of Adam and Eve, then it serves to reason that they, too, were anatomically and behaviorally modern humans.

H. sapiens sapiens stand apart from archaic *H. sapiens* in several significant ways. One of the most significant differences manifests in skull shape. Compared to other hominins (such as Neanderthals and *H. erectus*), modern humans have oddly shaped skulls. Hominin skull shape was elongated along the anterior-posterior axis. But the skull shape of modern humans is globular, with bulging and enlarged parietal and cerebral areas. The modern human skull also has another distinctive feature: the face is retracted and relatively small. Anthropologists believe that the difference in skull shape (and hence, brain shape) has profound significance and helps explain modern humans' advanced cognitive abilities. The parietal lobe of the brain is responsible for

- perception of stimuli

- sensorimotor transformation (which plays a role in planning)

- visuospatial integration (which provides hand-eye coordination)

- imagery

- self-awareness

- working and long-term memory

These sophisticated features make modern humans exceptional and unique compared to other hominins and account for the significant cognitive differences between modern and archaic humans.

When Did Adam and Eve Exist?

When do anatomically and behaviorally modern humans appear in life's history? Three independent lines of evidence help establish the scientific date for the appearance of *H. sapiens sapiens* (hence, Adam and Eve) in life's history.

Molecular Anthropology

Scientists working in molecular anthropology study the genetic variations of people around the world. Variations in DNA sequences reveal important clues about the origin of modern humans, such as the time window and location for humanity's beginnings and the pattern of humanity's spread around the world. Two genetic markers used by anthropologists, mitochondrial DNA and Y-chromosomal DNA, have yielded provocative findings. Analysis of mitochondrial DNA indicates that every human traces his or her origin back to a single ancestral sequence that could be interpreted as a single woman (dubbed mitochondrial Eve). Likewise, characterization of Y-chromosomal DNA indicates that all men trace their origin back to a single ancestral sequence that could be interpreted as a single man (dubbed Y-chromosomal Adam). These analyses also indicate that mitochondrial Eve and Y-chromosomal Adam lived about 150,000 years ago (±50,000 years).[19]

Fossil Record

The skull shape of modern humans is globular, with bulging and enlarged parietal and cerebral areas and a relatively small, retracted face. A detailed study of the fossil record reveals that modern human skull morphology (and brain shape) appeared between 130,000 to 100,000 years ago.[20] It stands to reason that this period would be the window of time that modern humans and, hence, Adam and Eve appeared on Earth.

Archaeological Record

Results from molecular anthropology and the fossil record align with recent archaeological finds that place the origin of symbolism around 125,000 to 150,000 years ago.[21] Symbolism—the capacity to represent the world and abstract ideas with symbols—appears to be uniquely human and is most likely a manifestation of the modern human brain shape, specifically an enlarged parietal lobe.

Bottom line: Three distinct approaches for dating the origin of modern humans converge to place the origin of humanity safely between 125,000 to 175,000 years ago.

Did Humanity Emerge from a Primordial Pair?

The discovery of mitochondrial Eve and Y-chromosomal Adam is provocative, to say the least. This astounding finding opens the

possibility that the traditional theological view of human origins and the science of human origins can harmonize. Thus, it gives credence to the idea that Adam and Eve likely existed as real persons who gave rise to all of humanity.

However, those committed to the evolutionary paradigm have challenged this interpretation. They argue that the genetic diversity data shows that humanity arose from thousands of individuals, not two. The chief basis for this claim comes from estimates of the ancestral population size of humans based on genetic diversity using mathematical models.

In the face of this challenge, it is important to recognize that population sizes generated by these mathematical tools are merely estimates, not hard-and-fast values. The reason: the mathematical models are highly idealized, generating differing estimates based on a variety of factors, including varying sample size and number of locations in the human genome that were studied.

Moreover, studies in conservation biology raise serious questions about the validity of these methods. When a species is on the verge of extinction, conservationists often know the number of individuals that remain. And because genetic variability is critical for species recovery and survival, conservation biologists monitor genetic diversity of endangered species. In other words, conservation biologists have the means to validate population size methods that rely on genetic diversity.

By employing these means, researchers have measured the genetic diversity generations after the initial populations are established. Results show that the genetic diversity is often much greater than expected—again, based on the models relating genetic diversity and population size. In other words, these methods often fail to accurately predict population size. If researchers use the measured genetic variability to estimate original population sizes, the sizes would have measured larger than they actually were.[22] Again, these findings cast aspersions on the results of any population size estimate based on genetic diversity. Thus, it is not clear that the genetic diversity of humanity observed today is incompatible with humanity's inception as a primordial pair.

In light of these concerns, it is worth noting how biologist Ann Gauger and mathematician Ola Hössjer have accounted for genetic diversity. Using a mathematical modeling approach, they demonstrated that the human genetic diversity observed today could hypothetically be accounted for even if humanity began as two individuals.[23] Prior to this landmark study, molecular anthropologists merely assumed that a two-person start could not account for the human genetic diversity observed today. But nobody had tested this idea. And when it *was* tested, it turns out that two individuals living about 100,000 years ago, under certain scenarios, could give rise to a modern-day population of humans with the observed genetic diversity.

Bottom line: The scientific evidence doesn't rule out the existence of Adam and Eve. And, while there is no direct proof that humanity came from two individuals, it is still within the realm of scientific possibility.

Is There Scientific Evidence for the Image of God?

To appreciate the scientific case for the image of God, it is necessary to know what the image of God entails. This is no simple undertaking. Scripture doesn't explicitly state what the image of God is; nevertheless, theologians have offered three leading views: the resemblance view, the representative view, and the relational view.

The resemblance view has been the historic Christian view. Accordingly, the image of God describes humanity's spiritual—though finite and limited—resemblance to God. Two New Testament passages (Colossians 3:10 and Ephesians 4:24) support the resemblance view. In these passages, Paul encourages the Christians at Colossae and Ephesus to allow the Holy Spirit to transform them into the image of their creator (God). These passages imply that God's image includes our capacity for knowledge, understanding, love, holiness, and righteousness. In other words, according to Paul, the image of God refers to attributes we possess as humans that distinguish us from all other creatures. Specifically, the image of God entails the following facets:

- *A moral component.* Humans inherently understand right and wrong and have a strong, innate sense of justice.

- *A spiritual component.* Humans recognize a reality beyond this universe and physical life. We intuitively acknowledge the existence of God and have a propensity toward worship and prayer and desire to connect to the transcendent.

- *A relational component.* Human beings relate to God, themselves, other people, and other creatures.

- *An intellectual component.* Humans possess the ability to reason and think logically. We engage in symbolic thought. We express ourselves with complex, abstract language. We are aware of the past, present, and future.

We display intense creativity through art, music, literature, science, and technical inventions.

Because humans uniquely bear God's image, Christians have historically regarded humans as exceptional. We display unique attributes and stand apart from all other creatures on the planet. Our exceptional nature results from being made in the image of God and entails that humans have infinite worth and value.

Yet because of the evolutionary paradigm's influence, anthropologists have historically rejected the notion of human exceptionalism. Instead, they maintain that humans are nothing more than animals—differing in *degree*, not *kind*, from other creatures.

Today, a growing minority of anthropologists and primatologists now argue that humans are indeed exceptional because we uniquely possess the following combination of qualities:[24]

- capacity for symbolism
- open-ended generative capacity
- theory of mind
- capacity to form complex social networks

Humans effortlessly represent the world and abstract concepts with discrete symbols. Our ability to represent the world symbolically has interesting consequences when coupled with our abilities to combine and recombine those symbols in nearly infinite ways. Our imagination is limitless. The human capacity for symbolism is manifest in language, art, music, and even body ornamentation. And we desire to communicate with other humans the scenarios we construct in our minds.

In many respects, these qualities stand as scientific descriptors of the image of God.

For anthropologists and primatologists who think that humans differ in kind—not degree—from other animals, these qualities demarcate us from the great apes and Neanderthals. No convincing evidence exists that leads us to think Neanderthals shared the qualities that make us exceptional.[25]

The separation becomes most apparent when we consider the remarkable technological advances we have made during our tenure as a species. As a case in point, Neanderthals—who first appear in the fossil record around 250,000 to 200,000 years ago and disappear around 40,000 years ago—existed on Earth longer than modern humans have. Yet our technology has progressed exponentially, while Neanderthal technology remained largely static.

According to paleoanthropologist Ian Tattersall and linguist Noam Chomsky (and their coauthors),

> Our species was born in a technologically archaic context, and significantly, the tempo of change only began picking up after the point at which symbolic objects appeared. Evidently, a new potential for symbolic thought was born with our anatomically distinctive species, but it was only expressed after a necessary cultural stimulus had exerted itself. This stimulus was most plausibly the appearance of language…Then, within a remarkably short space of time, art was invented, cities were born, and people had reached the moon.[26]

In other words, the development of human technology signifies that there is something special—exceptional—about us as humans.

Scientific evidence for the image of God, for a primordial human pair, and for the timing of that pair all buttress the biblical accounts. In short, there has been no other time in human history when it has been more scientifically credible to hold to the traditional theological view of Adam and Eve.

On Science and Scientism:
What Insights Does C.S. Lewis Offer?

Terry Glaspey

C.S. Lewis, one of the greatest Christian thinkers of modern times, was emphatically not an opponent of science. But he drew an important distinction between science as a method of intellectual enquiry and scientism, which he saw as an illegitimate usurping of power by the scientific elite and their supporters. Science, as properly understood, offers experimental and observational techniques that can be used to draw reasonable conclusions about such matters as fall under its purview. But scientism takes a critical and problematic next step by holding to the perspective that science's methodologies provide the *only* reliable means for obtaining knowledge about human beings and the world in which they live. If science is the only way that we can discover truth of all kinds, then it follows that scientists should not only be the ultimate arbiters of public policy, but that the scientific elite are also privileged to speak out authoritatively on matters of morality and religious beliefs. For Lewis, such a perspective was a bridge too far.

Science and Faith in Conflict?

The idea that there is warfare between faith and science has largely been proposed and perpetrated by the most extreme voices of each side of that supposed rift. There are some persons of faith who are quick to reject every finding of science that doesn't easily fit into their predetermined theological system and use the most extreme rhetorical pronouncements to drive a wedge between their resolute beliefs and the latest scientific conclusions. Such individuals are unwilling to put much effort into an ongoing reassessment of their theological interpretations in the light of new scientific interpretations.

Similarly, for those who view any kind of supernatural reality as merely psychological wishful thinking and intellectual fantasy, there can be no ground for meaningful discussion with those they deem too ignorant to be an equal partner in any sort of reasonable conversation. Only their own materialistic presuppositions are allowed a place in the discussion, and the proponents of religious faith are treated with dismissive contempt. When operating under the warfare metaphor, it is easier to see representatives of the other view as enemy antagonists rather than conversation partners.

C.S. Lewis was not comfortable with the idea that the Christian faith is at war with science. In fact, he seemed unhesitant about accepting widely held scientific perspectives that were based on research, observation, and experimentation, even on subjects as controversial as the place of evolution as the process for explaining human origins (as long as this process was not envisioned as being the result of blind, unguided chance). He took science seriously and understood its value in offering answers to some important questions.

For Lewis, the application of the scientific method was *one of the ways* we could gather information, improve our way of life, and progress in knowledge. But he did not accept the idea that the scientific method was the only, or even the best, way that we could do those things. For Lewis, science has a place at the table, but it is not a privileged place. In fact, as he pointed out, science cannot answer some of our most important questions, nor can it help us understand how to properly use the knowledge it provides. Science offers us information, but it does not provide wisdom.

For the proponents of scientism, science takes the place of religion as a source for providing meaning for our lives. It is made the foundation for an optimistic version of naturalism, which promises not only to improve human lives, but, in fact, to improve human beings.

The Flaw of Scientism

The flaw of scientism is that it puts forth excessive claims about the ability of the scientific method to tackle the full array of human problems. The foundation of Lewis's critique of scientism is his evaluation of the philosophical position of naturalism. Naturalism is the conviction that the only realities that exist are those we can engage with our senses. This leaves no room for any supernatural or transcendent realities or values. What we can see, hear, smell, touch, and taste are the only avenues we have for being certain of what we know. Therefore, the naturalist believes that every statement about ethics that involves a referent to supernatural or transcendent values is really nothing more than a statement about the preferences or emotional state of the one who makes it. Naturalists assert that the material realm alone is open for scientific investigation, therefore it is the only realm we can talk about with certainty. That being the case, an empirical approach is the only one open to us.

Scientia and *Sapientia*

Lewis is part of an intellectual tradition that emphasizes the necessity of two different and distinct kinds of knowledge, each important for its own sphere. *Scientia* is a knowledge of the physical world that is available through scientific observation and experimentation. It is concerned with matters of fact, quantity, repeatability, and matter. *Sapientia*, on the other hand, is the knowledge that results from intuition, common sense, and revelation, which leads to transcendent knowledge and is involved with issues of meaning, value, and purpose.

If we confuse these two ways of knowing, we are liable to make catastrophic errors about reality. This is precisely the problem with scientism: It does not recognize that each of these forms of knowing should operate in its own sphere. When we decide to run a scientific experiment to gain deeper understanding of something in the physical sphere, we need not (and should not) rely on intuition or dogma. And when the scientific method cannot be deployed, as it cannot for many of life's most pressing issues, we should look instead to wisdom and revelation. *Scientia* is valuable in the pursuit of certain kinds of truth, but we must understand its limits, which is what the philosophy of scientism fails to do.

One might consider our thought process this way: Imagine coming upon a teakettle that is sitting on an electric stove, whistling merrily and insistently. We ask ourselves, "Why is the teakettle making that whistling sound?" Part of the answer is that it is sitting on an element on the stove that has reached 212 degrees Fahrenheit, and that this has agitated the water within the kettle and caused it to boil, releasing some particles as steam that are being forced through the small opening at the top and thus producing a whistling sound. That is a perfectly adequate answer to our question.

But another answer might be this: because someone has decided they want to make a cup of tea.

Both answers are correct, but each deals with a different realm. Science tells us how things happen, but doesn't tell us much about the purposes behind what happens. As Lewis suggested, "In science we are only reading the notes to a poem; in Christianity we find the poem itself."[1]

Lewis believed that when science attempts to intrude on the questions of purpose, it can become dangerous, a pretext for a power grab against the long tradition of *sapientia*, or wisdom. At that point it becomes difficult to distinguish science from magic. Magic is the attempt to manipulate reality by using spells and incantations, and perhaps scientism is not so different. Like magic, scientism sees the natural world as something we can manipulate for our own ends. The results—atomic weapons, pollution, climate change, genetic manipulation, etc.—may well prove catastrophic for humans. In the process we end up manipulating not just nature, but other people as well.

The Implications of Scientism

The implications of scientism are the theme of Lewis's novel *That Hideous Strength*, which chillingly illustrates what can happen when science runs amok, controlled by an elite who have wrongly convinced themselves that they have the best intentions for humanity in mind as they work to control and shape human destinies. The scientific elites in Lewis's story are more concerned with what they *can* do than with what they *should* do. This is a crucial difference. As Lewis said in a letter to Arthur C. Clarke: "I agree Technology per se is neutral, but a race devoted to the increase of its own power by technology with a complete indifference to ethics does seem to me a cancer in the Universe."[2] Or, as the great French writer Francois Rabelais proposed over a hundred years before Lewis, "Science without conscience is nothing but ruination of the soul."[3]

The fundamental question is this: Is the philosophical foundation of scientism adequate to entrust it with the increasingly difficult ethical issues of the modern age? In his *A Guide for the Perplexed*, E.F. Schumacher offers this evaluation of scientism:

> The maps provided by modern materialistic Scientism leave all the questions that really matter unanswered; more than that, they deny the validity of the questions...the ever more vigorous application of the scientific method to all subjects and disciplines has

destroyed even the last remnants of ancient wisdom…[4]

Scientism is a philosophy that can easily lead to dehumanization because it takes the questions of transcendent value off the table. As Lewis suggested in *The Weight of Glory*, this leads to reductionism:

> The critique of every experience from below, the voluntary ignoring of meaning and concentration on fact, will always have some plausibility. There will always be evidence, and every month new evidence, to show that religion is only psychology, justice only self-protection, politics only economics; love only lust, and thought only cerebral bio-chemistry.[5]

What we need, suggests Lewis, is not only to look at the facts, but to look for the meaning behind them. Even as we have pursued the facts, we have traveled only part of the distance down the long road to understanding. In Lewis's *The Voyage of the Dawn Treader*, the young Eustace Scrubb confidently states, "In our world a star is a huge ball of flaming gas." But the wise old man sees it a little differently. "Even in your world, my son," replied the man, "that is not what a star is, but only what it is made of."[6]

Lewis offers a helpful metaphor for how reductionism causes us to fail to correctly interpret even the facts themselves:

> The strength of such a [materialistic] critic lies in the words "merely" or "nothing but." He sees all the facts but not the meaning. Quite truly, therefore, he claims to have seen all the facts. There is nothing else there, except the meaning. He is, therefore, as regards the matter in hand, in the position of an animal. You have noticed that most dogs cannot understand pointing. You point to a bit of food on the floor: the dog, instead of looking at the floor, sniffs at your finger. A finger is a finger to him, and that is all. His world is all fact and no meaning.[7]

A world of bare facts is a world open to manipulation by those who hold the keys to what they see as the only legitimate form of knowing. And, according to Lewis, the ultimate result of scientism is to empty everything of significance and meaning:

> You cannot go on "explaining away" forever: you will find that you have explained explanation itself away. You cannot go on "seeing through" things for ever. The whole point of seeing through something is to see something through it.[8]

Scientism, Power, and Control

Tellingly, it is often not the scientists themselves who have carried the torch for such a view of science, but more often it's popularizers and simplifiers—especially those with an anti-metaphysical axe to grind. They see the authority of science as a replacement for God and traditional ethics rather than what it has been historically: an ever-shifting set of paradigms for understanding our world. Scientism is not really about the exercise of legitimate science, but about the exercise of power and control, which is why C.S. Lewis and other thinkers have sought to question its presuppositions and warn against its excesses.

How Has Evil Been Done in the Name of Science?

Richard Weikart

When Ota Benga, an African pygmy, was put on display in the primate house of the Bronx Zoo in 1906, controversy erupted at this gross injustice. African-Americans denounced the spectacle as degrading and dehumanizing. Many scientists, however, including one of America's leading paleontologists, Henry Fairfield Osborne, president of the American Museum of Natural History, regarded the display as scientifically justified. These scientists argued that science demonstrated the inferiority of black Africans, who, according to them, were closer to apes on the evolutionary scale. The scientific racism that resulted in Benga's incarceration with apes also justified and reinforced many other immoral deeds in the late nineteenth and early twentieth centuries, including colonial wars and even genocide.[1]

In what way can we say that science was to blame for these and other nefarious deeds committed by people using the aura of scientific expertise to legitimize evil conduct? Some will object that scientific racism and other allegedly scientific constructs that resulted in immoral practices were merely pseudoscience, rather than real science. After all, almost every scientist today rejects scientific racism as misguided and counterfactual, viewing it as the product of ignorance and prejudices of the past. This distinction between science and pseudoscience rescues science from the moral taint of activities conducted under the auspices of allegedly pseudoscientific ideas.

Most historians of science today, however, reject this distinction because it illegitimately applies today's standards of science to an earlier time. In the late nineteenth century, most biologists and anthropologists embraced scientific racism, and they often contemptuously dismissed racial egalitarians as ignorant and hopelessly unscientific. Standard biology and anthropology textbooks taught that blacks were intellectually and even morally inferior to whites. That was considered mainstream science at the time.

Another objection to the involvement of science in evil deeds is that science is an amoral enterprise, so it cannot tell us how to conduct our affairs; thus, it cannot be implicated in any evil. There are three big problems with this objection.

First, some fields of science, such as

sociobiology and evolutionary psychology, claim to study the origins of morality and human conduct. If their claims are accurate (and I do not think they are), this has huge implications for the meaning of morality and thus for how we should act. Evolutionary psychologists regularly regale us with pronouncements about how humans should conduct themselves.

Second, some scientific theories have implications for our understanding of who we are as humans. Many evolutionary scientists claim that we as humans are nothing more than a cosmic accident that arose without any purpose or goal. This understanding of humanity influences moral decisions about how people should be treated and thus can lead to immoral conduct. Indeed, the sociologist John Evans demonstrated that people who reject a theistic understanding of humanity in favor of a strictly Darwinian biological view of humanity also tend to abandon regard for human rights.[2] Third, in carrying out the scientific enterprise, scientists have themselves engaged in evil deeds and atrocities, especially through immoral kinds of human experimentation.

Another objection I have encountered when discussing the evil spawned by some kinds of science is the claim that the evil deeds in question were a misuse of science. After making this claim, my interlocuter sometimes brings up the example of nuclear energy. Those who discovered nuclear energy contributed to our scientific knowledge, but then others came along and misused this knowledge by constructing bombs to kill people. This analogy, however, does not work very well. Scientific knowledge about nuclear energy can be used in a variety of ways, both for good or for evil (or simply to promote understanding of the cosmos). Scientific racism is quite different because it cannot be

used for any good purposes, and the theory itself has direct moral implications about the way we treat our fellow humans in a way that nuclear energy does not.

Before proceeding further and chronicling some of the evil deeds perpetrated in the name of science, I want to stress that science is a wonderful enterprise that benefits humanity by helping us to understand our world. We should celebrate its many triumphs. However, scientists are not emotionless computers who stoically crank out completely impartial, logical ideas. They are beset by the same prejudices and passions that influence all other mere mortals. In some fields of science, this may not have much effect on the theories and practices of science, but the closer the scientific field comes to dealing with humanity, the greater the perils. Allegedly scientific theories about human behavior must be treated with more caution than theories about the behavior of oxygen. Indeed, the field of psychology is currently beset by a "replication crisis" because scientists trying to replicate many experiments—even famous textbook examples—are coming up with results that differ from those of the original experiments that served as the modern foundations of the field.

Another way that human passions can influence the scientific enterprise is by scientific dishonesty, which is more prevalent than many like to admit. Some scientists are tempted to gain honor, prestige, and research funding by fudging or ignoring data to come up with results they consider more impressive or more dramatic or more congenial to their preconceptions (or more likely to get them further funding).

In the remainder of this chapter, I will examine how some forms of science have contributed to evil in two key ways: by theories that justified and encouraged immoral behavior, and by immoral experimentation. With regard to scientific theories, I will focus on those that were widely accepted by scientists in that time, not ones that were embraced by only a minority of scientists. Not all scientists would agree about the horrific ways that some of these theories were applied, but the scientific theories did play a role in inspiring bad behavior. I will not examine more remote applications of technology spawned by science (such as nuclear energy) because this would be tantamount to blaming the knife for the murder rather than blaming the knife-wielding assailant and the ideas that inspired him. We will focus on the ideas that influenced the evil deeds at the time when those ideas were considered scientific.

One of the difficulties we encounter in discussing these issues is that in many cases, such as scientific racism, preexisting ideas—such as racial prejudices and European feelings of racial superiority—were integrated into scientific theories. Scientists took ideas that were already widespread and inserted them into their science. In these cases, it is difficult to tease out the responsibility of science for the subsequent evil deeds. After all, racial animosity, race-based slavery, and colonial wars of conquest all preexisted scientific theories of racism. Thus, it could be that science merely justified what Europeans would have done anyway. This would mean that science did not really contribute to evil in any significant way.

However, the problem with this view is that after scientists integrated racism into their scientific theories during the nineteenth century, the authority of science convinced many contemporaries of the "truth" of racism. Stephen Jay Gould, a historian of science, admits that racism increased considerably during the late nineteenth century, and in an era when scientific authority was

highly esteemed, the claims that racism was scientific swept away many doubts. Thus, it seems that science, even when it did not originate an idea, still may have contributed something to an idea's widespread popularity and its intensity.

Scientific racism first arose in the late eighteenth century because of the influence of Enlightenment rationalism. Thinkers of the radical Enlightenment, which embraced either Spinozist pantheism or outright materialism, often espoused either biological determinism or environmental determinism (thus rejecting free will). While environmental determinism often militated against racism, biological determinism promoted racism and spawned scientific racism. Over the course of the nineteenth century, scientific racism increased in intellectual influence so that by the end of the century, most biologists and anthropologists considered it settled science.[3]

The emergence and growing influence of a scientific theory—Darwinism—in the late nineteenth century was a key factor contributing to the increased acceptance of scientific racism. Darwin's theory stressed variation within species, so he emphasized racial variations within the human species to try to convince his contemporaries that humans had evolved. This was not a peripheral point to Darwin's theory, either. Darwin professed that one of the main purposes of his book *The Descent of Man* was to describe "the value of the differences between the so-called races of man."[4]

Even more ominously, Darwin interpreted the European extermination of aboriginal peoples in Australia, the Americas, and elsewhere as an illustration of the struggle for existence in action. The fitter races were prevailing, in his view, and this would contribute to evolutionary progress. He stated in *Descent of Man*, "At some future period, not very distant as measured by centuries, the civilised races of man will almost certainly exterminate and replace throughout the world the savage races."[5] Rather than expressing outrage at this wanton killing, Darwin justified European brutality and genocide, claiming it was beneficial because it would contribute to evolutionary advance. Two leading historians of Darwinism explain that in light of Darwin's position on race relations, "racial genocide was now normalized by natural selection and rationalized as *nature's* way of producing 'superior' races."[6]

Many other Darwinian biologists and anthropologists in the late nineteenth and early twentieth centuries shared Darwin's view about racial inequality and racial extermination. The leading Darwinian biologist in Germany, Ernst Haeckel, for instance, was even more racist than Darwin, arguing that races were so different that humans should be divided into ten distinct species. He also claimed that "the differences between the lowest humans and the highest apes are smaller than the differences between the lowest and the highest humans."[7]

Because of this, Haeckel zealously supported colonizing areas inhabited by "primitive" people because he thought that the "value of life of these lower wild peoples is equal to that of the anthropoid apes or stands only slightly above them."[8] He was indignant when, in World War I, Germany's foes sent black colonial troops into battle because, he asserted, "a single well-educated German warrior, though unfortunately they are now falling in droves, has a higher intellectual and moral value of life than hundreds of the raw primitive peoples, which England and France, Russia and Italy set against us."[9] Haeckel's intense racism won many disciples in Germany in the early twentieth century.

Darwin's and Haeckel's belief in a racial struggle for existence came to influence political and military decision-makers, who believed they had the sanction of science for their rapacious policies. For instance, around the turn of the twentieth century, a German missionary in German Southwest Africa lamented that because of racism,

> the average German here looks upon and *treats* the natives as creatures being more or less on the same level as baboons (their favourite word to describe the natives) and deserving to exist only inasfar as they are of some benefit to the white man…Such a mentality breeds harshness, deceit, exploitation, injustice, rape and, not infrequently, murder as well.[10]

When the native Hereros in that colony revolted in 1904, the German general Trotha issued an annihilation order in what he called "a racial war." He explicitly appealed to Darwinism to justify the atrocities the Germans visited on the natives: "We cannot dispense with them [the natives] at first, but finally they must retreat. Where the labor of the white man is climatically possible…the philanthropic disposition will not rid the world of the above-mentioned law of Darwin's, the 'struggle of the fittest [sic].'"[11] Many scholars consider the Herero Revolt the first twentieth-century genocide, for it resulted in more than 80 percent of the Herero people being wiped out.[12]

The ideas of scientific racism and the Darwinian struggle for existence among races also played a prominent role in the ideology of Hitler and his Nazi colleagues. As I demonstrated in my book *Hitler's Ethic: The Nazi Pursuit of Evolutionary Progress*, Hitler's policies were founded on evolutionary ethics. He believed that whatever promoted evolutionary progress was good, while anything leading to biological decline was evil. He hoped to improve the human race by advancing the interests of what he considered the superior race while eliminating those deemed inferior. The physician Walter Gross, who directed the Nazi's Racial Policy Office, explained in a 1938 article, "Only in agreement with the natural and organic laws of the struggle for existence can the German people evolve to the highest possible level. The 'elimination of every inadequate organic being in the struggle for existence' is necessary, in order to secure the 'preservation and improvement of life.'"[13]

Gross and many Nazis appealed to science and nature to justify the elimination of races (and individuals) they deemed inferior. An expert on the history of scientific racism in Germany, Benoit Massin, concludes from his studies, "In fact, on many occasions, the impulse for racial and eugenic policy came not from politicians but from scientists and the medical profession." He also states, "Not only did these scientists help the regime, sometimes they directly inspired its murderous policies."[14]

Gross, Hitler, and their Nazi comrades had learned these precepts from the leading biologists and anthropologists in Germany, some of whom directly participated in Nazi programs to improve the race biologically. For example, during the Third Reich, Eugen Fischer, director of the prestigious Kaiser Wilhelm Institute for Anthropology, Eugenics, and Human Heredity, served on government committees and tribunals dealing with population and racial policies. He spearheaded the Nazi program to compulsorily sterilize the so-called Rhineland bastards—i.e., Germans fathered by black French colonial troops during the occupation of the Rhineland after World War I.

After World War II broke out and the Nazis had conquered vast swaths of territory in eastern Europe, Otto Reche, who was appointed professor of anthropology at the University of Leipzig six years before the Nazis came to power, offered his scientific expertise on race to help the Nazis cleanse the eastern occupied territories of what they considered racial riff-raff. The Nazis recruited anthropologists to "sift" through the populations of eastern Europe to identify the Nordic elements that could be salvaged.[15] These racist views are so repulsive that many people today find it difficult to comprehend how serious scientists could have embraced this as science. Yet they did. A leading human geneticist in Nazi Germany, Otmar von Verschuer, wrote to Fischer, "It is important that our racial policy—also the Jewish question—receives an objective scientific foundation that is recognized in wider circles."[16]

Another way that science contributed to inhumane policies in the early twentieth century was through the eugenics movement. There are some similarities between scientific racism and eugenics—both relied on biological determinism, and many scientists and physicians in the late nineteenth and early twentieth centuries embraced both ideas. However, eugenics generally focused more on improving human heredity within a given population. Eugenics advocates tended to disparage people with disabilities, often calling them inferior, weeds, burdens, or other derogatory terms.

The father of the eugenics movement, Francis Galton, was cousins with Darwin, and came up with the idea while reading Darwin's book *On the Origin of Species*. Most of the leading figures in the eugenics movement were scientists, psychiatrists, and physicians who marshalled scientific arguments to promote their programs of hereditary improvement. One of the leading figures in the American eugenics movement in the early twentieth century was Charles Davenport, a zoology professor at Harvard University, who in 1904 became the founding director of the Cold Spring Harbor Laboratory—a facility dedicated to investigating heredity and promoting eugenics. He insisted that eugenics was scientific, which was reflected in the title of his 1910 book *Eugenics: The Science of Human Improvement by Better Breeding.* In that work he explained that society "may annihilate the hideous serpent of hopelessly vicious protoplasm."[17] While this inflammatory statement suggests violence, he was not implying that we should kill people. However, he and many other eugenics advocates did promote compulsory sterilization for people whom they deemed unfit or inferior. At the behest of scientists and physicians, Indiana passed the first compulsory sterilization law in 1907, and soon many other states followed suit. By the 1960s, these laws had resulted in the forced sterilization of about 60,000 people in the United States.

In their pursuit of eugenics, Scandinavian countries also legislated compulsory sterilization in the late 1920s and early 1930s. However, nowhere was eugenics more radically pursued than in Nazi Germany. In July 1933, Hitler issued a decree mandating compulsory sterilization for people suffering from some hereditary illnesses. The Nazis applied this so fanatically that they sterilized about 350,000 Germans (about one of every 200). Then in 1939, Hitler radicalized his eugenics policies further by introducing a secret euthanasia program that, in a little more than five years, killed about 200,000 Germans with disabilities, plus tens of thousands of other people with disabilities in occupied countries. Hitler had no trouble finding psychiatrists and physicians willing and even zealous to carry out his murderous program. (Many

of the personnel involved in this were later transferred to the death camps in Poland to carry out the Final Solution.)

Hitler and his scientist and physician accomplices were not the only ones to appeal to science to justify euthanasia, both involuntary and voluntary. Peter Singer, one of the most famous bioethicists in the world today (he holds an endowed chair at Princeton University), and his philosophical compatriot James Rachels (in his book *Created from Animals: The Moral Implications of Darwinism,* published by Oxford University Press) have both argued that Darwinism undermines the Judeo-Christian sanctity-of-life ethic. They believe that this has profound moral implications, making abortion, infanticide, and euthanasia morally acceptable. Indeed, historians who have studied the origins of the euthanasia movement have agreed that Darwinism played a major role in helping launch the euthanasia movement in the late nineteenth and early twentieth centuries.[18]

In addition to scientific ideas influencing evil deeds, scientists have participated directly in atrocities by carrying out gruesome human experiments. The Nazi regime became notorious for this, and rightly so. SS physicians in Auschwitz carried out painful sterilization experiments on Jewish women, which brought misery and sometimes death to the subjects.[19] In his zeal to investigate human heredity, Josef Mengele carried out ghastly experiments on twins, sometimes infecting them with diseases to see their reactions. Whenever one of the twins in his custody died, he immediately killed the remaining twin by injection so he could scientifically compare the two corpses.[20] In Dachau, the air force physician Sigmund Rascher used inmates as guinea pigs to determine how much oxygen deprivation and how much exposure to cold water a person could endure. Most of his experimental subjects either died in the course of the experiment or were put to death soon thereafter.[21]

The Nazis were not the only ones to indulge in inhumane human experimentation. The United States, Britain, and many other countries have carried out barbaric experiments, both before and after the Nazi era. In the United States, one of the most notorious of these was the Tuskegee syphilis experiment of the mid-twentieth century. Medical scientists in the 1930s wanted to better understand the progression of syphilis, so they selected 400 black men in Macon County, Georgia, who were suffering from the disease. They deceived these men by not informing them that they had syphilis and by promising treatment for their malady. As treatment became available for these patients, the scientists refused to allow their test subjects to be fully treated solely because they wanted to gain greater knowledge of the disease. The patients, of course, were suffering miserably, but they were not informed that they were subjects of a scientific investigation. This experiment continued until 1972, when journalists uncovered it and provoked a public outcry against such immoral behavior.[22]

British medical researchers in the mid-twentieth century also conducted some experiments on subjects without gaining informed consent. In 1951, biochemical warfare experts at Porton Down began giving test subjects small doses of the deadly chemical sarin. They assured the participants that these tests were safe, but in 1953, one subject went into a coma and then recovered. Even though the scientists reduced the dosage thereafter, later that year another participant died. Nevertheless, the experiment continued, showing the callous disregard these scientists had for their test subjects' health and welfare.[23]

Lest one wrongly conclude that immoral scientific experiments are confined to the past, a Chinese researcher, He Jiankui, aroused controversy when he announced that he had genetically engineered some embryos that were later implanted into a woman, who gave birth in 2018. The purpose of his experiment was to see if he could block HIV from infecting the embryo and baby by knocking out a gene. This was the first time that a scientist had genetically engineered a human in such a way that the person would pass on the altered gene to their offspring. Most scientists—as well as nonscientists—were shocked and disappointed with He's human experiment, which they considered very premature because we do not know the consequences of this kind of gene editing on people. He was later convicted and imprisoned for not following proper research protocols.

What should we make of these dark episodes in the history of science? We certainly should not conclude that science is evil in and of itself. Most of what we call science is beneficial to humanity. However, one lesson we should draw is that not everything that scientists tout as science is necessarily true and accurate. Scientists are fallible humans, and some scientific theories in the past have had to be modified or even discarded. Undoubtedly some scientific theories today will be revised or replaced in the future. A second lesson is that the closer that scientific studies come to dealing with humans, the greater the dangers. Scientific theories explaining humans and human behavior can easily be tainted by the scientists' own preconceptions, prejudices, and subjectivity. Once these subjective ideas are incorporated into scientific theories, the results can be troubling—and sometimes even catastrophic—for humanity.

How Can We Use
Science in Apologetics?

Jay W. Richards

One of the most fruitful sources for Christian apologetics is natural science. That claim might seem odd because so many Christians and skeptics view science as a key source of disbelief. Indeed, the controversy over intelligent design in science might make it seem that Christian apologetics, which goes far beyond design arguments, should avoid science altogether. But this is to mistake certain bad ideas that travel under the name of science, with the evidence of science and science itself.

The most prominent bad ideas that masquerade as science are scientism and materialism. For science to be useful to apologetics, apologists must distinguish science from both scientism and materialism.

So, first, let's get clear what we mean by science, and in particular, natural science.

Natural Science

Natural science is the effort to understand the natural world systematically, to follow the evidence of nature where it leads, and to find, where possible, general events, rules, principles, and laws that underlie and explain what we observe. It always involves feedback between hypotheses (or theories) about the world and observations of the world, though the way hypotheses and observations relate differs from field to field.

In some cases, natural science involves controlled experiments in, say, a chemistry lab—experiments that, in principle, can be replicated by other experimentalists. In other cases, it involves not experiments per se, but rather, careful observations using equipment such as an electron microscope or a radio telescope.

At other times, the practice of science involves piecing together clues to explain some past event and trying to determine which, among competing hypotheses, best explains those clues. This includes almost any science related to the history of life, the universe, and our planet.

Scientism

Natural science, then, is a family of related empirical methods and institutions formed around these methods for learning systematically about the natural world. And over the last few centuries, it has given us unprecedented knowledge and insight about the world around us.

Science involves no claim that its methods are the *only* reliable sources of knowledge. That further claim comes from scientism, which draws on the success and prestige of science. As a result, many confuse it with science itself. But clearly, science and scientism are different things. One involves the methods for learning about nature. The other is a statement about science and knowledge itself.

Not only is science not the same as scientism, but scientism isn't a plausible view. First, if scientism were true, then no one would have known much of anything until science came along. That can't be right. For thousands of years, humans have known all sorts of things about themselves and their surroundings.

Second, we clearly know many things by methods other than science. Think of the knowledge you have from introspection, for instance, or your knowledge that it's wrong to torture a small child for the fun of it. You don't need science to know that, or to know what you ate for breakfast.

Third, scientism fails its own test. Clearly, one cannot show that scientism is true by the methods of science. So, it hoists itself on its own petard. It is self-refuting.

Materialism

The other bad idea that often travels under the banner of natural science is materialism. Whereas scientism is a belief about knowledge, materialism is a belief about the world. Materialism is the view that, ultimately, *the* fundamental reality—the thing that explains everything else and on which everything else depends—is matter (or matter and energy). There are no minds with real causal powers, no angels, no gods or God, no forms. Materialism finds its most famous liturgical expression in Carl Sagan's famous claim that "the Cosmos is all that is or ever was or ever will be."[1]

Some materialists, such as philosopher Alex Rosenberg, take materialism to its logical conclusion: There are no thoughts, he claims—at least not as we imagine them. "The notion that thoughts are about stuff is illusory," he contends in his book *The Atheist's Guide to Reality*. "Think of each input/output neural circuit as a single still photo. Now, put together a huge number of input/output circuits in the right way. None of them is about anything; each is just an input/output circuit firing or not."[2] Yes, Rosenberg wrote a book presumably trying to express his thought that thoughts aren't about anything. Materialism, like scientism, has this tendency to destroy itself. But let's not pursue that here. For our purposes, we should just remember that science is not materialism.

There is a more modest form of materialism, however, that many confuse with science. I'm referring to methodological materialism (or naturalism). This is the view that when doing science, one must act as if materialism is true. Proper scientific explanations can appeal *only* to material causes. Partisans use this supposed rule of science to dismiss teleological and design arguments from the domain of science. It's especially popular as a rear-guard defense of weak Darwinian claims in biology. The argument takes this basic form: Sure, we have no direct evidence that natural selection, sifting through random genetic mutations, gave rise to [some biological form], but the alternative—design—isn't science. Therefore, selection and mutation is the best explanation by default.

Critics such as Stephen C. Meyer have offered several arguments against methodological materialism, especially as a weapon against intelligent design. First, as Meyer has shown, any definition of science broad enough to encompass the diversity of science will also encompass ID arguments.

Darwinism has always presupposed that design arguments in biology are coherent, specific, and both falsifiable and falsified. Thus they are already part of the content of science. The special creation of individual species is the constant foil in Darwin's *Origin of Species*, but that isn't the only possible alternative to Darwinism. Rather, design broadly construed is the non-Darwinian alternative to which the mind naturally inclines. Even arch-Darwinist and atheist Richard Dawkins has admitted this. Contemporary design arguments have added heft and precision to this universal intuition.

Second, ID arguments are based not on religious texts or even broad philosophical concepts, but on publicly available empirical evidence drawn from nature. ID doesn't focus on agents themselves, but on the empirically detectible traces of intelligence in nature. ID arguments are not, in themselves, arguments for the existence of God. They are arguments that the effects of intelligence are detectible in nature and explain at least some aspects of nature.

As a result, ID arguments offer answers to questions that scientists ask *as scientists*. While a complete list of criteria for what constitutes science remains elusive, it's generally agreed that scientific arguments ought to be based on public evidence from nature, and ought, as philosopher Del Ratzsch has said, "to be put in empirical harm's way."[3] Normally that means they can be verifiable, falsifiable, or testable against competing hypotheses. Given these reasonable criteria—which fit the core definition of natural science as presented earlier—a number of contemporary ID arguments qualify.

Third, ID follows argumentative forms well attested in natural science. For instance, Stephen C. Meyer's argument for intelligent design as the best explanation for biological information follows the same canons of reasoning that Charles Darwin followed in *On the Origin of Species*.[4]

This means we have sound reasons for claiming that ID should be considered within science itself, even if, like so many scientific ideas, including Darwinism, it has larger philosophical implications.

Of course, even if design arguments drawing on evidence from science go beyond science itself, that would tell us nothing about whether those arguments are valid, sound, or persuasive. In particular, it would give us no good reason to reject my claim here, which is that science is a ripe source for Christian apologetics.

Apologetics

Now that we have clear definitions, we can turn to apologetics. Apologetics is the giving of a rational defense of the faith and can be divided into positive and negative forms. Positive apologetics includes arguments for God's existence, for the reliability of Scripture, and for the historicity of Christ's resurrection. In negative apologetics, the apologist critiques non-Christian ideas, such as materialism, and responds to objections to the faith, such as the problem of evil or the claim that miracles are impossible.

A less-direct way to do apologetics is to show the fruitfulness of the Christian worldview in other intellectual endeavors. "I believe in Christianity as I believe that the sun has risen," C.S. Lewis wrote, "not only because I see it, but because by it I see everything else."[5]

In which ways can apologists draw on science? Answer: All of the above. Apologists may draw on publicly available evidence from science in defense of some Christian belief and against non-Christian beliefs. They may point to the success of science itself as evidence for the truth of some Christian belief, such as the rationality of the Creator. And they may cite evidence for the role of Christianity in the rise of science for those who view science as a good thing. So when I say science is useful for apologetics, what I mean is that both the evidence from natural science and the existence of science itself provide positive evidence for Christian apologetics.

In particular, apologetic arguments that appeal to the evidence of science often show the tension between the evidence of science and the materialism and scientism with which science is often confused.

None of these arguments need to prove, say, the truths of the Nicene Creed, or even the existence of God. But they may still be good arguments. In some cases, this evidence, when framed with the right argument, may get us to a transcendent creator—to deism or perhaps to theism. In other cases, it may point simply away from materialism and toward purpose or intelligent design. That's okay. What evidence and arguments based on science lack in certainty and specificity, they often gain in accessibility.

Finally, in most cases, when dealing with evidence from natural science, the best we can do is to make cumulative case arguments or "an inference to the best explanation." We might like deductive certainty. But the nature of scientific evidence means we will often have to settle for less. (More on this below.)

Science and Good Arguments

What is a good apologetic argument? As noted above, it need not be a deductive proof, though of course it should be logically sound and valid. Yet it's possible for an argument to have a valid structure and be sound because it has true premises, but still not persuade the target audience. A good apologetic argument

should not just be sound and valid; it should also be rationally persuasive. But to whom? Does it have to convince a committed skeptic such as Alex Rosenberg, who wrote a book saying that thoughts aren't about anything?

No. That's far too high a bar. If your skeptical next-door neighbor doesn't want to believe that intelligent life exists on other planets, then probably no argument will convince him otherwise. Sure, if ET showed up at his front door, that might do the trick. But that would be an empirical demonstration, not an argument.

Let's say a good argument should be able to persuade what we might call "the rational fence sitter." This is the person who is open to arguments and evidence in favor of some conclusion. An argument that the universe has a finite age, for instance, is good if it can rationally persuade, or at least move someone toward, the conclusion that the universe has a finite age.

I say rationally persuade to rule out sophistry, hypnosis, or manipulation. Consider that next-door neighbor again. If you give him drugs that make him gullible, deprive him of sleep, and then hire actors in costumes to show up in the middle of the night looking like otherworldly aliens, you haven't rationally persuaded him—no matter what he comes to believe about aliens.

Of what does a good argument consist? To rationally persuade a rational fence sitter, you want to offer an argument that starts with a point he already believes, and move him rationally toward the conclusion you're interested in. In other words, you want to appeal to evidence that your intended audience, sitting in rows along the fence between belief and unbelief, counts as evidence.

This is where science can be especially useful for apologetics in the early twenty-first century. Unfortunately, we live in an age that is, for the most part, philosophically and metaphysically illiterate—even, if not especially, among those who enjoy many years of formal education. As a result, there are many powerful traditional philosophical arguments for God's existence, such as Thomas Aquinas's Five Ways, that fall on deaf ears. It's not just that they do not persuade. They don't even make sense except to those specialists who have the time to study their premises and structure. This is distressing because, in my view, some of these arguments establish the existence of God.

Persuasive arguments in every era must connect with some *belief* or *source of authority* that the target audience recognizes or accepts. For millions of people in our era—at least those who have not abandoned the notion of truth altogether—science provides that. It provides publicly available evidence.

Keeping all that in mind, here is some evidence from natural science that is especially fruitful for apologetics.

Good Evidence

1. A Cosmic Beginning

One of the most fundamental questions for apologetics is whether the material world is the ultimate reality. Perhaps, as Aquinas thought, it's possible that the material world could be eternal and still be contingent. But it can't be necessary and fundamental unless it is also eternal. (I trust this is obvious to you.) Determining whether the universe has always existed or has a finite age is quite important.

The evidence discovered over the last century in favor of big bang cosmology—the redshift of distant galaxies, the cosmic microwave background radiation, and the like—points toward a universe with an age. That is, it implies the universe began to exist in the finite past.

This evidence gives empirical force to the traditional Kalam cosmological argument:

1. Whatever begins to exist must have a cause for its existence.

2. The universe began to exist.

3. Therefore, the universe has a cause for its existence.

And that cause must, of course, be separate from the physical universe, since that cause itself is what needs explaining.

Until the twentieth century, the evidence for the second premise was mainly negative. That is, the alternative would be an infinite regress of sequential causes, or the passing of an infinite amount of time. And such a sequence of infinite seconds, minutes, hours, or whatever leads to paradoxes that can (arguably) only be avoided by affirming that the universe had a beginning.[6]

Only in the twentieth century did mainstream cosmology provide such strong empirical support for a cosmic beginning. Moreover, everyone understood that a cosmic beginning flatly contradicted the materialism of the nineteenth century. Indeed, the staunch resistance to big bang cosmology in the scientific community was due largely to the fact that it pointed the mind in a Godward direction.[7]

2. Fine-Tuning

The second, related line of evidence is the apparent fine-tuning of the universe. *Fine-tuning* refers to various features of the universe that are necessary conditions for the existence of complex life. Such features include the initial conditions and "brute facts" of the universe as a whole, the laws of nature or the numerical constants present in those laws (such as the gravitational force constant), and local features of habitable planets (such as a planet's distance from its host star, etc.).

The basic idea is that these features must

fall within a very narrow range of possible values for chemical-based life to be possible, and the overwhelming bulk of the possible universes within our "universe neighborhood" would be incompatible with any kind of life. Because a habitable universe has an intrinsic interest and value, this fine-tuning provides strong evidence for design.

There is controversy about the complete list of fine-tuned cosmic parameters. And philosophers argue about how to calculate the probabilities involved. Even so, there are many well-established examples of fine-tuning that are widely accepted even by scientists who are generally hostile to theism and design. For instance, Stephen Hawking made this admission in *A Brief History of Time*, the very book in which he tried (and failed) to avoid the need for a cosmic beginning: "The remarkable fact is that the values of these numbers [the constants of physics] seem to have been very finely adjusted to make possible the development of life."[8]

Here are the most celebrated and widely accepted examples of fine-tuning for the existence of life. For a more detailed discussion, see the references in the endnotes.

Cosmic Constants
1. Gravitational force constant
2. Electromagnetic force constant
3. Strong nuclear force constant
4. Weak nuclear force constant
5. Cosmological constant
Initial Conditions
6. Initial distribution of mass energy
7. Ratio of masses for protons and electrons
8. Velocity of light
9. Mass excess of neutron over proton[9]

Design arguments based upon fine-tuning are quite compelling and have led some prominent fence sitters toward theism or at least deism. But for some Christians, this represents a weakness in these arguments. These Christians object that if an argument stops short of full Christian truth—the Trinity, incarnation, resurrection, message of salvation, and so forth—then it could do more harm than good. It could leave someone a mere theist who believes in a God but holds all sorts of false ideas about him. They might even become a Sikh or Bahá'í or Muslim.

This is an odd objection, and I mention it only because it comes up in Christian circles more often than it ought. Most Christian apologists would love to be able to show that if you accept the law of non-contradiction, then you have to believe all the truths of the Nicene Creed. But alas, that's just not how God has chosen to reveal himself. God's eternal power and divine nature are clearly seen, Paul says in his letter to the Romans, "in the things that have been made" (1:20). That's God's general revelation of himself. But there's no hint that we can learn of what God has done through special revelation without turning to history.

Psalm 19 suggests that God reveals himself in this twofold way. The psalm begins by telling of the heavens declaring the glory of "God," of *El* in Hebrew. This is the word the Jews and other peoples in the ancient Near East used to refer to a creator. The psalm tells us that the heavens—which are visible to everyone—pour forth speech and display knowledge day and night throughout the world. But when the psalm begins to speak of the law, which God gave specially and personally to the Jews, it speaks of God not as *El*, but as the "LORD"—that is, *Yahweh*. That's

God's personal name. The psalmist knows God's name only because God has revealed it specially.

So, the God who is revealed clearly, but in a limited way, as the Creator through the natural world also reveals himself particularly in history, in Scripture, and especially in Jesus Christ.

Thus, if evidence from physics and cosmology points in the direction of a creator, then it's going as far as it's supposed to go. We should rejoice if it moves people from blind materialism to theism, even if they still have a way to go in learning that the Creator also loves them and has died for them.

3. The Correlation of Life and Discovery

The third and related body of evidence points to an eerie correlation between the requirements for complex life on a planet and the conditions for scientific discovery. Guillermo Gonzalez and I lay out this evidence in our book *The Privileged Planet*.[10] We make a cumulative case argument that the same narrow range of conditions that allows for complex observers like ourselves also provides the best overall place for making a wide range of scientific discoveries.

If you compare Earth to other less habitable planets, you find that the very factors that make it more habitable than its peers also make it more suited to doing science. This includes its geology, atmosphere, moon, host star, planetary neighbors, location in the galaxy, and location in cosmic time. If you could pick only one place for doing diverse kinds of natural science, you would pick Earth, or an Earth twin.

This pattern, we argue, makes much more sense if the universe is designed for discovery than if it is not. So, far from being antiscience, our design argument implies that the world is made for scientific discovery.

4. Origin of Life

The fourth and very impressive line of evidence is the origin of life on Earth. The central question with regard to the origin of life is the origin of biological information—information that, by definition, can't be reduced to the principles of chemistry and physics. This evidence is powerful because the type of information needed for life is so intimately connected with intelligent agency. The "creation of new information," discerned information theorist Henry Quastler, "is habitually associated with conscious activity."[11]

Indeed, in our experience, intelligent agents acting for a purpose are the *only* known and observed source of such information. In *Signature in the Cell*, Stephen C. Meyer makes the detailed case that the information in the biological world points to intelligence.[12] Biologist Douglas Axe develops his own related argument for design in biology in his highly readable book *Undeniable*.[13]

5. Biological Forms

The next category of evidence is in biology itself—subsequent to its origin. In this area, we have to take account of the neo-Darwinian process of natural selection and random mutation, which can mimic the effects of intelligence. Note that even if the Darwinian and related processes could fully account for biological complexity, this fact would not touch the previous lines of evidence 1. – 4. Arguments for purpose and design would simply need to stay focused on the larger context—just as we might point to the fine-tuning of physics to explain the structure of a snowflake, even if snowflakes don't constitute independent evidence for design.

But in fact, there is plenty of solid evidence for the strict limits of Darwinian explanations.

Biochemist Michael Behe provides the details in his three important books, *Darwin's Black Box*,[14] *The Edge of Evolution*,[15] and *Darwin Devolves*.[16] But more than that, there is positive evidence of design in biological forms. Behe tends to focus on tiny "molecular machines" that are "irreducibly complex," such as the bacterial flagellum. In other words, they lack the sort of functions that might be accessible to gradual evolution by the Darwinian mechanism. Instead, their function is largely the result of being in a complete or near-complete state. This is just the sort of thing you'd expect from intelligent agency.

Stephen C. Meyer focuses on the sudden appearance of animal body plans in the geological record in his book *Darwin's Doubt*.[17]

6. The Origin and Practice of Science Itself

The final line of evidence is the origin and practice of science itself. If materialism were true, we would have no reason to expect that the universe should be, from the human perspective, rationally transparent. Just in Darwinian terms, we could expect that our behavior in the world is survival-enhancing. But why should we be able to figure out what is happening at the atomic scale, or inside stars? Why should we be able to develop

formal mathematical models that capture material reality? None of this has anything to do with survival.

Science itself, then, suggests an underlying purpose, and almost all the founders of modern science understood the scientific enterprise in this way. Oxford mathematician and apologist John Lennox puts it this way: "Of course, I reject atheism because I believe Christianity to be true. But I also reject it because I am a scientist. How could I be impressed with a worldview that undermines the very rationality we need to do science?"[18] There is a growing literature on this topic, but a good place to start is sociologist Rodney Stark's delightful book *For the Glory of God*.[19]

Together, these lines of evidence drawn from science confirm that the universe is the result of purpose and design—a design in which human observers have a key role—and is not a mere conglomeration of blind forces and particles that exists of their own accord. While scientific arguments have limits, they are useful in apologetics in demonstrating these empirically supported truths. Indeed, this evidence from science points toward the existence of a creator.

What About the Historical Relationship Between Christianity and Science?[1]

H. Wayne House

The term *science* is from the Latin word *scientia*, meaning "knowledge," with the latter referring to all acquisitions of knowledge, including God and the creation by God. Generally, the modern use of the word is very specific and refers to the discipline of study relating to investigation of the physical world. So thoroughly has this discipline monopolized the term *science* that to say that one is studying something scientifically is tantamount to saying that truth is related to the study. Consequently, science is viewed as truth or knowledge itself. Even areas of study that are open to various interpretations or are based on subjective analysis, such as psychology, sociology, etc., have sought to associate with natural sciences to enjoy the credibility attached to the word *science*.

Three broad models. Ideologically, the relationship between Christianity and science may be viewed in three distinct models: Christianity and science as competitive or conflicting realms of knowledge; Christianity and science as complementary realms of knowledge; and, Christianity and science as distinctive and separate realms of knowledge. Throughout the history of the relationship between Christianity and science, there have been periods when each of the three has been the prevailing idea, and today, there are theologians and scientists who champion each model. In recent years, there has been increased interest in the study of the models and the relationship between the two disciplines. Institutions such as the Templeton Foundation have done much to foster discussion and study of the relationship and raise public awareness of it from a nonsectarian but religious perspective.

Within Christianity more specifically, individuals such as Sir John Polkinghorne (1930–2021) and Alister McGrath (b. 1953), who are dually and highly trained in science and theology, have done much to raise awareness of the study of science and religion. Polkinghorne was a British theoretical physicist, theologian, writer, and Anglican priest. McGrath is a theologian, Christian apologist, and former molecular biophysicist with doctorates in both science and theology. Knowledge and certitude with respect to the natural world is viewed as something over which there is an ideological battle between science and Christianity. The groups and individuals mentioned above all tend to adopt the "consensus" view on topics such as evolution and origins, but other groups, such as Discovery Institute and Reasons to Believe, have provoked thoughtful discussion over science and religion that credibly asks whether modern evolutionary theory is the best explanation for the complexity of life.

In the first model, science and theology are in conflict; the two disciplines are seen to be competing for truth in the natural world. Within Christianity, one often sees this in things such as the creation-evolution debates regarding human origins. The second model sees science and Christianity as complementary, understanding truth as something to which both disciplines can contribute without there being infringement or conflict between them. The third model of distinctive realms contends that the world of science and the world of Christianity share no common ground and address no common interests. One may participate personally or professionally in either, neither, or both disciplines without conflict because science addresses one set of questions and Christianity addresses a second but separate set of questions. One can find proponents of each of the models in every manifestation of Christian belief—Protestantism, Roman Catholicism, Orthodoxy, and other expressions of Christianity.

Within Christianity, theology was historically viewed as the "queen of the sciences" because knowledge could be acquired through the study of revelation as well as via study of the natural world. However, in the minds of many modern scientists, religion may not be empirically tested, and it no longer may be considered a science.

Early debates. Three key debates in the history of science and Christianity dominated previous interactions. The first was over answers to the question "What is the nature of the solar system?" Key figures within this debate were Copernicus and Galileo. The second debate was centered on the question "What is the nature of the universe?" In this discussion, the key individual was Isaac Newton, and this debate focused on the additional question of "What was and is the role of God in the universe?" This latter issue centered on the theological perspective of deism, where a supernatural deity creates the universe but then ceases to have direct involvement or intervention with it. The final debate focused on the question "What are the biological origins of life and the human species specifically?" The primary person in this field was Charles Darwin. In each of these debates there were Christian voices on both sides.

Recent debates. In the last century, and to the present day, debates about Christianity and science have largely addressed a different set of questions, although the human biological origins question is still prominent. The most recent debate regards what is often termed the *new atheism* or *scientific atheism.* It addresses the question, "Does science deny God?" A prominent advocate of the yes position is Richard Dawkins. A second question pertains to cosmology and asks, "Does the anthropic principle mean anything?" In essence, is the fine-tuning of the universe laden with any religious significance? Key figures in this question include John Barrow, Frank Tipler, B.J. Carr, and M.J. Rees. A third question pertains to quantum theory and asks, "Is there complementarity in science and religion?" The work of Niels Bohr and Albert Einstein figures prominently in this question.

A fourth question pertains to those who accept evolutionary biology and asks within that framework of acceptance, "Is there a religious end and significance to evolution?" Within this discussion are individuals such as Henri Bergson, Pierre Teilhard de Chardin, Ernst Mayr, and Simon Conway Morris. A fifth question relates to those accepting evolutionary psychology and asks, "What are the origins of religious belief, and to what extent might they be rooted in biological and sociological factors?" Here, one thinks of the work of Edward O. Wilson, Stephen Jay Gould, and David Sloan Wilson. "Can religion and religious experience be explained solely by psychology?" is the sixth question. Prominent in this discussion is the work of William James, Sigmund Freud, and Ralph W. Hood. A final question is, "To what extent is religion 'natural'?" This question arises from the new field of what is termed the *cognitive science of religion*. The prominent thinkers here include scientists such as Pascal Boyer, Justin Barrett, and Robert N. McCauley.

The questions above illustrate the breadth and depth of the "science and Christianity" (or "science and religion") field of inquiry. In so doing, one realizes that a challenge for any person seeking to enter the dialogue is the need to have a good general knowledge of the sciences and of at least one religion. The complexities, intricacies, and nuances can be significant. That, in part, is why an understanding of the three models of the nature of interaction between science and Christianity is important.

Demise of the Conflict Model

Under the first model—conflict—when dealing with religion and science, one must determine whether these categories are truly inimical to each other and truly separated into opposite poles. Science covers a large number of disciplines, each having their own procedures, and does not fit within one definition. The same is true of religion. Religions differ, so that religions like Judaism and Christianity, and to a lesser extent, Islam, are more amicable to scientific investigation because their theological claims are connected to events within the historical world, such as the exodus of the Hebrews from Egypt in Judaism, the resurrection of Jesus from the dead in Christianity, or the revelation to Muhammad spoken of in the Qur'an. The religions of Hinduism and Buddhism are different, in that they focus on suprahistorical myths and are more concerned with moral teaching than historical events.

The term *religion*, too, needs to be understood as broad because it may include rituals that are often not central to the discussion of truthfulness. Thus, it would be more accurate to speak of the compatibility of science and theology than science and religion.

A challenge for scientists in speaking of their discipline as being concerned only with the physical or natural world is that unlike religion, which is concerned with the spiritual world or superstition, scientists also use nonempirical means to achieve their ends. Mental processes and ideas are not physical or empirical. What may not be recognized or realized by scientists who have typically studied neither theology nor philosophy is that their assured results coming from certain areas of modern scientific investigation have their foundation in several early Greek philosophers, and later in Christian theologians. Thus, science and religion, or more specifically, science and Christianity, share common truths. There are historical parallels between science and religion. Some of the ideas within modern science conform to ancient Near Eastern nature religions—such as the steady state theory—that had a closed system in which even the gods were captive to nature.

Some ancient philosophers, such as Anaximander, believed in a view similar to evolution, in which he saw life arising from the watery depths. On the other hand, the view of origins articulated in the book of Genesis, in which the universe had a beginning, seems more in agreement with the big bang view of the universe accepted by the majority of contemporary scientists. Under this view, there is no obvious reason why the conflict model must be the correct one. Rather, a better model is that science and religion relate to one another in what philosopher Stephen C. Meyer calls "qualified agreement."[2] This model proposes that although there is the theoretical potential for conflict between science and religion (because they sometimes speak to the same subjects, such as the origin of life and the universe), when the evidence is evaluated fairly, we find much agreement between these two methods of seeking truth.

Origins of Science

Knowledge of the universe has slowly developed over thousands of years. At first, the study of nature was largely within the domain of religion, particularly the pagan myths of the ancient Near East. All of nature was understood as being under the sway of forces of nature that had no discernible beginning. Even the gods emerged (*theogony*) from this seemingly eternal watery chaos so that nothing existed outside this reality. With the teaching of the Hebrew people hundreds of years before Greek philosophers expressed their view, the author of Genesis offered a radically different perspective of creation compared to those found in cultures like Assyria, Babylonia, and Egypt. In

the Hebrew creation story, given in the first chapter of Genesis, the Creator God existed distinct from his creation. In the Mesopotamian myths, the various creatures were involved in producing chaos, but in the biblical account, these "creatures" are presented as inanimate objects (sea and sky) that emerged from the Creator's creative acts. God and nature were no longer the same, and the spiritual and physical worlds that were involved with each other also were not the same. God influenced the created order by being its creator, but he was also its sustainer. Because of this interest in understanding the world to understand God, nature itself took on its own importance.

Many years after the writers of the Hebrew Scriptures had pondered creation in books such as Genesis, Job, and Ecclesiastes, Greek philosophers began to approach the study of the physical world. In the early stages of this approach, they relied on deductive reasoning to understanding how the world operated. This sole approach to studying the creation had limited success. Because of this attempt to explain, entirely through deduction, how the world worked, this encouraged "the idea that we can anticipate nature's course through metaphysical analysis rather than by observation and experiment. Without testing ideas against nature, many erroneous ideas were developed such as the idea that falling rocks 'want' to reach their natural resting place."[3] Later Greek philosophers followed the model of Aristotle, who believed that philosophers were to examine the essential nature of things inductively, and from that determine how they work, often through serendipity.

The joining of deductive thinking with inductive investigation served to move the study of the world from mental constructions of truth to hands-on examinations of nature. With the advent of Christianity, which was influenced by the Hebrew Scriptures' view of the world that recognized the importance of nature, rather than adhering to nature religion, Christian theologians and apologists provided the impetus for the development of science that matured during the medieval period.

Origins of Modern Science

The nature of scientific inquiry before the Enlightenment. The empirical method of investigating nature arose in an unexpected time, with the emergence of early Christianity beginning in the third and fourth centuries, and due to an unexpected reason. As early Christian theologians sought to understand how God and nature related, they developed the theological doctrine of divine voluntarism—that is, God is a free and transcendent being. He governs the universe as he sees fit. This caused them to study nature more to understand God.

Later medieval theologians continued this quest and developed scientific views of the world long before the time of the early scientist Isaac Newton. In reality, Christian theology spawned modern science, and the Middle Ages were not dark times of superstition, but served as the basis of later science.[4] This understanding has not always been the case, because scientific study by many scientists over the last 150 years has established an orthodoxy of naturalism, in which any mention or use of religious, theological, or revelational claims are excluded from consideration and deemed to be "mere faith" claims that cannot be trusted, as science is, and cannot be considered in evaluating the world around us.

Thus, as mentioned earlier, theology as "queen of the sciences" was dethroned by naturalistic science because revelation was rejected as a legitimate and proper source of information for determining truth. With the rise of scientism and naturalism, a new orthodoxy arose that has excluded God and religion from the conversation regarding truth and facts.

The conversation in the past had been between Christianity (from which modern science arose) and science, but acceptance of naturalism caused many scientists to adopt a view of the world that excludes any nonempirical sources of truth. Opponents of Christianity, holding to a conflictual model of Christianity and science, contend that these early scientists departed from this dictum in seeking to explain their ideas because they included arguments often buttressed by nonempirical data that was based on philosophical assumptions and reasoning. Consequently, most scientific disciplines today are unaffected by the fact that a scientist may be a theist or an atheist because the investigations proceed from empirical data that can be tested through observation, are falsifiable, and are capable of making predictions.

The most contentious area in which Christianity and modern science have collided over the last 150 years is the subject of the origins of the universe and life. With the rise of naturalistic Darwinism and its progeny neo-Darwinism, naturalism permeated the sciences, and the claims of Christianity regarding the creation of the universe and humanity by a direct act of the God of the Bible have been largely rejected—particularly in the discipline of biology, though generally in most areas of scientific research today.

Whether this present-day animosity is necessary is an important discussion. What is often missed by some of the most vitriolic opponents of Christian thought having an impact on scientific questions is that it is the development of Christian views of the Creator and creation that gave rise to the modern scientific study of the world in the West. As T.F. Torrance explains in his book *Theology and Science at the Frontier of Knowledge* (2000), the classical way of thinking that developed in Western culture owes a debt to how much it was influenced by theology and philosophy. Western culture has a Christian foundation, with the classical way of thinking most easily seen in the work of the great Alexandrian thinkers of the third and fourth centuries.

These philosophers had developed the Christian doctrines relating to the creation of the world out of nothing as well as the incarnation of the Logos, which they had inherited from the Greek world that preceded them, such as Platonic, Aristotelian, and Stoic thought in relation to concerns about cosmology, epistemology, and logic. Christian philosophers sought to understand the tools and methods of Greek science and apply them to analysis regarding the nature of God and the universe. This is especially true of the great philosopher Origen, who adhered to both Platonic and Stoic ideas regarding the nature of comprehension and limitation. Origen held that though God is incomprehensible, he nonetheless can be considered despite not being finite. Reversing the ideas of Stoicism, Origen insisted that God makes the universe comprehensible to humanity by temporally limiting it with a beginning and an ending.[5]

The Alexandrian theologians, particularly Origen, sought to establish, in the words of Torrance, "a dynamic interaction between God and the world which gave the universe a cohesion and unity under God."[6] This is important to the rise of science because, in contrast to Greek polytheism and pluralism,

a unitary view of the universe was developed, which is necessary for the existence of science.

The development of science in the Middle Ages. It is common for people to believe that due to religious resistance and even persecution by the medieval church, the development of science was inhibited by Christianity during the Middle Ages. Often this has been tied to the myths that have arisen in connection with the controversy between Galileo and the Roman Catholic Church. Such is not the case, and, in fact, the Catholic Church supported scientific research. During the period between AD 500 to 1500, applications of science in the form of many inventions occurred, such as the printed book, cannons, and compasses. In their own way, each of these inventions revolutionized human progress.

Beyond the value of these and other inventions is the fact that this period laid the foundations for the rise of modern science. With the approval of the Roman Catholic Church, a scientific manner of combining reason and faith underpinned the study of science. In fact, the accomplishments that happened by men like Sir Isaac Newton would not have occurred had it not been for a twelfth-century theologian named Bernard of Chartres. Thus, the advances of Newton would have not occurred except for the achievements of medieval science.

Contemporary Conflicts Between Christianity and Naturalistic Science

Some who study Christianity and science contend that the perceived conflict between Christianity and science is really not about science but about philosophy. When a theist and an atheist examine the same physical facts, they should come to the same conclusion. When Darwinism was first introduced to scientists in the Western world, individuals who were creationists practiced their discipline no differently than those who were evolutionists, though the former not only were committed, generally speaking, to belief in a creator, but also to belief in the biblical Creator.[7]

Also upholding a conflictual model, many Darwinists argue that it is not possible to practice a scientific discipline as a creationist, though in fact it is done regularly. Creationists counter that there is no inherent reason why belief in a creator impinges on the capability of trained scientists to examine evidence and come to the same conclusions regarding the operation of science than that which is done by an evolutionist. Only when the discussion centers on philosophical considerations is there a difference, as there also is when the discussion turns to nonobservable historical assumptions. Such debate also illustrates the role that philosophy and presuppositions play for both camps as they address Christianity and science. The reason for a different interpretation relates to the theist and the atheist having different paradigms through which they interpret the facts.

In recent years, an additional concept, intelligent design, has been championed by some both inside and outside of Christianity. In addition, the shift away from belief that human beings are special creations of an intelligent creator resides not in some overwhelming evidence for a Darwinian view of human origins, but rather, in a philosophical change as to the nature of science.[8] Philosopher J.P. Moreland quotes Neal Gillespie's *Charles Darwin and the Problem of Creation* when he says, "Darwin's rejection of special creation was a part of the transformation of biology into a positive science. That is, one committed to a thoroughly naturalistic explanation of life based on material costs and the uniformity of nature."[9] The point that Gillespie is making, according to Moreland, is that the move from a creationist perspective of science to that of Darwin's naturalistic view related to the shift from a supernaturalistic view of creation to a naturalist perspective.

Current conflicts between Christianity and science. In contemporary expressions of the conflictual model, two major conflicts exist between Christianity and modern science: the creation of the universe and life, and the existence of miracles. Naturalistic scientists believe that the Christian worldview is inferior to the scientific worldview. They believe that science has provided the explanation for the universe, that theological answers are either false or superfluous, that the scientific method is considered objective while Christian faith is entirely subjective, and that science ostensibly deals with facts while religion deals with faith and what cannot be measured.

Christian proponents of the conflictual model (and in some instances, the distinctive realms model) contend that science is concerned with how nature works, while theology is concerned with why an orderly universe exists. Under these views, the contradictions between science and theology are apparent, rather than real, and are contradictions of interpretation rather than fact. In addition, they argue, scientists, no less than theologians, use faith as they approach their task, and have presuppositions just as theologians do. Scientists operate with paradigms, even when the scientific facts undermine the paradigm. Scientists hold on to their paradigms until they can find another one that works. In this model, the major contemporary struggle between science and theology is that of the acceptance or rejection or

Darwinism—essentially the assumption that only naturalistic answers are allowed to scientific questions.

Proponents of the distinctive realms model argue that there is no conflict because science and religion—specifically, Christianity—address different matters. Science is understood to deal with the world and its behavior, and religion is understood to relate to issues about God, morality, and the afterlife. Opponents of this model contend that it is not possible to compartmentalize religion and naturalistic science. After all, Christianity makes specific empirical and historical claims about reality, such as the fact that Christ rose from the dead. Religion is not limited to claims that have nothing to do with the real world.

In the complementary model, and to some degree in the distinctive realms model, there is no reason to resist or fear the conversation between Christianity and modern science. The problem arises when there is an exaltation of science to a philosophy of *scientism* (the view that science is the only source of true knowledge), and the unwillingness of some Christians to interact with the findings of "true" science. Nearly all matters relating to science would be uncontroversial to the vast majority of Christians, and the primary matters of disagreement are restricted to claims of a solely materialist universe and a Darwinian naturalistic perspective of the origin and history of life. Science is not inimical to Christian thought, but at times the interpretations of scientists may be opposed to it.

Future interaction between Christianity and science. Science was born in the cradle of Christian theology. The dilemma is between modern naturalist science and traditional science, which has recognized that God can be involved in his world without taking anything away from the natural order. Is knowledge only material in nature, or may it be both material and immaterial?

Many scientists are discovering that their research regarding the origin of the universe requires them to answer some of the same questions normally thought to be the domain of theologians. Astrophysicist Robert Jastrow has posed this oft-quoted dilemma for his fellow scientists:

It is not a matter of another year, another decade of work, another measurement, or another theory; at this moment it seems as though science will never be able to raise the curtain on the mystery of Creation. For the scientist who has lived by his faith in the power of reason, the story ends like a bad dream. He has scaled the mountains of ignorance; he is about to conquer the highest peak; as he pulls himself over the final rock, he is greeted by a band of theologians who have been sitting there for centuries.[10]

In recent years cosmology has been an active area of debate and discussion in the Christianity-and-science realm. Not only has a steady-state view of the universe been abandoned by many scientists, in which it was viewed as eternal, but many scientists now believe the physical universe had a beginning at a point in time. The latter view sounds more like the Hebrew Moses than the Greek philosophers who held to an eternal universe.

In addition, two arguments continue to gain ground in the scientific community, albeit more slowly than the big bang view—namely, the anthropic principle and intelligent design. According to the former, the conditions of life in our universe, galaxy, and planet are specially suited for advanced life. This suggests the universe was intended as a habitable environment for beings like us. Intelligent design, which is advocated by philosophers and scientists such as William Dembski (*The Design Inference*), Antony Flew (*There Is a God: How the World's Most Notorious Atheist Changed His Mind*), Stephen C. Meyer (*Signature in the Cell*), and Michael Behe (*Darwin's Black Box*), argues that the universe and life were brought about by an intelligent cause. This is another powerful area of agreement between Christianity and science.

Much of the tension between science and religion relates to mistrust because of myths about religion persecuting or inhibiting scientific discoveries, and because of the distrust among religious believers that scientists have generally overstated discoveries that support Darwinism, or viewed religious people as ignorant, if not stupid, because they question science and scientism on various matters. Not only are some scientists at fault for being dogmatic and refusing to even consider non-Darwinian views, but theologians are at fault for adopting interpretations of biblical texts that clearly counter widely accepted perspectives. For example, theologians have promoted a denotative understanding of

certain events, such as the stationary nature of the Earth, or have capitulated to the eternal nature of the universe, now rejected by science in favor of the big bang view.

At one time, science, in the ancient Near East religions, was associated with myths and interaction with the forces of a deified nature. Under the teaching of various Greek philosophers, nature began to be an object of study rather than something that reigned over the lives of people through the capricious gods and lower deities who themselves were caught up in control of this impersonal nature religion. But only with the emergence of the Hebrew and Christian worldviews was there a clear distinction between the creation and the Creator. Only then could a science be born that saw nature in nonreligious terms.

What the evidence demonstrates in all this is that scientific conclusions are not final, and that theological pronouncements may be properly reevaluated for accurate hermeneutics and bias. Science and theology will both benefit from a willingness of scientists and theologians to shed dogmatism and be willing to listen to each other, to recognize that the physical universe is not absolute and is not all that is, and that both parties have much to learn from the other.

Part II:

SCIENCE AND DESIGN

What Is the Evidence for Intelligent Design and What Are Its Theological Implications?

Stephen C. Meyer

Biologists have long recognized that many organized structures in living organisms—the elegant form and protective covering of the coiled nautilus; the interdependent parts of the vertebrate eye; the interlocking bones, muscles, and feathers of a bird wing—"give the appearance of having been designed for a purpose."[1] Before Darwin, biologists attributed the beauty, integrated complexity, and adaptation of organisms to their environments to a powerful designing intelligence. Consequently, they also thought the study of life rendered the activity of a designing intelligence *detectable* in the natural world.

Yet Darwin argued that this appearance of design could be more simply explained as the product of a purely undirected mechanism—namely, natural selection and random variation. Modern neo-Darwinists have similarly asserted that the undirected process of natural selection and random *mutation* produced the intricate designed-like structures in living systems. They affirm that natural selection can mimic the powers of a designing intelligence without itself being guided by an intelligent agent. Thus, living organisms may look designed, but on this view, that appearance is illusory and, consequently, the study of life does not render the activity of a designing intelligence detectable in the natural world. As Darwin himself insisted, "There seems to be no more design in the variability of organic beings and in the action of natural selection, than in the course in which the wind blows."[2] Or as the eminent evolutionary biologist Francisco Ayala has argued, Darwin accounted for "design without a designer" and showed "that the directive organization of living beings can be explained as the result of a natural process, natural selection, without any need to resort to a Creator or other external agent."[3]

Interestingly, some contemporary physicists also now make similar arguments about the origin of what physicists call the "fine-tuning" of the universe. Since the 1950s and 1960s, physicists have discovered that the laws and constants of physics and the initial conditions of the universe have been finely tuned to make life in the universe (and even basic chemistry) possible. To many physicists, this discovery has suggested the activity of a fine-tuner or super-intellect—i.e., an actual designing intelligence. Yet other physicists now argue that the fine-tuning of the physical parameters of the universe manifests the appearance, but not the reality, of design. For example, physicist Lawrence Krauss has argued that cosmological fine-tuning does not provide evidence of intelligent design, but instead, "the illusion of intelligent design."

So did Darwin explain away all evidence of apparent design in life? Have contemporary physicists explained away the evidence of design in the universe?

Proponents of the theory of intelligent design answer that question with an emphatic *no*. We argue that there are specific features of life and the universe that are best explained as the result of an *actual* designing intelligence as opposed to an undirected materialistic process (such as natural selection and random mutation) that merely mimics the powers of a designing intelligence. Moreover, we argue that the superior explanatory power of the design hypothesis makes the activity of a designing intelligence in the history of life and the universe *scientifically detectable*. This commitment to the detectability of intelligent design not only distinguishes the theory of intelligent design from materialistic evolutionary accounts of the origin of life and the universe, but it also distinguishes the theory of intelligent design from the idea of theistic evolution. Indeed, though most versions of theistic evolution affirm the existence of God, they deny that God's designing activity is detectable in the natural world.

So the theory of intelligent design asserts that evidence of design is detectable in nature. But what evidence do proponents of the theory cite to justify this claim? Let's consider two classes of such evidence and a scientific

method of design detection that can be used to detect intelligent design in nature.

The Origin of Life and the Information Enigma

As noted, Darwin attempted to explain the origin of new living forms starting from simpler preexisting forms of life. Nevertheless, his theory of evolution by natural selection did not attempt to explain the origin of life—the origin of the simplest living cell—in the first place. Yet there now is compelling evidence of intelligent design in the inner recesses of even the simplest living one-celled organisms. Moreover, a key feature of living cells—one that Darwin knew nothing about—has made the intelligent design of life scientifically detectable.

In 1953, when Watson and Crick elucidated the structure of the DNA molecule, they made a startling discovery. The structure of DNA allows it to store information in the form of a four-character digital code. Strings of precisely sequenced chemicals called *nucleotide bases* store and transmit the assembly instructions—the information—for building the crucial protein molecules and machines the cell needs to survive.

Francis Crick later developed this idea with his famous "sequence hypothesis," according to which the chemical constituents in DNA function like letters in a written language or symbols in a computer code. Just as letters of the English alphabet may convey a particular message depending on their arrangement, so too do certain sequences of chemical bases along the spine of a DNA molecule convey precise instructions for building proteins. The arrangement of the chemical characters determines the function of the sequence as a whole. Thus, the DNA molecule has the same property of "sequence specificity" that characterizes codes and language.

Moreover, DNA sequences do not just possess information in the strictly mathematical sense described by pioneering information theorist Claude Shannon. Shannon related the amount of information in a sequence of symbols to the *im*probability of the sequence (and the reduction of uncertainty associated with it). But DNA base sequences do not just exhibit a mathematically measurable degree of improbability. Instead, DNA contains information in the richer and more ordinary dictionary sense of alternative sequences or arrangements of characters that produce a specific effect. DNA base sequences convey instructions. They perform functions and produce specific effects. Thus, they not only possess "Shannon information," but also what has been called *specified* or *functional information*.

Like the precisely arranged zeros and ones in a computer program, the chemical bases in DNA convey instructions by virtue of their *specific* arrangement—and in accord with an independent symbol convention known as the genetic code. Thus, biologist Richard Dawkins notes that "the machine code of the genes is uncannily computer-like."[4] Similarly, Bill Gates observes that "DNA is like a computer program, but far, far more advanced than any software we've ever created."[5] Biotechnologist Leroy Hood likewise describes the information in DNA as "digital code."[6]

After the early 1960s, further discoveries revealed that the digital information in DNA and RNA is only part of a complex information processing system—an advanced form of nanotechnology that both mirrors and exceeds our own in its complexity, design logic, and information-storage density.

Where did the information in the cell come from? And how did the cell's complex information processing system arise? These questions lie at the heart of contemporary origin-of-life research. Clearly, the informational features of the cell at least appear designed. And, as I show in extensive detail in my book *Signature in the Cell*, no theory of undirected chemical evolution explains the origin of the information needed to build the first living cell.[7]

Why? There is simply too much information in the cell to be explained by chance alone. And attempts to explain the origin of information as the consequence of prebiotic natural selection acting on random changes inevitably presuppose precisely what needs explaining—namely, reams of preexisting genetic information. The information in DNA also defies explanation by reference to the laws of chemistry. Saying otherwise is like saying a newspaper headline might arise from the chemical attraction between ink and paper. Clearly something more is at work.

Yet the scientists who infer intelligent design do not do so merely because natural processes—chance, laws, or their combination—have failed to explain the origin of the information and information-processing systems in cells. Instead, we think intelligent design is detectable in living systems because we know from experience that systems possessing large amounts of such information invariably arise from intelligent causes. The information on a computer screen can be traced back to a user or programmer. The information in a newspaper ultimately came from a writer—from a mind. As the pioneering information theorist Henry Quastler observed, "creation of information is habitually associated with conscious activity."[8]

This connection between information and prior intelligence enables us to *detect* or infer intelligent activity even from unobservable sources in the distant past. Archaeologists infer ancient scribes from hieroglyphic inscriptions. SETI's search for extraterrestrial intelligence presupposes that information embedded in electromagnetic signals from space would indicate an intelligent source. Radio astronomers have not found any such signal from distant star systems. But closer to home, molecular biologists have discovered information in the cell, suggesting—by the same logic that underwrites the SETI program and ordinary scientific reasoning about other informational artifacts—an intelligent source.

DNA functions like a software program and contains specified information just as software does. We know from experience that software comes from programmers. We know generally that specified information—whether inscribed in hieroglyphics, written in a book, or encoded in a radio signal—always arises from an intelligent source. So the discovery of such information in the DNA molecule provides strong grounds for inferring (or detecting) that intelligence played a role in the origin of DNA, even if we weren't there to observe the system coming into existence.

The Logic of Design Detection

In *The Design Inference,* mathematician William Dembski explicates the logic of design detection. His work reinforces the conclusion that the specified information present in DNA points to a designing mind.

Dembski shows that rational agents often detect the prior activity of other designing minds by the character of the effects they leave behind. Archaeologists assume that rational agents produced the inscriptions

on the Rosetta Stone. Insurance fraud investigators detect certain "cheating patterns" that suggest intentional manipulation of circumstances rather than a natural disaster. Cryptographers distinguish between random signals and those carrying encoded messages, the latter indicating an intelligent source. Recognizing the activity of intelligent agents constitutes a common and fully rational mode of inference.

More importantly, Dembski explicates criteria by which rational agents recognize or detect the effects of other rational agents and distinguish them from the effects of natural causes. He demonstrates that systems or sequences with the joint properties of "high complexity" (or small probability) and "specification" invariably result from intelligent causes, not from chance or physical-chemical laws.[9] Dembski notes that complex sequences exhibit an irregular and improbable arrangement that defies expression by a simple rule or algorithm, whereas specification involves a match or correspondence between a physical system or sequence and an independently recognizable pattern or set of functional requirements.

By way of illustration, consider the following three sets of symbols:

"nehya53nslbywl`jejns7eopslanm46/J"

"TIME AND TIDE WAIT FOR NO MAN"

"ABABABABABABABABABABAB"

The first two sequences are complex because both defy reduction to a simple rule. Each represents a highly irregular, aperiodic, improbable sequence. The third sequence is not complex, but is instead highly ordered and repetitive. Of the two complex sequences, only the second, however, exemplifies a set of independent functional requirements—i.e., it is *specified*.

English has many such functional requirements. For example, to convey meaning in English, one must employ existing conventions of vocabulary (associations of symbol sequences with particular objects, concepts, or ideas) and existing conventions of syntax and grammar. When symbol arrangements "match" existing vocabulary and grammatical conventions (i.e., functional requirements), communication can occur. Such arrangements exhibit "specification." The sequence "Time and tide wait for no man" clearly exhibits such a match, and thus performs a communication function.

Thus, of the three sequences, only the second manifests both necessary indicators of a designed system. The third sequence lacks complexity, though it does exhibit a simple periodic pattern, a specification of sorts. The first sequence is complex, but not specified. Only the second sequence exhibits *both* complexity and specification. Thus, according to Dembski's theory of design detection, only the second sequence implicates an intelligent cause—as our uniform experience affirms.

In my book *Signature in the Cell*, I show that Dembski's joint criteria of complexity and specification are equivalent to "functional" or "specified information." I also show that the coding regions of DNA exemplify both high complexity and specification and, thus not surprisingly, also contain "specified information." Consequently, Dembski's scientific method of design detection reinforces the conclusion that the digital information in DNA indicates prior intelligent activity.

So, contrary to media reports, the theory of intelligent design is not based upon ignorance or gaps in our knowledge, but on scientific discoveries about DNA and on established scientific methods of reasoning in

which our uniform experience of cause and effect guides our inferences about the kinds of causes that produce (or best explain) different types of events or sequences.

Anthropic Fine-Tuning

The evidence of design in living cells is not the only such evidence in nature. Modern physics now reveals evidence of intelligent design in the very fabric of the universe. Since the 1950s and 1960s, physicists have recognized that the initial conditions and the laws and constants of physics are finely tuned, against all odds, to make life possible. Even extremely slight alterations in the values of many independent factors—such as the expansion rate of the universe, the speed of light, the masses of quarks, and the precise strength of gravitational or electromagnetic attraction—would render life impossible. Physicists refer to these factors as "anthropic coincidences," and to the fortunate convergence of all these coincidences as the "fine-tuning of the universe."

Many physicists have noted that this fine-tuning strongly suggests design by a pre-existent intelligence. Physicist Paul Davies has said that "the impression of design is overwhelming."[10] Fred Hoyle argued, "A commonsense interpretation of the facts suggests that a superintellect has monkeyed with physics, as well as chemistry and biology."[11] Many physicists now concur. They would argue that—in effect—these parameters appear finely tuned to make life possible because some*one* carefully fine-tuned them.

To explain the vast improbabilities associated with these fine-tuning parameters, some physicists, such as Lawrence Krauss and Leonard Susskind, have postulated not a fine-tuner or intelligent designer, but instead, the existence of a vast number of other parallel universes. This multiverse concept posits the existence of many other universes, each with different sets of physical parameters. In so doing, it attempts to show that a set of fine-tuning parameters necessary for life would—in all probability—inevitably arise somewhere in some universe, since this multiplicity of new universes would vastly increase the number of opportunities for generating a life-friendly universe.

Multiverse advocates not only posit a great multiplicity of other universes, they also posit the existence of some universe-generating mechanism to explain where these other universes came from. It's important to understand why they must do this. Most proponents think of the different universes that they postulate as causally isolated or disconnected from each other. Thus, they do not expect to have any direct observational evidence of universes other than our own.[12] Consequently, nothing that happens in one universe should have any effect on things that happen in another universe. Nor would events in one universe affect the *probability* of events in another universe, including the probabilities of whatever events were responsible for setting the values of the fine-tuning parameters in another universe—such as ours. As science writer Clifford Longley explains the concept: "There could have been millions and millions of different universes created each with different dial settings of the fundamental ratios and constants, so many in fact that the right set was bound to turn up by sheer chance."[13]

Yet if all the different universes were produced by the same underlying causal mechanism, then it would be possible to conceive of our universe as the winner of a cosmic lottery, where some winning universe with just the right laws, constants, and/or initial conditions, would eventually

emerge. Postulating a "universe-generating machine" could conceivably render the probability of getting a universe with life-friendly conditions quite high, and, in the process, explain the fine-tuning as the result of a randomizing element—like the action of a giant slot machine or a roulette wheel turning out either life-conducive winners or life-unfriendly losers with each spin or pull on the handle.

But, as I explain in my new book *Return of the God Hypothesis*[14] in much more detail, advocates of these multiverse proposals have overlooked an obvious problem. The speculative cosmologies (such as inflationary cosmology and string theory) they propose for generating alternative universes invariably invoke mechanisms that *themselves* require fine-tuning, thus begging the question as to the origin of that prior fine-tuning. Indeed, all the various materialistic explanations for the origin of the fine-tuning—i.e., the explanations that attempt to explain the fine-tuning without invoking intelligent design—invariably invoke prior unexplained fine-tuning.

Moreover, the fine-tuning of the universe exhibits precisely those features—extreme improbability and functional specification—that invariably trigger an awareness of, and justify an inference to, intelligent design.[15] Because the multiverse theory cannot explain fine-tuning without invoking prior fine-tuning, and because the fine-tuning of a physical system to accomplish a recognizable or propitious end is exactly the kind of thing we know intelligent agents do, it follows that intelligent design stands as the best explanation for the fine-tuning of the universe. And that makes intelligent design detectable in *both* the physical parameters of the universe and the information-bearing properties of life.

Intelligent Design: Theistic Implications

So why is a discussion of the theory of intelligent design important in a book about science and faith? After all, proponents of intelligent design have often argued that the method of design detection outlined in this chapter does not necessarily make it possible to determine the identity of the intelligent agent responsible for any particular designed system or artifact—only that such a system or artifact was designed by an intelligent agent *of some kind*. In addition, proponents of intelligent design, such as myself, insist that the case for intelligent design is based upon scientific evidence and upon established methods of scientific reasoning—not religious belief or authority.

All that is true. Nevertheless, as I've also argued, although the case for intelligent design depends upon scientific evidence and methods of reasoning, it may well have larger theistic implications. And, as I argue in a new book *Return of the God Hypothesis*, the evidence for intelligent design in life and in the universe—when considered together—does point strongly to a transcendent designing intelligence—i.e., God—rather than an immanent designing agent within the cosmos itself.

Of course, some scientists, such as Francis Crick,[16] Fred Hoyle,[17] and even Richard Dawkins,[18] have postulated that an intelligence elsewhere *within* the cosmos might explain the origin of the first life on Earth. Crick proposed this idea after candidly acknowledging the prohibitively long odds against life arising spontaneously here on Earth.[19] He consequently proposed that life first arose by some undirected process of chemical evolution somewhere else in the universe and then continued to evolve, eventually producing an intelligent form of alien life. This immanent intelligence—an extraterrestrial agent rather than a transcendent God—designed and then "seeded" a simpler form of life on Earth. Hence, the term *panspermia* (from the Greek *pan*, "all," and *sperma*, "seed").

Though logically possible, I've never found this explanation for the origin of life or the origin of biological information satisfying. For one thing, any theory of the origin of life, whether purporting to explain the origin of the first life here on Earth or elsewhere in the cosmos, must account for the origin of the specified information necessary to configure matter into a self-replicating system—something that most biologists take as a *sine qua non* of a genuinely living organism. Yet those who propose panspermia have not explained, or even seriously grappled with, the problem of the origin of specified biological information.[20]

Simply asserting that life arose somewhere else out in the cosmos does not explain how the information necessary to build the first life, let alone the first intelligent life, could have arisen. It merely pushes the explanatory challenge farther back in time and out into space. Indeed, positing another form of preexisting life only presupposes the existence of the very thing that all theories of the origin of life must explain—the origin of specified biological information.

Beyond that, the panspermia hypothesis certainly does not explain the origin of the cosmological fine-tuning. Because the fine-tuning of the laws and constants of physics (and the initial conditions of the universe) date from the very origin of the universe itself, the designing intelligence responsible for the fine-tuning must have had the capability of setting the fine-tuning parameters and initial conditions from the moment of creation. Yet, clearly, no intelligent being *within* the cosmos that arose after the beginning of the cosmos could be responsible for the fine-tuning of the laws and constants of physics that made its existence and evolution possible. Such an intelligent agent "inside" the universe might reconfigure or move matter and energy around in accord with the laws of nature. Nevertheless, no such being subject to those laws could possibly change the constants of physics simply by changing the material *state* of the universe. Similarly, no intelligent being arising after the beginning of the universe could have set the initial conditions of the universe upon which its later evolution and existence would depend. It follows that an immanent intelligence (an extraterrestrial alien, for instance) fails to qualify as an adequate explanation for the origin of the cosmic fine-tuning.[21]

Instead, the fine-tuning of the universe as a whole is better explained by an intelligent agent that transcends the universe, one that has the attributes that religious believers typically associate with God. Indeed, because theism conceives of God as an intelligent agent having an existence independent of the material universe—either in a timeless eternal realm or in another realm of time independent of the time in our universe—theism can account for (1) the origin of the universe in time (i.e., at a beginning), (2) the fine-tuning of the universe from the beginning of time, and (3) the origin of the specified information that arises after the beginning of time that is necessary to produce the first living organism.

Thus, deeper philosophical deliberation about the evidence of intelligent design in life and the universe may well lead to a theistic conclusion. And that suggests, as many authors of this book do, that science, properly understood, may well have faith-affirming implications.

Is Our Intuition of Design in Nature Correct?

Douglas Axe

According to the secular account of reality, God is an invention of the human imagination—a myth that persists either because it's an inevitable by-product of human mental capacities or because it benefits our survival.[1] Either way, he is a recent insertion into a flow of history that for eons got along just fine without him.

It goes without saying that many people disagree with that account of God. But if atheists are right, the God question is fundamentally a human thing—one of many propositions over which humans argue, and nothing more. If atheists are wrong, however, then this question goes much deeper. For atheists, facing arguments for God's existence is one thing. Facing God himself is another.

The design question likewise has two levels—one merely human, and the other divine. The first elevates the human intellect by putting smart people on the stage to debate the big questions. Is the universe designed? Is life designed? Are humans specially designed? Scholars argue the positions back and forth, hoping to persuade observers to join their camp, or (short of that) at least

to reassure fellow campers that they shouldn't pack up and leave.

All this changes if and when God chooses to end the pretense that his existence was ever really in doubt. This is the point where the shallow gives way to the deep, where those who took pride in their intellect realize their folly, where words fail those who relied too heavily on them.

The book of Job captures this divine invasion beautifully. The debate there wasn't over God's existence, but over something equally beyond doubt: his goodness. After 35 chapters of Job and his companions exchanging their wisdom on the matter as if God were too distant to hear (or to care), God himself made his closeness known.

He did this not by displaying his power anew, but rather by reminding Job that his creation already *is* the display. No earthquakes or lightning strike for Job, then—just a long series of humbling questions of this nature:

If you're so smart, Job, can you tell me how *this* works?

If you're so capable, can you do *this*?

It was the opposite of a test of knowledge, where correct answers tend to inflate the ego. In this case, the answers were all obvious—*no*—the point being to get Job's ego (and ours) out of the way.

It worked. "I have uttered what I did not understand," Job confessed, "things too wonderful for me, which I did not know."[2] The God who had seemed theoretical to humans wrapped up in their debate proved his reality not by arguing but by reclaiming the attention he has always deserved.

Arguing the Obvious

For millennia, philosophically minded people have argued that the universe requires a transcendent designer. Of the beginning, Plato wrote: "[I]n those days nothing had any proportion except by accident; nor did any of the things which now have names deserve to be named at all—as, for example, fire, water, and the rest of the elements. All these the creator first set in order, and out of them he constructed the universe…"[3] Order implies design.

Technology brought ever more compelling comparisons between the handiwork of humans and the handiwork of God—from Plutarch's shoe,[4] to Paley's watch,[5] to Behe's mousetrap,[6] and to Meyer's computer.[7] And the design argument was refined along the way, progressing from its initial focus on mere order to more sophisticated notions of complexity and information.

Interestingly, though, while some of these arguments predate the New Testament, there's no hint of them in Scripture. Indeed, both Old and New Testaments seem to prefer divine invasion over argument. Perhaps Paul's letter to the church in Rome explains this preference. In the first chapter, he says that those who deny God must actively "suppress the truth." Why? Because "what is

known about God is plain to them, because God has shown it to them. For his invisible attributes, namely, his eternal power and divine nature, have been clearly perceived, ever since the creation of the world, in the things that have been made."[8] This leaves deniers of God "without excuse."[9]

It's as if Scripture recognizes there's no point in arguing the obvious. Those who choose to deny what is obvious are being led by something other than reason, and this makes them immune to arguments. Spiritual conditions call for spiritual medicine, and rational argumentation (in itself, anyway) isn't that.

Of course, there are plenty of examples of argumentation in Scripture. Solomon, in the book of Ecclesiastes, masterfully demonstrates the futility of life lived entirely "under the sun." Jesus, even more masterfully, argues against the legalism of the Pharisees. Peter argues that the words of King David point forward to Jesus the Christ.[10] Paul and the author of Hebrews argue that we are saved by grace through faith, and not by works of the law. We could go on and on.

But none of these truths are patently obvious, such that anyone claiming not to know them would have to be disingenuous. The most elementary truths, on the other hand, *are* that obvious. Philosophers can spend their time debating whether existence, or consciousness, or God, or the self, or the universe are real, but Scripture wastes no time on such things. That stars and mountains and trees and humans could not have come to exist apart from God is another such thing.

These truths aren't really in doubt, however badly some insist otherwise. Young children are more honest about this than adults, not having learned to worry about what others think of them. "By elementary-school age," writes Berkeley psychology professor Alison Gopnik, "children start to invoke an ultimate God-like designer to explain the complexity of the world around them—even children brought up as atheists."[11]

Of course, Gopnik and others see this as childish ignorance that calls for education. To me, it has the look of artless honesty. In any case, it's a picture of simplicity, which is a virtue more often than a vice.

There's a key lesson here. In my experience, the thing that most robs people of confidence in expressing the certainty that God made us is the misconception that this should be complicated. Many of us tend to brush simple arguments aside in search of more complex arguments, thinking these should be taken more seriously. I've done this.

It's a big mistake.

After spending decades on technical scientific arguments for the design of living things, I've come to realize that, as valid as these may be, the most powerful defense of any obvious truth is the obvious one. Since the claim that we were made by a "God-like designer" is one of these obvious truths, our preferred defense should be simple, commonsensical, and obvious.

Accepting Cookies

Use your imagination to transport yourself back to age four. The moment a wonderfully familiar smell wafts into your bedroom, you race out to the kitchen to see a cookie sheet on the countertop with a dozen fresh-baked cookies lined up in rows. Risking a minor burn and a pre-dinner scolding, you carefully pry one off the sheet and trot back to your room, blowing on it to cool it off before you devour the evidence.

You don't for a moment wonder how those cookies came into existence. You know one of your older siblings or your parents must have made them. You would be

perplexed if someone pressed you on that intuition, as it seems too obvious to require justification. If they did, you would say something like: "You have to mix everything together to make cookies, so someone has to make them, and I don't know how."

And you would be absolutely correct. If we really wanted to make that simple deduction sound more erudite, we could bring in a team of experts with PhDs in food chemistry, physics, rheology, and probability. If someone is willing to pay the bill, this team could spend years filling volumes with technical studies and calculations that, in the end, affirm what you already knew at age four.

You may not have been able to pronounce *probability* when you were four. You certainly didn't know that the probability of multiple independent events all occurring equals the product of their individual probabilities. Nor would you have grasped the mathematical reason that this quickly turns improbabilities into impossibilities. The experts could use these facts to work up some numbers for the improbability of those cookies having been a fluke of nature: assume a mixing bowl on the counter; estimate the probability of a sack of flour accidentally falling out of the cupboard and dispensing three cups into the bowl at one in a trillion; same for the cup of granulated sugar, and for the cup of brown sugar; one in a septillion (generously) for two eggs to make their way out of the fridge and into the bowl without their shells, all by chance.

We can stop there. Without being anywhere close to a fluke batch of baked cookies yet, we already have an improbability of 1 in $10^{12} \times 10^{12} \times 10^{12} \times 10^{24}$. That equates to 1 in 10^{60}, or (writing that huge number out), 1 in:

1,000,000,000,000,000,000,000,000,
000,000,000,000,000,000,000,000,
000,000,000,000.

Your four-year-old self was spot on. Someone definitely made the cookies (and whoever did will be similarly confident about the cause of the missing one).

Intuition over Calculation

Now that you're considerably older than four, would you say that this calculation, or the scary number it produced, makes you more confident than you were back then? I think many of us would say it does. We shouldn't, though. After all, the probabilities we used in the calculation were nothing more than intuitive guesses on my part. I don't have time to waste watching refrigerators in the vain hope of witnessing the kind of fluke that would be needed for eggs to come out by chance, much less for the yolks and whites to land in the bowl without the shells. Instead, knowing that nothing remotely comparable to this is likely, I assigned a probability that seems sensibly low.

In other words, we don't really need numbers to know that a stocked kitchen can't produce cookies by accident. We simply need to know two things from experience: (1) *many* things have to happen correctly for cookies to be made, and (2) *none* of those things are apt to happen if someone doesn't make them happen.

The reason we don't need a number for (1) is that we naturally underestimate how many things need to be done. When we think of making cookies, we think of a recipe with around half-a-dozen steps. One of these might be *drop dough by large spoonfuls into rows on an ungreased cookie sheet*. Now, for a person with a bit of experience in the kitchen, there's no need to break that step down further. But what's easy for a human with a clear set of instructions is *impossible* if we take the human away. That's the whole point!

Because of (2), we're intuitively confident that mere physical chaos in a kitchen has no chance of transferring cookie dough from a bowl onto a cookie sheet—12 little mounds in rows of four. We feel no need to justify this because it seems so obvious. But we certainly *can* justify it. To do so, we simply imagine the unlikelihood of purely accidental events causing just *one* little mound of dough to land in the right place without making a hopeless mess of the rest of the dough or of the cookie sheet. Pick your version of chaos. The cats quarreling on the countertop? A large earthquake? A tornado? A car crashing into the house? We have enough experience with messes and disasters to know that if any of these were forceful enough to disturb the dough, the result wouldn't be pretty.

We can't put a number on the odds of one piece of dough getting transferred correctly to the cookie sheet, but we know it's very low. If every kitchen in the world were cleared of people after being supplied with a bowl of cookie dough and a cookie sheet, and all these kitchens were monitored by video for 24 hours, we would be very surprised if chaos were to come through in any of them. If we put the odds for success in a single kitchen in one day at one in a billion, no one would have grounds for saying we've grossly underestimated. Of course, people who don't like where an argument is going can always complain. But for their complaint to rise to the level of a serious rational objection, they would have to be willing to set up a smaller-scale version of the experiment to prove their point. More power to them if they do!

In all likelihood, they won't. Their own intuition is apt to tell them this would be a waste of time (and cookie dough), which would mean they actually agree with us, however reluctantly.

And if getting one little mound of

dough placed correctly is that improbable, then 12 is out of the question. If we assume each mound requires a separate disruptive event, independent of the others, then the likelihood of accidental causes properly completing that one recipe step in one kitchen in a day is no better than 1 in 10^{108}. To write that number out, we would put 108 zeros after a 1. There is no need to do that, however, because the point is clear. There haven't been nearly enough kitchens or days for an improbability of this magnitude to be surmounted.

Now, a case can be made that the 12 mounds of dough wouldn't actually require 12 independent events. That's fine. Understand, though, that if we ask more work of a single accidental event—say, two or three properly placed mounds of dough—then we make that event much more unlikely than it already was. There are many reasonable ways to do the analysis, and by all of them, chance is hopelessly inadequate for dropping the dough correctly onto the cookie sheet. And remember, we've only been dissecting one of the recipe's steps. We would reach the same conclusion for creaming the sugar with the butter, for cracking and beating the eggs, for transferring and mixing in the flour, for coordinating the oven controls with the opening and closing of the oven door, for transferring the unbaked cookies to the hot oven, and so on. All these already-impossibly-low probabilities multiply to give a probability so fantastically small as to be indistinguishable from zero, practically speaking.

What should we conclude from all this? Two things: First, our immediate automatic sense that certain things (baking cookies, for example) have to be done intentionally can, at least in some instances, be shown to be correct. Second, there seems to be a simple and reliable explanation for a whole class

of these instances. Namely, whenever it's intuitively obvious that some significant outcome—some big thing or result—required lots of smaller things to happen in the right way, and it's equally obvious that none of these would have happened unless someone who knew what they were doing made them happen, we can be confident that purposeful action was the cause. Our imagined four-year-old recognized that know-how is the key, and that's absolutely correct.

We aren't doing any mental math when we have these intuitions. They're too immediate to involve calculation. Nevertheless, we *can* do math to check them. And when we do, we invariably confirm our intuitions. There's no mystery here. Many things are both obvious and true, and this is one of them. There are heights from which we obviously shouldn't leap. No need for careful measurements or calculations to figure that out (though these could be done). There are tools obviously inadequate for the task at hand—ones we don't even consider. And there are causes obviously insufficient to explain effects at hand—ones we don't even consider. When it comes to the things that trigger our design intuition, all causes apart from purposeful action are obviously insufficient, which is why we don't even consider them—usually.

Spotting Genius

Even though our design intuition seems so reliable, we're told this intuition misleads us when it comes to life. The four-year-olds who instantly know cookies and dragonflies both had to be made are eventually given an alternative explanation—not for cookies, but for dragonflies. Despite their much greater sophistication, these living marvels came about without any purposeful action, we're told. Cookies have to be made, but dragon-flies just happened.

Really?

That many things had to be arranged correctly is no less obvious in the case of a dragonfly than it is in the case of a cookie, though this conclusion is arrived at some-what differently for the two. The cookie and the dragonfly are both recognizable as signif-icant things, but we're much more familiar with what it takes to make the one than the other. After all, people make cookies.

Nevertheless, the four-year-old who sees a drone in operation for the first time instantly recognizes it as something that was purposefully made, despite knowing nothing at all about how it was made. Indeed, most of us have never seen how drones are manu-factured, and yet we all know, intuitively, that the process must be purposeful in a much more sophisticated sense than the making of a cookie is. And again, we're absolutely right.

How can we arrive at this conclusion with no understanding of how the thing is made?

It all boils down to how we instantly know that many things had to be arranged correctly in order for the drone to work. The answer, surely, is that we see the drone doing something purposeful, and purposeful action always requires many things to be done in a way that aligns with the purpose. Instead of drifting with the air currents, the way ashes do, the drone moves in a controlled way, as though it *wants* to go somewhere and refuses to let the currents get in its way. There is effort on display here—not only seen in the drone's movements, but also heard in the undulating pitch of the whirring blades.

The same goes for the dragonfly, only more so.

From an early age, we realize that clever inventions like these—things that behave as though they have a purpose—always require

know-how. I call a thing of this kind a *busy whole*, meaning "an active thing that causes us to perceive intent because it accomplishes a big result"—i.e., a *clever* result. We know, intui-tively, that this can only happen by insightful action—"by bringing many small things or circumstances together in just the right way."[12]

In other words, the way we recognize clever inventions effectively guarantees that they meet the two conditions we identified earlier, which assures these things didn't hap-pen by accident. We wouldn't see something as a clever invention unless it were obvious (1) that many things had to happen correctly for it to work, and (2) that none of these things were apt to happen by accident. Both are obvious just from the fact that the thing rises above unguided processes like wind and erosion to accomplish a big result that pro-cesses like those can't accomplish.

Our eye for cleverness starts developing as soon as we begin to experiment with clev-erness. We start by recognizing that physical things have sizes and shapes, textures and colors, and sometimes sounds, smells, and tastes. We soon learn to distinguish these things according to their various kinds. While perfecting that skill, we learn to move and manipulate things. We learn to imagine how various things can be combined and arranged, and then challenge ourselves to do the combining and arranging. In learning to speak, we become proficient in associating physical symbols—phonetic sounds—with meanings, opening to us the infinite world of creating and communicating ideas.

Accompanying all this learning is a strong desire to master what we haven't yet mastered, made possible by a keen sense of the kinds of things that require mastery. To the voracious young explorer, dragonflies fit that description every bit as well as drones do. And for good reason.

The Science of Rebellion

Here's a one-paragraph summary of what we've accomplished so far. Starting with cookies, we realized that our innate ability to recognize things that require know-how serves us well from an early age. Without any calculation at all, we instantly conclude that some things come about only through the deliberate actions of people who know how to make them come about. Not only that, but we're right! In cases where we can use our know-how to do the probability calculations, we invariably find that the many things that had to come together couldn't possibly have done so without know-how. And if that's true for simple things like cookies, then it has to be even truer for more complex things where we lack the know-how to list the requirements. Indeed, we don't need to know how some clever invention works to spot instantly that *someone* knew what they were doing when they put it together. All we have to see is that it does something distinctly unnatural by achieving a significant end in a clever way. That kind of genius requires mastery over the natural elements, which can never be achieved by the natural elements themselves. A simple batch of cookies showed us why the odds against this are always insurmountable.

Everyone gets this line of reasoning—at least as it applies to things like cookies and drones. Why, then, do so many intelligent people insist that it doesn't apply to things like dragonflies? Anyone with even a bit of exposure to biology knows that the inner workings of dragonflies are much more sophisticated than the inner workings of drones. Equally obvious is that the design of dragonflies displays a much higher level of mastery over the physical elements than the design of drones does. Both zoom and hover, but only the dragonfly does this in

a completely autonomous way. The drone needs a human to direct it. Only the dragonfly takes care of its own energy needs. The drone needs a human to charge its battery. Only the dragonfly repairs itself—from the molecular scale on up. The drone needs humans to do the repairs. And, most profoundly, only the dragonfly makes more of its kind. Drones need humans to make them.

Dragonflies are as far above drones as drones are above cookies. How, then, can anyone believe, contrary to their own intuition, that these living marvels are unintended accidents of blind, natural processes?

The answer from Scripture (back to Romans 1) is that deep down, nobody really does believe this. The choice to not believe is more about rebellion than reason. Smarts are an asset for anyone genuinely seeking the truth, but for those actively suppressing the truth, they're just a way to make folly look respectable. When Richard Dawkins wrote that "Darwin made it possible to be an intellectually fulfilled atheist,"[13] he was expressing the same idea from his God-denying perspective. No one wants to look foolish. Though Scripture describes God-deniers in those terms,[14] they themselves want their denial to look respectable.

But that's a tall order. If young children automatically ascribe dragonflies to a "God-like designer" for all the right reasons—reasons that are both obvious and valid—then for educated adults to insist otherwise is, as Paul put it, inexcusable.[15]

Deconstructing Natural Selection

Dawkins was partly right. Darwin did indeed give atheists their favored explanation for things like dragonflies and humans. But while certain aspects of Darwin's explanation are undoubtedly true (the reality of natural selection, for example), it suffers the same implausibility problem as all other attempts to remove God from the story.

Natural selection is the centerpiece of evolutionary theory. And yet by its very definition, this process doesn't explain how things originate. In Darwin's own words, natural selection is the "preservation of favourable variations and the rejection of injurious variations."[16] That is, natural selection can preserve helpful traits in a species, but only after something else first brings those traits into existence. The problem is that evolutionists have never been able to identify an adequate "something else." I've referred to this deficiency as "the gaping hole in evolutionary theory"[17]—my term for something many others have described for well over a century. Dutch botanist Hugo De Vries famously gave this account back in 1904: "Natural selection may explain the *survival* of the fittest, but it cannot explain the *arrival* of the fittest."[18]

People who are determined to believe that blind evolutionary processes invented dragonflies must try to satisfy themselves with the hope that all the remarkable inventions on display in these insects came about through a long series of coincidences involving genetic mutation and natural selection. Masters of storytelling, like Dawkins, almost make it sound plausible, if you don't allow yourself to think about it too deeply.

Let's walk through a scenario to see why it doesn't work.

Evolutionists believe that all life on Earth came from an early form of bacterial life. None of them know how this early life appeared, but they think the Darwinian process explains how all subsequent life evolved in an unguided way from that humble beginning. Contrary to what our intuition tells us, no know-how was needed, they insist.

But if a stocked kitchen can't produce a batch of cookies by accident, how could a sea of bacteria produce things like brains and eyes and hearts, as parts of things like dragonflies and geckos and humans—all by accident? People who haven't run the numbers might think anything is possible in a vast ocean over a vast period of time. Nobel Prize-winning biologist George Wald expressed this sentiment well when he wrote that "time itself performs the miracles."[19] But we know better. We reached an improbability of 1 in 10^{60} before we were even close to factoring in everything that would have to happen for a batch of cookies to happen by accident. And that's *cookies*, mind you. Brains and eyes and hearts are in another league, making their improbabilities even more extreme. When we bear this in mind, the Earth's oceans become insignificantly small, at a meager 10^{21} liters, and pitifully young, at a paltry 10^{17} seconds.

Astute evolutionists get this. As Richard Dawkins put it in *The Blind Watchmaker*:

> ...however many ways there may be of *being alive*, it is certain that there are *vastly more ways of being dead*, or rather not alive. You may throw cells together at random, over and over again for a billion years, and not once will you get a conglomeration that flies or swims or burrows or runs, or does anything, even badly, that could remotely be construed as working to keep itself alive.[20]

But if they know *time* is not the answer to these astronomical improbabilities, what do they think the answer is? Though a growing number of evolutionary biologists have misgivings about its adequacy,[21] *natural selection* continues to be the favored candidate despite the fact that it can't explain the arrival of the fittest. The thinking is that long before an invention as complex as the insect eye existed, much simpler inventions were being generated by mutations and locked in by natural selection. When a functional compound eye finally did appear, it was (supposedly) the result of hundreds of these simpler inventions acting in concert.

This is nothing but storytelling. No scientist can specify the steps that would produce a new compound eye. We don't even have a scientific basis for thinking transitions like this are possible. But even if they are, they certainly can't happen by accident. The whole problem is that unguided processes have no way to bring all the right things together in the right way to produce something that rises to the level of a clever invention. Mutation and selection can lock in changes that fall well short of that, but these are invariably evolutionary dead ends.

The hemoglobin defect responsible for sickle cell anemia in humans is a good example. Babies who inherit the responsible mutation from both parents have substantially reduced life expectancies, but those inheriting it from only one parent are healthy. Now, it so happens that these healthy carriers of the mutation benefit from resistance to malaria. Because of this, natural selection has caused the defective hemoglobin gene to persist in regions where malaria is prevalent. This is natural selection at work for sure, but nobody thinks it's an evolutionary springboard to something remarkable. It's merely the lesser of two evils—something harmful that happens to reduce the risk of death from something even more harmful.

There are many similar examples of situations (usually atypical ones) where impairing genes can provide a selective advantage. Living things normally work hard to preserve their genes intact, but desperate times can call for desperate measures, including loss of genetic information. Biochemist Michael Behe has recognized this as a major recurring theme in adaptive evolution: "break or blunt any functional gene whose loss would increase the number of a species' offspring."[22]

This only underscores the profound difference between *adaptation* (tweaking something in a way that provides immediate benefit) and *invention* (bringing an ingenious thing into existence for the first time). When a ship is struggling to stay afloat, all kinds of things get thrown overboard. But that's adaptation, not invention. You don't get a new ship design that way.

Spooky Coincidence Versus Humble Truth

In the end, this is where Darwin's project sinks—and not only his, but *all* projects that attempt to ascribe life to unplanned, unguided natural processes. The mere fact that ingenious inventions always require highly exceptional arrangements—that when all raw possibilities are considered there are always *vastly* more useless arrangements than ingenious ones—means that someone clever has to do the arranging. All insistence to the contrary amounts to nothing more than an appeal to spooky coincidence.

Since nobody believes in spooky coincidences, these appeals are always disguised to look more respectable, but that doesn't change the underlying facts.

Cookies illustrate this nicely. The thought that an earthquake might cause things to become arranged in a kitchen in such a way that an accidental batch of cookies becomes probable has no more merit than the thought that an earthquake might cause an accidental batch of cookies. At best, the former is just a recasting of the latter, shifting the emphasis from the improbability of the outcome to the improbability of the circumstances needed to achieve it. At worst, it's even less plausible.

Consider, for example, the idea that an earthquake might just happen to throw things together to make a cookie-making machine—one that routinely cranks out batches of cookies. If that were to happen, then batches of cookies would be the expected outcome. But of course, the odds of accidental processes inventing such a machine are *far worse* than the already-impossible odds of them making a single batch of cookies. This avenue of thinking accomplishes nothing productive. In a desperate attempt to dodge the original spooky coincidence, it concocts an even spookier one.

Attempts to explain life as a product of accidental causes are even more unbelievable, because while we can specify in detail how an autonomous process can crank out cookies,[23] attempts to specify an autonomous process that cranks out species in all their glorious variety amount to nothing more than vague storytelling. Suffice it to say that if such a process were even possible, it certainly wouldn't come about by accident.

The children have always had this one right. And I suppose that makes everything I've said here the opposite of an impressive argument. I'm saying that what four-year-olds recognize with no deliberation at all is demonstrably correct, whereas the vague and convoluted story handed down to us by the experts—endlessly revised in a futile attempt to make it coherent—is high-end nonsense.

This humble, unpretentious truth happily takes its place alongside a few others of notable significance.

What Is Intelligent Design and How Should We Defend It?

Casey Luskin

Each year, the scientific case for intelligent design (ID) becomes more compelling as peer-reviewed pro-ID scientific studies, as well as papers by non-ID scientists, reveal increasing evidence of design in nature. Yet at the same time, prominent cultural and intellectual voices become louder in both their disdain for the ID position and their certainty that it is wrong. How can the nonexpert sort through these starkly contrasting voices and viewpoints and decide (1) whether ID is true, and (2) if ID is true, whether it is important enough to be worth expending the effort to defend it against seemingly powerful opposition? Or, to put the question another way, *Does ID matter?* If ID does matter, this leads to a practical question: *How should we navigate this complex and intense debate to defend intelligent design?*

This chapter will attempt to answer these questions while outlining the strong scientific case for ID and explaining why ID is an effective and vital tool for anyone who wishes to scientifically demonstrate that nature is full of evidence for design.

Is Intelligent Design True?

Countless students are taught that there is no evidence for intelligent design, and neo-Darwinian evolution is an unassailable fact. Unless a student proactively researches the topic, she has little chance of disputing her professors' doctrinaire declarations. Is there any hope?

In fact, there is much. With moderate study, one can articulate the case for ID and answer most common objections. Many excellent books have been written that cover these topics, and this present volume, or the resources listed in the bibliography at the end, are also great places to start. But we can cover some of the basics in this chapter.

The most common objection to ID is that it is not science, but religion. To understand why this objection is flawed, we must first appreciate how ID theorists argue for design.

Intelligent design is a scientific theory that holds that many features of the universe and living things are best explained by an intelligent cause rather than an undirected process like natural selection. ID aims to discriminate between objects generated by material mechanisms and those caused by intelligence.

ID theorists start by observing how intelligent agents act when they design things. By studying human intelligent agents, we learn that when intelligent agents act, they generate high levels of information. The type of information that indicates design is generally called *specified complexity*, or *complex and specified information* (CSI for short). Let's briefly unpack this term.

Roughly speaking, something is complex if it is unlikely. But complexity or unlikelihood alone is not enough to infer design. To see why, imagine that you are dealt a five-card hand for a poker game. Whatever hand you receive is going to be very unlikely. Even if you get a good hand, like a straight or a royal flush, you're not necessarily going to say, "Aha! The deck was stacked." Why? Because unlikely things happen all the time. We don't infer design simply because of something's being unlikely. We need more—according to ID theorist William Dembski, that is *specification*. Something is specified if it matches an independent pattern.

To understand specification, imagine you are a tourist visiting the mountains of North America. First, you come across Mount Rainier, a huge, dormant volcano in the Pacific Northwest. This mountain is unique; in fact, if all possible combinations of rocks, peaks, ridges, gullies, cracks, and crags are considered, its exact shape is extremely unlikely and complex. But you don't infer design simply because Mount Rainier has a complex shape. Why? Because you can easily explain its shape through the natural processes of erosion, uplift, heating, cooling, freezing, thawing, weathering, etc. There is no special, independent pattern to the shape of Mount Rainier. Its complexity alone is not enough to infer design.

Now you visit a different mountain— Mount Rushmore in South Dakota. This mountain also has a very unlikely shape, but its shape is special. It matches a pattern—the faces of four famous presidents. With Mount Rushmore, you don't just observe complexity; you also find specification. Thus, you would infer that its shape was designed (Figure 1).

 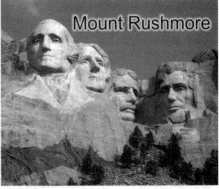

Figure 1. Which of these two mountains has a shape that allows us to detect design? Mount Rainier (left) has an unlikely (complex) shape, but it's not specified, so we do not detect design. In contrast, Mount Rushmore (right) has a shape that is both complex and specified, so we detect design. Credits: Mount Rainier: Casey Luskin. Mount Rushmore: Dean Franklin, CC BY 2.0 (https://creativecommons.org/licenses/by/2.0), via Wikimedia Commons.

We can further see that ID is science and not religion because ID uses the scientific method to make its claims. The scientific method is commonly described as a four-step process involving observation, hypothesis, experiment, and conclusion. ID uses this precise method:

- *Observations:* ID theorists begin by observing that intelligent agents produce high levels of CSI.

- *Hypothesis:* ID theorists hypothesize that if a natural object was designed, it will contain high CSI.

- *Experiment:* Scientists perform experimental tests upon natural objects to determine if they contain high CSI. For example, mutational sensitivity tests show enzymes are rich in CSI: they contain highly unlikely orderings of amino acids that match a precise sequence-pattern that is necessary for function.[1] Another easily testable form of CSI is irreducible complexity, wherein a system requires a certain core set of interacting parts to function. Genetic knockout experiments show that some molecular machines are irreducibly complex.[2]

- *Conclusion:* When ID researchers find high CSI in DNA, proteins, and molecular machines, they conclude that such structures were designed.

Contrary to popular conceptions, however, ID is much broader than biology. The laws of physics and chemistry show evidence of design because they are finely tuned to allow life to exist (see chapters 20, 21, 22, 23 for details). Universal laws are complex in that they exhibit unlikely settings—cosmologists have calculated that our universe is incredibly finely tuned for life to less than one part in $10^{10^{123}}$.[3] (That's 1 in 10 raised to the exponent of 10^{123}; we don't even have words or analogies to convey numbers this small!) Yet these laws are specified because they match an extremely narrow band of values and settings required for the existence of advanced life. This again is high CSI, and it indicates design. As Nobel Laureate Charles Townes observed:

> Intelligent design, as one sees it from a scientific point of view, seems to be quite real. This is a very special universe: it's remarkable that it came out just this way. If the laws of physics weren't just the way they are, we couldn't be here at all.[4]

To summarize, scientific discoveries of the past century have shown life is fundamentally based upon:

- A vast amount of CSI digitally encoded in a biochemical language in our DNA.

- A computerlike system of information processing where cellular machinery reads, interprets, and executes the commands programmed into DNA to produce functional proteins.

- Irreducibly complex molecular machines composed of finely tuned proteins.

- Exquisite fine-tuning of universal laws and constants.

Where, in our experience, do language-based digital code, computerlike programming, machines, and other high CSI structures come from? They have only one known source: intelligence.

The argument for design briefly sketched here is entirely empirically based. It offers positive evidence for design by finding, in nature, the types of information and complexity that we know, from experience, derive from intelligent causes. (This positive case for design is further elaborated in chapter 16.) One might disagree with the conclusions of ID, but one cannot reasonably claim that this argument is based upon religion, faith, or politics. It's based upon science.

Answering Serious Objections

An ID supporter could easily master the aforementioned arguments for design, showing not only that ID is science, but also that it is strongly supported by the scientific evidence. However, after making this argument, the budding ID defender will likely encounter many second-order objections. Answering these additional objections might seem like a potentially bottomless pit of issues that must be studied. Is there any hope to master the issue?

Again, the answer is yes. The key issues are usually less complicated than one might expect, and one can generally rely on the following rule: most of the time, serious* objections to ID fail to address the crux of ID's argument—*Where does new information come from?* Staying focused on that key scientific question will prevent distraction by red-herring objections.

When critics do offer serious, evidence-based rebuttals to ID, the objections typically fall into one of three categories: (1) inadequate,

(2) wrong, or (3) they end up confirming ID arguments. Even when sincerely made objections fall short, it's important to take them seriously and respectfully. The way in which you answer objections is as important for reaching people as the substance of your arguments. Let's examine some common responses that are raised to ID and discuss some answers.

A classic example of an inadequate response is found in how critics commonly reply to Michael Behe's arguments for irreducible complexity (see chapter 27 for details). ID critics have virtually never even attempted to provide step-by-step Darwinian accounts of how irreducibly complex features might evolve. Rather, they typically respond merely by citing sequence similarity (often called *homology*) between proteins, and then claim this is sufficient to establish an evolutionary history.[5] Aside from the fact that such explanations ignore that reusage of functionally similar genetic sequences and biological parts could be the result of common design rather than common descent, citing homology does nothing to demonstrate a Darwinian evolutionary pathway.

Another common rebuttal from ID critics has been to cite junk DNA. In his Oxford University Press book *Living with Darwin*, Philip Kitcher claims the "masses of genomic junk" and "apparently nonfunctional DNA" that "litters the genome" refute ID because "if you were designing the genomes of organisms, you would certainly not fill them up with junk."[6] Likewise, for years Richard Dawkins has argued that "creationists might spend some earnest time speculating on why the Creator should bother to litter genomes with untranslated pseudogenes and junk

---

* In this context a "serious" objection means one that is posed sincerely, not one that is necessarily difficult to answer. Tips for addressing nonserious (i.e., insincere) objections will be discussed later in this chapter.

tandem repeat DNA."[7] To a large extent I agree with the framing of the issue by critics that junk DNA could pose a potential challenge to ID. If our cells are in fact full of useless garbage DNA, then that would not be expected under an ID paradigm. However, in this case, the evidence has overturned this anti-ID argument, making it scientifically wrong.

The past decade has witnessed a scientific revolution in molecular biology, where mass functionality has been discovered for so-called junk DNA. For example, in 2012, an international consortium of molecular biologists working on a project called ENCODE (Encyclopedia of DNA Elements) reported that the vast majority of our "non-gene-coding" DNA is biochemically functional.[8] Literally thousands of scientific papers have been published reporting function for what was once called *junk DNA*[9]—a development ID successfully predicted.

In contrast, many evolutionary scientists bet that noncoding DNA would be useless junk—a gamble that didn't pay off. A striking example is seen in ID critic Kenneth Miller's "exhibit A" argument during the 2005 *Kitzmiller v. Dover* trial. He testified that a particular pseudogene was "broken" and "non-functional"—i.e., junk DNA—but then function for this pseudogene was reported in a 2013 paper in the journal *Genome Biology and Evolution*.[10] Even Dawkins has been forced to backtrack on junk DNA. When confronted with ENCODE's results, he claimed, "it is exactly what a Darwinist would hope for—to find usefulness in the living world."[11]

Finally, critics often claim that ID arguments have no scientific merit, but high-profile exchanges have shown that when leading ID proponents and critics go head to head, ID arguments end up being confirmed. Consider Stephen C. Meyer's 2013 *New York Times* bestselling book *Darwin's Doubt*. There, Meyer argues that the best explanation for the origin of the genetic information required to build new animal body plans in the Cambrian explosion is intelligent design. His book received a critical yet serious and respectful review in the journal *Science* by UC Berkeley paleontologist Charles Marshall.[12]

In a radio debate soon thereafter, Meyer pressed Marshall to explain the origin of the biological information necessary for animal body plans in evolutionary terms. Surely if such an explanation existed, this prominent evolutionary scientist could provide it. But Marshall did not explain how that genetic information evolved. Instead, he conceded that he assumes the existence of the information for the genes necessary to build animals, and never explains their origin.[13] This confirmed Meyer's argument that intelligent design alone can account for the origin of new information. Subsequent scientific papers confirmed Meyer's prediction that the origin of Cambrian animals would have required thousands of novel genes.[14]

In *Darwin's Doubt*, Meyer also cites mainstream scientific authorities who criticize neo-Darwinian evolution, leading biology into a post-Darwinian era. When responding to Meyer at BioLogos, the leading theistic evolutionist biologist Darrel Falk evaluated an accusation from another ID critic that Meyer had exaggerated the extent to which evolutionary biologists are rethinking or even rejecting neo-Darwinism. Did Meyer commit this error?

"I don't think so," Falk wrote. "Many evolutionary developmental biologists think that we are on the verge of a significant re-organization in our thinking about the mechanics of macro-evolution." Falk, who is a supporter of mainstream evolutionary science, further admitted that "the rapid generation of body plans *de novo*" in the Cambrian explosion is a "big mystery," and that "none" of the current evolutionary models can yet explain it.[15]

Many similar examples could be given, but these high-profile exchanges confirm that despite the bluster from critics, the arguments and evidence are trending in ID's direction: Neo-Darwinism is a struggling paradigm, and evolutionary biologists lack good explanations for the origin of new biological information. When ID arguments are put to the test by serious critics, ID is frequently confirmed.

ID's Research Program

Another common retort to ID is that it has published no peer-reviewed research. This is entirely false. The positive case for design was outlined above (and is elaborated further in chapter 16), and it is buttressed by an active ID research community that has published more than 100 peer-reviewed scientific publications to date.[16] There are multiple hubs of ID-related research. Biologic Institute, led by protein scientist Douglas Axe, has developed and tested the scientific case for intelligent design in biology via laboratory and theoretical research on the origin and role of information in biology.

Another ID research group is the Evolutionary Informatics Lab, founded by senior Discovery Institute fellow William Dembski along with Robert Marks, professor of electrical and computer engineering at Baylor University. Their lab has attracted graduate-student researchers and published multiple peer-reviewed articles in technical science and engineering journals showing that computer programming "points to the need for an ultimate information source qua intelligent designer."[17]

Other scientists around the world are also publishing peer-reviewed scientific papers supportive of intelligent design. These include biologist Ralph Seelke at the University of Wisconsin-Superior; Wolf-Ekkehard Lönnig, who recently retired from the Max Planck Institute for Plant Breeding Research in Germany; biochemist Michael Behe at Lehigh University; and mathematician Ola Hössjer at Stockholm University.

These labs and other researchers have published their work in a variety of appropriate technical venues, including peer-reviewed scientific journals, peer-reviewed scientific books (some published by mainstream university presses), and peer-reviewed philosophy of science journals and books. Pro-ID peer-reviewed papers have appeared in quite a few mainstream scientific journals, including *Protein Science, Journal of Molecular Biology, Journal of Theoretical Biology, Journal of Advanced Computational Intelligence and Intelligent Informatics, Quarterly Review of Biology, Cell Biology International, Rivista di Biologia/Biology Forum, Physics of Life Reviews, Quarterly Review of Biology, Journal of Theroetical Biology, Annual Review of Genetics*, and others. At the same time, pro-ID scientists have presented their research at conferences worldwide in fields such as genetics, biochemistry, engineering, and computer science. A list of many of these peer-reviewed pro-ID scientific papers can be found at https://www.discovery.org/id/peer-review/.

Collectively, this research is converging on a consensus: Natural selection and random mutation can produce minor changes, but many complex biological features—like new protein folds—cannot arise by unguided evolutionary mechanisms. These information-rich structures require an intelligent cause.

Dealing with New Objections

When it comes to handling new objections you've never heard, there is bad news and good news. The bad news is that every ID advocate, from the guru to the newbie, will encounter unfamiliar objections. The good news is that there are a finite number of objections out there, and over time, the number of objections you haven't heard will become fewer and fewer.

And there's more good news: There's no need to fear being asked a question to which you don't know the answer. Yes, it will happen, but when it does, don't get too stressed—there's no shame in honestly admitting when you don't know an answer, and then promising to go research the matter. For these situations, there are many good resources available. Try scouring credible ID websites listed in the bibliography at the end of this book to see if the objection has already been tackled.

Handling Heated Objections

Unfortunately, in the ID-evolution debate, it's not uncommon for objections to become emotionally heated, accompanied by insults and ridicule. It should go without saying that it's wrong to engage in this behavior. No "side" in this debate is completely blameless, but it's nonetheless an empirically observable fact that when harsh words are exchanged, ID proponents are usually on the receiving end. This is never fun to experience, but it does offer opportunities to learn and make a positive impact.

For example, nonexperts sometimes wonder how they can evaluate complex scientific arguments to determine which side in the ID debate has the "better argument." I've always been a firm believer that with a little study,

nonexperts *can* appreciate most arguments for and against ID. But there are other ways to approach the question. In addition to evaluating scientific arguments, however, we can ask: *Does one side make serious arguments and welcome dialogue, while the other persistently resorts to personal attacks that stifle conversation?* This can help reveal which camp is behaving like the evidence is on their side, and which is compensating for a weak position.[18]

When I first became involved with this debate, I went through this very process of evaluating arguments for and against ID, and I found that critics would often not address ID claims. Rather, they would employ labels (such as *creationist, antiscience,* or *pseudoscience*) and ad hominem or personal attacks (such as calling people "dishonest" or "liars for Jesus," etc.). Critics would also use straw-man definitions of ID and focus on irrelevant issues like personal religious beliefs or motives. Such tactics are typically deployed to avoid addressing ID arguments, shut down discussion, and intimidate ID advocates into silence. It's safe to say that ID proponents were being "cancelled" by critics long before "cancel culture" was widely known as a thing. In stark contrast, I found ID proponents kept their focus on the substantive questions, inviting dialogue and treating critics with respect.

On the one hand, such observations can help encourage us that ID's arguments may have merit. If ID opponents had good rebuttals, they would not resort to such tactics. On the other hand, it can be discouraging to know that when we defend ID, we may face personal attacks and opposition. This is exactly how those making the personal attacks want us to feel—intimidated, so we'll keep our mouths shut.

Those who defend ID should therefore be emotionally and spiritually prepared to face heated words and unfair rhetorical tactics. And those who experience ad hominem attacks can take some solace knowing they are in good company. Virtually every single scientist and scholar who has criticized the evolutionary viewpoint—even those with the most impeccable credentials—has faced personal attacks on his or her character. One of two things are true: Either (1) virtually every single critic of Darwinism is dishonest, immoral, etc., or (2) evolutionists are responding to scientific challenges with personal attacks.

After your first experience with making a strong evidence-based case for design in good faith but then receiving back nasty insults instead of substantive rebuttals, you will begin to appreciate the unpleasant but unforgettable lesson that the answer is (2). Even scholarly commentators writing in mainstream academic venues have observed the harsh and uncivil treatment to which evolutionists often subject their critics.[19] But how do we respond to such attacks?

On a logical level, it's always fair game to point out that these personal attacks commit logical fallacies—they represent the very definition of the genetic fallacy or the ad hominem fallacy because they attack the person or the source making an argument rather than the argument itself.

On an internal level, we can be encouraged. Usually when people use ad hominem arguments it's because they don't feel intellectually confident in the strength of their position, so they resort to emotionally charged rhetoric. Thus, ironically, the ID defenders who face personal attacks might be encouraged that they have the better argument, and the incendiary rhetoric might signal an underlying opportunity.

But how can you reach someone who is clearly angry and lashing out? The answer is

by following the teachings of Jesus and the Bible.

Jesus and the apostles taught us to love our enemies (Matthew 5:44), to repay evil with good and bless those who persecute us (Romans 12:14-21), and to not retaliate when insulted (1 Peter 3:9). Sometimes this can be difficult. But when we respond in a loving, respectful, and informed manner, malicious critics will sometimes be surprised and suddenly open to hearing our views. Christ's commands aren't just the right thing to do—they can also be extremely effective in reaching people during heated dialogue.

It's also important to remember that your audience is often much bigger than a few vocal, mean-spirited opponents. While heated critics may be resistant to your arguments, audience members in the "undecided middle" who are watching your response will see that you are responding in a kind, respectful, and informed fashion, and they may also be persuaded by your rhetorically winsome and intellectually credible replies. Harsh personal attacks are never pleasant to experience, but when handled correctly, they offer an opportunity to demonstrate that ID arguments are strong—and to show a love and grace that attracts people to your viewpoint.

The Straw-Man Definition

One common tactic among critics is to deploy a straw-man definition of ID, which claims, "ID says life is so complex it couldn't have evolved, therefore was created by a supernatural God." They may cite the 2005 *Kitzmiller v. Dover* case, where a federal judge adopted this straw-man definition and ruled that ID is a form of religion—ignoring how pro-ID expert witnesses defined ID in his own courtroom, and ignoring ID's peer-reviewed research. This critics' definition is false for two reasons.

First, it wrongly frames ID as a strictly negative argument against evolution, ignoring the positive case for design discussed previously (and elaborated in chapter 16).

Second, it wrongly claims ID appeals to the supernatural. All ID scientifically detects is the prior action of intelligence. While the complex and specified information in DNA points to an intelligent cause, that information by itself cannot scientifically tell you whether the intelligence is Jehovah, Allah, Buddha, Yoda, or some other intelligent source. ID respects the limits of scientific inquiry and does not attempt to address religious questions about the identity of the designer.

Some may rejoin by saying ID defenders are hiding their belief that the designer is God. But ID proponents don't hide their personal religious views. I am intentionally open about my Christian faith, whether speaking to religious audiences or secular ones. But I make clear that my belief that the designer is the God of the Bible is not a conclusion of ID; rather, it's a religious belief I hold for separate reasons.

Indeed, the ID movement includes people of many worldviews, including Christians, Jews, Muslims, people of Eastern religious views, and even agnostics. What unites them is not some religious faith, but a conviction that there is scientific evidence for design in nature.

The *Kitzmiller v. Dover* Case

Unserious critics will often try to end discussion on ID by citing the 2005 *Kitzmiller v. Dover* ruling, where a single district court federal judge ruled that ID is religion and not science, and therefore, it is unconstitutional to teach ID in a rural public school district

of central Pennsylvania. What's the best way to respond to this attempt to shut down conversations about ID?

The argument assumes that judges are inerrant, but spend a day in law school, and you'll learn that judges make mistakes all the time. Moreover, judges cannot settle scientific debates, and a court ruling has no ability to negate the evidence for design in nature. In fact, the *Kitzmiller v. Dover* ruling includes numerous false claims about law and science that make it a highly unreliable analysis. For example, the ruling did the following:[20]

- Adopted the straw-man definition of ID, wrongly claiming ID requires "supernatural creation"—a position refuted by ID proponents who testified during the trial.

- Ignored the positive case for design and falsely claimed that ID proponents make their case solely by arguing against evolution.

- Wrongly claimed that ID had been refuted when the court had been presented with credible scientific witnesses and publications on both sides showing evidence of a scientific debate.[21]

- Used poor philosophy of science by taking the level of support for a theory as a measure of whether it is scientific.

- Blatantly denied the existence of pro-ID peer-reviewed scientific publications and research that were documented before the court.

- Adopted an unfair double standard of legal analysis in which religious implications, beliefs, motives, and affiliations counted against ID, but never against Darwinism.

A later analysis found that the ruling's celebrated section on whether ID is science was largely copied directly from an ACLU brief,[22] and a leading anti-ID legal scholar called it "unnecessary, unconvincing, not particularly suited to the judicial role, and even perhaps dangerous both to science and to freedom of religion."[23]

Federal judges cannot settle scientific debates. The day after the Dover ruling, our cells were still full of language-based digital codes and miniature factories that produce micromolecular machines, and the universe remained exquisitely fine-tuned to sustain complex life. This debate can be settled only by the evidence, not legal declarations.

Is ID Worth Defending?

The forgoing discussion shows that ID can be defended using compelling scientific arguments. We might say ID is "true." But just because something is defensible doesn't mean it is important enough to put in the effort to defend. Does ID matter enough to be worth defending? Yes, but to address this question, we must appreciate the strategic value of the ID argument.

In his book *How We Believe*, psychologist Michael Shermer discusses a study that surveyed why self-described skeptics do or do not believe in God. Issues related to scientific evidence for God and design in nature ranked as the number one reason given for why they believe in God (29.2 percent), and also why they doubt (37.9 percent).[24]

Those who want to demonstrate—especially to scientifically minded skeptics—that life and the universe were designed will find ID is a vital tool. To defend ID, it's important to appreciate the strategic value of ID compared to other viewpoints—creationism and theistic evolution.

ID's Advantage over Creationism

The modern theory of ID was developed in the 1980s and 1990s by scientists seeking a strictly scientific approach to studying origins. Before that time, creationism, which always mixed theology into its arguments, dominated the debate.

Many of ID's founders, such as chemist Charles Thaxton, biochemist Michael Behe, mathematician William Dembski, and philosopher and historian of science Stephen C. Meyer realized that the study of information could provide a scientific basis for detecting design in nature—an approach that could appeal to the scientifically minded skeptic. Herein lies ID's advantage over creationism.

While young-Earth and old-Earth creationists debated the age of the Earth, or whether Noah rode a dinosaur or a camel onto the ark, elite materialists dominated our culture. UC Berkeley law professor Phillip Johnson, another cofounder of the ID movement, helped reframe the origins debate to reach that cultural landscape.

As a legal scholar, Johnson saw that the most important issue was not the age of the Earth or differing interpretations of Genesis. Rather, it was a fundamental question asked by everyone: *Are we the result of blind, undirected material causes or purposeful intelligent design?* By using scientific arguments and evidence to show that we are the result of design, ID directly answers this crucial worldview question.

Of course, various types of creationism can encourage people who want to reconcile the Bible with science. But for those wanting to reach skeptics and scientifically demonstrate design, ID should be of supreme interest.

However, ID doesn't answer every important question. As a science, ID does not attempt to address religious or theological questions about the designer's identity or the proper meaning of Genesis. Of course many good philosophical, theological, and historical arguments may address those important issues, but they go beyond ID.

ID must be taken for what it's worth: It's a compelling scientific argument showing that life and the universe were designed. Yet recently, another camp has arisen within the Christian community that claims science does not support ID and ID doesn't matter. This camp is called *theistic evolution* (TE).

ID's Advantage over Theistic Evolution

Theistic evolutionists (who sometimes call themselves *evolutionary creationists*) are Christians who believe that God used material evolutionary mechanisms to create life. To understand how ID interfaces with theistic evolution—and why ID is a better approach—we must first define *evolution*.

The term *evolution* can have different meanings, but it is generally used in three different ways. It can mean something as benign as (1) "Life has changed over time," or it can entail more controversial ideas, like (2) "All living things have a universal common ancestry," or (3) "Natural selection acting upon random mutations drove the evolution of the complexity of life."

ID does not conflict with evolution defined as "change over time" (1), or even the idea that some living things are related by common ancestry (2). However, the dominant theory of evolution today is neo-Darwinism (3), which contends that an apparently blind and unguided process of natural selection acting on random mutations produced the complexity of life. It is this specific claim made by neo-Darwinism that ID directly challenges. To the extent that theistic evolutionists accept neo-Darwinian evolution, ID conflicts with their view.

However, theistic evolution can mean different things to different people. The average "theistic evolutionist on the street" probably does not have neo-Darwinian evolution in mind when they say they believe "God used evolution." What they actually believe is that life has a long history (millions of years), but God supernaturally intervened at various points to direct the course of life. They accept evolution (1), and maybe (2) as well, but they're not so sure about (3). Many people who adopt the "theistic evolutionist" moniker therefore actually doubt neo-Darwinism and hold a view much closer to ID, where an intelligent agent has actively intervened—in a meaningful and detectable manner—to guide the development of life. The best way to reach such people is to give them accurate and clearly thought-out definitions of evolution so they can appreciate that their personal viewpoint is more like ID and differs dramatically from the standard neo-Darwinian paradigm.

But not all TEs think like this. Some are direct supporters of neo-Darwinism and/or other materialistic models of evolution. Neo-Darwinism—as defined by its proponents—entails a blind process of natural selection acting upon random mutations without any discernable guidance by an external agent. According to the architects of the theory, this process has no goals, no predetermined outcome, and is, by definition, "unguided."[25] Many theistic evolutionists adopt this view and are wedded to essentially materialistic scientific accounts of life, rejecting intelligent design. For example, the leading TE group BioLogos holds that design cannot in principle be scientifically detected, or that design could be scientifically detected, but isn't.[26] Thus, an important dividing line between full-blown TE and ID is this:

- TE accepts materialistic evolutionary explanations (e.g., neo-Darwinism) where the history of life appears unguided, and denies that we scientifically detect design.

- ID claims hold we may scientifically detect design as the best scientific explanation for many aspects of biology.

Do these differences matter? That is a key question. TEs tend to think the differences aren't very important, but ID proponents believe these differences are crucial.

According to textbooks and leading evolutionary biologists, neo-Darwinian evolution is defined as an unguided or undirected process of natural selection acting upon random mutation. Thus, when leading theistic evolutionists say that "God guided evolution," they mean that somehow God guided an evolutionary process that, for all scientific intents and purposes, appears *unguided*. As Francis Collins put it in *The Language of God*, God created life such that "from our perspective, limited as it is by the tyranny of linear time, this would appear a random and undirected process."[27] Whether it is theologically or philosophically coherent to claim that "God guided an apparently unguided process" is a matter I will leave to the theologians and the philosophers. ID avoids these problems by maintaining that life's history *doesn't* appear unguided, and

that we can scientifically detect that intelligent action was involved.

Theistic evolutionists sometimes try to obscure these differences, such as when BioLogos claims "it is all intelligently designed."[28] But when pressed, they'll admit this is strictly a blind faith-based theological view for which they can provide no supporting evidence. ID proponents wonder how one can speak of intelligent design if it's always hidden and undetectable. "We're promoting a scientific theory, not a theological doctrine," replies ID, "and our theory detects design in nature through scientific observations and evidence."

Some theistic evolutionists will then further reply by saying, "Since we both believe in some form of intelligent design, the differences between our views are small." ID proponents retort, "Whether small or not, these differences make all the difference in the world."

And there's the rub. By denying that we scientifically detect design in nature, theistic evolutionists cede to materialists some of the most important territory in the debate over atheism. Biologically speaking, TE gives no reasons to believe in God. This gives new atheists and their allies great pleasure, and dramatically weakens the ability of theists to defend belief in God.

To be clear, I'm *not* saying that if one accepts Darwinian evolution then one cannot be a Christian. Accepting or rejecting the grand Darwinian story is a disputable or secondary matter, and Christians (or other theists) may hold different views on this issue. But while it may be possible to claim God used apparently unguided evolutionary processes to create life, that doesn't mean Darwinian evolution is theologically neutral.

In chapter 8, Fazale Rana explains how evolutionary biology leads many people to abandon traditional Christian beliefs about

Adam and Eve, but the challenge goes much deeper. According to orthodox Darwinian thinking, human beings are accidents of history—apparently undirected processes created not just our bodies, but also our brains, our behaviors, our deepest desires, and even our religious impulses. Under theistic Darwinism, God guided all these processes such that the whole show appears "random and undirected," as Francis Collins put it. Similarly, two BioLogos authors wrote in response to ID, "On the evolutionary creationist account, the work is signed using invisible ink."[29] So God's involvement is "invisible" and life's origins appear "random and undirected"? This view stands in direct contrast to Romans 1:20, where the apostle Paul taught that God's handiwork is "clearly perceived" in nature.

TE may not be absolutely incompatible with believing in God, but it offers no scientific reasons to support faith. Perhaps this is why the atheist evolutionary biologist William Provine wrote, "One can have a religious view that is compatible with evolution only if the religious view is indistinguishable from atheism,"[30] and one BioLogos article admitted, "Evolutionary creationism does not necessarily add apologetic value to the Christian faith."[31] TE is a losing strategy for Christians. It capitulates to the "consensus," and opposes a robust scientific search for evidence of design in nature.[32] One medical physician compellingly explained why he abandoned theistic evolution in favor of an ID viewpoint: "Richard Dawkins famously said that Charles Darwin made it possible for him to be an intellectually fulfilled atheist, but I found that ID made it possible for me to be an intellectually fulfilled Christian."[33]

The ID Advantage

People hunger for compelling scientific arguments that can bolster their faith and persuade skeptical friends. In meeting this demand, ID offers the best of both worlds:

- Unlike creationism, ID accepts the best evidence offered by mainstream science and makes a strictly scientific argument.

- Unlike theistic evolution, ID doesn't challenge orthodox Christian theology and embraces the use of science to argue for design.

True, the ID argument has limits: It's a scientific theory that holds that some aspects of nature are best explained by an intelligent cause. If one desires to introduce a person to Christ on the cross, ID is not enough. But if one seeks convincing scientific arguments to show the universe and life require a designer, then, as ID's motto says, we must simply "follow the evidence wherever it leads."

The public—including skeptics—is eager to hear about this evidence, and with some study, patience, and grace, the task of informing them shouldn't be too difficult.

What Is the Positive Case for Design?

Casey Luskin

In chapter 15, we saw that intelligent design (ID) is a historical scientific theory that uses the scientific method to make testable claims about the origin of various features of nature. But on a scientific level, ID is much more than that. The positive case for design allows the theory of ID to also serve as a heuristic—a paradigm that can inspire scientific research and help scientists make new discoveries. This chapter will elaborate on how the case for design in nature uses positive arguments in multiple scientific fields, based upon finding in nature the type of information and complexity that, *in our experience*, comes only from intelligence—and explain how these positive arguments are turning ID into a fruitful paradigm to guide twenty-first-century scientific research.

What's a Positive Argument?

To understand how ID makes a positive argument, it's helpful to first appreciate what positive and negative arguments look like in historical sciences. Simply put, negative arguments in science proceed by saying, "Theory X is false; therefore, Theory Y is true." This form of argument only gets you so far because evidence against one theory does

not, in and of itself, necessarily therefore constitute positive evidence for another theory. A positive argument proceeds by saying, "Theory X predicts Y. Y is found. Therefore, we have evidence that is inferred to support Theory X." Such a positive argument uses abductive reasoning, where one infers a prior cause based upon findings its known effects in the world around us. As paleontologist Stephen Jay Gould put it, historical sciences use this kind of reasoning to "infer history from its results."[1]

Some might claim that such a positive, abductive argument commits the logical fallacy of affirming the consequent, where one wrongly infers a particular cause from its known effects because there might also be other causes that can potentially account for the data. The solution is to compare known causes which have the potentiality to explain the data and determine which one explains the most data. This is what ID theorist Stephen C. Meyer and other philosophers of science call making an "inference to the best explanation."[2]

But where do historical scientific explanations come from in the first place? Another important method of historical sciences is the principle of uniformitarianism, which holds that "the present is the key to the past."

Historical scientists apply this principle by studying causes at work in the present-day world in order to, as the famous early geologist Charles Lyell put it, explain "the former changes of the Earth's surface" by reference "to causes now in operation."[3] To put it more simply, historical scientists study causes at work in the present-day, and through these investigations can then make testable and falsifiable predictions about what we should expect to find today if a given cause was at work in the past. When these predictions are fulfilled, we have positive evidence that a particular cause was at work. The cause that accounts for the most data is inferred to be the most likely to be correct. This is how historical scientists make an inference to the best explanation.

Let's consider an everyday example.

Imagine that you took your 4x4 truck off-roading and you returned home with the truck covered in mud. You drop the truck off at a car wash to have it cleaned, and an hour later, return to pick it up. This may seem like a silly exercise, but how could you apply the scientific method of historical sciences to determine whether the truck was washed? Well, you could use your past experiences with car washes to make predictions about what you would expect to find if the truck was washed, and then you could test those predictions.

For example, your experiences with car washes have taught you that after a car goes through a car wash, it's completely free of dirt and mud, and has soapy residue on its paint.

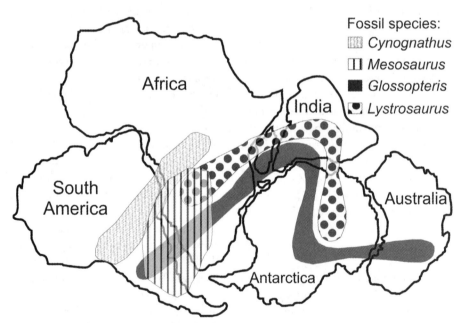

Figure 1. This map shows Gondwana, the southern portion of the Pangea supercontinent. Plate tectonics successfully predicts that locations of fossil species (shaded zones) found on continents that today are widely separated by oceans will match up when the continents are fit back into their ancient locations. Credit: Modified by Casey Luskin after "Rejoined Continents," *This Dynamic Earth: The Story of Plate Tectonics*, online edition, https://pubs.usgs.gov/gip/dynamic/continents.html (accessed March 16, 1996), public domain.

Thus, if the truck was washed, then you might predict that there will be no mud left on the exterior and would even be spotless. This prediction could be tested by a simple visual analysis. If you see chunks of mud remaining, then you refute your hypothesis that the truck was washed. You could also undertake a more technical analysis, predicting that if the truck was washed, then there should be small amounts of soap residue left on the paint surface. You could scrape material off the surface of the truck and perform a chemical analysis to confirm or refute this hypothesis. If you find that there are no chunks of mud on the truck, and soap residue is present on the truck's paint, you would have positive evidence that the truck was washed.

But is a car wash the *best explanation*? A competing hypothesis, the "rain washed the car" hypothesis, might explain a general lack of mud, but would not leave the car spotless and could not explain the presence of the soapy residue. We use this positive argument to infer that the *best* explanation for the observed data is that the truck went through a car wash.

Let's now try a scientific example from my field of geology. The theory of plate tectonics predicts that continents were once joined as a single supercontinent, often called *Pangea*. Plate tectonics predicts that continents that are now widely separated by oceans might show similar rocks and fossils—especially along the edges where they were once thought to be linked. This is in fact what we find, with plate tectonics making a successful prediction that provides evidence for the theory (Figure 1). No other theory made this prediction, making plate tectonics the best explanation for the evidence. This is a positive argument for plate tectonics.

As a historical scientific theory, ID works

in much the same way, making predictions that can be tested to provide positive evidence for the theory.

Outlining ID's Positive Argument

The theory of ID employs scientific methods commonly used by other historical sciences to conclude that certain features of the universe and living things are best explained by an intelligent cause, not an undirected process such as natural selection. To borrow Lyell's words, intelligent agency is a cause "now in operation" that can be studied in the world around us. Thus, as a historical science, ID employs the principle of uniformitarianism by beginning with present-day observations of how intelligent agents operate, and then converts those observations into positive predictions of what scientists should expect to find if a natural object arose by intelligent design.

For example, mathematician and philosopher William Dembski observes that "[t]he principal characteristic of intelligent agency is directed contingency, or what we call choice." According to Dembski, when an intelligent agent acts, "it chooses from a range of competing possibilities" to create some complex and specified event.[4] Thus, the type of information that we observe results from intelligent design is called *specified complexity* or *complex and specified information*—or CSI for short.

We explained CSI in more detail in chapter 15, but in brief, something is complex if it's unlikely, and specified if it matches an independently derived pattern. In using CSI to detect design, Dembski says ID is "a theory of information," where "information becomes a reliable indicator of design as well as a proper object for scientific investigation."[5]

ID theorists then positively infer design by studying natural objects to determine whether they bear the type of information that, in our experience, arises from an intelligent cause.

ID thus seeks to find in nature the types of information—to be precise, complex and specified information—that we know from experience is produced by intelligent agents. Human intelligence provides a large empirical dataset for studying what is produced when intelligent agents design things. For example, language, codes, and machines are all structures that contain high CSI, but in our experience, these things always derive from an intelligent mind. By studying the actions of humans, we can understand what to expect to find when an intelligent agent has been at work, allowing us to construct positive, testable predictions about what we should find if intelligent design is present in nature. High CSI thus reliably indicates the prior action of intelligence.

This positive argument for design follows the standard scientific method of observation, hypothesis, experiment, and conclusion. To be more specific, the positive case for design begins with observations of intelligent agents and what they produce when they design things. This leads to hypotheses (predictions) about what we should expect to find if intelligent agency was involved in the origin of a structure. These predictions are testable via studies of nature—often called *experiments*—but in this case meaning any empirical study of what exists in the natural world. Depending upon the outcome of the experiments and the nature of the data, the hypothesis/prediction is either confirmed or not. This leads to a (tentative) conclusion about whether design has been detected in nature.

At its simplest level, the positive case for design is thus a two-step process:

1. Study intelligent agents to understand what kind of information is produced when they act.

2. Study natural objects to determine whether they contain the type of information known to be produced when intelligent agents act.

Investigating the Evidence for Design

We'll now use this basic method to investigate the positive evidence for design in five fields: (1) biochemistry, (2) paleontology, (3) systematics (the relationships between organisms), (4) genetics, and (5) physics. Each example will begin with observations about how intelligent agents act based on previous studies by ID theorists. Then a testable hypothesis/prediction is made, followed by a discussion of what the data reveal (experiment), and finally, a conclusion.

1. The Positive Case for Design in Biochemistry

Observation (from previous studies): Intelligent agents think with an end goal in mind, allowing them to solve complex problems by taking many parts or symbols and arranging them in intricate patterns that perform a specific function—i.e., they generate high levels of complex and specified information:

- "Intelligence is a goal-directed process that is capable of thinking with will, forethought, and intentionality to achieve some end-goal."[6]

- "[W]e have repeated experience of rational and conscious agents—in particular ourselves—generating or causing increases in complex specified information, both in the form of sequence-specific lines of code and in the form of hierarchically arranged systems of parts... Our experience-based knowledge of information-flow confirms that systems with large amounts of specified complexity (especially codes and languages) invariably originate from an intelligent source—from a mind or personal agent."[7]

- "In all irreducibly complex systems in which the cause of the system is known by experience or observation, intelligent design or engineering played a role [in] the origin of the system."[8]

Hypothesis (prediction): Finely tuned high-CSI structures will be found in biology, including irreducibly complex systems that require multiple components to function.

Experiment (data): Natural structures contain many parts arranged in intricate patterns that perform a specific function (e.g., they contain high CSI). These include language-based codes in our DNA, irreducibly complex molecular machines like the bacterial flagellum,[9] and highly specified protein sequences. Mutational sensitivity tests have shown that the amino acid sequences of many functional proteins must be highly complex and specified in order to function.[10]

Conclusion: Irreducible complexity and high CSI systems are found, indicating these systems were designed.

2. The Positive Case for Design in Paleontology

Observation (from previous studies): Intelligent agents can rapidly infuse large amounts of information into systems:

- "Intelligent design provides a sufficient causal explanation for the origin of large amounts of information, since we have considerable experience of intelligent agents generating informational configurations of matter... We know from experience that intelligent agents often conceive of plans prior to the material instantiation of the systems that conform to the plans—that is, the intelligent design of a blueprint often precedes the assembly of parts in accord with a blueprint or preconceived design plan."[11]

- "Intelligent agents sometimes produce material entities through a series of gradual modifications (as when a sculptor shapes a sculpture over time). Nevertheless, intelligent agents also have the capacity to introduce complex technological systems into the world fully formed. Often such systems bear no resemblance to earlier technological systems—their invention occurs without a material connection to earlier, more rudimentary technologies. When the radio was first invented, it was unlike anything that had come before, even other forms of communication technology. For this reason, although intelligent agents need not generate novel structures abruptly, they can do so."[12]

Hypothesis (prediction): Novel biological forms requiring large amounts of new genetic information will appear abruptly in the fossil record, "fully formed," and without similar precursors.

Experiment (data): Biological novelty commonly appears in the fossil record suddenly, fully formed and without similar precursors or evolutionary intermediates.[13] As one zoology textbook states:

Many species remain virtually unchanged for millions of years, then suddenly disappear to be replaced by a quite different, but related, form. Moreover, most major groups of animals appear abruptly in the fossil record, fully formed, and with no fossils yet discovered that form a transition from their parent group.[14]

The dominant pattern in the fossil record is explosions of new biological forms. The Cambrian explosion is a prime example where most of the major animal phyla appear "fully formed" in a geologically abrupt manner,[15] but there are many other examples in the fossil record, including a bird explosion,[16] an angiosperm explosion,[17] and a mammal explosion.[18] Even our genus *Homo* appears abruptly.[19]

Conclusion: Many higher taxa arose by intelligent design.

3. The Positive Case for Design in Systematics (the Relationships Between Organisms)

Observation (from previous studies): Intelligent agents reuse functional components in different systems (e.g., wheels for cars and airplanes, or keyboards on cell phones and computers):

- "An intelligent cause may reuse or redeploy the same module in different systems, without there necessarily being any material or physical connection between those systems. Even more simply, intelligent causes can generate identical patterns independently."[20]

- "According to this [evolutionary] argument, the Darwinian principle of common ancestry predicts such common features, vindicating the theory of evolution. One problem with this line of argument is that people recognized common features long before Darwin, and they attributed them to common design. Just as we find certain features cropping up again and again in the realm of human technology (e.g., wheels and axles on wagons, buggies and cars), so too we can expect an intelligent designer to reuse good design ideas in a variety of situations where they work."[21]

- "[I]f different forms of life were intelligently designed, with a mosaic of characteristics, some of which they share in common with some organisms and others of which they share in common with different organisms, then we would expect 'phylogenetic' analyses to generate conflicting trees depending on which character was chosen. Indeed, phylogenetic analyses of different characters present in several different human-designed technological objects have been shown to generate precisely such conflicting trees."[22]

Hypothesis (prediction): Genes and other functional parts will be reused in different and unrelated organisms in a pattern that need not match a "tree," or nested hierarchy.

Experiment (data): Similar parts have been found reused in widely different organisms where even evolutionists believe the common ancestor did not have the part in question. Examples include similar genes controlling eye or limb growth in different organisms whose alleged common ancestors are not thought to have had such forms of eyes or limbs.[23] There are numerous examples of extreme convergent genetic evolution, including similar genes used in whales and bats for echolocation.[24] Genes and functional parts are frequently not distributed in a "tree-like" pattern or nested hierarchy predicted by common ancestry, but rather, show reusage in a non-nested pattern.[25] One mainstream scientific paper acknowledges:

> Incongruence between phylogenies derived from morphological versus molecular analyses, and between trees based on different subsets of molecular sequences has become pervasive as datasets have expanded rapidly in both characters and species...phylogenetic conflict is common, and frequently the norm rather than the exception.[26]

Conclusion: Common design is prevalent throughout life. The re-usage of highly similar and complex parts in widely different organisms in non-treelike patterns is best explained by the action of an intelligent agent.

4. The Positive Case for Design in Genetics

Observation (from previous studies): Intelligent agents generate structures that have a purpose or function:

- "Since non-coding regions do not produce proteins, Darwinian biologists have been dismissing them for decades as random evolutionary noise or 'junk DNA.' From an ID perspective, however, it is extremely unlikely that an organism would expend its resources on preserving and transmitting so much 'junk.'"[27]

- "[Intelligent] design is not a science stopper. Indeed, design can foster inquiry where traditional evolutionary approaches obstruct it. Consider the term 'junk DNA.' Implicit in this term is the view that because the genome of an organism has been cobbled together through a long, undirected evolutionary process, the genome is a patchwork of which only limited portions are essential to the organism. Thus on an evolutionary view we expect a lot of useless DNA. If, on the other hand, organisms are designed, we expect DNA, as much as possible, to exhibit function... Design encourages scientists to look for function where evolution discourages it."[28]

Hypothesis (prediction): Cellular components were originally designed for a purpose, and "junk" DNA will generally turn out to perform valuable functions.

Experiment (data): Numerous studies have discovered function for "junk DNA,"[29] including evidence of widespread biochemical function in the human genome discovered in the ENCODE project.[30] *Discover Magazine* summarized ENCODE's 2012 breakthrough report writing, "The key point is: It's not 'junk.'"[31] Specific examples include functionality in pseudogenes, microRNAs, introns, endogenous retroviruses, and repetitive LINE, SINE, and *Alu* elements.[32] Examples of unknown DNA functions persist, but ID encourages researchers to investigate functions, whereas neo-Darwinism has discouraged seeking such function.[33]

Conclusion: Functionality for junk DNA is prevalent, and was successfully predicted by intelligent design.

5. The Positive Case for Design in Physics

Observation (from previous studies): Intelligent agents can quickly find extremely rare or highly unlikely solutions to complex problems:

- "Agents can arrange matter with distant goals in mind. In their use of language, they routinely 'find' highly isolated and improbable functional sequences amid vast spaces of combinatorial possibilities."[34]

- "Intelligent agents have foresight. Such agents can determine or select functional goals before they are physically instantiated. They can devise or select material means to accomplish those ends from among an array of possibilities. They can then actualize those goals in accord with a preconceived design plan or set of functional requirements. Rational agents can constrain combinatorial space with distant information-rich outcomes in mind."[35]

Hypothesis (prediction): The physical laws and constants of physics will take on rare values that match what is necessary for life to exist (i.e., fine-tuning).

Experiment (data): Multiple physical laws and constants must be finely tuned for the universe to be inhabited by advanced forms of life. These include the strength of gravity (gravitational constant), which must be fine-tuned to 1 part in 10^{35} (ref. [36]); the gravitational force compared to the electromagnetic force, which must be fine-tuned to 1 part in 10^{40} (ref. [37]); the expansion rate of the universe, which must be fine-tuned to 1 part in 10^{55} (ref. [38]); the cosmic mass density at Planck time, which must be fine-tuned to 1 part in 10^{60} (ref. [39]); the cosmological constant, which must be fine-tuned to 1 part in 10^{120} (ref. [40]); and the initial entropy of universe, which must be fine-tuned to 1 part in $10^{10^{123}}$ (ref. [41]). The Nobel Prize-winning physicist Charles Townes observed:

> Intelligent design, as one sees it from a scientific point of view, seems to be quite real. This is a very special universe: it's remarkable that it came out just this way. If the laws of physics weren't just the way they are, we couldn't be here at all.[42]

Conclusion: The cosmic architecture of the universe was designed.

While the evidence points strongly to design in these five scientific fields, like all scientific theories, the conclusion of design is always held tentatively, subject to future scientific discoveries.

Using the Positive Case for Design to Answer Common Objections

The positive case for design allows us to answer some of the most common objections to intelligent design—including the claim that ID isn't science, the charge that ID is only a negative argument against evolution, and "God of the gaps" or "argument from ignorance" accusations.

First, ID's use of the scientific method in each of the scientific fields elaborated above provides strong evidence that ID is science.

Second, critics often claim that ID is merely a negative argument against evolution—that it has no positive content and merely critiques other viewpoints. During the *Kitzmiller v. Dover* trial, biologist Kenneth Miller from Brown University testified that ID "is always negative, and it basically says, if evolution is incorrect, the answer must be design."[43] In the court's ruling in that case, a federal judge relied upon Miller's testimony to similarly claim that "ID proponents primarily argue for design through negative arguments against evolution."[44] The positive case for design outlined above forcefully shows that ID is based upon positive evidence and is not merely a negative argument against Darwinian evolution or other material causes.

As for the "God of the gaps" charge, the basic objection is that ID is an argument from ignorance, based upon what we don't know (gaps in knowledge) rather than what we do know, and that future discoveries will likely close those gaps in our knowledge with material explanations and eliminate the argument for design.

For example, UC Berkeley paleontologist Charles Marshall calls ID "a (sophisticated) 'god of the gaps' approach, an approach that is problematic in part because future developments often provide solutions to once apparently difficult problems."[45] University of Alberta biologist and theologian Denis O. Lamoureux similarly charges that ID is a "God-of-the-gaps approach to divine action," which holds that "there are 'gaps' in the continuum of natural processes, and these 'discontinuities' in nature indicate places where God has miraculously intervened." In his view, this approach fails because "these purported gaps have always been gaps in *knowledge* and not actual gaps in *nature*."[46] An article with the theistic evolutionist BioLogos Foundation argues that "pragmatically" the argument for design "is in fact an argument from ignorance" because "it seems like you need to test for (lack of) natural explanations to discover irreducible or specified complexity."[47]

These accusations bear little resemblance to the actual theory of intelligent design as put forth by ID proponents. Indeed, if the positive case for design shows anything, it's that this objection is incorrect.

ID's positive arguments are based precisely upon what we have learned from studies of nature about the origin of certain types of information, such as CSI-rich structures. In our experience, high CSI or irreducible complexity derives from a mind. If we did not have these observations, we could not infer intelligent design. We can then go out into nature and empirically test for high CSI or irreducible complexity, and when we find these types of information, we can justifiably infer that an intelligent agent was at work.

Thus, ID is not based upon what we *don't know*—an argument from ignorance or gaps in our knowledge—but rather, is based upon what we *do know* about the origin of information-rich structures, as testified by the observed information-generative powers of intelligent agents.

Does Darwinian Theory Make the Same Predictions?

One potential objection to the positive case for design is that Darwinian evolution might make some of the same predictions as ID, making it difficult to tell which theory has better explanatory power. For example, in systematics, ID predicted reusage of parts in different organisms, but neo-Darwinism also predicts different species may share similar traits either due to inheritance from a common ancestor, convergent evolution, or loss of function. Likewise, in genetics, ID predicted functionality for junk DNA, but evolutionists might argue noncoding DNA could evolve useful functions by mutation and selection. If neo-Darwinism makes the same predictions as ID, can we still make a positive argument for design? The answer is yes, and there are multiple responses to these objections.

First, not all the predictions generated by positive arguments for design are also made by Darwinian theory. For example, Michael Behe explains that irreducible complexity is predicted under design but predicted to *not* exist by Darwinism:

> [I]rreducibly complex systems such as mousetraps and flagella serve both as negative arguments against gradualistic explanations like Darwin's and as positive arguments for design. The negative argument is that such interactive systems resist explanation by the tiny steps that a Darwinian path would be expected to take. The positive argument is that their parts appear arranged to serve a purpose, which is exactly how we detect design.[48]

The same could be said of high-CSI features like protein sequences, which require rare and finely tuned sequences of amino acids to function. These are predicted by ID but are not expected under a blind trial-and-error process of mutation and selection.[49]

Second, the fact that a different theory can explain some data does not negate ID's ability to successfully make positive predictions. After all, a "positive case" means that the arguments for design stand on their own and do not depend merely on refuting other theories. While refuting competing hypotheses can certainly help solidify a theory's status as the best explanation, a positive argument must be able to stand on its own. ID's fulfilled predictions show there is positive evidence for design, regardless of what other models may or may not say.

Third, it's not clear that in any of these cases neo-Darwinian evolution (or other materialistic models) make *exactly* the same predictions as ID. For example, in systematics, neo-Darwinism may predict the reusage of parts in different organisms, but it predicts that the distribution of parts will generally conform to a treelike pattern (or a nested hierarchy). Intelligent agents are not bound to distribute parts in a tree, and thus reusage of similar parts may be found even among very distantly related organisms. We can test between these different models. A 2018 paper by software engineer Winston Ewert in the journal *BIO-Complexity* proposed a model of common design called a *dependency graph*, which was "based on the technique used by software developers to reuse code among different software projects."[50] He compared the distribution of gene families reused in different organisms to a treelike pattern predicted by neo-Darwinism versus a dependency graph distribution used by computer programmers and predicted by ID. After analyzing the distribution of gene families in nine diverse types of animals, Ewert's preliminary analysis found that a common design-based dependency graph model fit the data 10^{3000} times better than a traditional Darwinian phylogenetic tree.[51] His ID-based dependency graph model predicted reusage of parts much better than neo-Darwinism.

Ewert tested the data against common descent. But even convergent evolution struggles to explain reusage of parts. Richard Dawkins acknowledges "it is vanishingly improbable that exactly the same evolutionary pathway should ever be travelled twice,"[52] yet we often find striking similarities across distantly related organisms, such as the camera-like structure of the vertebrate eye and the octopus (cephalopod) eye. What evolution biology calls extreme convergence is better explained by common design.

With junk DNA, it's true that neo-Darwinian evolution predicts that functionality could sometimes evolve for noncoding DNA, and that finding function in a given case does not necessarily refute that model. Yet a major prediction of modern evolutionary theory is that neutral (neither harmful nor beneficial) mutations occur frequently and accumulate as useless genetic junk in genomes. For example, in 1972 the pioneering molecular evolutionary biologist Susumu Ohno published an article titled "So much 'junk' DNA in our genome" in a volume titled *Evolution of Genetic Systems.* He argued that "at the most, only 6% of our DNA" entails functional genes, with the rest being "untranscribable and/or untranslatable DNA" representing "extinct genes" or "nature's experiments which failed"—akin to "fossil remains of extinct species."[53] Biologists soon envisioned additional evolutionary mechanisms for filling our genomes with junk. In his influential 1976 book *The Selfish Gene,* Richard Dawkins predicted that "a large fraction" of our genomes has no function, because "the true 'purpose' of DNA is to survive, no more and no less. The simplest way to explain the surplus DNA is to suppose that it is a parasite, or at best a harmless but useless passenger, hitching a ride in the survival machines created by the other DNA."[54]

In 1980, *Nature* published two papers by influential biologists furthering the concept of "selfish" junk DNA. The first article, "Selfish Genes, the Phenotype Paradigm and Genome Evolution," by W. Ford Doolittle and Carmen Sapienza, maintained, "Natural selection operating within genomes will inevitably result in the appearance of DNAs with no phenotypic expression whose only 'function' is survival within genomes."[55] A second

paper, "Selfish DNA: the ultimate parasite," was by Francis Crick, who won the Nobel Prize for determining the structure of DNA, and the eminent origin-of-life theorist Leslie Orgel. They concluded that "much DNA in higher organisms is little better than junk," and "it would be folly in such cases to hunt obsessively for" its function.[56] Since this time, Darwinian thinkers have been seduced by the idea that "parasitic" DNA and random mutations will spread junk throughout our genomes. In 1994, Kenneth Miller published an article claiming that "the human genome is littered with pseudogenes, gene fragments, 'orphaned' genes, 'junk' DNA, and so many repeated copies of pointless DNA sequences that it cannot be attributed to anything that resembles intelligent design."[57] Many similar quotes could be given showing that the idea of junk DNA was born, bred, and flourished from within an evolutionary paradigm.

As might be expected from such statements, the literature admits that evolutionary thinking has hindered research into functions for junk DNA. A 2003 article in *Scientific American* noted that "introns," a type of noncoding DNA found within genes, "were immediately assumed to be evolutionary junk"—a view that the article later called "one of the biggest mistakes in the history of molecular biology."[58] That same year, a paper in the journal *Science* observed that "[a]lthough catchy, the term 'junk DNA' for many years repelled mainstream researchers from studying noncoding DNA."[59] A striking admission came in a 2020 paper in *Nature Reviews Genetics* titled "Overcoming challenges and dogmas to understand the functions of pseudogenes," which argues that "dogma" in biology causes "demotivation into exploring pseudogene function by the a priori assumption that they are functionless." According to the paper, "[t]he dominant limitation in advancing the investigation of pseudogenes now lies in the trappings of the prevailing mindset that pseudogenic regions are intrinsically non-functional" and "there is an emerging risk that these regions of the genome are *prematurely dismissed* as pseudogenic and therefore regarded as void of function."[60]

The ID community's view of junk DNA stands in stark contrast to the typical evolutionary view. Going back to some of ID theory's early days in the 1990s, ID theorists have been predicting that noncoding DNA would turn out to have functions. In 1994, pro-ID scientist Forrest Mims submitted a letter to *Science* that warned against assuming that junk DNA was "useless."[61] In 1998, William Dembski wrote that "on an evolutionary view we expect a lot of useless DNA. If, on the other hand, organisms are designed, we expect DNA, as much as possible, to exhibit function…Design encourages scientists to look for function where evolution discourages it."[62] Many other ID theorists have made similar predictions over the years. What might have happened if their predictions had been heeded?

In 2021, the journal *Nature* acknowledged that prior to the Human Genome Project (HGP), which completed in 2003, there was "great debate" over whether "was it worth mapping the vast non-coding regions of genome that were called junk DNA, or the dark matter of the genome?" The article noted that over 130,000 "genomic elements, previously called junk DNA" have now been discovered, and highlighted how important these "junk" segments have turned out to be:

[I]t is now appreciated that the majority of functional sequences in the human genome do not encode proteins. Rather, elements such as

long non-coding RNAs, promoters, enhancers and countless gene-regulatory motifs work together to bring the genome to life. Variation in these regions does not alter proteins, but it can perturb the networks governing protein expression With the HGP draft in hand, the discovery of non-protein-coding elements exploded. So far, that growth has outstripped the discovery of protein-coding genes by a factor of five, and shows no signs of slowing. Likewise, the number of publications about such elements also grew in the period covered by our data set. For example, there are thousands of papers on non-coding RNAs, which regulate gene expression.[63]

Under an ID paradigm, debates over whether to investigate junk DNA would have ended much sooner with an emphatic "Yes!," furthering our knowledge of genetics and medicine. When it comes to junk DNA, ID has made superior predictions.

Predictions or Retrodictions?

Another potential objection to the positive case might be that we're not making positive predictions for design, but rather, looking backwards to make after-the-fact retrodictions. With junk DNA, this is clearly not the case—ID was predicting function years before biologists discovered those functions. The same could be said for the discovery of finely tuned CSI-rich biological sequences, something that design theory-inspired ID theorists like Douglas Axe and Ann Gauger investigate, and which they indeed have found.[64] Winston Ewert's dependency-graph model promises that ID can bear good fruit as we learn more about the gene sequences of organisms. Indeed, as we'll see in the last section of this chapter, ID makes useful predictions that can guide future research in many scientific fields.

But I would argue that a chronology of exactly what was predicted when is not dispositive as to whether a positive argument can be made. The purpose of this chapter is to show that a positive case can be made for intelligent design, and what matters is that the predictions of ID that are listed here flow naturally from observations about how intelligent agents operate, and that they successfully explain the observed data. This is what gives ID explanatory value to predict what we should find in nature. Exactly what happened when is less important than the raw explanatory power of intelligent design.

Can Materialistic Models Accommodate the Data?

Yet another potential objection is that even if materialistic models did not initially predict the data, materialists can still find ways to accommodate it. Even if this is true, a positive argument for design stands because it does not require refuting Darwinian evolution. Nonetheless, many materialistic explanations for the data that is predicted by design are unconvincing.

For example, in paleontology, evolutionary biologists expected to find gradual transitions rather than abrupt explosions of new life forms. As Stephen Jay Gould put it: "The extreme rarity of transitional forms in the fossil record persists as the trade secret of paleontology."[65] Because of this difficulty, in the 1970s, Gould and his colleague Niles Eldredge developed punctuated equilibrium as a model where evolution takes place in small populations over relatively short geological time periods that are too rapid for

transitional forms to become fossilized.[66] But this model has many problems.[67]

Punctuated equilibrium compresses the vast majority of evolutionary change into small populations that lived during shorter segments of time, allowing too few opportunities for novel, beneficial traits to arise. Punctuated equilibrium is also unconvincing in that it predicts that with respect to the fossil record, evidence confirming Darwinian theory will not be found. Would you believe someone who claimed that fairies and Leprechauns exist and were caught on video, but when asked to produce the film, declares, "Well, they are on camera, but they are too small or too fast to be seen"? That doesn't make for a compelling theory.

Analogous problems plague attempts to account for the life-friendly fine-tuning of physical laws by appealing to a multiverse (see chapter 42 by Bruce Gordon in this volume). Materialists hope that if there are a near-infinite number of universes, then perhaps the extreme unlikelihood of obtaining the precise parameters required for life is less difficult to overcome. Aside from the fact that the multiverse is not observable,[68] the mechanisms proposed to generate a multiverse *themselves require fine-tuning*, thus exacerbating rather than addressing the challenge of fine-tuning.[69] Even worse, appealing to a multiverse destroys our ability to do science.

Imagine that 100 percent of the people in a town of 10,000 get cancer in a year, and that the odds of this occurring by chance are 1 in $10^{10,000}$. Normally, scientists would reason that such great unlikelihood would rule out chance and suggest some physical agent is causing the cancer cluster. Under multiverse thinking, however, one might as well say, "Well, imagine there are $10^{10,000}$ universes, and our universe just happened to be the one where this unlikely cancer cluster arose—purely by chance!" Should scientists seek a scientific explanation for the cancer cluster, or should they just invent $10^{10,000}$ universes where this kind of event becomes probable?

The multiverse advocate might reply, "Well, you can't say there aren't $10^{10,000}$ universes out there, right?" That's the point: There's no way to test multiverse claims, and science should not seriously consider untestable theories. Multiverse thinking makes it impossible to rule out chance, which essentially negates our basis for drawing scientific conclusions. If this is the answer to cosmic fine-tuning, then intelligent design has not been given an adequate rebuttal.

The point is this: Simply because materialists make outlandish proposals to explain away data that was positively predicted by intelligent design does not mean that those materialistic ideas actually work. What matters is that intelligent design is making useful predictions, allowing it to become a fruitful paradigm for guiding scientific research, as we'll see in the next section.

ID as a Paradigm for Fruitfully Guiding Science

There's one final common objection to ID that the positive case helps us to answer. In his *Kitzmiller v. Dover* testimony, Kenneth Miller referred to intelligent design as a "science stopper."[70] Similarly, in his book *Only a Theory*, Miller stated, "The hypothesis of design is compatible with any conceivable data, makes no testable predictions, and suggests no new avenues for research. As such, it's a literal dead end…"[71]

Yet as we've already seen, ID makes a variety of testable and successful predictions. This allows ID to serve as a paradigm guiding scientific research to make new scientific discoveries. The list below shows various fields where ID is helping science to generate new knowledge. For each field, multiple ID-friendly scientific publications are cited to provide examples of where ID is inspiring the progress of science:

- *Protein science:* ID encourages scientists to do research to test for high levels of complex and specified information in biology in the form of fine-tuning of protein sequences.[72] This has practical implications not just for explaining biological origins, but also for engineering enzymes and anticipating/fighting the future evolution of diseases.

- *Physics and Cosmology:* ID has inspired scientists to seek and find instances of fine-tuning of the laws and constants of physics to allow for life, leading to new fine-tuning arguments such as the Galactic Habitable Zone. This has implications for proper cosmological models of the universe, hints at proper avenues for successful "theories of everything" that must accommodate fine-tuning, and other implications for theoretical physics.[73]

- *Information theory:* ID leads scientists to understand intelligence as a scientifically studiable cause of biological complexity and to understand the types of information it generates.[74]

- *Pharmacology:* ID directs both experimental and theoretical research to investigate limitations of Darwinian evolution to produce traits that require multiple mutations in order to function. This has practical implications for fighting problems like antibiotic resistance or engineering bacteria.[75]

- *Evolutionary computation:* ID produces theoretical research into the information-generative powers of Darwinian searches, leading to the discovery that the search abilities of Darwinian processes are limited, which has practical implications for the viability of using genetic algorithms to solve problems.[76]

- *Anatomy and physiology:* ID predicts function for allegedly "vestigial" organs, structures, or systems whereas evolution has made many faulty predictions of nonfunction.[77]

- *Bioinformatics:* ID has helped scientists develop proper measures of biological information, leading to concepts like complex and specified information or functional sequence complexity. This allows us to better quantify complexity and understand what features are, or are not, within the reach of Darwinian evolution.[78]

- *Molecular machines:* ID encourages scientists to reverse-engineer molecular machines—like the bacterial flagellum—to understand their function like machines, and to understand how the machine-like properties of life allow biological systems to function.[79]

- *Cell biology:* ID causes scientists to view cellular components as "designed structures rather than accidental by-products of neo-Darwinian evolution," allowing scientists to propose testable hypotheses about cellular function and causes of cancer.[80]

- *Systematics:* ID helps scientists explain the cause of the widespread feature of conflicting phylogenetic trees and "convergent evolution" by producing models where parts can be reused in non-treelike patterns.[81] ID has spawned ideas about life being front-loaded with information such that it is designed to evolve, and has led scientists to expect (and now find!) previously unanticipated "out-of-place" genes in various taxa.[82]

- *Paleontology:* ID allows scientists to understand and predict patterns in the fossil record, showing explosions of biodiversity (as well as mass extinction) in the history of life.[83]

- *Genetics:* ID has inspired scientists to investigate the computerlike properties of DNA and the genome in the hopes of better understanding genetics and the origin of biological systems.[84] ID has also inspired scientists to seek function for noncoding junk DNA, allowing us to understand development and cellular biology.[85]

Critics wrongly charge that ID is just a negative argument against evolution, that ID makes no predictions, that it is a "god of the gaps" argument from ignorance, or that appealing to an intelligent cause means "giving up" or "stopping science." As this chapter has shown, these charges are misguided. Ironically, when critics claim that research may not detect design because that would stop science, it is they who hold back science by preventing scientists from investigating the scientific theory of intelligent design. When scientists are allowed to infer intelligent agency as the best explanation for information-rich structures in nature, this opens up many avenues of research that are bearing good fruit in the scientific community.

Why Does Intelligent Design Matter?

William A. Dembski

Intelligent design, or ID, identifies patterns in nature that are best explained as the product of intelligence. Some patterns are best explained by invoking intelligence, others are not. The face of Mount Everest is best explained by natural forces of wind, erosion, and the like. The face of Mount Rushmore requires an intelligent agent.

Many types of human inquiry fall under this definition of intelligent design, including forensic science, archaeology, and the search for extraterrestrial intelligence (SETI). With forensic science and archaeology, the intelligence in question would be human. Thus, if forensic science identified a death by foul play or if archaeology identified a mound as a burial mound, in both cases, a human intelligence would be responsible.

With SETI, by contrast, the intelligence in question would be nonhuman. SETI is interesting because it assumes we don't need to directly observe the actual intelligent aliens to know that they exist. It's assumed that radio transmissions from outer space could exhibit patterns that would serve as clear markers of intelligence. English messages sent via bits in ASCII or Unicode would certainly be best explained as the result of intelligence, though we assume an alien intelligence would not be speaking English. But certain numerical patterns, such as prime numbers (recall the Carl Sagan novel *Contact*, which was turned into a movie), could also clearly signify intelligence.

Yet with such examples, the intelligence would still be some sort of embodied intelligence that could have evolved by natural means, or so the story goes. Thus, a blind, purposeless evolutionary process driven by natural selection could, it is thought, give rise to intelligent agents like us, who then leave behind patterns that can be distinguished from naturally formed patterns and that would be best explained as the result of intelligence.

The rub for intelligent design—and this is what makes it controversial—is to focus on patterns in nature that could not result from an evolved intelligence. Take a living cell. Things consisting of living cells, like humans, exhibit intelligence. But what formed the first living cell? Is the pattern of information in the cells on Earth best explained as the result of intelligence? If so, what sort of intelligence created it? An intelligence arising in another part of the universe? If so, how did that intelligence arise elsewhere in the universe? Unless the universe is eternal (which current cosmological theories reject), you can't just keep kicking the problem of intelligence in the universe back in time indefinitely. Either the intelligence responsible for life on Earth evolved without the need for intelligence, or it ultimately derives from an unevolved intelligence.

Atheists (and I'll include here most agnostics) are perfectly fine with products of evolution producing patterns best explained as the result of intelligence. This becomes the ultimate free lunch, getting complexity from simplicity, intelligence from unintelligence. The atheistic cosmos begins as a chaotic mix of material forces that initially has no intelligence. Intelligence, for the atheist, is something that evolves from non-intelligence. But if we find patterns in nature that require intelligence for their explanation and yet could not be the result of an evolved intelligence, then we refute atheism (as well as agnosticism, which then no longer has the luxury of basking in ignorance about the ultimate cause of intelligence in the universe).

The refutation of atheism here then becomes a simple matter of logic, the inference rule in question being *modus tollens*. The rule runs like this:

If A, then B. Not B. Therefore, conclude not A.

For instance: If it's more than 100 degrees outside, I'll be wearing shorts. I'm not wearing shorts. Therefore, it must not be more than 100 degrees. So, too, if there is no God or intelligence outside nature, then intelligence is something that evolved by undirected means from nature. But intelligence is not something that evolved by undirected means from nature (as evident in the patterns signifying intelligence in the first

living cell, whatever its place of origin, be it Earth or some other location in the universe). Therefore, we conclude that there is a God or intelligence not reducible to the undirected forces of nature.

In reply, atheists will counter with one of two lines: (1) argue that the patterns in all living cells are such that no God or external intelligence is required to explain them; or (2) argue that the patterns are such as to require intelligence, but not an intelligence that is external to the material world, not an intelligence such as God. In the case of (2), this would have to be an evolved intelligence. And if we take it that life could not have arisen without intelligent input on Earth, such an intelligence responsible for life on Earth can't be an intelligence that evolved on Earth. Rather, it would have to be an extraterrestrial intelligence that evolved elsewhere in the universe.

Such a view is seriously entertained (Francis Crick, the atheist co-discoverer of the structure of DNA was one such advocate), it being dubbed *panspermia*. Panspermia comes in two varieties: one in which spores randomly migrate through outer space and happen to seed Earth, thereby allowing for life to originate. And then there's directed panspermia, in which spaceships sent by intelligent and technologically advanced aliens intentionally seed Earth with the first life.

Now the evidence for panspermia in both varieties, directed and undirected, is nil. The only thing that commends it is that it removes God from the equation of life on Earth. This removal of God, in the view of its advocates, is, however, enough to commend panspermia and make it superior to any "God Hypothesis." Nonetheless, panspermia is instructive in showing the lengths to which some atheists will go to explain the "appearance of design" in the living cell.

The other approach by atheist scientists is simply to stick their hands into the fire-ant mound and, despite the pain, claim that there's nothing to see here that would exhibit the need for prior design. And so, rather than invoke panspermia, they invoke chemical and biological evolution. Chemical evolution refers to the interactions of chemicals on the primordial Earth that predate the first living cell and give rise to it. Many speculative scenarios purport to provide some insight into how such chance chemistry could bring about something so complex and marvelous as the cell. But hey, if there's no God or external intelligence, then chance chemistry is all you have.

Once chemical evolution brings about the first living cell, the handoff is made to biological evolution. Now you've got an entity (the cell) that can replicate and be subject to natural selection acting on random variations, the changes so produced being transmissible by heredity to future generations. This is biological evolution, and because of the preeminent role that natural selection tends to play for most evolutionary biologists, this is a Darwinian form of biological evolution. For the atheist, the beauty of Darwinian biological evolution is that it requires no input from intelligence. It is supposed to produce living things that display intelligence, but that are not themselves the product of intelligence.

Although atheists are happy to pretend that chemical and biological evolution have successfully closed off any need for an external designing intelligence, in fact, they are mistaken. It's actually quite easy to see that chemical evolution cannot account for certain patterns in the cell and that these are best explained by intelligence. The argument here is incredibly strong—it is a steel trap. With regard to biological evolution, the argument is also very strong, but not quite as decisive.

Here's the problem for chemical evolution: chemical evolution must explain the origin of life, but the only example of life that we have is the cell, and even the simplest cell is immensely complex. Viruses don't count because they depend on cells to replicate (they are thus causally downstream from cells). Idle speculations about possible simpler precursor life forms don't count either, which exist in no labs or in any detailed testable form but only in the imaginations of materialists who yearn for their existence. It's not just the sheer complexity of cells that's at issue for chemical evolution. There's one pattern in particular that must be explained, a pattern that is best explained by intelligence and that remains totally unexplained in terms of natural processes.

I'm talking about the genetic code. The genetic code is a code, in other words, a way of mapping one set of digital characters to another. A decoder ring provides such a mapping. Unicode, which takes blocks of bits and maps them to alphanumeric characters, is another example. The genetic code takes triplets of nucleotide bases (64 possibilities) and maps them to amino acids (20 possibilities). It does this to create proteins, which are the workhorses in the cell. But the point to note is that the genetic code is a *code*, and it requires immensely complex machinery in the cell to work, including messenger and transfer RNAs as well as ribosomes.

So, what explains a code? All codes, where we know how they originated, are the result of intelligence. The mathematician Claude Shannon, in the late 1940s, laid out the mathematics underlying such codes in what became known as *information theory*. And then in 1953, Francis Crick and James Watson discovered the structure of DNA, which in the next decade was used to show that all life depends on a code, the genetic code. All life as we know it requires the full genetic code. Life exhibits no simpler codes out of which the full genetic code might have evolved. It's all or nothing, and every living thing we know of has it. Codes, in the way they coordinate two digital character sets, are best explained as the result of intelligence. We may not know how the intelligence did it. But that's true even of human inventors when we don't know them or their methods of invention. Yet we can be confident that intelligent agency was responsible for the result.

Given that chemical evolution has proven an utter failure in explaining the genetic code (this code being a precondition for life), it follows that biological evolution begins with a gaping hole at the very place it is supposed to start. Namely, there is no clean naturalistic handoff from chemical evolution to biological evolution because there is no convincing, fully naturalistic account of the origin of life. The genetic code evinces a pattern that requires an intelligence for its explanation, and conversely, the chemical processes of chemical evolution don't suffice to explain this pattern.

But what about biological evolution? If we're willing to grant the existence of the first living cell, perhaps everything will take care of itself thereafter via the wonderworking power of natural selection. But this is a vain hope as well. Irreducibly complex cellular structures such as those described by Michael Behe in this book provide harsh obstacles to Darwinian biological evolution. And then there is the work on evolutionary informatics (see evoinfo.org) that shows how evolutionary processes need to be infused with external information in order to produce novel functions and the biological structures needed

to support them. The details here are fascinating and need to be waded through to see clearly that Darwinian biological evolution really is in the sad shape that I'm claiming.

And this brings me, after all this stage setting, to the title of this chapter: *Why does intelligent design matter?* It matters because it shows that there are clear patterns in nature that are best explained as the result of intelligence and where such an intelligence itself is not part of nature as conceived by the materialist atheist. Moreover, the inference to this intelligence outside nature is not a matter of religion but of science. Intelligent design, in standing against chemical and biological evolution, is therefore not making a religious but a scientific case.

Intelligent design makes the identification of intelligence in nature not a religion versus science question but a science versus science question. This is huge, especially in our day when science has all the clout and religion is largely regarded as passé in our culture. Creationism, especially of the young-Earth variety, has looked to science to challenge evolutionary theory and make the design in nature seem plausible, but at the end of the day, creationism is always a religious doctrine about where everything ultimately came from (namely, from the eternal infinite creator God of Judaism and Christianity). Moreover, young-Earth creationism is adamantly committed to an interpretation of Genesis in which the universe was created about 6,000 years ago in six 24-hour days. Any science it invokes therefore always gives way to religious requirements.

Intelligent design, by contrast, offers no doctrine about the designing intelligence ultimately responsible for the patterns in nature that are best explained as the product of intelligence. As far as the theory of intelligent design is concerned, it could be the deity of ethical monotheism that is the object of worship of the great monotheistic faiths. But it could also be a pantheistic of panentheistic deity. It could even be Nature, writ large, where nature is conceived not in its materialistic guise as a medley of material particles interacting by unbroken laws of attraction and repulsion, but rather where Nature is imbued with an inherent teleology capable of giving rise to embodied intelligences via teleological (non-Darwinian) processes of evolution. The "atheist" philosopher Thomas Nagel holds such a view. It's worth noting that his brand of atheistic teleological naturalism is as unacceptable to hardcore materialist Darwinists like Daniel Dennett and Richard Dawkins as is the theism of Christians like me.

Because intelligent design makes the question of design in nature not a religion-versus-science but a science-versus-science question, it constitutes much more a threat to the reigning atheistic materialists than creationism ever did. For this reason, atheists are quick to misrepresent intelligent design as creationism, hoping by this expedient to cast intelligent design as the religious player in a religion-versus-science controversy, thereby ensuring that intelligent design can be dismissed as religion without having to be engaged on its scientific merits and its challenge to chemical and biological evolution. Despite such efforts by atheists to misrepresent intelligent design, the theory of intelligent design has, over the last 25 years, been sufficiently developed and articulated that its scientific underpinnings are now clear and defensible.

At the same time that intelligent design faces opposition from materialistic atheists, it also faces resistance from some quarters of the young-Earth creationist community. The level of resistance from young-Earth

creationists to intelligent design was perhaps less in times past than now. I recall a conference at Biola University in 2004 at which Duane Gish got to the microphone and gushed how much he "loved intelligent design." John Morris, the son of Henry Morris, acknowledged to me, when I visited his creation museum in 2001, how even though he had differences with me and the intelligent design community, he saw intelligent design as the key to unseating Darwinism and natural selection.

In subsequent years, however, once it became clear that intelligent design would not directly support young-Earth creationism but was also compatible with other views of creation, and even non-Christian religious options for the origin of life and the universe, support for intelligent design among young-Earth creationists became more ambivalent. Young-Earth creationism, insofar as it gives pride of place to a literalistic interpretation of Genesis, cannot avoid framing the creationist challenge to materialistic atheism as a religion-versus-science controversy. Young-Earth creationism knows, on theological grounds, that Earth is only a few thousand years old and that the design in nature must therefore be real irrespective of scientific evidence. This view can result in a loss of urgency to demonstrate design in nature scientifically, which is the goal of intelligent design.

Intelligent design, by contrast with creationism, is at root a scientific position and gains its rhetorical edge from this fact. To be sure, intelligent design has theological implications. Indeed, intelligent design is friendly to theism and unfriendly to atheism. Some young-Earth creationists appreciate this point and therefore support intelligent design, seeing its ability to displace atheism as invaluable despite its theological limitations. Intelligent design shows that an unevolved intelligence is responsible for certain patterns in nature, notably in the formation of life, but it's not in a position to identify that intelligence with the Judeo-Christian creator God.

Why, then, does intelligent design matter? It matters because it poses the only effective challenge to materialistic accounts of chemical and biological evolution. These forms of evolution have in our culture become prime engines for atheism and for a secular worldview. In his video *The Case for a Creator*, Lee Strobel describes how he lost his faith in high school when in biology class he learned of the Miller-Urey experiment conducted in the 1950s at the University of Chicago, in which random chemical reactions produced certain building blocks of life. Of course, getting building blocks doesn't get you a building, and Strobel didn't think through the implications of this origin-of-life experiment very carefully at that time. Yet the conclusion he drew from it, which derailed his faith for many years, is one that many others have drawn from it: that life is easily formed without design, and therefore without God…and therefore God need not exist…and therefore God probably doesn't exist. Others have read defenses of biological evolution and followed Richard Dawkins in concluding that Darwin made it possible to be an intellectually fulfilled atheist.

Intelligent design, by contrast, in showing that the universe as a whole, and biology in particular, requires an unevolved intelligence, and by making its case scientifically rather than religiously, counters such attempts by atheist scientists to use science to bolster their atheism. Chemical and biological evolution, conceived as undirected materialist processes, have been notoriously successful engines for advancing atheism. No more. In overturning these engines, intelligent design becomes an engine for theism, or at least for worldviews friendly toward some form of purposiveness underlying nature.

This friendliness of intelligent design to theism is the prime reason that intelligent design matters. It's not just that intelligent design defeats the atheism and agnosticism that are rampant in our culture. It's not just that intelligent design is an interesting scientific enterprise in its own right. Rather, it gets at the most important truth underlying the natural world—namely, that the natural world is the product of a mind, an intelligence. Intelligent design's ultimate significance, therefore, is not as a tool of controversy for defeating views hostile to Christian faith, but rather for getting at the deepest truth of nature—that it exhibits real intelligence. Intelligent design shows that nature is here on purpose and that we are therefore here for a purpose.

Intelligent design matters because it reveals that nature radiates purpose!

Have Science and Philosophy Refuted Free Will?

Michael Egnor

D o we have free will? This is one of the most important philosophical and scientific questions of modernity. Our answer to it determines our fundamental understanding of human nature, the basis for moral accountability, and the framework for our culture, law, and politics. Much of the modern debate on free will is predicated on a materialist understanding of man and on a determinist understanding of nature—both of which, as we will see, find little support in logic or in science.

Current Debate on Free Will

The current debate may be summed up simply as this: materialists deny that free will is real because they assert that all of nature is determined by physical laws, and because human beings are wholly a part of nature, this leaves no room for libertarian free will.

Those who affirm free will argue that its reality is obvious and that free will is a fundamental aspect of moral accountability and of what it is to be human. In fact, the reliability of human reason and of the scientific process itself is fatally undercut if we deny free will.

In this chapter, I will argue that an objective and informed understanding of physics and neuroscience is entirely compatible with, and in fact supports, the reality of free will.

Definition of Free Will

What do we mean by free will? This is a surprisingly difficult question, to which philosophers have offered many definitions and countless nuances. Do we mean choices that are utterly uncaused, or do we mean only choices that are caused materialistically (by matter and its processes)? Is a choice free merely if we make it ourselves, without external compulsion? There are even profound theological questions: How is freedom possible if all that exists is created by God? How is it possible if God knows the future, and by doing so, fixes the future and thus fixes our choices? Is God free, for example, to choose evil?

For clarity, we can consider three standard definitions of free will.

The first definition, which philosopher Alfred Mele terms the "modest conception of free will,"[1] is that free will is just the ability to make rational informed decisions if not subjected to undue force. This is a common definition, and many philosophers (usually compatibilists) endorse this definition or some variant of it.

The second definition is termed "ambitious free will," which is, in Mele's words, a deep openness to alternative decisions. In this view, free will presupposes alternatives—we *could* have chosen otherwise, if free will is real.

The third definition of free will is that of Thomas Aquinas (which is the view that I endorse). Aquinas's definition of free will[2] differs in subtle but important ways from modest or ambitious free will. Aquinas believed that free will is real, but while he agreed that we usually *do* have alternative choices, it is not *necessary* to have alternative options for free will to be real. We can be free, according to Aquinas, even if we cannot choose other than what we do.

What did Aquinas mean? First, for Aquinas, the term *free will* is a misnomer. The rational powers of our soul are intellect and will, and freedom is true of us as a whole—both freedom of intellect as well as freedom of will. The will cannot be designated "free" in isolation. I can freely choose, but it is *me*, not just my will, that is free.

By freedom Aquinas meant that I am fully the source of my choices, without compulsion. In classical metaphysical terms, I am a prime mover (in a sense) for my choices. My choices, of course, can be influenced or conditioned by other factors (my emotions, my past, etc.) but in the final analysis, I am wholly the source of my choices.

The affirmation of freedom of will raises additional profound theological questions: If God is the creator of nature, and I am part of nature, how can I freely choose? Doesn't God create my choices? If God knows the future, how can my will really be free, since God already knows what I will do, and I cannot do otherwise? Aquinas answered that God creates in accordance with the essences of things, and that I am created as a creature with free will. My freedom is created, just as my body and soul are created, and created freedom is wholly free while at the same time being wholly created. God knows the future and knows what I will choose, but my lack of alternative choices implied by this does not in itself make my will unfree.

Aquinas defined free will as freedom from constraint but not as necessarily entailing alternatives for two reasons: First, if alternative choices are necessary for freedom, then God himself is not free, because it is self-evident that it is in the nature of God that he does not and *cannot* choose evil. Yet, in Aquinas's definition, God is free because freedom is not dependent on alternative choices. Moreover, human freedom remains despite divine omniscience because in Aquinas's understanding, freedom is not dependent on alternative choices. Second, in Thomistic theology, our souls in heaven cannot choose to sin because we have the beatific vision of God and thus we have in our knowledge and possession the ultimate good. Yet even in heaven, even when we *cannot* sin, we remain free because our choices remain wholly from us and caused only by us and we are not constrained.

Aquinas's definition of free will as uncoerced prime causation without the necessity of alternative choices corresponds to our experience of free will without introducing intractable theological problems.

The Framework for the Debate

Leaving the theological questions aside for a moment, let's address questions about free will from a secular philosophical, scientific, and social perspective. The modern framework for the understanding of free will is based on the concepts of determinism, compatibilism, and libertarianism.[3]

Determinism is the view that all nature, including human choice, is determined wholly by the laws of physics. Its most famous expression was given by French polymath Pierre-Simon Laplace in 1814. Laplace asserted (based on Newtonian insights) that if he knew the precise location and momentum of every particle in the universe at a given moment, he could know all past and future events with certainty. Nature is causally determined by physical laws, without remainder.

From the Newtonian perspective—prior to quantum theory—such hubris seemed at least plausible. Quantum mechanics has challenged Laplace's determinism—it seems that nature is fundamentally probabilistic, not deterministic (although many pre-quantum-theory philosophers and physicists in the early twentieth century argued that the determinism is an illusion). In 1935, Einstein and coworkers published a seminal paper (the EPR paper[4]) in which they proposed that quantum mechanics is incomplete in the sense that there must be hidden laws (called *local hidden variables*) that completely determine the apparently statistical outcome of quantum processes. Pure chance, in physics, is an illusion, Einstein argued. To reject determinism leads to "spooky action at a distance" and a host of counterintuitive phenomena. In this view, the probabilistic nature of the subatomic world is merely apparent and quantum mechanics is an incomplete description of nature. The debate between the determinists and the indeterminists was about this question: Is there *any room* for freedom independent of the laws of physics?

Was Laplace correct? For much of the twentieth century, this seemed to be a theoretical question only—there seemed to be no way it could be tested experimentally.

Remarkably, in 1964, Irish physicist John Bell published a relatively simple experimental design by which the existence of hidden variables could be tested decisively.[5] The theoretical and experimental details are beyond the scope of this discussion; they involve measuring the ratio of properties of emitted subatomic particles. Over the past half-century, more than a dozen experiments based on Bell's theorem have been conducted and each time, the *nondeterminism* of nature has been confirmed. Einstein was wrong. In technical terms, there are no local hidden variables, which means that quantum processes are *not* determined by their local state prior to the collapse of the quantum waveform. Determinism in this sense is *not* true. There remain subtleties to the debate, but the evidence in physics strongly implies that nature is *not* deterministic in the sense that Laplace and Einstein believed it was.

Thus, the modern debate about free will, which generally presupposes the deterministic nature of physics, is thus predicated on a scientific error. Determinism does not exist in nature, and thus, scientifically speaking, determinism is no impediment to the reality of free will. Nonetheless, there are still those who seek to oppose Aquinas's strong version of free will.

Compatibilism is the view that free will is compatible with determinism. As noted, physicists have demonstrated that nature is not deterministic, but this research is generally unknown to philosophers who discuss free will, so a discussion of compatibilism is useful. Compatibilism's precis is Schopenhauer's assertion, "Man can do what he wills but he cannot will what he wills."

Compatibilism has had many defenders (e.g., stoic philosophers, Hume, and Hobbes) and has many modern defenders (e.g., Strawson, Dennett, and Fischer) and it remains a popular view among philosophers and scientists. The philosophical defenses of compatibilism are subtle and many, but ultimately, they fail for three reasons:

1. Determinism in nature is false, as demonstrated by Bell's Inequality and experiments that follow on it, so compatibilism, which is predicated on the truth of determinism, is falsified.

2. If determinism were true, there can be no freedom in any meaningful sense. If determinism is true, then our choices are caused by physical laws and past events, neither of which we control.

3. Compatibilism is self-refuting in the sense that the argument that deterministic compatibilism is true is itself invalid because material processes are not in themselves propositions, and therefore, the argument that deterministic compatibilism is true is not capable of having truth value. If (physical) determinism is true, then all human action, including the expression of philosophical arguments, are chemical reactions, which have no truth value in themselves. If you are a meat machine, your statement "I am a meat machine" is a secretion, not an argument. You may be a meat machine, but your assertion that you are is meaningless, and there is no reason to pay attention to it as a proposition.

Incompatibilism is the view that free will is incompatible with determinism and generally entails the (mistaken) view that determinism is true. Incompatibilists argue that compatibilism—the view that will can be free even if all human acts are determined by physics and chemistry—is incoherent. They insist that genuine free will implies libertarianism—genuine freedom from physical determinism.

Those who deny free will are incompatibilists who deny the existence of any natural phenomena—including free will—other than material processes governed wholly by the laws of physics. This view is widely held today by many philosophers and scientists (e.g., Richard Dawkins, Sam Harris,[6] and Jerry Coyne).

Libertarianism is the view that free will is incompatible with determinism but that it is determinism, not free will, that doesn't exist. Libertarianism is the common understanding of free will held by most people who are not invested in materialism and determinism, and has been held by many theologians and philosophers, most notably by the Franciscan priest and philosopher Duns Scotus and by Jesuits such as Molina and Suarez. Even Lucretius—a materialist—held to libertarianism, believing that the randomness of the motion of atoms provided for the indeterminism necessary to ground free will.

Libertarianism comes in many iterations, but it is essentially the view that the will (or the whole person) is free in the sense that choices are not compelled. That is, choices originate in the person, not in the laws of physics or in extra-personal processes of nature. The prime mover analogy is helpful: for our choices, we are prime movers, in the sense that we are wholly the origin of our choice. Of course, our choices are influenced by innumerable factors, many of which are material—our emotions, our digestion, external circumstances, etc. But our choices are ours and originate wholly in us. Our will is *influenced* by agents other than ourselves, but not compelled.

Science and Free Will

Although I believe that metaphysical arguments alone establish the reality of free will (see below), there is a large and growing body of scientific research that bears on the free will question. These scientific perspectives can be classified as those of physics, of neuroscience, and of social psychology.

Physics

First, as noted above, contemporary experimental evidence in quantum physics rules out determinism—at least determinism based on the existence of local hidden variables. The experiments demonstrating the nondeterministic nature of reality at the quantum level are numerous and reasonably decisive, and any denial of free will predicated on the truth of determinism in nature is refuted by modern physics. Denial of free will based on determinism—which is the usual basis for denial—betrays ignorance of modern physics. *If* free will is not real, it is *not* because nature is deterministic.

Neuroscience

Two classes of experiments seem to address the question of free will—one not well known to the scientifically astute public, and one that is generally well-known. While the results of these experiments are commonly interpreted as refuting free will, this denial, as we will see, is not supported by the evidence.

Wilder Penfield

The first modern free will experiments in neuroscience were those of Wilder Penfield.

Penfield was the pioneer in brain surgery for epilepsy. He worked in Montreal in the early and mid-twentieth century. He performed many brain operations on awake patients—they were given local anesthesia so they felt no pain—and Penfield mapped their brains by stimulating the cortex with an electrical probe. He was able to remove areas of the cortex from which seizures arose while protecting vital brain tissue.

In his book *The Mystery of the Mind*,[7] Penfield noted a remarkable finding in his brain mapping of hundreds of patients over many decades: He was able to stimulate many different kinds of mind states—movements of limbs, sensations, perceptions, emotions, memories—but he was never able to stimulate the *will*. For example, during an operation, Penfield would ask the awake patient to move his arm randomly at will, and also during the operation, Penfield could (without telling the patient) stimulate the area of the brain that causes the arm to rise. He would then ask the patient, "Why did your arm rise? Did you do it, or did I?" The patient always knew whether they had willed their arm to rise or Penfield caused it to rise apart from their will. Penfield noted that he could evoke (by electrical stimulation of the brain) movement and many other neurological functions (such as memories), but the patients always knew that they had not willed it—*they always could tell the difference between willed acts and forced acts.* Penfield interpreted this remarkable result as evidence that the will is not determined by the brain, because it could not be evoked by stimulating the brain. He believed he had demonstrated experimentally the reality of free will—the fact that the will is not a material function of the brain.

Penfield began his scientific career as a materialist, but he ended as a passionate dualist and defender of the reality of free will.

Benjamin Libet

The most remarkable modern experiments on free will—and the research that is most widely known to the scientifically engaged general public—were conducted by physiologist Benjamin Libet in the mid-twentieth century.[8] Libet was fascinated by the timing relationship between thoughts and brain activity. In his book *Mind Time,* he provides a synopsis of his work for the general public.

Libet's most famous experiments involved the timing between brain waves, decisions, and acts. He used the simple model of making a decision to push a button. He studied hundreds of normal volunteers and asked them to freely decide to push a button while he monitored electrical activity from their brains. He developed a system whereby he could measure the exact moment a volunteer decided to push a button (to a resolution of about 10 milliseconds, using a clock visible to the volunteer) and correlate it with the exact moment the button was pushed and the exact timing of electrical activity in the volunteer's brain. He found consistently that three things happened in sequence:

1. At about 300 milliseconds before a volunteer was aware of his decision to push the button, a spike in brain waves appeared in the brain (called the *readiness potential*).

2. The volunteer consciously decided to push the button.

3. Then 200 milliseconds later, the volunteer actually pushed the button.

Initially it seemed that free will was an illusion, because the brain generated the decision to push the button—i.e., brain activity *preceded* conscious awareness of the decision by about a third of a second. The preliminary interpretation was that we are driven by unconscious material processes in our brains and only afterward do we have the illusion that our will is freely chosen.

But Libet continued his research and added a twist: He told the volunteers to occasionally *veto* the decision to push the button after it is made—to decide to push, and then to immediately decide not to. Remarkably, he found the same results. Whether or not the decision was vetoed, he found a readiness potential that preceded the conscious decision by 300 milliseconds, *but he found no brain activity that corresponded to the veto.* The temptation to push the button was a material product of the brain, but the veto was not material—that is, *it was not from the brain.* Our acceptance or veto of the impulse provided by the brain is, in a sense, spiritual—not from the brain tissue. We remain free in a libertarian sense to accept or veto the decision.

Libet famously quipped that he demonstrated the reality of "free won't"—the reality that we can accept or veto preconscious motives generated by our brains. He noted the consilience between this result and the religious view of temptation and sin: We cannot voluntarily eliminate temptations, but we can choose freely—without material compulsion—whether to comply with or resist the temptation.

Libet's research has been widely discussed and Libet-type experiments have been performed using other technologies, such as functional MRI imaging. It is noteworthy that Libet is often misrepresented as having demonstrated that free will is an illusion, when within the limits of his experimental technique (which we'll discuss below), he demonstrated the *reality* of free will. Indeed, even in those cases where brain activity was detected prior to conscious awareness of the choice ("preconscious intent"), the implications are not necessarily any different from those of the veto experiment—i.e., it's still possible that an immaterial will caused the initial brain activity (preconscious intent) that preceded conscious awareness of the choice. Thus, interpretations that these experiments support free will seem more robust than the interpretations that the experiments refute it.

Twenty-First-Century Libet-Type Experiments

In the early twenty-first century, several neuroscientists have replicated aspects of Libet's work using functional MRI (fMRI) imaging of the brain. fMRI imaging uses regional changes in blood flow to measure brain activity. It does not have the fine-time resolution of electrodes on the scalp, but it does provide a picture of brain activity correlated to the anatomy of the brain—brain activity shows up as a bright spot on the fMRI brain image.

In 2008, Soon et al.[9] used fMRI imaging to study the brain activity in volunteers who were given the option to choose to push a button with their right or left hand. Because the right hemisphere controls the left side of the body and vice-versa, consistent brain activity in the contralateral hemisphere corresponds to a decision to push the button with that hand.

Soon found brain activity that correlated with the decision to move and that this brain activity preceded the decision, just as happened in Libet's experiments, but Soon found activity as much as 10 seconds before the decision. Furthermore, Soon found only a 60-percent correlation by side—that is, only 60 percent of the trials showed brain activity on the side that would correspond to

the hand that pushed the button. Because 50 percent is random, the correlation between unconscious brain activity and the decision to move was quite poor, only slightly better than pure chance. Unlike Libet, Soon did not test brain activity with a veto, which would have been very difficult using fMRI technology, given the poor time resolution of fMRI.

In 2011, Fried et al.[10] repeated Libet's experiments using electrodes actually implanted in the brain in patients undergoing evaluation for epilepsy (Libet used electrodes on the scalp in normal volunteers). They found that 80 percent of the time, a readiness potential could be recorded in the brain approximately 700 milliseconds before the decision to push the button.

Synopsis of Penfield and Libet-Type Experiments

Penfield's brain stimulation experiments, Libet's experiments using electrodes on the scalp, and the results of researchers who have used fMRI imaging and electrodes implanted in the brain can be summarized as follows: The will is not material and cannot be evoked by brain stimulation (Penfield). There appears to be brain activity, of variable consistency, to choose a simple act like pushing a button, *prior* to conscious awareness of the decision (Libet and others). However, this evidence need not refute free will, and is consistent with the presence of material predispositions that our free will can accept or reject; indeed, our free will may be causing this preconscious intent. Importantly, these decisions could be vetoed—or presumably accepted—*without evidence for brain activity* (Libet), which supports the existence of an independent mind, or will.

By Penfield and Libet's own interpretations, *neuroscience confirms the reality of free will* in the sense that we retain the free (immaterial and not caused by brain activity) choice to accept or veto a predisposition. The predisposition is material and caused by the brain. Our choice to accept or reject the predisposition is immaterial—spiritual in this sense. Neuroscience clearly supports the reality of free will.

It is important to note that many neuroscientists and philosophers have questioned the relevance of Penfield and Libet-type experiments to our understanding of free will. Aside from the vagaries of experimental technique, it's not clear that such simplistic reduction of free choice—to a mere whim to raise an arm or push a button—has real relevance to the intricate interplay of habit, emotion, experience, and rational contemplation characteristic of complex real-world moral decision making, which is the real issue in the free will debate. It is safe to say that Penfield and Libet-type neuroscience experiments show that some aspect of will appears to be *independent* of brain activity, and that there is often *some* brain activity of which we are unaware that precedes simple choices. But even the latter is not incompatible with free will. We retain the free choice—free in the sense that (in Libet's experiments) it is not associated with measurable brain activity—to accept or veto our predilection.

Neuroscience supports, rather than refutes, the reality of libertarian—even of spiritual—free will.

Research Supporting the Immateriality of the Will

In addition to the rather direct investigations of free will by Penfield, Libet, Soon, and Fried, there is another large body of neuroscience that points to that immateriality of the will, and that supports the libertarian perspective that the will is not determined by matter or physics. While a rigorous description of these experiments is beyond our scope, more detailed discussion of these experiments is given in chapter 19.

Wilder Penfield

In addition to his research on the inability to invoke the will using electrical stimulation, Penfield also found that he could not invoke abstract thought or intellectual seizures using electrical stimulation.[11] All neurological functions invoked by electrical probes during surgery or by seizures involve concrete material acts such as moving limbs, having sensations, or having emotions or memories. Penfield found no *abstract* thought in *any* experience evoked by surgery or seizure—i.e., he couldn't evoke calculus or philosophy. Abstract thought is never evoked by stimulation of the brain. Penfield asked, famously, "Why are there no intellectual seizures?" His answer was that the abstract, rational part of the mind (which includes the will) is an immaterial power of the soul. Penfield's insight supports the reality of free will in the sense that the will is immaterial and, by its nature, free from materialistic determinism.

Roger Sperry

Roger Sperry[12] was a neuroscientist in the mid-twentieth century who studied patients who had the two hemispheres of their brain surgically disconnected. Although he found subtle but fascinating neurological disabilities associated with cutting the brain (more or less) in half, he did not find any "splitting" of the unity of consciousness nor of the self, nor any cleavage of the intellect or will. Cutting the brain in half did not split the will or cut the capacity for will in half, which implies that will is an immaterial and thus free power of the soul.

Awareness in the Persistent Vegetative State

In 2006, Cambridge neuroscientist Adrian Owen[13] and his coworkers published a study of the brain activity of a young woman who had suffered massive brain damage from an accident. Despite having brain damage so severe that she was diagnosed by bedside examination as having no mind at all—the deepest possible level of coma—Owen showed that the woman had a high level of awareness of her surroundings and could even answer questions (using patterns of brain activation on fMRI). Owen's work has been replicated with many other severely brain-damaged patients (more than 40 percent of deeply comatose patients in a persistent vegetative state can understand and answer questions), and supports the view that rational parts of the mind—including the will—are not caused wholly by material processes. Near-destruction of brain matter appears to leave abstract thought and will relatively intact in many patients, which implies that the will is not caused by the brain, and in that sense, is free.

Near-Death Experiences

There is a large body of scientific literature on near-death experiences that involve persistence, awareness, and mental experience in the absence of a heartbeat or brain function.[14] While the exact nature of these experiences remains controversial and under investigation, there is ample scientific evidence that there are aspects of the mind, including the will, that can function independently of brain matter. This evidence lends support to the view that the will is not determined by matter.

Neuroscience is frequently misrepresented as disproving free will despite the fact that, when considered objectively, neuroscience clearly *supports* the reality of free will.

Social Psychology and Free Will

As with neuroscience, social psychology experiments have been invoked (without justification) to deny the reality of free will. For example, social psychologist Daniel Wegner, in his book *The Illusion of Conscious Will*,[15] argues that many of our everyday acts to which we ascribe free will are really deeply ingrained habits over which we exercise no real control. He refers to small unconscious movements we make during ordinary life that seem to occur independently of conscious will, and to the phenomenon of "facilitated communication," in which strong, unconscious suggestibility leads people to act in ways that appear to be voluntary but are in fact caused by others who are "assisting" them. He also describes a rare medical syndrome that follows brain damage in which patients exhibit "utilization behavior." People with this disorder automatically carry out tasks such as filling a glass with water—without a conscious intention to do so—when an empty glass is placed in their hand. Wegner asserts that the automatic nature of such behavior implies that all behavior arises from reflex-like activity taking place in our nervous system, which is inconsistent with free will.

Other phenomena in social psychology that some researchers invoke to deny free will are implementation intentions, or are specific plans to do a specific task at a specific time in a specific place rather than doing something "whenever you can." It is clear that deliberate implementation intentions increase the likelihood that the task will be accomplished—there is much psychological research to support this. Some researchers draw from this work the conclusion that the neural correlates of implementation intentions—the specific neurochemicals and firing patterns of neurons associated with these intentions—are the actual cause of our completing of the tasks, and that this demonstrates that our will is determined by our neurochemistry and is thus not free.

The common inference here is that many of our behaviors are automatic or "programmed" in some sense, and that because it is reasonable to infer that *all* of our behavior is caused in the same way, then all of our behavior must be programmed by the brain, and free will is an illusion. Our behaviors are, in this paradigm, complex reflexes caused wholly by deterministic material processes.

I believe that Wegner's and other social psychologists' assertions that such automatic phenomena contradict the reality of free will is unfounded. It is undeniable that much of our behavior is unconscious and habitual. I don't think about every detail of every movement I make. Much of what I do is automatic in this sense, and I am unaware of the specifics unless I consciously intend to be aware, and even then there's much of what I do that I don't know. But merely because I do a task habitually and without explicit deliberation—like walking to the cafeteria to buy lunch—doesn't mean that I do not really choose to do so. It merely means that it is in my biological nature that my (free) choices are *facilitated* by efficiencies in my nervous system. When I choose to buy lunch, my brain and nerves and muscles habitually undertake tasks (walking, verbally ordering food, taking out my wallet, etc.) that allow me to carry out the task I have freely chosen to do, and to do so with a minimum of unnecessary effort.

In fact, the classical understanding of virtue, as described by Aristotle, is "the habituation of freely chosen good acts." The mere fact our bodies make the implementation of those choices efficient does not mean that the choices do not freely originate within us. There is no logical basis for the assertion that unconscious or habitual aspects of behavior imply that the behavior does not originate within us without determinism or compulsion.

Evidence in social psychology in no way refutes free will, and neuroscientific evidence is most reasonably interpreted as providing support for free will. The invocation of science to deny free will is the result of a profound misunderstanding of science.

Logical and Metaphysical Arguments for Free Will

Even if the weight of the scientific evidence favoring freedom of will is not taken into account, there are strong reasons to infer that the will is free based on logic and metaphysics. These reasons include:

1. As noted above, virtually all denial of free will is predicated on the belief that nature is deterministic. It is clear from modern quantum mechanics that nature is not deterministic, and thus there is genuine room in natural processes for libertarian free will as a hallmark of human choice. The deterministic denial of free will is logically indefensible because one of its predicates—that nature is deterministic—is not true.

2. A long line of metaphysical reasoning, going back at least to Aristotle, demonstrates that the human capacity for reason and abstract thought is an immaterial power of the human soul. This inference is well supported by logic and evidence (please see my other chapter in this volume, chapter 19). The will is the power of the soul that follows

on the intellect, and thus is also an immaterial power of the soul. By virtue of its immateriality, the will is not determined by any materialistic process. It is in its immateriality that the will is free.

3. Denial of free will based on materialism is self-refuting. If the will is wholly determined by matter and its interactions, then the act of denying free will, which itself is an act of will, is a wholly material process. However, only propositions can have truth value. Material processes—such as the secretions of neurochemicals, the transmission of action potentials, etc.—are not propositions, and therefore have no truth value. If free will is a wholly material process, then denial of free will is merely a secretion of brain tissue, so to speak, and is incapable of being true (or false). Furthermore, material processes are governed by physical laws, not by laws of logic, so if there is no free will, logic cannot be ascribed to the denial of free will. Thus, if it is true that free will is not real, then the assertion by a free-will denier that free will is not real is neither true (nor false) nor logical. A meaningful discussion of free will—even the denial of free will—*presupposes* free will.

Determinism and the End of Innocence

Despite the logical fallacies and misunderstandings of science that underly the denial of free will, the social and political consequences of free-will denial are quite real and are extraordinarily dangerous. I believe that the denial of free will is one of the *most* dangerous metaphysical positions we can take. This is because a respect for human dignity and moral innocence depends critically on a recognition of free will.

Moral Culpability and the End of Innocence

A common argument for the social benefit of free-will denial is that it humanizes the criminal justice system by eliminating the inference to moral culpability in criminal law and redirecting criminal justice away from retribution and toward the prevention of crime and the rehabilitation of the offender. This is a profound misunderstanding of the societal implications of free-will denial.

Retributive justice and respect for free will *increases* respect for human dignity because it acknowledges rational human agency and human exceptionalism. Rational moral agency is among the highest attributes of man. Unlike animals, we can choose to act morally or immorally. Even when we choose moral wrongs, we demonstrate our human dignity in the sense that the ability to freely choose moral actions—or to freely choose immoral actions—is a hallmark of the dignity of the human soul. Man's dignity is that he can be saintly or wicked—neither attribute is attributable to an animal or even to a human being who is not in possession of his rational faculties. We do not ascribe culpability to cattle or to infants or to schizophrenics—not because they are innocent, but because they are incapable of either guilt or innocence.

And in the incapacity for guilt or innocence lies the danger of free-will denial. If we have no free will, we have the moral standing of a cow, a newborn, or a psychotic—that is, we have no moral standing at all. And while we cannot be guilty, *we also cannot be innocent.* Society controls children's lives, restrains madmen, and pens, herds, and slaughters cattle because they lack moral agency—they lack free will. Lacking moral standing, children and madmen and cattle cannot be neither morally guilty nor innocent, so we do with them as we think best without respect for their moral agency. We *control* infants and madmen and livestock. We don't ask their opinions. We do what we choose to them, for a panoply of reasons, but we don't deal with infants or madmen or cattle according to our respect for their freedom because they have no moral freedom.

To incarcerate or even to execute a guilty man out of retribution is to acknowledge his moral dignity—to *respect* him—and is no less an act of respect than the acquittal of an innocent man. Guilt and innocence are two sides of the same coin. Moral guilt and innocence—moral agency—is the hallmark of human dignity and depends critically on the reality of free will.

Without free will, criminal "justice" and all social and political policy becomes management and control, not discourse and justice. Justice, without free will, has no real meaning because justice depends on respect for moral agency. Justice is not a principle of livestock management.

What happens to a criminal justice system when free will is denied? Consider, for example, incarceration for *predisposition* to commit a crime rather than for retribution for having committed a crime and thus having committed a moral transgression. Of course, free-will deniers would abhor such "precrime" policing, as would those of us who acknowledge free will and human dignity, but free-will deniers lack a solid rational basis for opposing incarceration for precrimes. A precriminal cannot claim, in a system devoid of respect for free will, "I'm innocent" because without free will, no one is

guilty and no one is innocent. We don't take guilt or innocence into account in livestock management. Inevitably, a denial of free will leads to the untethering of criminal justice from moral (and actual) guilt or innocence, and to a shift in criminal "justice" policy to maximization of efficiency and effectiveness. Why wait until a person in a high-risk demographic (say, a black teenage son of an unwed mother living in a high-crime neighborhood in an inner city) commits a crime when crime can be avoided by *preventative* incarceration?

No reasonable person would support such a policy, but denying free will leads to a slippery slope where such things become possible. We institutionalize schizophrenics and pen cattle for reasons of efficiency and effectiveness, not because they are *guilty* of anything. We constrain them in order to accomplish economic and societal goals without respect for their moral agency. Without respect for free will, there is no actual criminal justice system; justice presupposes moral accountability. There is a just system of crime management, for which actual moral guilt and innocence are, in the final analysis, not only irrelevant but incoherent.

In this vital sense, criminal justice as *retribution*—the bugbear of free-will deniers—is the supreme acknowledgment of human dignity. Retribution can be properly directed only to rational moral agents—one does not hold a cow or a neonate or a madman accountable. While crime prevention and the rehabilitation of offenders are laudable policies, they are akin to livestock management—the prevention and correction of bad behavior by controlling environment and by training. In a genuine system of justice that respects human dignity, crime prevention and rehabilitation must be predicated on respect for actual moral guilt and innocence, which depend on the reality of free will.

Arendt and Totalitarianism

The centrality of respect for free will and moral agency for human rights is not merely a theoretical proposition. Philosopher Hannah Arendt, who is the preeminent scholar of totalitarianism, has pointed out that totalitarian systems differ from all other political systems in that laws in totalitarian systems are unconcerned with actual moral guilt or innocence.[16] Hitler did not consign Jews to death because they were individually morally guilty of specific crimes—the trains to Auschwitz were not filled with miscreants convicted of statutory felonies. The felons were, in fact, the people operating the trains. The people in the trains—the Jews—were incarcerated and killed for *what they were*, not for what they did. The Holocaust was human livestock management—the cold, calculated decision to cull the human herd of unwanted individuals.

Stalin did not starve millions of Ukrainians because each kulak was guilty of breaking Soviet law. The Soviets consigned kulaks to death not for retribution for capital crimes but because kulaks were an impediment to state policy. For the Soviet government, kulaks were not morally different from the livestock they owned. Kulaks were consigned to oblivion just as diseased cattle were culled to free up resources for agricultural improvement. Neither a doomed Jew nor a starving kulak could plead innocence because in the totalitarian state, there is no innocence just as there is no real guilt. There is no moral culpability in totalitarian systems because no one is free in any sense. There are just historical currents and races and classes to be managed, herded, and culled.

The cornerstone of totalitarian rule is the denial of free will—the denial of moral guilt and innocence.

Dangers of Denying Free Will

The most reasonable understanding of free will is that of Thomas Aquinas, who asserted that man is free in the sense that his rational choices are caused entirely by himself without determination or compulsion. The reality of libertarian free will so defined is well supported by logic, metaphysics and physics, neuroscience and social psychology.

Determinism, which is the cornerstone of modern free-will denial, has been discredited by quantum mechanics, and the materialism on which much of free-will denial is predicated has little metaphysical credibility or scientific support. Criminal justice based on free-will denial entails the repudiation not merely of morality and of the concept of guilt, but also the repudiation of the possibility of innocence. The denial of free will is the denial of justice itself and is an essential predicate for totalitarianism. Denial of free will is a metaphysical error, a scientific misunderstanding, a repudiation of human dignity, and an existential threat to humanity.

Can Materialism Explain Human Consciousness?

Michael Egnor

odern neuroscience is done from a materialistic perspective. By the term *materialist* I mean that neuroscientists and most philosophers of mind assume that the mind is generated by the brain, without remainder, and that mental powers will ultimately be explained entirely in terms of matter and its interactions—nothing but physics and chemistry. Just how this happens is (necessarily) vague, but it is within this framework that ordinary neuroscience research is conducted and interpreted. However, this materialist inference is a stipulation, not an inference derived from evidence, and not an inference justified by experiment. Nor is the materialist inference, as we will see, even justified *logically*. Materialism is a tenuous—indeed, a highly dubious—predicate for neuroscience.

The Mind-Brain Relationship and the History of Philosophy of Mind

The materialist interpretation of the mind-brain relationship is an artifact of modernity—in its modern form, it began in the seventeenth century. Ancient materialism was of a somewhat different sort. Ancient Greek and Roman materialists like Epicurus, Leucippus, Democritus, and Lucretius proposed that the soul (by which they meant roughly what we mean by mind) is a composite of fundamental particles called *atoms*. The atomists' theory bore little resemblance to modern quantum mechanics. Then, "atoms" were tiny indivisible particles with various powers, not nondeterministic particle-wave dualities as we understand them today. Ancient Indian philosophers like Kanada and Chinese philosophers like Wang Chong proposed understandings of the soul not unlike those of the Greek and Roman atomists.

Materialism declined in the West several centuries before Christ in large part because of Platonism and its Aristotelian descendants. Most post-Socratic philosophers understood the soul in terms of ideas rather than particles of matter, and various iterations of Platonic idealism and Aristotelian hylomorphism ("matter-form") came to dominate the understanding of the soul. Christianity drew extensively from Platonism (in addition to revelation) in its philosophical understanding of the soul, and from the thirteenth century to the modern era, the Aristotelian perspective of Thomas Aquinas was the mainstream Christian understanding of the soul. In Thomistic psychology,[1] which was shared in its broad outlines by Jewish and Islamic philosophers, the soul is the form of the body, and the powers of the mind—sensation, perception, imagination, memory, and appetites—are the substantial form of the body. Plants and animals have vegetative and sensitive souls, respectively, but man has a rational soul that differs radically from lower souls because man has the capacity for reason and free will, which are immaterial powers. These immaterial powers were understood as spiritual gifts—the "image of God" in which man was created—and the immateriality of the rational soul renders the human soul immortal.

In the seventeenth century, modernist philosophers broke with the classical Thomistic understanding of man, and materialism in its modern form can be traced, ironically, to the substance dualism of Rene Descartes. Descartes proposed that man is a composite of two separate substances: a thinking substance and a material substance. Descartes ascribed all powers of the mind to the thinking substance and all aspects of the physical body to the material substance, which he understood as a biological machine. In the Cartesian view, man is a ghost in a machine. The bane of the Cartesian understanding of the mind-body relationship was the interaction problem—how can a thinking substance connect to a physical substance? Over time, the thinking substance was discarded as superfluous and philosophers came to understand the human soul purely in terms of its matter—that is, of its nonthinking properties that were extended in space.

Thus, ancient materialism rose again in modernist Cartesian terms.

The Metaphysics of Modern Neuroscience

By the early twentieth century, neuroscientists had understood enough about the correspondence between the brain and the mind that the systematic scientific study of the brain-mind connection became feasible. This required a coherent theory of mind, which in the modernist era was implicitly materialist. Materialistic theories of the mind essentially amount to ghostless Cartesianism. Materialists accept Descartes's understanding of matter as substance extended in space and discard Descartes's thinking substance, which leaves modern materialists with a conundrum. How can materialism explain the mind when matter is defined (in the Cartesian scheme) as nonthinking extension in space? When materialists exorcised Descartes's ghost from the human machine, they drove out thinking as well. The history of neuroscience in the modern era is implicitly the materialist struggle with self-refutation. How can the mind be explained by nonmental matter?

Behaviorism

The earliest modern neuroscientific approach to the study of mind was the behaviorist approach. Behaviorism[2] has a certain efficiency: Behaviorism is the theory that the only testable (and thus relevant) correlate of mind is behavior. Behaviorist neuroscience was greatly streamlined by ignoring the mind. Some behaviorists denied even the existence of mental states, while others admitted them, but all ignored them because behaviorism was functionally materialist in the sense that the behavior of material bodies was the only object of behaviorist neuroscience. A problem with behaviorism (there were many) was that a theory of mind that ignored the mind was, you might say, off to a bad start. A flowering of good science in the mid-twentieth century—not the least being Noam Chomsky's groundbreaking research on the uniqueness of human language[3]—exposed the sterility of the behaviorist project. By 1960, cognitive science (which at least acknowledges the existence of mind) was born, and behaviorism was shoved back into its box, so to speak.

Identity Theory

Behaviorism was eclipsed in the mid-twentieth century by identity theory, which at least brought mind to the study of the mind. Whereas behaviorism was efficient, identity theory was simple: thoughts, feelings, etc. were *identical* to brain states. For example, if you have a toothache, the pain you feel in your tooth is *the same thing* as the simultaneous physiological state of your tooth nerves, etc. The experience of pain is neurochemicals and nothing more. This radical reductionist take on neuroscience simplified the scientific study of the mind because as a scientist, all you had to do was sort out the chemicals, and you explained the mind for free.

The problem with identity theory—that your thoughts are chemicals—is that it violates Leibniz's Law, which is a fundamental principle of logic. Leibniz's Law states that for things to be identical, they must be exactly the same in all respects. To claim that two *different* things are identical is nonsense. Two trees of unequal height can be similar, or analogously treelike, but they are not identical because one tree is taller than the other. To state that your pain is identical to your neurotransmitters is a fallacy because your pain and your nerves can be distinguished

from one another in the sense that your pain hurts and your nerves are long stringy things. A surgeon can see your nerves, but he can't see your pain.

Thoughts and feelings aren't the same as nerves or chemicals, so they aren't identical, and identity theory is wrong. There's agreement on that today, even among materialists.

Eliminative Materialism

The latest iteration of materialism is eliminative materialism. Eliminative materialists acknowledge that the mind can't be explained in terms of matter, which is a reasonable acknowledgment given that matter as understood by modernists is essentially Cartesian nonmental substance. Eliminative materialists, faced with irreconcilable contradictions between materialism and the mind, eliminated the mind. Please understand: Eliminative materialists don't claim that the mind is irrelevant (behaviorism) or that it is the same thing as matter (identity theory). They claim that there are no minds (that's the eliminative part). Thoughts and feelings don't exist. It's all matter, all the way down, and we merely make a category error by referring to the mind, which they call folk psychology (as in folk tale). In the eliminative materialist view, we are just meat robots who haven't yet abandoned our quaint delusion that we have minds.

Holding this view presents a problem because it can't be held. If you believe that eliminative materialism is true, then belief in eliminative materialism isn't a belief, so you don't believe in eliminative materialism. You may see the problem here: How can you have the delusion that there are no minds and hence have no delusions? How can you believe there are no beliefs? For the eliminative materialist, a thought is a physical state, a certain concentration of brain chemicals or whatever that we (the uninitiated) foolishly call a thought. So a difference of opinion between an eliminative materialist and a dualist isn't really a difference of opinion at all. It's just two different concentrations of brain chemicals in different skulls. How puddles of skull chemicals get into a "disagreement" is vague, but "vagueness" only applies to thoughts, which we don't have anyway. If you aren't getting a headache over this, you don't understand eliminative materialism.

One presumes that eliminative materialists still request Novocain at the dentist's office—not, mind you, because they don't want pain (which does not exist), but because they don't want (for some reason) the physical state of their brain chemicals that we plebeians erroneously call pain.

At this point, you may feel (if we can use that word) a bit uncomfortable. Eliminative materialism, aside from being logical nonsense, has a real flavor of crazy, except that an eliminative materialist would say that there is no crazy; there are just chemicals we foolishly call crazy.

There is one sense in which eliminative materialism is more rational than other materialist theories of mind. Whereas other materialist theories in one way or another conceal their illogic, eliminative materialism makes no pretense of logic or of anything mental at all. Honesty is a virtue.

Computer Functionalism

Materialists who are still tethered to reality eschew the above iterations of their ideology, and many embrace functionalism. Functionalism is variably considered a form of materialism or of dualism (I consider it dualist), and it is defined colloquially as "the mind is what the brain does." Functionalism defines the mind according to the role it plays rather than by the substance it is

made of. Nearly all modern functionalists are computer functionalists, by which they mean that they understand the mind as a kind of computation. A common formulation is the assertion that the mind is software and the brain is hardware. The mind runs on the brain like a program runs on a computer.

Although popular, computer functionalism is beset with intractable problems. The brain can (and has been) described using the terms of computation, but *the mind is not any form of computation.* The salient critique of mind-as-computation was given a few decades ago by John Searle in his Chinese Room analogy.[4] I'll provide my own critique here, which is an adumbration of Searle's.

In the nineteenth century, philosopher Franz Brentano asked this salient question: What is unique to mental states that is never found in matter? How can we discern a thought from a physical thing? He answered *intentionality.* Intentionality is a technical philosophical term that means "aboutness." Intentional things are inherently *about* something (other than themselves), whereas nonintentional things are not really about anything. All thoughts are intentional in that they have an object to which they point. I think *about* lunch or *about* my vacation or *about* mercy. Physical objects are never about anything, except when we artificially attribute intentionality to them. A rock on the seashore isn't about anything. A tree isn't about anything. A pencil isn't about anything. Physical objects are just what they are.

Intentionality has been a profound problem—I think *the* fundamental problem—for materialist theories of the mind. Volumes have been written on materialist efforts to explain intentionality[5] of the mind if the mind is material, and all have failed.

We can apply the problem of intentionality to computer functionalism. Computation is algorithmic. Computation is a rule-based mapping of an input to an output. However, computational mapping is never intentional. That is, it is never inherently *about* anything. It's a mechanical process that is blind to meaning. Consider, for example, the Word program I used to type this chapter. The program doesn't care what I type. I can type a chapter supporting materialism or opposing materialism. The computer program is blind to meaning—it isn't *about* anything. In fact, it is the lack of intentionality, the blindness to meaning, that makes computation so useful. Imagine having to buy a different word-processing program for each opinion that you write about!

The mind is not computation because it is always intentional, and computation is never intentional. In fact, the mind, in this sense, is the *antithesis* of computation. All that the mind is, is what computation is not. Computer functionalism is an apophatic theory of mind: it is what the mind is not.

Dualism and Materialism in Modern Neuroscience

One might think that the logical problems with materialism would insulate twenty-first-century neuroscience from its influence, but that is not so. Most contemporary neuroscientists work from an implicitly materialist perspective—in part because they're unreflective, in part because materialism is the metaphysical correlate of the atheistic scientism that infests modern science, and in part because public admission of a dualist perspective is perceived (correctly) to be a career impediment in neuroscience. I recently had a friend (a tenured and accomplished neuroscientist) who is a devout Christian tell me privately that if he ever publicly questioned materialism, he would never get another grant.

Materialism is the framework for modern neuroscience for ideological reasons, not logical or scientific ones. Materialism is stipulated, not demonstrated, in neuroscience. The reason this stipulation has, despite logic and evidence, been easy to pull off is that it is rather difficult to test materialistic and dualistic theories with unambiguous rigor. How do we know, experimentally, if a thought is material or immaterial? How do you test this *metaphysical* question? It can be tested and has been tested. There are here two kinds of inferences we can make (discounting abduction): deductive and inductive. All natural science is inductive because, as Thomas Aquinas observed in the thirteenth century, deduction cannot prove the existence of a thing because essence (the formal structure of a deductive argument) is absolutely distinct from existence. We cannot deduce neuroscience. We gather evidence, apply explanatory frameworks, and see which inference best fits that evidence and the logic. Neuroscience, like all science, is inductive—it is inference to the best explanation drawn from evidence.

To test materialism and dualism, we need to define them. These are the working definitions I use:

1. Materialism entails the inference that the brain causes all aspects of the mind, without remainder.

2. Dualism entails the inference that the brain causes some aspects of the mind, with remainder.

All parties agree that the brain causes some aspects of the mind. The materialist claim is radical: The brain causes *all* aspects of the mind; there are no immaterial thoughts.

The dualist claim is less radical: There are some aspects of the mind that are not caused by the brain.

The philosophical issues get subtle here, involving supervenience, epiphenomenalism, etc., which are beyond our scope. But the basic claims of materialism and dualism can be tested with some rigor, as we will see.

As noted, materialist and dualist theories of mind differ on the immaterial remainder. That is, they differ on whether all mental states are caused *entirely* by brain states, or whether there are some mental states that are not caused by the brain (i.e., are immaterial). The evidence, of course, is incomplete—there are countless mental states that are uncorrelated with brain states because no effort has been made to study them, or (perhaps) because the scientific methods employed are insensitive. With these provisos in mind, we can still make reasonable inferences about materialism and dualism based on this fundamental concept: If all mind states are caused by brain states, then every mind state can be evoked by stimulating the brain in some fashion, every mind state can be suppressed by ablating (damaging) the brain in some fashion, and every mind state can be correlated with a brain state in some fashion. If some mind states are *not* caused by brain states, then these mind states will be recalcitrant to evocation, ablation, and correlation.

The Neuroscience Evidence on Materialism and Dualism

Most of the seminal experiments in neuroscience that address the mind-brain issue can be categorized as evocative (stimulate the brain), ablative (suppress the brain), or correlative (compare brain states to mind states). It is instructive to review the major experiments in neuroscience that speak to the mind-brain question.

Evocative Experiments
PENFIELD—INTELLECTUAL SEIZURES AND FREE WILL

Wilder Penfield[6] was a leading neurosurgeon and neuroscientist in the mid-twentieth century who pioneered the surgical treatment of epilepsy using stimulation and recording from the surface of the brain in awake patients undergoing brain surgery. This was possible because the brain itself feels no pain, and the scalp can be anesthetized with Novocain-like drugs to render the surgery painless. This surgery is still being done today.

Penfield was especially interested in the relationship between the brain and the mind. He began his career as a materialist and he ended it as a passionate dualist. He based his dualism on two observations:

1. *There are no intellectual seizures:* Seizures are sporadic electrical discharges from the brain and they cause a variety of symptoms, from complete loss of consciousness to focal twitching of muscle groups, sensations on the skin, flashes of lights or noises, smells, and even intense memories or emotional states. Penfield could record these electrical discharges from the surface of the brain. Penfield noted that *there are no intellectual seizures.* That is, there has never been a seizure in medical history that had specific intellectual content, or abstract thought. There are no mathematics seizures, no logic seizures, no philosophy seizures, and no Shakespeare seizures. If the brain is the source of higher intellectual function, as is widely believed, why in medical history has there *never* been a seizure that evoked abstract thought? This fascinated Penfield,

and he inferred quite reasonably that the reason there are no intellectual seizures is that abstract thought does not originate in the brain.

2. *Free will cannot be simulated by stimulation of the brain:* Part of Penfield's research was to stimulate the motor areas of the brain, which caused patients' limbs to move during surgery. He was the first surgeon to map the motor areas of the brain in this fashion. In doing this, he noticed that patients always knew the difference between stimulated movements and movements that they freely caused themselves. Penfield would ask patients to move their limbs freely whenever they chose, and he would (without telling them) stimulate their limbs to move. Patients always knew the difference between movements freely chosen and movements caused by the surgeon. Penfield could never find a region of the brain that simulated free will. He concluded that free will is not in the brain—it is an immaterial power of the mind.

Ablative Experiments
ROGER SPERRY AND SPLIT-BRAIN EXPERIMENTS

Some of the most remarkable experiments in modern neuroscience were carried out by Roger Sperry,[7] an American neurophysiologist and Nobel Laureate who did his work during the mid-twentieth century. He studied patients who had undergone a radical operation to treat epilepsy. In split-brain surgery, commonly known as corpus callosotomy, neurosurgeons cut the large bundle of nerve fibers that connect the two hemispheres

of the brain to prevent the propagation of seizures through the brain. This lessens the severity of seizures in certain epilepsy patients.

Sperry conducted meticulous research on people who had undergone this radical surgery. He found that after having the cerebral hemispheres disconnected—leaving the brain more or less cut in half—*that patients were, for all ordinary purposes, normal.* For his Nobel Prize-winning research, Sperry found that there were subtle perceptual abnormalities of which these patients were unaware that revealed different functions of the right and left halves of the brain. Speech was usually in the left cerebral hemisphere and spatial reasoning and orientation was usually in the right. His experiments were detailed and elegant and showed perceptual disabilities that were not evident during everyday life.

What is most remarkable about Sperry's research is how little the patients' minds were changed by this radical surgery. People whose brains were cut in half remained completely normal in everyday life. Specifically, they remained one person, with no splitting of consciousness or personality. The splitting that Sperry *did* find involved subtle perceptual disabilities that were fascinating, but of which the patients were completely unaware.

The most important inference to be drawn from Sperry's work is that the sense of unity of personhood and the capacity for abstract thought are completely unaffected by cutting the brain in half.

Sperry rejected reductive materialism, and over his career, he changed his view from behaviorist materialism to anti-mechanistic and nonreductive mentalism.

ADRIAN OWEN AND PERSISTENT VEGETATIVE STATE

In the September 2006 issue of the journal *Science*, Adrian Owen and his colleagues at the University of Cambridge published a study titled "Detecting Awareness in the Vegetative State."[8] Owen studied a woman who had been diagnosed as being in persistent vegetative state (PVS), which is a condition that is characterized by severe brain damage and complete and (usually) permanent lack of consciousness and awareness of any sort. People in PVS are assumed to have no internal mental processes whatsoever—the deepest level of coma. Owen's experiment involved a functional MRI of the woman's brain. Functional MRI imaging measures blood flow in regions of the brain that appears to correlate with brain activity. When this patient in PVS was in the MRI machine, Owen asked her to "imagine that you are playing tennis" and "imagine that you are walking across the room." Remarkably, functional MRI imaging showed that she was understanding and capable of cooperating mentally with the requests.

Owen compared the woman's functional MRI results with normal volunteers and found no difference, and he repeated the experiment and asked her the same words in a different order so they made no sense. Her brain showed no response. Owen concluded that despite having massive brain damage and being in a coma so deep she was diagnosed as having no mind at all, she was actually able to think abstractly and to understand and respond to detailed questions. Since Dr. Owen's original experiment, many patients in PVS have been shown, by functional MRI imaging, to be capable of complex abstract thought.

This shows a rather marked dissociation between very severe brain damage and high levels of abstract thought. The most parsimonious interpretation of this research is that these patients' capacity for abstract thought is to some degree *independent* of the functioning of the brain.

NEAR-DEATH EXPERIENCES

In modern times, some people who suffer cardiac arrest are resuscitated and can report on their experiences during the time their heart stopped functioning. These experiences are surprisingly common, and there is a growing medical literature of scientific studies of near-death experiences (NDEs).[9]

A famous example was Pam Reynolds,[10] who underwent radical surgery for a brain aneurysm in 1991. Her body was cooled and her heart was deliberately stopped so a surgeon could repair a dangerous aneurysm at the base of her brain. For 30 minutes she was clinically dead and documented to have no brain waves by EEG measurements.

Reynolds reported that when her heart stopped, she heard a humming sound and left her body and floated to the ceiling of the operating room and watched her own surgery. She reported that her senses were remarkably acute, and after the operation, she reported the details of the operation, including specific characteristics of the instruments and the conversations of surgeons that occurred during the period when she was clinically dead. She subsequently saw a beautiful light and felt pulled toward it, and she saw deceased relatives—including her grandmother and uncle—welcoming her to a beautiful and peaceful place. She was then told she had to return to her body, and reluctantly did so. She described the return as like jumping into ice water.

Tens of millions of people worldwide have had NDEs. About 20 percent of NDEs are veridical, meaning that people have experiences that can be checked by things that they could not have been aware of ordinarily.

Near-death experiences are typically

coherent and clear "hyper-real" experiences that are unlike hallucinations or hypoxia that might be the consequence of physiological events in a dying brain. A nearly universal quality of NDEs is that they entail an enhancement of awareness and clarity of insight rather than the stupor and confusion characteristic of brain impairment and hypoxia. Materialist neuroscientists have offered a plethora of materialist explanations for NDEs, none of them convincing, that tend to be either psychological or physiological. The weaknesses of materialist explanations include the inability to explain the complete experience as well as the inability to explain the intense realism, the vivid thoughts and perceptions, and the radical life-changing effects of NDEs. For example, materialist explanations cannot account for the fact that children who have NDEs encounter relatives whom they had never met because the relatives had died before their birth.[11]

All materialist explanations for near-death experiences are implausible. They simply do not explain the facts of NDEs. Succinctly, were it not for materialist bias, no materialist explanation for near-death experiences would be taken seriously in the scientific community.

Correlative Experiments
CEREBRAL LOCALIZATION AND PHRENOLOGY

In the nineteenth century, scientists discovered that certain regions of the brain were associated with very specific neurological functions. For example, a small region of the left frontal lobe is commonly associated with the ability to speak. Injury to this region impairs speech. The posterior parts of the frontal lobes on both sides contain neurons that mediate movement on the opposite side of the body. The arrangement

of these neurons is quite precise, and a map of the contralateral side of the body can be generated by the study of these neurons. Corresponding regions in the anterior parietal lobes are associated with tactile sense, and the cerebral cortex between the occipital lobes at the back of the brain correlates with vision. There are many quite specific regions in the deeper parts of the brain as well as in the brainstem that correspond to physiological functions such as the regulation of heart rate, the regulation of breathing, and the secretion of hormones.

During the nineteenth and early twentieth centuries, most neuroscientists believed that all aspects of mental function, including higher-level abstract thought, are caused by specific regions of the brain—in the same way that speech, movement, sensation, vision, and basic physiological functions are. This followed quite naturally from the materialist perspective—that the mind is caused by the brain, without remainder. This scientific perspective, which was called *phrenology*,[12] led to the attempt to correlate higher intellectual functions, such as the abstract capacity for judgment and reason, with specific gyri of the brain. In this era, doctors lacked the ability to image the brain during life (this was before CAT scans and MRIs and even before X-rays), so neuroscientists used the then-reasonable inference that the shape of the skull would correlate with the development of the brain underlying it.

Phrenologists, who were neuroscientists, attempted to correlate higher intellectual function with head shape, and phrenology was quite influential in neurology and psychiatry in the late nineteenth and early twentieth centuries. It was used extensively in criminology to identify individuals prone to crime and in psychiatry to flag predisposition to psychiatric illness. However, it became clear by the early twentieth century that phrenology was misguided. Abstract thought was not localized in the brain in the same way that other powers of the mind such as movement, sensation, and vision were, and head shape provided no clue to character.

Phrenology was not a pseudoscience. Rather, phrenology was a science that used systematic reasoning and experiment, the predictions of which were shown to be wrong. They were wrong because the *materialist metaphysical assumptions* of phrenology were wrong. While indeed it is true that certain powers of the mind, such as the ability to move limbs, to sense touch, and to see, are localized in specific regions of the brain, higher intellectual powers cannot be localized in this way. Chastened neuroscientists should have questioned, but didn't, the materialism on which phrenology was predicated.

The failure of phrenology is primarily a failure of the materialist understanding of the mind. Unfortunately, while phrenology was tossed into the dustbin of neuroscience, materialism continues to be the metaphysical framework by which modern neuroscience is done.

BENJAMIN LIBET AND FREE WON'T

Benjamin Libet[13] was a neurophysiologist who worked during the mid-twentieth century. He was fascinated with what he called *mind time*. He studied the correlation between the electrical activity of the brain and the simultaneous content of thought. Libet performed seminal experiments on the correlation between thought and brain activity, and his most famous experiments involved the study of free will.

Libet asked normal volunteers to participate in an experiment in which brain waves were recorded by electrodes taped to the volunteers' scalp. The volunteers were asked to sit in front of a screen that displayed a clock and were asked to push a button whenever they freely decided to do so. The clock recorded the exact time they made the decision to push the button (to the resolution of milliseconds) and allowed the subjects to precisely time the moment that they freely decided to push the button. The time of the volunteers' decision to push the button could be correlated precisely in time with their brain waves.

Libet's initial experiments showed a very consistent relationship between brain waves, voluntary decision, and execution of the decision to push the button. He found that approximately 400 milliseconds before the volunteers were aware of a conscious decision to push the button, there was a spike in brain activity that seemed to correspond with the thought that the button would be pushed. It is noteworthy that this spike in brain activity always occurred *before* the conscious decision.

Libet's early experiments seemed to support the view that free will does not exist. It seemed that the volunteers' conscious decision to push the button was caused by unconscious brain waves that preceded conscious awareness by almost half a second. This would imply that our sense of free will is mistaken; we are driven—forced—to make decisions by brain activity of which we are not conscious and over which we exert no control.

However, Libet was a careful scientist, and he explored the question of free will in more detail. He asked the volunteers to repeat the experiments in which they decided to push the button and noted the time of their decision, but he then asked the volunteers to *veto the decision to push the button* and thereby not push it. Remarkably, Libet found that while the decision to push the button was invariably preceded by an unconscious spike in brain activity, *the veto was electrically silent in the brain.* Libet interpreted this as meaning that the brain causes a steady stream of unconscious motivations—he compared them to temptations—of which we are freely (i.e., spiritually and immaterially) capable of refusal or compliance.

Libet noted the remarkable similarity between his experimental findings and the traditional Christian understanding of temptation and sin. He concluded that we are beset with unconscious temptations over which we have no control, but to which we can give assent or refusal to act freely. Our free will is real, Libet concluded, if we understand it as *free won't*—as our freedom to accept or reject temptation.

Chomsky and Universal Grammar

The insights of linguist Noam Chomsky[14] are of great importance to our understanding of language and of the mind and the brain. In the 1950s, Chomsky proposed a revolutionary theory of human language. Prior to his research, linguists generally favored a behavioral model of human language acquisition. They believed that young children learned language via a process of trial and error and reinforcement—that children spoke nonsensical words and phrases and learned meaning and grammar by parental reinforcement of proper language. Chomsky showed that while the childhood learning of the meaning of words did follow this paradigm, the childhood acquisition of grammar did not. He showed that human language contains a universal grammar that is present in all individual languages and that the capacity for grammar is an inborn trait and is not acquired by children after birth by simple trial and error. In other words, the specific words that we use are learned, but our knowledge of the structure of human language is not; it is innate.

Human beings are born with a remarkable capacity for syntax that allows human language to express an infinite range of ideas. In the words of nineteenth-century linguist Wilhelm von Humboldt, grammar is what enables us "to make infinite use of finite means."[15] Chomsky demonstrated that human beings are born with a "language organ"[16]—an inherent capacity for grammar—that all other animals lack.

Chomsky's view supports the inference that humans differ profoundly from animals in their capacity for unlimited abstract thoughts organized by grammar. Human beings use grammar to make infinite use of finite means. Chomsky stresses that in man, language is not primarily a means of communication. It is essential to our ability to think as human beings—to our ability to conceive an unlimited range of concepts. Language is the tool by which human beings think in a uniquely human way, and (in my view) it is a hallmark of the immaterial nature of human reason.

Thomistic Dualism and Materialism in Modern Neuroscience

The seminal research in neuroscience in the past century clearly supports a dualist view of the mind-brain relationship. Materialism fails logically, and it fails as a framework for understanding experiments. The brain can be split in half (Sperry), but abstract thought and one's sense of self are not split in half. Abstract reasoning cannot be evoked from the material brain by electrical stimulation or by seizures (Penfield). Higher abstract thoughts are not localized to particular regions of the brain (phrenology), but language is an abstract, inborn ability of human beings (Chomsky), and severe brain damage and even death do not ablate the mind (Owen and NDE research). Free will can be demonstrated experimentally (Libet and Penfield). A useful synopsis of the dualist interpretation of neuroscience suggests three experimentally demonstrated properties of the human mind:

1. The mind is metaphysically simple (it cannot be split).

2. The intellect is immaterial.

3. Free will is real.

There are several varieties of dualism, and I believe the type of dualism most consistent with logic and with modern neuroscience is Thomistic dualism. Thomistic dualism is the theory of the soul developed by Thomas Aquinas in the thirteenth century. It is Aristotelian in outline. The soul is the Aristotelian form of the body, and the mind is several powers of the soul. Some powers of the mind are closely linked to matter, such as sensation, perception, imagination, memory, and emotion. This explains the significant dependence of many mental abilities on brain function.

But some powers of the mind—the intellect and the will—are not generated by the brain, although ordinarily they depend on the brain for their normal function. For example, our ability to do mathematics is an immaterial power of the mind, but if we cannot see or hear or remain alert (which are material powers), we are not likely to acquire or exercise mathematical skills. The brain is necessary and sufficient for perception, imagination, emotion, and memory, and the brain is ordinarily necessary, but not sufficient, for intellect and will. Our material and immaterial powers of mind work in concert.

Thomistic dualism is the cornerstone of the Christian understanding of the human soul, and it is an excellent framework for modern neuroscience. It is noteworthy that Thomistic dualism naturally points to the spiritual nature of the human soul—immaterial reason and will, which are created in God's image—and points to the immortality of the soul. Immaterial intellect and will are not destroyed by death, for the dissolution of the matter of the body does not dissolve the immaterial soul, which is not ineluctably linked to matter. In Thomist terms, the human soul is subsistent and survives the death of the body.

The application of Thomistic dualism to neuroscience can be understood by considering Aristotle's four causes (formal, final, material, and efficient). To explain the neuroscience of my typing this chapter, my intellect is the formal cause, my will to do this work is the final cause, my brain and nerves and muscles are the material cause, and the physiology of my nervous system (neurotransmitters, action potentials, etc.) are the efficient cause. Neuroscience ordinarily studies material and efficient causes. Psychology and ethics study formal and final causes. While a materialistic theory of mind forces us to limit our understanding to mere matter, Thomistic dualism allows us to understand not only matter and its relations but also abstract thought and purposes inherent to human acts.

Thomistic dualism allows us to study all aspects of neurobiology without denying the reality of reason and free will. Unlike materialism, Thomistic dualism allows us to study human nature comprehensively without forcing impoverished materialist theories on neuroscience.

Does the Big Bang
Support Cosmic Design?

Brian Miller

A large percentage of physicists and astronomers at the beginning of the twentieth century assumed that the universe was eternal. Many preferred the idea of an eternal universe for reasons that were articulated by science historian Simon Singh:

> An eternal universe seemed to strike a chord with the scientific community, because the theory had a certain elegance, simplicity and completeness. If the universe has existed for eternity, then there was no need to explain how it was created, when it was created, why it was created or Who created it. Scientists were particularly proud that they had developed a theory of the universe that no longer relied on invoking God.[1]

Then a series of discoveries shattered this belief. Scientists came to realize that the universe had a beginning, a point at which all of time, space, matter, and energy came into existence—a point referred to as the big bang. This new understanding of our origins sent shockwaves not only through the sciences,

but they reverberated throughout the hallowed halls of departments of philosophy and even theology. The jarring revelation to many scientists was that an absolute beginning to our universe pointed to something outside of physical reality that started everything, and that something had the characteristics of the God of the Bible.[2]

Historical Background

The story begins with a German physicist named Albert Einstein, who developed the theory of special relativity. Einstein discovered that time and lengths measured in some frame of reference depend on the velocity of the observer making the measurement with respect to the reference frame under study. A classic thought experiment demonstrating this principle is an astronaut traveling in a spaceship approaching the speed of light. The astronaut could travel to a distant planet and back in a duration of time that he perceived to last for only a year. In contrast, those on Earth could have perceived his trip to have lasted for decades or even centuries.

Einstein expanded his initial insights

into the theory of general relativity, which started from the assumption that an observer of mass m in some frame of reference, such as an elevator, cannot distinguish between the reference frame accelerating with an acceleration of a and his experiencing a gravitation force of magnitude ma. This assumption led to the famous equations of general relativity, known as the field equations, that depicted gravity not as a force but as the effect of the curvature of space resulting from a mass.

A common illustration is that of a trampoline with a heavy bowling ball in the middle. The mass of the ball causes an indentation in the trampoline. The resulting curvature of the trampoline causes lighter objects moving near the bowling ball to bend toward it. For instance, the indentation could direct a moving ping-pong ball to circle around the bowling ball. In like manner, a large mass curves space in such a way that objects move toward it. A famous description of such interaction between space and matter was given by physicist John Wheeler: "Space tells matter how to move, and matter tells space how to curve."[3]

Einstein applied his theory to the entire universe and discovered that the universe should tend to expand or contract. Yet, he, like many of his contemporaries, assumed that the universe was eternal and static—it always stayed the same. Based upon this belief, he postulated a "cosmological constant" that acted against the force of gravity. In modeling the universe, he assigned the constant just the right value to counterbalance the effect of gravity, which would otherwise have caused the universe to shrink and eventually collapse into a dense ball. He thus provided a mathematical description for a static (unchanging) universe. Other physicists solved Einstein's field equations for the universe, and they pointed out how

the mathematics mandate that the universe should either expand or collapse. A static universe is too unstable to persist.

Einstein was at first reluctant to abandon his static universe, but decisive evidence discrediting a static universe came through the research of American astronomer Edwin Hubble. Hubble discovered that galaxies farther away from the Earth traveled away (receded) with greater velocities, and the recession velocity was directly proportional to the distance from Earth. This observation implied that our universe was in fact expanding. Consequently, the universe in the increasingly distant past must have been smaller and smaller until it began in an infinitely dense, vanishingly small volume. Both the beginning of time and the infinitely dense, vanishingly small volume are referred to as singularities. Einstein eventually described his proposal of the cosmological constant maintaining a static universe as the greatest blunder of his life, for he could have been the first to predict the big bang.

Responses to Big Bang

The theological implications of the universe having a beginning were immediately recognized and resisted. For instance, physicist Arthur Eddington responded:

> Philosophically the notion of a beginning of the present order is repugnant to me. I should like to find a genuine loophole. I simply do not believe the present order of things started off with a bang…it leaves me cold.[4]

In fact, the term *big bang* was coined by physicist Fred Hoyle to express his derision at the idea of a beginning. As a quirk of history, the term stuck. Hoyle acknowledged that his resistance was due to the theory's religious implications.

For others, the evidence dramatically affected their entire view of reality. For instance, it led famed astronomer Allan Sandage to embrace belief in God. At a conference featuring dialogues between atheistic and theistic scientists, Sandage explained the big bang theory's religious implications:

> Here is evidence for what can only be described as a supernatural event. There is no way that this could have been predicted within the realm of physics as we know it…science, until recently, has concerned itself not with primary causes but, essentially, with secondary causes. What has happened in the last fifty years is a remarkable event within astronomy and astrophysics. By looking up at the sky, some astronomers have come to the belief that there is evidence for a "creation event."[5]

Physicist Robert Jastrow described the ramifications of the evidence for a beginning in ever starker terms:

> This is an exceedingly strange development, unexpected by all but the theologians. They have always accepted the word of the Bible: In the beginning God created heaven and earth…The development is unexpected because science has had such extraordinary success in tracing the chain of cause and effect backward in time. For the scientist who has lived by his faith in the power of reason, the story ends like a bad dream. He has scaled the mountains of ignorance; he is about to conquer the highest peak; as

he pulls himself over the final rock, he is greeted by a band of theologians who have been sitting there for centuries.[6]

Early Attempts to Escape the Beginning

Unsurprisingly, many academics have attempted to overturn the conclusion of a beginning through the most creative of means. One of the first attempts was by Fred Hoyle, physicist Thomas Gold, and mathematician Hermann Bondi, who constructed a "steady-state" model for an eternal universe. They acknowledged that the galaxies were receding from each other, but they circumvented the need for a beginning by proposing that matter was constantly being created between the galaxies. The new matter eventually coalesces into galaxies so that the density and other features of the universe never change. The trio believed that this process could have been occurring indefinitely into the past. Ultimately, their model was rejected by the mid-1960s because it predicted that galaxies of all ages should be observed throughout the universe, but astronomers have only identified galaxies of middle age or older.

The next attempt to eliminate the beginning was the oscillating-universe model. This model postulates that the universe expands until it reaches a maximum size. Then it contracts due to gravity until it shrinks to a sufficiently small size that it, through some unknown mechanism, undergoes a cosmic bounce and then expands again. This cycle is said to repeat itself eternally. The oscillating model was eventually rejected due to the problem of entropy.

Specifically, the entropy (disorder) of the universe increases continuously. Consequently, each cycle would end up lasting a longer period of time. Looking backward in time, earlier cycles would last for shorter and shorter periods until the period would reduce to zero, which is physically impossible.

Eternal Inflation and String-Landscape Models

Another approach to avoiding a beginning is proposing that our universe is just one of an infinite multitude that are being continuously generated. Different theories propose distinct universe-generating mechanisms. For instance, physicists Alan Guth, Paul Steinhardt, and others envisioned a vacuum energy permeating space that drives a very rapid expansion. A miniscule volume of space could expand over a trillion trillion times in the tiniest fraction of a second. Different patches of space stop expanding due to a drop in the vacuum energy, and they then emerge as "bubble universes" that expand according to the traditional big bang model. Our universe is believed to be merely one of these bubble universes, and its beginning was the point where one particular patch of space stopped inflating. In principle, the inflation process could have continued indefinitely into the past, thus removing the need for a beginning.

Another mechanism is based on the application of string theory to the origin of our universe in what are termed *string-landscape* models. String theory attempts to unify all the forces and subatomic particles into a collection of strings vibrating with different frequencies. In one model, multidimensional "branes" are envisioned to inhabit a higher-dimensional space, and these branes occasionally collide, generating big bang events. This process might also have been occurring eternally into the past.

The hope that such universe-creating

mechanisms could bypass a cosmic beginning came to a crashing end due to a theorem developed by physicists Arvind Borde, Alan Guth, and Alexander Vilenkin. The Borde-Guth-Vilenkin (BGV) theorem states that any universe that has been expanding, on average, throughout its history must have had an absolute beginning. This constraint applies to inflationary, string landscape, and any other plausible model that could possibly generate our universe. The theorem's conclusiveness was best explained by Vilenkin:

> With the proof now in place, cosmologists can no longer hide behind the possibility of a past-eternal universe. There is no escape; they have to face the problem of a cosmic beginning.[7]

A Universe from Nothing

Atheist cosmologists had one last trick to play from the field of quantum cosmology. This domain of physics attempts to apply quantum mechanics to general relativity (i.e., gravity) in the context of the early universe. In standard quantum mechanics, a mathematical expression known as the *Schrödinger equation* is derived, and it can be solved to generate an equation known as the *wave function*. The latter equation yields the probability that some particle or system of particles has a particular value or range of values for such properties as momentum or position.

In quantum cosmology, a more complex expression known as the *Wheeler-Dewitt equation* is derived, and it can be solved to generate a universal wave function. This function, in like manner, yields the probability for a universe appearing with particular gravitational and mass properties. Physicists such as Stephen Hawking and Laurence Krauss have asserted that the mathematics behind solving the wave function demonstrate how our universe did not necessarily have a beginning, and they argue that our universe could have appeared from "nothing." Yet both of these claims are incorrect.

With regard to the first claim, Hawking solved the wave function using a mathematical trick where the time variable was replaced with imaginary time. The exact details are not crucial to understand. This substitution not only enabled him to solve the wave function, but it also eliminated the beginning of time in his analysis because the original time variable was replaced. In describing his work, Hawking declared that he had eliminated the need for God to explain the origin of the universe:

> So long as the universe had a beginning, we would suppose it had a creator. But if the universe is really completely self-contained, having no boundary or edge, it would have neither beginning nor end; it would simply be. What place, then, for a creator?[8]

In reality, Hawking's mathematical trick altered the equations in such a way as to disassociate the new time variable from anything real[9] in the physical universe. More importantly, at the end of his calculation, he transformed back into real time, at which point the beginning of the universe reemerged. Hawking even admitted this point:

> When one goes back to the real time in which we live, however, there will still appear to be singularities...In real time, the universe has a beginning and an end at singularities that form a boundary to space-time and at which the laws of science break down.[10]

His boast about eliminating the need for God is entirely duplicitous.

With regard to the second claim, Hawking stated:

> Because there is a law such as gravity, the universe can and will create itself from nothing...Spontaneous creation is the reason there is something rather than nothing, why the universe exists, why we exist. It is not necessary to invoke God to light the blue touch paper and set the universe going.[11]

Krauss made a similar assertion based on the work of Alexander Vilenkin: "The laws themselves require our universe to come into existence, to develop and evolve."[12]

In reality, neither Hawking's nor Vilenkin's research demonstrates that the universe could emerge out of nothing purely due the laws of physics. The underlying mathematics actually presuppose an already-existing universe. In other words, the "nothing" from which the models start is not a literal nothing but a universe that has already begun, albeit with zero spatial volume, after the big bang event. Both Hawking and Krauss are like a magician who is attempting to amaze the crowd by pulling a rabbit out of his hat, but observant viewers can see the rabbit's tail sticking out of the hat's bottom before the trick begins.

At a more philosophical level, the very claim that a physical law can do anything represents a serious error. Physical laws simply describe what occurs in an existing universe when existing masses and existing forces interact. Before a universe begins, the physical laws can be described by mathematical expressions, but the mathematics have no power to cause any event to take place. For instance, the law of gravity can be represented by the equation force equals mass times the gravitational acceleration: F=mg. This relationship describes how the gravitational attraction of a planet pulls an object toward itself, but the equation representing the interaction does not cause the planet, the object, or the gravitational pull to come into existence. The same holds true for the wave function and a universe it could potentially describe.

Mind Before Matter

The cause of the universe must have existed before the beginning of matter, energy, space, and time. Therefore, it must be immaterial, timeless, and immensely powerful. As mentioned, it could not have been some physical law in a higher-dimensional physical space, for that space would also require a beginning, as determined by the BGV theorem. It also could not have been some eternal lawlike force. In such a case, the circumstances that would cause the impersonal force to create a universe would have occurred infinitely far in the past because all chance circumstances would have transpired in an eternal reality. Therefore, the universe should be infinitely old, which it is not. The only remaining universe-generating cause is a mind, for only a mind with free will can choose to act (e.g., create a universe) at a given moment without any previous set of conditions causing it to do so.[13]

The necessity of a mind is further supported by physicists who have pondered the possibility of the mathematics describing the laws of nature having the capacity to create reality. Proponents of this hypothesis have had to eventually face the question of what grants math such creative power. For

instance, Hawking famously asked, "What is it that breathes fire into the equations and makes a universe for them to describe?"[14] Vilenkin also pondered this question, which led him toward the conclusion of a mind:

Does this mean that the laws are not mere descriptions of reality and can have an independent existence of their own? In the absence of space, time, and matter, what tablets could they be written upon? The laws are expressed in the form of mathematical equations. If the medium of mathematics is the mind, does this mean that mind should predate the universe?[15]

Identifying the Creator

In the following chapters, we will look at how the laws of physics and the initial conditions of the universe were carefully fine-tuned to allow for life. In addition, the origin of life can only be explained by an intelligent agent infusing matter with the needed information to construct the first cell. And, our planetary system appears designed not just for life but for scientific investigation. Consequently, our planet was particularly crafted for the benefit of humans. The evidence that the universe had a beginning, coupled with this additional evidence, points to the cause of the universe being an immaterial, infinitely powerful, personal being that must exist outside the universe. And, this being created the universe, our local environment, and life with the ultimate intent of allowing humans to flourish and to advance scientifically. This description is in perfect harmony with the God of biblical tradition.[16]

21

How Does Fine-Tuning Make the Case for Nature's Designer?

Hugh Ross

The evidence for design in the natural realm has long been a favorite argument for God's existence. Though in the past it has been criticized for its lack of rigor and thoroughness, the design argument has consistently remained among the most compelling scientific arguments for God. Evidence for design is clear, concrete, and measurable. Furthermore, design evidence has been increasing at an exponential rate throughout the past several decades. Nowhere is this design evidence more straightforwardly visible and expansive than in the "simple" sciences—specifically, astronomy and physics.

Cosmic Fine-Tuning Evidence

At least 50 books describing and documenting what is often called *the cosmological anthropic principle* have been published, most of them written by research scientists with no declared religious commitment. These books all identify and affirm a host of characteristics of the universe, including that the laws of physics themselves must be narrowly fine-tuned for any possibility of physical life in the universe, much less any possibility of *human* life. This fine-tuning implies not only unfathomable intellect and specific intent, but also inestimable power.

Although I would gladly recommend that you read these 50-plus books, the following paragraphs offer a sampling—a baker's dozen—of the more striking discoveries they describe. The first 12 lead up to one of the most profound measurable lines of evidence to come forth in the past few decades and the implications of the ongoing research surrounding it. Here they are, in no particular order:

1. In the first minuscule fraction of a second of cosmic existence, the universe contained a ratio of about 10 billion plus 1 nucleons (protons and neutrons) to every 10 billion antinucleons. Immediately, the antinucleons annihilated the nucleons, generating an enormous burst of energy. Eventually, the nucleons left over and dispersed from this blast formed all the galaxies and

stars that make up the universe today. If this initial slight excess of nucleons over antinucleons were any smaller, the universe would lack the ordinary matter (matter comprised of protons, neutrons, and electrons) necessary for the subsequent formation of galaxies, stars, and heavy elements essential for life. If the excess were any greater, any galaxies that formed would condense and trap radiation so efficiently and tightly as to prevent the formation of individual stars and planets.

2. A neutron is 0.138 percent more massive than a proton. This extra mass means neutrons require slightly more energy to be produced than do protons. So, as the universe cooled from the intensely energetic initial burst, it produced fewer neutrons than protons—in fact, about seven times fewer. If the neutron were just another 0.1 percent more massive, so few neutrons would remain from the cooling down of the early event that there would be too few available to serve as part of the nuclei of all the heavy elements essential for life.

3. The extra mass of the neutron relative to the proton also determines the rate at which neutrons decay into protons and protons combine with electrons and antineutrinos to form neutrons. If the neutron were 0.1 percent less massive, so many protons would end up as neutrons that all the stars in the universe would have rapidly collapsed into either neutron stars or black holes.[1] Thus for life to be possible in the universe, the neutron mass must be fine-tuned to better than 0.1 percent.

4. Another decay process involving protons must also be fine-tuned. Protons are believed to decay into mesons (a type of fundamental particle). I say "believed" because the decay rate is so slow that experimenters have yet to record a single decay event (average decay time for a single proton exceeds 4×10^{32} years). Nevertheless, theoreticians are convinced that protons do decay into mesons, and at a rate fairly close to the current experimental detection limits. If protons decayed any more slowly into mesons, the universe of today would have too few nucleons to make the galaxies, stars, and planets essential for our existence.[2]

Here's why: The factors that determine this decay rate also determine the ratio of nucleons to antinucleons in the earliest moments of the universe. Thus, if the decay rate were slower, the number of nucleons would have been too closely balanced by the number of antinucleons, thus too few nucleons would have been left to fulfill their essential role. If, however, the decay rate of protons into mesons were faster, the amount of energy released in this particular decay process would either seriously harm or destroy life. Thus, the decay rate must be fine-tuned.

5. The universe must be fine-tuned to provide not only a sufficient quantity of nucleons, but also a precise number of electrons. Unless the number of electrons is equivalent to the number of protons to an accuracy of one part

in 10^{37} or better, electromagnetic forces in the universe would have so overcome gravitational forces that galaxies, stars, and planets would never have formed. One part in 10^{37} is such an incredibly small number that it may be hard to visualize. So, consider this scenario: Cover the entire North American continent in dimes stacked to the distance to the moon, roughly 239,000 miles. Now repeat this stacking of dimes on a million additional continents the size of North America. Only one of all these dimes has been painted red and mixed in with the rest. Finally, blindfold someone and ask that person to pick out just one dime from the entire stack. The odds that this person will pick the red dime are one in 10^{37}.

6. Life molecules cannot exist unless sufficient quantities of the elements that comprise them exist. For these elements to exist requires a delicate balance among the constants of physics governing the strong and weak nuclear forces, gravity, and the nuclear ground state energies (quantum energy levels important for the forming of certain elements from protons and neutrons). For the strong nuclear force—the force governing the degree to which protons and neutrons stick together in atomic nuclei—the balance is easy to recognize. If this force were too weak, multi-proton nuclei would not hold together, and, thus, hydrogen would be the only element in the universe. But if the strong nuclear force were slightly stronger, protons

and neutrons would have such an affinity for one another that none would remain alone. They would all attach to other protons and neutrons, meaning no hydrogen would exist. Life chemistry is impossible without hydrogen and impossible if hydrogen is the only element.

How delicately balanced is the strong nuclear force? If it were just 4 percent stronger, diprotons (atoms with two protons and no neutrons) would form. Diprotons would cause stars to exhaust their nuclear fuel so rapidly as to make any kind of physical life impossible. On the other hand, if the strong nuclear force were just 10 percent weaker, carbon, oxygen, and nitrogen would be unstable, and, again, any conceivable kind of physical life would be impossible.[3]

7. As strange as it may seem, the strong nuclear force is both the strongest attractive force in nature and the strongest repulsive force. The fact that it is attractive on one length scale and repulsive on another length scale makes it highly unusual and counterintuitive. Nevertheless, without its weird properties, life would be impossible. For life to be possible, the strong nuclear force must be attractive over lengths no greater than 2.0 fermis and no less than 0.7 fermis (one fermi = a quadrillionth of a meter) and maximally attractive at about 0.9 fermis.[4] At lengths shorter than 0.7 fermis, the strong nuclear force must be strongly repulsive. Here's why: Protons and neutrons are packages

of more fundamental particles called *quarks* and *gluons*. If the strong nuclear force were not strongly repulsive on length scales below 0.7 fermis, the proton and neutron packages of quarks and gluons would merge. Such mergers would mean no atoms, no molecules, and no chemistry would ever be possible anywhere or any time in the universe. As with the attractive effect of the strong nuclear force, the repulsive effect must be exquisitely precise, both in its length range of operation and its strength of the repulsion.

8. In the case of the weak nuclear force—the force that governs, among other things, the rates of radioactive decay—if it were much stronger than what we observe, all ordinary matter in the universe would quickly be converted into heavy elements. But if it were much weaker, the only remaining ordinary matter in the universe would be just the lightest elements. Either way, the elements essential for life chemistry (carbon, oxygen, nitrogen, phosphorus, etc.) would either be nonexistent or would exist in amounts far too small for the assembly of life chemistry. What's more, unless the weak nuclear force were delicately fine-tuned (to better than one part in 10,000), those life-essential elements produced only in the cores of supergiant stars would never escape the confines of those cores because supernova explosions would be impossible.[5]

9. The strength of the gravitational force determines how hot the nuclear furnace at the core of a star burns.

If the gravitational force were any stronger, stars would be so hot as to burn up relatively quickly—too quickly and too erratically to provide what life on a planet orbiting a star requires. Any planet capable of sustaining life must be supported by a star that is both stable and long-burning. Then again, if the gravitational force were any weaker, stars would never become hot enough to ignite nuclear fusion. In such a universe, no elements heavier than hydrogen and helium would ever be produced.

10. In the late 1970s and early 1980s, cosmologist Fred Hoyle discovered that the nuclear ground state energies for helium, beryllium, carbon, and oxygen required exquisite fine-tuning for any kind of physical life to exist. If the ground state energies for these elements were any higher or lower with respect to each other by more than 4 percent, the universe either would be devoid of carbon and oxygen or would yield insufficient quantities of either oxygen or carbon for life.[6] Hoyle, who has in the past expressed his opposition to theism[7]—and to Christianity in particular[8]—nevertheless concluded on the basis of this quadruple fine-tuning that "a superintellect has monkeyed with physics, as well as with chemistry and biology."[9]

11. In 2000, a team of astrophysicists from Austria, Germany, and Hungary demonstrated that the level of design for electromagnetism and strong nuclear force actually exceeds by far what physicists

had previously determined.[10] The team began by noting that for any conceivable kind of physical life to be possible in the universe, certain minimum abundances of both carbon and oxygen must be present. Next, they pointed out that the only astrophysical sources of significant quantities of carbon and oxygen are red giant stars. (Red giant stars are large stars that, through nuclear fusion, have consumed all their hydrogen fuel and subsequently engage in the fusion of helium into heavier elements.) The team mathematically constructed models of red giant stars using slightly different values for the strong nuclear force and electromagnetic force constants. By this means they discovered that tiny changes in the value of either constant leads to insufficient production of carbon, oxygen, or both. Specifically, they determined that if the value of the coupling constant for electromagnetism were 4 percent smaller or 4 percent larger than what we observe, the carbon and oxygen essential for life would not exist. In the case of the coupling constant for the strong nuclear force, if it were 0.5 percent lesser or greater, life would be impossible.

12. These new limits on the strength of the electromagnetic and strong nuclear forces provide much tighter constraints on the quark mass and on the value of the Higgs vacuum expectation.[11] Without getting into the technical details of the Higgs vacuum expectation value and

quarks, I can say that the new limits demonstrate an ever greater degree of fine-tuning not only in the physics of stars and planets but also in the physics of fundamental particles.

The Most Spectacular Fine-Tuning Yet Measured

Several of the 140 fine-tuned features of the universe listed in the online "design compendium" I've been assembling for a few decades now reveal a level of precision that exceeds by far what we humans have been able to achieve in our design efforts.[12] One feature that stands out above the rest is a mysterious entity called *dark energy*. According to astronomer Lawrence Krauss, the very existence of dark energy (the energy embedded in the cosmic space surface) "would involve the most extreme fine-tuning problem known in physics."[13] Of course, dark energy represents a *problem* only for those who choose to reject the reality of a fine-tuner.

To make life possible, and especially to meet the needs of advanced life, the universe must expand at just-right rates throughout the entirety of its history. For any galaxies, stars, and planets to appear, ever, requires exquisite fine-tuning of the cosmic expansion rate. If the universe were to expand too rapidly from its origin event, gravity would be insufficient to pull together the primordial gas into the clumps that eventually become galaxies, stars, planets, asteroids, and comets. If the universe were to expand too slowly, gravity would gather all the primordial gas into black holes and neutron stars. A just-right cosmic expansion rate for life requires, in turn, exquisite fine-tuning of both the cosmic mass density and the dark energy density.

The cosmic mass density has a doubly significant impact. It factors into the rate at which the universe expands, and it establishes what kinds of elements will come to exist in the universe. It determines how much of the universe's primordial hydrogen converts into helium through nuclear fusion during the first few minutes of cosmic history. If this process yields too little helium, future stars will fail to produce any elements heavier than helium. If it yields too much helium, future stars will convert all their hydrogen and helium into elements as heavy as iron or heavier. In both scenarios, the carbon, nitrogen, oxygen, phosphorus, and potassium that physical life requires will be nonexistent.

Only through the exquisitely precise fine-tuning of the cosmic mass density can future stars produce all the elements life requires—and in the just-right ratios of abundance that life requires. Cosmic mass, by itself, however, cannot generate the exacting and variable cosmic expansion rates that produce, with just-right timing, all the galaxies, stars, planets, asteroids, and comets with the appropriate spatial distribution necessary for the possible existence and survival of advanced life. Dark energy must come to the rescue, filling in a critical gap in the process.

For this dark energy to play its part, however, it must be fine-tuned with vastly greater precision than even the cosmic mass density requires—not once, but twice! Its initial value must be extremely high, and then, within a split second after the cosmic origin event, drop to an extremely low value. By any accounting, the original density of dark energy must have been at least 122 orders of magnitude greater than the density that is detected today. Its original density must be this large to explain the cosmic inflation event that occurred between 10^{-35} and 10^{-32} seconds after the cosmic origin event, the moment when the universe expanded at millions of times the velocity of light. (Cosmic

inflation results from the symmetry breaking that occurs when, due to cosmic cooling, the strong-electroweak force of physics separates into strong nuclear force and the electroweak force. If the symmetry breaking had not occurred with its accompanying cosmic inflation, life would be nonexistent.[14]) This difference between the initial and current dark energy density implies that during the first tiny fraction of a second of cosmic existence, all but just one minuscule part (one in 10^{122}) was somehow canceled out. (This cancellation event is roughly analogous to the annihilation of nucleons by antinucleons, a cancellation effect that left a just-right quantity of nucleons in the universe.)

The precision represented by one part in 10^{122} ranks as one of the most spectacular fine-tuning detections to date. If one were to compare the fine-tuning of dark energy density with the greatest fine-tuning yet achieved by humans, the Laser Interferometer Gravitational-Wave Observatory (LIGO),[15] the fine-tuning of dark energy would rank 10^{99} times superior—a factor of one thousand trillion trillion trillion trillion trillion trillion trillion times superior.

What does this capacity for the precision of both design and implementation tell us about the source of our universe? First, design and implementation require intelligence, knowledge, and intention and capacity (power and resources), all of which are attributes of personhood, not of some nebulous force or forces. Second, this "person" is, at a minimum, some 10^{99} times more intelligent and capable, not to mention intentional, than any human or collaboration of humans. The fine-tuning observed and measured in dark energy, not to mention the fine-tuning of well over 100 other cosmic features and the laws of physics, leads to the reasonable conclusion that an omniscient, omnipotent, personal being caused the universe for a purpose that includes human existence.

Responding to Familiar Challenges

Pushback to this conclusion of the cosmic fine-tuner as a personal being often comes from those who question the necessity of such an immense and ancient universe if, indeed, its purpose is to provide a habitat for humans. Physicist Stephen Hawking wrote in his bestselling book *A Brief History of Time*, "Our solar system is certainly a prerequisite for our existence…But there does not seem to be any need for all these other galaxies."[16] Particle physicist Victor Stenger, in *God: The Failed Hypothesis*, quipped, "If God created the universe, he wasted huge amounts of space, time, matter, and energy."[17]

As the research previously described reveals, in a less massive universe, the periodic table would consist of just one or two elements: hydrogen and helium. Life would be impossible. In a less massive universe, supernovae and neutron star merger events would be absent, and yet from these come all elements heavier than iron, elements essential to life, such as cobalt, copper, zinc, arsenic, selenium, and iodine.

A reasonable explanation likewise can be offered to the question about age, given that billions of years pass between the origin of the universe and the appearance of life on Earth, and even more years before the appearance of human life. The answer, again, goes back to how the elements essential for life's formation and protection come into existence, and in sufficient quantities. At least three generations of star "birth" and "death" are required. Stellar furnaces must yield not only the components of life's buildings blocks but also such elements as uranium and thorium,

which make possible the formation of an enduring, powerful magnetic field around a planet. Only such a shield can protect life on a planet from deadly radiation and prevent the planet's atmosphere from sputtering away. Only a planet super-endowed with uranium and thorium can sustain plate tectonics sufficient to meet the complex needs of advanced life.[18]

The universe today also happens to be neither too young nor too old to provide astronomers with a clear view of the totality of cosmic history. If the universe were slightly younger, radiation from the cosmic origin event wouldn't have traveled far enough yet along the cosmic space surface to reach our telescopes. If the universe were slightly older, the accelerating expansion caused by dark energy would have sent us beyond the reach of visible radiation from the cosmic origin event.

It might seem a wondrously improbable coincidence that we humans exist at the one time in cosmic history that allows us to witness 100 percent of cosmic history. But it seems too far a plausibility stretch that we humans also reside in the one advanced-life conceivable location in the vast cosmos where all of cosmic history can be observed and studied.[19] Apparently, the universe has been designed not only to make our existence possible but also to reveal the glorious attributes of the One who so carefully and caringly formed it for our specific benefit.

Fine-Tuning at All Size Scales

Whether observing through the most powerful deep-space telescope or the most exacting electron microscope, we see compelling evidence of fine-tuning. Our Laniakea supercluster of galaxies is like no other yet known and uniquely appropriate for the needs of advanced life. Our Local Group of galaxies is the only known galaxy cluster in which an advanced-life-supporting galaxy could reside. Our Milky Way Galaxy is the only known galaxy possessing each of the 100+ characteristics essential for advanced life to exist for any significant length of time. And, astronomers have yet to find another star sufficiently like the sun such that it could possibly host a planet on which advanced life can not only survive but also thrive.[20]

Ongoing research reveals that each of our neighboring solar system planets and our moon feature the precise—and rare—characteristics that make advanced life possible on Earth.[21] More than 400 distinct attributes of Earth must be fine-tuned for the sake of human life, and each of the millions of species of nonhuman life on Earth plays a role in enhancing the well-being of humans. Even if the observable universe holds as many as ten billion trillion planets, the probability of finding just one that's suitable for life appears absurdly—and increasingly—remote.

The Accumulation of Evidence

Over a 20-year period, Reasons to Believe staff and volunteer scientists have cataloged fine-tuning evidence for the survival and support of various kinds of life, but even with an expanding team, we cannot keep pace with the rapidity by which the published findings accumulate. In one ten-year period alone, from 1996 to 2006, conservative probability estimates for spontaneous incidence and/or coincidence of microbial-life-essential characteristics ranging from the gross features of the universe to the features of Earth shrank by a factor of 10^{525} times![22]

Research has yet to devote significant attention to how dramatically more precise

the fine-tuning must be for life to survive through millions or billions of years as opposed to a few months or years. Even less attention has been given to how much finer the tuning must be for the sake of human life as compared with unicellular life.

The fine-tuning already observed certainly suggests that the universe has been shaped not just for any kind of life, but for human life in particular. This idea can be more rigorously established through comparison of the fine-tuning required for the support of

1. ephemeral simple life (unicellular life that persists for 90 days or fewer),

2. permanent simple life (unicellular life that persists for 3 billion years or more),

3. intelligent physical life (human beings or their functional equivalent), and

4. intelligent physical life capable of launching and sustaining a global high-technology civilization.

Although the numbers have grown since the list was last updated, here is what the published journals surveyed tell us about the life requirements of any potential life site in the observable universe:

1. Fine-tuned characteristics recognized as essential for briefly existing bacteria = 501

2. Fine-tuned characteristics recognized as essential for long-lasting bacteria = 676

3. Fine-tuned characteristics recognized as essential for pre-Neolithic humans = 816

4. Fine-tuned characteristics recognized as essential for global high-tech civilization = 922

With the number of fine-tuned characteristics essential for life, especially human life, increasing, the probability of finding a potential life site without the intervention of an omniscient, omnipotent, personal being is shrinking. That probability figure is now so tiny as to represent a virtual impossibility.

A Closing Consideration

As I survey the accumulating evidence that a fine-tuner is responsible for the existence of intelligent physical life, an impression of purpose comes into sharper focus. In the last few years of researching the scientific literature for evidence that points to a creator, it has become increasingly apparent to me that every component of and event in the universe, Earth, and Earth's life plays some role in a grand plan to allow billions of human beings to not only exist but also recognize and respond to a gracious offer of redemption from all that causes devastation and death.

Nontheists may attempt to argue that the evidence of fine-tuning has no philosophical or theological significance. And yet they cannot help but admit that everything *appears* designed at some level. Even scientists who flatly reject the notion of a designer continue to write books and articles about this apparent design. With time, its source becomes increasingly difficult to ignore or explain from a naturalistic perspective.

The mystery of nature's exquisite fine-tuning draws us ever more deeply into researching its details. New instruments allow us to probe with ever-increasing precision the complexities we had previously missed. And, at every step along the way, the impression of purposeful design grows stronger, not weaker, giving us more reasons to believe our existence is no accident. We humans are creatures of destiny.

Do We Live on a Privileged Planet?

Guillermo Gonzalez

In his book *Pale Blue Dot*, Carl Sagan reflected on an image of Earth taken by the Voyager 1 space probe in 1990 as it turned its cameras back to take a "family portrait" of the planets from the edge of the solar system. He wrote,

> Look again at that dot. That's here. That's home. That's us. On it everyone you love, everyone you know, everyone you ever heard of, every human being who ever was, lived out their lives… Our posturings, our imagined self-importance, the delusion that we have some privileged position in the Universe, are challenged by this point of pale light. Our planet is a lonely speck in the great enveloping cosmic dark. In our obscurity, in all this vastness, there is no hint that help will come from elsewhere to save us from ourselves.[1]

Never mind that size and importance have no logical connection. Miners risk life and limb for a tiny precious diamond in tons of relatively worthless rock. When considering the status of Earth within the wider universe, a person can come to this question from one of two directions. Either take Sagan's view that we are an insignificant speck floating aimlessly in a universe that doesn't care about us, or take the view that we are a privileged species living on a privileged planet. The materialist takes the former view; everyone else takes the latter view.

Can the evidence of nature help us to choose between these views? If it is possible for observations of nature to uncover evidence of purpose, then, yes, it is possible. Many scientists today, like the late Carl Sagan, are of the view that science cannot address questions of design or purpose in nature. They believe any purported evidence of design in nature is illusory. As Richard Dawkins stated, "[b]iology is the study of complicated things that give the appearance of having been designed for a purpose."[2]

This is a departure from the views of the great scientists leading up to the twentieth century, including Johannes Kepler, Isaac Newton, Robert Boyle, Michael Faraday, James Clerk Maxwell, John and William Herschel, and Alfred Russell Wallace, among many others.[3] They argued that evidence of design and purpose can be discerned in the patterns of nature, either in specific instances such as the solar system, geology, living things or in the physical laws themselves.

The modern intelligent design movement has adopted a position very similar to this.

The Evidence

The gratuitous beauty of the starry heavens above. The chimerical appearance of rainbows. The austere splendor of solar eclipses. These inspire people in every time and place. For most of human history, they were also mysteries.

In a sense, we've now removed many of the mysteries. Scientists routinely measure the distances to the stars. We know how sunlight passing through millions of suspended water droplets in the atmosphere produces a rainbow. We can predict the timing and location of solar eclipses years in advance anywhere on Earth to within one second.

Yet these discoveries point to even deeper mysteries. *Why* is our world, from our local and galactic environment to the constants of physics, set up so that *we* can see the stars, rainbows, and solar eclipses? After all, our ability to see these things is not logically required for our existence. Surely, the universe could have been otherwise.

This is the question I set out to answer with my coauthor, Jay W. Richards, in our book *The Privileged Planet: How Our Place in the Cosmos Is Designed for Discovery.*[4] The subtitle gives away our argument. We argued not just that the universe is vaguely habitable and open to science, but that those rare places in the universe best suited for complex life and observers are also the best places, overall, for scientific discovery (hereafter, the PP thesis). Moreover, we argued that this is evidence of a cosmic conspiracy rather than a mere coincidence. We set out to test this hypothesis against the best evidence from natural science. Together, these discoveries—that is, observations—form a cumulative

case in favor of our hypothesis. We didn't try to immunize our argument against future discoveries, but rather, sought to put it at risk. That's why we predicted that relevant new observations (or new analyses of existing observations) would confirm our argument.

In the present book, the chapter "How Do Solar Eclipses Point to Intelligent Design?" describes a small bit of the evidence for the PP thesis. The topic of total solar eclipses was the first instance of the PP thesis that I discovered, and it is the topic of the first chapter in *The Privileged Planet*. Since the PP thesis is a cumulative case argument, it becomes stronger with each instance that is added. Given the limited space available, I will discuss only two additional instances in this chapter: the visibilities of rainbows and the stars.

In *The Privileged Planet*, I ended the chapter on solar eclipses with a teaser on rainbows. How are rainbows like total solar eclipses? They're both beautiful visual spectacles (eye candy!) requiring no special optical aids to enjoy. We don't need to be able to see either one for our survival. Rainbows are far more common than total solar eclipses, but rainbows are rare in regions of the world with infrequent rainfall. From my experience, I see several rainbows a year on average; whenever I'm with others when I do see one, I hear that they never tire of the sight. Both rainbows and total solar eclipses are fleeting phenomena, typically lasting minutes.

For a rainbow to appear, you need suspended water droplets in the atmosphere and the direct sunlight that results from the sun being between the horizon and 42 degrees altitude. This typically occurs just after a thunderstorm has passed and small droplets are still in the atmosphere, and the sky is clearing in front of the sun. Seems like a simple setup. This must be a common phenomenon in the cosmos, right?

Let's consider the other major bodies in the solar system—the planets and large moons. Mercury, the closest planet to the sun, and the moon, look superficially similar. This is because they lack atmospheres. Can't have rainbows there. Same goes for Mars. Even though it has a thin atmosphere, it's too dry. It never rains. On the other hand, Venus has too much atmosphere. It is perpetually cloud-covered, and its surface is bone-dry. The giant planets aren't options because they have no solid surfaces. What about their moons?

The only moon in the solar system with a substantial atmosphere is Titan. Its atmosphere is dominated by haze. From the ground, the sun would be faint and the sky orange and hazy. Hypothetical Titans would not see rainbows either.

Life requires liquid water. Liquid water in the form of precipitation is required to make rainbows. A rich biosphere powered by photosynthesis needs not only abundant water, but also abundant sunlight. Precipitation is important, too, in order to permit life to flourish on the continental interiors. The link between rainbows and life is obvious.

Yes, rainbows are beautiful, but are they good for anything? Indeed, they have been very important for science, but in subtle ways. Rainbows have long intrigued casual and careful observers alike, and they've prompted many questions. How are they produced? Are the colors of the rainbow already present in the sunlight? Is it possible to make an artificial rainbow in the lab? Over the centuries, these and other questions have spurred research in optics and the physics of light.[5]

Descartes and Newton performed early experiments with prisms to produce artificial rainbows. These were the first spectroscopes—instruments that spread light into its constituent colors, its *spectrum*. These

could be made to spread out the spectrum of any light source over a larger range of angles than a natural rainbow, permitting more detailed examination. Indeed, when astronomers pointed their spectroscopes at the sun, important new discoveries were made. In 1802, English chemist William Hyde Wollaston discovered a series of dark lines in the solar spectrum when viewed at high resolution; Joseph von Fraunhofer rediscovered them in 1814. We now know them as Fraunhofer lines.

The application of the spectroscope to the fields of astronomy and chemistry led to enormous leaps in our understanding of the universe. Spectroscopists learned that each chemical element has a unique spectrum, a kind of fingerprint. This powerful tool permitted astronomers to identify the chemical elements present in the sun's atmosphere. In 1864, William Huggins married a spectroscope to his telescope and obtained spectra of distant stars. They measured their compositions and other properties as well, such as temperature and density. Jules Janssen discovered the second most abundant element in the universe, helium, in the spectrum of the sun in 1868 during a total solar eclipse! Astronomers also learned that they could measure the velocity of a star along the line of sight using the Doppler effect.

It's as if someone has been trying to get our attention with a pretty shiny object writ large across the sky, saying, "Look here, stupid. This is important!" In other words, the beauty of rainbows attracted our attention, and when smart people thought carefully about them, our knowledge of nature expanded greatly. A partly cloudy atmosphere with precipitation gives us rainbows and permits us to observe solar eclipses. Our kind of atmosphere also permits us to view the distant stars.

Think of all that needs to go right for us to be able to see the stars. In addition to clear-to-partly-cloudy skies, the atmosphere must transmit to the ground the kind of light we can see. This is not only true for us, but it is precisely the same kind of light that plants need and that the sun produces in abundance. The sun's energy output peaks where our sight is most sensitive, where the biosphere is most productive, and where most of the energy is present that's needed to warm the Earth. It didn't have to be this way.

Although living in a brightly lit environment doesn't mean we *must* have eyes, vision has definite survival benefits. We have eyes because we were born creatures of the sun-lit world. However, if we lived in the dark underground, not only would we not see the sunlight, but we wouldn't have eyes. We would be like the blind cave critters that lost their sight long ago. We can see the stars because we can see sunlight. But that's only part of the story. If the sun was unlike other stars, we might not be able to see them. If life were only able to exist around an exotic type of star that emits different light than most other stars, we wouldn't see most other stars.

The Earth rotates much faster than it revolves around the sun. This means we can see the stars potentially every 24 hours for nearly 12 hours at a time from a typical location on Earth. But, we know that there are planets around other stars that are *tidally locked*. The situation is very much like the moon. Because its rotation period equals its period of revolution, the moon always presents the same face to us. Planets that orbit close to their host stars are likely to be tidally locked. The starlit side of a planet is more likely to be habitable (assuming the planet is habitable at all). It is unlikely that any inhabitants of such a world would know the stars.

The moon brightens our night skies and interferes with our observations of fainter celestial objects, but only for about two weeks each month. It could have been much worse had we lived in a multiple star system, or on a world with several large moons, or in a tightly packed multiple planet system around a red dwarf.[6] In such scenarios, again, we might never see the stars.

The stars themselves must also be bright enough for us to see them. Even if all the other conditions are met (e.g., atmosphere is transparent, stars emit the right kind of light, few nearby interfering bright light sources), we might still find ourselves with a rather boring night sky. We can see stars at night because enough stars are close enough and luminous enough and our vision is sensitive enough.

We reside in a giant structure known as the Milky Way galaxy. It is a flattened structure containing stars, gas, and dust. The Milky Way is about 100,000 light-years across and its disk, where we reside, is about 1,000 light-years thick. We currently live in a rather busy region of the Milky Way galaxy, near the mid-plane of the disk and about half-way out from the center to edge of the visible disk. The solar system orbits about the center of the galaxy in a slightly noncircular path and also bobs up and down, somewhat like a carousel horse on a merry-go-round. As we continue to move away from the mid-plane over the next few millions of years, fewer and fewer stars will grace Earth's night skies.

We also reside in a relatively dust-free region of the Milky Way. Interstellar dust blocks some of the light from distant objects within and beyond the galaxy. While stars are packed more tightly in the inner galaxy, observers there would have to deal with the greater amounts of dust. Their skies would have lots of bright stars, but the dust would block views of the more distant stars and other galaxies. Ironically, the brightness of the night sky on an inner-galaxy planet from all the bright stars would also block the views of fainter objects, somewhat like scattered light from the full moon in our sky interferes with astronomical observations.

What Does It Mean?

Does it matter that we can see the stars? Apart from aiding in navigation, being able to see the stars (and planets, moons, asteroids, nebulae, and galaxies) has not been crucial for our survival. It has, however, been of inestimable importance in our scientific understanding of the cosmos. Not only have we been able to place the Earth within its proper astronomical context, but we have discovered that there was a beginning to the universe. This has profound metaphysical implications.

Children often ask simple yet profound questions—for example:

- "How high do I have to jump before I won't come back down?"

- "Why is the sky blue?"

- "Why does ice float?"

- "Did time have a beginning?"

- "How do I know I'm not dreaming now?"

- "How can fish breathe underwater?"

Note that when a child asks *why* questions, we often respond as if they were *how* questions. For instance, when I answer the second question above, I begin by explaining the nature of light and the concepts of wavelength and how light interacts with molecules of nitrogen and oxygen in the air. But this explanation is treating the question as if it were a *how* question. Responding to

the inquiry as a *why* question might get into more philosophical ideas. Rephrasing the question would make this clearer: "Why do we live in the kind of universe where observers experience blue skies?" There might not be a unique answer we can give to these sorts of questions.

Similarly, to this list we can add the question "Why can we see the stars?" In this case, I literally do mean it as a *why* question. Scientists understand *how* it is that we can see the stars. But I'd like to know why we exist in a universe that permits us to see the stars. The same can be asked of rainbows and total solar eclipses. As I see it, there are only two possible species of answers to these and similar questions.

The first species begins with the philosophical stance called *methodological naturalism*.[7] In this view, only those explanations that follow from the physical laws are permitted in science. Intelligent agency is permitted as an explanation only insofar as it is understood to have itself arisen strictly naturally. But what happens when the physical laws themselves need explaining, when they display a surprising pattern? Most physicists and philosophers agree that the physical constants, laws, and cosmological initial conditions could have been otherwise. So why do these particular laws that permit us to enjoy views of rainbows, eclipses, and stars exist and govern our universe?

Proponents of methodological naturalism would argue, for instance, that the fine-tuning evident in physics and cosmology can be explained if our universe is but one of a huge number of universes (the multiverse).[8] The reason we observe the particular values of this universe is because they are necessary for our existence. Other universes don't have the right combination of constants, laws, and initial conditions for life. So they don't have observers. Multiple criticisms have been leveled against this idea,[9] but for the sake of argument, let's say it can explain the fine-tuning problem. Notice how the argument goes: invoke a vast array of unseen explanatory resources to account for us observing those conditions.

Now, back to our need to explain why there is this close linkage between observers and the visibility of rainbows, total solar eclipses, and stars. We *don't need* to observe these things; they are *not logically necessary* for our existence. They are merely physically necessary. Given this, appealing to the multiverse to explain the properties of this universe fails. The multiverse's observer self-selection bias only works in principle for properties that are necessary for our existence. Surely, there are vastly more universes that permit intelligent beings but that don't offer them the opportunities for scientific discovery that our special universe does.

The second species of answer to my question requires something different than just "more of the same." A natural explanation just won't cut it. How do we explain our universe? The three examples of the PP thesis I described earlier are but the tip of the proverbial iceberg. There are multiple additional examples in *The Privileged Planet*, and at least one additional example published since.[10] Together they form a simple pattern. The most habitable places for intelligent beings like us are also the best places for scientific exploration and discovery. Or, to put it more succinctly, the universe is set up for observers to discover its properties. To me, this sounds like the answer to a *why* question.

Philosophers of science would frame the argument within confirmation theory. The PP pattern is what we would expect if the universe is designed for scientific discovery, but not otherwise. The PP pattern confirms design. Not only that, but we can say the PP pattern is evidence of purpose. And, we know from uniform experience that artifacts that have purposes are uniquely the products of intelligent agency.

Once we accept this conclusion, other mysteries start to make sense. For instance, why are mathematics so effective at describing the universe? Eugene Wigner noted that "mathematical concepts turn up in entirely unexpected connections...they often permit an unexpectedly close and accurate description of the phenomena in these connections."[11] Why can we trust our minds to discover and understand the physical laws? Melissa Cain Travis provides an answer to this question with her Maker Thesis: "If there is a Maker behind the cosmos in whose image we are made, then the astounding success of the sciences is no mystery; the Christian theist should expect that his or her mind has a special kinship with a Mind who desires to be known through his creation."[12]

Once we take ownership of this ancient idea, then the practice of science makes perfect sense. Otherwise, it is just a happy accident. After all, the naturalist has no good warrant to believe in the scientific enterprise. Naturalists do science because it works, not because they expect it to work based on their naturalistic assumptions. If you don't believe this, consider the lessons from the history of science.

Auguste Comte, the French Enlightenment philosopher and founder of positivism, wrote in 1835 about what we can know about other planets and stars:

> We understand the possibility of determining their shapes, their distances, their sizes and their movements; whereas we would never know how to study by any means their chemical composition, or their mineralogical structure, and, even more so, the nature

of any organized beings that might live on their surface…I persist in the opinion that every notion of the true mean temperatures of the stars will necessarily always be concealed from us.[13]

In Comte's vision of a materialistic universe, there was no reason to assume a rationality of laws or conditions that suggested those laws applied uniformly throughout the universe. His failed predictions are especially interesting coming from the Enlightenment culture with its contempt for traditional religion. It is understandable why Comte failed in his predictions. Don't you have to stick a thermometer in something to determine its temperature and take it apart in a lab to determine its composition? After all, why would the information needed to determine the compositions and temperatures of stars and planets be encoded in their light? Comte was not a scientist, even though he wrote about the philosophical aspects of science. It is doubtful that science would have arisen as early as it did had Enlightenment philosophy taken hold earlier. Enlightenment philosophers inherited science from deeply religious people who came before them.

Modern science grew out of deeply religious men applying theological ideas about the natures of God, man, and nature. Historian of science James Hannam writes,

The starting point for all natural philosophy in the Middle Ages was that nature had been created by God. This made it a legitimate area of study because through nature man could learn about its creator. Medieval scholars thought that nature followed the rules that God had ordained for it. Because God was consistent and not capricious, these natural laws were constant and worth scrutinizing… The motivations and justification of medieval natural philosophers were carried over almost unchanged by the pioneers of modern science. Sir Isaac Newton explicitly stated that he was investigating God's creation, which was a religious duty because nature reflects the creativity of its maker.[14]

We've come full circle. The success of the institution of modern science tends to confirm its founding assumptions. And, now the PP pattern provides additional confirmation. This also gives us more confidence in identifying the designer. To summarize, belief in the biblical God motivated people to discover science, and modern practitioners of science uncovered evidence that science was intended from the beginning.

While many examples of the PP pattern given in *The Privileged Planet* had to await modern science for their discovery, the three examples I described above are accessible to anyone. No special equipment is required. People have marveled about them at least since ancient times. David wrote in Psalm 19:1 that nature reveals truths about God continuously to all peoples: "The heavens declare the glory of God, and the sky above proclaims his handiwork. Day to day pours out speech, and night to night reveals knowledge. There is no speech, nor are there words, whose voice is not heard."

We have a God-given mandate to learn about his creation, and we should have confidence that we can discover important truths about it. There are practical aspects about what we learn through science, but there are also sublime truths. We have discovered that the cosmos points beyond itself to its Creator.

23

How Do Solar Eclipses Point to Intelligent Design?

Guillermo Gonzalez

On Monday, February 26, 1979, my best friend and I stayed home from school. We weren't exactly playing hooky, because on that day we observed a solar eclipse in Miami, Florida. We used a small refracting telescope with eyepiece projection, which provided a safe means of viewing the eclipse. We watched as the moon covered about 40 percent of the sun. While we were excited to witness the partial solar eclipse, we knew that diagonally opposite from us in the Pacific Northwest, observers were also enjoying views of the totally eclipsed sun. For them, the day turned to almost night, and in the sky, they saw a black disk surrounded by the sun's otherwise invisible corona. In contrast, everything around us looked normal. If you stepped outside and looked around, you wouldn't have known the sun was in eclipse.

I knew from my astronomy books that the next total solar eclipse visible from the continental US would occur on August 21, 2017. I began my countdown! I did enjoy a few annular and partial eclipses as "snacks" while waiting for the big one. Thankfully, I didn't have to wait that long.

On October 24, 1995, I was able to time a research trip to India to coincide with a total solar eclipse. On that occasion, I joined a scientific eclipse expedition from the Indian Institute of Astrophysics in Bangalore and observed the eclipse from northern India. Prior to this experience I had read descriptions, looked at photos, and watched videos of solar eclipses, but nothing compared with the actual experience. Everyone at the site—visiting scientists and locals alike—reacted with shouts of joy and applause. More than 20 years later, I would witness my second total solar eclipse, the one I had waited for since 1979, in Missouri with my family. The spectacle doesn't get old. If given another opportunity, I would eagerly observe another total solar eclipse.

There is something about experiencing a total solar eclipse that touches us deep inside. Many people feel awe, even a spiritual connectedness to the cosmos. One question I was inspired to ask after watching my first total solar eclipse was "Why this?" In other words, "Why are we privileged to witness such a glorious phenomenon?" It seems so out of place with our usual daily mundane needs for survival. After all, people can get by just fine without ever seeing a solar eclipse.

Just creating a total solar eclipse also seems like a lot of trouble to go through. What, exactly, do you need for a total solar eclipse to occur? On the one hand, you don't need much—just three bodies. In our case, you need the Earth, the sun, and the moon. In general, you need an observer platform, an eclipsing body, and a luminous body. If all you wanted to observe was just any solar eclipse, then you'd be done here.

But if you want to observe a total solar eclipse, you need to tune the sizes and separations of the three bodies. More specifically, you need to satisfy the following condition: As seen from the surface of the Earth, the angular size of the moon must be larger than the angular size of the sun. Or, more technically, the size-to-distance ratio of the moon must be larger than the size-to-distance ratio of the sun. This way, the moon completely covers the sun whenever the two bodies are aligned on the sky. These are the minimal requirements for a generic total solar eclipse—which ours isn't (more on this below).

How often do eclipses occur? It depends on the properties of a moon's orbit. If it orbits on the same plane as the planet orbits its host star, then total solar eclipses will occur every month. If the moon's orbital plane is tilted relative to the planet's orbital plane, like ours is, then solar eclipses will occur during two intervals each year separated by half a year. Some of them will be partial, and some will be total.

The average angular sizes of the sun and moon in our sky are a close match, differing by only a few percent. I say "average" because the Earth's orbit is very slightly noncircular and the moon's orbit a little more so. The moon appears 14 percent larger when it is closest compared to its farthest point. When the moon appears a little larger than the sun, we get a total solar eclipse. When the moon appears a little smaller, we get annular eclipses. During an annular eclipse, the sky does not noticeably darken much, and the various phenomena associated with a total eclipse are not visible. Our "perfect" total solar eclipses allow us to view the full outer atmospheric layers of the sun (the chromosphere and corona) while blocking the blindingly bright photosphere. If the moon appeared modestly larger, it would block the thin chromosphere and part of the corona.

It is not merely a theoretical exercise to consider how eclipses might appear from other planets. In 1999, I explored this question using data on all the moons known at the time in the solar system.[1] Far from showing Earth's eclipses are average, the data demonstrated, by multiple measures, that we enjoy the best solar eclipses in the solar system! First, as I already noted, the angular sizes of the sun and moon are almost a perfect match, as viewed from the surface of the Earth. Only one other moon of the 64 I examined has a similarly close match—Prometheus, which orbits Saturn. However, differences between it and the moon illustrate why our eclipses are much better.

Saturn is nearly ten times farther from the sun than the Earth, making the eclipsed sun appear only one-tenth as big. Indeed, Earth is the closest moon-bearing planet to the sun. That means we can see more detail in the eclipsed sun without visual aid than any hypothetical observer on any other planet. What's more, Prometheus is a small, potato-shaped moon. Its shape is not a good match to the round disk of the sun. Being a much larger body, our moon has stronger gravity, which has shaped it into a pretty good approximation to a sphere. Its profile on the sky is a good match to the sun's profile.

Prometheus orbits close to Saturn, within its ring system. Prometheus is one of Saturn's shepherd moons, located just inside the F ring. Because of its small orbit around such a massive planet, Prometheus moves fast, resulting in very short-duration eclipses lasting about one second. In contrast, our moon moves at a much more leisurely pace, giving us eclipses lasting up to seven-and-a-half minutes. The ring material between Prometheus and Saturn also poses a challenge to viewing the eclipse clearly. Not only would it block clear views of the eclipsed sun, but sunlight scattered off the ring particles would interfere with the view.

There is a less-common type of solar eclipse produced when a moon casts its shadow on another moon. These are called *mutual eclipses*. In an article in *Astronomy & Geophysics* published in 2009, I wrote about these eclipses and showed that they are not nearly as good as the eclipses we experience on Earth.[2] Only a small fraction of the eclipses display a close match between the sun and a given moon, and those that do have very short duration (less than a second).[3]

I'm not the first to take note of this remarkable coincidence in the close match between sun and moon. Most observers attributed this to chance and did not pursue it further. Others were genuinely bothered by these observations. For instance, the popular British science writer and astronomer John Gribbin commented on solar eclipses in his 2011 book *Alone in the Universe: Why Our Planet Is Unique,*

Just now the Moon is about 400 times smaller than the Sun, but the Sun is 400 times farther away than the Moon, so that they look the same size on the sky. At the present moment of cosmic time, during an eclipse, the disc of the Moon almost exactly covers the disc of the Sun. In the past the Moon would have looked much bigger and would have completely obscured the Sun during eclipses; in the future, the Moon will look much smaller from Earth and a ring of sunlight will be visible even during an eclipse. Nobody has been able to think of a reason why intelligent beings capable of noticing this oddity should have evolved on Earth just at the time that the coincidence was there to be noticed. It worries me, but most people seem to accept it as just one of those things.[4]

Gribbin alludes to a time dimension that makes the coincidence even more special. The moon is receding from us at 3.82 centimeters per year, and the sun is gradually growing in size. Taken together, these factors mean that earthlings will cease enjoying total solar eclipses in about 250 million years.[5] This is only about 5 percent of the age of Earth. As Gribbin noted, we are living during the best time in Earth's history to observe solar eclipses.

Caleb A. Scharf, an astronomer and director of the Astrobiology Center at Columbia University, opines,

So is there some great significance to the fact that we humans just happen to exist at a time when the Moon and Sun appear almost identically large in our skies? Nope, we're just landing in a window of opportunity that's probably about 100 million years wide, nothing obviously special, just rather good luck.[6]

In other words, "Move along, nothing to see here." But I didn't stop asking questions after noticing these coincidences and comparing our solar eclipses to others in the solar system. Why? Because my definition of the scientific method is to search for the best explanations for the patterns we observe, not just the best naturalistic explanations. Are these *just* coincidences, or are they pointing to a deeper truth?

Let's consider again the basic ingredients for producing solar eclipses, in general terms: host star, moon, and observer platform. Now, let's consider the basic requirements for Earth to be habitable to observers like us. First, the observer's home must be within the circumstellar habitable zone (CHZ). A planet within the CHZ can maintain liquid water on its surface; it's not too hot and not too cold. The location of the CHZ around a star depends only on its luminosity. The CHZ around a feeble red dwarf star is only about one-tenth the size of ours. Turning it around, an observer in the CHZ of their star will see their host star as having approximately the same brightness as we see from our sun.[7]

In addition, habitability is maximized when a planet orbits a main sequence (dwarf) star. About 90 percent of nearby stars are dwarfs, as is our sun. A dwarf derives its luminosity from the fusion of hydrogen to helium in its core. Since most of a star's mass is hydrogen, it will spend most of its life as a dwarf. Dwarfs differ in mass, luminosity, size, and surface temperature; all these parameters correlate. Thus, the most massive dwarfs are also the most luminous and the largest. The luminosity of a dwarf is a very sensitive function of its mass; for example, doubling the mass of the sun would increase its luminosity by just over a factor of 10.

What does all this have to do with solar eclipses? Very simply, the necessity of observers being within the CHZ of their host star determines how big it will appear in their sky.

In other words, the fact that our sun subtends an angle of half a degree in our skies is a consequence of our need to be within the CHZ. It is also a consequence of sunlike stars being more habitable than either more massive stars or less massive dwarfs. Now that we've made a connection between one aspect of solar eclipses (the eclipsed body) and our existence, we can ask, "What needs to be special about the eclipsing body?"

The moon is important for life on Earth. The moon contributes about twice as much as the sun to the ocean tides. The tides, in turn, quickly mix mineral and organic nutrients from the continents to the oceans. This is very important for marine life. The tides create an intertidal zone, with a rich diversity of creatures. The tidal energy from the moon is also important in powering the circulation of the world's oceans, which, in turn, helps mollify the climate.[8]

The moon helps to minimize climate swings another way—by stabilizing the tilt of Earth's rotation axis.[9] A small moon, perhaps like one of the two tiny moons around Mars, will not do. It must be a relatively large moon to have enough gravitational force to do its job of stabilizing. As it is, the small variations in the tilt of Earth's rotation axis cause significant climate variations on timescales of thousands of years. Complex creatures, like mammals and reptiles, would be less likely to survive on a planet with larger climate swings.

The moon, then, is also necessary for our existence.[10] Let's summarize where we are: How big the sun appears in our skies is closely tied to Earth's habitability for complex life, as is the presence of a large moon in orbit around Earth. Put these together, and it means that the occurrence of solar eclipses on Earth is tied to our presence here. In other words, if you satisfy the conditions required for complex life on a planet, then you also satisfy the requirements for producing solar eclipses on that planet.

The necessary conditions for complex life that I listed above are far from exhaustive. For instance, there are factors related to the properties of Earth, such as its size, water fraction, and iron content. Of these properties, only Earth's size is relevant to the appearance of solar eclipses to observers on its surface, but that is secondary in importance. Some fraction of planets within the CHZ and with a large moon will not be habitable.

Notice that the habitability conditions I discussed earlier only guarantee us solar eclipses, not total solar eclipses and not "perfect eclipses." These conditions for habitability allow us to say that it is more probable than not that observers on a habitable planet will enjoy total solar eclipses and even perfect eclipses. There might be another habitability constraint that could even account for the relation between life and perfect eclipses.

The currently favored theory for the formation of the moon involves a giant collision between the proto-Earth and a Mars-size planetesimal.[11] During the collision, Earth's structure and atmosphere would have been altered, possibly in ways that increased its habitability.[12] A small impact would not have produced the habitability-enhancing outcome of our moon's formation. The moon would have quickly receded from Earth and Earth's rotation would have slowed as it gave some of its angular momentum to the moon (and to Earth's orbit too via the solar tides). Had the moon's mass been only 10 percent greater, its rotation axis-stabilizing effects would have ended by now.[13] In terms of size, this corresponds to a mere 70 miles greater in diameter. In short, we have very nearly the largest possible moon that can still stabilize the tilt of Earth's rotation axis after 4.5 billion years, and there are plausible reasons to prefer a larger rather than a smaller moon for habitability. The moon is within a "just right" range to allow habitability of the Earth.

It is estimated that the moon will continue to stabilize the tilt of Earth's rotation axis for roughly another 1.5 billion years.[14] Models of the evolution of Earth's interior, atmosphere, and biosphere together yield a similar future maximum lifetime of the biosphere, which will become inhospitable to complex (animal) life probably less than 500 million years from now.[15] This is about the same timescale that total solar eclipses will continue to be visible from Earth. Another "coincidence"!

Designed for Us

The best current evidence shows that the eclipse coincidences on Earth are no accident. We enjoy the best solar eclipses in the solar system. Our existence on Earth at this time is closely tied to our ability to observe perfect solar eclipses. These coincidences are largely traceable to physical laws.

Doesn't this imply that eclipses like ours can happen in other planetary systems where the same conditions are met? Yes. Indeed, if there are extraterrestrial intelligent beings, then they are also likely to enjoy total solar eclipses. Since the coincidences are explained by the laws of nature, does this mean they are not designed? Not necessarily. All it means is that any design did not originate from a source within the universe. Instead, the design originated with the physical laws themselves at the origin of the universe.

As I noted in the introduction, Earth's perfect eclipses are anything but a dull phenomenon. Perhaps we might doubt whether Earth's solar eclipses were designed if they weren't so awe-inspiring. But total solar eclipses are even more than awe-inspiring— they are also important to scientific discovery.

I will only give a brief summary here of the discoveries total solar eclipses have enabled scientists to make. For a detailed treatment, the reader is encouraged to read chapter 1 of *The Privileged Planet*.

Albert Einstein knew that his theory of general relativity was going to be controversial before he published it in 1915. For this reason, he proposed a novel observational test. After consulting with a leading astronomer at the time, George Ellery Hale, he proposed carefully measuring the positions of stars near the eclipsed sun. Then, he compared the stars' relative positions to what they are when the sun is not in that part of the sky. Einstein predicted that the two sets of measurements would differ by a certain amount according to his theory; this was one observable way that general relativity made different predictions than Newtonian physics.

British astronomer Arthur Stanley Eddington led a successful scientific expedition to observe the 1919 total solar eclipse as a way to test Einstein's prediction. He confirmed the predictions, which led to the rapid acceptance of Einstein's theory.[16] Criticisms within the scientific community continued, but subsequent eclipse expeditions confirmed the initial findings. Total solar eclipses proved to be the key to the rapid acceptance of what has turned out to be the most important theory in physics. However, if total solar eclipses were not possible, this discovery could not have been made. Total

solar eclipses also have helped us to learn about the chemical composition of the sun.

It's as if we have been drawn to observe solar eclipses by their beauty and discovered something important when we looked closer. The awe-inspiring beauty and scientific fruitfulness of total solar eclipses make a powerful case for design. The physical laws are extravagant in providing us with total solar eclipses. It won't help to appeal to observer self-selection bias combined with the multiverse because there is no logical connection between our existence and total solar eclipses. In other words, it is not *logically* necessary that we have to be able to observe total solar eclipses, it is only *physically* necessary (or at least likely).

Put in more philosophical terms, a set of physical laws that links observers with total solar eclipses is what one would expect if the universe is designed for observers to experience beauty and engage in scientific discovery but not otherwise. Since the conditions required to produce total solar eclipses are in addition to the conditions required for life, the number of habitable universes that don't have total solar eclipses will outnumber those that do have them. Given this, picking a universe at random from the subset of habitable universes will very rarely include one that has observers who can experience total solar eclipses. Once you accept this conclusion, your eyes will be opened to seeing solar eclipses in a different light. And your eyes will be opened to seeing other many examples, like those I elaborated in chapter 22.

How Does the Intelligibility of Nature Point to Design?

Bruce L. Gordon

lbert Einstein (1879–1955) famously remarked that "the eternal mystery of the world is its comprehensibility...[t]he fact that it is comprehensible is a miracle,"[1] and the mathematical physicist Eugene Wigner (1902–1995) opined that "[t]he miracle of the appropriateness of the language of mathematics for the formulation of the laws of physics is a wonderful gift which we neither understand nor deserve."[2] As these remarks highlight, the intelligibility of the universe to the human mind requires explanation in two respects. The first is ontological: Why is nature ordered in such a way that it can be understood? The second is epistemological: Why is the human mind able to gain understanding of the natural order? In the past, these questions did not provoke the puzzlement they do today. Let's get some historical perspective on the rise of modern science and the current milieu before we examine why a metaphysically naturalistic worldview provides no good answers to these questions, and why theism, which understands the universe as the product of intelligent design, is the only metaphysical context in which the existence and intelligibility of nature has an explanation.

Historical Perspective

For science to be possible, there must be order present in nature, and it has to be discoverable by the human mind. But why should either of these conditions be met? Historically, while there were temporary manifestations of systematic research into nature in ancient Greece and early Islam, and isolated discoveries elsewhere, the seeds of modern science first came to concentrated and sustained fruition in Western culture before its methodologies and achievements were disseminated throughout the world. This lasting and world-changing development emerged in the context of the Judeo-Christian worldview that permeated medieval Europe.[3] What drove it was a deeply entrenched society-wide conception of the universe as the free and rational creation of God's mind so that human beings, as rational creatures made in God's image, were capable of searching out and understanding a divinely ordered reality. The freedom of God's creative will meant this order could not be abstractly deduced—it had to be discovered through observation and experiment—but God's stable and faithful character guaranteed it had a rational structure that diligent

253

study could reveal. This theological foundation gave solid answers to ontological and epistemological questions concerning the intelligibility of the universe, but as the quotes from Einstein and Wigner make clear, this foundation had been lost by the middle of the twentieth century. Why?

Some see it as the outworking of the seventeenth-century mechanical philosophy that sought to explain all natural phenomena in terms of material contact mechanisms.[4] On this view, mechanical philosophy conceptually reduced scientific causality to efficient and material causes, purging Aristotelian notions of formal and final causality from science. This is perhaps plausible methodologically, but not metaphysically. The conception of mechanism in the mechanical philosophy retained formal causes in their *design* and final causes in the *purpose* they were created to serve. The break with Aristotle arose from the fact that, in the conception of the theistic and deistic mechanical philosophers, design and purpose were *transcendently* imposed rather than *immanently* active, so the search for scientific explanations turned to the intelligent implementation of efficient material mechanisms. The purge of any sense of design and purpose from the "scientific" conception of nature is due to the late-nineteenth-century rise of Darwinian philosophy, which sees the mechanisms of nature as brute facts and the course of their development as completely blind and purposeless.[5]

It is Darwinism, so conceived, that renders the existence of mathematically describable regularities in nature and their intelligibility to the human mind (itself conceived as the accidental result of blind processes) as such a surprise, for it assumes naturalism—the self-contained character of nature and the denial of supernaturalism—as the context for science. Under the aegis of

naturalism, there can be no expectation that nature is regular in a way that allows presently operative causes to be projected into the past to explain the current state of the universe or into the future to predict its development. The absence of *any* sufficient cause to explain why nature exists leaves the philosophical naturalist with no reason to think that what does exist should be ordered, or that any order he finds should be projectable into the past or the future. By denying transcendence and defaulting to a conception of the universe as a closed and ultimately arbitrary system of causes and effects, naturalism makes science the uncanny enterprise on which Einstein and Wigner remarked.[6] On the other hand, the Judeo-Christian worldview recognizes that nature exists and is regular not because it is closed to divine activity, but because (and *only* because) divine causality is operative. It is only because nature is a creation and thus *not* a closed system of causes and effects that it exists in the first place and exhibits the regular order that makes science possible. God's existence and action is not an obstacle to science; it's what makes it possible.

The Ontological Problem

What are the ontological preconditions for the very possibility of knowledge, including science? There are at least three that we will explore.

The first, and most basic, is that there exists an explanation for the way everything is. If there is *no* explanation that exists for some things being the way they are, then there is no reason for them to be that way rather than another. This plays havoc not just with science, which is in the business of providing explanations, but also with the possibility of knowledge in general.

Second, nature has to possess an

intelligible order. It has to be *regular* in a way that can be understood and serve as the basis for expectations about the future. If nature were not regular, or its regularity were not discernible, then we couldn't know anything about it, and science would be impossible.

Third—and this follows from the first two points—insofar as the causal structure of the natural world is *not* sufficient unto itself, it must be completed by something that *transcends* it. If there is no *physical* explanation for something, then there must be a *metaphysical* explanation for it.

Let's consider these three points by examining (1) the principle of sufficient reason, (2) the metaphysical basis of physical regularities, and (3) the causal incompleteness of the natural realm in quantum physics and its implications.

The Necessity of Explanation: The Principle of Sufficient Reason

For present purposes, we may define the Principle of Sufficient Reason (PSR) as follows: every contingent state of affairs, without exception, has an explanation. (In philosophy, something is "contingent" if it is possible that it could be either true or false.) It is not hard to see why this principle has to be true. What would happen to our knowledge of reality if we were to suppose some contingent states of affairs had *no* explanation? If some states of affairs could lack explanation, the possibility of there being *no explanation* would become a competing "explanation" for *anything* that occurred. If something can lack explanation, we are prevented from separating things that have an explanation from those that do not, for there is no basis to conclude that something having an explanation was not, in fact, something occurring for no reason at all. Because no objective probability is assignable when

there's no explanation, there being no explanation becomes an inscrutable competitor to every proposed explanation, undermining our ability to decide whether *anything* has an explanation, scientific or otherwise. This means our current state of awareness might *also* lack explanation, irremediably severing our perceptions from reality. So denying that every contingent state of affairs has an explanation undermines scientific access to truth and opens the door to an *irremediable universal skepticism*. The principle of sufficient reason is thus a broad logical truth we know *a priori*; it's a precondition for knowledge and the intelligibility of the world.

We can demonstrate the broad logical necessity of the PSR straightforwardly. Consider any contingent state of affairs—for instance, the state of affairs consisting in the existence of our universe—and let p be a proposition that represents this state of affairs. Every state of affairs is representable by a (suitably complex) proposition and every contingently true proposition represents a contingent state of affairs that is actual. We now argue:[7]

1. For all p, if p is a contingently true proposition, then it's *possible* there's a proposition q such that q completely explains p.

2. (1) is uncontroversial: given any contingently true proposition, it is merely *possible* that there exists an explanation for its being true.

3. The fact that q explains p entails both p and q, since q cannot explain p if q is not true, and p must be true if it is explained.

4. For a contradiction, assume that p, in fact, has no explanation.

5. Let p^* be the following proposition: p is true and there is no explanation for p.

6. Since p is contingently true, so is p^*.

7. By (1), there is some possible world W at which p^* has a complete explanation, q.

8. If a conjunction has been completely explained, so has each conjunct.

9. Since p is a conjunct of p^* and q completely explains p^* at W, q explains p at W.

10. But q also explains p^* at W, so p^* is true at W, in which case there is no explanation for p at W.

11. Hence, p both has and lacks an explanation at W, which is contradictory.

12. The supposition that p has no explanation leads to a contradiction and therefore is false.

13. Thus, for any contingently true proposition p, p has an explanation.

Since every contingently true proposition represents a contingent state of affairs that is actual, and every contingently true proposition has an explanation, every actual contingent state of affairs has an explanation. This principle of sufficient reason is *necessarily* true. Since what is contingent cannot *ultimately* be explained by another contingent thing—for infinite regresses of contingent explanations (were such possible) would themselves be contingent and thus both need and have an explanation— ultimate explanations must terminate upon something noncontingent, i.e., necessary. It's a *necessary* truth, therefore, that some

necessary being is the *ultimate* explanation for contingent states of affairs that are actual, including our universe's existence.

What can we say about this necessary being? It transcends and explains the contingent existence of space, time, matter, and energy; so it's logically and ontologically prior to the universe, sufficient unto itself, nonspatiotemporal, and thus immaterial in nature. It cannot be an abstract object like a mathematical equation, however, for such things are causally inert. As Stephen Hawking asked in a rare moment of metaphysical lucidity: "What is it that breathes fire into the equations and makes a Universe for them to describe?"[8] The answer is a timeless, immaterial, and necessarily existent being capable of a timeless act of creation—in short, an immensely powerful and knowledgeable personal being possessed of a rational will. This requisite explanatory entity is, of course, more than merely suggestive of God's perfections.

Whether our universe—or multiverse,[9] if you grant credence to fashionable speculation—had an absolute beginning or is postulated to emerge from some timeless quantum state, the fact that something exists that *did not have to exist* is a contingent state of affairs requiring explanation. We thus arrive on God's doorstep as that being necessary both to ground knowledge and to explain why anything exists to be known.[10]

That such a being acted to create our universe is evidenced by features of nature that indicate design or purpose. This is, of course, the province of *teleological* arguments. As Aquinas recognized and Hume belabored, the teleological argument doesn't get you to theism or the uniqueness of the designer. Nonetheless, it has an important place. We will not dwell on it here.[11] We will turn instead to examining the fact that nature, as we experience it, *is* ordered, and subject to mathematical description.

Explaining the Regularity of Nature

What do we mean by "laws of nature"? Some philosophical naturalists argue they are *metaphysical necessities* similar to statements like "*No mammals are mathematical propositions.*" This cannot be right. Take Coulomb's law, for example: The fact that two like (or different) charges repel (or attract) each other with a force proportional to the magnitude of the charges and inversely proportional to the square of the distance between them gives *no* hint of being metaphysically necessary. The world could have been different. Other philosophers have suggested that laws of nature are *contingently necessary* relationships among universals mirrored at the level of the corresponding particulars. Oxymoronic appearances aside, no coherent account of this claim has been given either. Merely calling something "necessary" doesn't make it so.

Finally, other would-be necessitarians propose that physical laws derive from innate causal powers grounded in the essential natures of things and inherent in their material substance. These laws are manifested through forces or fields that necessarily emanate by nature from associated material substances, mediating physical interactions in a necessary way. But it's difficult to see why *this* causal power necessarily flows from *that* material substance. Things could have worked differently.[12]

All this goes to show that the idea of primitive metaphysical necessity in nature is unconvincing and problematic. Its theistic defenders generally maintain that to disallow it is to confuse metaphysics with epistemology or, more specifically, the necessary with the analytic or a priori. They say that it is not required for us to see why this causal power

flows essentially from that essential nature as long as the matter is clear from God's point of view. I am considerably less sanguine and inclined to think that causal powers, when not conceptually manifest in the essential nature of a natural kind, only coincide with that kind in virtue of God *freely making it so.* If there is no conceptual reason why a *different* causal behavior might not be associated with a natural kind, then it remains a metaphysical possibility that it *could* have been, had God so decided. To deny that conceivability is a reasonable guide to metaphysical possibility in this context is to constrain God's power in an unjustifiable way and strip us of the only analytical tool we have, leaving us defenseless against ungrounded assertion. This is not to say that divine decisions in this regard are arbitrary, however, for there may be *other* very good reasons that certain causal behaviors and not others are manifest in respect of the natural kinds that exemplify them.

From a metaphysical standpoint, however, this problematizes secondary causation models of providence. For secondary causation to be a nonvacuous explanation for the behavior of nature, material substances must possess an identity as part of a natural kind that generates essentially the lawlike behavior manifested by that kind, and we have just seen that there is *no* necessary metaphysical connection between the material identity of a substance and the causal powers it supposedly generates. God could have associated a *different* causal power with that material substance. The only reason any material substance is observed to function causally in the way that it does, therefore, is that God *chooses* to manifest that causal power through that substance. The power is therefore *not intrinsic* to the identity of that substance, so the material substance is *not* genuinely functioning in a secondary causal capacity.

What is really happening is that God is *continuously acting extrinsically* to produce that behavior by that substance. When we realize that material substances possess no intrinsic causal powers, however, we move from secondary causation to *occasionalism* as a better account of divine providence: Every event in nature that isn't subject to the influence of freely acting creatures—that is, all apparent causality in inanimate nature—is an *occasion* of direct divine action.

What if the metaphysical naturalist, in order to avoid this conclusion, rejects necessitarian (i.e., deterministic) accounts, maintaining that laws of nature are *mere* regularities, as David Hume did? Perhaps, like David Lewis, he develops a sophisticated regularist theory of physical laws.[13] The key question then becomes, *Why is nature regular at all*, and what keeps it so? A stable universe of *mere* regularities is a perpetual miracle. But if we say that there is no *physical* reason for the regularity of nature, is it tenable to hold that *no reason for nature being regular is required*? Clearly, embracing the viewpoint that no explanation is required entails denying the principle of sufficient reason—that is, denying the principle that *every contingent state of affairs has an explanation*—indeed, in this case, it requires denying this principle on a universal scale. But this would be a mistake of disastrous proportions, as we have already seen. The absence of a *physical* reason for the regularity of nature thus suffices to show that there is an immaterial *metaphysical* ground for its regularity. And the best account of this immaterial metaphysical ground is found in the mind of God and the mental causation constitutive of divine action.

Dealing with the Causal Incompleteness of Nature: Quantum Physics and Design[14]

Even if naturalistic accounts of physical law weren't *metaphysically* inadequate, however, their *physical* inadequacy manifests itself in the behavior of the world at the microscopic scale. Quantum physics, which describes this microphysical world, sets aside familiar conceptions of motion and the interaction of bodies and introduces acts of measurement and probabilities for observational outcomes in an *irreducible* way not ameliorated by appealing to our limited knowledge. The state of a quantum system is described by an abstract mathematical object called a *wave function*. As long as the system is unobserved, the wave function develops deterministically, but it only specifies the *probability* that various observables (like position or momentum) will have a particular value when measured. Furthermore, these probabilities can't all equal zero or one, and measurement results are *irreducibly* probabilistic.

This means that no sufficient physical condition exists for one value being observed rather than another permitted by the wave function—or, if you prefer, there is no sufficient condition for experiencing one reality rather than another included in the wavefunction. The physical reality behind our experience is causally incomplete in a way that shows that it *cannot* be autonomous. Causal closure is achieved by divine action, not by way of natural secondary causation, as if reality could otherwise function autonomously. God must *always be active*, and quantum physics is thus a *ceteris paribus* (all other things being equal) description of divine action and divine freedom in creation. The metaphysical lesson of quantum physics is that every instance of apparent causality in inanimate nature simply *is* the result of divine action. Divine design and action grounds the natural order. This point needs further elucidation.

The absence of sufficient material causality in quantum physics has a variety of experimentally confirmed consequences that preclude a world of mind-independent material substances governed by efficient material causation. The most rudimentary is that quantum reality does not exist until it is observed. A straightforward demonstration of this is provided by the delayed-choice quantum eraser experiment.[15] This experiment measures which path a particle took *after* wave function interference inconsistent with particle behavior has already been created. The interference can be turned off or on by choosing whether or not to measure which way the particle went *after* the interference already exists. Choosing to look *erases* wave function interference and gives the system a particle history. This experiment has been performed under conditions guaranteeing that *no physical signal* could connect the choice to look with the erased interference.[16] The fact that we can make a causally disconnected choice whether wave or particle phenomena manifest in a quantum system demonstrates that no measurement-independent, causally connected substantial material reality exists at the microphysical level.[17]

Consider two more results supporting this conclusion. First, the physically reasonable assumptions that an individual particle cannot serve as an infinite source of energy or be in two places at once entail that particles have *zero* probability of existing in any bounded spatial region, no matter *how* large.[18] Closing loopholes extends this result to more general conditions that include nonstandard interpretations of relativity.[19] Unobserved quanta don't exist *anywhere* in space, and thus have no existence apart from measurement. This consequence has been experimentally confirmed.[20] In short, there is no intelligible notion of microscopic material objects: Particle talk has pragmatic utility in relation to measurement results

and macroscopic appearances, but *no* basis in unobserved (mind-independent) reality.

Second, microphysical properties do not require a substrate. The Cheshire cat in *Alice in Wonderland* disappeared, leaving only its grin, which prompted Alice to say she'd seen "a cat without a grin, but never a grin without a cat." Quantum physics has its own Cheshire cat in which quantum systems that behave like their properties are spatially separated from their positions.[21] For example, an experiment using a neutron interferometer has sent neutrons along one path while their spins follow another.[22] In macroscopic terms, this would be like sending the redness of red balls along one path and their sphericity along another, or the spin of tops along one path and their positions along another. Under appropriate experimental conditions, quantum systems are decomposable into disembodied properties—a collection of Cheshire cat grins.

This leaves us in a situation in which there is no substantial material reality possessing properties that could ground necessitarian or counterfactual material relations at the microphysical level. So necessitarian theories of natural law cannot gain a purchase point in our most fundamental theories of matter and radiation. At this level, necessitarianism (and secondary causation) must be set aside, leaving us with a *regularist* account in which there are regularities in the natural world on a universal scale, but no real *laws* of nature. In short, microphysical reality behaves in ways we can count on and mathematically describe, but it does so for no discernible *physical* reason. This also points to occasionalism for its needed metaphysical explanation.

And what about the macroscopic world of our experience? How should we understand the transition between the microscopic and macroscopic worlds? Every quantum wave function is expressible as a superposition of different possibilities (states) in which the thing it describes *fails to possess* the properties those possibilities specify. No quantum system *ever* has simultaneously determinate values for *all* its associated properties.[23] This applies to macroscopic as well microscopic systems, since, under special laboratory conditions, we can create *macroscopic* superpositions (*coherent* macroscopic states). Large organic molecules have been put into superposition,[24] and Superconducting Quantum Interference Devices (SQUIDs) have superposed a billion electrons moving clockwise around a superconducting ring with another billion electrons moving anticlockwise, so that *two incompatible macroscopic currents are in superposition*.[25]

Contrary to some philosophers of physics who argue this provides evidence of dynamically interacting parallel realities,[26] I would contend that *none of* the mathematical-structural components of these quantum states are *materially* real and note, in the case of laboratory-created macroscopic superpositions, that our conscious self is *not* in the superposition, but rather, *observing* it. What such superpositions demonstrate is that quantum reality *is not materially substantial*; it is *merely* perceptual. Quantum reality goes no deeper than multimodal percepts superimposed in conscious awareness. The macroscopic stability we observe is the product of what physicists call *environmental decoherence*—the destructive interference of probability waves as quantum systems interact. So what does *this* imply about the world we experience?

Neither the phenomena of quantum physics nor their mathematical descriptions are consistent with the view that the causal basis for our experience is found in a mind-independent world of material substances. The physical world of our perception is a world of *mere* phenomena, the necessary explanation for which is found in the mental causation constitutive of divine action. In the absence of material substances and the presence of quantum-mechanical probabilities for *observables*, divine action is thus best conceived in terms of an occasionalist quantum idealism.[27] In this context, *physical laws* (so-called) *become regularities of divine action*. A general form for the physical regularities of our experience is thus:

> If collective conditions C were observed, all other things being equal, with quantum-mechanical probability p, God would cause state of affairs S to be observable.

For example, if the temperature of fresh water at sea level were observed to be raised to 100 degrees centigrade, all other things being equal, with high quantum-mechanical probability, God would cause the boiling of that water to be observable. The mathematically describable regularities of nature are thus *active* expressions of God's perpetual faithfulness (Psalms 33:4; 119:90; 2 Timothy 2:13). God is the one in whom we live and move and have our being (Acts 17:28), the one who is before all things, and in whom all things hold together (Colossians 1:17). From our standpoint within time, then, there is no distinction to be made between creation and providence, since reality, *in toto*, is continually realized through divine action (mental causation) as an expression of *creatio continua*. God's design of reality is continuously woven into the very fabric of existence.[28]

The Epistemological Problem

What does is it mean to have *knowledge* of something? For present purposes, I offer Dallas Willard's rough-and-ready definition:

We have knowledge of something when we are representing it (thinking about it, speaking of it, treating it) as it actually is, on an appropriate basis of thought and experience. Knowledge involves truth or accuracy of representation, but it must also be truth based upon adequate evidence or insight. The evidence or insight comes in various ways, depending on the nature of the subject matter. But it must be there.[29]

What assures us, though, that we're representing something as it *actually* is? We've already seen that the principle of sufficient reason provides an ontological basis for knowledge. Every contingent state of affairs must have a metaphysical explanation if we are to avoid an irremediable skepticism. But there is an explicitly *epistemic* condition for knowledge as well, demonstrating that the intelligibility of nature to the human mind rests on the *design* of our cognitive faculties. The intelligibility of the universe presupposes that our cognitive faculties are *not* the result of blind natural forces with no end in view, but instead have the goal, when functioning properly, of producing true beliefs. Let's elaborate.

If metaphysical naturalism were true, there would be little basis for supposing the human mind was capable of doing science at all.[30] Under naturalism, our cognitive faculties would have been produced by natural selection-sifting chance variations. What would matter about them is their fitness for ensuring survival and reproduction, *not* their ability to represent reality as it *actually* is. But if our minds don't aim at *truth* as correspondence with reality, and if our conception of *logical coherence* is an accidental byproduct of evolution expressing how our minds just

happen to work, then the *veridicality* of our perceptions and the *validity* of our reasoning processes is, at best, inscrutable. And the further removed from immediate survival our beliefs are, the less confidence we should have that they bear any relationship to reality at all. This means that evolutionary naturalists have no real warrant for believing anything—including evolutionary naturalism—to be *true*; in short, belief in evolutionary naturalism is self-defeating.

While the modern version of this "evolutionary argument against naturalism" has provoked critical discussion,[31] it is compelling to note that computational evolutionary psychology has drawn the same conclusion. Computational experiments using evolutionary game theory demonstrate that organisms acting in accordance with the true causal structure of their environment will be out-competed and driven to extinction by organisms acting in accordance with arbitrarily imposed species-specific fitness functions.[32] Ironically, this demonstration can only be trusted if we are *not* organisms with nonveridical fitness functions; otherwise, we have no confidence that conclusions drawn on the basis of experiments we have devised bear any relationship to reality. Without cognitive faculties aimed at true beliefs, all human knowledge, including science, is just a fitness-driven survival mechanism with an inscrutable connection to reality.

Note, however, that the probability that properly functioning cognitive faculties are reliable guides to truth is high *when conditioned on theism*, especially Christian theism, since God not only brings about our existence, but also wants us to *know* and have a relationship with him, and he intends that we *understand* the world well enough to be its stewards. Christianity gives us access to the world and a basis for thinking that science can lead us to the truth; naturalism takes this away. When the prerequisites of knowledge are considered, therefore, the game seems over for the philosophical naturalist before it has started: He lacks a basis for thinking any of his beliefs are true. So theism not only catalyzed the rise of modern science, it remains the only worldview on which the origin, order, and intelligibility of nature makes any sense.

Concluding Thoughts: The Ends of Methodological Naturalism

In recognizing that the ontological and epistemological foundations of science are better served by theism than naturalism, and that theism smoothed the way historically for the development of modern science, we are led to ask why we should acquiesce to a methodology that would require us to proceed *as if* metaphysical naturalism were true when it cannot be, especially when a broader scientific methodology that includes intelligent causation avoids this charade. Nonetheless, this is precisely what advocates of methodological naturalism ask us to do. While maintaining—rightly—that methodological naturalism doesn't entail metaphysical naturalism, they insist that scientists prescind from anything that might go beyond immanent natural causes when doing science, thereby acting *as if* metaphysical naturalism were true for the purposes of scientific explanation. But it's *not* true and, as we have seen, science is only possible *because* it's not true. So why persist with this pretense?

Methodological naturalism's current ascendancy is an artifact of a conception of nature that not only isn't intrinsic to the task of scientific explanation, it's *inimical* to it. What science *is* depends largely on how

human beings understand *the nature of nature itself.* Whether methodological naturalism is reasonable depends on nature's *nature*: If nature *isn't* a causally closed system and is subject to detectable nonmaterial influences, then adhering to methodological naturalism precludes science from getting a grip on *all* the aspects of reality it's equipped to investigate. Natural science has its name *not* because of methodological restrictions on *how* nature is studied, but because it's the natural world that is the *object* of study. And here's the thing: How nature behaves is an *empirical* question, not one decided a priori. If our goal is an *integrated and adequate scientific understanding of reality* rather than a compartmentalized and deficient one, we need to hold methodological questions loosely and follow the evidence where it leads. Since, as we have seen, science is *impossible* foundationally under the aegis of *metaphysical* naturalism, it is not unreasonable to expect that it will be *incomplete* heuristically under the restrictions of *methodological* naturalism.

How should we think about scientific explanation under these conditions? It's not hard to see that systematic abductive inferences in a *uniformitarian* framework provide a broader and more adequate approach than methodological naturalism.[33] Modern uniformitarianism explains past developments on the basis of presently operative causes under the assumption that the causal structure of the world has remained constant and permits reliable inferences. Our ability to do this notably depends on the principle of sufficient reason. The uniform mathematical hallmark of intelligent causation is *specified complexity. Complexity* is measured by improbability and *specification* by conformity to an independently given pattern. In the context of physics and biology, where detection of intelligent causes would evince nonmaterial influences indicative of the falsity of causal closure, these patterns often have *functional* significance. Since structures and processes that exhibit a degree of complex-specified information *exceeding* the probability bounds of the observable universe[34] are uniformly associated with intelligent activity, intelligent causation is part of the causal structure of the world and falls within the purview of scientific investigation. Design-theoretic explanations are a species of uniformitarian analysis.

Recently, different models of specified complexity involving semiotic,[35] algorithmic,[36] functional,[37] and irreducibly[38] complex specified information have been shown to have a common underlying mathematical form that, with additional constraints, allows the construction of generalized (canonical) specified complexity models demonstrating that systems exhibiting a large degree of specified complexity are exceedingly improbable under any relevant probability distribution.[39] These canonical models can be used to create statistical hypothesis tests[40] for specified complexity. Inferring design from a specified-complexity test that assigns large degrees of specified complexity only to features of designed structures takes the form of a likelihood-ratio test in which the alternative hypothesis is that the structure was most likely produced by intentional design.[41] Under these conditions, rejecting the null hypothesis in favor of the alternative provides good, but defeasible, evidence of particular design (extraordinary providence), and the more evidence of this kind that is gathered, the more certain this conclusion becomes, just as one would expect for any scientific procedure.[42]

Christians distinguish between *extraordinary* and *ordinary* providence. What the science of intelligent design enables us to do is to distinguish those developments in the history of the universe that require *extraordinary* providence for their proper explanation from those that can be explained by its *ordinary* course. As we have seen, science is impossible foundationally under the aegis of metaphysical naturalism, which leads us to suspect its methodological incompleteness under the restrictions of methodological naturalism. The science of intelligent design remedies these deficiencies by recognizing the only possible metaphysical foundation for science and putting an end to the pretensions of methodological naturalism.

Part III:

SCIENCE AND EVOLUTION

25

Did Life First Arise by Purely Natural Means (Abiogenesis)?

Walter L. Bradley and Casey Luskin

Major scientific magazines and journals often feature articles on the "Biggest unsolved mysteries in science"[1]—and the origin of life is almost always on that list, sometimes as the number one mystery.[2] In this chapter we will explore key challenges to a natural, chemical origin of life. We'll examine the formation of the essential functional polymers of life—proteins, DNA (deoxyribonucleic acid), and RNA (ribonucleic acid). How might these extraordinarily complex molecules have formed in oceans, lakes, or ponds from simple, naturally occurring molecular building blocks like sugars and amino acids? What is life? How does it operate? Could life originate by strictly natural means?

Darwin's theory of evolution and the development of the second law of thermodynamics by Boltzmann and Gibbs are two of the three major scientific discoveries of the nineteenth century. Maxwell's field equations for electricity and magnetism are the third. The second law of thermodynamics has had a unifying effect in the physical sciences much like the theory of evolution has had in the life sciences. What is intriguing is that the predictions of one seem to

contradict the predictions of the other. The grand story of evolution teaches that living systems have generally moved from simpler to more complex over time.[3] The second law of thermodynamics teaches just the opposite, a progression from order to disorder, from complexity to simplicity in the physical universe. Your garden and your house, left to themselves, go from order to disorder. But you can restore the order if you do the necessary work. In the winter, when it is cold, the interior of your house will gradually drop in temperature toward the outside temperature. But a gas heater can reverse this process by converting the chemical energy in natural gas into thermal energy in the house.

This simple analogy illustrates what is true of all living systems: they can only live by having access to energy and a means of converting this energy into the alternative forms of energy or work required to oppose the pull toward thermodynamic equilibrium, from complexity to simplicity. Living systems are much more complex than nonliving systems. Like a lawnmower with gasoline as a source of energy and an engine to convert that energy into movement of a blade to cut the grass, living systems must have access to

Figure 1: A backyard showing a dark area without grass, and a grass-filled area that receives sunlight. Image credit: © Walter Bradley.

sources of energy and systems to convert the energy into the needs of plants and animals. Nonliving objects in nature exist without any complex functional systems or any energy flow requirements. They are generally made of simple crystalline or amorphous materials. A picture of my (Walter Bradley's) backyard (Figure 1) shows a region in the foreground that is completely shaded by a large oak tree; it receives no sunlight, and consequently, has no grass. Adjacent to this shadowy, bare section is a region where sunlight is present about 50 percent of the daytime and consequently has a beautiful, grassy cover. The second law of thermodynamics is a law of nature (like gravity, everyone is subject to it). Living plants and animals can survive only with energy flowing through their systems. Nonliving objects such as mountains, rocks, sand, rivers, and soil have no need for energy flow, nor do they have the complexity to utilize energy toward some goal.

To summarize, plants can utilize solar energy to levitate above thermodynamic equilibrium. Nonliving objects such as mountains, oceans, rocks, sand, and soil have no need for such complexity; they do not store chemical energy like plants do; nor can they process solar or other forms of energy. Living matter is much more complex (e.g., RNA, DNA, protein, etc.), needing as it does to be able to utilize and store available energy from the sun or from the consumption of plants and animals.

All Living Systems Must (1) Process Energy, (2) Store and Utilize Information, and (3) Replicate

Aristotle posited the idea of spontaneous generation of life from nonliving matter (abiogenesis) that held sway for two millennia. But in 1859, Louis Pasteur

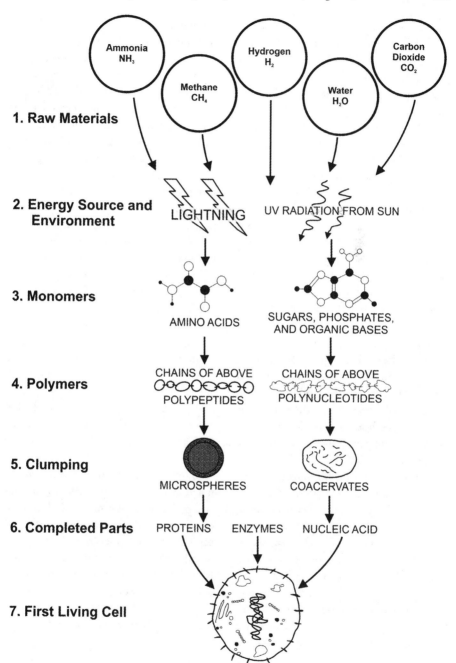

Figure 2: Major steps involved in the origin of life. All prebiotic evolutionary scenarios contain many hypothetical steps. Credit: Casey Luskin, modified with permission after Committee for Integrity in Science Education, *Teaching Science in a Climate of Controversy: A View from the American Scientific Affiliation* (Ipswich, MA: American Scientific Affiliation, 1986), 31.

showed persuasively—with a clever set of experiments—that what appeared to be life springing forth from nonliving matter was actually life emerging from exceedingly small living organisms, not lifeless matter. Pasteur's experiments were widely seen as having settled the question of whether life could *only* come from preexisting living matter, a process called *biogenesis*. In 1864, Pasteur triumphantly predicted to the science faculty at Sorbonne in Paris, "Never will the doctrine of spontaneous generation [of life coming from non-living matter (abiogenesis)] recover from the mortal blow of this simple experiment."[4] Pasteur's view remained dominant for almost a century.

In 1924, after 60 years of virtual silence since Pasteur's experiments, the Russian biochemist Alexander Ivanovich Oparin proposed that the complex molecular arrangements and associated functions of living systems evolved from simpler molecules that preexisted on the lifeless, primitive Earth. With this bold speculation and a recognizably modern hypothesis of how life might have arisen, Oparin reopened the discussion of abiogenesis.[5]

In 1929, the British biologist J.B.S. Haldane published a paper in the *Rationalist Annual* speculating on what initial conditions might be most favorable for a naturalistic origin of life.[6] He imagined an early Earth atmosphere rich in gases that was acted upon by lightning that caused chemical reactions to produce various building blocks for life—such as sugars and simple amino acids. In Haldane's view, these molecules might become sufficiently

concentrated in oceans, or more likely in lakes and ponds, such that they could chemically react to form long polymer chains that today we know are the key components in living cells (i.e., protein, DNA, and RNA).[7] In 1944, the noted quantum physicist Erwin Schrödinger observed that living systems are characterized by highly ordered, aperiodic structures that survive by continually utilizing (chemical or radiant) energy from their surroundings.[8] In 1952, Harold C. Urey proposed that the Earth's early atmosphere was rich in hydrogen, ammonia, and methane—chemicals that both provided the elemental building blocks and the energy to facilitate the chemical reactions necessary to make primary biopolymers, the chemical building blocks of life.[9]

The review above outlines early theories for generating the building blocks of life on Earth. But many additional steps would be needed for the origin of life to occur, which are sketched out in Figure 2.* In the next section, these various steps in a hypothetical origin-of-life scenario will be reviewed so that you can judge for yourself whether current theories are plausible.

First, it is vital to define the problem. As noted earlier, all living systems (1) process energy, (2) store information, and (3) replicate. In nature, these processes are performed primarily by molecules from three families of large biopolymers: proteins, DNA, and RNA. The mystery of how life began is essentially the mystery of how these three types of biopolymers formed and congregated within a cell with a barrier made of lipids as a self-replicating system.

---

* For a good understanding of the mainstream origin-of-life thinking, the reader is invited to watch the video "How did life begin? Abiogenesis. Origin of life from nonliving matter" by Arvin Ash, available at youtube.com/watch?v=nNK3u8uVG7o. This video describes, with helpful graphics, the basic components of a cell and the steps necessary for a living cell to emerge from the prebiotic "soup." The authors of this chapter do not agree with all the speculative narration in the video, but the graphics and commentary are helpful.

Steps 1, 2, and 3: Prebiotic Synthesis of Simple Organic Monomers

The Miller-Urey experiments were conducted in 1952–1953[10] and were celebrated as a great breakthrough in the search for a chemical pathway from gases assumed to be present in the early Earth's atmosphere to chemical reactions that produced amino acids, the building blocks for protein molecules. This experiment (see the apparatus depicted in Figure 3), and other similar experiments, have produced additional simple monomers—certain building blocks of life.

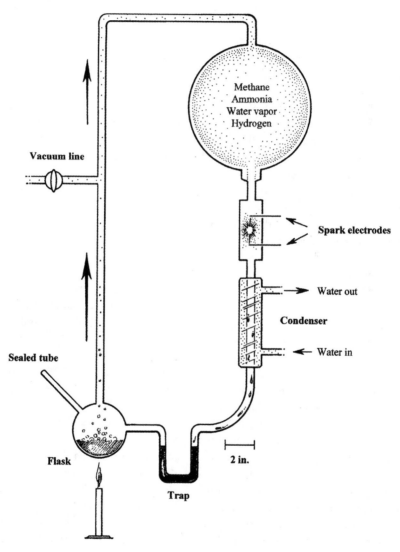

Figure 3: Experimental apparatus Miller–Urey used to test his hypothesis. Credit: © Jody F. Sjogren 2000, as used in Figure 2-1 in Jonathan Wells, *Icons of Evolution: Science or Myth?* (Washington, DC: Regnery, 2000). Used with permission.

Subsequently, careful critiques of the Miller-Urey[5] experiments and similar experiments created great doubt in their significance, though they are still taught in some high school textbooks as if they were scientifically sound. The atmosphere used in their experiments assumed a very energy-rich primordial atmosphere of methane, ammonia, and hydrogen, none of which would have been chemically stable in an early-Earth atmosphere. Studies of the early Earth's atmosphere by NASA during the 1980s confirmed that the mix of atmospheric gases used in the groundbreaking Miller-Urey experiments was wrong. The journal *Science* summed up the discoveries in 1980 by noting, "No geological or geochemical evidence collected in the last thirty years favors an energy rich, strongly reducing primitive atmosphere (i.e., hydrogen, ammonia, methane, with no oxygen). Only the success of the Miller laboratory experiments recommends it."[11] Later articles put it equally bluntly—in 1995, *Science* stated that "the early atmosphere looked nothing like the Miller-Urey situation."[12] Again in 2008, an article in *Science* reported, "Geoscientists today doubt that the primitive atmosphere had the highly reducing composition Miller used."[13]

There are good reasons to understand why the Earth's early atmosphere did not contain high concentrations of methane, ammonia, or other reducing gases. Earth's early atmosphere is thought to have been produced by outgassing from volcanoes, and the composition of those volcanic gases is related to the chemical properties of the Earth's inner mantle and core. Geochemical studies have found that the chemical properties of the Earth's interior would have been very similar in the past as they are today.[14] But today, volcanic gases do not contain methane or ammonia, and are not generally reducing.

Instead, an atmosphere dominated by carbon dioxide is preferred, but this poses a problem for prebiotic synthesis experiments, as prominent origin of life theorist David Deamer observed: "Carbon dioxide does not support the rich array of synthetic pathways leading to possible monomers, so the question arose again: what was the primary source of organic carbon compounds?"[15]

Another problem with Miller-Urey type prebiotic synthesis experiments is that when amino acids are synthesized from energy-rich gases, a racemic mixture of amino acids is created with 50 percent L-amino acids and 50 percent D-amino acids, sometimes called *left-handed* and *right-handed*. Protein molecules created in living systems must have 100 percent L-amino acids. If there are any D-amino acids in the chain, it would prevent the chain of amino acids from folding up into the proper three-dimensional protein structures associated with this amino acid string, preventing it from performing its function.

There are many additional problems with Miller-Urey-type research that seeks to identify plausible chemical pathways for the synthesis of proteins, DNA, and RNA molecules—the molecules of life. So drastic is the evidence against prebiotic synthesis of life's building blocks that in 1990, the Space Studies Board of the National Research Council recommended a "reexamination of biological monomer synthesis under primitive Earth-like environments, as revealed in current models of the early Earth."[16] Because of these difficulties, many leading theorists have abandoned the Miller-Urey experiment and the "primordial soup" model it is claimed to support. In 2010, University College London biochemist Nick Lane stated the primordial soup theory "doesn't hold water" and is "past its expiration date."[17] Instead, he proposes that life arose in undersea hydrothermal vents where water circulates through hot volcanic rock at the bottom of the ocean. But both the hydrothermal vent and primordial soup hypotheses face another major problem.

Step 4: Forming Polymers

Assume for a moment that there was some way to produce simple organic molecules on the early Earth. Perhaps these molecules did form a primordial soup, or perhaps they arose near some high-energy hydrothermal vent. Either way, origin-of-life theorists must then explain how amino acids or other key organic molecules linked up to form long chains (polymers), thereby forming proteins or RNA through a process called *polymerization*. A problem for the primordial soup version of this model is that it would be at chemical equilibrium, without any free energy for organic monomers to react further.[18] Indeed, chemically speaking, the last place you would want to link amino acids or other monomers into chains would be a vast, water-based environment like the primordial soup or in the ocean near a hydrothermal vent. As the US National Academy of Sciences acknowledges, "Two amino acids do not spontaneously join in water. Rather, the opposite reaction is thermodynamically favored."[19] Origin-of-life theorists Stanley Miller and Jeffrey Bada similarly acknowledged that the polymerization of amino acids into peptides "is unfavourable in the presence of liquid water at all temperatures."[20] In other words, water breaks protein chains of monomers back down into amino acids (or other constituents), making it very difficult to produce proteins (or other polymers like RNA) in the primordial soup or underwater near a hydrothermal vent.

The hydrothermal vent model is popular among origin-of-life theorists because

it represents a high-energy environment, but this model faces additional problems. Hydrothermal vents tend to be short-lived, lasting perhaps only hundreds of years[21]—timescales so short that the origin of life at undersea vents has been said to be "essentially akin to spontaneous generation."[22] It is also difficult to envision how prebiotic chemicals could become concentrated in such a chaotic, unbounded oceanic environment.[23]

But perhaps the biggest obstacle to the origin of life at hydro*thermal* vents is implied in their name: extremely high temperatures. According to *Scientific American*, experiments by Miller and Bada on the durability of prebiotic compounds near vents showed that the superheated water would "destroy rather than create complex organic compounds."[24] In the view of Miller and Bada, "organic synthesis would not occur in hydrothermal vent waters," indicating that vents are not an option for the origin of life because "[a]ny origin-of-life theory that proposes conditions of temperature and time inconsistent with the stability of the compounds involved can be dismissed solely on that basis."[25] Some might reply that certain alkaline thermal vents have lower temperatures,[26] but the high pH present near alkaline vents tend to precipitate carbon into carbonate minerals, with very little carbon remaining in the seawater for prebiotic chemical reactions,[27] and such a high pH is highly destructive to RNA.[28] As one paper put it, "the evolution of RNA is unlikely to have occurred in the vicinity of an alkaline deep-sea hydrothermal vent."[29]

Step 5: Clumping

Assuming that prebiotic organic polymers could be created under some set of natural conditions, the origin of life still cannot occur unless the requisite molecules can be concentrated or "clumped" together in some protective container where necessary chemical reactions can take place. In living organisms, such environments are the basic unit of life—the cell. But could something like a cell membrane arise naturally before life existed? In the 1970s, biochemist Sidney Fox and colleagues believed they had uncovered primitive cell membrane-like structures called *protenoid microspheres*.[30] Other structures called *coacervates* were proposed, first by Oparin, as potential precursors to modern cell membranes.[31] Because these structures lack any metabolism and the ability to self-reproduce,[32] they clearly could not constitute life. But even if these structures could do those things, they are unable to perform the most basic protective function of cell membranes: discriminate among nutrients, waste products, and toxic chemicals.

Campbell's Biology, a prominent college-level biology textbook, explains this requirement:

> One of the earliest episodes in the evolution of life may have been the formation of a membrane that enclosed a solution different from the surrounding solution while still permitting the uptake of nutrients and elimination of waste products. The ability of the cell to discriminate in its chemical exchanges with its environment is fundamental to life, and it is the plasma membrane and its component molecules that make this selectivity possible.[33]

Undoubtedly the textbook is correct: Without this extremely important protective barrier, the earliest forms of life would be unable to obtain food and be vulnerable to harmful molecules and chemical reactions in the outside environment, such as

oxidation. The membrane also keeps the cell's components together to allow for necessary cellular processes to take place. But the "lipid bilayer" of modern cells is no mere passive wall—it's a smart, active gatekeeper capable of allowing water and nutrients in, and letting waste products out. Specialized machines embedded in this smart membrane discriminate between helpful and harmful substances through a variety of biochemical pathways and molecular pumps. Hence the problem for origin-of-life theorists—as synthetic chemist James Tour of Rice University explains, no origin-of-life experiments have ever created "the required passive transport sites and active pumps for the passage of ions and molecules through bilayer membranes."[34] Tour elaborates on the daunting complexity of cell membranes that remains unexplained by origin-of-life theorists:

- Researchers have identified thousands of different lipid structures in modern cell membranes. These include glycerolipids, sphingolipids, sterols, prenols, saccharolipids, and polyketides. For this reason, selecting the bilayer composition for our synthetic membrane target is far from straightforward. When making synthetic vesicles—synthetic lipid bilayer membranes—mixtures of lipids can, it should be noted, destabilize the system.

- Lipid bilayers surround subcellular organelles, such as nuclei and mitochondria, which are themselves nanosystems and microsystems. Each of these has their own lipid composition.

- Lipids have a nonsymmetric distribution. The outer and inner faces of the lipid bilayer are chemically inequivalent and cannot be interchanged.[35]

Despite modest progress with the synthetic production of microspheres, coacervates, and similar structures, the lack of any discrimination ability means the clumping step in the origin of life has not been explained.

Amino Acid Sequence

Primary Structure

Tertiary Structure

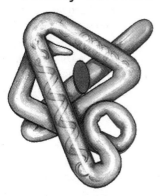

Quaternary Structure

Valerie Gower

Figure 4. Proteins have multiple levels of structure, where a chain of amino acids can fold up into a three-dimensional shape with a surface topography that can attract and hold atoms or molecules in place to facilitate their chemical reaction and release. Credit: Valerie Gower, © Discovery Institute. Used with permission.

Step 6. Completed Parts—the First Self-Replicating Molecules

In an undergraduate seminar taught by Stanley Miller that I (Casey Luskin) took as a student at the University of California, San Diego, Dr. Miller taught us that "making compounds and making life are two different things."[36] Many variants of Stanley Miller's experimental setup pictured in Figure 3 have been used in attempting to demonstrate the conversion of energy-rich, gaseous-phase chemicals into amino acids and other biomolecular monomers. But this is not nearly sufficient to generate life. Any origin-of-life explanation must include plausible biochemical paths from individual bio-building blocks like amino acids or nucleic acids to functional polymers such as proteins and DNA. The origin-of-life explanation must also include ways to speed up chemical reactions that are naturally slow. In living cells, long chains of amino acids fold up into 3-D structures that allow them to function as enzymes that greatly accelerate chemical reactions, as seen in Figure 4. How could these arise before life existed? More importantly, any origin-of-life model must account for the very particular sequencing of the molecules—i.e., the ordering of amino acids in proteins and nucleotide bases in RNA and DNA that allows them to function properly. This means explaining a crucial aspect of life: the origin of its information, or what proponents of intelligent design (ID) call the "information sequence problem."

For some theorists, the origin of life is defined as the natural origin of a self-replicating system capable of undergoing Darwinian evolution.[37] The most popular proposal for the first self-replicating molecule is RNA—where life was first based upon RNA carrying both genetic information (akin to modern DNA) and performing catalytic functions (akin to modern enyzmes), in what is termed the *RNA world*. Before we delve deeply into that, it is instructive to use the proceedings of a conference organized by the International Society for the Study of the Origin of Life (ISSOL) at the University of California, Berkeley in 1986 to measure the progress that has been made in origin-of-life research from 1952–1986.

I (Walter Bradley) attended this conference and watched one of the plenary sessions devoted to a spirited debate between scientists who believed that the first life was made of DNA ("DNA-first") and those who believed that the first biomolecules were proteins ("protein-first"). Neither group had yet been able to synthesize under plausible conditions either protein or DNA. Proteins can act as a chemical catalyst. DNA is the repository of information that is used to make functional protein. One of the outcomes from the conference was the sense that neither protein-first nor DNA-first were promising pathways to explaining the origin of life. But the difficulty demonstrating a plausible biochemical pathway for the origin of life that went through DNA-first or protein-first created an openness to new alternative possibilities. In 1986, the RNA world was just emerging as a popular alternative to protein-first or DNA-first models.

At the concluding plenary session, leading origin-of-life researcher Robert Shapiro addressed the RNA world and traced citations in the biochemical literature of the synthesis of RNA molecules under conditions thought to represent the early Earth conditions. The results were shocking. He cited a 1986 paper indicating RNA synthesis under prebiotic conditions had been demonstrated repeatedly, citing a 1985 paper and alluding to others. But that 1985 paper did not present original work—rather, it cited a 1984 paper and went all the way back to 1968 without any original work cited. A close reading of the 1968 paper indicated that the authors thought that they might have synthesized RNA molecules under prebiotic conditions but had not actually found any.

Shapiro's talk subsequently presented five huge barriers to this biochemical pathway from prebiotic chemistry to the first living systems. At the end of his dramatic presentation, the room of most of the world's most active origin-of-life researchers fell silent. The chair of the session, who was also the editor of the premiere journal *Origins of Life and Evolution of Biospheres*, repeatedly invited questions from the stunned audience. It was the only time in my (Walter Bradley) professional lifetime that I attended a plenary session of scientists and engineers where there were no questions. The chair closed the session without any questions offered, and he closed with the comment, "Robert, do you have to be so pessimistic?" Robert did not reply, but might have said he was letting the data do the talking, and the data told a very pessimistic story.

History has confirmed Shapiro's pessimism. Despite these difficulties, to this day, the RNA world remains the most popular model for the origin of life. But there are major problems with the RNA world hypothesis and claims that a self-replicating RNA molecule appeared by pure chance.

First, RNA has not been shown to assemble in a laboratory without the help of a skilled chemist intelligently guiding the process. Origin-of-life theorist Steven Benner explained that a major obstacle to the natural production of RNA is that "RNA requires water to function, but RNA cannot emerge in water, and does not persist in water without repair" due to water's "rapid and irreversible" corrosive effects upon RNA.[38]

In this "water paradox," Benner explains that "life seems to need a substance (water) that is inherently toxic to polymers (e.g., RNA) necessary for life."[39]

To overcome such difficulties, Benner and other chemists carefully designed experimental conditions that are favorable to the production of RNA. But Robert Shapiro explains that these experiments do not simulate natural conditions: "The flaw is in the logic—that this experimental control by researchers in a modern laboratory could have been available on the early Earth."[40] Reviewing attempts to construct RNA in the lab, James Tour likewise found that "[t]he conditions they used were cleverly selected," but in the natural world, "the controlled conditions required to generate" RNA are "painfully improbable."[41] Origin-of-life theorists Michael Robertson and Gerald Joyce even called the natural origin of RNA a "Prebiotic Chemist's Nightmare" because of "the intractable mixtures that are obtained in experiments designed to simulate the chemistry of the primitive Earth."[42] In the end, these experiments demonstrate one thing: RNA can only form by intelligent design.

Today, RNA is capable of carrying genetic information, but RNA world advocates claim that in the past, it also fulfilled the kinds of catalytic roles that enzymes perform today. A second problem with the RNA world is that RNA molecules do not exhibit many of the properties that allow proteins to serve as worker molecules in the cell. While RNA has been shown to perform a few roles, there is no evidence that it could perform all necessary cellular functions.[43] As one paper put it, proteins are "one million times fitter than RNA as catalysts" and "[t]he catalytic repertoire of RNA is too limited."[44]

The most fundamental problem with

the RNA world hypothesis is its inability to explain the origin of information in the first self-replicating RNA molecule—which experts suggest would have had to be at least 100 nucleotides long, if not between 200 and 300 nucleotides in length.[45] How did the nucleotide bases in RNA become properly ordered to produce life? There are no known chemical or physical laws that can do this. To explain the ordering of nucleotides in the first self-replicating RNA molecule, origin-of-life theorists have no explanation other than blind chance. As noted, ID theorists call this obstacle the information sequence problem, but multiple mainstream theorists have also observed the great unlikelihood of naturally producing a precise RNA sequence required for replication. Shapiro puts the problem this way:

> A profound difficulty exists, however, with the idea of RNA, or any other replicator, at the start of life. Existing replicators can serve as templates for the synthesis of additional copies of themselves, but this device cannot be used for the preparation of the very first such molecule, which must arise spontaneously from an unorganized mixture. The formation of an information-bearing homo-polymer through undirected chemical synthesis appears very improbable.[46]

Elsewhere, Shapiro notes, "The sudden appearance of a large self-copying molecule such as RNA was exceedingly improbable" with a probability that "is so vanishingly small that its happening even once anywhere in the visible universe would count as a piece of exceptional good luck."[47] A 2020 paper in *Scientific Reports* similarly notes, "Abiotic emergence of ordered information stored in the form of RNA is an important unresolved problem concerning the origin of life" because "the formation of such a long polymer having a correct nucleotide sequence by random reactions seems statistically unlikely."[48] Steven Benner refers to the "Information-Need Paradox," where self-replicating RNA molecules would be "too long to have arisen spontaneously" from available building blocks.[49] Benner raises an additional logical difficulty in that generating an RNA molecule capable of catalyzing its own replication is *much less likely* than generating RNA molecules that catalyze the destruction of RNA. This suggests a grave theoretical difficulty where RNA world theorists are faced with a "chemical theory that makes destruction, not biology, the natural outcome."[50]

The paper in *Scientific Reports* proposed a solution to these quandaries that showed just how intractable this problem is: It concluded that because the formation of a single self-replicating RNA molecule is prohibitively unlikely in the observable universe, therefore the universe must be far larger than we observe—an "inflationary universe" that increases the probabilistic resources until such an unlikely event becomes likely. This is just like the materialist response to the fine-tuning of physics: When the observed specificity of nature appears to indicate design, they invent multiverses to overcome probabilistic difficulties. When RNA world theorists are appealing to the origin-of-life's version of the multiverse to avoid falsification, it's clear that their project has fatal problems.

Step 7: The First Living Cell

In recent years, MIT physicist Jeremy England has gained media attention for proposing a thermodynamic energy-dissipation model of the origin of life. England's view was summarized when he famously said that the origin and evolution of life "should be as unsurprising as rocks rolling downhill." "You start with a random clump of atoms, and if you shine light on it for long enough, it should not be so surprising that you get a plant."[51] Another physicist, ID theorist Brian Miller, has responded to England's research.

Miller points out that the kind of energy that dissipates as a result of the sun shining on the Earth or other natural processes cannot explain how living systems have both low entropy (disorder) and high energy. As Miller puts it: "These are unnatural circumstances. Natural systems never both decrease in entropy and increase in energy—not at the same time." Living cells do this "by employing complex molecular machinery and finely tuned chemical networks to convert one form of energy from the environment into high-energy molecules"—things that cannot be present *prior* to the origin of life because they must be explained *by* the origin of life. Without this cellular machinery to harness energy from the environment and drive down entropy, England's energy-dissipation models cannot do the task they've been handed. As Miller said, England's model cannot account for the origin of biological information, which "is essential for constructing and maintaining the cell's structures and processes."[52]

Miller has highlighted a crucial deficiency in origin-of-life models: What is the origin of the cellular machinery, and the information that encodes the machinery that undergirds even the simplest cell? Forming a self-replicating RNA molecule, a seemingly impossible task under natural Earth conditions, is still a far cry from producing all the vast machinery required by cells to exist. A final obstacle for the RNA world—and any naturalistic account of the origin of life—is

therefore its inability to explain the origin of the genetic code and the molecular machinery of life.

There is an important distinction between the genetic code and the information in DNA or RNA: The genetic code is essentially the language in which the genetic information in the DNA or RNA is written. In order to evolve into the DNA/protein-based life that exists today, the RNA world would need to evolve the ability to convert genetic information into proteins. However, this process of transcription and translation requires a large suite of proteins and molecular machines—which themselves are encoded by genetic information. This poses a chicken-or-egg problem, where essential enzymes and molecular machines are needed to perform the very task that constructs them.

To appreciate the obstacle this poses to materialistic accounts of the origin of life, consider the following analogy. If you have ever watched a DVD, you know that it is rich in information. However, without the machinery of a DVD player to read the disk, process its information, and convert it into a picture and sound, the disk would be useless. But what if the instructions for building the first DVD player were only found encoded on a DVD? You could never play the DVD to learn how to build a DVD player. So how did the first disk and DVD player system arise? The answer is obvious: Intelligent agents designed both the player and the disk at the same time, and purposefully arranged the information on the disk in a language that could be read by the player.

In the same way, genetic information could never be converted into proteins without the proper machinery. Yet the machines required for processing the genetic information in RNA or DNA are encoded by those same genetic molecules—they perform and direct the very task that builds them. This system cannot exist unless both the genetic information and transcription/translation machinery are present at the same time, and unless both speak the same language. A functional living cell therefore can't evolve in a piecemeal fashion, but the likelihood of it arising all at once by unguided natural processes is far too low to be considered a viable model.

Biologist Frank Salisbury explained this problem in *American Biology Teacher* in 1971, not long after the workings of the genetic code were first uncovered:

> It's nice to talk about replicating DNA molecules arising in a soupy sea, but in modern cells this replication requires the presence of suitable enzymes…[T]he link between DNA and the enzyme is a highly complex one, involving RNA and an enzyme for its synthesis on a DNA template; ribosomes; enzymes to activate the amino acids; and transfer-RNA molecules…How, in the absence of the final enzyme, could selection act upon DNA and all the mechanisms for replicating it? It's as though everything must happen at once: the entire system must come into being as one unit, or it is worthless. There may well be ways out of this dilemma, but I don't see them at the moment.[53]

The same problem confronts modern RNA world researchers, and it remains unsolved. As two theorists observed in a 2004 article in *Cell Biology International*:

> The nucleotide sequence is also meaningless without a conceptual translative scheme and physical "hardware" capabilities. Ribosomes, tRNAs, aminoacyl

tRNA synthetases, and amino acids are all hardware components of the Shannon message "receiver." But the instructions for this machinery is itself coded in DNA and executed by protein "workers" produced by that machinery. Without the machinery and protein workers, the message cannot be received and understood. And without genetic instruction, the machinery cannot be assembled.[54]

Unless origin-of-life theorists can account for (1) the molecular machinery of the cell, (2) the information which encodes that machinery, and (3) the ability of cells to process that information to construct this machinery via a genetic code, the origin of even the simplest cell remains unexplained. Perhaps these seemingly intractable fundamental problems have an out: lots of time.

Saved by Time?

Many materialists believe that the severe unlikelihood of the series of events required for the origin of life is not a serious problem because there is essentially unlimited time for these events to occur. George Wald expressed this sentiment in 1954, writing in *Scientific American*, "Time is in fact the hero of the plot." Since he believed there were billions of years available for the origin of life on Earth, Wald poetically hoped, "Given so much time, the 'impossible' becomes possible, the possible probable, and the probable virtually certain. One only has to wait; Time itself performs the miracles."[55] But time isn't unlimited.

First, the early Earth was a hostile environment for any nascent biomolecules and even early life. While the Earth formed at about 4.54 billion years ago, the crust did not begin to solidify until about 4.4 to perhaps as late as 4 billion years ago.[56] Second, large bolide impact events occurred during the "heavy bombardment period" which lasted on Earth until about 3.8 billion years ago[57]— impacts large enough to vaporize the oceans and sterilize Earth's surface of any early life or prebiotic molecules.[58] Third, there is now good evidence of cellular life existing as early as 3.77 billion years ago based upon the presence of microfossils in jasper cherts in the Nuvvuagittuq belt in Quebec, Canada.[59]

Does this evidence imply less than 30 million years from the point at which Earth became habitable to the evidence of the first life? That may seem like a long time, but on geological timescales it is considered short.

Indeed, decades after Wald, such fossil evidence of early life led theorists to say things like "we are left with very little time between the development of suitable conditions for life on the Earth's surface and the origin of life"[60] and "we are now thinking, in geochemical terms, of instant life…"[61] While the precise dates of the earliest life and estimates of the onset of Earth habitability vary and these issues are debated vigorously in the literature, the point is clear: There is not unlimited time for the origin of life.

Time is *not* the hero of the plot; rather, it is the antagonist. The Herculean feats required by origin-of-life models are matched only by the poverty of resources available on the early Earth in terms of time and available chemical reactants. No wonder Francis Crick, the Nobel Prize-winning biochemist who co-discovered the structure of DNA, lamented, "An honest man, armed with all the knowledge available to us now, could only state that in some sense, the origin of life appears at the moment to be almost a miracle."[62] Based upon current knowledge, the first life could not have arisen by purely natural means.

An Optimistic Solution to the Mystery of Life's Origin

One might think that we have been overly pessimistic in our analysis of the current status of origin-of-life research. But consider what five prestigious origin-of-life thinkers say about the current status of origin-of-life research:

• *Nobel Prize-winning biologist Jack Szostak:* "It is virtually impossible to imagine how a cell's machines, which are mostly protein-based catalysts called enzymes, could have formed spontaneously as life first arose from non-living matter…Thus, explaining how life began entails a serious paradox."[63]

• *Harvard chemist George Whitesides:* "Most chemists believe, as do I, that life emerged spontaneously from mixtures of molecules in the prebiotic Earth. How? I have no idea… We need a really good new idea."[64] "I don't understand how you go from a system that's random chemicals to something that becomes, in a sense, a Darwinian set of reactions that are getting more complicated spontaneously. I just don't understand how that works."[65]

• *"Origin of Life" entry in the Springer* Encyclopedia of Astrobiology *by Mexican theoretician Antonio Lazcano:* "A century and a half after Darwin admitted how little was understood about the origin of life, we still do not know when and how the first living beings appeared on Earth."[66]

• *Richard Dawkins, leading evolutionary biologist and new atheist:*

"The universe could so easily have remained lifeless and simple…The fact that it did not—the fact that life evolved out of nearly nothing, some 10 billion years after the universe evolved out of literally nothing—is a fact so staggering that I would be mad to attempt words to do it justice."[67]

- *Eugene Koonin, a prestigious biologist at the National Center for Biotechnology Information:* "The origin of life is one of the hardest problems in all of science, but it is also one of the most important. Origin-of-life research has evolved into a lively, inter-disciplinary field, but other scientists often view it with skepticism and even derision. This attitude is understandable and, in a sense, perhaps justified, given the 'dirty' rarely mentioned secret: Despite many interesting results to its credit, when judged by the straightforward criterion of reaching (or even approaching) the ultimate goal, the origin-of-life field is a failure—we still do not have even a plausible coherent model, let alone a validated scenario, for the emergence of life on Earth. Certainly, this is due not to a lack of experimental and theoretical effort, but to the extraordinary intrinsic difficulty and complexity of the problem. A succession of exceedingly unlikely steps is essential for the origin of life, from the synthesis and accumulation of nucleotides to the origin of translation; through the multiplication of probabilities, these make the final outcome seem almost like a miracle."[68]

But there is an alternative solution to the information sequence problem and the mystery of life's origin—and it has the benefit of being based upon our uniform experience with how information arises. I (Walter Bradley) and my coauthors hinted at this solution in the original edition of *The Mystery of Life's Origin*, published in 1984, wherein we observed, "We know by experience that intelligent investigators can synthesize proteins and build genes" and concluded that "intelligence is the authentic source of the information in the biological world."[69] In 2020, Discovery Institute published an updated edition of *The Mystery of Life's Origin*, and all involved in the project were struck at how few changes were needed, owing to the fact that little meaningful progress had been made in the field of origin-of-life research over the previous 35 years. Stephen C. Meyer, James Tour, Brian Miller, and other scientists also contributed chapters updating the arguments.

As the baton is passed to the next generation of ID theorists, it's worth giving Meyer, a Cambridge-trained philosopher of science, the last word as his chapter in the 2020 edition of *Mystery* expanded our arguments that ID is the only known cause for the information-rich biomolecules required for the origin of life:

> [O]ur uniform experience affirms that specified information—whether inscribed in hieroglyphs, written in a book, encoded in a terrestrial radio signal, or produced in an RNA-world "ribozyme engineering" experiment—always arises from an intelligent source, from a mind and not a strictly material process. So the discovery of the functionally specified digital information in DNA and RNA provides strong grounds for inferring that

intelligence played a role in the origin of these molecules. Whenever we find specified information and we know the causal story of how that information arose, we always find that it arose from an intelligent source. It follows that the best, most likely explanation for the origin of the specified, digitally encoded information in DNA and RNA is that it too had an intelligent source. Intelligent design best explains the specified genetic information necessary to produce the first living cell.[70]

ID theorists thus propose that the action of an intelligent agent was required for the origin of the first living cell. In keeping with their materialistic outlook, origin-of-life theorists maintain that a self-replicating cell arose naturally, and then Darwinian evolution took things the rest of the way and allowed the grand diversity of living organisms to evolve. Other chapters in this book will evaluate claims about the Darwinian evolution of life as well.

26

What Are the Top Scientific Problems with Evolution?

Jonathan Wells

In 1973, biologist Theodosius Dobzhansky wrote that "nothing in biology makes sense except in the light of evolution."[1] In 1989, biologist Richard Dawkins wrote, "It is absolutely safe to say that if you meet someone who claims not to believe in evolution, that person is ignorant, stupid, or insane (or wicked, but I'd rather not consider that)."[2]

But what *is* evolution?

What Is Evolution?

The word *evolution* has many meanings. In one sense, it simply means change over time. In another it refers to the history of the cosmos, or the progress of technology, or the development of culture. No sane person believes that nothing changes over time, or that the cosmos, technology, and culture have no history. In these senses, *evolution* is uncontroversial.

In biology, *evolution* can refer to the fact that many plants and animals now living are different from those that lived in the past. It can also refer to the fact that minor changes occur within existing species; we

see such changes in our own families. But these uncontroversial meanings of biological evolution were not what Dobzhansky and Dawkins had in mind when they used the word. They meant *Darwinian* evolution.

Charles Darwin called his theory *descent with modification*, by which he meant that all living things are descended from one or a few common ancestors that lived in the distant past. The ancestors were then modified by unguided processes such as small variations and natural selection (survival of the fittest). Darwin wrote in *On the Origin of Species,* "I view all beings not as special creations, but as the lineal descendants of some few beings which lived long" ago, and that "Natural Selection has been the main but not exclusive means of modification."[3] He also wrote in his *Autobiography,* "There seems to be no more design in the variability of organic beings, and in the action of natural selection, than in the course which the wind blows."[4]

In the modern version of Darwin's theory, often called *neo-Darwinism,* accidental DNA mutations are considered to be the primary source of new variations. I will use *evolution* to mean neo-Darwinism throughout the rest of this chapter.

Like *evolution*, *science* has several meanings. In this chapter, I will use *science* to refer to *empirical* science: the enterprise of searching for the truth by comparing hypotheses with evidence. So the question is: What are the top problems with the evidence for neo-Darwinism?

Following Darwin's term *descent with modification*, I will first consider the evidence for descent (the hypothesis that all living organisms are descended from common ancestors). Specifically, I will focus on homology, fossils, and molecular phylogeny. Then I will consider the evidence for modification (the hypothesis that organisms have evolved by strictly unguided natural processes). I will focus on natural selection, mutation, and speciation (the origin of new species).

Homology

Classically, *homology* meant similarity of structure and position: for example, in the bones in the human hand and the wing of a bat. Darwin considered homologies to be evidence of common ancestry. He wrote in *On the Origin of Species*,

> What can be more curious than that the hand of a man, formed for grasping, that of a mole for digging, the leg of the horse, the paddle of the porpoise, and the wing of the bat, should all be constructed on the same pattern, and should include the same bones, in the same relative positions?[5]

Darwin regarded this as inexplicable if all species were separately created: "On the ordinary view of the independent creation of each being, we can only say that so it is;—that it has so pleased the Creator to construct each animal and plant."[6] (In the fourth edition of *On the Origin of Species*, Darwin added "but this is not a scientific explanation."[7]) Instead, he argued, homologies were explicable by his hypothesis of descent with modification:

> If we suppose that the ancient progenitor, the archetype as it may be called, of all mammals, had its limbs constructed on the existing general pattern, for whatever purpose they served, we can at once perceive the plain signification of the homologous construction of the limbs throughout the whole class.[8]

Yet animals and plants possess many features that are similar in structure and position but are clearly *not* derived from a common ancestor with those features. The camera eye of a vertebrate and the camera eye of a squid or octopus are remarkably similar, but no one thinks they were inherited from a common ancestor that possessed a camera eye. The spines of South American echidnas and North American porcupines are remarkably similar, yet echidnas give birth by laying eggs, while porcupines give birth to live babies after nurturing them in a womb, like human beings. This fundamental difference means that echidnas and porcupines had very different origins, and they did not inherit their spines from a spiny common ancestor. The folds of skin between the forelimbs and hind limbs of Australian flying phalangers and North American flying squirrels are very similar. Yet the former give birth to fetuses that crawl into a pouch to complete development, like kangaroos, while the latter nurture their fetuses in a womb, like human beings. Again, they had very different origins.[9]

Examples also include the pits of sticky nectar in carnivorous plants, which apparently originated separately six different times.[10] Plants of the euphorbia genus in Africa have thickened, fleshy stems to store water and prickly spines instead of leaves, like plants of the cactus family in the Americas, yet they originated separately under very different conditions.[11]

So similarity of structure and position is evidence for common ancestry, except when it isn't. Modern biologists call similarity *not* due to common ancestry *convergence*, and they have redefined *homology* to mean similarity due to common ancestry. Berkeley evolutionary biologist David Wake wrote in 1999, "Common ancestry is all there is to homology."[12] But according to philosopher of biology Ronald Brady, "By making our explanation [common ancestry] into the definition of the condition to be explained [homology], we express not scientific hypothesis but belief."[13]

Furthermore, once homology is defined in terms of common ancestry, it cannot logically be used as *evidence* for common ancestry. To do so would be to reason in a circle: How do we know that feature A and feature B are descended from a common ancestor? Because they are homologous. How do we know that A and B are homologous? Because they are descended from a common ancestor.[14]

Another problem with using homology as evidence for common ancestry is that examples of convergence are widespread. Cambridge paleobiologist Simon Conway Morris wrote in 2003 that "convergence is ubiquitous." He concluded, "Not only is the Universe strangely fit to purpose, but so, too…is life's ability to navigate to its solutions."[15] So similarities in structure and position do *not* provide clear-cut evidence for the common ancestry aspect of evolution.

Fossils

A fossil is "a remnant, impression, or trace of an organism of past geologic ages."[16] The study of fossils (called *paleontology*) started long before Darwin. They provide our best glimpse of the history of life before the present. Assuming that fossils in one layer of rock are younger than fossils in layers below them, pre-Darwinian paleontologists had already grouped them according to their relative ages. The result is known as the fossil record.

Darwin wrote this about the fossil record in *On the Origin of Species*:

> By the theory of natural selection all living species have been connected with the parent-species of each genus, by differences not greater than we see between the varieties of the same species at the present day; and these parent-species, now generally extinct, have in their turn been similarly connected with more ancient species; and so on backwards, always converging to the common ancestor of each great class. So that the number of intermediate and transitional links, between all living and extinct species, must have been inconceivably great.[17]

But the "inconceivably great" numbers of transitional links postulated by Darwin have never been found. Indeed, one of the most prominent features of the fossil record is the Cambrian explosion, in which the major groups of animals (called *phyla*) appeared around the same geological time in a period called the *Cambrian*, fully formed and without fossil evidence that they diverged from a common ancestor.

Darwin knew about this evidence in 1859, and he acknowledged it to be a serious problem that "may be truly urged as a valid argument" against his theory.[18] He hoped that future fossil discoveries would help to fill in many of the blanks, but more than 150 years of additional fossil collecting has only made the problem worse. In 1991, a team of paleontologists concluded that the Cambrian explosion "was even more abrupt and extensive than previously envisioned."[19]

The abruptness seen in the Cambrian explosion can also be seen on smaller scales throughout the fossil record. Species tend to appear abruptly in the fossil record and then persist unchanged for some period of time (a phenomenon called *stasis*) before they disappear. In 1972, paleontologists Niles Eldredge and Stephen Jay Gould called this pattern *punctuated equilibria*.[20] According to Gould, "every paleontologist always knew" that it is the dominant pattern in the fossil record.[21] In other words, the "inconceivably great" numbers of transitional links postulated by Darwin are missing not just in the Cambrian explosion, but throughout the fossil record.

Even if we *did* have a good fossil record, we would still need our imagination to produce narratives about ancestor-descendant relationships. Here's why: If you found two human skeletons buried in a field, how could you know whether one was descended from the other? Without identifying marks and written records, or perhaps in some cases DNA, it would be impossible to know. Yet you would be dealing with two skeletons from the same recent, living species. With two different, ancient, extinct species— often far removed from each other in time and space—there would be no way to demonstrate an ancestor-descendant relationship.

Decades ago, paleontologist Gareth Nelson wrote, "The idea that one can go to the fossil record and expect to empirically recover an ancestor-descendant sequence, be it of species, genera, families, or whatever,

has been, and continues to be, a pernicious illusion."[22] In 1999, evolutionary biologist Henry Gee wrote that "it is effectively impossible to link fossils into chains of cause and effect in any valid way." He concluded, "To take a line of fossils and claim that they represent a lineage is not a scientific hypothesis that can be tested, but an assertion that carries the same validity as a bedtime story— amusing, perhaps even instructive, but not scientific."[23]

Molecular Phylogeny

The word *phylogeny* refers to the evolutionary history of an organism.[24] The word was coined by German Darwinian biologist Ernst Haeckel several years after the publication of *On the Origin of Species*. Evolutionary biologists have proposed phylogenies based on homologies in fossils, but as we have seen, there are problems with both fossils and homology. With the rise of modern molecular biology, evolutionary biologists have increasingly sought to base phylogenies on molecules such as proteins and DNA.

Proteins consist of sequences of subunits called *amino acids*, and DNA consists of subunits called *nucleotides*. Different species may contain similar proteins or DNA molecules that exhibit slight differences in the sequences of their subunits. If three different species contain a similar DNA molecule, and its sequence in species A is more similar to its sequence in species B than in species C, then an evolutionary biologist might infer that A is more closely related to B than it is to C.

But the meaning of *related* is ambiguous. In one sense it can refer to genealogy, as in "Charles Darwin was more closely related to Erasmus Darwin (his grandfather) than either was to Geronimo." In another sense it can refer to similarity, as in "iron is more closely related to aluminum than either is to a daffodil."[25] Phylogenetic inferences assume that molecular relatedness (the second sense) is equivalent to genealogical relatedness (the first sense). This premise is based on the assumption of common ancestry.

Molecular comparisons are complicated by the problem of alignment. DNA sequences in living things typically contain repeated and/or deleted segments, so it is often unclear where to line them up. If two sequences can be aligned in more than one way, then any comparison will depend heavily on what alignment the investigator chooses. And when many sequences are compared, as they are in molecular phylogenies, the problem becomes much worse.[26]

Darwin thought that the history of living things could be represented as a "great Tree of Life," with common ancestors as the trunk and modern organisms as the tips of the branches.[27] If the history of life is tree-like, one would expect that the data from molecular phylogeny would eventually converge on a single tree, and that as more data were found, the fit would improve. Yet from the very beginning, molecular phylogenetics has been plagued with discrepancies among trees based on different sequences and different alignments.

And the problem has only grown worse as more data have accumulated. In 2005, three biologists who compared 50 DNA sequences from 17 animal groups concluded that "different phylogenetic analyses can reach contradicting inferences with [seemingly] absolute support."[28] In 2012, four evolutionary biologists reported "incongruence between phylogenies derived from... different subsets of molecular sequences has become pervasive."[29]

So the idea of common ancestry remains an assumption. It does not follow from

homology, except by circular reasoning. The fossil record remains (as Darwin acknowledged) a serious problem. And common ancestry does not emerge from the inconsistent findings of molecular phylogenetics.

Natural Selection

In the Introduction to *On the Origin of Species* Darwin wrote, "I am fully convinced that species are not immutable." He continued, "Furthermore, I am convinced that Natural Selection has been the main but not exclusive means of modification."[30]

But Darwin had no evidence for natural selection. In *On the Origin of Species*, the best he could offer was "one or two imaginary illustrations."[31] So instead of direct evidence for natural selection, Darwin (who himself bred pigeons) based his argument on domestic breeding, or what is often called *artificial selection*. He noted that "the breeding of domestic animals was carefully attended to in ancient times," and that "its importance consists in the great effect produced by the accumulation in one direction, during successive generations, of differences absolutely inappreciable by an uneducated eye."[32]

Yet in all the years of domestic breeding, no one ever reported the origin of a new species, much less a new organ or body plan. In the 1930s, neo-Darwinian biologist Theodosius Dobzhansky used the word *microevolution* to refer to changes within existing species (such as those observed by domestic breeders), and the word *macroevolution* to refer to the origin of new species, organs, and body plans. He wrote,

> There is no way toward an understanding of the mechanisms of macroevolutionary changes, which require time on a geological scale, other than through a

full comprehension of the microevo-lutionary processes observable within the span of a human lifetime and often controlled by man's will. For this reason we are compelled at the present level of knowledge reluctantly to put a sign of equality between the mechanisms of macro- and microevolution, and pro-ceeding on this assumption, to push our investigations as far ahead as this work-ing hypothesis will permit.[33]

But a "working hypothesis" is not evi-dence. It wasn't until the 1950s that British naturalist Bernard Kettlewell discovered what appeared to be the first evidence for natural selection. Peppered moths in the UK exist predominantly in two varieties: dark ("melanic") and light. Before the nineteenth-century industrial revolution, melanic forms were rare or absent, but when smoke from industrial cities darkened nearby tree trunks, the melanic form became much more com-mon. This phenomenon, called *industrial melanism*, was attributed to melanic moths being better camouflaged than light moths and thus less visible to predatory birds: in other words, to natural selection.

Kettlewell captured some of each vari-ety and marked them with a tiny spot of paint. Then he released them onto dark- or light-colored tree trunks. When he recap-tured some the next day, he found that a significantly greater proportion of better-camouflaged moths survived. Kettlewell termed this "Darwin's missing evidence."[34] The story, usually illustrated with photos of light- and dark-colored peppered moths on light- and dark-colored tree trunks, was fea-tured for decades in many biology textbooks as compelling evidence for evolution.[35]

By the 1980s, however, it had become clear that peppered moths don't normally rest on tree trunks in the wild. They fly by night and rest during the day in upper branches where they can't be seen. By releasing moths onto tree trunks in the daytime, Kettlewell's experiment failed to simulate natural con-ditions. It turned out that most textbook photographs had been staged by pinning dead moths on tree trunks or by placing live moths in unnatural positions and photo-graphing them before they moved away.[36]

Better evidence for natural selection came from finches in the Galápagos Islands in the 1970s. The islands were home to what biologists listed as 13 different species of finches, and biologists Peter and Rosemary Grant and their colleagues studied one of these on a single island. The Grants and their colleagues kept detailed records of each finch species' anatomy, including the length and depth of their beaks. When a severe drought in 1977 killed many of the islands' plants, about 85 percent of the birds died. The Grants and their colleagues noted that the survivors had beaks that were, on average, 5 percent larger than the population average before the drought, presumably because the surviving birds were better able to crack the tough seeds left by the drought. In other words, the shift was due to natural selec-tion. The Grants estimated that if a similar drought occurred every ten years, the birds' beaks would continue to get larger until they would qualify as a new species in 200 years.[37]

When the drought ended and the rains returned, however, food was plentiful, and the average beak size returned to normal. No net evolution had occurred.[38] Nevertheless, "Darwin's finches" found their way into most biology textbooks as evidence for evolution by natural selection.[39]

So there *is* evidence for natural selection, but like domestic breeding, it has never been observed to produce anything more than microevolution. As Dutch botanist Hugo de Vries wrote in 1904, "Natural selection may explain the survival of the fittest, but it cannot explain the arrival of the fittest."[40]

For the arrival of the fittest, most modern evolutionary biologists rely on mutations.

Mutation

Darwin insisted that new variations—the raw materials for natural selection—originated without purpose or direction, but he did not know their source. It wasn't until 1953, when James Watson and Francis Crick discovered the molecular structure of DNA, that many biologists thought the source had been found.

Watson and Crick inferred that DNA consists of two complementary strands, each composed of a string of four subunits. In 1958, Crick proposed that the sequences of subunits specify sequences of RNA molecules that function as intermediaries in the synthesis of proteins. The RNA sequences then specify the sequences of amino acids, the subunits of proteins.[41]

Some modern biologists think that the sequence of amino acids specifies the final form of a protein, and that proteins specify the final form of an organism. This line of reasoning is sometimes called the *central dogma* of molecular biology, and it can be crudely summarized as "DNA makes RNA makes protein makes us." In 1970, molecular biologist François Jacob wrote that an organism is the realization of a "genetic program" written in its DNA.[42] Under this view, changes (mutations) in DNA sequences would change the genetic program and thus modify the organism in any number of ways. Molecular biologist Jacques Monod (who shared a 1965 Nobel Prize with Jacob) wrote that with this realization, "and the

understanding of the random physical basis of mutation that molecular biology has also provided, the mechanism of Darwinism is at last securely founded. And man has to understand that he is a mere accident."[43]

But can DNA mutations really be the source of the variations needed for macroevolution? Certainly they can cause changes in an organism, but biologists have long recognized that most DNA mutations are either neutral (that is, they produce no observable changes) or harmful. To lead to the sort of evolution that could produce plants and animals from lower forms of life, we need mutations that cause *beneficial* variations. Otherwise, natural selection would either ignore them or tend to eliminate them.

Rare beneficial mutations have been found, but all of them produce only small biochemical changes—not new organs or body plans. Frequently these advantageous changes involve the loss or diminishment of function at the biochemical level.[44] Many biologists have concluded that the idea of a genetic program was wrong, and that DNA does *not* control the development of an organism. DNA is necessary, but not sufficient; other factors are also involved. One of these is spatial information in membrane patterns.[45] According to evolutionary biologist Thomas Cavalier-Smith, the idea that DNA contains all the information needed to make an organism "is simply false." Membrane patterns play

a key role in the mechanisms that convert the linear information of DNA into the three-dimensional shapes of single cells and multicellular organisms. Animal development creates a complex three-dimensional multicellular organism not by starting from the linear information in DNA...but always starting from an already highly complex three-dimensional unicellular organism, the fertilized egg.[46]

Since the 1970s, molecular biologists have performed comprehensive screens for mutations affecting embryo development in fruit flies, roundworms, zebrafish, and mice. Hundreds of mutations have been identified, but none of them change development in the fundamental ways needed for macroevolution. All the available evidence leads to the conclusion that no matter how much we mutate a fruit fly embryo, only three outcomes are possible: a normal fruit fly, a defective fruit fly, or a dead fruit fly. Not even a house fly, much less a roundworm, a zebrafish, or a mouse, can be produced via mutations.

Speciation

We know that speciation has occurred because many new species have appeared in the history of life. Evolutionary biologist Ernst Mayr wrote, "Darwin called his great work *On the Origin of Species*, for he was fully conscious of the fact that the change from one species into another was the most fundamental problem of evolution."[47] According to evolutionary biologist Douglas Futuyma, speciation "is the sine qua non of diversity" required for evolution. Speciation "stands at the border between microevolution—the genetic changes within and among populations—and macroevolution."[48]

But how does speciation occur? Part of the problem is that the term *species* is notoriously difficult to define. A definition applicable to plants and animals won't necessarily work for bacteria, and definitions applicable to living things won't necessarily work for fossils. As of 2004, several dozen

definitions were in use among biologists and paleontologists.[49] The definition most often used by evolutionary biologists is the "biological species concept," according to which species are groups of interbreeding natural populations that are reproductively isolated from other such groups.[50]

If species are defined this way, then in one sense speciation has been observed in the laboratory. Normally when two different species hybridize, either naturally or artificially, the hybrids are sterile because the maternal and paternal chromosomes are too dissimilar and cannot pair up in cell division. Occasionally, however, the hybrid undergoes chromosome doubling, or *polyploidy*. With matching sets of chromosomes that can undergo cell division, the hybrid may then be fertile and constitute a new species under the biological species concept. In the first decades of the twentieth century, Swedish scientist Arne Müntzing used two plant species to make a hybrid that underwent chromosome doubling to produce hemp-nettle, a member of the mint family that had already been found in nature.[51]

Speciation by polyploidy is called *secondary speciation* to distinguish it from *primary speciation*—the splitting of one species into two. According to Douglas Futuyma, polyploidy "does not confer major new morphological characteristics…[and] does not cause the evolution of new genera" or higher levels in the biological hierarchy.[52] So although secondary speciation by polyploidy has been observed in flowering plants, it is not the solution to Darwin's problem. The solution would be primary speciation by variation and selection, which has not been observed.

In 1940, geneticist Richard Goldschmidt argued that "the facts of microevolution do not suffice for an understanding of macroevolution." He concluded, "Micro-evolution does not lead beyond the confines of the species, and the typical products of microevolution, the geographic races, are not incipient species."[53]

Darwin used the term *incipient species* to refer to a variety of one species he thought was in the process of becoming a new species: "I believe a well-marked variety may be justly called an incipient species."[54] But how can we possibly know whether two varieties (or races) are in the process of becoming separate species? Saint Bernards and Chihuahuas are two varieties of the dog species (*Canis lupis familiaris*) that, for anatomical reasons, do not interbreed naturally. Are they on their way to becoming separate species? The Ainu people of northern Japan and the !Kung of southern Africa are members of the human species (*Homo sapiens sapiens*). Although people from both groups could undoubt-edly interbreed, without modern technology, which affords mass movement of people around the globe, they would be (for all practical purposes) reproductively isolated geographically, linguistically, and culturally. Are they therefore incipient species? Clearly, Darwin's term *incipient species* is a theoretical prediction, not evidence.

We sometime read in the news media that scientists have finally observed the ori-gin of a new species. Such cases, however, are invariably either examples of incipient speciation, or cases in which scientists have inferred from already-existing species how they might have split in the past.[55] Observa-tional evidence for primary speciation is still missing.

In 1992, evolutionary biologist Keith Stewart Thomson wrote, "A matter of unfinished business for biologists is the iden-tification of evolution's smoking gun," and "the smoking gun of evolution is speciation,

not local adaptation and differentiation of populations." Before Darwin, Thomson explained, the consensus was that species can vary only within certain limits; indeed, centuries of artificial selection had seemingly demonstrated such limits experimentally. "Darwin had to show that the limits could be broken," wrote Thomson, and "so do we."[56]

In 1996, biologists Scott Gilbert, John Opitz, and Rudolf Raff wrote,

> Genetics might be adequate for explaining microevolution, but microevolutionary changes in gene frequency were not seen as able to turn a reptile into a mammal or to convert a fish into an amphibian. Microevolu-tion looks at adaptations that concern the survival of the fittest, not the arrival of the fittest.

They concluded, "The origin of species—Darwin's problem—remains unsolved."[57]

English bacteriologist Alan Linton went looking for evidence of primary speciation and concluded in 2001,

> None exists in the literature claim-ing that one species has been shown to evolve into another. Bacteria, the simplest form of independent life, are ideal for this kind of study, with gener-ation times of twenty to thirty minutes, and populations achieved after eigh-teen hours. But throughout 150 years of the science of bacteriology, there is no evidence that one species of bacte-ria has changed into another…Since there is no evidence for species changes between the simplest forms of unicel-lular life, it is not surprising that there is no evidence for evolution from pro-karyotic [e.g., bacterial] to eukaryotic

[e.g., plant and animal] cells, let alone throughout the whole array of higher multicellular organisms.[58]

In 2002, evolutionary biologists Lynn Margulis and Dorion Sagan wrote, "Speciation, whether in the remote Galápagos, in the laboratory cages of the drosophilosophers [those who study fruit flies], or in the crowded sediments of the paleontologists, still has never been directly traced."[59] So evolution's smoking gun is still missing.

One Wrong Argument

Darwin called *On the Origin of Species* "one long argument."[60] It was an argument opposing the doctrine that species had been individually created, and an argument proposing the hypothesis that all living things are the modified descendants of one or a few common ancestors. But the hypothesis was unsupported in 1859, and the evidence for it is still insufficient. Homology has become circular reasoning. The fossil record remains at best inconclusive (and likely opposed to Darwinian gradualism), and molecular phylogeny is shot through with inconsistencies. Natural selection and mutation produce nothing more than changes within existing species. And the origin of species—Darwin's central problem—remains unsolved.

On the Origin of Species may have been one long argument, but from the standpoint of empirical science, continued claims that the evidence for evolution is "incontrovertible"[61] (as Richard Dawkins put it) might be better termed one long bluff.

27

How Does Irreducible Complexity Challenge Darwinism?

Michael Behe

I will praise thee; for I am fearfully and wonderfully made." So wrote King David in Psalm 139 (KJV) three millennia ago. Although the ancient Hebrews had barely any scientific understanding of biology, the psalmist apparently thought he didn't need much to be grateful. He knew that he had eyes to see and ears to hear, a tongue to speak and hands to grasp. Such marvels compellingly bespoke purposeful design—whether David knew or not how they worked at a fundamental level.

David's conclusion was continually reaffirmed as the knowledge of biology increased, painfully slowly, with time. More than 1,000 years after the psalms were written, the Roman physician Galen—an expert on anatomy and arguably the best scientific mind of the classical era—noted that the human body is the result of a "supremely intelligent and powerful divine Craftsman"—that is, "the result of intelligent design."[1] One of the earliest microscopists, Anton van Leeuwenhoek, remarked in the seventeenth century that the fantastic world of tiny creatures revealed by his then-new instrument could not be the result of a haphazard

process: "[T]his most wonderful disposition of nature with regard to these animalcules for the preservation of their species; which at the same time strikes us with astonishment, must surely convince all of the absurdity of those old opinions, that living creatures can be produced from corruption of putrefaction."[2] In the early nineteenth century, the abolitionist, apologist, and Anglican clergyman William Paley proclaimed the requirement of "an intelligent designing mind for the contriving and determining of the forms which organized bodies bear."[3] Yet in 1859, this historical unanimity of judgment was suddenly shattered by Charles Darwin.

Born in 1809, Darwin grew up during an age of exploration, when great ships sailed out from English harbors to reconnoiter mysterious, distant lands. At the age of 23, he signed on as ship's naturalist for a five-year, round-the-world exploratory voyage aboard the HMS *Beagle*. The expedition took him to South America, the Galápagos Islands, Australia, and more. During the trip, Darwin encountered many curiosities that few Englishmen had seen. In particular, he was impressed that there were species in those faraway places that clearly resembled, but were different from, European species. What's more, even within the new lands themselves there seemed to be odd distributions of some species. For example, the mainland of South America held just one species of mockingbird. Yet in the Galápagos archipelago (situated about 600 miles due west of Ecuador), three islands each had their own individual mockingbird species that were quite similar to, but distinctively different from, the mainland variety. For years after his return to England, Darwin puzzled over such relationships. Then he had an idea.

Darwinism

In his magnum opus, *On the Origin of Species*, Darwin proposed that such minor differences between mockingbird species—and even the amazing, apparently designed features of life over which David, Galen, van Leeuwenhoek, and Paley had exclaimed—all could be explained as the workings of chance and necessity over untold ages. In a nutshell, Darwin's argument went like this: First, he observed that there was variation within all species. Some members of a species might be larger than average, some members might be faster, others might have brighter plumage. He knew that not all the offspring of a species that were born would survive to reproduce. (If they did, the Earth would quickly be overrun and food supplies depleted.) So he reasoned that a member of a species whose chance variation gave it an edge in the struggle to survive would, on average, survive longer than those without it, and thus would have more opportunities to produce more offspring. Darwin called that simple chain of reasoning the *theory of natural selection*.

If the offspring inherited a parent's beneficial variation, then, other things being equal,

they, too, would have a better chance of survival than those without it. Consequently, over succeeding generations, the percentage of individuals of a species that had the beneficial variation would increase until eventually most or all individuals had it. At that point it would no longer be a variation; it would be a standard or "fixed" characteristic of the species. Then (Darwin's idea continued) perhaps another, different, helpful variation might come along, and the same scenario would replay itself. And then another variation, and another, until after very many generations the species had changed into something altogether different from its ancestor. If at some point a portion of the ancestor species separated from the main group—by, perhaps, crossing over a mountain chain or a desert—then different random variations might be favored in the separated lineage, eventually yielding a species different from both the ancestor and from the other lineage. Imagining the branching of species over eons, Darwin proposed that all species on Earth had ultimately come from one or a very few common ancestors.

This was a clever idea, and many scientists quickly saw that it could explain a number of features of biology, including changes in size, coloration, and more. But most scientists of the day were skeptical that Darwin's evolutionary mechanism could account for major differences between groups of organisms, or for complex features of life.[4] Darwin had given a vague, rhetorical account in *On the Origin* about how "Organs of Extreme Perfection and Complication" such as the eye might evolve, but few of his peers then were persuaded. A prominent biologist who was a contemporary of Darwin, St. George Mivart, wrote a book in 1871 entitled *On the Genesis of Species*, in which he argued that variation and selection might well explain small changes in preexisting biological systems but were powerless to account for "the incipient stages of useful structures."[5]

Although Darwin's peers were skeptical of his proposed *mechanism* of evolution (i.e., random variation plus natural selection), the great majority did quickly sign on to what is sometimes called the *fact* of evolution—that is, simply the proposal that organisms can change over time and that all likely did descend from some common ancestor, but with the crucial question of *What in the world might have caused such fantastic transformations?* left unresolved. And that's where matters stood for nearly 70 years.

Neo-Darwinism

Then in 1930, a mathematically inclined biologist named Ronald Fisher performed an elementary calculation to estimate how much potential variety there might be in a species for natural selection to sift. Building on the earlier work of the Austrian monk and founder of the science of genetics Gregor Mendel, which showed that some physical traits of an organism were inherited as discrete characteristics (such as the yellow or green color of the peas that he grew in his monastery garden), Fisher reasoned that if one kind of discrete gene came in two different varieties (like, say, that for normal hemoglobin and for sickle-cell hemoglobin) and a second kind of discrete gene also came in two varieties (like, say, that for brown versus blue eyes), then there would be $2 \times 2 = 4$ possible combinations of those genes that an organism could inherit, one from each parent. If a third gene came in two varieties, there would be $2 \times 2 \times 2 = 8$ possible combinations, and for four genes, $2 \times 2 \times 2 \times 2 = 16$ possible combinations. Now, organisms can have thousands of genes. Fisher noted

that if only 100 of them came in two varieties, there would be 2^{100} possible combinations; expressed as a normal number, that is a bit greater than a billion billion trillion. Fisher concluded that there was more than enough variation in life to evolve organisms in whatever direction natural selection beckoned.[6]

With Fisher's work in hand, other biologists strove to reconcile Darwin's theory with genetics. The amalgam was dubbed the *Evolutionary Synthesis* or *neo-Darwinism*. Neo-Darwinism was increasingly accepted by biologists in the 1930s and 1940s and remains the default explanation to this day. Nonetheless, there was a huge hole in Fisher's thinking—no one, including Fisher, knew then what a "gene" was.

To help put in perspective the state of science in the era when neo-Darwinism was accepted, let's take a look back at its state when Darwin first proposed his theory. In the mid-nineteenth century, molecules—which are now known to form the physical basis of life—were hypothetical entities; no one knew if they actually existed or not, let alone what their properties might be. The cell—which is now known to be the foundation of life—was thought to be a rather simple blob of jelly called *protoplasm*. In fact, it seemed so simple that some scientists—including the eminent biologists and fervent Darwin boosters Thomas Huxley and Ernst Haeckel—thought that cells might form spontaneously from sea mud.[7] No one of the time could even imagine such things as a genetic code, molecular machines, or any of the other elegant molecular systems of life. Thus, even though Darwin's theory depended critically upon the inheritance of variations, neither he nor anyone of that era understood genetics. Darwin was flying blind. In the absence of an understanding of genetics, Darwin proposed a theory he called *pangenesis*, in which nondescript particles dubbed *gemmules* from all areas of the body somehow were collected in the reproductive organs. The theory was completely wrong.

Although some progress had been made by the 1930s (when neo-Darwinism began to harden into orthodoxy), the molecular basis of life was still largely unknown. Whether a gene was composed of DNA, protein (the favored, but incorrect, candidate), carbohydrates, or something else was as yet a mystery. The nature of the uncanny proteins called *enzymes*, which catalyze the chemical reactions of metabolism, was anyone's guess. Thus Fisher was nearly as much in the dark when fashioning neo-Darwinism as Darwin himself had been when proposing the original idea.

Fathomless Elegance

After World War II, progress in biology accelerated at an ever-increasing rate. It was determined in the late 1940s that DNA—not protein—was the hereditary material. In the early 1950s, Watson and Crick solved the riddle of the elegant double-helical structure of DNA, immediately suggesting the manner in which its structure allowed genetic information to be copied and transferred to offspring. In the late 1950s, the genetic code was cracked, showing how a sequence of linked DNA subunits (dubbed *nucleotides*) instructs the cell to make a corresponding chain of linked protein subunits (called *amino acids*). Around the same time, British scientists first solved the structure of a protein, myoglobin, which helps store oxygen in muscles for later use. They were flabbergasted to find that its shape was highly irregular—unlike the pretty double-helical structure of DNA. Scientists later learned that the irregular structure of myoglobin—and of proteins

in general—is needed for them to do their jobs. Proteins are *machines* and, as for a lawnmower, their ability to perform their jobs is inextricably bound to their shapes.

The steady development of a slew of new laboratory techniques allowed a host of experiments to be done that were previously impossible and kicked research into ever-higher gears. Organic chemists learned how to make synthetic fragments of proteins and DNA from chemicals. Other methods were invented to determine the exact sequence of units in both natural proteins and nucleic acids. With the parallel advancement in computing power, very large databases of sequences could be compared to each other and searched for hidden patterns. The 400-year-old technique of microscopy was tricked out with new late-twentieth- and twenty-first-century gizmos that improved resolution astoundingly. In short, we are living through an age of astonishing progress in biology, one that (among much else) allows Darwin's theory, as well as modern modifications of it, to be tested at a level that was inaccessible until beginning only a few decades ago.

Arguably the most fundamental overarching discovery of the past three-quarters century is that cells are run not by some mysterious force, but by a bevy of enormously complex molecular *machines* whose work is orchestrated by astoundingly intricate instructions coded into molecules. The nanotechnology touted in sci-fi programs such as *Star Trek* is crude in comparison to that discovered at the foundation of life. There are machines in cells that act as outboard motors, others that behave as molecular trucks and buses to ferry supplies throughout the cell to their various proper destinations, more that act as chemical factories synthesizing the necessary molecular components of life, and—just as important—breaking them down once their jobs have been completed.

As I write this chapter, it is the summer of our coronavirus discontent, so to illustrate just the mechanical complexity of the molecular basis of life, I will showcase a virus—not the coronavirus, but one that infects bacteria. Viruses that infect bacteria are called *bacteriophages*. The bacteriophage shown in Figure 1, T4, looks something like the alien walking-crafts in the 2005 Steven Spielberg movie *War of the Worlds*. It has a hollow geometric structure that stores the viral DNA as its "head." Attached to the head is a long, hollow tube through which the DNA passes to enter the cellular prey. At the bottom of the tube is a structure called the *baseplate*. Attached to the baseplate are what are termed *long-tail fibers*. When the long-tail fibers touch the right place on a bacterial cell,

other shorter attachment fibers, which had been folded up inside the baseplate, quickly extend and firmly grab onto the cell. This triggers the tube to rapidly shorten, thrusting a needle-like structure through the bacterial cell membrane.[8] The viral DNA is then injected into the hapless cell, where it begins to hijack the cell's machinery to make copies of the virus. In brief, bacteriophage T4 is a smart, fully automated, nanoscale, hypodermic needle.

Fatal Difficulties

The modern discovery of the elegant molecular basis of life has exposed multiple ruinous problems for Darwin's mechanism as the main driver of the unfolding of life. One of the major difficulties is that the molecular machinery of the cell· strongly resists explanation by the many slow, gradual, incremental steps that Darwin postulated his mechanism required. In a discussion of the possible evolution of the vertebrate eye in the *Origin of Species*, Darwin remarked, "If it could be demonstrated that any complex organ existed, which could not possibly have been formed by numerous, successive, slight modifications, my theory would absolutely break down," adding hopefully, "But I can find out no such case." In the past 75 years, biology has discovered numerous such cases at the molecular foundation of life.

For example, let's consider bacteriophage T4 again. How might it have arisen by multiple tiny, functional steps, each almost always an improvement over the previous one, as Darwin's theory would require? All the components of T4 serve the end of delivering its DNA into a cell. Considering just T4's gross mechanical features, none of its multiple components works by itself. The head, which contains the genetic material, is helpless to

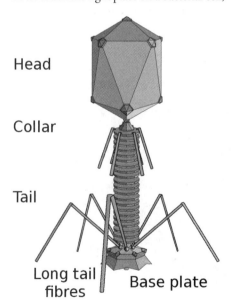

Head

Collar

Tail

Long tail fibres   **Base plate**

Figure 1. The structure of bacteriophage T4. Credit: Image modified from Adenosine (original); en:User:Pbroks13 (redraw), CC BY-SA 2.5 <https://creativecommons.org/licenses/by-sa/2.5>, via Wikimedia Commons.

enter the cell. The tube is useless without the head, as is the baseplate without either of the other parts. Ditto for the tail fibers. In fact, the bacteriophage is what I have termed *irreducibly complex*. Briefly, that means that all the parts of a structure are needed for it to work. And if that's the case, then there simply is no slow, gradual route to its evolutionary assembly by an unguided Darwinian process.

T4 is not some freak of nature; the example of its sophisticated structure is the rule at the cellular and molecular level of life, not the exception. As I have written previously,[9] the molecular machinery that fills the cell is commonly irreducibly complex in the same way as the bacteriophage. That should not surprise us. A machine is defined as "an apparatus consisting of interrelated parts with separate functions, used in the performance of some kind of work."[10] Thus (other than one of the few simple machines, such

as a lever or inclined plane), a machine is *expected* to be irreducibly complex. As I have shown, even such ordinary, everyday machines like a mousetrap have that property (Figure 2). And if the construction of a common mousetrap is strongly resistant to explanation by "numerous, successive, slight modifications," the sophisticated machinery of the cell is exponentially more so.

Although the discussion above paints a bleak picture for Darwin's mechanism, the actual situation is far worse. The "parts" *themselves* are extremely complex, made of multiple proteins that usually consist of hundreds of amino acid components. The parts themselves don't work unless a very large number of distinct components are in place.[11]

But even that's not all. A completely separate intractable difficulty for Darwin's explanation has come to light only within the

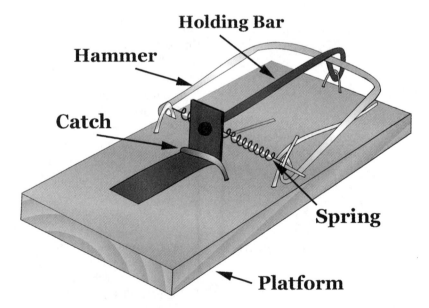

Figure 2. A common mechanical mousetrap is irreducibly complex. Image credit: "Mousetrap Rat Trap." Illustration by Christian Dorn (Commongt, 2020) Pixaby. Pixaby license. Modified by Amanda Witt, who granted permission.

past several decades. Prior to then, the gross effect of a mutation in a plant or animal could often be observed, but the exact change in the DNA sequence underlying the mutation could not easily be tracked down—simply because no laboratory techniques were available that could do so. But that has changed in recent years, and the ability to discover the DNA alterations underlying observable mutations has blossomed. As I have written,[12] it has, surprisingly, turned out that the great majority of *helpful* mutations are ones that *degrade* or *destroy* preexisting genes.

For example, of the 17 most highly selected (that is, most helpful) genes that differentiated the polar bear from its ancestor (the brown bear), fully 75 percent had suffered one or more mutations that were expected to damage the gene. Most mutations in dogs that underlie the traits of various breeds (such as, say, the Chihuahua and the French Poodle) are similarly due to mutations that break or blunt the respective genes for the traits. Even with bacteria it's the same story. A decades-long experiment by Richard Lenski, a microbiologist at Michigan State University, has shown that after more than 50,000 generations of evolving the bacterium *E. coli* in his laboratory, and after the arrival of dozens of helpful mutations that made the bug grow faster, the very large majority of those mutations are ones that, again, are expected to degrade or destroy the genes in which they occur.

The bottom line from such recent discoveries is that not only can't Darwin's mechanism *build* complex functional molecular machines, but it actively *breaks* them whenever that would help the organism in any specific circumstance. Due to the very progress of science, we now know that *Darwin's mechanism works chiefly by squandering genetic information for short-term gain.*

Apprehending Intelligent Activity

The discussion to this point has shown why Darwin's mechanism of evolution, which had seemed so promising, turned out to be so woefully inadequate to the task it had set itself once the molecular foundation of life was uncovered. But why should we think that tiny, elegant machinery—or, for that matter, the larger features of life that caused David, Galen, Paley, and countless others to marvel—were purposely designed? What is the *positive* case that intelligence was required to produce them?

The pertinent definition of *design* is "the purposeful or inventive arrangement of parts or details."[13] Let's simplify that to "the purposeful arrangement of parts." That means we apprehend design—we conclude that an intelligent agent has acted—whenever we see an arrangement of parts that strongly appears to have been arranged for a purpose. For example, as William Paley pointed out in his famous watchmaker example from his 1809 book *Natural Theology*:

> In crossing a heath, suppose I pitched my foot against a *stone*, and were asked how the stone came to be there, I might possibly answer, that for any thing I knew to the contrary it had lain there for ever…But suppose I had found a *watch* upon the ground…I should hardly think of the answer which I had before given…For this reason, and for no other, namely, that when we come to inspect the watch, we perceive that its several parts are framed and put together for a purpose…[T]he inference we think is inevitable, that the watch must have had a maker.[14]

Although Paley also made some off-target

arguments in *Natural Theology*, here he is correct. It is *precisely* because "we perceive that its several parts are framed and put together for a purpose" that we conclude "the watch must have had a maker." We reach one of our most basic rational conclusions—that another mind has acted, that an intelligent agent has been at work—by perceiving *a purposeful arrangement of parts*. That is *the* way—the *only* way—that we can recognize the action of another mind.

The reason is easy to see. We know that minds can have purposes. Thus, to the extent that an intelligent, mindful agent can manipulate parts, it can arrange them to achieve its purposes. And because we ourselves are intelligent beings, we can often perceive the purpose behind an arrangement of parts. On the other hand, unintelligent matter and processes do not have purposes. Thus, any juxtaposition of various parts brought about by an unintelligent process will not show a clear purpose. Occasionally, intriguing coincidences can arise at a very low level of specificity (such as when we see faces in the clouds), but nothing like what we find in life. Much more can be said about this, and I have written extensively on the topic.[15] Yet the basic conclusion is simple: As intelligent beings ourselves, we recognize when another intelligence has acted solely by perceiving a purposeful arrangement of parts.

That feature is what unites conclusions of design over the ages with our conclusion today. David could easily see that his fingers and muscles were coordinated for grasping. Galen knew that the wonderful arrangement of the larynx allowed production of the human voice. Anton van Leeuwenhoek observed the amazing, coordinated features of microscopic life. William Paley understood how the many functioning parts of the human eye allowed for sight. And in our day, we can see that all those marvelous features are underlain by even more spectacularly purposeful arrangements of molecular parts at the very foundation of life. Thus, from the knowledge accumulated in our own time, we can say with more certainty than ever that we are indeed fearfully, wonderfully, deliberately designed.

28

Can New Proteins Evolve?

Douglas Axe

I think that everyone who is interested in a critical examination of the creative power of evolutionary processes should be interested in proteins. But this needs some explanation.

First, what *are* proteins? We're all familiar with protein as an essential component of a healthy diet, but beyond a few rules we've heard for health or weight loss or strength training, most of us have only a vague notion of what proteins are. Even that plural form of the word may sound unfamiliar if you think of protein only as a substance, like fat or sugar.

To a chemist, all three of these words—proteins, fats, and sugars—refer to classes of molecules, each having its own defining chemical features. Moreover, all three of these classes encompass a huge variety of distinct chemical species. It makes sense, then, to speak not only of sugar as a substance but also of *sugars*—referring to the many chemically distinct forms that come under this heading. Likewise for fats and proteins.

Cellular production of a particular fat or sugar molecule involves the activities of many genes. By comparison, the making of a particular protein molecule is simple, at least conceptually. Cells read the DNA sequence of a single gene and translate this into a

sequence of linked amino acids, forming a long chain—a protein. The whole translation process is pure genius. We won't go into all the details here, but the fact that it could equally be called *decoding* gives you an idea of how sophisticated it is.

Figure 1 shows how the biological genetic code works. Genes, which are long chains of DNA bases represented by the letters A, C, G, and T, are read one *codon* at a time, this being the term for a group of three consecutive bases. Each of the 64 possible codon sequences specifies one of the 20 amino acids, or a stop signal to mark the end of the protein chain. Having "cracked" this code back in the 1960s, biologists can correctly predict the amino-acid sequence of the protein chain specified by any given gene.[1]

This simple relationship between genes and proteins makes proteins attractive for studying the effects of mutations (accidental sequence changes) to a gene. This is one of several reasons I say that everyone who's interested in a critical examination of the creative power of evolution should be interested in proteins. Mutations provide the genetic variation that's thought to fuel evolutionary change, and proteins offer a perfect opportunity to see how promising this fuel is.

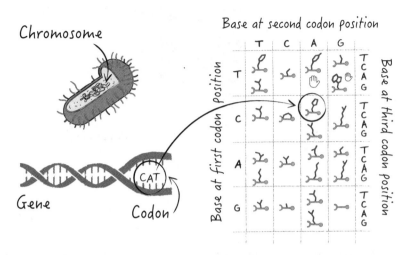

Figure 1: Genes and the genetic code that cells use to translate them. Bacterial cells carry a single chromosome—a long loop of DNA on which all the genes reside. Most living cells use the same basic set of 20 amino acids depicted (with artistic license) in the table on the right. The amino acids differ in the appendages, shown projecting upward. A highly sophisticated molecular system involving about 100 specialized proteins is used within cells to interpret each of the 64 possible codon sequences as specifying one of the 20 amino acids (or the end of the protein chain, represented by the hand symbol). The end result is the set of codon "meanings" that we refer to as the *genetic code*, often represented in table form, as here. (Reprinted from Douglas Axe, *Undeniable* [New York: HarperOne, 2016], used with permission.)

A second reason has to do with natural selection, which is the other main component of the evolutionary mechanism. Natural selection preserves traits in a species that are helpful to reproduction and suppresses traits that hinder reproduction. That's not controversial. The controversial claim is that mutation and selection in combination have the power to invent remarkable new traits and to combine them to invent radically new forms of life.

Proteins have something to offer here as well. Because proteins perform highly specific and important functions within cells—from structural scaffolds to chemical reactors and rotary motors—they bear a clear connection to traits. This connection is particularly direct in bacteria, where organismal functions are never far removed from molecular functions. In summary, then, proteins—bacterial proteins in particular—provide an ideal opportunity to test whether the combination of genetic mutations with natural selection really does have the inventive power to build the molecular systems that produce new biological traits.

Protein Folding Makes Remarkable Functions Possible

What enables a long chain of linked amino acids to perform highly specific molecular functions with machine-like precision? The answer is machine-like *structure*. Figure 2 shows one example (among thousands) of these remarkable structures. The multipart machine depicted is an ATP synthase—an assembly of 22 protein molecules that produces the energy molecule ATP. Biochemists

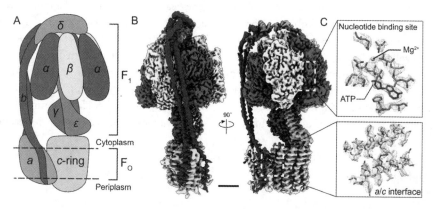

Figure 2: Structure of a bacterial ATP synthase. (A) Schematic diagram showing the protein parts. (B) Images of the ATP synthase, from different angles, obtained with a method called *cryogenic electron microscopy*. (C) Molecular details (not of concern here). This figure is taken from H. Guo, T Suzuki, J.L. Rubinstein (2019) "Structure of a bacterial ATP synthase." *eLife* 2029;8:e43128, https://elifesciences.org/articles/43128#fig1 (accessed August 24, 2020), under CC BY 4.0 license.

refer to this as an *enzyme* because it accomplishes a chemical conversion (making ATP from ADP). But it's no exaggeration to call it a molecular machine as well. Operating as a sophisticated nano-generator, the ATP synthase has a rotor (consisting of the parts labeled *c*-ring, γ, and ε) that spins at 8,000 rpm![2]

But how do long chains of linked amino acids form stable parts to make machines like this? The short answer is that it's *possible* for the different amino acids to be arranged along the chain in such a way that the whole thing locks into a specific three-dimensional form. The process is called protein *folding*, with the term *fold* referring to the overall form. Figure 3 illustrates the basics.

I emphasized the word *possible* there for a reason. A random gene would specify a random sequence of amino acids, which would flop around without folding. Chains like that are rapidly broken back down into amino acids to keep them from interfering with cellular processes. Very special amino acid sequences are needed for protein chains to fold into stable structures.

Measuring the Remarkable

It's possible to measure just how special these sequences must be. Because random genes are hopeless, the best way to do this is to start with a natural gene that specifies a selectable trait by producing a protein that imparts this trait. Laboratory methods exist for introducing random mutations into the chosen gene. This can be done in a controlled way by restricting how many changes occur or by confining the mutations to a small portion of the gene. By starting with a gene that specifies a working protein and by controlling in this way the extent to which the gene is mutated, experimenters can produce large numbers of mutant versions of the gene, some of which will almost certainly still work.

Building on earlier work of this kind,[3] I applied the method to a natural gene that enables bacteria to inactivate penicillin-like antibiotics.[4] The trait in this case is antibiotic resistance, which is very easy to select for in the lab. First, one simply puts the bacteria on petri dishes with a small amount of penicillin. Cells carrying a mutant version of the gene specifying an enzyme that's still able to break penicillin will form visible colonies on the petri dishes, whereas cells with inactive genes will die.

I chose four clusters of amino acids, ten each, for my experiments. In each experiment, I heavily mutated the gene locations for one cluster, resulting in many mutant genes, each of which specified a mutant version of the enzyme with a jumble of amino acids in those ten locations. Of roughly 100,000 mutant genes tested per cluster, some were found to work in three of the clusters. None in the fourth. Testing more mutants presumably would have turned up some that worked in that fourth cluster. In

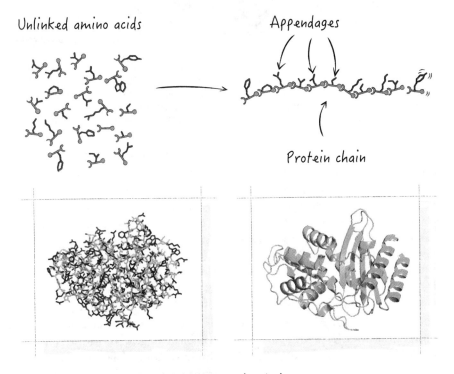

Unlinked amino acids

Appendages

Protein chain

Folded protein chain

Figure 3: The construction of proteins from amino acids. Amino acids are linked one by one in the precise sequence specified by a gene, to form a long, flexible chainlike molecule (upper right). The amino-acid sequences specified by most natural genes have the highly special property of causing the whole chain to fold into a well-defined three-dimensional structure, an example of which is shown in the lower left. Scientists use simplified representations to make it easier to see the features of these folded protein structures, the most common one being the "ribbon" diagram, shown for the same protein (called *beta-lactamase*) in the lower right. Each coil in a ribbon diagram represents an element of structure called an *alpha helix*, and each arrow represents a *beta strand*. These two elements make up most of the structures of all proteins, with the connections between the elements called *turns* or *loops*. Although the loops look floppy, like spaghetti, they usually have a firmly fixed structure just like the rest of the protein. (Reprinted from *Undeniable*, with permission.)

any case, I was able to estimate from these results a fraction of mutants that work for each cluster, though this fraction is really an upper-bound estimate for the fourth cluster.

From these experimentally based estimates, I calculated the fraction of mutants that would be expected to work if the entire gene had been mutated in a similar way. This fraction is closely related to another fraction that carries a great deal of significance for protein evolution. But before we talk about what this important fraction turned out to be, let's take some time to understand evolutionary thinking well enough to put this fraction in its proper context.

The Big Question for Atheists

For just a minute or two, put yourself in the place of someone who is determined to deny God. To make that easier, take yourself back to a moment when you yourself were resolved to assert your sovereignty over your life. In those moments (we all have them), God's presence seems highly inconvenient, so we push him out of our minds. I'm suggesting that atheists are merely people for whom this momentary lapse becomes a consciously adopted way of life. We aren't nearly as different, in terms of our basic human wiring, as some people think.

Atheism is a commitment, not a deduction.

Imagine, for a moment, that you've made this commitment. Having taken that step, your next step is to order your beliefs around that commitment. Whether the commitment was rational or not, you want to think of yourself as being rational, so you seek a way to make sense of the world and yourself without exposing your fundamental commitment to critique.

We call these ways of making sense of things *worldviews*. What we're picturing, then, is the process of choosing or constructing a worldview, specifically an atheistic one. Your (hypothetical) atheistic worldview doesn't have to provide answers for everything. No worldview does. It merely needs answers for the most pressing questions—the ones that will leave you feeling very uncomfortable if you have no answers for them.

Topping the list is the question *How did we get here?* How did we humans come to exist? Thinking as an atheist, what makes this one particularly urgent is that we all automatically sense that we couldn't exist apart from God having made us. I unpack this claim in the chapter titled "Is Our Intuition of Design in Nature Correct?" If we accept it here, we see that the atheist needs a way to quell his or her own most nagging doubts before fielding criticisms from anyone else.

What would an adequate atheistic answer to the question of our origin look like? It's helpful to turn that around. What would make a God-free answer *inadequate* to an atheist? In other words, what kind of atheistic explanation for our existence would self-respecting atheists find profoundly unsatisfactory?

The most obvious answer is a *non*explanation. No atheist worth his or her salt would be content to say, "I have no explanation whatsoever for life; I'm merely committed to rejecting any divine role." Or at the least, any atheist who *is* content with such a statement can expect criticism along the lines of "That sounds like a *you*-thing, not a rational thing." I haven't met any such atheists. Like everyone, atheists value reason highly enough to be uncomfortable leaving something so fundamental completely unexplained.

A plausible atheistic explanation for life, if there is such a thing, must start with a recognition that rejection of divine causation presents a large probabilistic hurdle.

Smart atheists get this. One of the smartest, Thomas Nagel, writes that "for a long time I have found the materialist account of how we and our fellow organisms came to exist hard to believe, including the standard version of how the evolutionary process works."[5] However much his fellow atheists would like to blame his incredulity on ignorance, the fact that scientific advances only make the problem worse seems to contradict that interpretation: "The more details we learn about the chemical basis of life and the intricacy of the genetic code, the more unbelievable the standard historical account becomes."[6] Nagel recognizes that "doubts about the reductionist account of life go against the dominant scientific consensus," but, he continues, "that consensus faces problems of probability that I believe are not taken seriously enough, both with respect to the evolution of life forms through accidental mutation and natural selection and with respect to the formation from dead matter of physical systems capable of such evolution."[7]

Even the famously brash atheist Richard Dawkins has described the probabilistic hurdle well. In *The Blind Watchmaker*, he put the problem this way:

> ...however many ways there may be of being alive, it is certain that there are vastly more ways of being dead, or rather not alive. You may throw cells together at random, over and over again for a billion years, and not once will you get a conglomeration that flies or swims or burrows or runs, or does anything, even badly, that could remotely be construed as working to keep itself alive.[8]

Dawkins is exactly right here, and the point is very general: *The accomplishment of any high-level function requires an arrangement of things or circumstances that chance is obviously unable to deliver.*

Of Hurdles and Impasses

In other words, this isn't merely a probabilistic hurdle—it's a probabilistic *impasse*. Misplacing the bike-rack key somewhere in your apartment presents a probabilistic hurdle. After you check the obvious places without success, each subsequent place you check starts to seem less promising. Nevertheless, you persist with the confidence that a thorough search must ultimately be successful. In this way, you overcome the unlikelihood that any single randomly chosen place to look is the right place by your willingness to keep looking. Eventually, the place you look *is* the right place.

Probabilistic impasses are different. If instead of having lost the key in your apartment, you lost it on a road trip—somewhere along the 400 miles you've driven since you inadvertently left it on the back bumper—you can safely kiss it goodbye. The key exists somewhere on that vast stretch of interstate asphalt, but since you can't afford to spend the rest of your life searching, it may as well not exist.

Nagel and Dawkins agree that life presents a probabilistic impasse for random searches, but they disagree on what options this leaves for atheists. For those who subscribe to materialism—the belief that the physical universe is all there is—Nagel doesn't think there are any plausible options. As a hardcore materialist, Dawkins thinks otherwise.

The Principle of Conservation of Coincidence

Nagel has the clearer grasp of this. Dawkins's perennial point of confusion is that he thinks probabilistic impasses don't apply to *blind* processes as long as they aren't *random* (in the mathematical sense of a uniform probability distribution). He's just plain wrong about this. In fact, in thinking blind natural selection solves the problem, he's *doubly* wrong.

His second error (I'll get to the first one next) is evident in his own words quoted above. Whatever Dawkins thinks does a better job of putting cells together than randomness, it *can't* be natural selection. Why? Because natural selection only happens *after* a conglomeration exists that works to keep itself alive! That is, the probabilistic impasse that Dawkins describes so vividly has to be overcome *before* natural selection comes into play, because natural selection is the favoring of things that *already do* work to keep themselves alive! Somehow this man made a lucrative career promoting natural selection, and neither he nor his supporters seem to have appreciated this basic point.

Dawkins's first error is his neglect of something so comprehensive that it rules out *all* materialistic explanations of life. It's a principle I call the *conservation of coincidence*. The basic idea is that the spooky coincidence that Dawkins acknowledges would be needed for the random throwing together of cells to produce, say, a hummingbird, can't actually be made less spooky merely by proposing an account of events—not, that is, unless the account includes intelligent causation. Absent intelligence, these accounts merely shuffle the coincidence around. A good story might give the impression that the spooky coincidence has been converted into something more respectable, but this is a false impression. In the end, it's just snake oil.

Dawkins's simple, correct observation that hopelessly nonviable arrangements (whether of cells or molecules) so vastly outnumber the ingenious arrangements needed

for any form of life means that life is, in effect, an infinitesimally small target—a target *far too small* to be hit by accident. Dawkins recognizes this when he pictures certain accidental processes, such as cells being thrown together at random, but he somehow loses sight of it when he thinks of Darwin's accidental process. Somehow, vague and incoherent stories of hummingbirds appearing over billions of years by accident seem less obviously implausible to him than sudden appearance by accident. But the magnitude of the coincidence comes straight from the simple observation that the target is too small to be hit by accident. The precise details of any supposed accident are of no consequence whatsoever. What makes them all tantamount to spooky coincidences is the mere fact that accidents are being credited with something accidents can't do.

Nagel is right when he says, "It is no longer legitimate simply to imagine a sequence of gradually evolving phenotypes, as if their appearance through mutations in the DNA were unproblematic—as Richard Dawkins does for the evolution of the eye."[9] Atheists have no escape from this.

For his own part, Nagel acknowledges that life can't be an accident, which makes him an exceptional atheist. In his words:

> The inescapable fact that has to be accommodated in any complete conception of the universe is that the appearance of living organisms has eventually given rise to consciousness, perception, desire, action, and the formation of both beliefs and intentions on the basis of reasons. If all this has a natural explanation, the possibilities were inherent in the universe long before there was life, and inherent in early life long before the appearance

of animals. A satisfying explanation would show that the realization of these possibilities was not vanishingly improbable but a significant likelihood given the laws of nature and the composition of the universe. It would reveal mind and reason as basic aspects of a nonmaterialistic natural order.[10]

According to his analysis, a satisfactory account of reality "will have to include teleological elements"[11]—goal-directedness that's nowhere to be found in photons or quarks or electromagnetic fields. That is, "in addition to physical law of the familiar kind, there are other laws of nature that are 'biased toward the marvelous.'"[12]

I admire the lucidity of his thought and his unreserved frankness. If ever there has been a more earnest and clear-thinking atheist, I haven't encountered him or her.

Nor can I think of a version of atheism that's closer to theism. By Nagel's way of thinking, the universe has, from the beginning, been directed toward marvelous goals, and this is what removes what would otherwise be a probabilistic impasse.

The only thing missing from Nagel's view is a way of making this overarching purposiveness coherent. Purpose is, in all our experience, inextricably linked to personhood. When we say that a hammer has a purpose, we mean not that it has its own intent but that it was fashioned with an intended use in mind. And when we speak of mind, we likewise take personhood to be the context. Nagel speaks of mind as though it can be detached from personhood, but I have no way to make that notion coherent, much less correct. I don't think he does either.

Indeed, he openly admits to searching for a coherence that escapes him. "A plausible picture of how we fit into the world" is

not anything he claims to have. It is, rather, his "*hope*."[13]

Why does Nagel not allow himself to acknowledge that God is at the very center of this plausible picture? Here again, I admire the frankness of his own account. Atheists of the Dawkins variety like to pretend that scientific progress has made God less plausible. They love that conclusion. What they lack is an argument to support it.

Nagel refuses to play that game. He's an atheist because he dislikes the thought that he comes under God's authority, which I suspect is true of most atheists, including Dawkins.[14] Nothing intellectual or scientific in that. Just very human.

Here's how Nagel put it in *The Last Word*:

I want atheism to be true and am made uneasy by the fact that some of the most intelligent and well-informed people I know are religious believers. It isn't just that I don't believe in God and, naturally, hope that I'm right in my belief. It's that I hope there is no God! I don't want there to be a God; I don't want the universe to be like that.

My guess is that this cosmic authority problem is not a rare condition and that it is responsible for much of the scientism and reductionism of our time. One of the tendencies it supports is the ludicrous overuse of evolutionary biology to explain everything about life, including everything about the human mind.[15]

In summary, the fact that the very best thinking among atheists acknowledges that life can't be an accident underscores just how obvious, commonsensical, and ultimately undeniable this is.[16] Atheists have their motives for denying the obvious, but these have never come from science or reason, however much they wish to pretend otherwise.

Back to Proteins

Returning now to proteins, what do they add to the picture? The answer is that they exemplify the commonsensical principles by which we rightly reject purpose-free accounts of life. We can fully grasp and affirm these principles without turning to the subject of proteins (or any other technical subject), but in proteins we find elegant confirmation.

The important fraction I referred to above is the likelihood that a random chain of amino acids would have the special physical properties needed to fold into a stable structure that's suitable for a particular function. Putting that in evolutionary terms, assuming a particular new capability that can be achieved with a new protein fold would benefit an organism, and that a genetic mistake in that organism has produced a gene sequence that differs substantially from what existed before, this fraction is the probability that the new gene happens to encode a new protein that performs the desirable new function. Using the results of my experiments on the penicillin-resistance enzyme,[17] I estimated this probability to be in the ballpark of:

1/100,000,000,000,000,000,000, 000,000,000,000,000,000,000,000, 000,000,000,000,000,000,000,000, 000,000,000

Clearly, what Richard Dawkins knew intuitively to be true of random conglomerations of cells—that they have no hope of doing anything useful—is equally true of random conglomerations of amino acids. Indeed, according to the above general principle ("the accomplishment of any high-level function requires an arrangement of things or circumstances that chance is obviously unable to deliver"), random conglomerations must *always* be hopeless messes. What Dawkins overlooked is that this is equally true of *accidental* conglomerations. The smallness of the target assures us these things can't be stumbled upon by accident. And the above fraction, small as it is, merely reflects the target for a single modest-sized enzyme. When we consider larger inventions—everything from the ATP synthase to the hummingbird—the targets become mind-bogglingly smaller.

Stories of these marvels being stumbled upon through the accrual of thousands of lucky accidents should be rejected for the same reason we reject all spooky coincidences. Luck never delivers on such grand scales. It can't.

And again, in the tiny protein marvels we see elegant proof of this. None of the remarkable protein functions we've talked about happen without the remarkable folded structures that enable them to happen. And the folding of a protein chain is, for basic physical reasons, very much like an all-or-nothing process. Either the arrangement of amino acids has the very special properties needed to stabilize a three-dimensional structure or it doesn't. If it doesn't, the chain remains floppy and none of the innumerable functions that require structure are possible. There is no halfway. The gradual steps that Darwin imagined and Dawkins embraces are pure fiction.

So, as in many other cases, scientific discoveries have fully affirmed what common sense has already told us—loud and clear and from an early age.[18] Some would say that this reinforces common sense. I believe it reinforces science.

29

Does the Evidence Support
Universal Common Ancestry?

Casey Luskin

The *central* claim of the theory of evolution, as laid out in 1859 by Charles Darwin in *The Origin of Species*, is that living species, despite their diversity in form and way of life, *are the products of descent (with modification) from common ancestors*,[1] declared an article in the leading journal *Science* in 2005. Yet as other chapters in this book have explained, the "theory of evolution" makes different claims and the term *evolution* can have different meanings. For many, *evolution* simply means "change over time." Both evolutionary scientists and Darwin-skeptics affirm this definition.

A second definition—spotlighted in the *Science* article quoted above—is common ancestry, the hypothesis that living organisms are genetically related through descent with modification. The standard modern interpretation of this definition is "universal common ancestry," where every single living organism shares a common ancestor. Under this view, not only are all living humans related, but we also share a common ancestor with apes, and going back further, we're related to everything from horses to tuna fish to broccoli to foot fungus and bacteria. This definition is

controversial among many (though not all) Darwin-skeptics and is increasingly debated among evolutionary biologists.

The third definition of *evolution*—which claims that natural selection acting upon random mutations was the driving cause of change—is the most controversial among scientists because it holds that the mechanisms that produced change over time (definition one) and common ancestry (definition two) were blind and undirected. The inadequacy of evolution's mechanism (definition three) has been amply addressed by Douglas Axe, Michael Behe, and others in this volume. The purpose of this chapter is to examine *only* the second definition—universal common ancestry.

At first blush, common ancestry may seem less important compared to natural selection. After all, the third definition directly addresses whether the history of life appears unguided—a central question in the debate over intelligent design (ID). Indeed, common ancestry is compatible with ID. For example, one possible way to view ID is that God actively guided the history of life, but did so where organisms are related.

Nonetheless, for a variety of reasons, common ancestry is an important topic.

First, the pursuit of truth is of the utmost importance. If universal common ancestry is true, we should want to know that. If not, we should modify our views accordingly.

Second, although ID theorists may not see universal common ancestry as the most important issue in the debate, many evolutionists apparently feel otherwise. As the *Science* article quoted above notes, common ancestry is viewed as "[t]he central claim of the theory of evolution"—and consequently evolutionists devote a great deal of energy to arguing for it. Indeed, the article maintains not only that "evolutionary trees serve…as the main framework within which evidence for evolution is evaluated," but that common ancestry is a crucial pillar of evolutionary apologetics:

> Phylogenetic trees are the most direct representation of the principle of common ancestry—*the very core of evolutionary theory*—and thus they must find a more prominent place in the general public's understanding of evolution…[W]e can hope that a wider segment of society will come to appreciate the overwhelming evidence for common ancestry and the scientific rigor of evolutionary biology.[2]

However, in their eagerness to take Darwin to the public, evolutionists often make a mistake, confusing evidence for common ancestry as evidence for the full-blown Darwinian story, thus conflating the second and third definitions of evolution.[3] While this rhetorical strategy is logically flawed (evidence for common ancestry is not necessarily evidence for blind natural selection[4]), if the evidence for universal common ancestry is weak, then their arguments for neo-Darwinism face not just a logical problem, but also a factual one.

Third, evolutionists have often used ridicule and outlandish rhetoric in attempts to bully people into accepting universal common ancestry, suggesting something might be awry about the strength of their case. We just saw this when the *Science* article trumpeted the "overwhelming evidence for common ancestry." But even some Christian evolution popularizers do this. In a 2011 InterVarsity Press book, *The Language of Science and Faith*, theistic evolutionists Francis Collins and Karl Giberson compared those who doubt common ancestry to geocentrists, writing that "virtually all geneticists consider that the evidence proves common ancestry with a level of certainty comparable to the evidence that the earth goes around the sun."[5] Elsewhere, Giberson employed similar rhetoric, stating that "biologists today consider the common ancestry of all life a fact on par with the sphericity of the earth"[6]—unsubtly implying that those who doubt common ancestry are no better than flat-Earthers. When people overstate their case and attempt to smear critics, this is often because the evidence for their viewpoint is weak—a red flag suggesting we should look more carefully at the evidence.

Finally, once the mechanism of evolution (definition three) comes under scrutiny, and life's history no longer appears unguided, then evolutionary scientists lose an important (though not necessarily exclusive) rationale for claiming that all life is genetically related. Conversely, if all life is not related, then this challenges the standard core of neo-Darwinian accounts of biological history.[7] Given the strong criticisms of natural selection that have been put forth, and the prominent role of common ancestry among arguments for evolution, universal common ancestry is worth putting on the table for further investigation.

Before we investigate the evidence, it's important to note that it's theoretically possible that common ancestry might be true, or false, at multiple levels of the taxonomic hierarchy. For example, universal common ancestry hypothesizes that all living organisms are related. That hypothesis might be false, but common ancestry could still be true at lower taxonomic levels, such as among all animals, all vertebrates, all mammals, or all primates, etc. Even if we question common ancestry among higher groups, everyone agrees that all humans share a common ancestor.

Ultimately, common ancestry must be evaluated case by case, but doing so within every single taxonomic grouping would require an impossibly lengthy inquiry. This chapter will thus evaluate the case for universal common ancestry as it is commonly articulated in textbooks and popular books by evolution advocates. The case for universal common ancestry is often said to be cumulative, based upon multiple lines of evidence including biogeography, fossils, DNA and anatomical similarities, and embryology.[8] Because common descent is said to be demonstrated by multiple independent lines of congruent evidence, those categories of evidence should be evaluated independently. Let's examine whether the evidence supports common ancestry in those different areas, starting with biogeography.

Biogeography

Biogeography is the study of the distribution of organisms in both time and space over Earth's history, and defenders of neo-Darwinism contend that biogeography supports their viewpoint. For example,

the National Center for Science Education (NCSE), a pro-Darwin advocacy group, cites a "consistency between biogeographic and evolutionary patterns" and argues "[t]his continuity is what would be expected of a pattern of common descent."[9]

Much biogeographical data, however, has little to do with Darwinian evolution, and does not provide special evidence for common ancestry. This data can be easily explained as the result of migration and continental drift—two conventional ideas accepted by virtually everyone in this debate. Moreover, the NCSE's arguments ignore the many biogeographical puzzles that have vexed evolutionary biologists because they show a marked *dis*continuity between biogeography and common ancestry.

Evolutionary explanations of biogeography fail when terrestrial or freshwater organisms appear in a location (such as an isolated island or continent) at which no standard migratory mechanism can explain how those species arrived from their proposed evolutionary ancestors. In other words, take any two populations of organisms, and evolutionary biology claims that if we go back far enough, they must be linked in space and time by descent. But sometimes it's virtually impossible to explain how two populations arrived at their current geographical locations from some common ancestral population.

For example, a severe biogeographical puzzle for common ancestry is the origin of South American monkeys, a large group called *platyrrhines*. Based upon molecular and morphological evidence, New World platyrrhine monkeys are thought to be descended from African Old World or *catarrhine* monkeys.

The fossil record shows that monkeys have lived in South America for about 30 million years.[10] But plate tectonics shows that Africa and South America separated around 100–120 million years ago (mya), and South America was an isolated island continent from about 80 to 3.5 mya.[11] If South American monkeys split from African monkeys around 30 mya, neo-Darwinism must somehow explain how monkeys crossed hundreds, if not thousands, of kilometers of open ocean to end up in South America.

This poses a major problem for common ancestry—one recognized by multiple experts. A textbook on human evolution states, "The origin of platyrrhine monkeys puzzled paleontologists for decades…When and how did the monkeys get to South America?"[12] Primatologists John Fleagle and Christopher Gilbert explain:

> The most biogeographically challenging aspect of platyrrhine evolution concerns the origin of the entire clade. South America was an island continent throughout most of the Tertiary… and paleontologists have debated for much of this century how and where primates reached South America.[13.]

For those unfamiliar with the explanations of evolutionary scientists, their responses to such puzzles can be almost too incredible to believe. They propose *not that common descent might be wrong,* but that monkeys must have *rafted* across the Atlantic Ocean, from Africa to South America, to colonize the New World. The textbook explains: "The 'rafting hypothesis' argues that monkeys evolved from prosimians once and only once in Africa, and…made the water-logged trip to South America."[14]

Of course, there can't be just one seafaring monkey, or it will die leaving no offspring. Thus, at least two monkeys (or perhaps a single pregnant monkey) must have made the rafting voyage.

Fleagle and Gilbert admit the rafting hypothesis "raises a difficult biogeographical issue" because "South America is separated from Africa by a distance of at least 2,600 km [~1,600 miles], making a phylogenetic and biogeographic link between the primate faunas of the two continents seem very *unlikely*."[15] (South America and Africa were closer in the past due to continental drift, but it is still estimated that they were 1,500–2,000 km [~1,000–1,200 miles] apart at the time monkeys made the journey—a huge distance to travel by ocean.[16]) But they are wedded to common ancestry and are obligated to find such a "link," whether likely or not. Given no other evolutionary options, they conclude that "the rafting hypothesis is the most *likely* scenario."[17] In other words, the "unlikely" monkey-rafting hypothesis is made "likely" only because they *assume* common descent *must* be true.

Needless to say, the rafting hypothesis faces serious problems. Mammals like monkeys have high metabolisms and require large amounts of food and water.[18] Fleagle and Gilbert thus concede that "over-water dispersal during primate evolution seems truly amazing for a mammalian order," and conclude that "[t]he reasons for the prevalence of rafting during the course of primate evolution remain to be explained."[19] Or, as another expert puts it, "the mechanical aspect of platyrrhine dispersal [is] virtually irresolvable" because evolutionary models "must invoke a transoceanic crossing mechanism that is implausible (rafting) or suspect…at best."[20] Fleagle and Gilbert compare the monkeys' voyage to winning the lottery: "by a stroke of good luck anthropoids were able to 'win' the sweepstakes."[21]

This is not the only case where evolutionary biologists are forced to invoke rafting or other speculative mechanisms of "oceanic dispersal" to explain away difficult problems. Other biogeographical conundra include the presence of lizards and large rodents in South America,[22] the arrival of bees, lemurs, and other mammals in Madagascar,[23] the appearance of elephant fossils on various islands,[24] the existence of freshwater frogs across isolated oceanic island chains,[25] and many other examples.[26]

This problem exists for extinct species as well. A 2007 paper in *Annals of Geophysics* notes the "still unresolved problem of disjointed distribution of fossils on the opposite coasts of the Pacific."[27] However, this paper doesn't invoke rafting—instead, it proposes something even more unlikely: populations became separated due to an "expanding Earth"—a long-discarded geological hypothesis (different from well-accepted modern theories of plate tectonics) that could only be taken seriously when trying to rescue common descent from falsification. A review in *Trends in Ecology and Evolution* describes the problem:

> A classic problem in biogeography is to explain why particular terrestrial and freshwater taxa have geographical distributions that are broken up by oceans. Why are southern beeches (*Nothofagus spp.*) found in Australia, New Zealand, New Guinea and southern South America? Why are there iguanas on the Fiji Islands, whereas all their close relatives are in the New World?[28]

After considering several "unexpected" biogeographical examples, the review concludes that "these cases reinforce a general message of the great evolutionist [Darwin]: given enough time, many things that seem unlikely can happen."[29]

Indeed, that does appear to be the message here. If you're going to retain common ancestry, you must accept some extraordinary biogeographical claims. When evolutionary scientists are forced to appeal to fantastical "expanding Earth" hypotheses, or "unlikely" accounts of species rafting across oceans, common ancestry clearly faces a challenge.

The Fossil Record

A popular college-level biology textbook explains, "Fossils are the only direct record of the history of life."[30] This seems generally correct, making the fossil record an ideal place for testing universal common ancestry.

The textbook's author, geologist Donald Prothero, has elsewhere written, "The fossil record is an amazing testimony to the power of evolution, with documentation of transitions that Darwin could only have dreamed about."[31] If you feel otherwise, Prothero continues, then you're a "creationist" who shares "much in common with the Neo-Nazi Jew-hating Holocaust deniers."[32]

But what do fossils say about evolution? If all living organisms are related, then the fossil record should produce transitional organisms that show the intermediate stages between life's various groups. But the history of life bears a repeated pattern of explosions, where new fossil forms appear abruptly without clear evolutionary precursors. Perhaps the most famous example of a fossil explosion is the Cambrian explosion, where many of the major living animal groups (called *phyla*) appear in the fossil record in a sudden geological eyeblink—five to ten million years, and possibly less.[33] Before the Cambrian, very few fossils having anything to do with modern phyla are found in the record. As one invertebrate zoology textbook states:

Most of the animal phyla that are represented in the fossil record first appear, "fully formed" and identifiable as to their phylum, in the Cambrian some 550 million years ago…The fossil record is therefore of no help with respect to understanding the origin and early diversification of the various animal phyla…[34]

Some defenders of neo-Darwinism respond by denying that the Cambrian explosion occurred. However, biologists familiar with the evidence have concluded that the Cambrian explosion was real, as the eminent Dutch biologist Martin Scheffer explains:

> The collapse of the [Precambrian] Ediacaran fauna is followed by the spectacular radiation of novel life-forms known as the Cambrian explosion. All of the main body plans that we know now evolved in as little as about 10 million years. It might have been thought that this apparent explosion of diversity might be an artifact. For instance, it could be that earlier rocks were not as good for preserving fossils. However, very well preserved fossils do exist from earlier periods, and it is now generally accepted that the Cambrian explosion was real.[35]

The diversity of complex animals that appear in the Cambrian explosion is impressive, ranging from worms to arthropods to mollusks to even vertebrate fish. But some familiar animals—like dinosaurs, parrots, or camels—don't appear until much later. Some evolutionists claim this progression demonstrates common ancestry. It doesn't.

While the fact that life has "changed over time" doesn't bother ID, the fact that

reptiles, birds, and mammals don't appear until after the Cambrian period could be a major problem for neo-Darwinian evolution if, whenever these groups do appear, they do so in an abrupt fashion that betrays the predictions of neo-Darwinian common ancestry. For many of these subgroups of animals, again, they appear abruptly, in patterns of explosions.

"While during the Cambrian explosion numerous phyla and classes representing basic body plans originated," writes paleontologist Walter Etter, "the Ordovician radiation was manifested by an unprecedented burst of diversification at lower taxonomic levels."[36] He continues, "The almost exponential increase in diversity was much more rapid during this Great Ordovician Biodiversification Event (GOBE) than at any other time [from the Cambrian to the present]," noting the increase was "for the most part abrupt."

Regarding the origin of major fish groups, Columbia University geoscientist Arthur Strahler wrote, "This is one count in the creationists' charge that can only evoke in unison from paleontologists a plea of *nolo contendere* [no contest]."[37] We also see an "explosive" and rapid appearance of other marine organisms such as ammonites,[38] other hard-shelled marine invertebrates,[39] and mosasaurs.[40]

As for plants, a paper in *Annual Review of Ecology and Systematics* explains that the origin of land plants "is the terrestrial equivalent of the much-debated Cambrian 'explosion' of marine faunas."[41] Regarding angiosperms (flowering plants), scientists refer to a "big bloom" or "explosion"[42] event. As one paper states, "[a]ngiosperms appear rather suddenly in the fossil record…with no obvious ancestors for a period of 80–90 million years before their appearance."[43]

Land animals show similar patterns. The fossil record shows an "explosion" of tetrapods when terrestrial vertebrates appear.[44] A 2011 article in *Science* admitted that tracing the evolutionary origin of major dinosaur groups "has been a major challenge for paleontologists."[45] A prominent ornithology textbook observes the "explosive evolution" of major living bird groups.[46] Similarly, authorities cite an "explosion" or "explosive diversification" of major mammal groups in the Tertiary.[47] Paleontologist Niles Eldredge notes that "there are all sorts of gaps: absence of gradationally intermediate 'transitional' forms between species, but also between larger groups—between, say, families of carnivores, or the orders of mammals."[48] Additional examples of fossil explosions are discussed by paleontologist Günter Bechly in chapter 31 of this volume.

Eldredge and others attempt to explain the abrupt appearance of major fossil groups through punctuated equilibrium—a flawed model that essentially renders common ancestry untestable with regard to fossils.[49] Nonetheless, a literal reading of the fossil record consistently shows a pattern of abrupt explosions of new types of organisms that contradicts common descent—the opposite of a Darwinian process of small changes adding up to larger ones. As biologist Jeffrey Schwartz at the University of Pittsburgh explains:

> We are still in the dark about the origin of most major groups of organisms. They appear in the fossil record as Athena did from the head of Zeus—full-blown and raring to go, in contradiction to Darwin's depiction of evolution as resulting from the gradual accumulation of countless infinitesimally minute variations.[50]

Comparing those who recognize this non-Darwinian pattern to "Holocaust deniers" won't make the problem go away.

DNA- and Anatomy-Based Phylogenetic Trees

One of the most common arguments for universal common ancestry is the universality of the genetic code—the claim that all life uses the same nucleotide triplets to encode the same amino acids in DNA.[51] However, the genetic code *isn't* universal; many variants in the genetic code are known among organisms.[52]

If the universality of the genetic code provides evidence for universal common ancestry, does its nonuniversality count as evidence against? In his chapter in this volume, Paul Nelson argues it ought to. Nonetheless, despite the variants, the vast majority of organisms do use the same "standard code," and all life forms employ similar building blocks (such as amino acids and nucleotides) and other biomolecules, such as DNA, RNA, and proteins. Are such widespread biomolecular similarities evidence for common ancestry? A 2010 paper in *Nature*, "A formal test of the theory of universal common ancestry," argued yes:

> [T]he "universal" in universal common ancestry is primarily supported by two further lines of evidence: various key commonalities at the molecular level (including fundamental biological polymers, nucleic acid genetic material, L-amino acids, and core metabolism) and the near universality of the genetic code.[53]

The article's author, evolutionary biochemist Douglas Theobald, concluded that universal common ancestry is the "best" explanation for these biomolecular similarities. But "best" compared to what? Theobald tested universal common ancestry against the exceedingly unlikely hypothesis that living organisms independently evolved the same biomolecules by sheer "chance." Of course, this is impossible, so universal common ancestry only appeared compelling because it was being compared to a preposterous null hypothesis. As critics writing in *Biology Direct* put it, they "cringed when they saw the Theobald paper" because it tests the "trivial" hypothesis that "significant sequence similarity might arise by chance as opposed to descent with modification."[54]

It's true that universal common ancestry is one possible explanation for genetic similarities between organisms—but are there others? Intelligent agents frequently reuse the same parts in different designs to meet functional requirements, such as reusing wheels on cars and airplanes, or reusing the key computer codes in different versions of Microsoft Windows. As Paul Nelson and Jonathan Wells observe:

> An intelligent cause may reuse or redeploy the same module in different systems, without there necessarily being any material or physical connection between those systems. Even more simply, intelligent causes can generate identical patterns independently...If we suppose that an intelligent designer constructed organisms using a common set of polyfunctional genetic modules—just as human designers, for instance, may employ the same transistor or capacitor in a car radio or a computer, devices that are not "homologous" as artifacts—then we

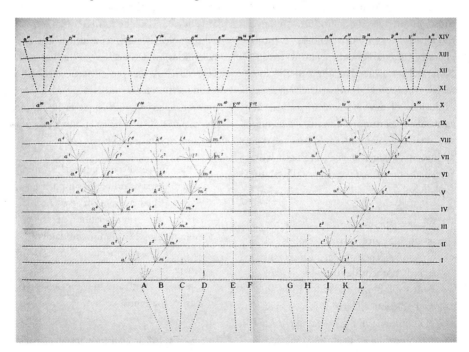

Figure 1. Darwin's tree of life. Credit: Public domain.

can explain why we find the "same" genes expressed in the development of what are very different organisms. [55]

Thus, common design—the intentional reuse of a common blueprint or components—is also a viable explanation for widespread similarities among biomolecules found in different types of organisms.

But it isn't just mere similarity among biomolecules that evolutionary biologists claim demonstrates universal common ancestry. (Contra Theobald, not all fundamental biomolecules are universal among organisms.[56]) Biologists often claim that *patterns* of similar nucleotide and amino acid sequences of genes and proteins allow organisms to be organized into a phylogenetic "tree of life."[57]

This "tree of life" was Darwin's only illustration in *On the Origin of Species* (Figure 1), and it has become the most famous icon representing his theory. But does the tree of life exist? In the 1990s, biologists made a startling discovery: Life falls into three basic domains that cannot be resolved into a treelike pattern. Thus, the prominent biochemist W. Ford Doolittle lamented,

> Molecular phylogenists will have failed to find the "true tree," not because their methods are inadequate or because they have chosen the wrong genes, but because the history of life cannot properly be represented as a tree.[58]

He explained that for many biologists, "It is as if we have failed at the task that Darwin set for us: delineating the unique structure of the tree of life."[59] The basic problem is that some genes yield one version of the "tree of life," but other genes require an entirely different tree. To put it another way, biological

similarity is constantly appearing in places where it wasn't predicted by common descent, leading to conflicts between phylogenetic trees.

Numerous papers have noted the prevalence of contradictory phylogenetic trees. A 1998 study in *Genome Research* plainly observed "different proteins generate different phylogenetic tree[s]."[60] A 2009 article in *Trends in Ecology and Evolution* acknowledged "evolutionary trees from different genes often have conflicting branching patterns."[61] A 2013 paper in *Trends in Genetics* reported that "the more we learn about genomes the less tree-like we find their evolutionary history."[62] A 2012 paper in a technical journal observed that "phylogenetic conflict is common, and frequently the norm rather than the exception."[63]

Candid admissions came in a 2009 article in *New Scientist* titled "Why Darwin Was Wrong About the Tree of Life."[64] It quoted researcher Eric Bapteste admitting that "the holy grail was to build a tree of life," but "today that project lies in tatters, torn to pieces by an onslaught of negative evidence."[65] According to the article, "[m]any biologists now argue that the tree concept is obsolete and needs to be discarded."[66] The paper recounted the results of a study by Michael Syvanen, which compared 2,000 genes across six animal phyla:

> In theory, he should have been able to use the gene sequences to construct an evolutionary tree showing the relationships between the six animals. He failed. The problem was that different genes told contradictory evolutionary stories.[67]

Syvanen succinctly explained the problem: "We've just annihilated the tree of life. It's not a tree anymore, it's a different topology entirely. What would Darwin have made of that?"[68]

Some defenders of common ancestry propose that these conflicts merely reflect gene-swapping among microorganisms at the base of the tree. But Carl Woese observes, "Phylogenetic incongruities [conflicts] can be seen everywhere in the universal tree, from its root to the major branchings within and among the various taxa to the makeup of the primary groupings themselves."[69]

Conflicts also exist between trees based upon molecules (DNA) versus those based upon body structure (called *morphology*). A review article in *Nature* titled "Bones, Molecules, or Both?" explained that "[e]volutionary trees constructed by studying biological molecules often don't resemble those drawn up from morphology," admitting that "[b]attles between molecules and morphology are being fought across the entire tree of life."[70]

A classic example involves attempts to construct a phylogenetic tree of the animal phyla. Traditionally, many phyla were grouped according to whether they have a central body cavity called a *coelom*. But molecular data contradicted that grouping, and instead, placed organisms that are morphologically distinct, such as nematodes and arthropods, very close—a result that *Nature* called "surprising."[71] Higher up the tree, conflicts persist. In 2014, the sequencing of various bird genomes showed that birds that were previously thought to be closely related—water birds, birds of prey, and songbirds—evolved their groups' defining traits convergently.[72] As *Nature* put it, "the tree of life for birds has been redrawn."[73]

Even when the data does fit a treelike pattern, that does not necessarily demonstrate common ancestry. It's important to appreciate that even the best-supported phylogenetic trees are still based upon the *assumption* that biological similarity results from inheritance from a common ancestor. A bioinformatics textbook admits this point:

> The key *assumption* made when constructing a phylogenetic tree from a set of sequences is that they are all derived from a single ancestral sequence, i.e., they are homologous.[74]

Evolutionary biology doesn't test whether organisms are related via a tree, but rather *assumes* that common ancestry is true. That assumption—*and it's merely an assumption*—is so deeply embedded in evolutionary thinking that theorists often forget it's there. One rare technical article admitted that this assumption exists, and often fails:

> [M]olecular systematics is (largely) based on the assumption…that degree of overall similarity reflects degree of relatedness…[T]he history of molecular systematics and its claims in the context of molecular biology reveals that *there is no basis for the "molecular assumption."*[75]

This *main assumption* of tree-building sounds nice, but the trade secret of evolutionary biology is that the assumption fails frequently, and when it does, evolutionary biologists employ various methods to force-fit the data back into a tree. An article in *Annual Review of Genetics* tacitly admits this fact:

> Because tree analysis tools are used so widely, *they tend to introduce a bias into the interpretation of results.* Hence, one needs to be continually reminded that submitting multiple sequences

(DNA, protein, or other character states) to *phylogenetic analysis produces trees because that is the nature of the algorithms used.*[76]

The point is that just because someone presents a nice, neat, tidy phylogenetic tree does not necessarily mean that the underlying dataset is treelike.

For example, when responding to Stephen C. Meyer's book *Darwin's Doubt*, evolutionary biologist Nicholas Matzke cited two phylogenetic trees to claim that the relationships of various Cambrian explosion animals could be explained by common ancestry.[77] Yet in these trees, 43.5 percent and 61.6 percent of the characters (Figure 2), respectively, were *not* distributed in a

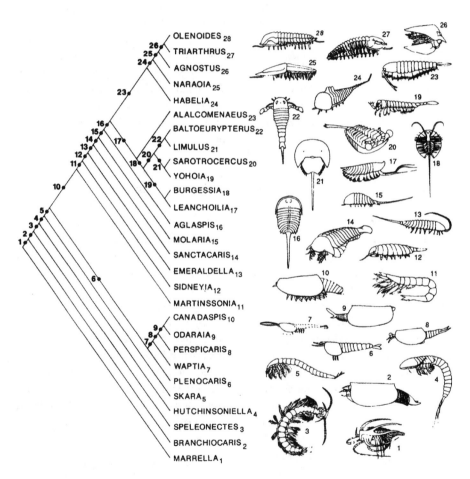

Figure 2. A tree cited by ID critic Nicholas Matzke to purportedly demonstrate how common ancestry explains evolutionary relationships between various arthropods. The *main assumption* failed 61.6 percent of the time, which the authors admitted was a "rather low" success rate. Credit: Figure 1, Derek Briggs and Richard Fortey, "The Early Radiation and Relationships of the Major Arthropod Groups," *Science* 246 (October 13, 1989), 241-243. Reprinted with permission from AAAS.

treelike pattern.[78] If the *main assumption* of tree-building fails 43.5 percent of the time, can one justify making it in the first place? Similar examples abound, with trees published in the literature reporting that the main assumption failed with respect to 57 percent,[79] 64 percent,[80] 65 percent,[81] or even 90 percent[82] of the data. Again, that means *more than 50 percent of the time a trait* did not have a treelike distribution. Perhaps the main assumption should be rewritten as *biological similarities indicate inheritance from a common ancestor, except for when they don't.*

Often, when the *main assumption* fails, evolutionary biologists force the data back into a treelike structure by invoking independent gain (called *convergent evolution*) or loss of traits. For instance, under the main assumption, the reason we have two eyes and dogs have two eyes is because humans and dogs shared a common ancestor with two eyes. That's fine. Yet cephalopods (octopi and squid) also have two eyes—and their "camera eye" design is almost identical to vertebrate eyes, even though standard evolutionary thinking doesn't hold that our most recent common ancestor with cephalopods had two eyes, much less a camera eye.

"Perhaps," the evolutionist replies, "both organisms independently evolved two highly similar camera eyes, just by chance." This is called *convergent evolution*, where two different organisms independently stumbled upon highly similar complex biological designs. Richard Dawkins admits there are "numerous examples…in which independent lines of evolution appear to have converged, from very different starting points, on what looks very like the same endpoint." Yet in the same breath he acknowledges that "it is vanishingly improbable that exactly the same evolutionary pathway should ever be travelled twice." Undaunted, Dawkins declares "it is all the more striking a testimony to the power of natural selection."[83]

Dawkins's dubious logic aside, convergent evolution shows that the main assumption has failed—that the data does not fit a tree, and biological similarity *does not* necessarily indicate inheritance from a common ancestor. This challenges the heart of the methodology used to infer common descent.

Now, some datasets are largely treelike. But in those happy cases, how can we know the main assumption was warranted, given how often it fails elsewhere, and given that there are other potential explanations for shared biological similarity such as common design? Computer scientist Winston Ewert applied the concept of "common design" to produce a "dependency graph"[84] model of organismal relationships based upon the principle that software designers frequently reuse the same coding modules in different programs. He tested his model by comparing the distribution of gene families in nine diverse organisms to a treelike pattern predicted by neo-Darwinism versus a dependency graph distribution used by computer programmers. The results showed that a common design-based "dependency graph" model fit the genetic data from these species 10^{3000} times better than a Darwinian evolutionary tree.[85] Common design is a superior explanation for many of the shared similarities among different organisms, and unlike common ancestry, it can account for shared traits that don't fit a treelike pattern.

Embryology

In an 1860 letter to the American botanist Asa Gray, Charles Darwin urged that embryology was "by far the strongest single class of facts in favor"[86] of his theory. Much has changed since Darwin penned those words, but embryology remains a favorite line of evidence cited to support common descent.

Most modern biology textbooks will portray the early embryos of different vertebrate species as highly similar, then claiming these similarities reflect common ancestry. Holt's *Life Science* provides typical language: "Early in development, the human embryos and the embryos of all other vertebrates are similar. These early similarities are evidence that all vertebrates share a common ancestor."[87]

For decades, students were also taught that "ontogeny recapitulates phylogeny." Called *recapitulation theory*, this idea was promoted by the German biologist Ernst Haeckel, who believed that the development of an organism (ontogeny) replays (recapitulates) its evolutionary history (phylogeny). Because the standard evolutionary view holds that humans evolved from fish, recapitulation theory taught that at one point between conception and birth, everyone went through a "fish stage."

Biologists now know that vertebrate embryos do *not* replay their supposed earlier evolutionary stages and have rejected recapitulation theory.[88] The concept has been removed from textbooks, but many still include inaccurate diagrams that overstate

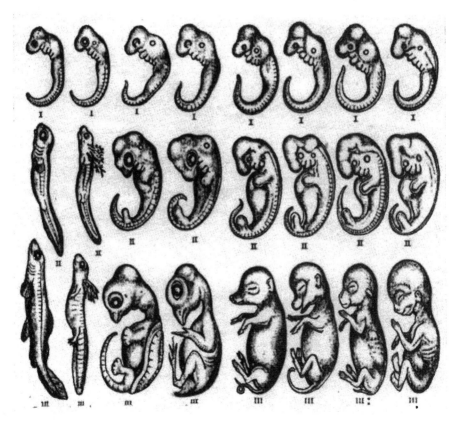

Figure 3. Haeckel's embryo drawings depicting (in this order): fish, salamander, tortoise, chicken, hog, calf, rabbit, human. Credit: George Romanes's 1892 book *Darwinism Illustrated* (public domain).

the degree of similarity between vertebrate embryos. Indeed, the journal *Science* observed that "[g]enerations of biology students may have been misled" by Haeckel's embryo drawings in textbooks.[89]

These drawings—commonly reprinted or adapted in textbooks[90]—overstate the degree of similarity between vertebrate embryos in their earliest stages. According to Stephen Jay Gould, Haeckel's methods were "fraudulent" because he "simply copied the same figure over and over again"[91] without adding differences between embryos of different species. This led embryologist Michael Richardson to call them "one of the most famous fakes in biology."[92]

In 2000, biologist Jonathan Wells published *Icons of Evolution,* which raised the public's consciousness about Haeckel's fraud, ultimately forcing many publishers to remove Haeckel's inaccurate drawings from most textbooks. Many textbooks, however,

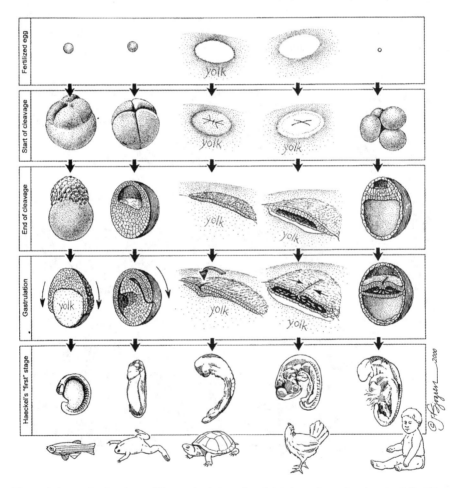

Figure 4. Accurate drawings of the early stages of vertebrate embryo development. Credit: © Jody F. Sjogren, 2000, as used in Figure 5-3 in Jonathan Wells, *Icons of Evolution: Science or Myth?* (Washington, DC: Regnery, 2000). Used with permission.

still claim the early stages of vertebrate development are highly similar, leading to textbook captions like the one quoted at the beginning of this section. But are those captions accurate?

No, they're not. Embryologists have found considerable differences among vertebrate embryos from their earliest stages onward, contradicting what we are told to expect from common ancestry.

Two of the earliest stages of vertebrate development are *cleavage* and *gastrulation*. During cleavage, a newly fertilized zygote undergoes rapid cell division until the embryo becomes a tiny ball of cells, laying out the basic axes that will define the body plan. Next, during gastrulation, the embryo increases in size while forming distinct germ layers that will later develop into individual organs. Yet a paper in *Systematic Biology* states that "early stages as initial cleavages and gastrula[tion] can vary quite extensively across vertebrates."[93]

Likewise, a 2010 paper in *Nature* stated, "Counter to the expectations of early embryonic conservation [i.e., similarity], many studies have shown that there is often remarkable divergence between related species both early and late in development."[94] Or, as another article in *Trends in Ecology and Evolution* stated, "Despite repeated assertions of the uniformity of early embryos within members of a phylum, development...[in those early stages] is very varied."[95]

Rather than looking highly similar in their early stages, vertebrate embryos look more like how they are depicted in Figure 4 on the previous page.

To their credit, some evolutionary biologists acknowledge that vertebrate embryos begin development differently, but then they claim that embryos pass through a highly similar midpoint stage, called the

pharyngular or *phylotypic* or *tailbud* stage. They propose an "hourglass model," where this converging midpoint stage of development reveals common ancestry, as depicted in Figure 5.

New atheist and developmental biologist PZ Myers named his popular blog *Pharyngula*, where he argues that "[v]ertebrate embryos at the phylotypic or pharyngula stage do show substantial similarities to one another that are evidence of common descent. That's simply a fact."[96] But does this pharyngula stage exist?

In a groundbreaking study published in *Anatomy and Embryology*, a team of embryologists noted that the phylotypic stage is sometimes treated as "a biological concept for which no proof is needed."[97] After

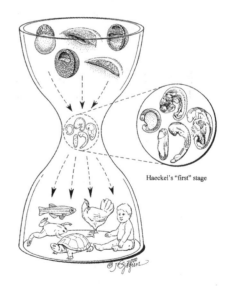

Figure 5. The "hourglass model" of development, where vertebrate embryos start development differently but are said to appear somewhat similar at a midpoint in development. Credit: © Jody F. Sjogren 2000, as used in Figure 5-4, Jonathan Wells, *Icons of Evolution: Science or Myth?* (Washington, DC: Regnery, 2000). Used with permission.

photographing vertebrate embryos during this purportedly similar stage, they found differences in major traits, including body size, body plan, growth patterns, and timing of development, concluding the evidence is "[c]ontrary to the evolutionary hourglass model" because vertebrate embryos show "considerable variability" during "the purported phylotypic stage." In their view, this "wide variation in morphology among vertebrate embryos is difficult to reconcile with the idea of a phylogenetically conserved tailbud stage."[98]

Likewise, a study in *Proceedings of the Royal Society of London* found that embryological data runs "counter to the predictions" of the phylotypic stage, since "phenotypic variation between species was highest in the middle of the developmental sequence." It noted a "surprising degree of developmental character independence argues against the existence of a phylotypic stage in vertebrates."[99]

Even PZ Myers has conceded that early vertebrate embryos can "vary greatly"[100] and that "there is wide variation in the status of the embryo."[101] Yet he tried to explain why these facts don't challenge common ancestry: "I wish I could get that one thought into these guys heads," he wrote. "[E]volutionary theory predicts differences as well as similarities."[102]

That's intriguing. Earlier Myers had cited the "substantial similarities" between vertebrate embryos as "evidence of common descent." But later, when forced to admit the "wide variation" among embryos, he argued that "evolutionary theory predicts differences" too. Perhaps so, but then how can he cite the "similarities" among embryos in the pharyngula stage as evidence for common ancestry?

In reality, Dr. Myers's comments reflect the fact that in practice, evolutionary theory predicts whatever it happens to find. In other words, common ancestry predicts nothing. His logic might help save the theory from falsification, but it doesn't construct a robust model that makes testable predictions. As the old adage says, "The theory that explains anything really explains nothing."

Is Universal Common Ancestry Testable?

At the beginning of this chapter, we noted that the case for common descent is often described as cumulative, based upon multiple lines of evidence, including biogeography, fossils, DNA, anatomy, and embryology. How is the theory faring?

In biogeography, evolutionists refused to let biogeographical conundrums become a challenge to common ancestry and resorted to unlikely explanations where species must raft across vast oceans in order to preserve common descent. Paleontology fails to reveal the continuous branching pattern predicted by common ancestry, and the fossil record is dominated by abrupt explosions of new life-forms. Yet rather than admitting a problem for common ancestry, paleontologists preferred weak models like punctuated equilibrium.

Regarding molecular and morphology-based trees, conflicting phylogenies have left the "tree of life" in tatters. But by invoking convergent evolution and other ad hoc explanations, evolutionary biologists have force-fitted the data into trees, tolerating frequent failures of their core assumptions in order to retain common ancestry. Similar problems confound embryology, where evolutionary biologists predict similarities will exist between vertebrate embryos, except for when we find differences, and then it predicts those too.

Much data contradicts the sometimes-made predictions of common descent, but what, if anything, does evolutionary biology actually predict?

As PZ Myers has shown us, common descent seems to predict whatever is expedient. If there's any clear pattern here, it's this: Too often the data fails to fit the predictions of universal common descent, but when that happens, proponents of common descent simply change their predictions, proposing dubious secondary explanations to retain their core model. This raises the question of the scientific status of common descent. At best, it's a scientific theory that is contradicted by much evidence. At worst, it's not even a scientific theory that makes concrete, testable predictions.

For these and many other reasons, even some mainstream evolutionary scientists are becoming increasingly skeptical of universal common ancestry.[103] A 2009 paper in *Trends in Genetics* wrote that breakdowns in core neo-Darwinian tenets like the "traditional concept of the tree of life" or that "natural selection is the main driving force of evolution" indicate "the modern synthesis has crumbled, apparently, beyond repair."[104] A paper that same year in *Nature Reviews Microbiology* maintained that "there is no such thing as a tree of life" because "the tree look[s] more like a forest."[105] A 2012 paper in *Annual Review of Genetics* explicitly doubted universal common ancestry, suggesting "life might indeed have multiple origins."[106] Another paper in *Biology Direct* noted that the "sudden emergence" of new complex life-forms contradicts a "tree pattern":

> Major transitions in biological evolution show the same pattern of sudden emergence of diverse forms at a new level of complexity. The relationships between major groups within an emergent new class of biological entities are hard to decipher and do not seem to fit the tree pattern that, following Darwin's original proposal, remains the dominant description of biological evolution.[107]

These authors, of course, support some form of unguided evolution. But the precise reason that they are critiquing the classical evolutionary model is because much data contradicts universal common ancestry. Twenty-first-century science seems to be taking biology beyond universal common ancestry and rejecting the neo-Darwinian "tree of life."

Can Universal Common Descent Be Tested?

Paul Nelson

C an universal common descent (UCD) be tested? The short answer is no. But also...*yes.*

The long answer, which will require this whole chapter to lay out, explains why this apparent contradiction exists in evolutionary theory. Stay with me: the long answer is worth your time. You will gain a fresh—and I hope surprising—perspective on evolution and its relation to biological evidence. A word of warning, however: This new perspective takes some getting used to, because saying that Darwin's tree of life is both testable *and* untestable sounds paradoxical or even crazy on its face. How could any theory that applies to every organism on Earth escape normal scientific scrutiny (that is the no answer), and yet also be vulnerable to refutation by evidence (that is the yes)?

Come and see. To make the counterintuitive and seemingly contradictory thesis of this chapter less daunting, we should start with the familiar. Then we will move on to the more difficult aspects, but into more rewarding territory as well.

Introduction: Universal Common Descent

Charles Darwin knew what was most important to him about the theory of evolution, and it was *not* the mechanism of natural selection. In May 1863, in a letter to Harvard botanist Asa Gray, Darwin described what he called the "turning point" of his theory. This logical axis represented the defining proposition of evolution, or its principal empirical claim: "change of species by descent." As Darwin elaborated, "Personally, of course, I care much about Natural Selection; but that seems to me utterly unimportant compared to the question of *Creation* or *Modification*."[1]

By "Creation," Darwin meant the separate origins of the species within the major groups of plants and animals; "Modification" meant their common origin (Figure 1). History has borne out his judgment of relative importance. While Darwin was still alive, natural selection fell under the critical scrutiny of some of his closest friends (e.g., Thomas Henry Huxley) and followers (e.g., St. George Mivart, who later turned against Darwin). Darwin himself diminished

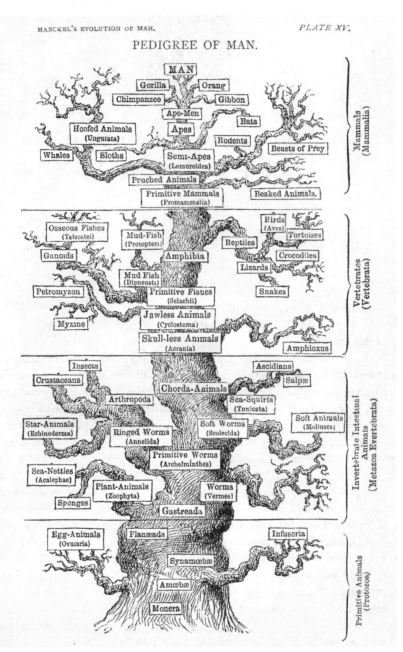

Figure 1. This figure was drawn by the German embryologist Ernst Haeckel, who was also an accomplished artist. It reflects the widespread scientific and popular perception that *evolution* meant the common ancestry of all life on Earth, as argued by Darwin in the *On the Origin of Species* (1859, 484). Credit: Ernst Haeckel, *The Evolution of Man: A Popular Exposition of the Principal Points of Human Ontogeny and Phylogeny* (New York: Appleton & Co., 1897), Plate XV, public domain, via Wikimedia Commons.

selection's explanatory role in favor of "use and disuse" in the expanded sixth edition of *On the Origin of Species* (1872). This skepticism about the efficacy of natural selection has waxed and waned, but in some form, it has been steadily counted as a significant opinion within evolutionary biology up to the present day.

The core proposition of descent with modification from a single common ancestor, by contrast, fared much better. This is the textbook historical geometry of the "tree of life"[2] (TOL): the theory of universal common descent. Throughout the twentieth century, Darwin's argument that all species on Earth find their place in the TOL, rooted in some "one primordial form, into which life was first breathed,"[3] ranked as the canonical geometry or topology of relatedness in historical biology. Doubters were few, and in every instance, stood outside the mainstream of biology. Even those who saw deep discontinuities among living things, such as the Scottish biologist and mathematician D'Arcy Thompson (1860–1948), were certain to affirm universal relatedness, with all organisms stemming ultimately from a single root.[4] Let us abbreviate this root, Darwin's "one primordial form," now known as the *Last Universal Common Ancestor*, as LUCA.

With the arrival of the twenty-first century, and the advent of rapid and increasingly inexpensive DNA sequencing producing an ocean of unexpected genetic findings, certitude about UCD and the real existence of LUCA has begun to fray (see discussion below). Nonetheless, UCD remains the default position of most working biologists, not to mention scientists, generally speaking. Organizations such as the Christian group BioLogos express this default acceptance of UCD as follows:

There is very little doubt in the scientific community about this broad characterization of evolution (anyone who claims otherwise is either uninformed or deliberately trying to mislead). The observational evidence explained by common ancestry is overwhelming.[5]

As we shall see, however, despite holding its default position for most biologists, UCD is now the subject of much more than "very little doubt." But that is a story for the end of the chapter.

UCD in its Standard Formulation

Consider five representative formulations of UCD from prominent evolutionary biologists:

Evolution asserts that the pattern of similarity by which all known organisms may be linked is the natural outcome of some process of genealogy. In other words, all organisms are related.[6]

The theory of evolution predicts that all organisms are related to one another by a unique, hierarchical genealogy.[7]

We shall take evolution to mean that all modern species have descended from a single common ancestral species; species have thus changed in appearance as they have descended and diversified through time. According to the theory of evolution, life originated only once and species are not immutable. The present diversity of forms has been produced from one ancestor by the splitting of species.[8]

All living organisms of the earth are connected, by reproduction, as ancestor and descendant in an uninterrupted nexus of relationship.[9]

It is widely accepted that all life on Earth today is descended from a common ancestral cellular organism that existed some time between 1.5 and 3.0 billion years ago.[10]

Observe that UCD is expressed using generic or common nouns—"life," "organism," "ancestor," "species," "descendant"—with a transitive verb denoting the uniquely biological relationship of *begetting* (i.e., parents producing progeny): "linked," "related," "connected," "descended." In this general formulation, UCD provides no mechanistic details about how such begetting occurs, nor about the specific biological features of organisms (e.g., their biochemistry, genetics, or physiology). Those details are not required, however—merely the postulate that LUCA existed, and that organisms come from other organisms (*omne vivum ex vivo*), with "like producing like," which humans since antiquity have understood to be distinctive of the living world. The global domain of UCD follows from the universal quantifier "all," and would be lost if we substituted the existential quantifier "some."

Thus: Is *x* an organism on Earth? If yes, then according to UCD, *x* shares its history of begetting relations with any other Earth organism *y*. By implication, the shared ancestors within the begetting lineages of *x* and *y*, respectively, must include (at least) the last universal common ancestor, LUCA. Given UCD, one can speak definitively about the history of any living thing—namely, that its ancestors must have included LUCA—but without having to supply any more

information about *x* than that *x* is, or was, an organism on this planet. LUCA must also have been an organism, of course, but again, the standard formulation of UCD requires one to say no more than that.

The standard formulation of UCD is therefore perfectly general. Indeed, UCD can be (and often is) stated as a single short sentence: *All organisms on Earth, extant or extinct, share common ancestry from LUCA.* The simplicity and universality of UCD are enormously attractive. Scientists want theories that compress as much as possible into as little as possible.

The Cost of Perfect Generality

In its standard formulation, however, UCD is untestable by observation. This is the *no* answer to the question posed in this chapter's title.

Look again at the five statements about UCD above. Consider that little, if anything, in the way of specific observational expectations or predictions follows from these statements. "Pattern of similarity by which all organisms may be linked" and "unique genealogical hierarchy" appear closest to being observable predictions. Yet without specifics, both notions remain naked abstractions. Which similarities are predicted by UCD? How is the unique hierarchy of all organisms to be identified?

When formulated using common nouns such as *organism* or *living thing*, and verbs denoting only a begetting relation, such as *connected* or *linked*, UCD rules out nothing. Actually ruling something out would mean that if we observed the biological state of affairs forbidden by UCD, the geometry of universal relatedness would be false. Nor does UCD predict what we *should* observe (such that failing to observe it would also threaten the theory). If UCD turns out to be consistent with whatever we see, the theory is not telling us anything.

Now, if the reader is like me, alarm bells should be going off. To say that UCD is untestable, or that nothing observable follows from it, seems counterintuitive to the point of absurdity, and moreover, contradicted by biological knowledge understood broadly. Surely the empirical (i.e., testable) content of UCD is readily available to anyone opening a biology textbook. What about the universal genetic code? Or universally shared molecular characters, such as ribosomes, found in all organisms? Or universally conserved genes? And so on.

Reflect for a moment on these widely cited predictions. The universal genetic code refers to a *particular molecular feature* of organisms that we can describe in minute detail (see below). Likewise, the ribosome is a well-defined molecular machine. And universally conserved genes have proper names that one can retrieve from GenBank or the EMBL database, along with their corresponding DNA sequences.

In other words, conspicuously absent from these ostensible predictions from UCD are the common or generic nouns of the standard formulation and its abstract notion of "linking" (i.e., material descent). The perfect generality of UCD lies at the opposite pole from biological specificity. For testing by observation, specificity is what we need. The data we will be checking will necessarily consist of specific instances: *these* populations of *Drosophila melanogaster, that* strain of *Escherichia coli, these* fossils of *Kimberella, that* gene or protein family.

To bring UCD into contact with data, therefore, specific features of organisms must be identified and embedded in a theory. *That*

theory cannot be UCD itself. While appealing, even elegant in its simplicity and universality, UCD floats free of anything observable. Whatever theory we use to connect UCD to specific lines of evidence must be an independent body of testable propositions about some biological state of affairs.

We may represent this schematically as follows:

UCD + independent or auxiliary biological theory → testable predictions

Now we have the *yes* answer to the chapter's question in hand. UCD only needed a helper theory, or two, or three, to connect its abstract geometry of universal relatedness to the data. Testability shows up after all.

Alas, life is much messier than that. A detailed case study (below) should illuminate the logic and the deeper problem we face. Here is a heads-up about where we will be going in the next two sections. What happens when a prediction generated by this theoretical coupling—that is, UCD + auxiliary biological theory à prediction—*fails*?

The Universal Genetic Code Prediction from UCD

One of the most widespread predictions cited as following from UCD is the universality of the genetic code, the 64-trinucleotide set of rules mediating information transfer from nucleic acids (DNA and RNA) to amino acids during protein assembly (see Figure 2). Consider three representative versions of this prediction from leading evolutionary or cell biologists:

> If organisms had arisen independently they could perfectly well have used different codes to connect the 64

trinucleotide codons to the 20 amino acids; but if they arose by common descent any alteration of the code would be lethal, because it would change too many proteins at once.[11]

Whatever code was used by the common ancestor [LUCA] would, through evolution, be retained. It would be retained because any change in it would be disastrous. A single change would cause all the proteins of the body, perfected over millions of years, to be built wrongly; no such body could live...Thus we expect the genetic code to be universal if all species have descended from a single ancestor.[12]

Consider what might happen if a mutation changed the genetic code.

Such a mutation might, for example, alter the sequence of the serine tRNA molecule of the class that corresponds to UCU, causing them to recognize UUU sequences instead. This would be a lethal mutation in haploid cells containing only one gene directing the production of tRNAser, for serine would not be inserted into many of its normal positions in proteins. Even if there were more than one gene...this type of mutation would still be lethal, since it would cause the simultaneous replacement of many phenylalanine residues by serine in cell proteins.[13]

Note that UCD is coupled in these passages with a theory about the necessary functional invariance (FI) of the genetic code.

Second Base				
U	**C**	**A**	**G**	
UUU ⎤ Phe UUC ⎦ UUA ⎤ Leu UUG ⎦	UCU ⎤ UCC UCA ⎥ Ser UCG ⎦	UAU ⎤ Tyr UAC ⎦ UAA ⎤ STOP UAG ⎦	UGU ⎤ Cys UGC ⎦ UGA — STOP UGG — Trp	U C A G
CUU ⎤ CUC CUA ⎥ Leu CUG ⎦	CCU ⎤ CCC CCA ⎥ Pro CCG ⎦	CAU ⎤ His CAC ⎦ CAA ⎤ Gln CAG ⎦	CGU ⎤ CGC CGA ⎥ Arg CGG ⎦	U C A G
AUU ⎤ AUC ⎥ Ile AUA ⎦ AUG — Met or Start	ACU ⎤ ACC ACA ⎥ Thr ACG ⎦	AAU ⎤ Asn AAC ⎦ AAA ⎤ Lys AAG ⎦	AGU ⎤ Ser AGC ⎦ AGA ⎤ Arg AGG ⎦	U C A G
GUU ⎤ GUC GUA ⎥ Val GUG ⎦	GCU ⎤ GCC GCA ⎥ Ala GCG ⎦	GAU ⎤ Asp GAC ⎦ GAA ⎤ Glu GAG ⎦	GGU ⎤ GGC GGA ⎥ Gly GGG ⎦	U C A G

First Base (U, C, A, G) — *Third Base*

Figure 2. The universal or standard genetic code. Credit: Sarah Greenwood, CC BY-SA 4.0, https://creativecommons.org/licenses/by-sa/4.0, via Wikimedia Commons.

That coupling yields the testable prediction that if UCD is true, then the genetic code, when sampled from any species, should be invariant (i.e., universally distributed when compared to any other species):

UCD + functional invariance (FI) of the code → universal genetic code (UGC)

Significantly, the FI theory about constraints on the code dates from the first formal publication of the UGC prediction. In November 1963, in the journal *Science*, biologists Ralph Hinegardner and Joseph Engelberg predicted that the genetic code would be universal. At the time, molecular biologists knew (1) that DNA carried genetic information in sequences of three-nucleotide codons, (2) that this information was transcribed into messenger RNA, and (3) that the nucleotide sequences in DNA and RNA corresponded, one to one, with the amino acid sequences in proteins.

Accurate information transfer from nucleic acids to proteins required a consistent set of sequence-conserving relations between codons and amino acids. "The abstract set of rules," wrote Hinegardner and Engelberg, "which associates a nucleotide triplet with a given amino acid is known as the genetic code."[14] Once established, the functional centrality of these rules for faithful information transfer, they reasoned, strongly constrained their possible variation:

> …there is no way in which an organism could gradually change from one code to another without passing through a random phase in which more than one amino acid can be placed in a locus… In fact, a change of this kind would almost certainly have large scale deleterious effects on any organism and

therefore the change would not be perpetuated. Thus, once established, the genetic code will never change, barring an incredible event, and all organisms descendant from a given organism having the complete code will have the same code.[15]

Before moving on, two important points must be made about the role of FI in the UGC prediction:

- It should be clear that the theory of functional invariance is what gives the inference from UCD its *empirical specificity*. That is, if the code were *not* functionally invariant, we might expect that over time any number of different codes could have evolved. FI, however, gives UCD genuine predictive strength. It is, as it were, a strong lever or medium of inference, projecting the abstract geometry of UCD into the molecular phenomena. FI does all the work of enabling a *specific* prediction about the code's structure and taxonomic distribution.

- FI, as an independent theory about the constraints on information transfer in cell and organismal function, falls within the set of biological generalizations collectively known as the *Principle of Continuity* (PrC). Continuity obtains as a theoretical rule whenever any transformation of form and function is posited to have occurred. Quite simply, any such transformation, from one generation to the next, *must be biologically possible*. Continuity (meaning "viability enabling reproduction") must be maintained.

At the end of this chapter, PrC will turn out to be critical to the abandonment of UCD by many evolutionary biologists. But that is getting ahead of the story.

When UCD Predictions Fail, Who Pays the Fine?

Prominently featured as a universally valid proposition in biology textbooks from the late 1960s until the mid-to-late 1980s, dressed confidently in a gray silk Armani suit, the UGC prediction (UCD + FI → UGC) now arrives at the restaurant's front entrance sniffling and bundled in the long underwear, fraying cardigan, lumpy overcoat, ski hat, woolen scarf, and mittens of numerous caveats and exceptions. The genetic code is no longer universal—or, to borrow the heavily pixelated, logically murky language of many textbooks, the code is "nearly universal," a phrase with all the crisp exactitude of "approximately pregnant" or "almost alive."

The first variants, discovered in 1979, occurred in the mitochondrial code, where "it was found that the code in vertebrate mitochondria differed from the universal code by using codons AUA for methionine and UGA for tryptophan."[16] As Cornell geneticist Thomas Fox argued in *Nature* in 1985, however, "mitochondria could be thought of as exceptions that prove the rule: their genetic systems produce only a very limited number of proteins and so might tolerate changes."[17]

Yet variants in the nuclear code discovered in the mid-1980s were, Fox continued, of a different order: "Some 'real' [nuclear] exceptions have come to light in both eukaryotic and prokaryotic free-living organisms, and the notion of universality will have to be discarded."[18] For instance, "in at least four species of ciliated protozoa, the codons UAA and UAG [stop codons in the universal code] occur in nuclear genes and are translated as Gln during cytoplasmic protein synthesis."[19] Similarly, in the bacterium *Mycoplasma capricolum*, UGA encodes Trp, rather than termination (stop) as in the universal code.[20]

"It seems obvious," opined French biologist François Caron of the École Normale Supérieure graduate school in 1990, "that the number of cases of deviations observed will increase rapidly in the future."[21] Caron's hunch was prophetic. The National Center for Biotechnology Information (NCBI) now maintains a regularly updated archive of variant codes, including the following:

- The Standard Code

- The Vertebrate Mitochondrial Code

- The Yeast Mitochondrial Code

- The Mold, Protozoan, and Coelenterate Mitochondrial Code and the Mycoplasma/Spiroplasma Code

- The Invertebrate Mitochondrial Code

- The Ciliate, Dasycladacean and Hexamita Nuclear Code

- The Echinoderm and Flatworm Mitochondrial Code

- The Euplotid Nuclear Code

- The Bacterial, Archaeal and Plant Plastid Code

- The Alternative Yeast Nuclear Code

- The Ascidian Mitochondrial Code

- The Alternative Flatworm Mitochondrial Code

- Chlorophycean Mitochondrial Code

- Trematode Mitochondrial Code

- Scenedesmus obliquus Mitochondrial Code

- Thraustochytrium Mitochondrial Code

- Rhabdopleuridae Mitochondrial Code

- Candidate Division SR1 and Gracilibacteria Code

- Pachysolen tannophilus Nuclear Code

- Karyorelict Nuclear Code

- Condylostoma Nuclear Code

- Mesodinium Nuclear Code

- Peritrich Nuclear Code

- Blastocrithidia Nuclear Code

- Cephalodiscidae Mitochondrial UAA-Tyr Code[22]

But what of the theory of functional invariance (FI), instrumental to the UGC prediction? Recognizing that, in the face of the disconfirming facts of variant codes, one cannot assume the truth both of UCD and FI, nearly all researchers working on the problem have either explicitly or implicitly dumped FI:

> Variations in codon assignments must arise as a result of mutations affecting the codon specificities of tRNAs or the interactions between tRNAs and aminoacyl tRNA synthetases. In either case the immediate result of such mutations in a genetic system must usually be wholesale changes in the proteins produced by that system, adversely affecting at least some and leading to a selective disadvantage or inviability. *Nevertheless, such variations have occurred during evolution.*[23]

Postulating that such fundamental variations occurred is, however, very far from knowing *how* they occurred.[24] "Direct replacements of one amino acid by another throughout proteins," argued Osawa et al. in 1990, "would be disruptive in intact organisms and even in mitochondria."[25] That is, we should not think that the body of molecular knowledge motivating functional invariance can be jettisoned at will. Yes, if UCD is true, and variant codes exist, FI has to go to the wall. Yet how do we know, independently of the failure of the UGC prediction, that FI is no longer the case?

Rather, taking UCD as given, we are now faced with another novel research problem: "How could non-disruptive code changes occur?"[26] As Caron notes:

> The scenarios have to answer the question: how, with our current knowledge of molecular mechanisms, can we imagine a termination codon becoming a glutamine codon or a leucine codon, a serine codon?[27]

The schema below shows how the discovery of variant codes upended the UGC logic:

UCD + FI → ? ← Variant
(non-universal) genetic codes

Note that the empirical content of UCD—now, vis-á-vis the genetic code—is indeterminate. That is, the predictive outcomes of the theory await the findings of the new research program (i.e., how can variant codes evolve?) to which the observational anomalies, the variant codes themselves, have been referred. UCD now predicts nothing in particular about genetic codes. Maybe they will resemble the erstwhile "universal" code, or maybe not; "nearly universal" rules out nothing. We'll just have to see what the data turn up.

Was UCD Tested, or Not, by the Failure of Its UGC Prediction?

Before we answer that question, look at Figure 3, which shows the distribution of variant genetic codes over the three major domains of life.[28] In 2007, systematist Lars Vogt (then at Harvard University) argued that such character distributions could not *in principle* refute the branching evolutionary tree on which they were arrayed:

A given tree hypothesis is logically congruent with any specific

observable evidence of character state distribution. In other words, a given tree, in combination with decent [sic] with modification as background knowledge, *does not prohibit* any specific character state distribution pattern. As there is no deductive link between any tree hypothesis and any specific character state distribution there exists no direct empirical test of hypotheses of monophyly (i.e., clades) *sensu* Popper—one cannot think of any observation, which, in case it would represent a true statement, would allow to conclude the falsity of a clade or a given cladogram through *modus tollens*.[29]

Modus tollens is the logical form *if P, then Q; not Q, therefore, not P*. Classically, *modus*

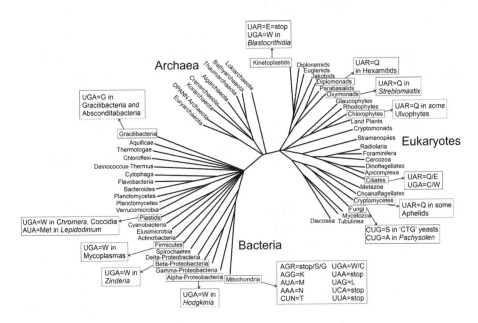

Figure 3. Phylogenetic distribution of variants of the genetic code. Any group not connected in this figure to a specific code variant uses the standard genetic code depicted in Figure 2. Credit: Figure modified after Figure 1 of Patrick J. Keeling, "Evolution of the Genetic Code," *Current Biology* 26 (September 26, 2016), R838–R858, used with permission from Elsevier.

tollens represents the logical structure of empirical testing. A theory predicts an outcome; we check to see if the outcome obtains. If the outcome (prediction) does obtain, the theory is confirmed; if it does not, the theory is imperiled.

Vogt argues, however, that the truth or falsity or phylogenetic hypotheses—descent with modification propositions—cannot be decided by *modus tollens* because such hypotheses rule out nothing. Any hypothesis that forbids no outcome to occur cannot be tested. Now, if UCD is anything, it is *the* phylogenetic hypothesis *non pareil*: the unity by descent, from LUCA, of all life on Earth. Thus, it seems that, indeed, UCD cannot be tested—once again, the paradoxical *no* answer in the opening paragraph.

A cynic might object that taking *modus tollens* to be an accurate description of scientific inquiry is, at best, touchingly naïve. Perhaps, he suggests, we should see UCD as an *axiom* of biology, adopted as a core commitment for the purposes of investigation and therefore protected from premature falsification by anomalous findings. This view—call it the axiom thesis—is not as outlandish as it may appear, and it helps to make sense of the scientific practice of evolutionary theorists, as in the UGC example given above. When reconciling a theoretical bundle (UCD + independent auxiliary theory) with apparently disconfirming observations, evolutionary biologists act to conserve the truth of UCD. As philosopher of science Harold I. Brown observes,

> In science, not all propositions are treated as testable empirical hypotheses. It is only because a large body of knowledge is taken as paradigmatic that we can isolate individual propositions for purposes of testing, and what

conclusions we draw from a particular test depends on what propositions we take as paradigmatic.[30]

Systematists Kevin de Queiroz and Michael Donoghue argue that "the principle of common descent" unifies the "patterns of living things in space, in time, and in form under a single general theory."[31] But, as they also argue,

> The theory of common descent…is "evolutionary" only in the most general sense, for it does not even refer to change. It certainly is not tied to any particular model of the evolutionary process, nor is it at odds with the results of systematic analysis.[32]

In other words, we need not worry that anything in our biological experience will ever run afoul of the theory. Except there is this problem: UCD *is* tested, and rejected, with increasing frequency. What's up with that? How can a supposedly untestable theory, found wanting on empirical grounds, end up discarded on the side of the road?

The Principle of Continuity to the Rescue

Keep the videotape idling on the year 2007. That year, a major paper from National Academy of Sciences molecular geneticist and NCBI lab director Eugene Koonin jettisoned UCD in its standard formulation:

> …it is generally assumed that, in principle, the TOL [tree of life, or UCD] exists and is resolvable…Here, I argue for a fundamentally different solution, i.e., that a single, uninterrupted TOL *does not exist*, although the evolution of

large divisions of life for extended time intervals can be adequately described by trees. I suggest that evolutionary transitions follow a general principle that is distinct from the regular clado-genesis [evolutionary branching]. I denote this principle the Biological Big Bang (BBB) Model.[33]

Evolution's core proposition of UCD, which Francisco Ayala said was as indubitable as "the roundness of the earth,"[34] was being tossed over the ship's railing, into the water. Why?

What happened was the birth of a new method of molecular data gathering in biology—one that is as historically important to that science, it turns out, as the invention of the telescope was to astronomy. From stage right during the mid-1990s, rapid and increasingly inexpensive DNA sequencing entered biology and quickly became a widely used research technology. For the first time, biologists could survey vastly more—and in many cases, genomically complete—DNA sequences from a multitude of species. Before 1995, molecular comparisons among species, based on single gene or protein sequences, were akin to trying to map the entirety of New York City by measuring ten inches of curb in Brooklyn and Manhattan, a foot or two in Staten Island, and another small section of curbstone in Queens and the Bronx.

In brief, whole genome DNA sequencing has overturned the genetic "unity of life" prominent in biology textbooks for the last 50 years. One remarkable finding in this respect has been termed *non-orthologous gene displacement* (NOGD), an awkward phrase that means the expected conservation (*orthology*) of genes and proteins, when assessed across the tree of life, and as predicted by UCD, is not observed. Rather, in the central information-processing and metabolic pathways inferred to have been present in LUCA, different (*non-orthologous*) genes and their protein products have "displaced" what biologists expected to find.

Consider a striking example of NOGD from a system at the heart of every free-living cell: the molecular structure of release factor, an essential player in protein assembly within the ribosome.[35] Ribosomes are unquestionably *the* most fundamental molecular machine in any free-living cell (viruses are defined by lacking ribosomes), the locus of action for turning DNA sequence information, via messenger RNA (mRNA), into functioning proteins. If you are a cell, or a collection of cells, on Earth, you must have ribosomes.

During the translation of mRNA in the ribosome, most codons are recognized by "charged" transfer RNA (tRNA) molecules, called *aminoacyl-tRNAs* because they connect to specific amino acids corresponding to each tRNA's anticodon, the three-nucleotide signal on the mRNA-binding stem of the tRNA. But "stop," or termination codons, are handled differently. In the so-called universal genetic code, there are three mRNA stop codons: UAG, UAA, and UGA. While these stop codons represent triplets, like ordinary amino-acid specifying codons, tRNAs do not decode them. Instead, in 1967, Mario Capecchi found that tRNAs do not recognize stop codons at all.[36] Rather, proteins he named *release factors* perform that task.

Briefly, when the ribosome moving along an mRNA strand arrives at a stop codon, release factor enters the ribosome and hydrolyzes (i.e., cuts) the chemical bond holding the last amino acid in the newly synthesized protein to its corresponding tRNA. This "releases" the nascent protein to exit the large subunit of the ribosome, and fold—hence, the name for the family of proteins performing this essential task.

Now, let's try a mini-thought experiment. If LUCA existed and possessed a DNA genome, ribosomes, and release factors—and if these features were functionally essential within LUCA, and therefore inherited by all its descendants—what should one expect to find when surveying the protein folds (three-dimensional structures) of release factors across the tree of life? In other words: assume UCD, couple that monophyletic geometry to our knowledge of ribosomal function, and make a prediction.

UCD + necessity for release
factors in translation → ?

Figure 4 shows release factor proteins in bacteria and eukaryotes. These are not the same proteins. They are non-orthologous: rotate them as you like; their three-dimensional structures remain topologically incongruent. Although the release factors perform the same *functional* task in the ribosome, they differ at the molecular level, in the same way a butterfly's wing differs from a bird's wing at the anatomical level: total absence of homology. This raises what Baranov et al. call "several unsolved puzzles":

> Since there is no strong evidence for an evolutionary relationship between bacterial class-I RFs [release factors] and their counterparts from archaea and eukaryotes, it is unknown how termination was mediated in the last common ancestor.[37]

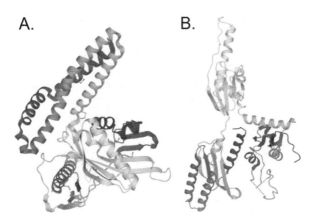

Figure 4. Release factor proteins in bacteria and eukaryotes. A. Three-dimensional structure of release factor protein from the bacterium *Escherichia coli* (domain Bacteria). Credit: Image from the RCSB PDB (rcsb.org) of PDB ID 1GQE (B. Vestergaard, L.B. Van, G.R. Andersen, J. Nyborg, R.H. Buckingham, and M. Kjeldgaard, "Bacterial Polypeptide Release Factor RF2 Is Structurally Distinct from Eukaryotic eRF1," *Molecular Cell* 8 (December 2001), 1375-1382). B. Three-dimensional structure of release factor protein in *Homo sapiens* (domain Eukarya). Credit: Image from the RCSB PDB (rcsb.org) of PDB ID 1DT9 (H. Song, P. Mugnier, A.K. Das, H.M. Webb, D.R. Evans, M.F. Tuite, B.A. Hemmings, and D. Barford, "The Crystal Structure of Human Eukaryotic Release Factor eRF1—Mechanism of Stop Codon Recognition and Peptidyl-tRNA Hydrolysis," *Cell* 100 (February 4, 200), 311-321). Images used under RCSB usage policies at https://www.rcsb.org/pages/usage-policy and https://www.rcsb.org/pages/policies.

Assuming, that is, that there *was* a last universal common ancestor (LUCA). For a growing cadre of evolutionary biologists, among them Eugene Koonin, Eric Bapteste, Didier Raoult, W. Ford Doolittle, and Michael Syvanen (none of whom are intelligent design proponents, incidentally), the wide extent of NOGD, when added to the other molecular anomalies revealed by whole-genome sequencing, render the hypothetical construct of LUCA a problematical or nonexistent entity, which historical biology is better off without.

The molecular unity of life, UCD and LUCA's original claim to fame, has been eroded away by unanticipated genetic findings:

> As the genome database grows, it is becoming clear that NOGD reaches across most of the functional systems and pathways such that there are very few functions that are truly "monomorphic," i.e. represented by genes from the same orthologous lineage in all organisms that are endowed with these functions. Accordingly, the universal core of life has shrunk almost to the point of vanishing.[38]

UCD and the real existence of LUCA remain the default position for most working biologists, if for no other reason than disciplinary inertia, or because they do not bother to think about the large-scale comparative questions that motivate the growing community of UCD skeptics.

But the *yes* answer to the testability of UCD is easy to understand, if one reflects (in closing) once again on this schema:

UCD + independent or auxiliary biological theory → testable predictions

Nothing forces the hand of an investigator when his prediction fails. When testing a complex theoretical bundle, such as UCD plus some auxiliary theory or theories, a failed prediction tells us only that a problem exists somewhere in the bundle—but not exactly where. Thus, when the fine for the failed prediction must be paid, the biologist can elect to pass the pink slip down the table to the auxiliary theory, or to UCD. Koonin slid the fine down to UCD. So have a growing number of other biologists.

Does the Fossil Record Demonstrate Darwinian Evolution?

Günter Bechly

The role of fossils for the theory of evolution has a quite checkered history. Famous evolutionary biologists like Willi Hennig and Colin Patterson questioned whether fossils are necessary or useful for establishing evolutionary relationships at all, while other scientists like Richard Dawkins, Donald Prothero, and Jerry Coyne have claimed that fossils rank among the most compelling evidence for evolution. On the other hand, many creationists have either claimed that fossils contradict Darwin and confirm a young-Earth creationist viewpoint and a global flood, or they have rejected certain published fossil evidence as forgeries. Is it possible to make interpretations somewhere in the middle (like intelligent design)—accepting that fossils are ancient (and real) but questioning the neo-Darwinian evolutionary spin? It is noteworthy that even some famous evolutionists (e.g., Stephen Jay Gould and Gerd Müller) have acknowledged that the pattern of the fossil record poses a problem for the Darwinian paradigm. What should an interested layman make of all this? Who is right, and what does the fossil record really say?

What Are Fossils, and What Is Paleontology?

Paleontology is the science of past life that is preserved as the fossil remains of extinct life forms that populated our planet millions of years ago. Usually fossils are either imprints in soft sediments or petrified or mineralized hard parts like bones or shells. More rarely, even soft tissues and colors can be preserved. Inclusions of insects and other small organisms in amber, which is a fossil tree resin, are three-dimensionally preserved in exquisite detail and look like they had just been still alive. The fossil record spans almost four billion years, from the first bacterial cells to Siberia's frozen mammoths of the last ice age. With the help of fossils, scientists attempt to reconstruct the life history of these organisms and their ecosystems, which were often quite different from today. Life very much changed over time in composition and appearance, with numerous new life forms and their distinct body plans abruptly appearing and disappearing over the eons. At least five major mass extinction events can be traced in the fossil record, and during each of these cataclysmic events, the majority of the species living at that time vanished from the scene.

Are Many Fossils Forgeries?

There have been some famous cases of forgeries that fooled even scientists, like the notorious Piltdown skull[1] or, more recently, the case of *Archaeoraptor*,[2] an alleged feathered link between dinosaurs and birds. However, all these cases were sooner or later recognized as forgeries by paleontologists. Today, the analytic techniques, like computed tomographic scans, have become much more sophisticated and make it much easier for experts to detect forgeries. Nevertheless, there still exist genuine industries for fossil forgeries (e.g., in Morocco, Russia, and China), but these rather exploit the large market of amateur fossil collectors. Generally, the modern published scientific evidence of the fossil record can be trusted and requires an adequate explanation. The debate should be whether the best explanation is provided by Darwinian evolution or rather by alternative approaches like intelligent design. Denying the authenticity of the empirical evidence provided by the fossil record, or even promoting conspiracy theories about the scientific community, is neither reasonable nor helpful for a serious scientific critique of unguided evolution.

How to Date a Fossil

Scientists confidently talk about millions of years, but do they really have evidence for these old ages? How do paleontologists determine the ages of stones and bones, which, after all, have no birth certificate stamped upon them? Some critics have accused scientists of circular reasoning because a common creationist argument claims that they date the geological layers (rocks) with so-called index fossils, and date the fossils with the geological layers (rocks) in which they are found. This is not correct, even though index fossils indeed play an important role in relative chronology (so-called biostratigraphy).

However, nowadays the absolute ages of most fossil localities can be determined with radiometric dating methods. These methods measure the relative abundance of different radioactive isotopes in the minerals. The dating works similar to an hourglass: When we know the rate of sand flow, we can easily say how long ago the hourglass was turned. Modern radiometric methods (called *isochrons* and the *concordia-discordia method*) do not require unproven assumptions about the initial amount of daughter isotopes or constant decay rates. When different dating methods are used for the same localities and rocks, they usually converge to the same dates, which gives us good reason to be confident that the dates are indeed correct. Thus, geology and paleontology indeed establish deep time and millions of years. Likewise, an overwhelming body of evidence from geology and paleontology, especially the systematically ordered pattern of the fossil record, cannot be convincingly explained by a single short-timed global catastrophic event.

Are the "Missing Links" Still Missing?

Evolutionists often say that there are many transitional fossils, while creationists often say that there are none. It seems that one side must be wrong, but actually both are right because they talk past each other and use the same term for two different things. When evolutionists talk about transitional fossils, they usually only mean transitional in the anatomical sense. This refers to fossils that possess a mosaic pattern of characters, with some primitive characters of the assumed ancestors still retained, while some (but not all) derived characters of the assumed descendants are already developed so that the fossil is anatomically intermediate. Evolutionists do not necessarily imply with the term *transitional fossil* that these forms are direct ancestors, as they could well be side branches from the ancestral lineage. Therefore, transitional fossils are not necessarily in the correct temporal sequence because such side branches could persist and even outlive more advanced forms.

When creationists and critics of Darwinism say that there is a lack of transitional fossils, they usually mean transitional in the sense of a gradual sequence of direct ancestor-descendent relationships, which implies not only a fine-graded directional anatomical transition but also a correct temporal order. While transitional fossils in the first sense are indeed very common and exist for most groups of organisms, transitional fossils in the second sense are extremely rare and mostly missing indeed. Overall, the best explanation of the available fossil data may be some form of common descent but certainly not an unguided gradual process of transition, but rather, abrupt transitions and saltational changes that require intelligent design.

The Discontinuous Fossil Record Refutes Darwinian Gradualism

Every theory makes certain predictions. The core prediction of Darwin's theory of evolution is gradualism, which means that all the transitional changes in the history of life are not supposed to have happened as sudden big changes, but by a continuous accumulation of small changes over vast periods of time. The simple reason is that Darwin wanted a naturalistic explanation and was fully aware that sudden big changes of organisms would require miraculous events. Therefore, Darwin mentioned not less than six times in his magnum opus *On the Origin of Species* the Latin phrase *natura non facit saltus*, which means that nature does not make jumps.

This claim is still made by Darwinians today. The most well-known modern popularizer of Darwinism, the infamous atheist Richard Dawkins, wrote in his 2009 bestselling book *The Greatest Show on Earth* the following remarkable statement: "Evolution not only is a gradual process as a matter of fact; it *has* to be gradual if it is to do any explanatory work."[3] This shows that gradualism is not just one optional element of Darwinism, but that it is very much essential for its success as a naturalistic explanation for the complexity and diversity of life. If gradualism is wrong, then Darwinism is refuted.

Indeed, the fossil record is highly discontinuous and strongly contradicts Darwin's prediction of gradualism. Even Darwin himself was already quite aware of this problem for his theory and therefore tried to explain it away as a mere artifact of undersampling of a very incomplete fossil record. The famous vertebrate paleontologist Philip Gingerich once snipingly remarked that "gaps of evidence are gaps of evidence and not evidence of gaps."

However, such appeals to the incompleteness of the fossil record are no longer tenable. Intelligent design theorist and philosopher of science Paul Nelson cogently explained why: Imagine you have a new hobby, beachcombing. Every day you walk along the shore and collect what the tide washes in. In the beginning you are surprised every day with new discoveries—shells of new types of snails and mussels, starfish, sand dollars, and driftwood, etc. But after a while you are finding mostly the same stuff over and over again, and you must be lucky to find something new that you have not seen before (like a stranded whale or a message in a bottle). When you have reached this point of mostly repetition, then you know that you have sampled enough to be sure that you have not missed much that is out there to find.

The same approach is used by paleontologists for a statistical test of the completeness of the fossil record; it is called the *collector's curve*. In most groups of fossils, we have reached this point of demonstrable saturation, where we can be pretty confident that the distinct discontinuities that we find are data to be explained and not just sampling artifacts. There is another reason why we know this: If the gaps and discontinuities in the fossil record were just artifacts, they should more and more dissolve with our greatly increasing knowledge of the fossil record. But the opposite is the case. The more we know, the more acute these problems have become. "Darwin's doubt"[4] did not get smaller over time but bigger, and if he were still alive, he would likely agree that the evidence simply does not add up, since he was much more prudent than many of his modern followers.

Of course, we have to consider the appropriate timescale in Earth history to estimate whether some event in the history of life is abrupt or not. In human history, we would not consider an event that lasts many years (say, decades or a century) to be abrupt. But in biological or geological terms, the appearance of a new group of organisms with a new body plan within, say, a window of time of 5–10 million years is very abrupt indeed. Why is this so? Because the average longevity of an invertebrate or vertebrate species (not

an individual organism) varies between 2.5–10 million years. This means that a transition that required 5–10 million years happened within the lifespan of a single species! This is much too short to allow for Darwinian evolution to explain the required changes. Here are some examples of such major discontinuities in the fossil record:[5]

The early origin of life: Planet Earth is 4.54 billion years old, but life could not arise until the end of a violent early period called the *Late Heavy Bombardment*, about 4.1–3.8 billion years ago.[6] During this period, our planet was hit by numerous gigantic meteorites that evaporated all of the world's oceans several times. The oldest uncontroversial fossils are 3.77 billion years old bacterial filaments from Quebec.[7] Thus, complex cellular life popped into existence right when environmental conditions allowed for its existence, as if it was planted there by a creator.

The early origin of photosynthesis: The first evidence for photosynthesis by blue-green algae appears abruptly and contemporaneously with the earliest life at about 3.8 billion years ago. Darwin's theory would predict that it required hundreds of millions of years of gradual evolution for this innovation to come into being because the conversion of sunlight into chemical energy certainly represents one of the most complex physiological achievements of life.

The Avalon explosion: The first larger fossil organisms appear suddenly 575–565 million years ago in the Ediacaran period with an event that has been named the *Avalon explosion*,[8] after an ancient continent Avalon. This ecosystem has been called the *Garden of Ediacara*[9] because there is no evidence for predation. It was a strange, alien world of microbial-algal mats covering the sea floor, and enigmatic large sessile organisms that looked very different from anything alive today. They had a glide symmetry (the left and right sides of the body were similar but somewhat offset along the median body axis), fractal growth, and a quilted air-mattress-like structure with no visible inner organs. These unique life forms appeared out of nowhere, without any intermediate precursors in the preceding geological layers.

The Cambrian explosion: The most well-known example of abrupt appearance in the fossil record is the Cambrian explosion about 535–515 million years ago, which was called *Evolution's Big Bang* on a *Time* magazine cover in December 1995.[10] In this event, 21 of the 28 known animal phyla appear suddenly without any precursors in the fossil record. Phyla are the highest groupings in animal classification, which denote the different body plans, like arthropods, mollusks, and vertebrates. Darwin would have predicted that we should find a succession of fossils that slowly diversify from an ur-multicellular animal into slightly different and then more and more different species into different genera, families, and orders, until finally, after a long period of diversification, we find the different body plans of the phyla.

However, instead of such a bottom-up diversification pattern, we find a top-down pattern with the big phyla differences appearing all at once and only later diversifying into different variations of these blueprints. The pattern looks very much like creation rather than Darwinian evolution. Of course, evolutionary scientists have tried to explain away this inconvenient truth with the claim that there were no suitable sediments in the older layers to preserve the postulated soft-bodied ancestors of the Cambrian phyla (the "artifact hypothesis"). But when such sedimentary layers were finally discovered in the preceding older strata in Mongolia and China, there were only fossil algae but no animals at all, while in the Lower Cambrian we find complex animals like trilobites, complete with their exoskeleton, articulated legs, and highly efficient compound eyes. Where are all the thousands of intermediate species that are required by the Darwinian story, and how could they fit into the very brief available window of time that only allows for two successive marine invertebrate species?[11]

The great Ordovician biodiversification event (GOBE): About 470 million years ago, the marine animal phyla very quickly diversified into numerous different families of marine invertebrates, such as brachiopods, gastropods, and bivalves. This "great Ordovician biodiversification event" was appropriately called *Life's Second Big Bang* by an article in the journal *New Scientist* in 2008.[12]

The Siluro-Devonian terrestrial revolution: Land plants appeared suddenly without aquatic precursors in the Late Silurian to Lower Devonian age about 420 million years ago. The oldest uncontroversial fossil land plant *Baragwanathia* was discovered in India and already belongs to the still-living group of club mosses. In the mid-Devonian age we already find fossil forests of large tree-like plants, of which some (the so-called cladoxylopsids) even had a more complex trunk anatomy than modern trees. This Siluro-Devonian terrestrial revolution was heralded as the "terrestrial equivalent of the much-debated Cambrian explosion of marine faunas" in a scientific publication on the early evolution of land plants by Bateman et al. (1998).[13] Such a Siluro-Devonian terrestrial revolution also happened among animals with the sudden appearance of land-living arthropods and vertebrates. The latter appear so suddenly that the oldest evidence for quadruped land vertebrates (the Zachełmie footprints in Poland) is actually

5–10 million years *older* than the oldest fossils of their presumed ancestors among the lobe-finned fish.[14] This is a genuine temporal paradox that gives proponents of Darwinism a reason for pause.

The Odontode explosion: This event refers to the abrupt appearance of tooth-like structures (odontodes) in different groups of jawed fish (sharks, lobe-finned fish, and ray-finned fish) about 425–415 million years ago.

The Devonian Nekton revolution: In 2010, my graduate-school colleague Christian Klug described a previously overlooked event that he named the Devonian Nekton revolution, which happened in the oceans 410–400 million years ago. Before this time, nearly all marine organisms lived either close to the sea floor or passively drifting as plankton. After this event, almost 80 percent of marine biodiversity was constituted by active swimmers like fish and cephalopods.

The Carboniferous insect explosion: In the next geological period, the Carboniferous, with its worldwide tropical swamp forests that generated our coal deposits, there were at least two sudden events. First, the reptile revolution in the Lower Carboniferous brought the transition from amphibians to the first reptiles with their completely different amniotic egg that allowed them to be independent from water. Later, in the Upper Carboniferous, flying insects appear suddenly without any transition that would show the stepwise evolution of insect wings. They appear fully formed and with great diversity. Many of those early insects belonged to primitive groups like mayflies, dragonflies, and roaches, which have larval stages that resemble the adults and simply grow larger with every larval molting and develop wings gradually.

Also, to the great surprise of evolutionists, different groups of holometabolous insects like beetles and wasps were discovered in Carboniferous layers. These insects already had the almost miraculous mode of development called *metamorphosis*, where the caterpillar-like larval stage is followed by a pupal resting stage, during which the complete body is dissolved into a kind of soup and reorganized into the adult flying insect. To find the "wondrous transformation of caterpillars," as insect metamorphosis was called by the seventeenth century naturalist Maria Sybilla Merian,[15] already present at the very beginning with the first flying insects is not at all what Darwin's theory would predict.

The Triassic explosions: After the big end-Permian mass extinction we find a genuine "carpet bombing" of explosive origins in the Triassic era. These include the sudden origin of modern tetrapods (the oldest turtles, crocodiles, lizards, dinosaurs, and genuine mammals all appear 251–240 million years ago), a sudden burst of marine reptiles abruptly going from 0 to 15 different families (248–240 million years ago), and a sudden origin of gliding and flying reptiles, including the first pterosaurs (230–228 million years ago). Apart from mammals, all these groups are without fossil precursors in the older geological strata.

In 2018, Massimo Bernardi of the University of Bristol lead-authored a study on the origin and diversification of dinosaurs,[16] and in light of his study, strikingly remarked that it is "amazing how clear cut the change from 'no dinosaurs' to 'all dinosaurs' was"[17] in the Upper Triassic. The evolutionists' explanation for this amounted to *it was raining a lot at this time.* You cannot make this stuff up!

The angiosperm revolution: The sudden appearance and diversification of modern flowering plants in the Lower Cretaceous era strongly conflicts with the gradualist perspective of Darwin's theory, which is why Charles Darwin considered this fact as an "abominable mystery."[18] Darwin hoped that with increasing knowledge of the fossil record this mystery might dissolve, but instead, it has been reinforced and has even become deeper in the 150 years since Darwin.

The explosive Tertiary butterfly radiation: The different modern families of diurnal butterflies and also the larger nocturnal butterflies appear abruptly at the Eocene-Oligocene transition during the Lower Tertiary period. From the preceding Mesozoic era, only fossils of small moth-like lepidopterans have ever been discovered, but no putative ancestors of real butterflies.

The explosive Tertiary bird radiation: Scientists were stunned by the sudden appearance of most major groups of modern birds (Neoaves) within a 10-million-year window of time in the Lower Tertiary about 60–50 million years ago. Ornithologist Alan Feducia appropriately called this phenomenon the "'big bang' model for bird evolution."[19]

The explosive Tertiary mammal radiation: Based on molecular clock data, evolutionists predicted that modern mammal orders all appeared in the Cretaceous era, but not a single fossil of modern (crown group) mammals has been found from these strata. Instead, most orders of modern placental mammals appeared abruptly in the Lower Tertiary between 62–49 million years ago. In many cases like bats, even the oldest fossil representatives of such highly sophisticated groups already are fully formed and look like their living relatives. Their abrupt appearance has been described as "an Eocene Big Bang for bats."[20]

The "Big Bang" of the genus Homo: Contrary to common belief, our own genus *Homo* does not gradually emerge from ape-man fossils in Africa, but instead, appears abruptly with its distinct differences and special adaptations for long-distance running. In the year 2000, renowned paleoanthropologist John Hawks copublished a study that documented the morphological gap between the australopithecines and *Homo*.[21] His results were celebrated in the press as a "Big Bang theory"[22] of the origin of the genus *Homo*.

The Upper Paleolithic human revolution: Even the various cultural achievements that involve complex symbolic thought—like jewelry, ivory carvings, and cave paintings—did not develop gradually. Instead, they appeared suddenly during the "Upper Paleolithic human revolution"[23] about 40,000 years ago, together with the first humans that have a globular braincase and a chin. Might this event correlate with the origin of real humans as the image bearers of God? It certainly looks like that. And contrary to common misconceptions, modern population genetics studies have shown that we could have come from a first couple.[24]

The Fossil Record Contradicts an Unguided Neo-Darwinian Mechanism

If we look at mainstream evolutionary biology, there are two areas of research that are generally considered to strongly support Darwin's theory of evolution: On the one hand, there is the fossil record that establishes deep time, change over time, and intermediate forms. On the other hand, there is population genetics, which calculates the change of gene frequencies in populations. Such microevolution can even be experimentally demonstrated in the lab, for example, with the origin and spread of drug resistance in microbes in a petri dish. Evolutionists then simply assume that the extrapolation and accumulation of microevolution over the long periods of time is sufficient to explain macroevolution.

So much for the theory, but in real life there emerges a genuine problem for the Darwinian theory of evolution if we combine the results from the fossil record with the mathematical apparatus of population genetics. The windows of time that are established by the fossil record for certain major transitions (e.g., the transition from quadrupedal pig-like animals to fully marine dolphin-like whales) prove to be much too short to accommodate the required waiting times for the origin and spread of the genetic changes necessary for the re-engineering of the body plans. This so-called "Waiting Time Problem" is well known and acknowledged and debated in the mainstream technical literature.[25] Even evolutionary biologists who attempted to refute the claims of Darwin critics like Michael Behe obtained results that confirmed prohibitive waiting times.

For example, Durrett and Schmidt (2008), in a widely discussed paper in the prestigious journal *Genetics*, calculated that for a single pair of coordinated mutations the waiting time in human evolution would have been 216 million years,[26] even though the fossil record shows that only about six million years were available since the assumed divergence of the chimp and human lineages. In-depth computer simulations and mathematical calculations, based on standard textbook methods of population genetics, refute the feasibility of an unguided neo-Darwinian mechanism of random mutation and natural selection within available geological timeframes. Therefore, the problem posed for Darwinism by the fossil record turns out to be even bigger than Darwin himself thought. It is for such reasons that a fellow paleontologist and friend of mine, who is neither a religious believer nor an intelligent design proponent, finally came to

doubt Darwinism himself. The facts simply no longer made sense to him.

Do Fossils Support Common Ancestry?

Apart from the proposal of an unguided mechanism of random mutation and natural selection, the other main element of Darwin's theory is common descent—i.e., the ancestor-descendent relationship of all organisms in a universal tree of life. This issue can be debated with respect to the fossil evidence, and we can explore "both sides" of the argument.

Lines of fossil evidence that may be cited to support common ancestry include the above-mentioned systematic order of the fossil record from simple to complex, and from less similar to more similar to the modern flora and fauna; the existence of transitional fossils that sometimes even form transitional series (e.g., in whales and horses); the fuzzy delimitation of groups in time (clearly distinct modern groups become less distinct when one looks at their earliest fossil representatives, respectively); the absence of extremely out-of-place fossils; and the often relatively good congruence between the reconstructed phylogenetic branching pattern and the sequence of appearances in the fossil record (so-called "stratigraphic fit").

Also, paleobiogeography often agrees with the assumption of common ancestry—for example, when fossil kangaroos are found only in Australia, the sole place where their modern relatives occur. Finally, the assumption of common ancestry allowed for successful predictions of how "missing links" should look, which were only later discovered and confirmed the theoretical predictions. A good example is the recent discovery of a primitive spider with an annulated tail-filament in 100-million-year-old Burmese amber. No known living or fossil spiders possess such tails, but it was predicted for fossil ancestors of spiders because such tails occur in the assumed phylogenetic relatives like whipscorpions. Such empirically confirmed predictions are always a good indication that a theory has explanatory power and increases the confidence in its general correctness.

On the other hand, there is also substantial fossil evidence that questions common descent. Examples include the general mismatch between the fossil record and molecular clock data; the fact that different data sets often produce conflicting phylogenetic trees; the many cases of poor stratigraphic fit (a conflict of stratigraphic and phylogenetic order of appearance) and temporal paradoxes of cases where the oldest fossils of assumed descendants are *older than all fossils* of the presumed ancestors. To accommodate such problems, paleontologists have had to postulate a plethora of so-called ghost lineages, which are merely theoretical lineages of organisms through time that are totally unsupported by the fossil record.

Furthermore, it is not true that there are no out-of-place fossils that appear in the wrong geological strata. Two striking examples are the land plant *Parafunaria* from the Cambrian in China and an advanced winged insect larva (thrips) from the Devonian Rhynie chert in Scotland. Both have been found in much-too-early periods of Earth history, where they definitely should not exist, according to the evolutionary narrative. Both cases are documented in the technical literature,[27] but have been consistently ignored by evolutionary scientists rather than considered as conflicting evidence and refutations of Darwin's theory.

In paleobiogeography, there are also numerous conundrums for common ancestry. A well-known example is the case of New World monkeys. The fossil record shows that New World monkeys would have separated from their assumed African ancestors and arrived in South America about 40 million years ago. This would have required the impossible scenario of a breeding group of monkeys to raft for 60 days about 870 miles across the south Atlantic Ocean.[28] For a genus of trapdoor spider on the South Australian Kangaroo Island, evolutionists even proposed a 6,000-miles raft journey from Africa across the wild Indian Ocean.[29] Many similar problems exist for flightless ratite birds, worm lizards, and for snails, iguanas, and boid snakes on Pacific islands.

Here is another problem for common ancestry: If common descent is true, we should find that different lines of evidence converge to a single correct evolutionary hypothesis. However, different sources of data, like phylogenomics and morphology, often produce highly incongruent trees for the reconstructed phylogenetic relationships. Furthermore, fossil evidence and evidence from embryology and genetics (evo-devo) often suggest very different conclusions about the origins of certain structures. Insect wings are a good example. Fossil evidence strongly supports their origin from stiff outgrowths of the dorsal breastplates of the exoskeleton (paranotal theory), while evo-devo data strongly support an origin from mobile leg appendages (exite theory).[30] Another example is the controversial issue of the identity of the three finger digits in feathered dinosaurs and birds: Fossil evidence suggests that it is clearly digits 1-2-3, while embryological evidence suggests that it is clearly digits 2-3-4 and genetic evidence suggests it is digits 1-3-4.[31] If common ancestry is true, then such conflicts cannot be explained without massive appeal to ad hoc hypotheses.

Yet another problem arises by the fact

that fossil evidence can severely conflict with the assumption that uniquely shared derived similarities (so-called synapomorphies) are good evidence for common ancestry. Take, for example, a group of mammals that has been named Tethytheria, which includes elephants, manatees, and the extinct hippo-like mammals Desmostylia. Based on various anatomical similarities, these three groups are considered to be close relatives. Also, they all share a horizontal mode of tooth replacement that is clearly derived and absolutely unique among all mammals. They lack permanent premolars, and as these cheek teeth wear down and fall out, they are replaced by new check teeth that slowly shift forward from behind. No scientist doubted that this shared derived similarity is due to common descent until paleontologists looked at the oldest fossil representatives of each of the three groups. Lo and behold, the earliest elephants, manatees, and desmostylians did not have horizontal tooth replacement, but the common vertical mode of other mammals. The unique horizontal tooth replacement obviously originated independently in each of these lineages and does not support their close relationship. Without fossils, we would have never suspected this. How many other similarities might mislead us to infer common descent?

Finally, Darwin's theory of common descent with modification suggests that ancestral species gradually "morph" into descendent species. But in the fossil record, we do not find these predicted gradual species-to-species transitions. The few textbook examples proposed in the past (e.g., the Steinheim snails[32] and *Globorotalia* foraminifera[33]) have all been refuted by new evidence. Paleontologists Niles Eldredge and Stephen Jay Gould formulated their famous hypothesis of "punctuated equilibria" to explain away this lack of fossil evidence for gradualism.

However, their slick idea of quick evolution in small isolated subpopulations that could hardly leave traces in the fossil record was never met with enthusiasm by mainstream paleontology and could, at best, explain some exceptional cases.

Overall, the fossil evidence is at least ambiguous, so that critique of the notion of universal common ancestry is far from being irrational or antiscientific. Different views on this topic exist within the intelligent design community, and it is my view that the cumulative evidence from geographical distribution, the hierarchical pattern of morphological and genetic similarities, atavistic organs, and endogenous retroviruses is still better explained by common descent. But of course, common ancestry does not necessarily imply unguided evolution. Considering all the other problems of an unguided evolutionary process, I personally believe that the core issue is not a conflict of special creation versus common descent, but rather, a synthesis of intelligent design with common ancestry in a form of guided evolution. If the history of life required multiple instances of infusion of new information, then the distinction between special creation and common ancestry becomes blurred anyway. There is nothing in the Bible that says God had to create plants and animals *de novo* or from dead matter. He could as well use living organisms as material for creation, somewhat analogous to the miraculous creation of Jesus's human body in the womb of his mother.

Living Fossils and Stasis in the Fossil Record

Finally, there is a considerable problem (for Darwinian evolutionists) posed by the numerous cases of evolutionary stasis and so-called living fossils. Living fossils are modern organisms—like the nautilus, the coelacanth, the tuatara lizard, or the gingko tree—that still look mostly unchanged compared to fossils that are hundreds of millions of years old. A formidable example includes horseshoe crabs that are still alive in several species. The Jurassic horseshoe crab *Limulus darwini* lived 148 million years ago and looked indistinguishable from a modern *Limulus*, and the same is true for the 244-million-year-old Triassic species *Limulitella tejraensis*. Even the oldest known horseshoe crab, *Lunataspis aurora* from the Ordovician period, which lived 445 million years ago, looks very similar to its modern relatives. If mutations occur constantly like clock ticks and natural selection is always at work, why did horseshoe crabs survive almost half a billion years without significant change, while other groups changed dramatically or went extinct, like the trilobites?

The usual Darwinian explanation is that natural selection is a transformative force only when environmental conditions change, but is conservative and stabilizing when environmental conditions stay the same. The claim is that living fossils like horseshoe crabs occupy an ecological niche that never changed. However, this explanation fails because during the almost half billion years of their existence, horseshoe crabs witnessed all five major mass extinction events, each of which eliminated 50–96 percent of biodiversity. They also witnessed several dramatic changes of all marine ecosystems, like the Devonian Nekton revolution. The convenient claim of a stable niche clearly does not make sense.

The only other ad hoc explanation available to Darwinians is dumb luck. They maintain that evolution is a contingent process like a lottery game that has many poor losers (like trilobites) and some lucky winners (like horseshoe crabs). The problem with this explanation is that the ubiquitous phenomenon of convergence (independent appearance of very similar traits) strongly contradicts the notion of evolution as a largely contingent process.[34] Evolutionary stasis shows that natural selection is an empty concept that explains everything and nothing, just like the old weather proverb: "If the cockerel crows from his favorite spot, the weather may change, or again it may not."

Information Infusions

We can safely conclude that the fossil record is far from proving the bleak materialist picture of unguided Darwinian evolution. It not only contradicts the gradualist predictions of Darwin's theory, but even shows that the available timeframes are much too short to allow the neo-Darwinian mechanism of random mutations and blind natural selection to account for the necessary genetic changes. This conflicting evidence can also no longer be explained away as an artifact of an incomplete fossil record. This is not at all the "crazy" idea of creationists, but more and more realized by mainstream evolutionary scientists. In December 2016, I attended a conference at the prestigious Royal Society in London titled "New Trends in Evolutionary Biology." The keynote speaker was the famous theoretical biologist Professor Gerd Müller from Austria. What he had to say was remarkable indeed and politically very incorrect. He listed several explanatory deficits of the modern evolutionary synthesis (also known as neo-Darwinism)—things that this theory fails to explain. Among these deficits he included phenotypic complexity (complex organs), phenotypic novelty (novel body plans), and nongradual forms of transition.

Many specialists agree: If these are what the theory fails to explain, then not much is left, and a new approach is needed.

An inference to the best explanation of all available evidence suggests that a top-down infusion of new information by an intelligent agent is more adequate than any mechanistic bottom-up process. Christians do not have to deny the scientific evidence from the fossil record to firmly reject the materialist delusion that we are nothing but the accidental byproducts of a random process. Instead, the evidence is absolutely compatible with a view of humans as intended by a loving God to be his image-bearers and good stewards of his creation.

Do Fossils Demonstrate Human Evolution?

Casey Luskin

Evolutionists commonly tell the public that the fossil evidence for the Darwinian evolution of our species, *Homo sapiens*, from ape-like creatures is incontrovertible. In 2009, Southern Methodist University anthropology professor Ronald Wetherington testified before the Texas State Board of Education that human evolution has "arguably the most complete sequence of fossil succession of any mammal in the world. No gaps. No lack of transitional fossils...So when people talk about the lack of transitional fossils or gaps in the fossil record, it absolutely is not true. And it is not true specifically for our own species."[1] According to Wetherington, human origins shows "a nice clean example of what Darwin thought was a gradualistic evolutionary change." But does the fossil record support such claims? Digging into the technical literature reveals a starkly different story.

Far from supplying "a nice clean example" of "gradualistic evolutionary change" that has "no gaps" or "no lack of transitional fossils," the record shows a dramatic discontinuity between ape-like and human-like forms. Human-like fossils appear abruptly in the record, without clear evolutionary precursors, contradicting Darwinian expectations. The fossil record does not show that humans evolved from ape-like precursors.

The Fragmented Field of Paleoanthropology

The discipline of paleoanthropology studies the fossil remains of ancient hominins and hominids. Paleoanthropologists face many daunting challenges in their quest to explain human evolution from this hypothetical human/ape common ancestor. Their field is fragmented in multiple senses, making it difficult to confirm evolutionary accounts of human origins.

First, the fossil record is fragmented, and long periods of time exist for which there are few hominin fossils. So "fragmentary and disconnected" is the data, according to Harvard zoologist Richard Lewontin, that "[d]espite the excited and optimistic claims that have been made by some paleontologists, no fossil hominid species can be established as our direct ancestor."[2]

A second challenge is the fragmented nature of the fossil specimens themselves. Typical hominid fossils consist of mere bone scraps, making it difficult to form defini-tive conclusions about their morphology, behavior, and relationships. As Stephen Jay Gould commented: "Most hominid fossils, even though they serve as a basis for endless speculation and elaborate storytelling, are fragments of jaws and scraps of skulls."[3]

Flesh reconstructions of extinct hominins are likewise subjective. They often attempt to diminish the intellectual abilities of humans and overstate those of apes. One high school textbook[4] caricatures Neanderthals as intellectually primitive even though they exhibited intelligence and culture, and casts

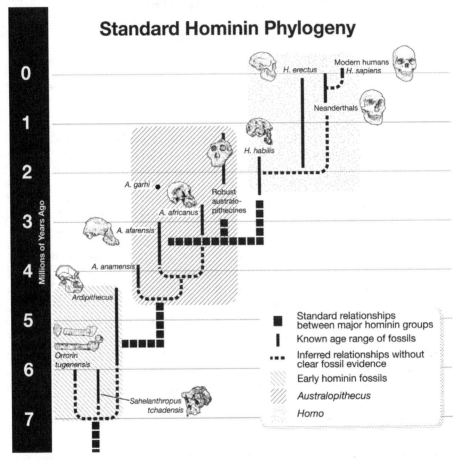

Figure 1. A typical phylogeny of hominins based upon information from multiple sources, especially Zimmer, *Smithsonian Intimate Guide to Human Origins*, 41; Meave Leakey and Alan Walker, "Early Hominid Fossils from Africa," *Scientific American* (August 25, 2003), 16; Potts and Sloan, *What Does It Mean to Be Human?*, 32–33; Ann Gibbons, *The First Human: The Race to Discover Our Earliest Ancestors* (New York: Doubleday, 2006); Ann Gibbons, "A New Kind of Ancestor: *Ardipithecus* Unveiled," *Science* 326 (October 2, 2009), 36–40; David Strait, Frederick E. Grine, and John G. Fleagle, "Analyzing Hominin Phylogeny: Cladistic Approach," *Handbook of Paleoanthropology: Principles, Methods, and Approaches*, 2d ed., eds. Winfried Henke and Ian Tattersall (Heidelberg, Germany: Springer, 2015), 1989–2014. Credit: Jonathan Jones. © Discovery Institute. Used with permission.

Homo erectus as a bungling, stooped form—even though its skeleton is extremely similar to that of modern humans. Conversely, the same textbook portrays an australopithecine (which, in reality, had a chimp-sized brain) with gleams of human-like intelligence and emotion—a common tactic in illustrated books on human origins.[5] The words of the famed physical anthropologist Earnest Hooton from Harvard University remain valid: "alleged restorations of ancient types of man have very little, if any, scientific value and are likely only to mislead the public."[6]

Third, the field itself is fragmented. The sparse nature of the data, combined with the desire to make confident assertions about human evolution, often betrays objectivity and leads to sharp disagreements.[7] After interviewing paleoanthropologists for a documentary, PBS NOVA producer Mark Davis recounted that "[e]ach Neanderthal expert thought the last one I talked to was an idiot, if not an actual Neanderthal."[8]

Even the most established and confidently promoted evolutionary models of human origins are based on limited evidence. *Nature* editor Henry Gee conceded that the "[f]ossil evidence of human evolutionary history is fragmentary and open to various interpretations."[9]

The Standard Story of Human Evolution

Despite the disagreements, there is a standard story of human evolution that is retold in countless textbooks, news media articles, and documentaries. Indeed, virtually all the scientists cited in this chapter accept *some* evolutionary account of human origins, albeit flawed. A representation of the most commonly believed hominin phylogeny is portrayed in Figure 1.

Starting with the early hominins and moving through the australopithecines, and then into the genus *Homo*, this chapter will review the fossil evidence and assess whether it supports this standard account of human evolution. As we shall see, the evidence—or lack thereof—often contradicts this evolutionary story.

Early Hominins

In 2015, two leading paleoanthropologists reviewed the fossil evidence regarding human evolution in a prestigious scientific volume titled *Macroevolution*. They acknowledged the "dearth of unambiguous evidence for ancestor-descendant lineages," and admitted,

> [T]he evolutionary sequence for the majority of hominin lineages is unknown. Most hominin taxa, particularly early hominins, have no obvious ancestors, and in most cases ancestor-descendant sequences (fossil time series) cannot be reliably constructed.[10]

Nevertheless, numerous theories have been promoted about early hominins and their ancestral relationships to humans.

Sahelanthropus tchadensis: The Toumai Skull

Although *Sahelanthropus tchadensis* (also known as the Toumai skull) is known only from one skull and some jaw fragments, it has been called the oldest-known hominin on the human line. When first published, articles in the journal *Nature* called it "the earliest known hominid ancestor"[11] and "close to the last common ancestor of humans and chimpanzees"[12]; as of 2020, the Smithsonian Institution still called it "one of the oldest known species in the human family tree."[13]

But not everyone agrees. Brigitte Senut, of the Natural History Museum in Paris, called Toumai "the skull of a female gorilla,"[14] and cowrote in *Nature* that "*Sahelanthropus* was an ape," not bipedal, and that many features "link the specimen with chimpanzees, gorillas or both, to the exclusion of hominids."[15] In 2020, nearly two decades after the fossil was first reported, the debate was seemingly settled when the femur of *Sahelanthropus* was finally described, confirming that it was a quadruped with a chimp-like body plan.[16] This evidence forced the researchers to suggest that if *Sahelanthropus* were a human ancestor, then that would mean bipedality is no longer a necessary qualification for status as a hominid[17]—an unorthodox view that would wreak havoc with the primate tree. More likely is the view of Madelaine Böhme at the University of Tübingen in Germany: "it's more similar to a chimp than to any other hominin,"[18] meaning, as another commentator put it, *Sahelanthropus* "was not a hominin, and thus was not the earliest known human ancestor."[19]

Precious Little Orrorin tugenensis

Orrorin, which means "original man" in a Kenyan language, was a chimpanzee-sized primate known only from "an assortment of bone fragments,"[20] including pieces of the arm, thigh, lower jaw, and some teeth. When initially discovered, *The New York Times* declared, "Fossils May Be Earliest Human Link,"[21] and reported that *Orrorin* "may be the earliest known ancestor of the human family."[22] *Nature* responded to such hype by warning that the "excitement needs to be tempered with caution in assessing the claim of a six-million-year-old direct ancestor of modern humans."[23]

That seems like wise advice. Paleoanthropologists initially claimed *Orrorin*'s femur indicates bipedal locomotion "appropriate for a population standing at the dawn of the human lineage,"[24] but a later Yale University Press commentary admitted, "All in all, there is currently precious little evidence bearing on how *Orrorin* moved."[25]

Ardipithecus ramidus: Irish Stew or Breakthrough of the Year?

In 2009, *Science* announced the long-awaited publication of details about *Ardipithecus ramidus*, a would-be hominin fossil that lived about 4.4 million years ago (mya). Expectations mounted after its discoverer, UC Berkeley paleoanthropologist Tim White, promised a "phenomenal individual" that would be the "Rosetta stone for understanding bipedalism."[26] The media eagerly employed the hominin they affectionately dubbed *Ardi* to evangelize the public for Darwin.

Discovery Channel ran the headline "'Ardi,' Oldest Human Ancestor, Unveiled," and quoted White calling Ardi "as close as we have ever come to finding the last common ancestor of chimpanzees and humans."[27] The Associated Press declared, "World's Oldest Human-Linked Skeleton Found," and stated that "the new find provides evidence that chimps and humans evolved from some long-ago common ancestor."[28] *Science* named Ardi the "breakthrough of the year" for 2009,[29] and introduced her with the headline, "A New Kind of Ancestor: *Ardipithecus* Unveiled."[30]

Calling Ardi "new" may have been a poor word choice, for it was discovered in the early 1990s. Why did it take some 15 years to publish the analyses? A 2002 article in *Science* explains the bones were "soft," "crushed," "squished," and "chalky."[31] Later reports similarly acknowledged that "portions of Ardi's skeleton were found crushed nearly to smithereens and needed extensive digital reconstruction," including the pelvis, which "looked like an Irish stew."[32]

Claims about bipedal locomotion require accurate measurements of the precise shapes of key bones (like the pelvis). Can one trust declarations of a "Rosetta stone for understanding bipedalism" when Ardi was "crushed to smithereens"? *Science* quoted various paleoanthropologists who were "skeptical that the crushed pelvis really shows the anatomical details needed to demonstrate bipedality."[33]

Even some who accepted Ardi's reconstructions weren't satisfied that she was a bipedal human ancestor. Primatologist Esteban Sarmiento concluded in *Science* that "[a]ll of the *Ar. ramidus* bipedal characters cited also serve the mechanical requisites of quadrupedality, and in the case of *Ar. ramidus* foot-segment proportions, find their closest functional analog to those of gorillas, a terrestrial or semiterrestrial quadruped and not a facultative or habitual biped."[34] Bernard Wood questioned whether Ardi's postcranial skeleton qualified it as hominin,[35] and cowrote in *Nature* that if "*Ardipithecus* is assumed to be a hominin," then it had "remarkably high levels of homoplasy [similarity] among extant great apes."[36] A 2021 study found that Ardi's hands were well-suited for climbing and swinging in trees, and for knuckle-walking, giving it a chimp-like mode of locomotion.[37] In other words, Ardi had ape-like characteristics which, if we set aside the preferences of Ardi's promoters, should imply a closer relationship to apes than to humans. As the authors of the *Nature* article stated, Ardi's "being a human ancestor is by no means the simplest, or most parsimonious explanation."[38] Sarmiento even observed that Ardi had characteristics different from both humans *and* African apes, such as its unfused jaw joint, which ought to remove her far from human ancestry.[39]

Whatever Ardi was, everyone agrees the fossil was initially badly crushed and needed extensive reconstruction. No doubt this debate will continue, but are we obligated to accept the "human ancestor" position promoted by Ardi's discoverers in the media? Sarmiento doesn't think so. According to *Time* magazine, he "regards the hype around Ardi to have been overblown."[40]

Later Hominins: The Australopithecines

Many paleoanthropologists believe the australopithecines were upright-walking hominins and ancestral to our genus *Homo*. Dig into the details, however, and ask basic questions like Who?, Where?, and When?, and there is much controversy. As one paper noted, "there is little consensus on which species of *Australopithecus* is the closest to *Homo*,"[41] if any. Even the origin of genus *Australopithecus* itself is unclear.

Retroactive Confessions of Ignorance

In 2006, *National Geographic* ran a story titled "Fossil Find Is Missing Link in Human Evolution, Scientists Say,"[42] reporting the discovery of what the Associated Press called "the most complete chain of human evolution so far."[43] The fossils, belonging to species *Australopithecus anamensis*, were said to link *Ardipithecus* to its supposed australopithecine descendants.

What exactly was found? According to the technical paper, the claims were based upon canine teeth of intermediate "masticatory robusticity."[44] If a few teeth of intermediate size and shape make "the most complete chain of human evolution so far," then the evidence for human evolution must be indeed quite modest.

Besides learning to distrust media hype, there is another lesson here. Accompanying the praise of this "missing link" were retroactive confessions of ignorance, where evolutionists acknowledge a severe gap in their model *only after* thinking they have found evidence to plug that gap. Thus, the technical paper reporting these teeth admitted, "Until recently, the origins of *Australopithecus* were obscured by a sparse fossil record" and noted, "The origin of *Australopithecus*, the genus widely interpreted as ancestral to *Homo*, is a central problem in human evolutionary studies."[45]

Evolutionists who retroactively confess ignorance risk that the evidence that supposedly filled the gap may not prove very convincing. This seems to be the case here, where a couple of teeth were all that stood between an unsolved "central problem in human evolutionary studies"—the origin of australopithecines—and "the most complete chain of human evolution so far." Moreover, we're left with admissions that the origin of australopithecines is "obscured."

Australopithecines Are like Apes

While early hominins are controversial due to their fragmentary remains, there are sufficient known australopithecine specimens to generally understand their morphology. *Australopithecus*, which literally means "southern ape," is a genus of extinct hominins that lived in Africa from about 4.5 to 1.2 mya. "Splitters" (paleoanthropologists who infer many different species) and "lumpers" (those who see fewer) have created a variety of taxonomic schemes for the australopithecines. The four most commonly accepted species are *afarensis*, *africanus*, *robustus*, and *boisei*. *Robustus* and *boisei* are larger-boned and more "robust," and are sometimes classified within genus *Paranthropus*.[46] They are thought to represent a later-living offshoot that went extinct

without leaving any living descendants. The smaller "gracile" forms, *afarensis* and *africanus*, probably lived earlier, and are classified within the genus *Australopithecus*.

The most well-known australopithecine fossil is Lucy (which belonged to *afarensis*), one of the most complete known fossils among pre-*Homo* hominins. She is often described as a bipedal ape-like creature that is an ideal precursor to humans. Yet only 40 percent of Lucy's bones were found, with a large percentage being rib fragments. Very little useful material from Lucy's skull was recovered, and yet she is one of the most significant specimens ever found. Bernard Wood refutes the misapprehension that she resembled some ape-human hybrid: "Australopithecines are often wrongly thought to have had a mosaic of modern human and modern ape features, or, worse, are regarded as a group of 'failed' humans. Australopithecines were neither of these."[47]

Others have questioned whether Lucy walked like humans or was significantly bipedal. An article in *Nature* observed that much of her body was "ape-like," especially with respect to the "relatively long and curved fingers, relatively long arms, and funnel-shaped chest."[48] It further reported "good evidence" from Lucy's hand-bones that her species "'knuckle-walked', as chimps and gorillas do."[49] A *New Scientist* article adds that Lucy appears well-adapted for climbing, since "[e]verything about her skeleton, from fingertips to toes, suggests that Lucy and her sisters retain several traits that would be very suitable for climbing in trees."[50] Richard Leakey and Roger Lewin argue that *A. afarensis* and other australopithecines "almost certainly were not adapted to a striding gait and running, as humans are."[51] They recount paleontologist Peter Schmid's striking surprise upon realizing Lucy's nonhuman qualities: "What you see in *Australopithecus* is not what you'd want in an efficient bipedal running animal."[52]

As for Lucy's pelvis, many claim it indicates bipedal locomotion, but Johanson and his team reported it was "badly crushed" with "distortion" and "cracking" when first discovered.[53] These problems led one paper to propose Lucy's pelvis appears "different from other australopithecines and so close to the human condition" due to "error in the reconstruction…creating a very 'human-like' sacral plane."[54] Another paper concluded that a lack of clear fossil data prevents paleoanthropologists from making firm conclusions about Lucy's mode of locomotion: "The available data at present are open to widely different interpretations."[55]

Other studies confirm australopithecine differences from humans, and similarities with apes. Their inner ear canals—responsible for balance and related to locomotion—are different from *Homo* but similar to great apes.[56] Traits like their ape-like developmental patterns[57] and ape-like ability for prehensile grasping by their toes[58] led a *Nature* reviewer to say that "ecologically they [australopithecines] may still be considered as apes."[59] Another analysis in *Nature* found the australopithecine skeleton shows "a mosaic of features unique to themselves and features bearing some resemblances to those of the orangutan," and concluded that "the possibility that any of the australopithecines is a direct part of human ancestry recedes."[60] A 2007 paper reported "[g]orilla-like anatomy on *Australopithecus afarensis* mandibles," which was "unexpected," and "cast[s] doubt on the role of *Au. afarensis* as a modern human ancestor."[61]

Paleoanthropologist Leslie Aiello states that when it comes to locomotion, "[a]ustralopithecines are like apes, and the *Homo* group are like humans. Something major occurred when *Homo* evolved, and it

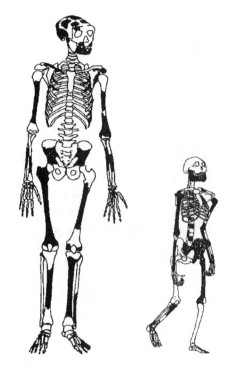

Figure 2. A comparison of Lucy (right) to early *Homo* (left). Black bones indicate those which have been discovered. The original caption states, "The first members of early *Homo sapiens* are really quite distinct from their australopithecine predecessors and contemporaries." Credit: Figure 1 in John Hawks, Keith Hunley, Sang-Hee Lee, Milford Wolpoff, "Population Bottlenecks and Pleistocene Human Evolution," *Molecular Biology and Evolution*, 17:2-22, © 2000 by Oxford University Press. Used with permission.

wasn't just in the brain."[62] The "something major" was the abrupt appearance of the human-like body plan—without direct evolutionary precursors in the fossil record.

Lacking Intermediates

If humans evolved from ape-like creatures, what were the transitional species between the ape-like hominins just discussed and the truly human-like members of the *Homo* genus found in the fossil record? There aren't any good candidates.

The Demise of Homo habilis

Many have cited *Homo habilis* (literally "handy man") as a tool-using species that was a transitional "link" between the australopithecines and *Homo*.[63] But its association with tools is doubtful and appears driven mainly by evolutionary considerations.[64] Anthropologist Ian Tattersall calls it "a wastebasket taxon, little more than a convenient recipient for a motley assortment of hominin fossils."[65] Ignoring these difficulties and assuming *habilis* was a real species, chronology precludes it from being ancestral to *Homo*: habiline remains *postdate* the earliest fossil evidence of the genus *Homo*.[66]

Morphological analyses further confirm that *habilis* makes an unlikely "intermediate" between *Australopithecus* and *Homo*—and show *habilis* doesn't even belong in *Homo*. An authoritative review in *Science* by Bernard Wood and Mark Collard found that *habilis* differs from *Homo* in terms of body size, shape, mode of locomotion, jaws and teeth, developmental patterns, and brain size, and should be reclassified within *Australopithecus*.[67] A study by Sigrid Hartwig-Scherer and Robert D. Martin in the *Journal of Human Evolution* found the skeleton of *habilis* was *more* similar to living apes than were other australopithecines like Lucy.[68] They conclude, "It is difficult to accept an evolutionary sequence in which *Homo habilis*, with less human-like locomotor adaptations, is intermediate between *Australopithecus afaren[s]is*…and fully bipedal *Homo erectus*."[69] Alan Walker and Pat Shipman similarly called *habilis* "more apelike than Lucy" and remarked, "Rather than representing an intermediate between Lucy

and humans, [*habilis*] looked very much like an intermediate between the ancestral chimplike condition and Lucy."[70] Hartwig-Scherer explains that *habilis* "displays much stronger similarities to African ape limb proportions" than Lucy—results she calls "unexpected in view of previous accounts of *Homo habilis* as a link between australopithecines and humans."[71]

Homo naledi versus Australopithecus sediba: The Link Resurrected?

The news media might be heavily biased toward evolution, but at least it is predictable. Whenever a new hominin fossil is discovered, reporters seize the opportunity to push human evolution. Thus it was no surprise when news outlets buzzed about the latest "human ancestor" after a new species, *Homo naledi*, was unveiled in 2015.

CNN declared, "*Homo naledi*: New Species of Human Ancestor Discovered in South Africa."[72] *The Daily Mail* reported, "Scientists Discover Skull of New Human Ancestor *Homo Naledi*."[73] PBS pronounced, "Trove of Fossils from a Long-Lost Human Ancestor."[74] And so on.

The find is striking because it represents probably the largest cache of hominin bones—many hundreds—ever found. In a field where a single scrap of jaw ignites the community, this is a big deal. But do we know that *Homo naledi* is a human ancestor, as news outlets declared? Dig into the details, and the answer again is no.

The primary claim about *Homo naledi* is that it was a "transitional form" or "mosaic"—a small-brained, upright-walking hominin with a trunk similar to the australopithecines, but with human-like hands and feet. But the technical material shows that even some of those supposedly human-like traits have unique features:

- The hands showed "a unique combination of anatomy"[75] including "unique first metacarpal morphology,"[76] and long, curved fingers that suggest *naledi* was, unlike humans, well-suited for "climbing and suspension."[77]

- Its foot "differs from modern humans in having more curved proximal pedal phalanges, and features suggestive of a reduced medial longitudinal arch," giving it an overall "unique locomotor repertoire."[78] The foot shows, again, that unlike humans, it was "likely comfortable climbing trees."[79]

The technical papers also reveal "unique features in the femur and tibia"—making a hindlimb that "differs from those of all other known hominins."[80] As for the head, "Cranial morphology of *H. naledi* is unique…"[81] Sound familiar? Whatever it was, overall *naledi* appears highly unique.

Indeed, the discoverers of *naledi* called it "a unique mosaic previously unknown in the human fossil record."[82] Such terminology should raise a red flag. In the parlance of evolutionary biology, "mosaic" usually means a fossil has a suite of traits that are difficult to fit into the standard evolutionary tree. This is the case here.

In 2010, some of the same scientists who discovered and promoted *naledi*—a team led by Lee Berger of the University of Witwatersrand—were promoting a *different* hominin species, *Australopithecus sediba*, as the intermediate *du jure* between the australopithecines and *Homo*. However, *sediba* and *naledi* differ in important ways that make them unlikely partners in an evolutionary lineage. Specifically, *sediba* (classified within *Australopithecus*) had an advanced "*Homo*-like pelvis,"[83] "surprisingly human teeth,"[84]

and a "human-like" lower trunk,[85] whereas *naledi*—placed within *Homo*—bears an "australopith-like" and "primitive" pelvis,[86] "primitive" teeth, and a "primitive or australopith-like trunk."[87] An australopithecine with apparently advanced *Homo*-like features seems a poor candidate to evolve *into* a member of *Homo* with primitive australopith-like versions of those same features. Thus, although both *sediba* and *naledi* have been said to be a human ancestor—by some of the same people, no less—evolutionarily speaking, traits are evolving in the wrong direction. As one news outlet put it: "Each [*sediba* and *naledi*] has different sets of australopith-like and human-like traits that can't be easily reconciled on the same family tree."[88]

In any case, *sediba* cannot be ancestral to *Homo* because, like *habilis*, it postdates the origin of our genus and has the wrong morphology.[89] Based upon fossil chronology, a 2019 study found that the likelihood that *sediba* is a human ancestor is less than 0.001.[90] Commenting on *sediba*, Harvard's Daniel Lieberman said, "The origins of the genus *Homo* remain as murky as ever,"[91] and Donald Johanson remarked, "The transition to *Homo* continues to be almost totally confusing."[92]

Another dubious claim about *naledi* is that it intentionally buried its dead—a testimony to its supposedly human-like intellect. Burying dead in the cave where it was found would require shimmying through a steep, narrow crevice while dragging a body a long distance in the dark—a physically challenging task for any hominin of any level of intelligence. For many reasons, multiple scientists—including two of Berger's colleagues at the University of Witwatersrand—dispute the intentional burial hypothesis.[93] Alison Brooks of George Washington University observed that claims of intentional burial are "so far out there that they really need a higher standard of proof."[94]

But the deathblow to claims for *Homo naledi* as an ancestral or transitional fossil is its age. When first published, *naledi*'s promoters suggested, on the basis of evolutionary considerations rather than geological evidence, that it lived 2–3 mya. But at that time the fossils hadn't been dated geologically. Carol Ward of the University of Missouri warned, "Without dates, the fossils reveal almost nothing about hominin evolution."[95] This didn't stop paleoanthropologists from speculating, predicting that *naledi* lived 2–3 mya and "represents an intermediate between *Australopithecus* and *Homo erectus*."[96] In 2017, *Homo naledi*'s remains were dated to the "surprisingly" and "startingly young" age of 236,000–335,000 years[97]—an order of magnitude younger than the age predicted by evolutionary considerations, and far too young to be ancestral to our species. Anthropologist James Kidder candidly admitted, "Nearly everyone in the scientific community thought that the date of the *Homo naledi* fossils, when calculated, would fall within the same general time period as other primitive early *Homo* remains. We were wrong."[98]

Many cautioned against the hype over *naledi*,[99] and its trajectory resembles other hominins for which hyped claims of transitional or ancestral status eventually failed. When evaluating media claims of the newest human ancestor, a dose of healthy skepticism is warranted.

A Big Bang Origin of *Homo*

After realizing that *habilis* could not serve as a link between *Homo* and *Australopithecus*, two paleoanthropologists lamented "this muddle leaves *Homo erectus* without a clear ancestor, without a past."[100] Indeed, it is difficult to find fossil hominins to serve as direct transitional forms between the ape-like australopithecines and the first human-like members of *Homo*. The fossil record shows abrupt changes that correspond to the appearance of our genus *Homo* about two million years ago.

From its first appearance, *Homo erectus* was very human-like, and differed markedly from prior hominins that were *not* human-like. Yet *Homo erectus* appears *abruptly*, without apparent evolutionary precursors. An article in *Nature* explains:

> The origins of the widespread, polymorphic, Early Pleistocene *H. erectus* lineage remain elusive. The marked contrasts between any potential ancestor (*Homo habilis* or other) and the earliest known *H. erectus* might signal an abrupt evolutionary emergence some time before its first known appearance in Africa at ~1.78 Myr [million years ago]. Uncertainties surrounding the taxon's appearance in Eurasia and southeast Asia make it impossible to establish accurately the time or place of origin for *H. erectus*.[101]

A 2016 paper likewise admits, "Although the transition from *Australopithecus* to *Homo* is usually thought of as a momentous transformation, the fossil record bearing on the origin and earliest evolution of *Homo* is virtually undocumented."[102] While that paper argues that the evolutionary distance between *Australopithecus* and *Homo* is small, it concedes that the lineage that led to *Homo* is "unknown."[103]

Early members of *Homo*, namely *Homo erectus*, show unique and previously unseen features that contributed to this "abrupt" appearance. The technical literature observes an "explosion,"[104] "rapid increase,"[105] and "approximate doubling"[106] in brain size associated with the appearance of *Homo*. Wood and Collard's major *Science* review found that *only one single trait of one hominin species qualified as "intermediate"* between *Australopithecus* and *Homo*: the brain size of *Homo erectus*.[107] However, this one trait of brain size does not necessarily indicate that humans evolved from less intelligent hominids. Intelligence is determined largely by internal brain organization, and is much more complex than the singular dimension of brain size.[108] Christof Koch, president of the Allen Institute for Brain Science, observes that "total brain volume weakly correlates with intelligence...brain size accounts for between 9 and 16 percent of the overall variability in general intelligence."[109] Because of this, brain size is not always a good indicator of evolutionary relationships (Figure 3).[110] In any case, *erectus* had an average brain size within the range of modern human variation (Table 1, see page 369). A few skulls of "intermediate" size do not demonstrate that humans evolved from primitive ancestors.

Much like the explosive increase in skull size, a study of the pelvis bones of australopithecines and *Homo* found "a period of very rapid evolution corresponding to the emergence of the genus *Homo*."[111] One *Nature* paper noted that early *Homo erectus* shows "such a radical departure from previous forms of *Homo* (such as *H. habilis*) in its height, reduced sexual dimorphism, long limbs and modern body proportions that it is hard at present to identify its immediate ancestry in east Africa."[112] A paper in the *Journal of Molecular Biology and Evolution* found that *Homo* and *Australopithecus* differ significantly in brain size, dental function, increased cranial buttressing, expanded body height, visual, and respiratory changes, and stated,

> We, like many others, interpret the anatomical evidence to show that early *H. sapiens* was significantly and dramatically different from...australopithecines in virtually every element of its skeleton and every remnant of its behavior.[113]

Noting these many differences, the study called the origin of humans "a real acceleration of evolutionary change from the more slowly changing pace of australopithecine evolution" and stated that such a transformation would have required radical changes:

> The anatomy of the earliest *H. sapiens* sample indicates significant

Figure 3. Got a big head? Don't get a big head. Brain size is not always a good indicator of intelligence or evolutionary relationships. Case in point: Neanderthals had an average skull size *larger* than modern humans. Moreover, skull size can vary greatly within a species. Given the range of modern human variation, a progression of relatively small to very large skulls could easily be obtained by using bones from living humans alone. This could give the false impression of some evolutionary lineage when it is no such thing. The lesson? Don't be too impressed when textbooks, media stories, or documentaries display skulls lined up from small to large sizes. Credit: Jonathan Jones. © Discovery Institute. Used with permission.

368 THE COMPREHENSIVE GUIDE TO SCIENCE AND FAITH

modifications of the ancestral genome and is not simply an extension of evolutionary trends in an earlier australopithecine lineage throughout the Pliocene. In fact, its combination of features never appears earlier.[114]

These rapid and unique changes are termed "a genetic revolution" in which "no australopithecine species is obviously transitional."[115]

For those not constrained by an evolutionary paradigm, it is not obvious that this transition took place at all. The stark lack of fossil evidence for this hypothesized transition is confirmed by three Harvard paleoanthropologists:

> Of the various transitions that occurred during human evolution, the transition from *Australopithecus* to *Homo* was undoubtedly one of the most critical in its magnitude and consequences. As with many key evolutionary events, there is both good and bad news. First, the bad news is that many details of this transition are obscure because of the paucity of the fossil and archaeological records.[116]

As for the "good news," they admit, "[A]lthough we lack many details about exactly how, when, and where the transition occurred from *Australopithecus* to *Homo*, we have sufficient data from before and after the transition to make some inferences about the overall nature of key changes that did occur."[117]

In other words, the fossil record shows ape-like australopithecines ("before"), and human-like *Homo* ("after"), but not fossils documenting a transition between them. In the absence of intermediates, we're left with inferences of a transition based strictly upon the assumption of evolution—that

an undocumented transition must have occurred somehow, sometime, and someplace. They assume this transition happened, even though we do not have fossils documenting it.

The literature thus admits the "abrupt appearance"[118] of early *Homo* and calls the origin of our genus "an enduring puzzle."[119] The great evolutionary biologist Ernst Mayr recognized these problems:

> The earliest fossils of *Homo*, *Homo rudolfensis* and *Homo erectus* are separated from *Australopithecus* by a large, unbridged gap. How can we explain this seeming saltation? Not having any fossils that can serve as missing links, we have to fall back on the time-honored method of historical science, the construction of a historical narrative.[120]

Another commentator proposed that the evidence implies a "big bang theory" of the appearance of *Homo*.[121] This large, unbridged gap between the ape-like australopithecines and the abruptly appearing human-like members of genus *Homo* challenges evolutionary accounts of human origins.

All in the Family

In contrast to the australopithecines, the major members of *Homo*—i.e., *erectus* and the Neanderthals (*Homo neanderthalensis*)—are very similar to us. Some paleoanthropologists have even classified *erectus* and *neanderthalensis* as members of our own species, *Homo sapiens*.[122]

Homo erectus appears in the fossil record a little more than two million years ago. Its name means "upright man," and unsurprisingly, below the neck, they were extremely similar to us.[123] An Oxford University Press

Table 1. Cranial Capacities of Extant and Extinct Hominids.[132]		
Taxon:	Cranial Capacities:	Taxon Resembles:
Gorilla (*Gorilla gorilla*)	340–752 cc	Modern Apes
Chimpanzee (*Pan troglodytes*)	275–500 cc	
Australopithecus	370–515 cc (Avg. 457 cc)	
Homo habilis	Avg. 552 cc	
Homo erectus	850–1250 cc (Avg. 1016 cc)	Modern Humans
Neanderthals	1100–1700 cc (Avg. 1450 cc)	
Homo sapiens	800–2200 cc (Avg. 1345 cc)	

volume notes *erectus* was "humanlike in its stature, body mass, and body proportions."[124] An analysis of 1.5-million-year-old *Homo erectus* footprints[125] indicates "a modern human style of walking" and "human-like social behaviours."[126] Unlike the australopithecines and habilines, *erectus* is the "earliest species to demonstrate the modern human semicircular canal morphology."[127]

Another study found that total energy expenditure (TEE), a complex character related to body size, diet, and food-gathering activity, "increased substantially in *Homo erectus* relative to the earlier australopithecines," approaching the high TEE value of modern humans.[128] While the average brain size of *Homo erectus* is less than the modern human average, *erectus* cranial capacities are within the range of normal human variation.[129] Intriguingly, *erectus* remains have been found on islands where the most likely explanation is that they arrived by boat. Anthropologists have argued this indicates high intelligence and the use of complex language.[130] Donald Johanson suggests that were *erectus* alive today, it could mate with modern humans to produce fertile offspring.[131] In other words, were it not for our separation by time, we might be considered interbreeding members of the same species.

As for Neanderthals, though they have been stereotyped as bungling and primitive, if a Neanderthal walked down the street, you probably wouldn't notice. Wood and Collard note that "skeletons of *H. neanderthalensis* indicate that their body shape was within the range of variation seen in modern humans."[133] Washington University paleoanthropologist Erik Trinkaus maintains that Neanderthals were no less intelligent than contemporary humans[134] and argues, "They may have had heavier brows or broader noses or stockier builds, but behaviorally, socially and reproductively they were all just people."[135] University of Bordeaux archaeologist Francesco d'Errico agrees: "Neanderthals were using technology as advanced as that of contemporary anatomically modern humans and were using symbolism in much the same way."[136]

Though controversial, hard evidence backs these claims. Anthropologist Stephen Molnar explains that "the estimated mean size of [Neanderthal] cranial capacity (1,450 cc) is actually higher than the mean for modern humans (1,345 cc)."[137] One paper in *Nature* suggested that "the morphological basis for human speech capability appears to have been fully developed" in Neanderthals.[138] Indeed, Neanderthal remains have been found associated with signs of culture, including art, burial of their dead, and complex tools[139]—including musical instruments like the flute.[140] While dated, a 1908 report in *Nature* reports a Neanderthal-type skeleton wearing chain mail armor.[141] Archaeologist Metin Eren said, regarding toolmaking, that "in many ways, Neanderthals were just as smart or just as good as us."[142] Morphological mosaics—skeletons showing a mix of modern human and Neanderthal traits—suggest "Neandertals and modern humans are members of the same species who interbred freely."[143] Indeed, scientists now report Neanderthal DNA markers in living humans,[144] supporting proposals that Neanderthals were a subrace of our own species.[145] As Trinkaus says regarding ancient Europeans and Neanderthals, "[W]e would understand both to be human. There's good reason to think that they did as well."[146]

Darwin skeptics continue to debate whether we are related to Neanderthals and *Homo erectus*, and evidence can be mounted both ways.[147] The present point, however, is this: *Even if* we do share common ancestry with Neanderthals or *erectus*, this does *not* show we share ancestry with any nonhuman-like hominins.

According to Siegrid Hartwig-Scherer, the differences between human-like members of *Homo* such as *erectus*, Neanderthals, and us reflect mere microevolutionary effects of "size variation, climatic stress, genetic drift and differential expression of [common] genes."[148] Whether we are related to them or not, these small-scale differences do *not* show the evolution of humans from nonhuman-like or ape-like creatures.

Culture Explosion

In 2015, two top paleoanthropologists admitted in a major review that "the evolutionary sequence for the majority of hominin lineages is unknown."[149] Despite the claims of evolutionary paleoanthropologists and unceasing media hype, the fragmented hominin fossil record does not document the evolution of humans from ape-like precursors. The genus *Homo* appears in an abrupt, non-Darwinian fashion without evidence of an evolutionary transition from ape-like hominins. Other major members of *Homo* appear very similar to modern humans, and their differences amount to small-scale microevolutionary change—providing no evidence that we are related to nonhuman-like species.

But there's more evidence that contradicts an evolutionary model.

Many researchers have recognized an "explosion"[150] of modern human-like culture in the archaeological record about 35,000 to 40,000 years ago, showing the abrupt appearance of human creativity,[151] technology, art,[152] and even paintings[153]—as well as the rapid emergence of self-awareness, group identity, and symbolic thought.[154] One review dubbed this the "Creative Explosion."[155] Indeed, a 2014 paper coauthored by leading paleoanthropologists admits we have "essentially no explanation of how and why our linguistic computations and representations evolved," since "nonhuman animals provide virtually no relevant parallels to human linguistic communication."[156] This abrupt appearance of modern human-like morphology, intellect, and culture contradicts evolutionary models, and may indicate design in human history.

Is Evolutionary Psychology a Legitimate Way to Understand Our Humanity?

Denyse O'Leary

Evolutionary psychologists analyze human behavior on the assumption that Darwinian natural selection in a nonhuman (or scarcely human past) accounts for our minds, attitudes, values, beliefs, and behavior. In this chapter, I will not consider whether their account of human origins is true. Rather, I will focus on whether, in any event, their attempts to discern human/not-quite-human behavior in a dark evolutionary past provide much insight into our behavior today.

ScienceDaily explains the discipline thus:

Evolutionary Psychology proposes that the human brain comprises many functional mechanisms, called psychological adaptations or evolved cognitive mechanisms designed by the process of natural selection.

Examples include language acquisition modules, incest avoidance mechanisms, cheater detection mechanisms, intelligence and sex-specific mating preferences, foraging mechanisms, alliance-tracking mechanisms, agent detection mechanisms, and so on.[1]

Evolutionary psychology (EP) claims to account in this way not only for big topics like philanthropy and religion but also countless less-momentous enterprises like shopping and gossip. Indeed, given the premise above, there is no human behavior that the evolutionary psychologist *cannot* account for.

The earliest well-known classics developing this line of thought were Charles Darwin's *On the Origin of Species* (1859) and *The Descent of Man, and Selection in Relation to Sex* (1871). In the latter book, Darwin proposed the theory of sexual selection—the way that the success or failure of the drive to mate determines which genes get passed on in a population.[2]

Many of Darwin's adherents saw his theories as evidence for Social Darwinism (unfortunate people are inferior in the struggle for life). Social Darwinism fell into disfavor after World War II because its theories justified colonialism, exploitation of labor, and eugenics.[3] In the 1970s, the basic idea was resurrected—without direct political implications—under the new name of *sociobiology*. Sociobiologists, for example, turned to insect colonies to explain traits like caring for others—our genes, like theirs, are shared in large part with relatives.[4] But sociobiology too became controversial when it attracted allegations of racism.[5]

Soon afterward, evolutionary psychology, a much broader and more widely accepted movement, burst onto the scene. Thus, everywhere in popular science culture today we run into EP explanations. Evolutionary psychology may be how most lay readers of popular science literature know Darwinism.

The Evolutionary Psychologist Explains the "True Reasons" for Our Behavior

We may not know why we do things, but, no matter, for the evolutionary psychologist does. Evolution explains, for example, why we shop: "Gatherers sifted the useful from things that offered them no sustenance, warmth or comfort with a skill that would eventually lead to comfortable shopping malls and credit cards."[6] Or gossip: "Back in the day, if you didn't care to find out what was going on, you were more likely to die and less likely to pass on your incurious genes."[7] Oh, and anger over trivial matters was once key to our survival.[8]

As the examples above illustrate, EP does not explain puzzling human behavior so much as it offers Darwinian explanations for conventional behavior, thus supplanting traditional ones. For example, why we are sexually jealous (not fear of abandonment, but "sperm competition"); why we don't stick to our goals (evolution gave us a kludge brain); why we developed music (to "spot the savannah with little Pavarottis" in order to catch the ear of ladies); why art exists (to recapture that lost savannah);[9] why many

women don't know when they are ovulating (if they knew, they'd never have kids);[10] why some people rape, kill, and sleep around (our Stone Age ancestors passed on their genes via these traits),[11] and why big banks sometimes get away with fraud (we haven't evolved so as to understand what is happening).[12]

EP also accounts for dreams (they increase reproductive fitness),[13] false memories (there might be a tiger in that tall grass),[14] menopause (men pursuing younger women),[15] monogamy (control of females or else infanticide prevention),[16] premenstrual syndrome (breaks up infertile relationships),[17] romantic love (a "hardwired" drive to reproduce),[18] rumination on hurt feelings (our brains evolved to learn quickly from bad experiences but slowly from the good ones),[19] smiling (earlier, a cringe reaction),[20] and wonder at the universe (explained by how early man lived).[21]

Accounts such as these can sit very oddly with accepted facts about human nature. For example, at *Psychology Today*, we might have learned, as #1 of "Ten Politically Incorrect Truths About Human Nature," that men prefer women with larger breasts because they make fertility easier to ascertain and primal man unconsciously tried to spread his selfish genes.[22] Indeed? It has nothing to do with a preference for more pleasure of many kinds—bigger paychecks, cars, and steaks? Don't many men who seek pleasure prefer to *avoid* more mouths to feed? The top explanation does not so much provide insight into men's behavior as map it onto Darwinian thinking—and that is the goal. Evolutionary psychology needs to be fully Darwinian much more than it needs to square with observations about contemporary human behavior.

EP explanations can also dispense with historical fact. For example, in *Delusions of*

Gender (2010), Cordelia Fine recounts an EP explanation of why little girls are dressed in pink (their brains evolved to process emotion differently). That must have been one of the most rapid instances of human evolution ever recorded. The practice of dressing girls in pink only took root in the early twentieth century.[23] But no matter. Give us Darwin; you can have history.

The accounts can even fail as parody. Neuroscientist Vilayanur Ramachandran tried parody with "Why Do Gentlemen Prefer Blondes?," but there is no reason not to take his explanations as seriously as all the others.[24] Give us Darwin, and you can have science too.

But the truly serious EP enterprises involve Darwinizing compassion, morality, and religion, as noted earlier. As before, this is done by substituting Darwinian explanations, treated as the "science" explanations, for traditional accounts.

Compassion: The Itch that the Darwinian Always Seems to Need to Scratch

In a Darwinian framework, if one person sacrifices for another, individual survival and reproduction must nonetheless be the true goal. Hence such behavior has been called "an anomalous thorn in Darwin's side" and a "conundrum that Darwinians would need to solve, given their view of the ruthless struggle among living beings for survival."[25]

A first step in reframing the group of qualities we often call compassion, self-sacrifice, or quiet heroism, etc., is renaming it. The group is now loosely termed *altruism*. It is a trans-species concept; it includes, for example, the mindless worker ants who pass on their genes by serving their queen instead of reproducing individually (kin selection).[26] As one writer puts it:

For decades, biologists have struggled to explain altruism in evolutionary terms. There's reciprocal altruism: a vampire bat regurgitates blood to another and receives a "payback" meal later on. There's sexual selection: male chimps look more attractive to females if they groom others or share food. There's kin selection. Altruism might not benefit the individual actor, but maybe it increases "inclusive fitness": passing on genes within a family or lineage. Hamilton's Rule puts kin selection into numbers: $c < r \times b$. The cost to the actor must be less than the benefit times the "relatedness" between the actor and the recipient. Mothers and children share half their genes, so r is ½. Same for siblings. An uncle-nephew share ¼, cousins share ⅛, and so on. Once, biologist J.B.S. Haldane was asked, "Would you lay down your life for your brother?" "No, but I would for two brothers or eight cousins," he replied.[27]

Evolutionist William Hamilton described the idea mathematically, calling it "inclusive fitness,"[28] and voila!, it was science. For a while.

In 2010, inclusive fitness exploded into a huge controversy. In a paper that year in *Nature,* coauthored with mathematicians, E.O. Wilson, the founder of sociobiology, suddenly rejected the kin selection he had earlier espoused. The paper declared, "Altruism can be explained by natural selection."[29] In his return to strict Darwinism (natural selection acting on random mutation produces evolution), Wilson insisted that Darwinism "provides an exact framework for interpreting empirical observations," rejecting theories he had promoted for decades. More than 140 top biologists subsequently denounced the new paper in a letter to *Nature*.[30] It must have been one of the few times in the history of modern biology that a return to strict Darwin-and-Darwin-*only* principles was angrily denounced.

New atheist evolutionary biologist Jerry Coyne weighed in, saying that Wilson et al. are "misguided" and "wrong—dead wrong." Oddly, he acknowledged, "The 'textbook' explanation, based on a higher relatedness of workers to their sisters than to their own potential offspring, no longer seems feasible…But we've known all this for years!" If so, most of them were very economical with the information.[31]

Most observers found the uproar hard to comprehend. Neuroscientist Michael Gazzaniga offered, in Wilson's defense, "Although he says his new theory opposes the idea of kin selection, in another sense he is simply maintaining that everybody is right."[32]

So then why is Wilson wrong? In *The New Republic,* John Gray called the episode "an exercise in sectarian intellectual warfare of the kind that is so often fought in and around Darwinism."[33] It certainly demonstrates the fragility of the attempt to Darwinize compassion and self-sacrifice.

A number of less rigorous Darwinian accounts of altruism are also available, usually riffing off Richard Dawkins's notion of the "selfish gene." Humans, along with social insects, bacteria, and viruses "share gene products and behave in ways that can't be described as anything but generous."[34] Elsewhere, we are told that altruism is just another form of manipulation[35] or that it may be hardwired into our genes ("in the end people are just biological organisms"[36]). Or, on another view, it's a mating signal;[37] indeed, it gets one more sex partners.[38] It enables cooperative breeding and raising of offspring[39] or otherwise serves "group needs."[40]

It feels like emptying Darwin's wastebasket.

Oxford physiologist Denis Noble observes that "selfish genes" have no empirical basis in science. Though it is "one of the most successful science metaphors ever invented," it does not coincide with how genes actually work.[41] David Dobbs made a similar point at *Aeon* magazine.[42]

On the ground, it's not clear that exceptionally altruistic people are governed by a need to spread their genes. But, as we have seen, an EP explanation need not be particularly informative; it can fail to account for obvious facts and still pass muster as long as it is fully naturalistic and does not stray from Darwinism. We will see this again when we look at EP's account of religion.

Religion Can Be Good, Bad, or Useless; the Only Thing It Can't Be Is True to Reality

That Darwinizing any aspect of human behavior is "scientific" is never doubted among evolutionary psychologists.[43] The conundrum is how religion functions as an evolutionary adaptation for survival to which truth claims are irrelevant.[44]

First, it could be a useful adaptation. Psychologist Jonathan Haidt speculates that "there was a long period in human evolution during which it was adaptive to lose the self and merge with others. It wasn't adaptive for individuals to do so, but it was adaptive for groups."[45]

Others consider religion to be a byproduct, a parasite on useful traits. Ilkka Pyysiäinen and Marc Hauser published a paper in 2010 arguing that view:

> Considerable debate has surrounded
> the question of the origins and

evolution of religion. One proposal views religion as an adaptation for cooperation, whereas an alternative proposal views religion as a by-product of evolved, non-religious, cognitive functions. We critically evaluate each approach, explore the link between religion and morality in particular, and argue that recent empirical work in moral psychology provides stronger support for the by-product approach.[46]

But some theorists see religion as a bad adaptation. Evolutionary biologist Jerry Coyne attributes Americans' doubts about Darwinian evolutionary theories to religious faith, which, he claims, correlates highly with social dysfunction.[47]

But again, the relationship between religion and adaptation that we can observe in the world around us is mixed. Christians, certainly, are warned against supposing that their faith is adaptive: "Blessed are you when others revile you and persecute you and utter all kinds of evil against you falsely on my account" (Matthew 5:11). Many persist in their faith anyway.

These theories about religion (useful, useless, or harmful) have one main thing in common: They typically spill forth with no real engagement with religion. For example, one recent study claimed that believers subconsciously endow God with their own beliefs on controversial issues.[48] But if it's that simple, why don't the many adherents whose religion requires them to do things they don't like (fast, give more to the poor) and give up what they really like (smoking pot, casual sex) just endow God with more permissive beliefs?

If one's research is in a hole that deep, why not stop digging? Well, in this case, the hole *is* the enterprise. To whatever extent religion is not merely the spread of selfish genes,

it is necessarily invisible to the evolutionary psychologist.

Why the Critics Have Never Made Any Dent So Far

Curiously, the best-known academic dissent does not come from Christian quarters. Agnostic philosopher David Stove and atheist philosopher Jerry Fodor have written books, respectively *Darwinian Fairytales*[49] and *What Darwin Got Wrong*,[50] assailing EP's evidence-challenged assertions. Social scientists Steven and Hilary Rose edited an anthology, *Alas, Poor Darwin,* featuring a number of notables in science and social science skewering the discipline.[51]

In 2009, science writer Sharon Begley noted at *Newsweek* evolutionary psychologists' characteristic backpedalling when challenged on extreme claims[52] and their comfort with indemonstrable hypotheses:

> From its inception, evolutionary psychology had warned that behaviors that were evolutionarily advantageous 100,000 years ago (a sweet tooth, say) might be bad for survival today (causing obesity and thence infertility), so there was no point in measuring whether that trait makes people more evolutionarily fit today. Even if it doesn't, evolutionary psychologists argue, the trait might have been adaptive long ago and therefore still be our genetic legacy. An unfortunate one, perhaps, but still our legacy. Short of a time machine, the hypothesis was impossible to disprove.[53]

In 2012, medical historian Andrew Scull, reviewing a book on psychiatry's legitimacy crisis, wrote that EP theories are "unnecessary, and get in the way of an argument that depends on no more than the self-evident proposition that all of us experience fears and anxieties, which are intensified in certain social situations and by large-scale trauma, but which cannot be termed 'mental illnesses.'"[54]

In any event, at the heart of EP is a searing contradiction: "Evolution" is supposed to be true beyond reasonable doubt. Yet evolutionary psychologists believe that humans have not fundamentally changed in any way in the last million years. But to even raise such objections has meant risking the label *antiscience*.

EP shares in Darwinism's global immunity from rational criticism from within the science disciplines. It fits so well with a prevailing naturalistic worldview that it need not recommend itself on either reason or evidence. That says far more about the culture that created it than about the history of the human race.

Science writer Hank Campbell offers a suggestion as to why so much sheer nonsense goes unchallenged: "Scientists are inclined to give it a break because they cleverly use the word 'evolutionary' in the name."[55] When you think about it, that has been a rather costly indulgence.

But for EP, the indulgence has been a perfect environment: Evolutionary psychology is a branch of Darwinism, therefore it is science; hence critics of the discipline, whatever their contentions, are antiscience. A decades-long flood tide of nonsense might flow on forever on those terms.

Or does something eventually have to give? A recent rash of criticism may damage the discipline more than the previous ones by focusing on two main points: EP simply doesn't have enough grounding in evidence to be a science, and treating it as a science leads to misleading interpretations of human behavior.

Leveling the Playing Field

In 2020, a new challenger to EP arose: University of New Hampshire philosopher Subrena E. Smith. Her article "Is Evolutionary Psychology Possible?"[56] received considerable sympathy with comparatively little "Attack on science!" pushback.

She starts with the fact that evolutionary psychologists believe that we inherited Stone Age minds, largely unchanged, from remote ancestors in the form of modules that govern our behavior for "predator avoidance, mate selection, and cheater detection."[57] But neuroscience has never identified any such modules. Even if it did, are they identical to those of humans who lived 50,000 years ago? Brains, after all, don't fossilize. Calling this the "matching problem," she says, "Unless the challenge can be overcome, evolutionary psychological explanations fail. Put more strongly, if the matching problem cannot be solved, evolutionary psychology is impossible."[58]

As is so often done, turning to chimpanzees for the rescue doesn't really help. Smith notes that wild chimpanzees today appear to behave as they did millions of years ago. We deduce that their present behavior is inherited from their past behavior. Human life, meanwhile, has vastly changed over the same period, so we simply can't make the same assumptions.

Smith offers, by way of illustration, a 2009 study of college students that attempted to show men care more about infidelity than women because of the biological cost: "Ancestral men…were susceptible to an additional and profound cost if they failed to detect a partner's infidelity: cuckoldry—the unwitting investment of resources into genetically unrelated offspring."[59] She offers a number of reasons for dismissing such conclusions as unjustified, including, "There is no good evidence that males' skepticism about their female partners' fidelity is under genetic control, and there is no dedicated neurological mechanism known to mediate it (the module controlling the attitude is entirely hypothetical)."[60]

Smith's arguments are cogent, but then so have been the arguments of many of her predecessors. Yet—unlike so many past critics of EP—she has found support in science media. That may signal a weakening of the hold of "all things Darwin" on the public imagination.[61] For once, the scientists and science writers who side with her don't seem to fear being labelled *antiscience*. Could many of them be starting to realize something? Evolutionary psychologists cannot claim to both represent science and refuse to be bound by its recognized standards.

Smith uses the term "subpersonal"[62] to describe evolutionary psychologists' explanations for human behavior, which make no distinction between human and animal cognition. And in doing so, she has identified one of the devastating weaknesses of the discipline.

Evolutionary Psychology as the Triumph of the Subpersonal

In EP, the role of abstract thought, including reason and moral concerns, is set aside. Human behavior is explained as if humans were nonrational animals. On that view, we may imagine that we do things for rational or moral reasons, but the true driving forces are those hypothesized modules that drive us to survive and reproduce.

Several recent examples of such subpersonal explanations come to mind.

At *Forbes* magazine, the question was asked, Why do men see themselves as "fathers" but male gorillas don't? The article explained:

> Being a "dad" is uncommon in the primate world. Most male primates have little to do with their offspring, especially apes. Other primate males are invested in mating. They typically don't take the time to care for their young. They donate their DNA, and then they are on their way to mate with as many females as possible. So why are human males different?[63]

Evolutionary psychologists offer an answer: "An oft-invoked explanation for the evolution of paternal provisioning in humans is that ancestral females started mating preferentially with males who provided them with food, in exchange for female sexual fidelity." But other evolutionary psychologists offer a different answer: Female preferences aren't the driving force. The second group has a knack for clever popular culture messaging: They compare "dads" to "cads." Dads provide food, thus leaving more offspring than the useless cads. So more children inherit provident "dad" traits.[64]

Leveling the playing field between humans and gorillas leaves out the human mind. Do gorillas even know that they are "fathers"? Fatherhood is an abstract concept. It involves knowing that babies don't just happen, like rainfall; sex produces them. Second, a given child arises from a specific sex act. Thus, a man is a child's father because his and his partner's actions caused the child to exist. Thus, the human mind *recognizes* fatherhood. And that is why men see themselves as fathers but gorillas do not.

Recently, ethics professor Arianne Shahvisi of Brighton and Sussex Medical School took aim at evolutionary psychologists over their claims about why pregnant women "nest" (bustle around during the third trimester, getting ready):

> Nesting is depicted not as a set of rational space-preparation activities for expectant parents, but as a set of irrational, hormonally compelled and evolved behaviours, unique to women. No scientific evidence is cited, but it's assumed that nesting in humans is an analogue of nesting in other animals—birds, mice, rabbits, rats—for whom it is literal: they build physical nests in which their infants will be born and housed.[65]

She found little science evidence supporting these EP claims; practical reasons explain better. For example, "…maternity leave typically doesn't begin until the third trimester, which is when nesting is reported to kick in." Human reason prompts a wise use of the time, not hormones or modules.[66]

Similarly, a recent column in *Psychology Today* debated whether women are more likely to cheat during the COVID-19 crisis. Martin Graff, a principal lecturer in psychology at the University of South Wales, explains:

> Genetic immunity from disease is obviously necessary for the wellbeing of a women's [sic] offspring, yet because there are no real obvious visual indications of genetic immune functioning in men, beyond perhaps appearing to be in good health, it may be difficult for women to select men on this criterion. The only way in which women can attempt to increase the probability of strong immunity in their offspring in a time of infection in the environment, is to engage in a multi-male mating strategy, which would provide the genetic diversity afforded when their children are fathered by different men, and make it more likely that at least some offspring will possess the necessary genetic immunity to protect them. Therefore, the threat of disease in the environment may increase the likelihood of women seeking sexually diverse mating strategies driven by such bet-hedging behaviour.[67]

Dr. Graff's assertions sound out of touch. During a pandemic, most human beings will try to avoid rather than cultivate strangers. Some women may very well cheat during the crisis, but few of them will be trying to get pregnant. If a woman does engage in indiscriminate sexual behavior during a pandemic (a risky proposition for her), a more obvious cause might be erratic behavior in the face of an unfamiliar threat.

But such explanations do not sound "sciencey." And EP needs to sound sciencey far more than it needs to conform to science findings or even commonsense observations, or provide useful explanations.

Why Does Anyone Take Evolutionary Psychology Seriously?

Evolutionary psychology is taken seriously because its subpersonal approach to human psychology—the assumption that humans no more understand their environment than chimpanzees or gorillas do—supports naturalism (nature is all there is). Lack of evidence, conflict with evidence, and superficiality are much more easily tolerated[68] than a departure from naturalism would ever be. As an outgrowth of Darwinism, EP will probably continue to enjoy popular science support. It is probably the means by which Darwinism, as a point of view about our world, is conveyed to a great many average readers. Possibly, however, critiques like that of Subrena E. Smith will enable discussion of its hopeless failure to be legitimate and not appear to be a mere attack on science.

Does Darwinism Make Theological Assumptions?

Cornelius Hunter

C harles Darwin's book on evolution, *On the Origin of Species*, was published in 1859. It is widely seen as a seminal work that establishes the foundation of the theory of evolution. This much is not controversial. But *On the Origin*, and more broadly, Darwin's thought in general, are also widely seen as empirically based or, simply put, scientific. In fact, Darwinism is often viewed as a sort of watershed event that transitioned the life sciences from an activity that typically incorporated religious assumptions to an activity that typically eschewed such thinking. In this view, Darwin signaled a shift of focus away from theology and toward a purer science.

This broad consensus view of Darwinism is perplexing given how badly it fails on the historical facts. Darwin's thought and work, including his theory of evolution, are deeply theological in every sense. Darwin uses theological language, he presents theological findings based on theological assumptions and beliefs, and he formulates theological arguments that are crucial to his thesis. This cannot be emphasized too strongly given the broad misconception about the role of theology in Darwinism. As this chapter demonstrates, Darwin and his theory of evolution are not merely motivated or influenced by theology; they incorporate, entail, and rely on theological positions.

Darwin's Theological Language

The first two chapters of *Origin* are on the topic of biological variation. The first chapter, "Variation Under Domestication," discusses what breeders had learned, and the second chapter, "Variation Under Nature," discusses biological variability in the wild. The two chapters summarize what was known at the time, but readers could be excused if they skipped over Darwin's long and sometimes tedious passages. Darwin was attempting to demonstrate that what we observe today as distinct species are, in fact, rather arbitrary. His point was that we are observing the species at a particular point in time and so they may seem to be distinct. But over long periods of time the picture is fluid as boundaries shift and new species emerge. How well Darwin did in introducing readers to his theory can be debated, but at the end of this two-chapter introduction, Darwin made an important pivot.

At the end of chapter 2, in a short final section entitled "Summary," Darwin did much more than merely summarize the material. He introduced a new idea that would be a fundamental theme of *Origin*: the doctrine of independent creation was failing. This was no mere summary but a powerful new argument:

> In genera having more than the average number of species in any country, the species of these genera have more than the average number of varieties. In large genera the species are apt to be closely, but unequally, allied together, forming little clusters round other species. Species very closely allied to other species apparently have restricted ranges. In all these respects the species of large genera present a strong analogy with varieties. And we can clearly understand these analogies, if species once existed as varieties, and thus originated; whereas, *these analogies are utterly inexplicable if species are independent creations.*[1]

After more than 40 pages of dense scientific material, Darwin briefly offers a summary with a rather abrupt statement about "independent creations." Clearly this observation was important for Darwin. Not only did he feel the need to include it in the chapter 2 summary, but it signaled a pattern. Darwin would provide lengthy passages of technical material to which he would attach theological interpretations. As Chris Cosans has observed, "Throughout the *Origin*, he [Darwin] usually contrasts his account not with that of other evolutionists

such as Lamarck or Chambers, but with that of someone we would now call a 'special creationist.'"[2]

Simply put, theology was an important theme in *Origin*, and theological language appeared from beginning to end. I have counted a total of 49 theological arguments made by Darwin.[3] To undertake this count I used the 1872 (6th) edition, although most arguments appear throughout all editions. In addition to the theological arguments, theological language is even more abundant. Darwin's favored theological key words included *creation* (25 appearances), *created* (19), *creator* (6), *independent* or *independently* (20), *independently created* (11), and *special* or *specially* (4). Note that these frequencies are not determined from automated searches over the entire book. Instead, I manually identified and counted these terms when used as key words in a theological argument. Furthermore, these arguments and key words do not appear more or less uniformly distributed across the *Origin*, but are biased toward the end of the book. Just as Darwin concluded lengthy technical discussions with theological interpretations, this pattern also appears across the entire book, where the majority of the theological language appears toward the end. Specifically, 65 percent of the theological arguments appear in the final 28 percent of the book, and 39 percent of the theological arguments appear in the final 10 percent of the book. These arguments, and their theological language, were obviously important to Darwin's presentation.

Although Darwin used the word *God* only once, otherwise theological language was common. Darwin also coined his own term: *ordinary*.[4] In the *Origin*, Darwin used it 66 times, or about 3.4 times per 10,000 words. For comparison, this is about twice the rate in Hugh Miller's *Testimony of the*

Rocks (1.7 per 10,000 words) and almost four times the rate in the *Bridgewater Treatises* (0.9 per 10,000 words). Of its 66 occurrences in the *Origin*, about half (35 occurrences) indicated "typical" or "common." In the remaining 31 occurrences, however, Darwin had a specific meaning that was specific to his presentation. These 31 occurrences fall into two categories depending on the context. In 19 of those occurrences, Darwin used *ordinary* to be roughly synonymous with *natural.* In these instances, *ordinary* may refer to heredity, genealogical relationships, or selection that are *natural* as opposed to miraculous or artificial.

But in the remaining 12 occurrences of *ordinary*, Darwin had a very different meaning in mind. Here Darwin used *ordinary* to refer to the theory of independent creation. In these 12 occurrences the word *ordinary* was always followed by *view* except the first occurrence, in which it was followed by *belief.* In that first occurrence, Darwin referred to the doctrine of the fixity of species:

> On the other hand, the ordinary belief that the amount of possible variation is a strictly limited quantity is likewise a simple assumption.[5]

In the remaining 11 occurrences Darwin referred, more generally, to the theory of independent creation. For instance, in the next three occurrences, Darwin explicitly defined this ordinary view as the doctrine of independent creation. He wrote, "On the ordinary view of each species having been independently created..."[6]

Following these, in the next three occurrences, Darwin included the words *creation, independent creation,* or *independent creation of each species* to remind the reader what is meant by the ordinary view. In the remaining

five occurrences, Darwin continued to provide reminders of the meaning in all except the one occurrence, where he simply referred to the ordinary view in stating that it is not obvious why, on the ordinary view, embryonic structures should be important in classification. So, readers need to understand that when Darwin states that the ordinary view has difficulty explaining why, for example, embryonic structures are important for classification, he is referring to independent creation. More generally, readers need to understand that when reading Darwin, theology is always in play.

Darwin's Theological Findings

If Darwin commonly used theological language, it is reasonable to ask what, exactly, was he saying *about* theology? This question is nontrivial and too often answered by projecting favored contemporary positions onto Darwin rather than understanding Darwin's actual writings and their historical context. Darwin does not fit comfortably into the stock categories he is sometimes assigned to, ranging from atheist and materialist to theological evolutionist. In fact, what is often not appreciated is that Darwin made a plethora of distinct theological statements in the *Origin*. Rarely are these theological statements critically examined, much less even recognized. This section looks at a representative sampling of these statements.

Darwin's many theological statements in the *Origin* followed the same general format; namely, that an observation in the natural world contradicted the ordinary view of independent creation. For the observations, Darwin stated that no explanation was possible under independent creation, that the observation was utterly inexplicable, that the observation could *only* be accounted for by his theory of inheritance and common descent, and so forth. In all cases, Darwin's many statements entailed theological claims about the creator and how he would create the world. In the following list I provide 20 representative observations that Darwin found to be theologically objectionable—i.e., he claimed they were difficult to explain under the "ordinary" view of divine creation. I have used Darwin's language but paraphrased for brevity, with page numbers given from the 6th edition of the *Origin* in 1872.

1. More varieties occur in a group having many species, than in one having few (p. 44)

2. Hybrids, produced by crossing distant species within a genus, may resemble, in some ways, other species of the genus (pp. 130-131)

3. Sterility (p. 422)

4. Blind cave animals share affinities with other inhabitants of the same continent rather than with blind cave animals in other continents (pp. 110-111)

5. Species on volcanic islands, hundreds of miles from a continent, may be similar to the inhabitants of that continent (pp. 322, 354)

6. Species have a close relationship with other species inhabiting distant lands (p. 334)

7. Species introduced by man into a region sometimes exterminate native species (pp. 347-348)

8. Whole orders of animals, such as batrachians (frogs, toads, and newts), are absent on oceanic islands that otherwise are favorable environments for them (p. 350)

9. There are bats and no other mammals on remote islands (p. 351)

10. Lower organisms range more widely than the higher organisms (p. 359)

11. For a given species, the part of the structure that differs from the same part in other species of the same genu is more variable than those parts that are closely alike in the several species (p. 122)

12. Nature has so much variety yet so little real novelty (p. 156)

13. The habits and structure of some animals are not in agreement—there are upland geese with webbed feet which rarely go near the water; and no one except Audubon has seen the frigatebird, which has all its four toes webbed, alight on the surface of the ocean (p. 142)

14. The similar bones in the arm of the monkey, the foreleg of the horse, the wing of the bat, and the flipper of the seal are clearly not of special use to these animals (p. 160)

15. Trifling characters that prevail throughout many and different species, especially those having very different habits of life, have high value in species classification (pp. 372-373)

16. Specific characters, or those by which the species of the same genus differ from each other, are more variable than generic characters in which they all agree; for instance, the color of a flower is more likely to vary in any

one species of a genus if the other species possess differently colored flowers than if all possessed the same colored flowers (pp. 415-416)

17. A part developed in a very unusual manner in one species alone of a genus, and therefore, as we may naturally infer, of great importance to that species, is eminently liable to variation (p. 416)

18. All the parts and organs of many species are commonly linked together by graduated steps; nature never takes a sudden leap from structure to structure (p. 156)

19. We never find the bones of the arm and forearm, or of the thigh and leg, transposed (p. 382)

20. Organs frequently bear the plain stamp of inutility, such as the teeth in the embryonic calf or the shrivelled wings under the soldered wing covers of many beetles (p. 420)

In this representative list, Darwin addressed a wide variety of observations. It would seem that, in his study of the natural world, Darwin unearthed a veritable treasure trove of theological findings. But while Darwin may have provided reasonably detailed explanations of the observations, his theological conclusions are usually more abrupt. Rarely does Darwin provide justification or rationalization for his claims. And even a casual glance at the list makes it rather obvious that these are by no means self-explanatory or trivial claims. We may rightfully ask, Why exactly is it true that the lower organisms ranging more widely than the higher organisms goes against divine intent?

Attaching theological interpretations to empirical observations was certainly not unique to Darwin. For more than a century, natural theologians had engaged in this practice. But they spelled out their theological reasoning and, in any case, it was plain to begin with. The many functional and complex designs found in nature were evidence of the creator's power and wisdom. And the aesthetics and patterns found in nature were signs of the creator's love of order, harmony, and beauty. The theological reasoning was straightforward, and it was explained.

Darwin now switched the argument around. Instead of finding signs of how the creator was manifest in the creation, he found signs of how the creator was *not* manifest in the creation. For the natural theologians, it seemed that whatever nature had to offer could reveal the creator. For Darwin, it seemed that whatever nature had to offer contradicted the creator. But in this case the theology was far from obvious. True, for observations such as natural evil (e.g., predation) and inefficiency, the negative theological interpretation are obvious enough. A good God would not make a bad world. But Darwin went far beyond such observations and arguments. Most of his theological claims, as can be seen in the list above, dealt with more nuanced aspects of nature. And typically, Darwin provided little explanation or justification for his claims.

What was obvious, however, is that his claims were theological. While they refer to observations, these claims are not empirical in the sense that no experimental result can confirm or refute them. It may be an accurate statement of scientific observation that the lower organisms range more widely than the higher organisms, but the claim that this is contradictory to divine intent is not. That can be arrived at only via theology.

Darwin's Theological Beliefs

It is sometimes said that Darwin's theological claims in the *Origin* were not statements of religious belief, but rather, arguments tailored for readers with religious sensibilities. Darwin had done the scientific work, arrived at a scientific finding, and in presenting it needed to address the reigning paradigm of independent creation. But the facts suggest this view is closer to a rational reconstruction than an accurate historical description. First, Darwin's theological statements appeared throughout the *Origin* in support of all his major arguments. They were far more extensive and foundational than would be the case if he had merely been throwing a bone to fundamentalists. Second, independent creation was hardly the reigning paradigm to merit the detailed attention Darwin gave to it. And third, these deeply theological claims did not suddenly appear in the *Origin*, but were a consistent theme in Darwin's earlier thought, going back to his earliest writings.

In 1844, Robert Chambers anonymously published *Vestiges of the Natural History of Creation*.[7] This work was more fanciful than scientific, but it presented an uncompromising naturalistic account of the origin of life, including the descent of humans from primates. It sold well and it popularized evolutionary ideas for lay audiences. Even evolution cofounder Alfred Wallace recalled the excitement 50 years later:

> I well remember the excitement caused by the publication of the *Vestiges* and the eagerness and delight with which I read it. Although I saw that it really offered no explanation of the process of change of species, yet the view that change was effected, not through any unimaginable process, but through

the known laws of reproduction, commended itself to me as perfectly satisfactory, and as the first steps towards a more complete and explanatory theory.[8]

By 1859, when Darwin first published the *Origin*, audiences were increasingly receptive to evolutionary ideas. Rather than refute independent creation, what would be important for Darwin would be to present a compelling, scientific account of the origin of species. This would not be easy. Darwin's presentation of his theory was more often a defense. And natural selection, a key component of his theory, was not well accepted. Darwin's theological arguments helped to fill the gap.

This helps to explain Darwin's wide use of theological argumentation in the *Origin*. We can also explain Darwin's wide-ranging theology simply as Darwin being Darwin. Consider this passage from an 1844 essay Darwin wrote, outlining his early ideas on evolution:

> Why on the theory of absolute creations should this large and diversified island only have from 400 to 500 (? Dieffenbach) phanerogamic plants? and why should the Cape of Good Hope, characterised by the uniformity of its scenery, swarm with more species of plants than probably any other quarter of the world? Why on the ordinary theory should the Galapagos Islands abound with terrestrial reptiles? and why should many equal-sized islands in the Pacific be without a single one or with only one or two species? Why should the great island of New Zealand be without one mammiferous quadruped except the mouse,

and that was probably introduced with the aborigines? Why should not one island (it can be shown, I think, that the mammifers of Mauritius and St Iago have all been introduced) in the open ocean possess a mammiferous quadruped? Let it not be said that quadrupeds cannot live in islands, for we know that cattle, horses and pigs during a long period have run wild in the West Indian and Falkland Islands; pigs at St Helena; goats at Tahiti; asses in the Canary Islands; dogs in Cuba; cats at Ascension; rabbits at Madeira and the Falklands; monkeys at St Iago and the Mauritius; even elephants during a long time in one of the very small Sooloo Islands; and European mice on very many of the smallest islands far from the habitations of man. Nor let it be assumed that quadrupeds are more slowly created and hence that the oceanic islands, which generally are of volcanic formation, are of too recent origin to possess them; for we know (Lyell) that new forms of quadrupeds succeed each other quicker than Mollusca or Reptilia. Nor let it be assumed (though such an assumption would be no explanation) that quadrupeds cannot be created on small islands...[9]

The passage continues, but in this quote alone Darwin introduces no less than six theological claims regarding the "ordinary theory." These theological claims are very much consistent with those we saw above in the *Origin*. Such claims also appear in Darwin's notebooks dating back to the 1830s. For instance, in one early entry, Darwin expressed disdain at the idea of the creator repeating a pattern: "Has the Creator since the Cambrian formation gone on creating animals with same

general structure—miserable limited view."[10] Here, Darwin was writing for no one but himself. There is no reason to believe that he was expressing anything other than his genuine religious beliefs—beliefs that could not be separated from his conclusions about origins.

Darwin's Theological Legacy

As we have seen, Darwin used theological language to express many theological findings about the natural world, and these arose from Darwin's long-standing theological beliefs. The result was a long list of arguments that juxtaposed his theory of evolution against the theory of creation. This has been noticed by scholars who concluded that while Darwin did present empirical evidences and arguments for natural selection in the *Origin*, these were not particularly strong, and that his theory instead was advanced by powerful contrastive arguments.[11] So important was this mode of argument that Sober has dubbed it "Darwin's principle."[12] To reiterate and elaborate what Cosans noted: "Throughout the *Origin*, he [Darwin] usually contrasts his account not with that of other evolutionists such as Lamarck or Chambers, but with that of someone we would now call a 'special creationist.' The position of Darwin's hypothetical creationist is the dialectical opposite of that endorsed in the *Origin*. The *Origin*'s creationist would seem in fact to be a younger less sophisticated version of Darwin himself."[13]

These are the arguments we have seen above, and while they are powerful, they rely on how creation should look if a creator had created it. Simply put, Darwin was making "God wouldn't do it that way" arguments. Darwin's powerful and enduring point, which continues to be repeated today, is that there are no viable explanations other than his. The observations may not be fully understood under evolution, but under creation, the story becomes downright impossible. As Ernst Mayr wrote:

> The greatest triumph of Darwinism is that the theory of natural selection, for 80 years after 1859 a minority opinion, is now the prevailing explanation of evolutionary change. It must be admitted, however, that it has achieved this position less by the amount of irrefutable proofs it has been able to present than by the default of all the opposing theories.[14]

For Darwin, the most important of those opposing theories was independent creation. On that theory, Darwin found the evidence to be "utterly inexplicable." It just did not make sense. As Stephen Jay Gould observed:

Odd arrangements and funny solutions are the proof of evolution—paths that a sensible God would never tread but that a natural process, constrained by history, follows perforce. No one understood this better than Darwin. Ernst Mayr has shown how Darwin, in defending evolution, consistently turned to organic parts and geographic distributions that make the least sense.[15]

All of this depends on how one chooses to define a "sensible" God, which is to say, all of this is theological. In other words, evolution relies on religious premises. Students of Darwin need to understand that ultimately, evolution is not about the scientific details. Ultimately, evolution is about God.

35

How Has Darwinism
Negatively Impacted Society?

John G. West

Ideas have consequences, and Charles Darwin's theory of evolution is an idea that has had momentous consequences for society. "Darwinian theory is a scientific theory...but that is not all it is," writes philosopher Daniel Dennett. "Darwin's dangerous idea cuts much deeper into the fabric of our most fundamental beliefs than many of its sophisticated apologists have yet admitted, even to themselves."[1]

According to the modern version of Darwin's theory, all living things ultimately evolved from one simple ancestral form through a process of natural selection acting on random genetic mutations and recombinations of genes.

Darwin's theory fueled three big ideas with significant consequences for humanity.

The first idea was that humans are not unique.

Darwin himself recognized that his theory diminished the case for human uniqueness, writing in one of his notebooks that "it is absurd to talk of one animal being higher than another."[2] He also complained that "people often talk of the wonderful event of intellectual Man appearing" when,

in fact, "the appearance of insects with other senses is more wonderful."[3]

Darwinian biologists today relish emphasizing that humans are just another animal. Biologist Charles Zuker says humans "are nothing but a big fly."[4] Geneticist Glen Evans claims that "the worm represents a very simple human."[5] A science journalist writes that "there isn't much difference between mice and men."[6] And the late Morris Goodman of Wayne State University argued that humans are "only slightly remodeled chimpanzee-like apes."[7]

Darwinian social theorists across the political spectrum make similar claims. John Derbyshire, formerly a writer with the conservative magazine *National Review,* argues approvingly that "the broad outlook on human nature implied by Darwinian ideas contradicts the notion of human exceptionalism...To modern biologists, informed by Darwin, we are merely another branch on Nature's tree."[8] Princeton University bioethicist Peter Singer, a political progressive and author of *A Darwinian Left,* agrees. In Singer's words, Darwin "showed...that we are simply animals. Humans had imagined we were a separate part of Creation, that there was some magical line between Us and Them. Darwin's theory undermined the foundations of that entire Western way of thinking about the place of our species in the universe."[9]

A second big idea fueled by Darwinism was that nature is the product of an unguided process. As Darwin himself made clear, natural selection is an unintelligent process that is blind to the future: "There seems to be no more design in the variability of organic beings and in the action of natural selection, than in the course which the wind blows."[10] Natural selection cannot select new features based on some future goal. It only favors traits that are beneficial to survival right now. Consequently, evolution by natural selection is "the result of an unguided, unplanned process," to cite the words of dozens of Nobel laureates who issued a statement defending Darwin's theory in 2005.[11]

According to Darwinism, amazing biological features such as the vertebrate eye, or the wings of butterflies, or the blood-clotting system are in no way the purposeful result of evolution. They are unintended byproducts of the interplay between chance (random mutations) and necessity (natural selection). The same holds true for higher animals such as human beings. In the words of late Harvard paleontologist George Gaylord Simpson, "Man is the result of a purposeless and natural process that did not have him in mind."[12] In the Darwinian worldview, human beings are accidents of natural history, not the purposeful creations of a loving creator.

A third big idea fueled by Darwin's theory is that the engine of progress in the history of life is mass death. Instead of believing that the remarkable features of humans and other living things reflect the intelligent design of a master artist, Darwin portrayed death and

destruction as our ultimate creator. As he wrote at the end of his most famous work: "Thus, from the war of nature, from famine and death, the most exalted object which we are capable of conceiving, namely, the production of the higher animals, directly follows."[13]

For more than 150 years, these three Darwinian ideas have shaped social beliefs and actions in virtually every sphere of human life, including race relations, medicine, environmentalism, criminal justice, ethics, and religion.

Darwinian Racism

Charles Darwin was not the world's first racist, and he was better than some racists in that he opposed slavery. Nevertheless, Darwin's theory exerted a powerful influence on the development and growth of scientific racism. As Harvard evolutionary biologist Stephen Jay Gould acknowledged, "Biological arguments for racism may have been common before 1859 [when Darwin published his book *On the Origin of Species*], but they increased by orders of magnitude following the acceptance of evolutionary theory."[14]

Darwin believed that his theory of natural selection provided a scientific explanation for why races should have unequal capacities and why there were "higher" and "lower" races. The specific traits that an animal needs to survive differ based upon its environment. Thus, there is no reason to expect that natural selection, acting on different populations, will produce the same traits in every population—or, in this case, in every race. That is why Darwinists *expected* to find differences in the capacities of different races.

Accordingly, Darwin declared that there are significant differences in the mental faculties of what he called "men of distinct races."[15]

He also argued that the break in evolutionary history between apes and humans came "between the negro or Australian [aborigine] and the gorilla," thus depicting blacks as the closest human beings to apes.[16] Darwin's supporters further popularized these ideas.

German scientist Ernst Haeckel was a correspondent of Darwin and one of the most celebrated champions of Darwin's theory in Germany in the late nineteenth and early twentieth centuries. Haeckel created a widely disseminated diagram of human evolution that portrayed the evolutionary gap between the highest human and the lowest human as being larger than the gap between the lowest human and the highest ape-like creature, which was given African features.[17]

The idea that nonwhite races represented a throwback to lower stages of evolution was widely embraced throughout the scientific community in the early twentieth century. Consider the views of American biologist Charles Davenport, a member of the National Academy of Sciences and regarded as one of the founding fathers of modern genetics. Davenport was obsessed with the idea that some races were still stuck in lower evolutionary stages. In his words, "it seems probable that in the same country we have, living side by side, persons of advanced mentality, persons who have inherited the mentality of their ancestors of the early Stone Age, and persons of intermediate evolutionary stages."[18]

Darwinian racism played a role in the justification of one of the first genocides of the twentieth century. From 1904–1908, the German military attempted to eradicate the Herero and Nama peoples in South West Africa. In October 1904, General Lothar von Trotha issued what became known as his extermination order declaring that the Hereros either had to leave German South West Africa or face extinction. Herero men would

be executed and Herero women and children would be driven into the desert, where they would die of starvation or dehydration. Von Trotha justified his extermination campaign by an explicit appeal to Social Darwinism, telling one newspaper that human feelings of philanthropy could not override the "law of Darwins, the 'struggle of the fittest.'"[19]

In the United States, meanwhile, thousands of indigenous people from around the world were put on public display at the St. Louis World's Fair in 1904 in what has become known as a "human zoo."[20] Those displayed were supposed to represent lower stages of human evolution. In 1906, the Bronx Zoo in New York City displayed a man from Africa, Ota Benga, in a cage with a monkey as an evolutionary missing link between humans and apes.[21] By the 1920s, new immigration restrictions had been imposed in cooperation with Darwinian biologists who decided that certain races were lower on the evolutionary scale and so they needed to be kept out.[22]

Since the civil rights movement, most scientists have abandoned Darwinian racism. But there are exceptions. In 2007, Nobel-prize winning biologist James Watson, the codiscoverer of the structure of DNA, sparked an uproar by suggesting that blacks are biologically inferior to whites. Watson further suggested that human evolution was the explanation for this biological inferiority.[23]

Outside the scientific community, white supremacists in the so-called alt-right are unfortunately attempting to resurrect Darwinian arguments for racism previously rejected by the scientific community.[24]

Darwinian Eugenics

One of the most far-reaching social impacts of Darwinism was in the field of medicine. It became known as eugenics, which was described by its proponents as the self-direction of human evolution. The inspiration for eugenics sprang directly from Darwin's theory. In Darwin's view, the key reason humans developed their outstanding capabilities was not because they were designed that way by a creator, but because natural selection ruthlessly weeded out the unfit. The problem, according to eugenists, was that in the name of humanitarianism, civilized societies were now doing their best to care for those who nature would have killed off. They thought that would lead to disaster.

In *The Descent of Man,* Darwin criticized modern society for methodically undermining natural selection's "process of elimination" by offering asylums for the mentally ill, homes for the handicapped, hospitals for the sick, and welfare programs for the poor. Darwin even warned about the dangers of vaccinating people against smallpox, which he thought "preserved thousands, who from a weak constitution would formerly have succumbed." Darwin's stark conclusion: "Thus the weak members of civilised societies propagate their kind. No one who has attended to the breeding of domestic animals will doubt that this must be highly injurious to the race of man…hardly any one is so ignorant as to allow his worst animals to breed."[25]

A kindly man, Darwin was torn by the implications of his theory, and he thought human sympathy would not allow people to follow his theory to its logical conclusion. Likewise, many of his followers believed it would be too cruel for humans to go back to the law of the jungle. They wanted to develop a kinder way to mimic natural selection through modern science. That was the goal of eugenics.

Charles Darwin's cousin, Francis Galton, is generally recognized as the official founder of eugenics, and he actually coined a term for this kinder way, adapted from a Greek root word meaning "good in birth."[26] "Positive eugenics" focused on encouraging those deemed the most fit to reproduce more, while "negative eugenics" focused on curtailing reproduction by those deemed unfit, including mental defectives and criminals. Eugenics became the consensus view of the scientific community for decades, and it was promoted by leading scientists around the world. In America, eugenics supporters included biologists at Harvard, Princeton, Yale, Columbia, Stanford, and the American Association for the Advancement of Science.[27]

Today the Darwinian roots of eugenics tend to be downplayed, but the Darwinian rationale for eugenics was explicit in the writings of eugenists themselves. For example, Harvard geneticist Edward East insisted that "eugenic tenets are strict corollaries" of "the theory of organic evolution."[28] Princeton biologist Edwin Conklin similarly advocated for eugenics as a way to undo modern society's violation of "natural selection, the great law of evolution and progress."[29]

The impact of eugenics on American public policy was far-ranging, including laws on who could marry, immigration restrictions on races considered lower on the evolutionary scale, and forced sterilization of those considered less fit in Darwinian terms. In America, 60,000 women were sterilized against their will in the name of eugenics.[30]

Elsewhere around the world, the impact of Darwinian eugenics was even more horrifying. Germany's eugenics movement sterilized hundreds of thousands of people and murdered 300,000 disabled children and adults, many in gas chambers dressed up as shower stalls. It was during the "eugenic" killings of the handicapped that the Nazis developed some of the methods they used to murder millions of Jews.[31]

After the general public learned about the atrocities committed in the name of eugenics in Nazi Germany, Darwinian eugenics was widely discredited. But Darwinism continued to influence bioethics debates in other ways.

Darwinian Abortion and Infanticide

Scientists and political activists alike have appealed to Darwinian theory to justify abortion by claiming that babies in the womb are not fully human.[32] Invoking an idea known as embryonic recapitulation, these proponents of abortion argue that human infants replay the history of evolution as they develop in the womb. They go through a fish stage, a lower mammal stage, and more before finally reaching the state of a human being. Thus, if someone aborts an infant while she is still in the fish stage, it is no more immoral than killing a fish.

Embryonic recapitulation is junk science and has been discredited for decades even among evolutionary biologists. That has not stopped this argument from being invoked repeatedly as a justification for abortion in public policy debates. In 1981, for example, University of Michigan geneticist James Neel testified to the US Congress that "[t]he early embryo appears to pass through some of the stages in the evolutionary history of our species. The scientific dictum is: 'Ontogeny recapitulates phylogeny,' which translates into: during embryological development we repeat in abbreviated form many aspects of our evolutionary past."[33] Neel told lawmakers that because of "these facts," he found "it most difficult to state, as a scientist, just when in early fetal development human personhood begins, just as I would find it

impossible to say exactly when in evolution we passed over the threshold that divides us from the other living creatures." Neel was a member of the National Academy of Sciences and one of America's top geneticists.

The recapitulation argument for abortion has continued to resurface. In 1990, celebrated astronomer Carl Sagan and his wife, Ann Druyan, published a defense of abortion that relied heavily on the idea of recapitulation.[34] In 2007, the late journalist Christopher Hitchens similarly defended abortion by claiming that "in utero we see a microcosm of nature and evolution itself… we begin as tiny forms that are amphibian."[35]

Another Darwinian justification for abortion focuses on natural selection. Abortion proponents have claimed that spontaneous abortions (i.e., miscarriages) developed through natural selection to weed out the unfit. In their view, medical abortions are merely an example of humans capitalizing on an evolutionary innovation through science. Planned Parenthood official Alexander Sanger has gone even further. He argues that human-directed abortion itself is a product of natural selection, saying that "humanity has evolved to take conscious control of reproduction and has done so in order to survive…We cannot repeal the laws of natural selection. Nature does not let every life survive. Humanity uniquely, and to its benefit, can exercise some dominion over this process and maximize the chances for human life to survive and grow."[36]

Darwinism has fed into debates over infanticide as well. University of Chicago evolutionary biologist Jerry Coyne has argued on his blog for legalizing infanticide for babies with biological defects. He wrote, "After all, we euthanize our dogs and cats when to prolong their lives would be torture, so why not extend that to humans?"[37]

Coyne recognizes that the reason we do not do so is because of a view of human beings that Darwinism has yet to completely overcome: "The reason we don't allow euthanasia of newborns is because humans are seen as special, and I think this comes from religion—in particular, the view that humans, unlike animals, are endowed with a soul... When religion vanishes, as it will, so will much of the opposition to both adult and newborn euthanasia."[38]

Darwinian Ecology

A similar Darwinian devaluation of human life can be found among some radical environmentalists. In *The War on Humans*, lawyer and bioethicist Wesley J. Smith has documented how the growing coercive utopianism of some environmentalists is grounded in a visceral hatred of humans and the denial that human beings are special or unique.[39] In the words of University of Texas evolutionary zoologist Eric Pianka, "Humans are no better than bacteria,"[40] and, "Other things on this earth have been here longer than us...and they have a right to this planet too—that includes wasps that sting you, ants that bite you, scorpions and rattlesnakes."[41] Pianka goes on to criticize humans for "sucking everything we can out of mother Earth and turning it into fat human bio-mass."[42] Pianka urges the reduction of the Earth's human population by up to 90 percent and calls on the government to confiscate all the earnings of any couple who has more than two children. "You should have to pay more when you have your first kid—you pay more taxes," he insists. "When you have your second kid you pay a lot more taxes, and when you have your third kid you don't get anything back, they take it all."[43]

Underlying the radical environmentalists' hatred for humans is the Darwinian rejection of human uniqueness. Christopher Manes, one of the early leaders of the environmentalist group Earth First!, explains: "Darwin invited humanity to face the fact that the observation of nature has revealed not one scrap of evidence that humankind is superior or special, or even particularly more interesting than, say, lichen."[44]

This kind of Darwinian misanthropy motivated ecoterrorist James Lee, who in 2010 took staff of the Discovery Channel hostage. Lee called on the Discovery Channel to "Talk about Evolution. Talk about Malthus and Darwin until it sinks into the stupid people's brains until they get it!" Lee's stated goal was to save "what's left of the non-human Wildlife by decreasing the Human population. That means stopping the human race from breeding any more disgusting human babies!"[45]

Darwinian Ethics

Darwin's theory has also exerted a corrosive influence on morality and human accountability. In a Darwinian view, morality is reduced to preprogrammed behavior in service of physical survival. In the words of Darwinian philosopher Michael Ruse and Harvard evolutionary biologist E.O. Wilson, "Morality...is merely an adaptation put in place to further our reproductive ends...In an important sense, ethics as we understand it is an illusion fobbed off on us by our genes to get us to cooperate."[46] As a result, "human free will is nonexistent," to cite the words of late Cornell University professor William Provine.

Attacks on personal responsibility have featured prominently in Darwinian accounts of human behavior during the past century and a half. For example, Darwinism played a key role in the development of the "new school of criminology" by Cesare Lombroso and others in the late nineteenth century.[47] These criminologists tried to find Darwinian explanations for why people engaged in crime, even labeling some persons "born criminals" because they were supposed to be throwbacks to an earlier stage in evolutionary history. Lombroso and his followers repudiated the traditional idea that "crime involved...moral guilt."[48] Italian jurist Enrico Ferri, one of Lombroso's most celebrated disciples, argued that it was no longer reasonable to believe that human beings could make choices outside the normal chain of material cause and effect given the advent of modern science, particularly the work of Charles Darwin. Ferri looked forward to the day when punishment and vengeance would be abandoned and crime would be treated as a "disease."[49]

The diminishment of free will is rampant among today's purveyors of sociobiology and evolutionary psychology. Evolutionary psychology proponent Robert Wright, for example, declares that "free will is an illusion, brought to us by evolution,"[50] and "in many realms, not just sex—we're all puppets."[51] Wright does add that "our best hope for even partial liberation is to try to decipher the logic of the puppeteer." But if "free will is an illusion," precisely how can we liberate ourselves from "the puppeteer"? And if human beings truly are "puppets" to their genes, puppets whose "emotions are just evolution's executioners"[52] (again quoting Wright), in what sense can people be blamed if they simply act according to their deepest impulses?

But the diminishment of free will and personal responsibility is not the only way Darwinism has influenced morality. Darwinism has also provided a robust biological justification for moral relativism. According to Darwin, specific moral precepts develop because, under certain environmental conditions, they promote survival.[53] Once those conditions for survival change, however, so too do the dictates of morality. That is why we find in nature both the maternal instinct and infanticide, both honoring one's parents and killing them when they become feeble. Natural selection "chooses" whatever behavioral traits best promote survival under the existing circumstances.

A Darwinian understanding of morality makes it very difficult to condemn as evil any human behavior that has persisted because every trait that continues to exist even among a subpopulation has an equal right to claim nature's sanction. Presumably even antisocial behaviors such as fraud, pedophilia, and rape must continue to exist among human beings because they were favored at some point by natural selection and therefore have some sort of biological basis. Of course, one could still justly condemn such behaviors if there existed a permanent moral standard independent of natural selection. But the existence of such a standard is precisely what orthodox Darwinism denies.

For the most part, Darwin himself did not press his relativistic analysis of morality to its logical conclusion, but he laid the groundwork for others who came after him, and his ideas helped reshape how people think about morality. In the United States, for example, 55 percent of adults now believe "evolution shows that moral beliefs evolve over time based on their survival value in various times and places."[54]

Nowhere has the Darwinian view of ethics had a severer impact than in family life and human sexuality. The thinker most responsible for the breakdown of traditional sexual ethics in Western culture was Harvard-trained evolutionary zoologist Alfred Kinsey.

Adopting a thoroughly Darwinian approach to sexual morality, Kinsey argued that any sexual practice that could be found somewhere among mammals could be regarded as "normal mammalian behavior" and therefore was unobjectionable.[55]

Today, many evolutionary psychologists have gone beyond mere sexual relativism and are affirmatively arguing against monogamy. They claim that we were bred by Darwinian evolution to have multiple sex partners, which means that we are programmed for promiscuity and infidelity. In their view, the very idea of a faithful and monogamous marriage contradicts our biology and must therefore be abandoned.

One prominent evolutionary psychologist to advocate this view is Christopher Ryan, coauthor of the *New York Times* bestseller *Sex at Dawn*. In the words of Ryan, "Marriage in the West isn't doing very well because it's in direct confrontation with the evolved reality of our species."[56] Ryan says he wants to save marriage by making it consistent with Darwinian biology. For him, that means redefining marriage to include multiple partners at the same time.

Darwinian Rejection of God

A final profound impact of Darwin's theory has been in the area of religion, where Darwinism has supplied a potent scientific rationale for atheism. Darwinism does not logically necessitate atheism, but it certainly does encourage it. If nature really supplies proof that the history of life is the product of an unguided process, then that would seem to make an atheist worldview much more credible.

That is why so many scientists have made unguided Darwinian evolution a cornerstone in their case for atheism. Oxford University biologist Richard Dawkins has famously claimed that "Darwin made it possible to be an intellectually fulfilled atheist."[57] Harvard evolutionary biologist E.O. Wilson has declared that the existence of a God "who directs organic evolution and intervenes in human affairs…is increasingly contravened by biology and the brain sciences."[58] University of Washington evolutionary psychologist David Barash makes a similar boast: "The more we know of evolution, the more unavoidable is the conclusion that living things, including human beings, are produced by a natural, totally amoral process, with no indication of a benevolent, controlling creator."[59]

In a nationwide survey of American atheists and agnostics, nearly seven in ten atheists and more than four in ten agnostics said that Darwin's unguided mutation/natural selection mechanism made the existence of God less likely for them personally. Similarly, more than seven in ten atheists and nearly four in ten agnostics agreed with evolutionary biologist Richard Dawkins that "the universe we observe has precisely the properties we should expect if there is, at bottom, no design, no purpose, no evil and no good, nothing but blind, pitiless indifference."[60]

But Darwinism's impact on religion has not been limited to just atheists and agnostics. It also has reshaped how many Christians view God, especially at Christian colleges and universities, where scientists and theologians have tried to revise Christianity to make it compatible with Darwinism. Because Darwinian evolution is by nature unguided, theistic Darwinists often downplay or reject the idea that God actively directs the development of life. For example, Anglican priest and physicist John Polkinghorne wrote that "an evolutionary universe is theologically understood as a creation allowed to make itself."[61] Roman Catholic biologist Kenneth Miller of Brown University insists that "mankind's appearance on this planet was *not* preordained, that we are here…as an afterthought, a minor detail, a happenstance in a history that might just as well have left us out."[62] Former Vatican astronomer George Coyne even claimed that because evolution is unguided "not even God could know…with certainty" that "human life would come to be."[63]

Other theistic Darwinists repudiate the biblical doctrine of a historic fall, the idea that human beings were originally created good and then fell into sin through a voluntary act of disobedience. According to Christian physicist Karl Giberson, because human beings were created through Darwinian evolution, they were essentially sinful from the start because "Selfishness…drives the evolutionary process."[64]

Still other theistic Darwinists challenge the idea that God's handiwork in nature is observable, especially in biology. Christian geneticist Francis Collins suggests that "evolution could appear to us to be driven by chance, but from God's perspective the outcome would be entirely specified. Thus, God could be completely and intimately involved in the creation of all species, while from our perspective…this would appear a random and undirected process."[65] In other words, God makes the history of life look "random

and undirected," even though it really is not. Contra Collins, for thousands of years, Jewish and Christian thinkers believed otherwise, maintaining that God's design could be clearly seen throughout nature.[66]

Darwinism as "Universal Acid"

Daniel Dennett has aptly described Darwinism as a "universal acid" that dissolves traditional ideas about morality, human responsibility, and God.[67] Without question, Darwinian theory has exerted a revolutionary impact on human society. Ironically, Darwinism's cultural prestige has continued to expand even while its scientific basis has sharply diminished in the face of withering critiques in recent years.[68]

Nobel Prize-winning physicist Robert Laughlin at Stanford University has gone so far as to declare, "Evolution by natural selection…has lately come to function more as an antitheory, called upon to cover up embarrassing experimental shortcomings and legitimize findings that are at best questionable and at worst not even wrong."[69] If Laughlin and others are right, it may be time to fundamentally challenge Darwinism's continuing hegemony on culture.

36

Do Scientists Have the Intellectual Freedom to Challenge Darwinism?

John G. West

Open debate is one of the corner-stones of scientific progress. In the words of physicist J. Robert Oppenheimer, "The scientist…must be free to ask any question, to doubt any assertion, to seek for any evidence, to correct any errors."[1] Unfortunately, when it comes to Darwinian evolution, the freedom to engage in robust debate and discussion is often sorely lacking. Italian geneticist Giuseppe Sermonti once wrote that "Darwinism…is the 'politically correct' of science."[2] His observation remains painfully accurate.

Modern Darwinian theory (often called *neo-Darwinism*) makes two main claims. First, all animals ultimately descended with modification (evolved) from a common ancestor in the deep past. Second, the primary mechanism driving this evolution of life is an unguided process known as natural selection. Randomly occurring mutations and recombinations in genes produce unplanned variations among individual organisms in a population. Some of these variations will help organisms survive and reproduce more effectively. Over time, these beneficial variations will come to dominate a population of organisms, and over even more time, these beneficial variations will accumulate, which is supposed to result in entirely new biological features and organisms.

According to Darwin's theory, natural selection is an unintelligent process that is blind to the future. It cannot select new features based on some future goal or potential benefit. As a result, evolution, in a Darwinian sense, is "the result of an unguided, unplanned process," to cite the words of 38 Nobel laureates.[3]

Although it is frequently claimed that no scientific critics of Darwinian evolution actually exist, in reality there have been serious scientific objectors to the theory from its inception. In fact, the cofounder of the theory of evolution by natural selection, Alfred Russel Wallace, doubted how much unguided natural selection could actually achieve. By the end of his life, Wallace championed what historian Michael Flannery has called *intelligent evolution*—the idea that the evolutionary process required guidance from an overseeing mind.[4]

Today there remains considerable internal debate among biologists over key parts of Darwinian theory. "The science journals right across the sub-disciplines of biology are full of criticisms of the Darwinian approach to the history of life," notes philosopher of science Stephen C. Meyer. "The Darwinian mechanism is held by many evolutionists to be insufficient to produce the new form and function in the history of life."[5] Dozens of peer-reviewed articles from mainstream science journals and books challenge, question, or raise problems for different aspects of Darwinian theory.[6]

When it comes to public discussions of evolutionary theory, however, the freedom for scientists to raise hard questions largely disappears. "There's a feeling in biology that scientists should keep their dirty laundry hidden because the religious right are always looking for any argument between evolutionists as support for their creationist theories," explains evolutionist and computer scientist W. Daniel Hillis. "There's a strong school of thought in biology that one should never question Darwin in public."[7]

As a result of this kind of groupthink, scientists today who publicly dissent from Darwin are likely to face intimidation, discrimination, demotion, or even removal from their jobs. Punishment can be especially severe for scientists who are skeptical of neo-Darwinism because they think nature displays clear evidence of intelligent design. Such scientists argue that the exquisite functionality of biological systems is better explained by an intelligent cause (i.e., a mind) rather than an unguided process such as natural selection.

Scientists Suppressed

The list of scientists, teachers, students, and others who have faced retaliation or discrimination for their public skepticism of Darwinism is long and growing.

At San Francisco State University, tenured biology professor Dean Kenyon was removed from teaching introductory biology classes. Once an influential proponent of Darwinian evolution, Kenyon had come to doubt key parts of Darwin's theory and expressed those doubts to students in class, including his belief that some biological features exhibited evidence of intelligent design. Kenyon was more fortunate than many academic critics of Darwin. After his plight was publicized by an article in *The Wall Street Journal*, the university was shamed into reinstating him.[8]

Biology professor Caroline Crocker at George Mason University was "barred by her department from teaching both evolution and intelligent design" after committing the crime of mentioning intelligent design in a course on cell biology. "It's an infringement of academic freedom," she told the journal *Nature*.[9] Subsequently her contract was not renewed.[10]

Oregon community college instructor Kevin Haley was terminated after it became known that he criticized evolution in his freshman biology classes. Haley's college refused to state why his contract was not renewed, but some of Haley's colleagues were upset that students who took his biology class were starting to challenge evolution in their classes.[11] Before the controversy over evolution, Haley had been regarded as an excellent teacher. Indeed, his former department chair had praised him in glowing terms, saying that students "perceive that he is interested in them. He generates curiosity and stimulates their thinking. Those are things that I think are not always there in a professor."[12]

Scientists outside of biology who express skepticism about Darwinism can also face discrimination and bullying. At Baylor University, mathematician William Dembski was fired as director of an academic center he had founded to explore the idea of intelligent design as an alternative to unguided Darwinian evolution. Eventually his faculty contract was not renewed as well, and he lost his job. Dembski, who holds doctorates from the University of Chicago and the University of Illinois at Chicago, had exemplary academic credentials and publications, but his research center had been strenuously opposed by Baylor's biology faculty.[13]

Chemistry professor Nancy Bryson was removed from her post as head of the science and math division of Mississippi University for Women after she delivered a lecture to honors students about some of the scientific weaknesses of chemical and biological evolution. "I was harshly attacked by Darwinist colleagues," she explained later. "…students at my college got the message very clearly, do not ask any questions about Darwinism."[14]

Sometimes scientists can find themselves blacklisted if they merely express openness or sympathy to a critical examination of Darwinism. Astronomer Martin Gaskell was a top applicant to become the head of an observatory at the University of Kentucky. In the words of one university faculty member there, "his qualifications…stand far above those of any other applicant."[15] But Gaskell was ultimately rejected for the job after the biology faculty waged an internal war against his hiring. Why did they want to prevent him from getting the job? First, Gaskell was perceived by other faculty to be "potentially evangelical."[16] Worse, although he identified himself as a supporter of evolution, in online notes for a science and faith talk, Gaskell respectfully discussed the views of intelligent design proponents and acknowledged that modern evolutionary theory had unresolved problems—just like any scientific theory. The Gaskell case illustrates how some

Darwinian biologists are not content to stop dissent over their theory within their own field. They want to censor disagreement with Darwin in other scientific disciplines as well. Indeed, sometimes they try to silence other scientists from raising the issue of intelligent design outside of biology without any reference to evolution. Eric Hedin was an assistant professor of physics at Ball State University. Like Gaskell, he had a long list of peer-reviewed science publications.[17] For many years, he taught an interdisciplinary honors class at Ball State called "The Boundaries of Science," which explored the limits of science.

During one small part of the course, Hedin discussed the debate over intelligent design in physics and cosmology—not biology.[18] Hedin's course received positive student reviews.[19] However, atheist evolutionary biologist Jerry Coyne at the University of Chicago and the Freedom from Religion Foundation filed complaints.[20] Ball State then violated its own procedures and appointed an ad hoc committee stacked with avowed critics of intelligent design, including two who spoke at a previous Darwin Day conference organized by the Ball State Freethought Alliance,[21] a group whose "original goal," according to its president, was "belittling religion."[22] Hedin's class was eventually cancelled by Ball State. In addition, the university president issued a campus speech code not only banning professors from covering intelligent design in science classes but also from expressing support for the concept in social science and humanities classes.[23]

Even nonscientists can face problems for suggesting that there might be a serious intellectual debate over Darwinism. At Baylor University, philosopher and legal scholar Francis Beckwith was initially denied tenure despite an outstanding record of academic research and publications.[24] Although Professor Beckwith was well known for his prolife views, he was most controversial for his law review articles and an academic book defending the constitutionality of teaching about intelligent design as an alternative to Darwinism.[25] It is important to note that Beckwith did not advocate that intelligent design *should* be taught in public schools—only that it was constitutional to teach it in an appropriate manner. But that nuanced position was too much for some of his colleagues, who were defenders of Darwin's theory. Fortunately for Beckwith, after a public outcry, the president of Baylor later granted him tenure.[26]

College professors are not the only targets in academia who face discrimination because of their skepticism of Darwinism. Students can be even more vulnerable. Ohio State University doctoral candidate Bryan Leonard had his dissertation defense put in limbo after three pro-Darwin professors filed a spurious complaint attacking Leonard's dissertation research as "unethical human subject experimentation." Leonard's dissertation project looked at how student beliefs changed after students were taught scientific evidence for and against modern evolutionary theory. The complaining professors admitted that they had not actually read Leonard's dissertation. But they were sure it must be unethical. Why? According to them, there is no valid evidence against evolutionary theory. Thus—by definition—Leonard's research must be tantamount to child abuse.[27]

Outside of academia, there have been similar cases of discrimination in government-funded science organizations. David Coppedge was a senior computer systems administrator for the Cassini Mission to Saturn at NASA's Jet Propulsion Lab in California. He faced demotion and discharge after he offended his supervisor by occasionally offering to loan colleagues DVDs about intelligent design.[28] No one had ever complained to Coppedge about his offers of DVDs, but when the supervisor found out, Coppedge faced a punitive investigation. His employment evaluations, which had been outstanding, suddenly became negative, and ultimately he lost his job. Coppedge's dismissal was justified as a budgetary reduction unrelated to his views on intelligent design, but that explanation was questionable given the facts of the case.

Evolutionary biologist Richard Sternberg faced similar retaliation by officials at the Smithsonian Institution's National Museum of Natural History (NMNH) after accepting for publication a peer-reviewed article supportive of intelligent design in a biology journal he edited. A research associate at the museum, Sternberg said that after the article was published, he was told to vacate his office space and was shunned and vilified by colleagues. Efforts were also made by administrators to discover Sternberg's personal religious and political beliefs.[29] Investigators for the US Office of Special Counsel concluded that "it is…clear that a hostile work environment was created with the ultimate goal of forcing [Dr. Sternberg]… out of the [Smithsonian]."[30]

Smithsonian officials denied any wrongdoing, but Sternberg was demoted from a research associate to a research collaborator without explanation.[31] A 17-month investigation by subcommittee staff of the House Committee on Government Reform subsequently confirmed and elaborated on the previous findings of the US Office of Special Counsel. In a detailed report released to the public, subcommittee investigators concluded that they had uncovered "substantial, credible evidence of efforts to abuse and harass Dr. Sternberg, including punitively targeting him for investigation in order to supply a pretext for dismissing him, and applying to him regulations and restrictions not imposed on other researchers."[32]

Congressional investigators further accused NMNH officials of conspiring "on government time and using government emails...with the pro-evolution National Center for Science Education (NCSE)...to publicly smear and discredit Dr. Sternberg with false and defamatory information."[33] The NCSE even provided a set of "'talking points' to...NMNH officials on how to discredit both Sternberg and the Meyer article." In addition, the NCSE was asked by senior museum administrator Dr. Hans Sues "to monitor Sternberg's outside activities...The clear purpose of having the NCSE monitor Dr. Sternberg's outside activities was to find a way to dismiss him."[34] Congressional investigators concluded that "the extent to which NMNH officials colluded *on government time and with government resources* with the NCSE to publicly discredit Dr. Sternberg's scientific and professional integrity and investigate opportunities to dismiss him is alarming."[35]

When asked about Sternberg's plight by *The Washington Post*, Eugenie Scott of the NCSE seemed to suggest that Sternberg was lucky more was not done to get rid of him: "If this was a corporation, and an employee did something that really embarrassed the administration, really blew it, how long do you think that person would be employed?"[36]

Science teachers in K-12 schools also face challenges if they criticize Darwinian theory. In Minnesota, high school teacher Rodney LeVake was removed from teaching biology after expressing doubts about Darwin's theory. LeVake, who holds a master's degree in biology, agreed to teach evolution

as required in the district's curriculum, but said he wanted to "accompany that treatment of evolution with an honest look at the difficulties and inconsistencies of the theory."[37]

In Washington State, longtime high school biology teacher Roger DeHart faced continuing harassment from pro-Darwin activists, who succeeded in getting his school district to prohibit him from discussing scientific criticisms of modern Darwinian theory with his students. DeHart was even banned from sharing mainstream science publications with students that corrected textbook errors about evolution. Although DeHart complied with his district's gag order, ultimately, he was removed from teaching biology. When he took a job in an adjoining school district so that he could continue to teach biology, the harassment continued. He was eventually reassigned from teaching biology in that district as well, even though there were no allegations by his new district that he was not following the prescribed curriculum. DeHart finally was driven from public education altogether.[38]

Dogmatic Science

In the popular mind, science is often thought of as a disinterested search for empirical truth. However, philosophers and historians of science have known for a long time that science, just like any other human profession, can be rife with egos, ideology, partisanship, and struggles for power. As three philosophers of science wrote in an Oxford University Press volume, "What science says the world is like at a certain time is affected by the human ideas, choices, expectations, prejudices, beliefs, and assumptions holding at that time."[39] In particular, when one scientific theory becomes enshrined as a reigning paradigm, dissenting views are often silenced for reasons other than lack of evidence. Dissenting views represent a threat to the ruling paradigm, and so those who have earned power and prestige from advancing it are reluctant to let their authority be eroded. For this and other reasons, scientists who have spent their lives working within one paradigm may have a difficult time acknowledging problems within that paradigm no matter how much contrary evidence accumulates.[40] Scientific paradigms can sometimes end up as dogmas where little dissent is tolerated.

Some scientists complain that this is exactly what has happened with Darwin's theory of evolution. "Darwinism has all the trappings of a secular religion," writes biologist Jonathan Wells. In Wells's view, Darwinism's "priests forgive a multitude of sins in their postulants…but never the sin of disbelief."[41] Wells is a critic of Darwinian evolution, but even some supporters of the theory have made similar comments. Leading Darwinian philosopher Michael Ruse, for example, argues that "evolution is promoted by its practitioners as more than mere science. Evolution is promulgated as an ideology, a secular religion."[42]

The dogmatic nature of Darwinism may help explain the toxic rhetoric some of its supporters employ to try to silence those who disagree. After 9/11, in many states, it became routine to apply the label of *Taliban* to anyone who supported teaching students about scientific criticisms of Darwinian theory.[43] Biology professor PZ Myers at the University of Minnesota, Morris, has even demanded "the public firing and humiliation of some teachers" who express their doubts about Darwin.[44] He further says that evolutionists should "screw the polite words and careful rhetoric. It's time for scientists to break out the steel-toed boots and brass

knuckles, and get out there and *hammer* on the lunatics and idiots."[45]

This kind of rhetoric chills open debate on many college campuses, making those who challenge Darwin the targets of hate speech. Consider the following vitriolic comments attacking Discovery Institute, a group that supports intelligent design, which were posted on the website of the University of Nevada's student newspaper: "…the theocratic morons of the Dishonesty Institute are traitors who want to destroy America's science education. If it was up to me they would be put in prison for treason. They are enemies of America, no better than terrorists, and they should be treated like terrorists."[46] Another commenter on the same page compared intelligent design proponents to Nazis.[47]

Some pro-Darwin activists are now resorting to what are essentially mob actions to muzzle critics of Darwinian theory. Stanley Wilson is a biologist who was scheduled to teach a course on the debate over Darwinian evolution and intelligent design at Amarillo College in Texas. The seminar was to be offered as a noncredit personal enrichment course for adult education. But then members of a local "freethinkers" group threatened the college, saying that if the course went ahead, they would physically disrupt the class. As a result, the college cancelled it.[48]

Censorship efforts have even extended to privately sponsored events. In Los Angeles, the California Science Center rented its theater to a private group to screen a film titled *Darwin's Dilemma*, which challenged Darwinian explanations of the Cambrian explosion in the history of life. After local pro-Darwin scientists complained, the Center broke the contract and cancelled the screening.[49] In another case, activists mounted a massive denial-of-service attack designed to shut down online registration for a conference sponsored by another pro-intelligent design group.[50]

Scientists Who Dare to Doubt

Despite the ongoing hostility and persecution, a growing number of scientists have been willing to pay the cost and publicly declare their skepticism of traditional Darwinism. In 2019, for example, Yale University computer scientist David Gelernter announced that he no longer accepted Darwinian explanations for the major innovations in the history of life. Darwinism can explain "the small adjustments by which an organism adapts to local circumstances," wrote Gelernter. "Yet there are many reasons to doubt whether [Darwinism] can answer the hard questions and explain the big picture—not the fine-tuning of existing species but the emergence of new ones. The origin of species is exactly what Darwin CANNOT explain."[51] Gelernter's apostasy attracted widespread coverage and sparked vigorous debate.[52]

Gelernter is far from Darwinism's only apostate. More than 1,000 doctoral scientists have signed their names to "A Scientific Dissent from Darwinism," which announces that they are "skeptical of claims for the ability of random mutation and natural selection to account for the complexity of life" and states that "careful examination of the evidence for Darwinian theory should be encouraged."[53] Signers of the declaration include members of the national academies of science in the United States, Russia, Poland, the Czech Republic, Brazil, and India (Hindustan), as well as faculty and researchers from a wide range of universities and colleges, including Princeton, MIT, Dartmouth, the University of Idaho, Tulane, and the University of Michigan.

"Some defenders of Darwinism embrace standards of evidence for evolution that as scientists they would never accept in other circumstances," said signer Henry Schaeffer, director of the Center for Computational Quantum Chemistry at the University of Georgia.[54] Other signers expressed similar concerns. "The ideology and philosophy of neo-Darwinism, which is sold by its adepts as a scientific theoretical foundation of biology, seriously hampers the development of science and hides from students the field's real problems," declared Vladimir L. Voeikov, a professor of bio-organic chemistry at Lomonosov Moscow State University.[55] Microbiologist Scott Minnich at the University of Idaho complained that Darwinian theory was "the exceptional area that you can't criticize" in science, something he considered "a bad precedent."[56]

According to signer and US National Academy of Sciences member Philip Skell, Darwinian theory is largely "superfluous" in biological explanations of how things work. Skell, the Emeritus Evan Pugh Professor at Pennsylvania State University before his death, argued that "Darwinian evolution…does not provide a fruitful heuristic in experimental biology" and "the claim that it is the cornerstone of modern experimental biology will be met with quiet skepticism from a growing number of scientists in fields where theories actually do serve as cornerstones for tangible breakthroughs." Skell reported that he "asked more than 70 eminent researchers if they would have done their work differently if they had thought Darwin's theory was wrong. The responses were all the same: No." Moreover, when Skell examined "the outstanding biodiscoveries of the past century," such as "the discovery of the double helix" and the "mapping of the genomes," he "found that Darwin's theory had provided no discernible guidance, but was brought in, after the breakthroughs, as an interesting narrative gloss."[57]

The growing number of scientists who are publicly skeptical of neo-Darwinism has made it harder to deny the existence of real scientific controversies over Darwinian theory. Such denials become even more implausible when peer-reviewed science journals regularly debate issues such as the role of microevolution in explaining macroevolution, the efficacy of natural selection, the origin of animal body plans during the Cambrian explosion, or flaws in standard lines of evidence for evolution such as vertebrate embryos and peppered moths.[58]

Despite the climate of intimidation that exists at academic institutions on the subject of evolution, the public supports free and open discussion on the topic. According to a nationwide survey in 2016, an overwhelming 93 percent of American adults agree that "teachers and students should have the academic freedom to objectively discuss both the scientific strengths and weaknesses of the theory of evolution." Another 88 percent agree that "scientists who raise scientific criticisms of evolution should have the freedom to make their arguments without being subjected to censorship or discrimination."[59] More broadly, they agree that dissenting views in science are healthy:

- 84 percent believe that "attempts to censor or punish scientists for holding dissenting views on issues such as evolution…are not appropriate in a free society."

- 94 percent believe "it is important for policymakers and the public to hear from scientists with differing views."

- 87 percent think that "people can disagree about what science says on a particular topic without being 'antiscience.'"

- 86 percent affirm that "disagreeing with the current majority view in science can be an important step in the development of new insights and discoveries in science."[60]

For the public, free speech in science is not a partisan issue. It is supported by decisive majorities across party lines, gender, religion, and age: 95 percent of Republicans believe that teachers and students should have the freedom to discuss the scientific strengths and weaknesses of evolution—and so do 93 percent of Democrats and 94 percent of independents. So do 96 percent of theists, 92 percent of agnostics, and 86 percent of atheists. Similarly, 87 percent of Republicans oppose attempts to punish or censor scientists who hold dissenting views on issues like evolution and climate change—and so do 84 percent of independents and 82 percent of Democrats, as well as 86 percent of theists, 83 percent of agnostics, 76 percent of atheists, 82 percent of women, and 86 percent of men.

Ironically, many of Darwin's current defenders are pursuing an intolerant approach to science that the person they are extolling would have repudiated. Darwin, to his credit, treated objections to his theory seriously and respectfully throughout his life. Supporters of his theory today might want to ponder his words in the introduction to *On the Origin of Species*, where he acknowledged that in evaluating his theory, "[a] fair result can be obtained only by fully stating and balancing the facts and arguments on both sides of each question."[61]

37

Is Darwinism a Theory in Crisis?

Jonathan Wells

What does it mean to say that a theory is "in crisis"? It's not enough to point out that a theory is inconsistent with evidence. Critics have been pointing out for decades that Darwinism doesn't fit the evidence from nature. Biologist Michael Denton published *Evolution: A Theory in Crisis* in 1986.[1] Thirty years later, he drove the point home with *Evolution: Still a Theory in Crisis*.[2]

But Darwinism is still with us, for two reasons. First, Darwinism is not just a scientific hypothesis about specific phenomena in nature, like Newton's theory that the gravitational force between two bodies is inversely proportional to the square of the distance between them (seventeenth century), Lavoisier's theory that things burn by combining with oxygen (eighteenth century), or Maxwell's theory that light is an electromagnetic wave (nineteenth century). As I wrote in chapter 26 of this book, Darwin called *On the Origin of Species* "one long argument," and a central part of it was a *theological* argument against the idea that species were specially created.[3]

Second, established scientific research programs such as Darwinism are never abandoned just because of some problems with the evidence. The idea that all species are descendants of one or a few common ancestors that have been modified by mutation and natural selection will maintain its dominance until large numbers of scientists embrace a competing idea. Currently, the major competing idea is intelligent design (ID), which maintains (contra Darwin) that some features of living things are better explained by an intelligent cause than by unguided natural processes. The shift, if and when it happens, will be a major scientific revolution. One way to approach this phenomenon is through philosopher of science Thomas Kuhn's 1962 book *The Structure of Scientific Revolutions*.[4]

I will begin by summarizing some of Kuhn's key insights. I will then apply those insights to the present conflict between Darwinism and intelligent design. As I do so, I point out some problematic aspects of Kuhn's work, but I conclude that recent events fully justify calling Darwinism a theory in crisis.

Kuhn's *Structure of Scientific Revolutions*

According to Kuhn, "normal science" is "research firmly based upon one or more past scientific achievements, achievements that some particular scientific community acknowledges for a time as supplying the foundation for its further practice." Those achievements were "sufficiently unprecedented to attract an enduring group of adherents away from competing modes of scientific activity." They were also "sufficiently open-ended to leave all sorts of problems" to be solved. Kuhn called achievements that share these two characteristics "paradigms."[5]

Once a paradigm becomes dominant, the normal practice of science is simply to solve problems within that paradigm. In the process, an "institutional constellation" forms that includes "the formation of specialized journals, the foundation of specialist societies, and the claim for a special place in the curriculum."[6] The last is very important, because one "characteristic of the professional scientific community [is] the nature of its educational initiation." In "the contemporary natural sciences...the student relies mainly on textbooks" until in the third or fourth year of graduate work, at which point the student begins to do independent research. "It is a narrow and rigid education, probably more so than any other except perhaps in orthodox theology."[7]

Kuhn wrote,

No part of the aim of normal science is to call forth new sorts of phenomena; indeed, those that will not fit the box are often not seen at all. Nor do scientists normally aim to invent new theories, and they are often intolerant of those invented by others.[8]

Yet "no paradigm that provides a basis for scientific research ever completely resolves all its problems." When anomalous evidence emerges, however, scientists' first line of defense is usually to "devise numerous articulations and *ad hoc* modifications of their theory in order to eliminate any apparent conflict." They never simply renounce the paradigm unless another is available to take its place. Thus "the decision to reject one paradigm is always simultaneously the decision to accept another," and "the judgment leading to that decision involves the comparison of both paradigms with nature *and* with each other."[9]

The most effective claim that proponents of a new paradigm can make is that "they can solve the problems that have led the old one to a crisis."[10] Even then, Kuhn wrote,

The defenders of traditional theory and procedure can almost always point to problems that its new rival has not solved but that for their view are no problems at all...Instead, the issue is which paradigm should in the future guide research on problems many of which neither competitor can yet claim to resolve completely. A decision between alternate ways of practicing science is called for, and in the circumstances that decision must be based less on past achievement than on future promise.[11]

How does a new paradigm originate? Kuhn wrote,

Any new interpretation of nature, whether a discovery or a theory, emerges first in the mind of one or a few individuals. It is they who first learn to see science and the world differently, and their ability to make the transition is facilitated by two circumstances that are not common to most other members of their profession.[12]

First, Kuhn wrote, "their attention has been concentrated upon the crisis-provoking problems." Second, these individuals are usually "so young or so new to the crisis-ridden field that practice has committed them less deeply than most of their contemporaries to the world view and rules determined by the old paradigm."[13]

According to Kuhn,

Paradigms differ in more than substance, for they are directed not only to nature but also back upon the science that produced them. They are the source of the methods, problem-field, and standards of solution accepted by any mature scientific community at any given time. As a result, the reception of a new paradigm often necessitates a redefinition of the corresponding science.[14]

Redefining Science

Kuhn noted that scientific revolutions are often marked by disputes over the "standard that distinguishes a real scientific solution from a mere metaphysical speculation." Newton's theory of gravity was resisted because "gravity, interpreted as an innate attraction between every pair of particles of matter, was an occult quality" like the medieval "tendency to fall." Critics of Newtonianism claimed that it was not science and "its reliance upon innate forces would return science to the Dark Ages."[15]

Centuries later, some scientists claimed that the big bang was not science. In 1938, German physicist Carl F. von Weizsäcker gave a lecture in which he referred to the relatively new idea that our universe had originated in a big bang. Renowned physical chemist Walther Nernst, who was in the audience, became very angry. Weizsäcker later wrote:

He said, the view that there might be an age of the universe was not science. At first I did not understand him. He explained that the infinite duration of time was a basic element of all scientific thought, and to deny this would mean to betray the very foundations of science. I was quite surprised by this idea and I ventured the objection that it was scientific to form hypotheses according to the hints given by experience, and that the idea of an age of the universe was such a hypothesis. He retorted that we could not form a scientific hypothesis which contradicted the very foundations of science.

Weizsäcker concluded that Nernst's reaction revealed "a deeply irrational" conviction that "the world had taken the place of God, and it was blasphemy to deny it God's attributes."[16]

Similarly, intelligent design has been criticized for not being science. In 2004, American Society for Cell Biology president Harvey Lodish wrote that intelligent design is "not science" because "the ideas that form the basis" of it "have never been tested by any scientific peer-scrutiny or peer-review."[17] In 2005, the American Astronomical Society declared, "Intelligent Design fails to meet the basic definition of a scientific idea: its proponents do not present testable hypotheses and do not provide evidence for their views."[18] And the Biophysical Society adopted a policy stating, "What distinguishes scientific theories" from intelligent design "is the scientific method, which is driven by observations and deductions." Since intelligent design is "not based on the scientific method," it is "not in the realm of science."[19]

As previous chapters in this volume have demonstrated, the claims about evidence and peer review in the statements quoted above are false. Nevertheless, the statements illustrate that critics of intelligent design, like the critics of Newtonianism and the big bang, claim that the new paradigm does not qualify as science.

Some pro-Darwin writers have argued that intelligent design is even *anti*science. In 2006, philosopher Niall Shanks wrote that "a culture war is currently being waged in the United States by religious extremists who hope to turn the clock of science back to medieval times." The "chief weapon in this war is…intelligent design theory."[20] In 2008, biologist and textbook writer Kenneth Miller claimed that "to the ID movement the rationalism of the Age of Enlightenment, which gave rise to science as we know it, is the true enemy." If intelligent design prevails, he wrote, "the modern age will be brought to an end." For Miller, what is at stake "is nothing less than America's scientific soul."[21]

It's true that intelligent design operates with a definition of science that differs from the definition used by pro-Darwin scientists. For the latter, science is the enterprise of seeking natural explanations for everything. Only material objects and the forces among them are real; entities such as a nonhuman mind (which would have to be the source of any intelligent design in nature) are unreal. In Darwinian science, any evidence that seems to suggest intelligent design is ignored or ruled out. In 1999, a biologist wrote in *Nature* that "even if all the data point to an intelligent designer, such an hypothesis is excluded from science because it is not naturalistic."[22] But in an intelligent design paradigm, science seeks to follow the evidence wherever it leads. According to Kuhn, disputes such as this over the nature of science are common in scientific revolutions.

Dissatisfaction and the Proliferation of New Articulations

A scientific revolution is fueled in part by growing dissatisfaction among adherents of the old paradigm. This leads to new versions of the theoretical underpinnings of the paradigm. Kuhn wrote:

> The proliferation of competing articulations, the willingness to try anything, the expression of explicit discontent, the recourse to philosophy and to debate over fundamentals, all these are symptoms of a transition from normal to extraordinary research.[23]

A growing number of biologists now acknowledge that there are serious problems with modern evolutionary theory. In 2007, biologist and philosopher Pigliucci published a paper asking whether we need "an extended evolutionary synthesis" that goes beyond neo-Darwinism.[24] The following year, Pigliucci and 15 other biologists (none of them intelligent design advocates) gathered at the Konrad Lorenz Institute for Evolution and Cognition Research just north of Vienna to discuss the question. Science journalist Suzan Mazur called this group "the Altenberg 16."[25] In 2010, the group published a collection of their essays. The authors challenged the Darwinian idea that organisms could evolve solely by the gradual accumulation of small variations preserved by natural selection, and the neo-Darwinian idea that DNA is "the sole agent of variation and unit of inheritance."[26]

In 2011, biologist James Shapiro (who was not one of the Altenberg 16 and is not an intelligent design advocate) published a book titled *Evolution: A View from the 21st Century*. Shapiro expounded on a concept he called *natural genetic engineering* and provided evidence that cells can reorganize their genomes in purposeful ways. According to Shapiro, many scientists reacted to the phrase "natural genetic engineering" in the same way they react to intelligent design because it seems "to violate the principles of naturalism that exclude any role for a guiding intelligence outside of nature." But Shapiro argued that

> the concept of cell-guided natural genetic engineering is well within the boundaries of twenty-first century biological science. Despite widespread philosophical prejudices, cells are now reasonably seen to operate teleologically: Their goals are survival, growth, and reproduction.[27]

In 2015, *Nature* published an exchange of views between scientists who believed that evolutionary theory needs "a rethink" and scientists who believed it is fine as it is. Those who believed that the theory needs rethinking suggested that those defending it might be "haunted by the specter of intelligent design" and thus want "to show a united front to those hostile to science." Nevertheless, the former concluded that recent findings in several fields require a "conceptual change in evolutionary biology."[28] These same scientists also published an article in *Proceedings of the Royal Society of London*, in which they proposed "an alternative conceptual framework," an "extended evolutionary synthesis" that retains the fundamentals of evolutionary theory "but differs in its emphasis on the role of constructive processes in development and evolution."[29]

In 2016, an international group of biologists organized a public meeting to discuss an extended evolutionary synthesis at the Royal Society in London. Biologist Gerd Müller opened the meeting by pointing out that current evolutionary theory fails to explain (among other things) the origin of new anatomical structures (that is, macroevolution). Most of the other speakers agreed that the current theory is inadequate, though two speakers defended it. None of the speakers considered intelligent design an option. One speaker even caricatured intelligent design as "God did it," and at one point another participant blurted out, "*Not* God—we're excluding God."[30]

The advocates of an extended evolutionary synthesis proposed various mechanisms that they argued were ignored or downplayed in current theory, but none of the proposed mechanisms moved beyond microevolution (minor changes within existing species). By the end of the meeting, it was clear that none of the speakers had met the challenge posed by Müller on the first day.[31]

A 2018 article in *Evolutionary Biology* reviewed some of the still-competing articulations of evolutionary theory. The article concluded by wondering whether the continuing "conceptual rifts and explanatory tensions" will be overcome.[32] As long as they continue, however, they suggest that a scientific revolution is in progress.

Circling the Wagons

Kuhn compared scientific revolutions to political revolutions. Like a political revolution, a scientific revolution typically divides people "into competing camps or parties, one seeking to defend the old institutional constellation, the others seeking to institute some new ones."[33] The camp defending the old paradigm uses every means at its disposal, including all of its professional societies and publications, to resist the challenger. Since the mid-twentieth century, established paradigms have also controlled enormous funding from foundations and taxpayers, and thus jobs in educational and research institutions. With careers at stake, things can get ugly.

And things have gotten ugly. In the late 1990s, in Burlington, Washington, high school biology teacher Roger DeHart taught evolution as required. But he also shared with his students a few articles from mainstream science publications that questioned some aspects of neo-Darwinian theory. Militant Darwinists intimidated the local school board with threats of a lawsuit, so DeHart was reassigned to another subject and his biology class was turned over to a physical education instructor. In 2002, DeHart left his career as a public high school teacher and eventually moved with his wife and children to another country.[34]

In 2003, Dr. Nancy Bryson was head of the Division of Science and Mathematics at the Mississippi University for Women. After she presented an honors forum titled "Critical Thinking on Evolution," a senior biology professor read to the audience a previously prepared statement calling her presentation "religion masquerading as science" and accusing her of being unqualified to talk about evolution. The next day, Dr. Bryson was informed that her contract as division head would not be renewed. She subsequently had to find work elsewhere.[35]

In 2004, biologist Caroline Crocker was a visiting professor at George Mason University. While covering a required section on evolution, she gave one lecture on evidentiary problems with Darwin's theory and briefly mentioned the controversy over intelligent design. At the end of the lecture, she told students to "think about it for yourself." For this reason, Crocker's contract was not renewed.[36]

In 2005, biology teacher Bryan Leonard was about to get his PhD in science education from Ohio State University. His dissertation, which was a quantitative study about how a group of students reacted to the critical analysis of evolution, had already been approved by his committee. At the last minute, however, three pro-Darwin professors (who admitted they had not read Leonard's dissertation) lodged a complaint against him. The complaint alleged that he had engaged in unethical behavior by implying to students that there were weaknesses in neo-Darwinism. As a result, the university blocked Leonard's PhD.[37]

Physicist David Coppedge began working at the Jet Propulsion Laboratory in California in 1996. For nine years he served as the team lead system administrator for the ambitious Cassini mission to Saturn. Then he was reprimanded and demoted for privately giving DVDs about intelligent design to co-workers who requested them. In 2011, he was let go.[38]

Internationally renowned paleontologist Günter Bechly directed the 2009 Darwin Day exhibit at State Museum of Natural History in Stuttgart, Germany. The exhibit was strongly pro-Darwin, but it included a critique of intelligent design that featured some books by intelligent design advocates. After reading some of the pro-intelligent design books, Bechly concluded that Darwinists had been misrepresenting intelligent design. He gradually changed his views and publicly declared his support for intelligent design in 2015. After that, Bechly reported, the museum told him he was "no longer welcome, and that it would be appreciated if I would decide to quit." He was eventually forced to resign.[39]

As Kuhn pointed out, mainstream scientific journals (like scientific societies) are also largely controlled by the dominant paradigm.

For this reason, articles about intelligent design, or even articles on other subjects that have been written by known advocates of intelligent design, have rarely been published in mainstream journals.

Some years ago, I submitted an article on cell biology to a prominent scientific journal. The article did not mention intelligent design. After I made some recommended changes, my article passed peer review, and the editor emailed to tell me he wanted to publish it. He had just one final question: Was I "the Jonathan Wells of intelligent design fame?" (His words exactly.) I answered that I was. Afterward he sent the article to yet another reader, whose "review" didn't really deal with its contents but sounded like an angry rant from a pro-Darwin blog. The editor then informed me he had decided not to publish my article.[40]

In this same scientific journal in 2020, biochemist Dave Speijer justified the prejudice against intelligent design. He recommended that Internet searches hosted by tech giants explicitly discriminate against intelligent design; if the tech giants resist, the government should "make them," he wrote. In particular, Speijer recommended "mandatory color-coded banners warning of consistent factual errors or unscientific content, masquerading as science."[41]

Some Cautionary Words

Kuhn was criticized for various inconsistencies in his argument, including his tendency to switch back and forth among several meanings of *paradigm* and *theory*. More seriously, he was criticized for his relativism because he sometimes wrote as though no paradigm is any closer to objective reality than any other. But it seems to me that Kuhn's biggest problem was that he himself

operated within a paradigm—Darwinism—without recognizing it as such. For example, in opposition to Karl Popper's view that theories cannot be verified but only falsified, Kuhn wrote that "verification is like natural selection: It picks out the most viable among the actual alternatives in a particular historical situation."[42]

Kuhn even concluded *The Structure of Scientific Revolutions* by calling his approach "the evolutionary view of science." At the end of his last chapter he wrote,

> The analogy that relates the evolution of organisms to the evolution of scientific ideas can easily be pushed too far. But with respect to the issues of this closing section [i.e., progress through revolutions] it is very nearly perfect...[T]he resolution of revolutions is the selection by conflict within the scientific community of the fittest way to practice future science. The net result of a sequence of such revolutionary selections, separated by periods of normal research, is the wonderfully adapted set of instruments we call modern scientific knowledge. Successive stages in that developmental process are marked by an increase in articulation and specialization. And the entire process may have occurred, as we now suppose biological evolution did, without benefit of a set goal, a permanent fixed scientific truth, of which each stage in the development of scientific knowledge is a better exemplar.[43]

Kuhn's defense against critics who thereafter called him a relativist was based on the analogy between biological evolution and the history of science. In a 1970 postscript to his 1962 book, he wrote:

Imagine an evolutionary tree representing the development of the modern scientific specialties from their common origins in, say, primitive natural philosophy and the crafts. A line drawn up that tree, never doubling back, from the trunk to the tip of some branch would trace a succession of theories related by descent. Considering any two such theories, chosen from points not too near their origin, it should be easy to design a list of criteria that would enable an uncommitted observer to distinguish the earlier from the more recent theory time after time...[If such a list can be compiled] then scientific development is, like biological, a unidirectional and irreversible process. Later scientific theories are better than earlier ones for solving puzzles in the often quite different environments to which they are applied. That is not a relativist's position, and it displays the sense in which I am a convinced believer in scientific progress.[44]

In the end, therefore, it seems that even Kuhn admitted that unguided processes do not solve problems or lead to truth; intelligent direction is necessary.

A Shift Toward Design

Despite these problems with Kuhn's argument, we can still benefit from his descriptions of what happens during scientific revolutions. These include (1) the focus on debates over the definition of science; (2) the proliferation of variant articulations of the existing paradigm, which represent growing dissatisfaction among its adherents; and (3) the way defenders of an existing paradigm use all the institutional means at their disposal—including professional journals, membership in professional societies, and funding for jobs and research—to resist the challenger.

All of these are evident in the current controversy between Darwinism and intelligent design. Whether intelligent design will be the paradigm that successfully replaces Darwinism remains to be seen. But without a doubt, the modern neo-Darwinian model of evolution is a theory in crisis.

Part IV:

HARD
QUESTIONS
ON SCIENCE
AND FAITH

38

Is Science the Only Means for Acquiring Truth?

David R. C. Deane

September 11, 2001 was a watershed moment in Western history. That many of us can recollect the details of where we were, what we were doing, and who we were with at the exact moment of receiving the news of the terrorist attacks is evidence of their far-reaching effects. Experiences of personal shock-horror wound the heart and enlist the mind to register certain autobiographical details like cognitive scar tissue. Two decades later, our recollection of these little details pays tribute to those who were killed in the name of a lie, and as we reflect, we are reminded of those unanswered and perennial questions that haunt all of us in moments of great tragedy: *Why* did this happen? *How* was it possible? *What* does it all ultimately mean?

As the US and its allies mobilized their forces to fight Islamic extremism in a new and political War on Terror, select voices from the scientific community banded together to push a rhetorically fueled *science-only* offensive against all things religious in a new atheist[1] War on Religion. Triggered by 9/11, evolutionary biologist and leading atheist Richard Dawkins was the first to attack

with an article titled "Religion's Misguided Missiles" on September 16, 2001.[2] Less than a week later he followed up with a second attack in an article titled "Time to Stand Up," where he declared, "It is time to stop pussy-footing around. Time to get angry. And not only with Islam...[atheists have] moderated our language for reasons of politeness...[but now we must] stand up and break this absurd taboo. My respect for the Abrahamic religions went up in the smoke and choking dust of September 11th." Concluding, Dawkins's call to arms read, "'Enough!' Let our tribute to the dead be a new resolve: to respect people for what they individually *think*, rather than respect groups for what they were collectively brought up to *believe*."[3] As atheist physicist Victor J. Stenger would later write, "Science flies us to the moon. Religion flies us into buildings."[4]

The new atheist War on Religion had begun. In 2006, Dawkins published his infamous salvo *The God Delusion*, attracting global attention and rallying together a new generation of science popularizers intent on swaying public opinion against religion. Along with a spate of other science versus religion bestsellers,[5] *Time* magazine ran a cover story on November 5, 2006, titled "God vs. Science," while news outlets published headlines such as "Future of science: We will have the power of the gods,"[6] and "Forget faith, only science can save us now."[7]

During 2008–2009, an advertising campaign was organized by the British Humanist Association and Richard Dawkins Foundation funding the decoration of public buses with the slogan "There's probably no god. Now stop worrying and enjoy your life." The 2012 Reason Rally, dubbed the "Woodstock for atheists and skeptics," saw up to 20,000 people descend on Washington, DC, to listen to a host of atheist scientists, comedians, and performers. Strangely enough, even skeptics found themselves proselytizing the public with quasi-religious hope-filled tracts like Michael Shermer's "Science Is My Savior."[8]

Ironically, in the years following 9/11, the new atheist War on Religion ended up bearing an uncanny resemblance to organized religion itself, complete with high priests, corporate fellowship, and a *God Delusion* "gospel" all of its own. Science thus became a pseudo-messianic figure heralding a post-religious hope in the aftermath of 9/11. "Science is the only way we know to understand the real world,"[9] wrote Dawkins. "There is no reason to suppose that science cannot deal with every aspect of existence,"[10] wrote a fellow Oxford don, chemist Peter Atkins (who had previously lauded the "universal competence" of science in his article "Science *as* Truth").[11] American evolutionary biologist Jerry Coyne went so far as to suggest that nonscientific "disciplines" can only "yield knowledge...to the degree that their methods involve what I'll describe as 'science broadly construed': the same combination of doubt, reason, and empirical testing used by *professional scientists*..."[12] In other words,

science isn't simply one means among many for acquiring truth, science is the very condition for truth and rationality itself, the *sine qua non* for knowing anything at all.

Defining Scientism

The totalizing or exaggerated view that science is the only means for acquiring truth has been labelled, somewhat pejoratively, as *scientism*. According to the *Oxford English Dictionary* (OED), scientism is:

> [1] A mode of thought which considers things from a scientific viewpoint.
>
> [2] Chiefly *depreciative*. [a] The belief that only knowledge obtained from scientific research is valid, and that notions or beliefs deriving from other sources, such as religion, should be discounted; [b] extreme or excessive faith in science or scientists. Also: [c] the view that the methodology used in the natural and physical sciences can be applied to other disciplines, such as philosophy and the social sciences.[13]

From this definition three primary features emerge. First, as an *ism*, scientism is a "belief" or philosophy *about* science and not a demonstration *of* science (note the irony of this with respect to Dawkins's "call to arms" above). As a "belief," scientism is therefore not a matter of science but a matter of philosophy (the significance of this point is often overlooked, as scientism has been largely popularized by scientists rather than philosophers).[14] Second, scientism is a belief about *how we know* (epistemology), which, in turn, leads to the third feature—namely, that scientism's claims about *how we know* guarantees certain conclusions about the fundamental nature of *what we know* (metaphysics). Perhaps no one expressed this as clearly as astrophysicist and astronomer Carl Sagan when he famously stated that "The Cosmos is all that is or ever was or ever will be."[15]

Learning from the Past to Move Forward

Currently, much of the heat and noise surrounding the more than decade-long new atheist War on Religion has subsided. While not without a public voice, "New Atheism" is all but dead and the reprieve of space and silence provides an opportune moment to reflect upon and critically evaluate the cardinal doctrine of New Atheism, namely, scientism: what it is, and how it has influenced thinking in both academia and popular opinion.

The remainder of this chapter will critically examine the above-mentioned primary features of scientism. *Criticism 1* will argue that there is a systemic contradiction at the core of scientism, the self-refuting consequences of which will be considered by *Criticism 2* regarding the way scientism methodologically eliminates all nonscientific knowledge. *Criticism 3* will consider the way scientism metaphysically reduces our view of the universe in general and of human beings in particular.

Importantly, this chapter is not a criticism of authentic science, its technological applications, nor the scientific competency of those scientists with whom it disagrees. The reason for this, again, is that questions *about* science are not questions *of* science. In addition, this chapter does not purport to criticize atheism per se, for just as religion comes in many varieties, so atheism comes in many varieties and, as it is, there are many articulate atheists (and agnostics) who are decidedly critical of scientism.[16] As the title suggests, what this chapter is criticizing is the philosophical contention—held by *some* atheists and represented by *some* scientists—that science is the *only* means for acquiring truth.

Scientism Is Logically Self-Refuting

Criticism 1: If the premise of scientism is valid, it is false by definition and consequently self-refuting.

As suggested from the outset, scientism is a belief *about* science not a demonstration *of* science. That scientism is self-refuting can be demonstrated by simply arranging the OED definitions of "science" and "scientism" into the following argument:

1. Science is "[t]he intellectual and practical activity encompassing those branches of study that relate to the phenomena of the physical universe and their laws…"

2. Scientism is "[t]he belief that only knowledge obtained from scientific research is valid…"

3. "The belief that only knowledge obtained from scientific research is valid" is not, itself, obtained from "[t]he intellectual and practical activity encompassing those branches of study that relate to the phenomena of the physical universe and their laws…"

4. Therefore, if the premise of "[t]he belief that only knowledge obtained from scientific research is valid…"[17] is, itself, valid, it is false by definition and consequently self-refuting.

But as this initial encounter with things is before the abstractive activity of measurement in the order of knowing, it follows that it is a *prescientific* form of knowledge. Science cannot be the only means for acquiring truth as this very statement cannot be obtained from scientific research. In this lies the irredeemable incoherence of scientism: by definition, scientism is self-refuting.

The charge that scientism is self-refuting can be traced back to criticisms that were leveled against a philosophy known as *positivism*.[18] Positivism has its modern roots in the writings of the French atheistic philosopher Auguste Comte (1798–1857), who was known as the father of sociology. In the aftermath of the French Revolution and Napoleonic Wars, Comte was driven to find a way to reorient and ground society (note the parallel with Dawkins et al. following 9/11).

In a series of texts known as the *Course of Positive Philosophy*, Comte proposed a "law of three stages" to describe the intellectual evolution of societies in search for truth.[19] The first was the primitive theological or fictitious stage, in which reality was understood by reference to God and divine will. The second was the slightly more sophisticated philosophical or metaphysical stage, in which reality was understood by abstract universal concepts (e.g., the universal rights of post-Enlightenment humanism). The third and most advanced was the positive or scientific stage, in which reality was understood according to the methods of science. A central theme of Comte's sociological schema was thus the idea of progress: To know the *process* of *how* human intelligence evolves is to know the goal or *purpose* for *why* human beings know at all—namely, to achieve a positive or scientific understanding of the world (Stage 3).

While there is no all-encompassing definition for what precisely constitutes the scope of science, there is a general agreement on at least one feature of science that guarantees the application of its methods—namely, *empirical verification*. Empiricism, coming from the Greek word translated "experience" (*empeiría*), is a modern theory of knowledge based on experience derived from the five senses (i.e., sight, taste, touch, hearing, smell). As it relates to science, empirical verification is typically achieved through experiments that observe and measure physically embodied "things" (matter) in order to draw out (abstract) their verifiable quantities to produce exact or positive descriptions. The reason scientists are interested in quantities is because quantitative information is susceptible to prediction and control according to the basic laws of mathematics. At a very elementary level, knowing the quantities of things means that we can add, subtract, multiply, and divide for the purposes of development and application (e.g., medicine and engineering). In the words of Comte, "from science comes prediction; from prediction comes action."[20]

For this reason, Comte viewed positive or scientific knowledge (Stage 3) as superior to all other kinds of knowledge. Theology (Stage 1) and philosophy (Stage 2) were inferior to science because they could not produce the same kind of *exact* descriptions.[21] Moreover, as the degree of numerical positivity or exactness is the degree to which a thing can be subjected to mathematical principles, Comte regarded mathematics as the fundamental gauge according to which all of the various sciences are determined (e.g., physics, biology, chemistry, etc.). Hence, positivism is the philosophical assertion that *what* we know about reality is ultimately determined by the *process* of *how* abstract mathematical principles *describe* reality as derived from the senses.[22]

Positivism was critically received across

Europe and North America during the nineteenth and twentieth centuries. During the 1920s and 1930s, philosophers developed positivism's central thesis of exactitude by bringing it up to date with developments in mathematics and logic,[23] as well as discoveries in the natural sciences.[24] The result was a new and more sophisticated version of positivism known as *logical positivism* (sometimes *neo-positivism*).[25] The central thesis of logical positivism was the principle of *verificationism* laid out by A.J. Ayer (1910–1989) in his 1936 publication *Language, Truth and Logic*. Simply stated, verificationism asserted that any and all propositions that could not be analytically or empirically verified were ultimately meaningless.

The force of verificationism made logical positivism stronger in its *science-only* affirmation compared to Comtean positivism. Whereas Comte had argued that only science can give exact or positive *descriptions* of reality, Ayer and the logical positivists argued that science constitutes the very rational paradigm of *explanation* itself. If a proposition could not be demonstrated through logical or mathematical principles or empirically based practices, it was simply meaningless (unintelligible, without truth value). Essentially, logical positivism carried Comtean positivism to its logical conclusion by postulating a synonymity between science and truth, insofar as truth, classically understood, is the end goal of knowledge—Stage 3 of Comte's law of three stages. "To know" the truth of a "thing" is to know the thing's "positive data," the measurable properties that describe the thing, as opposed to the cause and effect of the thing's actual substance. To the extent that disciplines such as theology and philosophy consider the fundamental substance of things and their ultimate causes beyond the domain of natural phenomena, they were to

be considered meaningless enterprises. In the words of David Hume (1711–1776), "Commit it then to the flames. For it can contain nothing but sophistry and illusion."[26]

As a philosophical movement, logical positivism is dead.[27] In its hopeless exploit of trying to eliminate metaphysics (anything meta- beyond the physical) it ended up positing a central thesis that was itself metaphysical, thus rendering it meaningless on its own terms. After all, what analytic principle or empirical practice could verify the truth of verificationism, itself?[28] Verificationism proved unverifiable. "Philosophy always buries its undertakers,"[29] wrote Étienne Gilson, and ironically, that the smell of positivism's decomposition lingers today under the new name of scientism is not suggestive of progress, as Comte supposed, but rather, indicates an uninformed and philosophically naïve regress. When so eminent a scientist as the late Stephen Hawking wrote sweeping statements like "philosophy is dead...Scientists have become the bearers of the torch of discovery in our quest for knowledge,"[30] he, and those of his ilk, display a philosophical ignorance symptomatic of their scientistic hubris. The statement "philosophy is dead" is not, itself, a statement of science, thus proving the contention that not everything said by scientists is scientific.

Scientism Is Methodologically Eliminative

Criticism 2: If the premise of scientism is valid, it eliminates all knowledge outside the scope of science including that which makes science possible to begin with.

A second criticism of scientism is that it radically restricts *what* we know to only *how* science *describes* physical things in terms of their quantities. As long as there are things that can be empirically verified (observed and measured to abstract their quantities), science is limitless in its possibilities. However, this limitless application of the scientific method is, at the same time, its most tangible limit.[31] Science cannot deal with the nonquantitative aspects of reality. It does not merely deny the reality of those aspects, it simply does not deal with them because the ultimate value of science is its practical application, which again, requires abstract quantitative information susceptible to prediction and control.[32] However, it is here that scientism deviates from science. Where science *intentionally excludes* nonquantitative aspects of reality from its descriptions, scientism *irrevocably eliminates* their existence altogether.

To unpack this contention, let's briefly reflect on the faculties and processes by which we routinely understand the world around us. When we awake from sleep, our sense faculties are inundated with all sorts of information, such as sights, sounds, smells, tastes, and the sensation of touch. We have an awareness of these aspects of reality because our body's sense faculties act like a gateway for our minds to access a worldview of things to think about. Everything we know is thus borne of our conscious experience through the joint exchange of sensation and intellection, for a mind absent a body to sense things would not have things to think about, while a body to sense things, absent a mind, would not be able to think about what it senses.[33]

Now, as science is an "intellectual and practical activity" (according to the OED definition above), it follows that the application of its methods necessarily depends upon the things given in conscious experience—even though its conclusions *intentionally*

exclude them. The reason for this is because in order to give a quantitative description of a thing, that thing has to be measured, but before that thing *can be* measured it *has to be* a measurable thing that we encounter in conscious experience (i.e., in order to describe a thing, that thing has to be such that it is describable).

But as this initial encounter with things is before the abstractive activity of measurement in the order of knowing, it follows that it is a *prescientific* form of knowledge. Another way of saying this is that science does not determine existence; rather, existence is the reality that governs the nature of all our thoughts—scientific or otherwise. Our conscious experience of existence generally and existing things in particular are the necessary preconditions explaining *why* science can describe how things are. This, in short, is the philosophical reason for scientific reason.[34]

The eliminativism of scientism is a non sequitur. It simply does not follow that there is no such thing as knowledge outside the scope of science on account of the descriptions that science gives to us. As Edward Feser notes, that is like supposing that because metal detectors only reveal metallic objects that metallic objects are all that really exist, which is not at all obvious given that metal detectors *intentionally exclude* everything that is nonmetallic.[35] That is their *purpose*, their very *why*, so to speak, which is presumed by the *how* of the mechanical processes keyed into the electromagnetic technology that makes a metal detector *what* it is: a detector of metal (only!). In the same way, the nonquantitative aspects of our knowledge do not appear in scientific descriptions because the methods of science *intentionally exclude* them—the very fact of which implies their existence.

To clarify, I am not suggesting that conscious experience is not open to scientific analysis. What I am suggesting is that the very activity of science presupposes conscious experience. The reason science cannot completely explain the fundamental truth of *what* things are is because science only explains the *effects* of *how* things are, not the *cause* of *why* they are in the first place, which is presupposed by the *how*. Science *describes* the *process* of *how* things are; it does not *explain* the *purpose* of *why* things are. For this reason, science cannot be the *only* means for acquiring truth. For if truth is the correspondence of thoughts to things,[36] and if science is only one aspect of our thoughts regarding *how* things are (quantitatively), then science is only *one means* among many contributing to the process of truth acquisition, not the end-goal of truth itself. Truth and science are categorically different. To assume their synonymity is akin to saying, "Mathematics tastes delicious!" It simply confuses different categories. Truth is primarily a moral issue, not a scientific one, which is why lying is generally considered wrong, not unscientific. As the act of corresponding our thoughts to the reality of things (or issues), truth bears upon all of our activities, "the intellectual and practical activity" of science notwithstanding. Truth is not a *how-only* judgment; it involves a *why* of intention, purpose, discernment, and commitment in the act of assent.

To bring all this together, consider a simple analogy offered by mathematician and philosopher of science John Lennox:

> Why is the water boiling? Well, the heat from the Bunsen burner flame is being conducted through the copper base of the kettle which is agitating the molecules which are moving faster and faster so that's why it's boiling. Hmm… Actually it's boiling because I would love a cup of tea.[37]

Notice that Lennox gives a *how* description of *process* before answering the question with a *why* explanation of *purpose*. Clearly, the *how* and *why* are not contradictory; they are complementary.[38] There is no rational objection to the fact that we can fully describe the *process* of a boiling kettle and yet at the same time provide a nondescriptive explanation in terms of *purpose*. The issue is not *degrees* of understanding but *kinds* of understanding. Only if we suppose that the *how description* and *why explanation* are of the same category or kind will we be faced with a contradiction and therefore pressed to choose between alternatives (which is what scientism does by categorically confusing science with truth, absorbing or dissolving the *why* of *purpose* by the *how* of *process*).

Moreover, notice that the *how description* is contingent on the *why explanation*. If there were no *purpose* for boiling the water in the kettle, we would have no grounds for knowing the thermal *process* of the boiling water because *how* water boils in a kettle presupposes an individual *purpose* for *why* water is boiling in the kettle to begin with. The *why explanation* carries with it the fundamental dimension of human agency assumed by the *how description* of mathematical principles.[39] Indeed, as Lennox goes on to explain, that the *why explanation* is foundational to the *how description* is only too obvious when we realize that people have been making tea for literally thousands of years before understanding the theory of heat conduction.

In short, as an "activity," science is first and foremost a conscious experience before it is a thought-through conclusion.[40] The inseparability of quantities from things (physical phenomena) justifies the very essence of *what* science is as an activity insofar as the reality of things affects our senses, providing our minds with objects of thought.

Thus, we can assert that *science is a means for acquiring truth*, and a most powerful one at that. However, as its descriptive justification is not by necessity of its own nature as an "intellectual and practical activity," it follows, once again, that science cannot be the only means for acquiring truth.

Scientism Is Metaphysically Reductionistic

Criticism 3: If the premise of scientism is valid, it conceptually reduces the existence of all things to their quantitative properties, including human beings.

A third criticism of scientism is one that looks at the other side of *Criticism 2* above, namely that if scientism radically restricts *what* we know to only *how* science describes physical things, then it follows that scientism also radically reduces *what* we know by confusing it with *how* we know. Whereas science is an activity that describes the quantities of things, scientism is the belief that those descriptions are the explanation for *what* actually exists; that is, the whole of *what* we know can be broken down (reduced) to the level of particular entities (things) and their elementary phenomena.

To begin with, recall the OED definition of science as an "intellectual and practical activity encompassing those branches of study that relate to the phenomena of the physical universe and their laws." Notice that, according to this definition, science is not a thing per se; rather, it is an activity. As such, the very definition of science is not scientific because the definition (Latin *dēfīniō*, "set bounds to") does not *limit* science as a thing to be measured but defines it as an activity for making sense of things by measuring them.

If we think about the activity of science

grammatically in terms of a verb, an active *doing*, we bring to mind the often-overlooked metaphysical preconditions necessary for the possibility of any activity, the activity of science notwithstanding. Take the activity of running as an example: When we reflect on this activity, we become aware in the first instance of the rather trivial fact that the act of running presupposes the existence of a runner. Reflection on the existence of a runner also makes us aware that the act of running presupposes the existence of a physical world with the right kind of environmental conditions in which the runner can exercise that act (e.g., a track, foot path, or park regulated by laws such as gravity, friction, etc.). Finally, as we reflect upon the existence of a runner and the existence of a suitable environment in which to run, we are aware that all that is required for the runner to run is the realization of the activity itself (i.e., as it is carried out through the functionality of legs, cognitive stability, general health, fitness, etc.).

Like the activity of running, the activity of science presupposes a number of preconditions. First, the existence of a scientist; second, "the phenomena of the physical universe and their laws;" and third, "the intellectual and practical" potential of the scientist to practice science. In summary, there thus appears to be at least three metaphysical preconditions presupposed by the activity of science: (1) a mind; (2) a world; (3) a rational intelligibility accounting for the reciprocal harmony between (1) and (2). In order to think, there has to be a mind capable of thinking, a worldview of things to think about, and some sort of rationale or intelligence (*logos*) common to both our minds and the world of things that *explains why* our thoughts can correspond to things (i.e., that grounds *truth*).[41] Without these metaphysical preconditions, the activity of science would not be possible.

What this brief excursus on the activity of science tells us above all else is that science is an intelligent human interaction with a rationally ordered world. How an individual understands these metaphysical preconditions will determine their view of science, the scope of its methods, its ultimate value and purpose, the kind of knowledge it can furnish, and the types of questions it can reasonably answer.[42] As the prefix *pre-* implies, these presuppositions are *before* science in the order of knowing; thus, by their very nature they cannot be subjected to scientific demonstration because they are *pre*scientific. They are *pre*supposed by science as the *pre*conditions of science. To presume otherwise is to argue in a circle as a scientific demonstration of its own possibility assumes its own existence, which is what it is seeking to demonstrate. But this is precisely what scientism insists: If science is the only means for acquiring truth, then the truth of *what* science is itself (i.e., as it entails those presupposed metaphysical preconditions) must be susceptible to scientific demonstration. So the question is, Has scientism found a way to escape this circularity?

Let's take theoretical physicist Lawrence Krauss as an example. In his 2012 publication *A Universe from Nothing: Why There Is Something Rather Than Nothing*, Krauss boldly declared that "surely 'nothing' is every bit as physical as 'something,' especially if it is to be defined as the 'absence of something.'"[43] Even out of context the patent absurdity of this sentence is self-evident such that it needn't warrant further elaboration. However, Krauss is not the only scientist to publish a proof-text of *Criticism 1*. The exact same contradiction is found in Stephen Hawking and Leonard Mlodinow's 2010 publication *The Grand Design*, the central thesis of which reads, "Because there is a *law* such as gravity, the universe can and will

create itself from nothing. *Spontaneous* creation is the reason there is something rather than nothing, *why* the universe exists, *why* we exist."[44] Like Krauss, Hawking and Mlodinow make an assertion of existence (gravity), imply its nothingness, and conclude that the something, which is nothing, is the reason there is something.[45]

What these and other such *nonscientific* statements reinforce is the systemic contradiction at the heart of scientism (*Criticism 1*). The logical force of scientism demonstrates the impossibility of trying to squeeze *why explanations* into *how descriptions*. If we were feeling charitable, we might say, "Okay, these guys may not have it all sorted out right now, but just give it some time. One day science will be able to close all the explanatory gaps."[46] But charity can be misplaced; the issue is not if the activity of science will one day provide an explanation for the existence of the universe. Rather, the issue is that today the existence of the universe provides an explanation for the activity of science. As quantum physicist Louis de Broglie pointedly wrote, "We are not sufficiently astonished by the fact that any science may be possible, that is, that our reason should provide us with the means of understanding."[47]

The activity of science in abstracting the quantities of empirically verifiable things presupposes a knowledge of existing things given in conscious experience. As abstractions, these quantities do not entail actual existence; they *belong* to the substance of the particular existing thing that grounds the reality of those quantities. Simply thinking about the quantity of $100 does not entail its actual existence in my wallet, after all![48] While not intending to be a statement of metaphysics, the famous line from Shakespeare's *Hamlet*, "To be or not to be?," strikes at the very core of existence in that "to know" is to implicitly register a knowledge of existence that is not, in and of itself, measurable. Can we measure particular existing things? Yes, that's called *science*. Can we measure our conscious experience of existence per se? No, and any attempt on the part of scientists to do so ends in a tangle of absurdities, contradictions, and self-refutations.[49]

Scientific descriptions do not explain the ultimate fact of existence; rather, they presuppose it. When scientists turn themselves into creators by supposing they can explain the existence of the universe with mathematical models, they are not doing cutting-edge physics, they are naively cavorting with the ancient Greco-pagan philosophy of Plato. Like Pythagoras, Plato believed that because numbers and mathematics were the best at dealing with quantities that they must also hold the explanatory key accounting for the existence of the universe.[50] On the surface the inductive argument makes sense: If the *effects* of things are mathematically quantifiable, then the *cause* of things must be mathematical quantities themselves. However, as we argued in *Criticism 2*, it is a non sequitur to suppose that because *some things* are quantitative that therefore *all things* are quantitative. But even beyond the theoretical objection, think about the practical consequences of a world in which everything can be given a quantitative formulation. If everything was reducible to *how descriptions* of quantities interacting according to mathematical principles, *what* would be left of the *why explanations* that carry the fundamental dimensions of human agency, such as truth? Consciousness? Free will? Love? Good? Evil?[51]

Although not wanting to belabor the point, let's consider just one example of a rather shocking science-only answer given by Dawkins because it takes us right back to where we began this chapter in our reflection upon 9/11. In his 1995 publication *River Out of Eden: A Darwinian View of Life*, Dawkins famously wrote:

> In a universe of blind physical forces and genetic replication, some people are going to get hurt, other people are going to get lucky, and you won't find any rhyme or reason in it, nor any justice. The universe we observe has precisely the properties we should expect if there is, at bottom, no design, no purpose, no evil and no good, nothing but blind, pitiless indifference... DNA neither cares nor knows. DNA just is. And we dance to its music.[52]

As though completing the circle of scientism's self-refutation, we cannot help but note that this statement completely undermines Dawkins's sense of moral outrage expressed in the aftermath of 9/11. If there is "no purpose," *why* a "call to arms"? If there is "no evil," *why* was 9/11 a "time to get angry"? If "DNA neither cares nor knows," *why* is Dawkins's DNA telling us something it "knows" and apparently "cares" about as though our DNA should also "know" and "care"? Indeed, *why* blame the blind fundamentalism of the terrorists and not the enlightened technology of aircraft itself if, as Dawkins elsewhere writes, "Each one of us is a machine, like an *airliner* only much more complicated"?[53] *Why, why, why, why, why?* Dawkins's seeming difficulty to answer these basic questions bears witness to the systemic self-refuting nature of scientism as outlined in *Criticism 1*.[54]

Scientism is not scientific. It is its own form of fundamentalism. The systemic consequences of scientism's self-refuting premise destroys all sense of meaning, including what it means to be human. As the old saying from Confucius goes, "Where words lose their meaning, people lose their lives." By erasing *why explanations*, scientism erases the fundamental human dimension assumed by *how descriptions*, thus abdicating the right of even scientists to engage in their scientific trade.

Summing It Up

Is science the only means for acquiring truth? No, at least not in the modern sense of "science" we have considered in this chapter. If the reader is looking for a quick way to refute scientism, my advice is simple: Let scientism speak for itself. There is really nothing more to it than that. The only defensible position of scientism is that it is the personal opinion of certain atheistic individuals who have otherwise committed themselves philosophically and methodologically to the assumption that the natural sciences are the only source of knowledge. But this simply begs the question as to *why* they feel compelled to do so.

I propose that the answer to this question has little, if anything, to do with authentic science and everything to do with the presuppositions of atheism, with which scientism goes hand-in-hand. Science has never spoken a word in its life for the fact that it is not a thing to have life; it is the activity of things that have life—namely, *scientists*. Science does not speak, *scientists do*. And that is why scientism does not logically end with a science that is limitless in its possibilities, but with *scientists* who are. This accords perfectly with atheism: The belief in the limitlessness of human potentiality makes sense in a world without God where humans are "gods unto themselves."

What this tells us is that scientism is fundamentally a worldview issue. Our understanding of science depends upon our understanding of the metaphysical preconditions of science, which, in turn, are founded upon our core convictions regarding the existence of God as the creator of all things. In this way, religious knowledge furnishes proposals for both philosophical reflection and scientific activity. The issue is not *science only*, or *philosophy only*, or *theology only*; the issue is *how do these distinct yet not-divorced sources of knowledge integrate with one another to establish a comprehensive worldview?* If scientists, philosophers, and theologians can work together toward this end—careful to acknowledge the limitations of their respective fields—then it might make amends to the increasing fragmentation and disintegration of our present society, which has only been exacerbated by the so-called new atheists' War on Religion. By facilitating a culture of relationality and reciprocity through collaboration and dialogue, such a worldview would dignify the fundamental differences that make each one of us uniquely human.

In closing, and as a start toward this end, I propose that the Judeo-Christian revelation contained in the Bible is capable of furnishing such a worldview. The Bible tells us that a rational God created an intelligible world with uniquely rational creatures known as human beings. And it tells us that God has given each one of us not only the capacity but the command to actively engage in the world around us that we might be fruitful and multiply.

But not only does it provide the reason for scientific reason, it also provides the reason for scientism's unreasonableness. The belief in the limitlessness of human potentiality is what the Bible calls idolatry—an attempt to replace the one true God with an alternative, which, in this case, is scientists. The tragedy of scientism is as old as Genesis 3, and it is this: We do not need to strive to "be like God" as though we could be *as* God, arbitrating truth and "good and evil" (Genesis 3:5). For

God has already made us *like* him in a *similar* way in that we have been made in his image (Genesis 1:26-27) as rational beings, which is borne out in the very fact that we can do science itself. The extent to which we have lost this truth today is what the Bible calls sin, yet therein lies good news—news that radically disrupts the dogma of scientism that says everything is susceptible to prediction and control. Through repentance and faith in Jesus Christ as the Savior of sins, we as human beings can *overcome* our natural predilections with a spiritual transformation that infuses life with *incalculable* meaning—incalculable because it is a life found in the infinite beatitude of God himself: in Christ Jesus.

Is Theistic Evolution a Viable Option for Christians?[1]

Jay W. Richards

When a fellow Christian asks me, "Can you believe in God and evolution?," I always respond, "That depends. What do you mean by *God*, and what do you mean by *evolution*?" No one seems to be very satisfied with this retort, which seems evasive; but it's an honest response because their question, as it stands, is ambiguous.

Asking whether one supports so-called theistic evolution has the same problem. Unless you define *theistic* and *evolution* very carefully, it might refer to views that, on closer inspection, are more different than they are alike. One version might be an oxymoron, one a trifle, one a curiosity, and another a complete muddle.

Besides being vague, these questions, and practically every answer to them, are controversial. Such social pressures don't encourage clear thinking or speaking. So when they encounter such questions, many people, especially academics, choose to obscure rather than clarify. If pressed, they may try to stake out a moderate both-and position: "Evolution is God's way of creating."[2] For the conflict-averse, using the term *theistic*

evolution, as we might call it, may be a reassuring response, but what does it mean?

In the century and a half since Charles Darwin first proposed his theory of evolution, Christians, Jews, and other religious believers have not only pondered its truth— or lack thereof—they have grappled with how to make sense of it theologically. So far, they haven't reached a consensus and tend, instead, to argue among themselves. It can all be quite confusing. In fact, the whole subject of God and evolution, and especially theistic evolution, is an enigma surrounded by a blanket of fog.

To answer the questions, we need to clear away the fog and the enigma.

A Range of Views

There are practically as many views of how God relates to evolution as there are people who have pondered the subject. Still, most views fall into one of several categories. Unfortunately, before defining the categories, you must first overcome a terminological hurdle. What do you do with the troublesome word *creationist*? The word is usually used pejoratively to bring to mind young-Earth creationists who believe that God created the universe in six 24-hour days sometime in the last 10,000 years. Critics assume that the young-Earth view is so disreputable that anyone associated with it will likewise be tarnished. However you judge that uncharitable assumption, you can't use the word *creationist* these days without carrying some of this baggage along for the ride.

This is an accident of history. In a less complicated world, the word *creationist* would not be a put-down but simply a way to refer to people—Christians, Jews, Muslims, Sikhs, and other theists—who believe in a doctrine of creation. Regrettably, that's not how the world or the word works. Like it or not, the discussion about God and evolution takes place in a context designed to misdirect and misrepresent certain views, especially those views that take God seriously.

Because we're stuck with the word *creationist*, though, we'll just have to slog ahead.

Besides young-Earth creationists, there are folks who refer to themselves as old-Earth creationists and still others who call themselves progressive creationists. Old-Earth creationists generally hold to mainstream scientific views of the age of the Earth and the universe but believe that God worked directly in nature (as a "primary" or "efficient cause"[3]) to create some things. These might include heavenly bodies like galaxies and the solar system, the first reproducing cell, various forms of life, human beings, human souls, and so forth. Old-Earth creationists disagree among themselves on the *loci*—the places—where they think that God acts directly, but all agree that, sometimes at least, God acts directly in natural history to bring about things that nature would not produce if left to its own devices.

Progressive creationists also believe that God acts directly at various points in cosmic history, but they tend to see more evolutionary development between the seams of God's specific acts.

Then there are those who don't fit simply on the creationist spectrum but do challenge materialistic theories of evolution. For example, intelligent design or ID theorists argue that nature, or certain aspects of nature, are best explained by an intelligent cause rather than blind, undirected mechanism. On this view, repetitive, lawlike, or mechanistic explanations that invoke, say, a gravitational force and natural selection explain some aspects of nature, but a full explanation of the natural world will include intelligent agency as well.

Moreover, ID theorists have argued that physical laws are themselves the result of intelligent design—even if they are not arguably adequate to explain everything in nature.

At the same time, ID theorists focus on the detectable effects of intelligence rather than on the specific locations or modes of design within nature. As William Dembski puts it: "Intelligent design (ID) is the study of patterns in nature that are best explained as the result of intelligence."[4] Because ID is minimal, it is logically compatible with almost any creationist or evolutionist view that allows for intelligent agency as an explanation within nature. (The contributors to this volume fall into the ID camp.)

Finally, there are theistic evolutionists who would appear to subscribe to a hybrid position that combines both theism and evolution. Most theistic evolutionists contrast their view with *special creationism*, which would include any view that suggests that God has acted directly in natural history. However, logically speaking, a theistic evolutionist could also be an ID proponent (in fact, there are many such people). Nevertheless, most self-described theistic evolutionists distinguish themselves from ID proponents, and are, in some cases, harsh critics of ID. So, like the word *creationist*, the term *theistic evolution* tends to have a meaning different from what its etymology alone would suggest.

So what exactly is theistic evolution? It would be nice to open *Webster's*, find the definitions of *theism* and *evolution*, bring the definitions together, and be done with it. Alas, it's not that simple. Behind the phrase *theistic evolution* lurks a lot of mischief and confusion.

A Dilemma

When dealing with God and evolution, most people have an intuitive feeling that there's some contradiction lurking in the neighborhood, some dilemma that has to be resolved. Even children, at some point, begin to sense this. Most probably ask their parents what my 11-year-old daughter once asked me: "So why did God make dinosaurs that all died out millions of years before Adam and Eve?" Several years earlier she had asked, obviously garbling the kindergarten evolution lesson, "Did we used to have tails?" Perhaps you already have answers to these questions. But if you're like millions of other parents, though, you will probably try to punt.

For punters, theistic evolution (or *evolutionary creationism* as it's sometimes called) might seem to promise some relief. But at some point, if you tell an attentive child that evolution is just God's way of creating, she's going to ask you what you mean. It would be nice to have something more than a pat answer accompanied by some hand-waving.

The trouble starts when we dig into the common textbook definitions of the term *evolution*. Here, evolution is often *defined* by its opposition to creation. Consider just two academic sources among legion: "That

organisms have evolved rather than having been created is the single most important and unifying principle of modern biology."[5] And here's the Harvard paleontologist George Gaylord Simpson: "Man is the result of a purposeless and natural process that did not have him in mind."[6] Darwin himself understood his theory this way. As he said, "There seems to be no more design in the variability of organic beings, and in the action of natural selection, than in the course which the winds blow."[7]

These descriptions of (Darwinian) evolution don't leave a lot of wiggle room. And notice that the idea of organisms evolving *rather than* having been created is not presented as a sidelight, as the private opinion of a few scientists. In the first quote, as in many others, evolution is described as the "single most important and unifying principle of modern biology." It would be hard to put the point any more strongly.

Surely it's better to ask and answer the follow-up questions than to avoid them. Presumably a theistic evolutionist is someone who claims that both theism *in some sense* and evolution *in some sense* are true—that both God and evolution somehow work together in explaining the world. But all the real interest is hidden behind the phrase *in some sense*. So, let's lay out the main senses of the two words in question.

Theism

Although different people understand God differently, the word *God* has a pretty stable meaning in ordinary conversation. If I tell Richard Dawkins, an atheist, that I believe in God, he has some sense of what I mean. In ordinary English and other Western languages, *God* usually refers to a creator, a personal being who has chosen to create the world, who is powerful and perfect in whatever ways such a being could be powerful and perfect, and who transcends the universe. That is, God would exist, would *be*, whether or not he had chosen to create the world. The world, in contrast, exists as the result of his free choice, for his purposes and at his discretion.

Of course, *God* doesn't refer to just any old being like a bunny resting on a down or the guy in Mumbai who answers your questions when you call Dell tech support. God, though a "being" in the sense that he "exists" (or, more precisely, *is*) is himself the source of other beings, and in that way, he is qualitatively different from all other beings. Classical theists often say that God is "Being itself." That way of speaking is a bit obscure to the uninitiated. At the very least, however, what this means is that God doesn't participate in some more fundamental reality called "being" along with everything else. He is the Source of all being. Moreover, unlike you, me, and the burrito I had for lunch, God necessarily exists. He exists in every possible world.

Technically, you could believe that such a God exists and be either a theist or a deist. A theist believes that God both created the world and continues to conserve and interact in and with it.[8] In fact, God is so intimately related to the world that, while being separate from the world, he still wholly pervades it. Christian theists speak of God as both transcendent and immanent.[9] What the theist will never do is identify God with the world.

A deist holds a more minimal view and believes that God created the world but doesn't really keep up with the day-to-day activities on the ground. Or even if he does keep up, he doesn't get directly involved. He maintains a strictly hands-off policy.

Besides theism and deism, and leaving aside polytheism, the other main options are pantheism and panentheism. *Pantheism* identifies God and nature. For the pantheist, God doesn't transcend the world nor is he independent of it. He's not even immanent in the world. Rather, God is the world and the world is God. For most pantheists, moreover, God is not really personal, either. After all, the universe just doesn't look much like an agent with purposes and a will. For the pantheist, God might be thought of as a rational principle or a life force that somehow pervades the universe; but God, for the pantheist, most certainly is not a transcendent creator.

Panentheism is a hybrid position that holds that God has some transcendent qualities but is nevertheless in the world. Or, to put it differently, the world is in him. The world, we might say, is part of God. God and nature may be distinct but they're inseparable. A panentheist might think of God as a creator, but not in an absolute sense. God might push or pull or persuade or cajole things to go in a certain direction. He might have purposes. But he won't call everything into existence from nothing simply by his free choice. God will evolve along with the world.

Though there are a few Christian academics who identify with panentheism,[10] the vast majority of Christians, Jews, Muslims, and historic thinkers in these traditions are theists. That's because the basic tenets of their religions hold that God is a transcendent creator who at least occasionally acts directly in the world. For instance, all three of these Abrahamic faiths believe that God specially communicated with Abraham and Moses.

In addition, Christians believe that God became a man, Jesus, at a particular time and place; that Jesus was conceived by the Holy Spirit in the Virgin Mary rather than by ordinary means; and that after Jesus died, he was raised from the dead and ascended into heaven. All this implies the Christian belief that God is triune. He exists eternally as three persons while still being one God. Though less central to Christian doctrine, most Christians also believe that Jesus worked certain extraordinary miracles, such as calming a storm and raising a girl from the dead.

Take away all beliefs about God acting in history, and you have, at best, only a shadow of theistic belief.

Of course, theists don't believe that God is aloof from the world except when he acts directly in nature. For theists, God transcends the world, is free to act directly in it, and always remains intimately involved with it.

At the same time, the theist need not believe that God always acts directly in the world. Traditionally, Christian theologians have argued that God can act in the world in two different ways. He can act directly or primarily, such as when he creates the whole universe or raises Jesus from the dead. It's God's world, so that's his prerogative. He's not violating the universe or its laws when he does this, or invading alien territory, because he's the source of both the universe and whatever laws it might have.

He also can act through so-called secondary causes. These include the choices or tendencies of the creatures he has made. For instance, he can work through the evil choices of Joseph's brothers to achieve a greater good of getting the descendants of Abraham to move to Egypt so that they don't die from famine.

God can also bring about his purposes through natural processes and laws that he has established, such as electromagnetic force. An event might be both an expression of a physical law and the purposes of God. It's not as if atheists appeal to gravity while theists appeal to miracles. Gravity is as consistent with theism as are miracles. But for the theist, gravity is a creature, or rather, it describes the behavior of creatures. It's like a mathematical description of how God has ordained physical objects to act in ordinary circumstances; it's not an eternal law governing God's behavior.

Christians, Jews, and other theists recognize that God can act through secondary causes when they thank God for their food. Still, they know that God normally provides our food not as manna from heaven, but through natural causes like rain, the spring season, and soil, and through human actions like sowing and reaping. God is so free and so powerful that he can act either directly or through secondary causes. He's like a doting gardener who creates his own sun, seeds, water, nutrients, and dirt. And he's perfectly happy to have "flowers" who can make their own decisions.

Therefore, for theists, God—who acts either directly or through secondary causes—continually upholds, oversees, and superintends his entire creation in "providence," even as he allows his creatures the freedom appropriate to their station.[11]

We've just scratched the surface, but we've probably said enough about theism for our purposes.

Evolution

While God is the grandest of all subjects, the meaning of the word *evolution* is a lot harder to nail down.

In an illuminating article titled "The Meanings of Evolution," Stephen C. Meyer and Michael Keas distinguished six different ways in which the term *evolution* is commonly used:

1. Change over time; history of nature; any sequence of events in nature.

2. Changes in the frequencies of alleles in the gene pool of a population.

3. Limited common descent: the idea that particular groups of organisms have descended from a common ancestor.

4. The mechanisms responsible for the change required to produce limited descent with modification, chiefly natural selection acting on random variations or mutations.

5. Universal common descent: the idea that all organisms have descended from a single common ancestor.

6. "Blind watchmaker" thesis: the idea that all organisms have descended from common ancestors solely through unguided, unintelligent, purposeless, material processes such as natural selection acting on random variations or mutations; that the mechanisms of natural selection, random variation and mutation, and perhaps other similarly naturalistic mechanisms, are completely sufficient to account for the appearance of design in living organisms.[12]

Meyer and Keas provide many valuable insights in their article. But here we're only concerned with the term *evolution* insofar as it's relevant to theology.

The first meaning is uncontroversial—even trivial. Everyone—from the most convinced young-Earth creationist to theistic evolution advocates—agrees that things change over time, and that the universe has a history.[13] Populations of animals wax and wane depending on changes in climate and the environment. At one time, certain flora and fauna prospered on the Earth, but later they disappeared, leaving mere impressions in the rocks to mark their existence for future generations.

Of course, "change over time" isn't limited to biology. There's also cosmic evolution, the idea that the early universe started in a hot, dense state and, over billions of years, cooled off and spread out, formed stars, galaxies, planets, and so forth. This includes the idea of stellar nucleosynthesis, which seeks to explain the production of heavy elements (everything heavier than helium) in the universe through a process of star birth, growth, and death. These events involve change over time, but they have to do with the history of the inanimate physical universe rather than with the history of life. While this picture of cosmic evolution may contradict young-Earth creationism, it does not otherwise pose a theological problem. The generic idea that one form of matter gives rise, under the influence of various natural laws and processes, to other forms of matter does not contradict theism. Surely God could directly guide such a process in innumerable ways, could set up a series of secondary natural processes that could do the job, or could do some combination of both.

In fact, virtually no one denies the truth of evolution in senses 1, 2, or 3. And pretty much everyone agrees that natural selection and random mutation explain some things in biology (sense 4).

What about the fifth sense of evolution, universal common ancestry? This is the claim that all organisms on Earth are descended from a single common ancestor (or common pool of ancestors) that lived sometime in the distant past. Universal common ancestry is distinct from the mechanism of change. In fact, it's compatible with all sorts of different mechanisms or sources for change, though the most popular mechanism is the broadly Darwinian one.

It's hard to square universal common descent with some interpretations of biblical texts, of course; nevertheless, it's logically compatible with theism. If God could turn dirt into a man, or a man's rib into a woman, then presumably he could, if he so chose, turn a bacterium into a jellyfish, or a dinosaur into a bird. Whatever its exegetical problems, an unbroken evolutionary tree of life guided and intended by God, in which every organism descends from some original organism, sounds like a logical possibility. As a result, there's some logical space where both ID and theistic evolution overlap—even if ID and theistic evolution often describe people with different positions.[14]

Besides the six senses mentioned by Meyer and Keas, there is also the metaphorical sense of evolution, in which Darwinian theory is used as a template to explain things other than nature, like the rise and fall of civilizations or sports careers. In his book *The Ascent of Money*, for instance, historian Niall Ferguson explains the evolution of the financial system in the West in Darwinian terms.[15] He speaks of mass extinction events, survival of the fittest banks, a Cambrian explosion of new financial instruments, and so forth. This way of speaking can sometimes be illuminating, even if, at times, it's a stretch. Still, no one doubts that there are examples of the fittest surviving in biology and finance. We might have some sort of "evolution" here, but not in a theologically significant sense.

Finally, there's evolution in the sense of progress or growth. Natural evolution has often been understood in this way so that cosmic history is interpreted as a movement toward greater perfection, complexity, mind, or spirit. A pre-Darwinian understanding of evolution was the idea of a slow unfolding

of something that existed in nascent form from the beginning, like an acorn eventually becoming a great oak tree. If anything, this sense of evolution tends toward theism rather than away from it because it suggests a purposive plan. For that reason, many contemporary evolutionists (such as the late Stephen Jay Gould) explicitly reject the idea that evolution is progressive and argue instead that cosmic history is not going anywhere in particular.

So clearly, theism, properly understood, is compatible with many senses of evolution. For most of the senses of evolution we've considered, in fact, there's little appearance of contradiction. Of course, this is a logical point. It doesn't tell us what is true—only what could be true.

But there's one clear exception: the blind watchmaker thesis. Of all the senses of evolution, this one seems, at least at first blush, to fit with theism like oil with water. It claims that all the apparent design in life is just that—apparent. That apparent design is really the result of natural selection working on *random* genetic mutations. (Darwin proposed variation. Neo-Darwinism attributes new variations to genetic mutations in DNA.)

The word *random* in the blind watchmaker thesis carries a lot of metaphysical baggage. In neo-Darwinian theory, *random* doesn't mean uncaused; it means that the changes aren't directed—they don't happen for any purpose. Moreover, they don't occur for the benefit of organisms, species, or ecosystems—even if, under the guidance of natural selection, an occasional mutation might redound to the benefit of a species.

Darwin, at least in *On the Origin of Species*, assumed a form of radical deism in which God establishes general laws that govern matter, but then leaves the adaptation and complexity of life up to random variations and natural selection. (Note that Darwin's personal views are a separate matter from the structure and rhetoric of his argument in the *Origin*.[16]) Nowadays, though, most evolutionary biologists are more thoroughgoing materialists, at least when it comes to their science. So the blind watchmaker thesis is more or less the same as the mechanism of neo-Darwinism as its leading advocates understand it.

The blind watchmaker thesis is usually wedded to some materialistic origin-of-life scenario, which isn't about biological evolution per se, though it is sometimes referred to as chemical evolution.

From the time of Darwin, who first proposed it, to the present, Darwinists have contrasted their idea with the claim that biological forms are designed. Here's how the late Darwinist Ernst Mayr put it:

> The real core of Darwinism, however, is the theory of natural selection. This theory is so important for the Darwinian because it permits the explanation of adaptation, the "design" of the natural theologian, by natural means, instead of by divine intervention.[17]

Notice that he says "instead of." Darwinists almost always insist that their theory serves as a designer substitute. That's the whole point of the theory. This makes it different from other scientific theories, like Newton's law of gravity. Newton didn't formulate the law to get God out of the planet business (in fact, for Newton, God was involved in every aspect of the business). And theories that invoke physical laws are determinate: They allow the scientist to make predictions about what will happen, all things being equal.

Darwin's theory isn't like that. It simply says that whatever has happened, and whatever will happen, the adaptive complexity we see in organisms is (primarily) the result of natural section and random variation, not design. From the very beginning, the theory was intended to rule out teleological (purposive) explanations. As William Dembski once said, "The appeal of Darwinism was never, That's the way God did it. The appeal was always, That's the way nature did it without God."[18]

That's why, even if not all agree with Richard Dawkins that Darwin "made it possible to be an intellectually fulfilled atheist,"[19] the vast majority of Darwinists claim that Darwin's mechanism makes God superfluous. It's their theory, so presumably they have a right to tell us what it means. Theists, in contrast to Darwinists, claim that the world, including the biological world, exists for a purpose—that it is, in some sense, designed. The blind watchmaker thesis denies this. So anyone wanting to reconcile strict Darwinian evolution with theism has a Grade-A dilemma on his hands. It's akin to reconciling theistic evolution with antitheistic evolution.

The easy way out of this dilemma is to drop or radically redefine the theistic part (dropping the Darwinian part is usually much riskier to one's career). Dissolving a dilemma, however, is not the same as resolving it. If the adjective *theistic* in the phrase *theistic evolution* is to be an accurate description, it should include a theistic view of God.

If you're new to the debate over God and evolution, you might already be anticipating how to be a theistic evolutionist. A theistic evolutionist, as suggested above, is someone who holds that God somehow sets up or guides nature so that it gives rise to everything from stars to starfish through a slowly developing process. Organisms share a common ancestor but reach their goal as intended by God. God works in nature, perhaps through cosmic initial conditions, secondary processes, discrete miracles, or some combination of these, to bring about his intended results rather than creating everything from scratch. Or perhaps God created the universe as a whole primarily, but everything else he "delegates," as it were, to natural causes. But whatever the details, by definition, the process of change and adaptation wouldn't be random or purposeless. It would implement a plan. It would have a purpose. So a theistic evolutionist, you might assume, would hold to a *teleological* version of evolution, which may include cosmic evolution, the origin of life, and biological evolution, and would for certain not endorse the Darwinian blind watchmaker thesis.

But it's rarely so straightforward. Consider the view of Presbyterian pastor Timothy Keller. In his popular book *The Reason for God*, he tells readers, "For the record I think God guided some process of natural selection, and yet I reject the concept of evolution as All-encompassing Theory."[20]

Earlier he says, "Evolutionary science assumes more complex life-forms evolved from less complex forms through a process of natural selection. Many Christians believe that God brought about life this way."[21] He also quotes approvingly from a Bible commentary that affirms evolution as a mere "scientific biological hypothesis," but rejects it as a "world-view of the way things are." Thus partitioned, Keller tells the reader "there is little reason for conflict."[22] Elsewhere Keller observes that he has "seen intelligent, educated laypeople really struggle with the distinction…Nevertheless, this is exactly the distinction they *must* make, or they will never grant the importance of" evolution as a biological process.[23]

But those "intelligent, educated, laypeople" struggle for a reason. What is the distinction he is proffering, and what does it distinguish? Is he saying that while it's okay to speculate about various evolutionary hypotheses, we should not affirm any? Surely not, because he seems to affirm a broad, semi-Darwinian evolutionary hypothesis. Is he saying that Darwinian evolutionary theory explains hearts and arms and ears and bacterial flagella, but not our love of music and our moral intuition? And if so, on what basis is he maintaining the distinction? After all, it's not as if we have solid empirical evidence that natural selection, acting on random genetic mutations, can give rise to an avian lung but not to our belief in the golden rule. At best, such a distinction would be ad hoc.

Or does Keller have something else in mind? He doesn't say. In any case, distinguishing evolution as a hypothesis from evolution as a "world-view of the way things are" offers scant guidance one way or another. To be useful, he would need to specify what he means by evolution, what he thinks it explains well, and what he thinks it leaves out that keeps it from constituting a "world-view of the way things are." Instead, we get a vague distinction without a difference. It's no surprise that the laypeople to whom he's commended the distinction don't find it very illuminating.

Notice that Keller speaks of God "guiding some process of natural selection," but does not mention *random variation*, which is as much a part of Darwinian theory as is natural selection. Perhaps that avoidance is intentional. But because he doesn't say outright that he rejects the idea that natural selection acts only on random genetic mutations, the careful reader is left guessing.

If we read him charitably, though, Keller seems to want to affirm that God guided the origin and development of life forms, all of which are linked by a chain of common ancestry, by coordinating his guidance with natural selection. The outcome isn't really random.

At the same time, Keller explicitly rejects the blind watchmaker thesis. So he's apparently not an orthodox Darwinist. He doesn't quite realize that to hold this view consistently, however, he needs to embrace teleology and reject orthodox neo-Darwinism and materialistic origin-of-life scenarios and not merely reject "evolution as a worldview of the way things are," whatever that means.

I don't intend to pick on Keller. I'm using him to illustrate how confusing this issue can be, and how even smart, orthodox religious thinkers often get into a muddle when they try to wed their Christian beliefs with Darwinian theory.

If we peel away these confusions and look for a straightforward, coherent position, however, we usually end up with the idea of God-guided common ancestry. This is probably what most people would think theistic evolution means. But they would be wrong, at least when it comes to describing the views of many who describe themselves as theistic evolutionists. These days, most theistic evolutionists seek to somehow reconcile theism with *Darwinian* evolution. They may affirm

design in some broad sense at the cosmic level, but things get patchy when it comes to biology. Though it's not always easy to understand what they're saying, many theistic evolutionists want to integrate the blind watchmaker thesis into their theology.

By now, it should be clear that this is like trying to square a circle. It just makes no sense to speak of God guiding an unguided process. Logically, something has to give. And to judge from the trajectory of Christian scholars who've tried to do this, what usually gives is theology. In other words, theistic Darwinists jettison some or another traditional Christian belief about God, such as his omnipotence or omniscience.

The price for adopting orthodox Darwinism—in either logical consistency or theological integrity—is quite high for Christians. As a result, it's surprising that more theistic evolutionists don't seem to bother with this prior question: Is there solid evidence that the Darwinian mechanism, aside from some low-level successes, can really account for biological complexity? If more Christian scholars would explore that question earnestly, then we would not need lengthy essays exploring the viability of theistic evolution.

Will Intelligent Machines Rise Up and Overtake Humanity?

Robert J. Marks II

A dvances in artificial intelligence (AI) are jaw-dropping. In 2018, a painting made by AI was auctioned at Christie's for $435,000.[1] AI has beat all human competition in games ranging from Go to Texas Hold'em.[2] China uses AI face recognition technology to monitor its citizenry. An unmanned armed drone was used to assassinate Iranian terrorist Qassem Soleimani.[3] Israel's Harpy missile operates using AI. It flies around a predefined kill zone waiting to be illuminated by radar, and then locates and destroys the source of the radar.[4] All this can be done autonomously without human oversight.[5]

Are these accomplishments of AI a red flag for the future? Will we someday be subservient to advanced AI as science fiction movies depict? In *The Terminator*, AI-based Skynet tries to destroy humanity. *The Matrix* depicts a future where AI exploits humanity while keeping people in a happy state of virtual reality distraction.

Some believe AI will usher in a dystopian future. The time at which the intelligence of computers surpasses that of humans is called the *singularity* by Google's Ray Kurzweil.[6] In 1999, he wrote, "Before the next century is over, human beings will no longer be the most intelligent or capable type of entity on the planet."[7]

Likewise, in his bestselling book *Homo Deus*, Yuval Harari posits that the main products of the twenty-first-century economy "will not be textiles, vehicles, and weapons but bodies, brains, and minds." He continues: "…the way humans have treated animals is a good indicator for how [AI] upgraded humans will treat us."[8]

Other top minds, in the roles of AI Chicken Littles, think likewise:

- Physicist Stephen Hawking warns that the emergence of artificial intelligence could be the "worst event in the history of our civilization."[9]

- Henry Kissinger, former Secretary of State under two US presidents, warns: "Philosophically, intellectually—in every way—human society is unprepared for the rise of artificial intelligence."[10]

- Entrepreneur extraordinaire Elon

Musk says AI is humanity's "biggest existential threat."[11]

If AI is to become this superintelligent threat, presumably it will need to become more intelligent than we are. To achieve this, AI will need to display human traits like understanding, sentience, and creativity. Not everyone agrees this is possible.

Other AI experts believe in human exceptionalism and claim that computer-based AI will never be able to duplicate the human mind. Noted mathematician and physicist Roger Penrose coauthored with Stephen Hawking the theory of black hole singularities.[12] He contends many human attributes, like creativity, are beyond the reach of AI. After discussing a computer-generated conversation between a computer "therapist" and its patient, he observes, "Though this may give an eerie impression that the computer has some understanding, in fact it has none, and is merely following some fairly simple mechanical rules."[13] Elsewhere Penrose is quoted as saying, "Intelligence cannot be present without understanding. No computer has any awareness of what it does."[14]

Gregory Chirikjian, director of the Johns Hopkins University robotics lab, agrees: "[AI does not display human traits] nor will robots be able to exhibit any form of creativity or sentience."[15] In his book *Hit Refresh,* Microsoft CEO Satya Nadella also agrees: "One of the most coveted human skills is creativity, and this won't change. Machines will enrich and augment our creativity, but the human drive to create will remain central."[16]

AI Limitations

Computer limitations support human exceptionalism.[17] These limitations do not mean AI will never be dangerous. Like

electricity, dangers in AI will need mitigating. (We'll address the dangers in AI later.) But AI will never try to take over like in the science fiction movies *The Terminator* and *The Matrix.*

Here is one major reason why: Fundamentally, anything computable must follow a step-by-step procedure written in computer code. Step-by-step procedures are called *algorithms.* Computers can only execute algorithms. There are many problems that are provably nonalgorithmic and therefore beyond the capability of computers.

An example of a nonalgorithmic and therefore noncomputable task is taken from Rice's theorem. Can properties of computer code be determined by an algorithm? In other words, can computer software be written to determine what an arbitrary computer code will or won't do? In many important cases, the answer is no. Rice's theorem says computer code cannot be written to examine an arbitrary computer program to determine whether the program at some time will print the number 3. If the first line in the software says "PRINT 3," then the software can easily be identified to print 3. The key to Rice's theorem is that the examining computer code must work for *all* possible programs. This is provably not possible.

A special case of Rice's theorem is the Turing halting problem.[18] It is not algorithmically possible to determine whether an arbitrary computer program will stop or run forever. No computer program can ever be written to determine whether another arbitrary computer program runs forever or halts.

Another nonalgorithmic operation is compression. Given a computer file of an arbitrarily large size, determining how much the file can be compressed is not algorithmic. The smallest a file can be compressed is called its *Kolmogorov complexity.* Above a certain file size, the Kolmogorov complexity cannot be computed. It is nonalgorithmic.[19]

This leads to an important question: Are there human traits that are nonalgorithmic? If so, these traits will never be computable. If this is true, the human mind will never be able to be uploaded to a computer. Computers can only handle the algorithmic. If you are stripped of your nonalgorithmic properties, you will be either very boring or, more accurately, not you.

Noncomputable human traits include sentience, understanding, and creativity. Each will now be addressed in more detail.

Sentience

A component of sentience is qualia. *Qualia* is an experience from the senses, including taste, smell, and touch.

Let's do a thought experiment. You bite into a segment from an orange. As you bite, the skin on the segment bursts and juice from the bite covers your tongue. You taste a sweet orange flavor as you chew and swallow.

You are now assigned the job of explaining your experience to a man with no sense of taste or smell since birth. How can this be done? You can provide explanations. You can present the man with the chemical components of the segment. He can understand the physics of chewing and the biology of the taste buds. But the true experience of biting and tasting of the juice exploding from the orange's juice vesicles is not possible to communicate to the sense-deprived man. Qualia is beyond description to those without a shared experience.

If the experience of biting a segment from an orange cannot be explained to a man without the senses of taste or smell, how can we expect to duplicate the qualia experience in a computer using computer code? If the true experience of biting into a segment from

an orange can't be explained, it is nonalgorithmic and therefore noncomputable.

Many computer operations, including the Turing halting problem and computation of the Kolmogorov complexity, cannot yet be solved by humans. Neither can describing the true experience of biting into an orange segment.

Duplicating qualia is beyond the capability of AI.

Understanding

Philosopher John Searle illustrated that computers do not understand what they do with his example of the "Chinese room."

Searle imagined himself in a room with many file cabinets. A question written in Chinese is slipped through a slot on the door. Searle does not read Chinese. But the file cabinets contain billions of easily searchable questions along with answers written in Chinese. Searle searches through the file cabinets until he finds a match to the question being posed. The file cabinet also contains the answer to the question. When he finds a match, Searle copies the answer in Chinese. He walks to the door and slips the answer through the slot to whoever is on the outside.

From the outside, it appears the occupant of the Chinese room is not only fluent in Chinese but is able to understand questions. Not so. The occupant of the Chinese room has no understanding of the Chinese language. In generating the answer to the submitted question, he is simply following the algorithm of pattern matching.

In 2011, IBM's Watson beat champion contestants in the quiz show *Jeopardy*. Processing natural language, Watson answered queries faster than the human contestants. Watson was the equivalent of a large Chinese room. Instead of file cabinets, Watson had access to all of Wikipedia and then some.

Like Searle in the Chinese room, Watson had no understanding of the meaning of the queries it fielded. It was following an algorithm written by computer programmers.

AI does not understand what it is doing. Understanding is nonalgorithmic and therefore noncomputable.

Creativity

AI has produced some astonishing results. AlphaGo software developed by DeepMind beat the world champion in the complex game of Go. A computer program dubbed Pluribus has beat poker pros in the game of no-limit Texas Hold'em poker.[20] OpenAI's program GPT-3 generates short bursts of astonishingly coherent prose given only a few prompts.[21] Don't these and other computer programs display creativity?

The answer depends on your dictionary. To answer the creativity question, the term *creativity* must first be defined.

The Lovelace test is a simple and easily understood definition of AI creativity offered by Selmer Bringsjord. Named after Ada Lovelace, the first computer programmer, the Lovelace test asks a simple question to determine software creativity: Did the AI in question perform a task that is beyond the explanation of the person who wrote the AI code (or someone with comparable expertise)? If so, the Lovelace test has been passed and AI has been shown to be creative.

Surprise results don't count. Computer programs generate surprising results all the time. They can be a result of faulty computing or generating results outside expectations.

In beating the world champion at Go, the AI program AlphaGo made an unusual move that surprised many, including the Go champion Lee Sedol. Was this creative? Not according to the Lovelace test. AlphaGo was written to play Go. And that's what it did. If AlphaGo could provide an explanation of the game of Go when queried to assist you with your taxes, it would be creative. But AlphaGo was not programmed to explain the game of Go or to fill out your taxes. It was written to play the game of Go and nothing else.

To achieve superintelligence based on AI writing better and better AI software, creativity is required. If computer programs cannot be creative as defined by the Lovelace test, they cannot write better programs beyond the explanations of the original programmer. For this reason alone, superintelligence is not achievable.

To date, no one has successfully claimed a computer program has passed the Lovelace test.

Arguing Outside Your Silo

The Future of AI

Despite the evidence just presented that AI will never display creativity, understanding, or sentience, there are a lot of people who believe that superintelligence will be achieved. Many see a dystopian future where AI will ruin our society. We'll call them AI dystopians. As we have quoted, Bill Gates and Stephen Hawking are in this highly populated camp. But consensus should not be taken as evidence in the status of science or its future. Michael Crichton, author of classic science fiction books like *Jurassic Park*, said as much in a talk given at CalTech:

> There is no such thing as consensus science. If it's consensus, it isn't science. If it's science, it isn't consensus. Period.[22]

But the question under discussion is not about the current state of AI, it is over the future of AI. The limitations of AI rest largely on computer science, so this foundational knowledge is essential in predicting what AI will accomplish.

Ideology is a contributing factor to opinions about AI of the future. According to technology prophet George Gilder, those who believe AI will exceed the intelligence of humans suffer from a "materialist superstition" that assumes the mind is a meat computer and can therefore be replicated by AI.[23] Ignorance of the computer science thus far presented is also a factor. Those who code and use computer software for a living are often unaware of the limitations placed on AI by established computer science theory. This background is not required to be an excellent coder.

Algorithmic information theory (AIT) is the study of algorithms in computer science.[24] AIT addresses what computer codes can and can't do.[25] Those without a background in AIT may not have solid ground for appreciating or defending the limitations of AI. AIT is a foundational tool for establishing AI limitations, and those who are trained to code for a living may not be familiar with AIT. Great coders do not need to know about AIT.

We have quoted Microsoft CEO Satya Nadella, Johns Hopkins University robotics scientist Gregory Chirikjian, and Oxford mathematician and physicist Sir Roger Penrose. All believe computers will never be creative. All have backgrounds in computer science and mathematics. But so do some of the AI dystopians.

Often, celebrity AI dystopians speak outside of their silo of expertise.

Confusion of Expertise

Ever wonder why actor Kevin Costner testified in front of Congress on the topic of oil spills, as did Ben Affleck on a children's project, and quizmaster Bob Barker on the Captive Elephant Accident Prevention Act?[26]

Many people, including some in Congress and the media, apparently equate celebrity in one area to across-the-board genius in everything.

News commentator Laura Ingraham disagrees; she has told clueless politically pontificating music celebrities to *Shut Up and Sing.*[27] Similarly, innovative comedian Ricky Gervais informed celebrity loudmouths, "You know nothing about the real world. Most of you spent less time in school than Greta Thunberg."[28]

One statistic to test the accuracy of prophetic statements is examination of credentials. Kevin Costner is no expert in oil spills and quizmaster Bob Barker has no credentials in elephant accident prevention. They are listened to because they are celebrities. The same analysis can be applied to business and science celebrities who make predictions about AI. Celebrity scientists and business tycoons are often not experts in AI.

In critiquing a person's background, care must be taken. In a debate, citing the lack of credentials of your opponent is called the *genetic fallacy.* The opponent's background is made the issue rather than the topic of debate. That is not the purpose here. The reasons underlying AI limitation have been established earlier in this chapter. The key debate issues have thus been addressed. We are now asking whether the AI dystopians are aware of the simple AIT-related limitations we have discussed.

Henry Kissinger, previously quoted, is alarmed about the impact of AI if not bridled. Kissinger is a gifted politician, diplomat, and geopolitical consultant. But his knowledge of AI apparently comes from reading the writings of others—many of whom adhere to materialist superstition. I suspect Kissinger is unaware of the deeper issues of computer science.

Elon Musk, whom we also quoted previously, is an AI dystopian. He has said, "With artificial intelligence we are summoning the demon."[29] Musk received undergraduate degrees in economics and physics from the University of Pennsylvania and then attended Stanford for a graduate degree in physics but dropped out. He is a hugely successful entrepreneur, having foundational roles in businesses such as PayPal, SpaceX, Tesla Motors, and OpenAI. During 2020 and 2021, Musk floated between being the seventh richest, the second richest, or even the richest man in the world with a net worth reportedly over $150 billion.[30] But does he have the computer science foundation to appreciate the limits of AI? Does expertise in business translate to expertise in AI?

Facebook CEO Mark Zuckerberg doesn't think so. He called Elon Musk's doomsday AI predictions "pretty irresponsible."[31] Business guru George Gilder questions Musk's opinions outside of the silo of business. He notes, "I think Elon Musk is a tremendous entrepreneur, yet he's a quite retarded thinker."[32]

Ray Kurzweil, a successful inventor and businessman, received a BS degree from MIT in computer science. Kurzweil is an avid supporter of superintelligence and has written books on the topic. Superintelligence assumes AI software will write better software that writes better software and on and on. According to Kurzweil, AI will soon be smarter than humans. But if AI writes better AI software not anticipated by the original programmer, then the AI is being creative. It would pass the Lovelace test. Kurzweil, with an undergraduate degree in computer science and vast experience in business and technical innovation, believes AI can be creative. Carefully defined by the Lovelace test, however, computer software cannot be creative.

Kurzweil once worked for Bill Gates at Microsoft. Does Bill Gates—another AI dystopian—have a background in computer science and AIT? Apparently not. As an undergraduate, Gates dropped out of Harvard University to pursue the founding of Microsoft. He is a talented entrepreneur whose success does not require deep studies in computer science. Gates was a knowledgeable programmer with early computer hardware. But much of his success came from his business instincts and his team of lawyers. Gates's father was a named partner in the Seattle law firm Preston Gates & Ellis. When I consulted for Microsoft, my first meeting was with Microsoft lawyers who told me, in no uncertain terms, my legal responsibilities. As expected, I was told Microsoft would own 100 percent of any intellectual property I created. I was instructed not to look at any patents associated with my assigned duty. This was new to me. If Microsoft was sued relating to my work and lost, punitive damages would kick in if I had looked at patents. Doing so could be construed as plagiarizing intellectual property. If I didn't look at patents, my contribution would be ruled a coincidental discovery by the courts and only monetary damages could be recovered.

In my experience, consulting typically requires the signing of documents like nondisclosures and specifying who owns what intellectual property. But Microsoft is my only consulting experience that started with a nose-to-nose meeting with a gaggle of lawyers.

Microsoft's success is due in large part to smart business dealings and not creativity. Their first historical coup was acquisition of MS-DOS. Microsoft did not write DOS. It was purchased by Microsoft in the early 1980s. Because of the rise in popularity of the IBM PC, MS-DOS became a cash cow for Microsoft.

Microsoft continued to expand, not necessarily by innovation but by acquisition, copying technology, and court battles. Flagship Microsoft software tools that were purchased or copied from other companies and not innovated by Microsoft include Windows, Word, PowerPoint, Excel, Internet Explorer, and Bing. Business practice, not innovation, is the secret of Microsoft's success.

Bill Gates must be celebrated as a gifted entrepreneur, businessman, and philanthropist. But his background in computer science, other than coding and its application, must be questioned. AI dystopian Gates opined, "I do think we do have to worry about [AI] but I don't think it's inherent that as we create super intelligence that it will necessarily have the same goals in mind that we do."[33]

So like Ray Kurzweil, Bill Gates believes in AI creativity that leads to superintelligence.

Great scientists risk similar overreach in expertise. Consider, for example, Stephen Hawking's fears of AI. Hawking, whose celebrity appearances include *Star Trek: The Next Generation*, is a genius in cosmology. With fellow genius Roger Penrose, he formulated the Penrose-Hawking singularity theorems, applying general relativity so as to better understand black holes. But artificial intelligence disturbed Hawking. He told BBC:

> The development of full artificial intelligence could spell the end of the human race…It [AI] would take off on its own, and re-design itself at an ever-increasing rate…Humans, who are limited by slow biological evolution, couldn't compete, and would be superseded.[34]

Like Gates and Kurzweil, Hawking bought into the idea of superintelligence. Despite his gifts in cosmology, Hawking was outside of his silo of expertise. He assumed that AI is creative.

Hawking's concern was all the more curious in light of his earlier abandonment of an ultimate Theory of Everything that would unify the physics of the universe in a nicely wrapped interconnected set of equations. He changed his mind about the viability of that project on account of Kurt Gödel's landmark theorems on incompleteness and inconsistency. No matter how much physicists discovered, he concluded, there would still be more to learn.[35]

Remarkably, the distance between Gödel's work and the limitations of computer creativity is not far. Indeed, Gödel's theorems form the foundation for AIT. Alan Turing, the father of modern computer science, built on Gödel's thesis, showing that some problems are nonalgorithmic and therefore cannot be captured by computer code. The nonalgorithmic nature of the Turing halting problem discovered by Turing is a manifestation of Gödel's work.

Roger Penrose, Hawking's coauthor on black hole physics, recognized this connection and wrote about it. His book *The Emperor's New Mind* (1989) wonderfully links Gödel to Turing and establishes the noncomputable nature of creativity. Penrose posits that the noncomputable characteristics of human thinking spring from quantum effects in microtubules in the brain. The quantum collapse of a wave function to a deterministic state, reasoned Penrose, is nonalgorithmic and thus might be the source of noncomputable creativity.

Hawking seems not to have considered Penrose's work when he offered his scary prediction about a dystopian AI future. One would not say to a man of his intellectual stature, "Shut up and do physics!" We can say, however, that Hawking had the intellect to learn about Penrose's insights into computers and AI via Gödel and thereby understand the limits of computers.

Yet apparently he didn't.

AI Dangers

The danger of AI lies not in its potential ability to become conscious and take over the world, but in its incomplete vetting prior to use. AI, like any system, will have its unexpected consequences. Contingencies can increase exponentially as the complexity is increased linearly.

Undesirable and unexpected contingencies have already been manifested in the deployment of AI systems. There are multiple examples of unexpected contingencies arising from complex systems ranking from the simply curious to the very serious:

1. A deep convolutional neural network was trained to detect wolves. After the trained neural network incorrectly classified a husky dog as a wolf, the programmers did some forensics and discovered there was undesirable bias in the training data. The pictures of wolves all contained snow. The picture of the misclassified dog also contained snow. In training, the neural network had learned the presence and absence of snow. The features of the animals were not considered in the classification problem.

2. An inconvenience for self-driving cars is the false classification of objects like plastic bags. A stationary plastic bag can be categorized as a large rock[36] while a wind-blown plastic bag may be mistaken for a deer.[37] These are unintended contingencies of the self-driving car's software.

3. A more serious problem with self-driving cars is fatalities. In 2018, an Uber self-driving car in Tempe, Arizona, struck and killed pedestrian Elaine Herzberg. Steven Shladover, a UC Berkeley research engineer, noted, "I think the sensors on the vehicles should have seen the pedestrian well in advance."[38] The death was a tragic example of an unintended contingency of a complex AI system. Unintended contingencies remain a major obstacle in the development of general (level 5) self-driving cars. Some developers, believing the problem is insurmountable, have given up.[39]

4. During the height of the Cold War, the Soviets deployed a satellite early-warning system called Oko to watch for incoming missiles fired from the United States. On September 26, 1983, Oko detected incoming missiles. At a military base outside of Moscow, sirens blared and the Soviet brass was told by Oko to launch a thermonuclear counterstrike. Doing so would result in millions being killed. The officer in charge, Lieutenant Colonel Stanislav Petrov, felt something was fishy. After informing his superiors of his hunch that Oko was not operating correctly, Petrov did not obey the order. Upon further investigation, Oko was found to have mistakenly interpreted the effects of the sun reflecting off of clouds.[40] There was no US missile attack. Petrov's skepticism of Oko's alarm likely saved millions of lives.

These examples of unintended contingencies deal with systems of broad complexity. Narrow AI systems are typically more error-free. Examples of a narrow AI system are anti-radiation missiles like the previously mentioned Harpy missile from Israel. The missile is launched and flies about (loiters) over a predefined kill zone. The missile can operate autonomously without human oversight. If fuel gets low, the missile returns home. Alternately, if illuminated by radar, the anti-radiation missile zeroes in on the location of the radar's source. The missile follows the radar beam and destroys the radar installation.[41] Whether or not one agrees with the mission of such a system, the anti-radiation missile is an example of relatively narrow AI that has historically worked without flaw. There are few if any unforeseen contingencies in anti-radiation missiles that would distract from their duties.

Self-driving cars require a more complex AI system. For tightly connected AI systems, the number of contingencies and therefore the number of unexpected consequences increase exponentially as the complexity increases linearly.

The dangers of AI can be mitigated through proper design practices:

1. *Domain expertise.* AI software should be developed by those with experiential knowledge of the problem being solved. Experts will better identify undesired contingencies during development of the AI software.

2. *Testing.* AI systems must be tested under different conditions and in different environments.

3. *Disjunctive design.* AI consisting of conjoined narrow AI systems is more easily testable than tightly designed systems. Smaller systems are easier to both conceive and test.

Exploding contingency count should give pause to anyone designing complex AI systems. Even if the development of superintelligence were possible, the required complexity of such systems would present programmers with the overwhelming problem of eliminating unexpected and undesirable performance.

Takeaways

The nonalgorithmic capacity of the human mind remains beyond the reach of computers and AI.[42] But, as with all new technologies like electricity, care must be taken in the development of AI to assure its safe and proper use. Frayed electrical wires still burn down houses and downed electric lines still electrocute. But the advantages of electricity far outweigh the dangers. The negative consequences have been mitigated by legislation and best practices. The hope is that AI dangers can be likewise contained.

AI will never be sentient, creative, nor have understanding, and it will never have dominion over humans because of its own initiative.[43] If AI becomes dangerous, it will be the fault of humans who develop and use AI. Following good ethics will likewise never be the responsibility of the AI itself, but will always belong to those who write and test AI systems and to the end users (humans).

The limitations of AI are further evidence of human exceptionalism.[44] We are "fearfully and wonderfully made."[45]

41

Can Panspermia
Explain the Origin of Life?

Guillermo Gonzalez

In 1971, Francis Crick (of DNA fame) and Leslie Orgel proposed the theory of *directed panspermia* at a Communication with Extraterrestrial Intelligence conference in Soviet Armenia. Two years later, they published their theory in the journal *Icarus*.[1] The idea is simply that an intelligent space-faring civilization seeded Earth with life. Implicit in the proposal of these scientists is that origin-of-life theories were inadequate. Crick famously wrote, "An honest man, armed with all the knowledge available to us now, could only state that in some sense, the origin of life appears at the moment to be almost a miracle, so many are the conditions which would have had to have been satisfied to get it going."[2] Thus, Crick and Orgel sought an external source for Earth life.

Directed panspermia is a species of the *panspermia* hypothesis for the origin of life. I will not be considering directed panspermia further in this article.[3] Lord Kelvin introduced panspermia theory in 1871 at the British Society for the Advancement of Science.[4] Swedish scientist Svante Arrhenius introduced a different version in 1908 in *Worlds in the Making*.[5] Kelvin's version

involved lithopanspermia, the transfer of life inside rocks, such as meteoroids. Arrhenius's version involved radiopanspermia, the transport of microorganisms (either naked or on small dust grains) by radiation pressure from the sun. There was little additional progress in the field until Crick and Orgel's work prompted renewed interest.

There are four broad steps involved in the transfer of viable organisms between planetary bodies: (1) launch from an inhabited planet's surface, (2) transit through interplanetary and/or interstellar space, (3) arrival at a habitable planet, and (4) colonization of the habitable planet. The first step is achieved when an asteroid or comet impacts the surface of a planet. Some fraction of the impact ejecta achieves escape velocity from the planet and enters the space environment. The second step is determined simply from application of Newtonian gravitational physics in the case of lithopanspermia as well as stellar radiation pressure forces in the case of radiopanspermia. Combined with probability calculations, then, it is possible to calculate the fraction of viable organisms that encounter the target body within a certain timeframe. The third step involves calculating the fraction of those organisms that survive reentry in the target body's atmosphere and reach its surface. The final step is not often considered, but it is essential. Some fraction of the arriving living organisms will not grow and multiply on the target planet because they lack some essential ingredient that was present on the home world. I will consider these steps in more detail below as applied locally to our solar system and then to other planetary systems.

Local Panspermia

Scientists found new motivation to work on panspermia following the announcement in 1996 of remains of ancient life in the Martian meteorite ALH84001. The claimed evidence for ancient Martian life has not held up well following more careful scrutiny.[6] However, the Martian meteorites did demonstrate to previously skeptical scientists that rocks from the surface of one planetary body can be transported at least partially intact to another. It is true that it is easier to get rocks off the surface of Mars, given its smaller escape velocity and thinner atmosphere, than the Earth. Does this mean that we should not expect to find pieces of Earth on other planetary bodies in the solar system?

This is a question that motivated me to conduct studies to calculate the amount of material blasted off Earth's surface early in its history.[7] As a way to test our calculations, we predicted the amount of Earth material that should be present in the lunar rock samples brought back by the Apollo astronauts (about three grams). In 2019, scientists found a two-gram piece of Earth's crust in an Apollo lunar sample called *Big Bertha*, confirming our predictions.[8] Although most of the ejected material will have suffered extreme alteration

from the high temperatures and pressures of the impact, a small fraction, located at Earth's surface, should have been launched relatively intact. Thus, we can have confidence that step 1 is plausible and even likely, at least for planets of Earth's size or smaller.

Earth has diverse surface environments. Most of its surface is covered by oceans, but a large impactor can easily slice through them and launch pieces of the ocean floor into space. Life is found on every type of surface on Earth. It is most abundant and diverse in the tropical forests. It is least abundant on the ice fields and the dry deserts. Many organisms will be alive when the impactor hits, but there will also be spores, seeds, and organisms in a dormant state. Smaller organisms are more likely to survive the impact shock than larger ones.

Scientists have studied the survival of many kinds of small organisms in the environment of space. Two intensively studied species are *Bacillus subtilis* spores and *Deinococcus radiodurans* bacteria, one of the most radiation-resistant organisms.[9] Experiments have been conducted by placing samples on the outer walls of the International Space Station, showing these organisms can survive the cold and vacuum of space.[10]

The transport of viable organisms ejected from planetary surfaces occurs via lithopanspermia and radiopanspermia. Since many of the ejecta fragments from large impacts are small, down to tiny dust grains, radiopanspermia should transport large numbers of organisms. Radiation pressure from the sun exerts a force on dust grains larger than gravity for grains smaller than about 0.2 microns.[11] Thus, radiation pressure would not be able to transport any organism larger than this size. Most bacteria are larger than this limit; the smallest of the ultramicrobacteria are close to 0.2 microns, as are the smallest plankton.[12] Interestingly, this size limit corresponds to the division between the viruses and other forms of life. While a few of the smaller species of bacteria and plankton are small enough to be propelled by radiation pressure in space, there would be no room for shielding from radiation. From this we can conclude that radiopanspermia can transport nothing larger than viruses while at the same time providing some protection. Radiopanspermia could potentially transport viruses to planets and moons in the solar system outside Earth's orbit and even beyond the solar system.

Lithopanspermia can transport larger organisms, but the typical transport times within a planetary system are much longer. For example, the typical transfer time of impact ejecta fragments from Earth to Mars is millions of years, but the first transfer occurs just after "only" about 100,000 years.[13] The first transfer to Venus is much faster, taking only about 8,000 years.[14] About 30 times more material is transferred from Mars to Earth than from Earth to Mars.[15]

The greatest threats to life in the space environment are solar high-energy electromagnetic radiation (UV, X-ray) and solar and galactic cosmic ray particle radiation. Microorganisms in the interiors of boulder-size impact ejecta fragments are protected from these hazards if they are larger than about 11 meters in diameter.[16] However, the natural radioactivities of uranium (including U-238, U-235 isotopes), thorium, and radioactive potassium-40 (K-40) in the boulder can pose additional threats to organisms deep in its interior.

Can such large (~11 meters in diameter) chunks of rock be ejected from planetary bodies? Today, very little material is being lost from the terrestrial planets, but the early solar system was different in at least a couple of ways relevant to our evaluation of panspermia. First, the vast majority of ejecta fragments launched from the planets happened within the first few hundred million years of the solar system's history. Large impacts were common, and some were large enough to cause planetary sterilizations. Ironically, the very impacts that caused the sterilizations were most effective in launching viable organisms into space. A significant fraction of the ejecta return to the parent body over a relatively short time period, possibly reseeding it after recovery from the impact.[17]

In addition, the abundances of the long-lived radioactive isotopes were higher on the early Earth (and the other planets). The dose rate for organisms in crustal rocks would have been about 2.5 times the present-day rate.[18] Thus, the timescale for the destruction of organisms in impact ejecta fragments would have been shorter than it is now. The young sun, too, was more dangerous, producing more frequent powerful flares. This would have exposed organisms in the less-protected regions of the ejected boulders to more intense radiation.

The third step poses the challenge of reentry. Ejecta boulders large enough to provide sufficient protection from external radiation reach the surface of a planet, even if the planet has a thicker atmosphere than Earth's. The atmospheric heating does not affect the interior of the boulder, given the short duration of the heating. The impact is what poses the greatest threat. However, survival is still possible for two situations. First, impactors entering at shallow angles less than about 30 degrees relative to the surface will have a sufficiently small vertical component of the impact velocity to permit a small fraction of microorganisms to survive.[19] Second, impact on water is much more gentle. It is likely Venus, Earth, and Mars all had oceans and lakes early on. From these considerations, then, a substantial fraction of viable organisms within impacting ejecta should survive on these worlds.

The final step requires the newly arrived organisms to begin colonization. In ideal laboratory settings, biologists can quickly grow microbial cultures starting with one or a few individual organisms. However, the conditions on the target world are likely to be far less than ideal for growth. The new environment might be very different from the organism's habitat on the home world. For example, a thermophile from a hot spring on Earth landing on one of the ice caps of Mars will be out of place and unlikely to survive for long. Or, perhaps the pH and/or salinity of the body of water an organism arrives in is outside its growth range.

A second important factor to consider is the ecological support community of an organism. An organism might have been part of a broader ecological community on its home world. Any arriving viable organisms in a given ejecta fragment are likely to be very few in number; they are effectively on their own. If the new arrivals are not able to have access to all the needed nutrients provided by the ecological support network on their home world, then they will not survive. While difficult to quantify, this would seem to be a major obstacle to infecting another world via panspermia. On the other hand, due to the short transfer times and amount of material present, bringing life back to a world sterilized by a giant impact is much more likely.

Today, Earth remains, to our best knowledge, the only inhabited planet in the solar system, and there is fossil evidence for going as far back as 3.7 billion years ago.[20] It is probable that large impacts on the Earth

during its first billion years delivered viable microorganisms to Mars and Venus. The early history of Venus is currently unknown, but we have learned much about Mars's history. There is no convincing evidence that life once existed on Mars or is presently living, although we continue to search for such. This is in spite of the facts that Mars is a close neighbor and has the most Earthlike climate of any other body in the solar system. This is an important lesson to keep in mind as we venture to consider more distant worlds.

Interstellar Panspermia

Recent discoveries have motivated some scientists to reconsider the possibility of transport of life between planetary systems (*interstellar panspermia*). For instance, we now know that planets orbit many nearby stars. As of this writing, just over 4,000 exoplanets are known.[21] The first exoplanet was found in 1995 around the sunlike star 51 Pegasi, a gas giant with an orbit very unlike those of the planets in our solar system. If this planet were in our solar system, it would orbit far inside the orbit of Mercury! Subsequent discoveries have only reinforced the diversity of planetary systems and how atypical the solar system is.

Another proposed aspect of interstellar panspermia moved from the theoretical to empirical category with the discovery of the first interstellar bodies passing through the solar system. On October 19, 2017, astronomers discovered Oumuamua as it passed relatively near the Earth.[22] It was estimated to be a few hundred meters in size and elongated. It did not display signs of cometary activity. On August 30, 2019, the first interstellar comet, 2I/Borisov, was discovered as it passed through the solar system.[23] Both objects are expected to pass through the solar system without being captured or colliding with another body. These discoveries proved once and for all that solid bodies can traverse the vast distances between planetary systems and thus serve as potential vehicles for the transport of life.

However, these two objects would not be carriers of microorganisms because they are probably not impact ejecta fragments. They presumably formed in other planetary systems like our solar system's asteroids and comets and were lost when they came too close to a massive planet.

I already noted earlier that radiopanspermia can easily accelerate small grains to escape the solar system. Only viruses are small enough to have modest protection in grains while they are being accelerated by sunlight. There are two severe problems with viruses riding along on small grains. First, upon reaching an exoplanet and entering its atmosphere, the grains and their viral contents will completely burn up. Second, viruses require a host to reproduce. By themselves, they are not going to spread life to another planet. For these reasons, in the following I will focus only on lithopanspermia as a means of the transport of microorganisms between planetary systems.

Lithopanspermia is effective at transporting viable microorganisms between close planets in a given system, but how can it transport them between planetary systems? The short answer is close encounters of the ejecta fragments with planets. Just like NASA uses the "gravity assist" maneuver to give a space probe a speed boost, close misses between ejecta fragments and planets alter their paths and can increase their speeds. In particular, close encounters between ejecta fragments and Jupiter can eject the fragments out of the solar system. To generalize the problem for any pair of "source" and "target"

planetary systems in the Milky Way galaxy, I will not just limit the discussion below to the solar system as the source of ejecta fragments but will also review the diversity of source systems.

It is not possible to come up with a general rule for all planetary systems because they are so diverse in their architectures. Still, there are some general trends that are relevant to our discussion. First, the exoplanet occurrence and multiplicity rates are greater for low-mass (red dwarf) stars. Also, red dwarfs are far more numerous than solar-mass stars. All else being equal, it would seem that red dwarfs would be the greatest sources of lithopanspermia. But all else is not equal. Gas giants like Jupiter are very rarely found around red dwarfs. What's more, the escape velocity from the habitable zone region of a red dwarf is larger. For these reasons, red dwarfs, although far more common than sun-like stars, should be much weaker sources of impact ejecta fragments for lithopanspermia.

Numerical simulations show that anywhere from 5 to 20 percent of the impact ejecta from Earth escape the solar system; for Mars the fraction escaping material is a little higher at 15 to 20 percent.[24] This process is almost entirely facilitated by Jupiter. This is a significant amount of material, but the more important metric is the travel time. It takes anywhere from just under 700,000 to 4 million years for the first Earth ejecta fragment to leave the solar system; the *typical* timescales for solar system exit after ejection from Earth is 10 to 50 million years.[25] I'll round it off to 30 million years.

Many planetary systems have gas giant planets in smaller orbits than Jupiter's. If Jupiter were moved to the present location of Mars, it would shorten the median time of ejection from the solar system to a few million years.[26] However, Jupiter's location has a strong effect on the formation and stability of the terrestrial planets as well as the asteroid belt. It is unlikely Earth would have been habitable if Jupiter had formed so close.[27] Nevertheless, to account for the possibility of a modestly smaller orbit for Jupiter, I'll adopt 10 million years for the typical loss timescale of a typical source planetary system with a Jupiter-mass planet.

The next step is the transfer of the fragments to a nearby planetary system. How far do they need to travel? In our region of the Milky Way galaxy, stars are typically separated by a few light-years. At typical departure velocities from the solar system of about 5 kilometers per second,[28] the trip to a star 5 light-years away will take about 300,000 years. This is optimistic, since it assumes a straight-line trajectory from the solar system to the target system. This will be the case for a few lucky pieces, but the typical trajectory will be a far more circuitous one.

Recall that most impacts on planets take place early on. Like most stars, the sun was very likely born in a star cluster. We don't know how long it remained there. However, it is known that most clusters disperse within 10–20 million years after their birth, and about 10 percent last up to 500 million years and a few last even longer.[29] The stars within a cluster are far more likely to exchange impact ejecta fragments with each other than field stars are. It's not so much the tighter packing of stars in a cluster that helps, but rather their relatively slow motions relative to each other.[30]

The strong bias of interstellar panspermia for young systems places strong constraints on the timing of the origin of life. The origin of life needs to take place early in a planet's history if it is to serve as a source of ejecta fragments for lithopanspermia. It did happen early enough in our solar system for it

to serve as a source. If life is rare in the Milky Way galaxy, and if it tends to start later in a planet's history than it did on Earth, then lithopanspermia sources will be few and far between. Even if life always arises early, there will be only a relatively short window of time available for impact ejecta with viable organisms to be useful for interstellar panspermia.

The next step is the capture of the ejecta fragments by the target system. The capture by a binary star system (i.e., two stars) is about 100,000 times more likely than it is for a single star with an orbiting Jupiter.[31] And, capture by a single star with a Jupiter is similarly more probable than a single star without a Jupiter. These facts have an important implication for the theory that the solar system was seeded by interstellar panspermia. For every planet orbiting a single star with a Jupiter that was seeded by interstellar panspermia, there are 100,000 planets in stellar *binary systems* that were seeded. In other words, *if Earth were seeded by interstellar panspermia, then we should be very surprised to find ourselves living around a single star.* Binary star systems are expected to be less habitable than single star systems, but some types of binaries should be compatible at least with single-celled life and perhaps even with complex life.

A direct hit from an interstellar body with a planet can occur, but its probability is very low. Otherwise, a collision with an interstellar body is a multistep process. It begins when an incoming interstellar object has a close encounter with a star or planet that significantly alters its path. If it is captured into a closed orbit, then the fragment will continue orbiting within the system until it collides with another body in the system. Its orbit is likely to be more eccentric and larger than the orbits of any habitable planets in the system. This means that a collision with a habitable planet will likely occur at relatively high velocity. The timescale for collision with a planet should be comparable to the timescale for the loss of the impact fragments from the source system, about 10 million years.

The total timescale from the moment of ejection of fragments to their capture by a planet in another system is, at minimum, 10 + 10 + 0.3 million years. For field stars, this timescale is true for only a tiny fraction of the ejecta; the vast majority of fragments will have much longer transit times. This timescale will be approximately true for the typical transfer between stars in a cluster.

It is now clear that seeding single-field stars via capture of viable impact ejecta fragments is very unlikely.[32] In recent years, panspermia research has focused instead on binary star systems and star clusters. Binary stars within star clusters are by far the most likely systems to capture impact ejecta fragments, followed by single stars with Jupiters in clusters. But, even if some fraction of the planets in a given cluster can be seeded via lithopanspermia, it is very unlikely that systems outside the cluster will be seeded. Certainly, as stars are lost from the cluster over time, they will disperse throughout the Milky Way, but they will not seed other planetary systems.

So, is interstellar lithopanspermia possible within the restricted confines of a star cluster? The transit times are a few tens of millions of years. Can organisms survive such a long journey and establish a new colony on an alien world? With every step, an ever-smaller fraction of the original organisms survive. A critical parameter is the survival time of the organisms. The number of delivered organisms depends on the survival time raised to the fourth power.[33]

There have been reports of viable organisms being revived after thousands or even

millions of years from ice, amber, or salt crystals. Such experiments are sensitive to contamination. Careful experiments with bacterial DNA show that it degrades on timescales of hundreds of thousands of years.[34] However, there are some credible studies demonstrating revival of ancient life. Frozen nematodes have been revived from permafrost dated between 30,000 to 40,000 years old.[35] Prokaryotes have been cultured from fluid inclusions in halite crystals from core samples in dry lake beds several tens of thousands of years old.[36] In this case, the authors suggest that the prokaryotes were part of a trapped microbial community with single-celled algae that mutually benefitted.

Aside from desiccated and dormant microorganisms, then, two other possible ways of keeping life viable are in a frozen state and as living micro-communities in halite crystals. While it is possible that permafrost can be blasted into space partly intact during a large impact, it is likely to be broken up into smaller bits given that its strength is less than that of rock. But the bigger challenge is enduring the space environment. Exposure to sunlight near Earth's orbit will cause the ice in the permafrost fragment to sublimate quickly. Even if a fragment survives its transit to another planetary system, it will undergo additional sublimation when it is within the habitable zone of that system. As it shrinks, a permafrost fragment will offer ever less shielding to imbedded organisms. Clearly, this is not a viable path for lithopanspermia.

On the other hand, fluid inclusions in halite require sufficiently warm temperatures to remain liquid. When a halite fragment leaves the vicinity of Earth's orbit, the inclusions will be too cold to remain liquid. Once frozen, the DNA will begin degrading. The range of ages of halite crystal deposits examined for viable organisms was 10,000 to

150,000 years; successful cultivation of the prokaryotes in the inclusions was only 0.4 percent.[37] The researchers concluded that "most ancient prokaryotes in halite are dead or viable but not culturable, or that our culturing conditions were simply not suitable."[38]

This gives us a glimpse of the difficulties of achieving step 4, the colonization of another planet. Even in a controlled laboratory setting with intelligently guided experiments, the cultivation of ancient microorganisms is barely possible.

It is probable that Earth delivered viable organisms to Mars and Venus early on. We know Mars had oceans, and we can speculate about Venus. We don't know if these interplanetary migrants were able to colonize these worlds, but we do know they are dead now. If they had life, it was short-lived. Given this, panspermia can increase the probabilistic resources for origin of life by a modest factor, to three planets, within a solar system. It is possible that panspermia can transfer life between neighboring stars very early in the development of a star system. But for the vast majority of the lifespan of the vast majority of stars, panspermia is not a viable model for transferring life. Thus, our solar system remains in quarantine relative to the rest of the universe.

Does the Multiverse
Refute Cosmic Design?

Bruce L. Gordon

In physics and cosmology, *MTV* doesn't refer to the cable TV channel—it means *Multiple Trials for Viability*.[1] Universes from nothing (quantum cosmology) and initial conditions for free (inflationary cosmology) laws and constants on the MTV (the string landscape) are the elements of the dire straits multiverse, the last refuge of metaphysical naturalism. No topic moves to the heart of whether matter or mind is more basic to understanding reality than the origin of the universe and its fine-tuning for the existence of life. Our task is to evaluate the work of cosmologists and theoretical physicists who are trying to explain cosmological origins and fine-tuning in purely naturalistic terms. We'll consider the explanatory adequacy of the many-worlds interpretation of quantum cosmology, innumerable bubble universes in inflationary cosmology, and the embarrassment of riches constituted by the solutions (vacua) of string theory. In the end, transcendent intelligent agency will be recognized as the only causally sufficient and metaphysically sound explanation—one that is more parsimonious, elegant, and resonant

with meaning than all the ad hoc machinations of multiverse cosmology.

Quantum Cosmology, Parallel Worlds, and Universal Origins

Quantum gravity and quantum cosmology investigate the intersection of general relativity and quantum physics. General relativity deals with the very large and quantum physics with the very small, but they overlap at the beginning of the universe and in black holes. In the late 1960s, Roger Penrose and Stephen Hawking showed that every path backward through spacetime leads to a beginning—a singularity—from which matter, energy, and spacetime itself emerged. Since singularities are spacetime regions of infinite curvature and matter density, they require quantum-mechanical treatment. *Quantum gravitational* theorists try to reconcile general relativity with quantum physics, but the incompatible mathematical structures of the two theories and the different ways they treat space and time make it difficult to do so. *Quantum cosmologists* treat the whole universe quantum-mechanically in the hope of removing singularities and understanding the "universal wavefunction" resulting from the quantum entanglement of physical reality as a whole emerging from the big bang.[2] Universal wavefunctions are solutions of the Wheeler-DeWitt equation, which serves heuristically as a general-relativistic analogue of the Schrödinger equation in ordinary quantum mechanics. In principle, universal wavefunctions represent all the information about the geometry and matter content of the universes they describe.[3] In addressing how quantum cosmology is bound to the many-worlds interpretation of quantum physics and what we should make of this, we'll examine quantum tunneling models for universal origins, loop quantum gravity and cosmology, and the idea that spacetime itself emerges from quantum entanglement.

Trying to explain the origin of the universe through quantum cosmology—as a precursor to explaining its initial conditions with inflationary cosmology and its laws and constants with the string landscape—presupposes the timeless preexistence of a quantum-gravitational vacuum where standard notions of space, time, energy, and entropy have no meaning.[4] This primordial vacuum did not need to exist—it is *contingent*. If our universe sprang from it, it most surely was not *nothing*, and an explanation for its existence is needed. That there would be such an explanation follows from the principle of sufficient reason.[5] The existence of a primordial vacuum would have to be explained by something that exists necessarily, transcends spacetime and physical reality, and is capable of bringing the foundations of any universe or multiverse into existence. We're well on our way to theism here, so note that Christians have *nothing* to fear from multiverse

cosmology. Nonetheless, it's both instructive and amusing to watch metaphysical naturalists trip while trying to lift themselves by their own bootstraps, so put on your thinking cap, grab some popcorn, and let's get started.

Tunneling Models

In tunneling models, universes pop into existence out of a primordial quantum vacuum. This is often described as tunneling from "nothing," but it obviously *isn't*. The models for this process are *semiclassical* because *quantum* matter and radiation fields are linked to *classical* gravitational fields. In the 1980s, Alexander Vilenkin developed a model of this sort, as did James Hartle and Stephen Hawking.[6] The mathematical space where universal wavefunctions live and out of which universes tunnel is the pairing of *all* curved three-dimensional spaces with every possible configuration of matter.[7] The universal wavefunction assigns a complex number to every point in this space.[8] Every path through this space describes a four-dimensional spacetime and its matter configuration, and any strictly increasing path provides a time measurement describing a possible "history" of that universe. Such descriptions deviate from paths representing classical spacetimes because the wavefunction must oscillate rapidly to make the "right" quantum-mechanical correlation between and development of spacetime curvature and matter variables highly probable. These correlations aren't unique but correspond to whole *families* of classical histories (paths) in superposition.[9] How do quantum cosmologists look for a suitable universal wavefunction? So far, either by inventing "natural" constraints on the "boundaries" of the mathematical space, or by constructing a "natural" algorithm for computing the universal wavefunction that would contribute to its uniqueness. The first approach is taken by Alex Vilenkin, and the second, in partnership with Jim Hartle, by Stephen Hawking (1942–2018), one of the most famous physicists since Einstein.

Vilenkin proceeds by inventing constraints on the mathematical space where the universal wavefunction lives. Roughly described, his proposal restricts the oscillatory behavior of the wavefunction to include only the paths of a family of classical solutions of Einstein's equations and orients them so they are moving *out* of the mathematical space. This means—by theoretician's fiat—that classical spacetimes can *end* in a singularity, but not *begin* in one. As for where the wavefunction *begins*, Vilenkin proposes it emerges from the rather ill-defined boundary between regions *in* the space where the wavefunction oscillates (and hence where an underlying classical picture is possible) and regions where it does not (which are purely quantum-mechanical in character). These regions where no oscillatory behavior is possible depend on parameterizing the wavefunction by a complex time variable (imaginary time) and a precise choice of matter distribution and interactions. The real time of possible universal histories therefore "begins" at a boundary inside the mathematical space between nonoscillatory and oscillatory regions where acausal "quantum tunneling" transitions the wavefunction from imaginary to real time. After this, the wavefunction, constrained to include only oscillations involving classical spacetimes, moves *outward* through the space to terminate on its boundary—that is, on a point *not* in the space. If you think this sounds rather arbitrary and vague, you're not wrong.

The famous Hartle-Hawking "no boundary" proposal is different. Rather than specifying conditions on the boundary of the mathematical space, Hartle and Hawking instead choose a specific algorithm to compute the universal wavefunction. This algorithm involves changing from real to imaginary time[10] so that finite answers result when the sum over possible histories of the universe, restricted by boundary conditions specifying certain paths over closed and bounded 4-geometries, is taken between a point and a finite 3-geometry.[11] As a first approximation, this involves restricting the number of degrees of freedom of the gravitational and matter fields in the mathematical space, then solving the Wheeler-DeWitt equation for this simpler case.[12] The resultant picture is that of a universe tunneling out of a "minimal temporal radius" in imaginary time, expanding in an inflationary and then matter-dominated fashion to a maximum radius, then recollapsing to a singularity.[13]

What should we say? Both of these approaches involve an *infinite winnowing* of mathematical structures through the precise selection of boundary conditions to establish the right relationship between matter variables and the curvature of space and render the geometry of our universe probable. Absent a priori reasons to privilege certain boundary conditions and algorithms as "natural," this infinite winnowing constitutes fine-tuning *after the fact* to create mathematical models that hold some chance of *describing* our universe, but no chance of *explaining* it. In addition to post facto fine-tuning, the whole procedure ignores the central issue of sufficient causality. Mathematical descriptions do *not* bring universes into existence. As Stephen Hawking remarked in a metaphysically lucid moment, "What is it that breathes fire into the equations and makes a universe for them to describe?"[14] In short, quantum cosmology is a metaphysical nonstarter if it lacks a transcendent information-winnowing catalyst.

Despite all the fine-tuning courteously provided by the theoreticians themselves, these models, by using the Feynman sum-over-histories approach to calculate the universal wavefunction, still require the many-worlds interpretation (MWI) of quantum physics and its associated multiverse. Interpreted realistically—which it must be if it is to have any explanatory value—this approach describes our reality as emerging by decoherence from the superposition of multiple parallel realities in the universal wavefunction.[15] Parallel worlds split from each other after they decohere because they are no longer capable of influencing each other. The MWI, also known as the Everett interpretation,[16] raises issues of technical concern in addition to the metaphysical worries it generates. The first technical concern is making sense of outcome probabilities. Everything that *can* happen, quantum-mechanically speaking, *does* happen in the MWI. How, then, do we make sense of different quantum outcomes having *different* probabilities if everything *must* happen? How could multiple mutually exclusive outcomes *each* have 100 percent probability? Efforts to resolve this issue usually propose that quantum probabilities arise from consciousness being bound to *one* world, then analyze quantum probabilities using decision theory.[17] A second technical issue is the "preferred basis" problem. Unless position is privileged in analyzing a quantum system, objects can have multiple simultaneous locations, something we do not observe. But there's no reason, from a purely quantum perspective, to privilege position this way. Defenders of MWI argue that this difficulty is resolved by decoherence, which quickly localizes the reality we experience.[18] Interpreted realistically,

the MWI also creates many problems about the nature of self and our knowledge of the world as reality divides exponentially. There is another way of looking at such questions in theistic metaphysics, however, that mitigates this concern. We'll discuss it when we deal with the idea that spacetime emerges from quantum entanglement.

Let us consider also the Hartle-Hawking procedure, which involves transforming to imaginary time. Once the procedure has been performed, the universe doesn't have a singularity at the beginning of time. Of course, the procedure was used as an *instrumental* expedient and Hawking speaks of it as such, then later interprets it *realistically*, remarking that if the universe has no beginning, "What place, then, for a Creator?"[19] This is disingenuous. First, when the transformation is reversed so the solution describes our reality, the singularity *reappears* and the universe *does* have a beginning. Second, even if the universe *didn't* have a beginning, its existence would still be contingent and need explanation. The theist can argue cogently that the best explanation would be God himself as the timeless and necessarily existent transcendent cause of a contingent universe with no temporal beginning.

Finally, the Hartle-Hawking model has recently come under severe criticism as new mathematical techniques for doing the requisite calculations without transforming to imaginary time have been discovered. Job Feldbrugge, Jean-Luc Lehners, and Neil Turok (FLT), who developed the new model, argue that any realistic approach to quantum cosmology must only consider paths over *real* time variables.[20] In the Hartle-Hawking but not the Vilenkin tunneling approach, the sums over paths in real time *always* blow up to infinity, something that doesn't happen with the new techniques. In light of this,

FLT argue the Hartle-Hawking approach fails miserably—it's just not possible for a universe to start quantum-mechanically in imaginary time. James Hartle, Jonathan Halliwell, and Thomas Hertog disagree, as Stephen Hawking presumably would too.[21] They argue no choice of path is more "physical" than any other; it's a mathematical tool that should be adjusted instrumentally to the greatest advantage.

Of course, consistently adopting this attitude prohibits any metaphysically realistic conclusions about the significance of quantum cosmology, negating all of Hawking's popular pronouncements on the theological significance of the no-boundary proposal. Indeed, without a realistic interpretation, semiclassical quantum cosmology has *no* explanatory value. Even so, it's not as though the FLT approach is unproblematic. Their model requires unsuppressed and unbounded primordial gravitational wave fluctuations, which doesn't match our universe either. Furthermore, semiclassical approaches assume the existence of a coherent theory of quantum gravity we do not yet possess. When and if we do have one, it may not be amenable to any of these procedures. As theoretical physicist Juan Maldacena has remarked, seeking the wavefunction of the universe "is the right kind of question to ask," but "whether we are finding the right wavefunction, or how we should think about the wavefunction, is much less clear."[22]

Loop Quantum Gravity and Cosmology

Loop quantum gravity (LQG) has produced some interesting results in cosmology that merit our attention.[23] Unlike semiclassical approaches, LQG searches for a coherent quantum theory of gravity before trying to apply quantum physics to cosmology. Some preliminary concepts are needed. Every quantum particle has a *spin angular momentum* that's peculiar to the interaction between quantum mechanics and relativity and has no classical analogue. It's represented by vectors whose movement in curved space generates linked systems called *spin connections*. These systems of connections can be used to replace the spacetime metric in general relativity, something discovered by Amitabha Sen in the late 1970s and brought to fruition in Abhay Ashtekar's spin-connection reformulation of general relativity in 1986. This reformulation introduced the *Ashtekar variables* that are central to LQG.

Roger Penrose invented *spin networks* in the early 1970s, a variant of which is also central to LQG. Given a collection of quanta with spin angular momenta, a spin network can be constructed as a set of nodes joined by links carrying an integral number of them. Penrose showed that networks with a large total angular momentum establish directions against which the orientation of other large networks can be measured, yielding an emergent space. LQG uses this procedure, interpreting the nodes as discrete quantized volumes of space and the links as discrete quantized areas where adjacent volumes touch. These LQG spin networks are static, so introducing time requires stacking the networks so evolving relationships between the nodes and links can be constructed. This turns the nodes into edges and the links into faces of quantized volumes of space, producing what LQG theorists call *spin foam*. Feynman's sum-over-histories approach to quantum physics is then used, as it was in the semiclassical models, to trace the development of spin foam through time, and spacetime emerges as a superposition of spin foams that are universal wavefunction solutions of the Wheeler-DeWitt equation.[24]

One of the predictions of LQG is the

existence of discrete quanta of area and volume at the Planck length (1.6×10^{-35} meters). This leads to some interesting results in *loop quantum cosmology* (LQC). Most interestingly for our purposes, LQG/LQC predict or entail: (1) The big bang singularity is avoided because space cannot be compressed smaller than the volume represented by one cubic Planck length, the smallest physically possible volume; (2) our universe resulted from a "big bounce" rather than a big bang; and (3) there was a short period of superinflation induced by the bounce that continues as the mass-energy density of the universe falls from its critical bounce density to half this value.

These consequences generate some pressing questions. One concern is whether there is any evidence that spacetime is *discrete*. In quantum gravity, normal geometry is quantized at distances shorter than the Planck scale, but this *doesn't* mean that the geometry *has* to become discrete. While some quantities can only have discrete values in quantum physics, not *all* observables have to possess a discrete spectrum. Whether *positions* or *distances* are discrete is an *empirical question*. Some argue this question has already been resolved in the negative. Notably, discrete theories of spacetime usually entail the time required for photons of different energies to reach the Earth from distant places will be different, violating special relativity.[25] Nonetheless, various measurements indicate such violations have to be much smaller than discrete theories of spacetime—including LQG—would require them to be.[26] Aside from the fact that LQG theorists, anticipating this consequence, have emphasized that violations of Lorentz invariance are merely *possible*, not necessary in LQG,[27] they can also show how LQG avoids violating special relativity,[28] indicating a potential connection to string theory in the process.[29]

Importantly, LQC is not derived directly from LQG. Rather, LQC *assumes* the universe is homogeneous and isotropic and thus symmetric *before* the loop quantization. If one began with LQG and sought homogeneous and isotropic solutions matching our universe, there's no guarantee one would find them. So fine-tuning is *presumed* rather than explained away by the tidy results of LQC (a big bounce, superinflation, and so on). What's more, even though LQC begins with a fine-tuned homogeneous and isotropic universe that, *if* it contracts, will bounce and undergo bounce-related superinflation, it must still transition to slow-roll inflation to address the fine-tuning present in the horizon and flatness problems.[30] As we'll see when we discuss inflationary cosmology, inflation *increases* the fine-tuning problem. In particular, the initial entropy fine-tuning of our universe (one part in 10 to the 10 to the 123rd power) is made exponentially worse by superinflation and slow-roll inflation. Inflation overcomes this by invoking *unlimited* resources guaranteeing *every* possible outcome and destroying scientific rationality in the process. It fills pothole-sized fine-tuning by creating Grand Canyon-sized fine-tuning, then tells us not to worry because it also creates an infinite amount of dirt to fill the hole.

Finally, aside from being fine-tuned, there could only be a finite number of bounces in LQC because thermodynamic constraints prevent an endless succession. The process had a beginning. Furthermore, our universe isn't going to collapse and bounce in the future. Indeed, if the intent of LQC is to avoid universal beginnings through an infinite cycle of bounces, it fails spectacularly. Inflation is *incompatible* with future collapse, but LQC bounces entail an early period of *superinflation*. Furthermore, while universes *would* bounce if they ever contracted, there's nothing in LQC that makes contraction remotely plausible.[31] LQG and LQC thus have a long way to go. And even if the research program eventually proves successful, the principle of sufficient reason still requires that it have a transcendent ground.

Spacetime Emergent from Quantum Entanglement

A new approach to quantum gravity, ER = EPR, connects wormholes in general relativity with quantum entanglement. Wormholes in spacetime, technically known as Einstein-Rosen (ER) bridges because they were first discussed in a 1935 paper written by Albert Einstein and Nathan Rosen,[32] are a staple of science fiction. In theory, if you fell into one side of a wormhole, you'd pop out the other side almost instantaneously, even if it were on the opposite side of the universe. And wormholes aren't just portals through space, they're portals to different times as well. Quantum entanglement, on the other hand, occurs when quanta interact or share spatial proximity, entwining their wavefunctions so the state of individual quanta cannot be described independently of the state of the others, even when separated by a large distance. This means that changing one part of an entangled system instantaneously changes all of it, even when the distance involved would *not* allow a signal subject to the limiting velocity of the speed of light to account for the correlation.

Albert Einstein, Boris Podolsky, and Nathan Rosen (EPR) discussed this phenomenon in another famous 1935 paper.[33] In 2013, Juan Maldacena and Leonard Susskind conjectured that nontraversable wormholes in general relativity were equivalent to a pair of maximally entangled black holes.[34] Extending this conjecture has led

to a research program in quantum gravity whereby spacetime structure emerges from the entanglement of quantum information.[35] The conjecture is indirectly evidenced in that certain characteristics of ER bridges have the same nontrivial mathematical properties as measures of degrees of entanglement in quantum systems.[36]

Before we discuss a model for time's emergence through entanglement in the universal wavefunction, let's talk about the fusion of space and time into *spacetime*. In spacetime, the measurement of time's passage is relativized to the reference frame (inertial frame) of the observer and affected by that frame's state of motion and the intensity of the surrounding gravitational field. Observers in *different* reference frames may experience the *same* spacetime event as present, past, or future; the experience of time is an *artifact* of one's inertial frame *within* spacetime. The standard relativistic interpretation of this is that *all* past, present, and future instants of time *coexist* in universal spacetime, their respective realities timelessly contained in the whole. From a global perspective, the universe is timeless and static; from the perspective of any local observer's reference frame, however, it is dynamic—time flows, and there is a past, a present, and a future.

How is the global view reconciled with the local measurement of time's passage or the human experience of time's flow and our decision-making freedom? Let's go back to the heuristic reconciliation of general relativity with quantum physics in the Wheeler-DeWitt equation, solutions of which give us the quantum-gravitational wavefunction of the universe. In *global* quantum-cosmological perspective, the Wheeler-DeWitt equation describes the universe as a static four-dimensional entity. How does the time we experience arise, then? Donald Page and William Wootters have argued that the time we experience is actually internal proper (clock) time rather than external coordinate time, so the universe can exist in a static quantum-entangled state while its subsystems evolve by an appropriate internal measure.[37] Various technical issues with this proposal have been resolved by Vittorio Giovannetti, Seth Lloyd, and Lorenzo Maccone.[38] What is more, Ekaterina Moreva et al. (2014, 2015) have performed an experiment showing how this mechanism works by creating a quantum state in the laboratory that an *external* observer perceives as static even while mathematically correlated subsystems *internally* measure each other's development.[39] This reconciliation of global and local perspectives produces two points of view: (1) that of a creator/observer *external* to the universe (God); and (2) that of *internal* observers (us). The transcendent creator-observer perceives the whole universe as a static system, but internal observers experience time as flowing from the past into the future. Physics reconciles an eternal static perspective with a temporal dynamic one.

The cosmic histories superposed in the universal wavefunction don't need to be materially substantial as they are in the Everett interpretation. The universal wavefunction can be regarded as a divine bookkeeping device that has its existence solely in the divine mind. Created consciousnesses, by the decisions they make, participate in the divine creation of reality, with divine assistance synergistically forging a unique path through the superposition of possibilities represented by the universal wavefunction. Divine omniscience is represented in the ramified universal wavefunction that exhaustively maps the possibilities inherent in created reality. God not only knows everything that has happened and can or will happen, he knows everything that might have happened and what could have resulted. But there is no need to join many-worlds theorists in the fantasy of genuine parallel realities with real parallel versions of ourselves, for divine direction keeps all consciousness on one synergistically forged path.

What is happening at the level of our psychological experience of freedom, however, especially with respect to its role in grounding moral responsibility? We can avail ourselves of Augustine's astute observation that time perception is a distention of the mind: the present of things past is memory, the present of things present is perception, and the present of things future is expectation.[40] Every moment of our conscious experience is *distended* to include memories of the past and anticipations of the future, all synthesized to provide a basis for our decisions. If we ask how we can have freedom of choice and moral responsibility when the future, with every decision we will make, exists timelessly as a creative act of God, the answer is that we have freedom *in each moment of consciousness*, and our lives are the collection of such moments. Each conscious moment gives us freedom of rational choice within the constraints of our nature, and God eternally perceives what that choice is. If we would have chosen differently—and we are created with the freedom to do so—then God, in his timeless creative act, would have brought a different universe into being. So God includes the chosen actions of all finite moral agents in his timeless creative act and, indeed, he can do so transhistorically so that the effect of a choice at a *given* point in space-time history can, if God so wills, affect *all* of spacetime history. We can conceive the cosmic impact of both the fall and Christ's redemption in this way.

Inflationary Cosmology, Bubble Universes, and Initial Conditions

The idea of cosmic inflation is that, a split second after the big bang, the universe underwent a short period of hyper-accelerated expansion that "smoothed out" our local cosmic environment by pushing any inhomogeneities beyond the boundary of what can be seen.[41] The most viable theoretical model of the inflationary process, chaotic eternal inflation, requires that once inflation starts it never ceases. Inflation thus produces a potentially infinite number of "bubble universes," each with *different* initial conditions, which suggests that bubble universes with initial conditions as fine-tuned as our own are bound to occur repeatedly.

The irony of this proposal is that inflationary processes actually *increase* rather than decrease the initial-condition fine-tuning of our universe. For instance, the energy of the inflationary field has to be shut off with tremendous precision in order for a universe like ours to exist, with inflationary models requiring shut-off accuracies ranging from one part in 10^{53} to as much as one part in 10^{123}, depending on the particular inflationary model in view. More devastatingly, achieving thermodynamic equilibrium in the cosmic microwave background radiation by inflation is an *entropy-increasing process*, yet even without it, as Roger Penrose has shown, our universe's initial entropy is fine-tuned to one part in 10 to the 10 to the 123rd power.[42] Tacking exponential inflationary growth onto the *already* hyper-exponentially fine-tuned entropy required by the big bang has the effect of *exponentially increasing* its already hyper-exponential fine-tuning!

Furthermore, Sean Carroll and Heywood Tam have shown that the chance of inflation actually occurring as part of any *realistic* cosmological history is only one in 10 to the 66,000,000th power.[43] Of course, endless chaotic inflation still gives metaphysical naturalists what they want: a scenario in which these staggering improbabilities don't matter because *every* initial condition, no matter how improbable, is not only realized sooner or later, but infinitely many times! This desideratum comes at a cost, however, that transfers to the *inflationary string landscape* as well. I'll postpone consideration of these costs to the next section and focus here on the peculiarities and explanatory inadequacies of inflation itself.

First, there is no physical theory predicting the form of inflation. Different potentials, and different initial conditions for the same potential, yield different predictions. In practice, the choice of potential and shut-off energies are *reverse-engineered* to fit the data, then put forward as an "explanation" for what is observed. As theoretical physicist William Unruh remarked, "I can fit any dog's hind leg with inflation." It has zero predictive value.[44] Second, Roger Penrose has noted that, quite independently of inflation, expansion from a *generic singularity* can become any irregular universe we please—inflation doesn't explain the universe's flatness *apart* from the independent choice of a fine-tuned metric for spacetime.[45] In other words, after the big bang, there must be a patch of space where quantum fluctuations have subsided so that Einstein's field equations apply. That space has to be flat enough, with a smooth enough distribution of energy, that inflationary energy can grow to dominate every other form of energy. But if that patch were "flat and smooth enough to start inflation, then inflation would not be needed in the first place."[46] Finally, we should note that grafting inflaton fields onto solutions of Einstein's field equations doesn't even guarantee that inflation will happen, as Stephen Hawking and Don Page have shown.[47]

Next, consider the supposed predictive strengths of inflationary models. We've already questioned whether inflation has *any* predictive value, but let's address its predictions about the cosmic microwave background radiation (CMBR). Inflationary models predicted that the CMBR would display a normal distribution of energy density fluctuations having the same spectrum at all scales, a prediction largely confirmed by observation. The difficulty is that this prediction is *not unique to inflation*. The normal distribution also follows from the Central Limit Theorem in statistics, which states that the mean of a sufficiently large iteration of random variables with well-defined means and variances will have a near-normal distribution.[48] A scale-invariant spectrum of energy fluctuations was also proposed for independent reasons by Edward Harrison[49] and Yakov Zeldovich[50] *prior* to the advent of inflationary cosmology. Furthermore, the CMBR data from the *Planck satellite* has created some significant problems.[51] The measured deviation from scale invariance is *tiny*, much smaller than one would expect from the simple inflationary dominating the literature. While the inflationary energy density can be adapted to produce whatever scale-invariant patterns of CMBR energy are observed—in this sense the theory is toothless—the *simplest* models predict a significantly *larger* deviation due to quantum fluctuations. Matching the satellite data requires making implausible ad hoc assumptions. Rather than starting at the Planck energy and slowly rolling downward, to match, the data inflation would have to begin on a flat plateau at an energy density a trillion times *less* than the big bang. Not only does this involve a very precise and

precipitous initial drop, it requires the plateau to give way to slow-roll inflation in an area without gravitational warping or unevening energy because these conditions would prevent inflation from getting started.

Finally, the simplest inflationary models predict cosmic gravitational waves, so if inflation had taken place, quantum fluctuations in the inflaton field would generate random warps in space propagating as spatial distortions across the universe when inflation ended. These gravitational waves would have distinctive polarizing effects on hot and cold spots produced in the CMBR. But these effects *have not been found* in the COBE, Planck, or BICEP2 data, despite an ongoing search since 1992. Excitement over the polarized CMBR detected by BICEP2 at the South Pole in March 2017 quickly dissipated when its proper explanation was found to be cosmic dust in our own galaxy.[52] The absence of CMBR polarization caused by cosmic gravitational waves, which should be readily detectable with modern instruments, "strongly disfavors" the simplest inflationary models, requiring them to be increasingly "arcane" to match the data.[53] In short, the inflationary mechanism has deep problems with arbitrariness and empirical tenability. As we will see, the costs it inflicts on scientific rationality are also prohibitive.

The String Landscape and the Laws and Constants of Nature

How do metaphysical naturalists address fine-tuning in the form of the *laws of nature* and the values of the *constants of nature*? Only one cosmological theory—the string landscape—offers mechanisms aimed at explaining this kind of fine-tuning. What naturalists need are blind universe-creating mechanisms producing universes with an endless variety of laws, constants, and initial conditions so our universe could be the chance outcome of these operations. An *observer selection effect* would then explain our existence: given that we exist, we must be in a region of the multiverse having conditions compatible with our existence. This is how the "anthropic string landscape"[54] functions as a "solution" to the naturalist's fine-tuning problem.

To understand the mechanisms, we need to note a few facts about string theory. String theory has received a lot of attention in the last 40 years as a potential "theory of everything" that could unite the four fundamental forces of nature—gravity, electromagnetism, the weak force, and the strong force—as manifestations of a single fundamental force: gravity. It postulates the fundamental constituents of nature are one-dimensional filaments instead of particles. These filaments are either open-ended or closed into loops and they vibrate in different ways to produce the different "particles" we observe. To allow for both radiation and matter while satisfying the rules of quantum mechanics, string theory must exhibit supersymmetry and strings must move in ten-dimensional spacetime.[55] The extra six spatial dimensions are "compactified" at each point of spacetime into structures so small they can't be observed. Unfortunately, there are *infinitely many ways* of folding extra spatial dimensions into unobservable structures. In string theory, the *shape* of each such compact structure dictates the *form* of the laws of nature in the three large dimensions, and the relative *sizes* of the curled dimensions dictates the *strength* of the physical constants. Every possible compactification represents a universe with different natural laws and constants that collectively form an infinite *landscape* of universes, each having a *different* physics.

How do you turn the *vice* of a theory with infinitely many solutions into a *virtue* explaining away the fine-tuning of our universe? It turns out there are somewhere between 10^{500} and 10^{1000} compactifications with a *positive* cosmological constant.[56] Since our universe's cosmological constant is positive and fine-tuned to one part in 10^{120}, the just-so story[57] told by string landscape theorists runs like this: The multiverse containing our universe started in the highest energy state for the cosmological constant (for the explanation to work) and, through random quantum decay of initial compactification features, cascaded down the energy landscape, each sequential branching decay launching inflationary bubble universes with *different* laws and constants. By such means, it is conjectured (without justification) that the *whole* landscape of different laws and constants might be explored. Our universe, with its properties fine-tuned for the existence of life, would then be explained by an observer selection effect: While there are infinitely many universes in the landscape incompatible with life, we exist, as we must, in a region compatible with our existence. So living in a universe with finely tuned conditions *necessary* to our existence is not a surprise.

Are string landscape theorists justified in pinning their hopes on this approach? Handling the fine-tuning of initial conditions, laws, and constants requires fusing inflationary cosmology with string theory, compounding the inadequacies and improbabilities of each. We have seen that inflationary cosmology requires fine-tuning that exceeds the fine-tuning it was invoked to explain, an irony dubiously mitigated by its unlimited resources. We now examine further difficulties with inflation before detailing the implausibilities of string theory and arguing that a naturalistic resolution of the fine-tuning issue is impossible in principle.

1. Arvind Borde, Alan Guth, and Alexander Vilenkin have shown that all inflationary multiverses have a finite history.[58] Chaotic inflationary models can have an endless future, but they must have a beginning, so their implications are the same as standard big bang cosmology. The inflationary string landscape obviously also satisfies this constraint. Furthermore, the timeless primordial quantum vacuum required to spawn all of this requires a transcendent explanation that puts us on the path to theism, making design hypotheses a natural extension of the evidence.

2. Swamping the fine-tuned improbabilities with infinite resources undermines scientific rationality. In a materialist multiverse resting on the hypothesis of an undirected and irreducibly probabilistic quantum inflationary mechanism, anything that *can* happen—no matter how improbable—*does* happen with unlimited frequency. This generates what physicists call the *measure problem*. It undermines confidence that the future will resemble the past in a way that subverts the inductive inferences that make science possible. In short, the inflationary multiverse *undermines* the very possibility of scientific rationality. MIT theoretical physicist Max Tegmark expresses the problem this way:

> When we try to predict the probability that something particular will happen, inflation always gives the same useless answer: infinity divided by infinity... [D]espite years of teeth-grinding in the cosmology community, no consensus has emerged on how to extract sensible answers from these infinities. So, strictly speaking, we physicists can no longer predict anything at all! This means that today's best theories need a

major shakeup by retiring an incorrect assumption. Which one? Here's my prime suspect: ∞.[59]

3. Beyond this, the measure problem reveals our place in reality to be very special, undermining the whole purpose of the multiverse. We see this in the Boltzmann Brain Paradox[60] and the Youngness Paradox.[61] Briefly, if the inflationary mechanism generates an infinite multiverse, then with probability indistinguishable from one, the typical observer in such a multiverse will be a spontaneous fluctuation with memories of a past that never existed (a Boltzmann brain) rather than observers like ourselves. Similarly, post-inflationary universes will overwhelmingly have *just* been formed, so a universe as old as our own has effectively zero probability. Either way, if our universe were part of an inflationary multiverse, it wouldn't be *typical*, but infinitely improbable in its age and compatibility with stable life forms. The fact that we're *not* Boltzmann brains and we live in a stable universe that is an estimated 13.7 billion years old severely undermines the multiverse theorist's goal of showing that we're not special. Ironic, much?

4. While the evidence for inflationary cosmology is contentious at best,[62] evidence for string theory is *nonexistent*.[63] String theory doesn't make any *unique* predictions testable by currently conceivable experiments, and its mathematical structure is so rich and all-encompassing that, if supersymmetry proves tenable (see the next point), there is virtually no experimental result it cannot accommodate. But a theory compatible with everything *explains nothing*.

5. String theory presupposes supersymmetry, a fundamental symmetry between matter particles (fermions) and radiation particles (bosons), allowing transformations

between them. If supersymmetry is false, then string theory fails and the string landscape hypothesis comes to nothing. Currently, the energy scale at which supersymmetry was expected to be discovered has been revised multiple times and it still hasn't been observed. The failure to observe it in experiments at the Large Hadron Collider (LHC) in Geneva, Switzerland, has contributed to the growing consensus that supersymmetry *is* false[64] and that, if further progress is to be made, physicists need some new ideas.

6. The string multiverse was invented to explain away fine-tuning, but it incorporates the mathematical structures of quantum theory, requiring both the quantization of energy and the exclusion principle, constraints that are *necessary* for life-supporting universes.[65] Obviously, then, the string landscape does *not* explain away the law structures necessary to life that it presupposes for its own function, so it *cannot* eliminate all fine-tuning considerations.

7. Last, any mechanism generating universes ad infinitum must have stable characteristics in order to remain functional. This means any "universe generator" will have *design parameters* that also require explanation. Postulating a random universe generator only bumps the need for design explanations up one level. Avoiding an infinite regress of explanatory demands requires termination in *actual* design by a necessarily existing intelligence that transcends spacetime, matter, and energy. So multiverse cosmologies make sense only in a theistic context,[66] but this makes them unnecessary for understanding the design parameters of our universe.

The Multiverse and the Ends of Naturalism

Mathematical descriptions have ontological implications, but they don't function as efficient causes. They are *causally inert* abstract objects. When quantum cosmology describes universes tunneling into existence from a highly structured faux-nothingness (a primordial vacuum state), its mathematical description doesn't provide an explanation, let alone an efficient cause, for such things. There can be no landscape of mathematical possibilities giving rise to an immeasurable realm of actualities that provides a mindless solution to the problem of cosmological origins and fine-tuning. Even an infinite arena of mathematical possibilities lacks the power to generate one solitary universe.

The mindless multiverse "solution" to the problem of fine-tuning is a metaphysical nonstarter. The absence of efficient material causality in fundamental physics and cosmology reveals the need for a deeper understanding of the world's rationality and orderliness. That explanation must always be mind over matter. When the logical and metaphysical necessity of an efficient cause, the demonstrable absence of a material one, and the realized implication of a universe both contingent and finite in temporal duration are all conjoined with the fact that we exist in an ordered cosmos—the conditions of which are fine-tuned beyond the capacity of any credible mindless process—the scientific evidence points inexorably toward transcendent intelligent agency as the only sufficient cause and the only reasonable explanation. A clarion call to intellectual honesty and metaphysical accountability reverberates throughout the cosmos: *In the beginning, God created the heavens and the earth; and the heavens tell his glory, the sky displays his handiwork*. If anyone has ears to hear, let him hear.[67]

What About Human Exceptionalism and Genetic Engineering?

Wesley J. Smith

Somewhere in the Great Beyond, Aldous Huxley is whispering, "I told you so." Huxley authored the great prophetic dystopian novel *Brave New World*, published in 1932, in which he conjured a society where people are genetically engineered into different castes. It is a sclerotic civilization with no chance of advancement or change, in which the people of each caste are enslaved by the naked power of raw biology to possess different capabilities and, furthermore, enjoy their degradation and dearth of free will. As Huxley put it in the 1946 republished edition foreword, "The theme of *Brave New World* is not the advancement of science as such," it "is the advancement of science as it affects human individuals."[1]

Ninety years ago, Huxley was already worried that science was leading "a really revolutionary revolution" to "be achieved, not in the external world, but in the souls and flesh of human beings."[2] In other words, human biology, and indeed, *human nature itself* could, Huxley feared, become the subjects and objects of scientific manipulation.

The First Genetically Engineered Babies Have Already Been Born

That time has now arrived as embryos are manipulated for eugenics purposes and the first human cloned embryos have already been manufactured. But the thrust toward *Brave New World* took a great leap forward in November 2018 when Chinese scientist He Jiankui announced in Hong Kong that he had created the first genetically engineered babies, twins with a gene deleted. *Fortune* magazine reported,

> The twins, named Lulu and Nana… were the result of in vitro fertilization (IVF). A few weeks old, they appear to be healthy. When they were a single cell, genetic surgery using a popular tool, CRISPR, "removed the doorway through which HIV enters to infect people."[3]

CRISPR—an acronym that stands for "clustered regularly interspaced short palindromic repeats"—is a tool by which biotechnologists can alter the genes of any cell or organism. When performed on an egg or sperm or early embryo—as with Lulu and Nana—it is known as *germline engineering* because the alterations pass down to subsequent generations. If done on a fetus or born organism, it is called *somatic cell engineering* because—scientists *think*—the alterations will only affect the edited individual.

For the purposes of this chapter, it is irrelevant *how* CRISPR is accomplished. What matters is that it *can* be done, *has been done* in humans, and that researchers are developing increasingly sophisticated techniques by which CRISPR can *change the essence of life* at its most fundamental levels. Indeed, biotechnology is moving at such breakneck speed that the term *brave new world* has transcended its literary roots and come to symbolize a particular mindset, nay, an *ideology*, a new bio-utopian mindset "committed to the process of human enhancement and self-directed evolution," which could not only embed "cultural distinctions...in our genetics," but would eventually "increase the biological differences among human populations."[4]

This is a profoundly portentous moment. We are in the "process of redefining ourselves as biological, rather than cultural and moral beings,"[5] a crusade furthered by an almost religious belief in biotechnology as the ultimate supplier of our future well-being. In the words of the great moral philosopher Leon Kass:

[I]t is a way of thinking and believing and feeling, a way of standing in and toward the world. Technology in its full meaning is the disposition rationally to order and predict and control everything feasible in order to master fortune and spontaneity, violence and

wildness, and leave nothing to chance, all for human benefit.[6]

If allowed free rein, where could such bio-absolutism take us? Kass warns evocatively of a "soft dehumanization of well-meaning but hubristic biotechnical 're-creationism.'"[7] He and others, such as the liberal social critic Jeremy Rifkin, predict that the ideal of human equality is not only at risk but that human life itself, may in some circumstances, be reduced to the status of a natural resource or a mere product ripe for exploitation and harvest. At the very least, I see it as leading toward the deconstruction of the Judeo-Christian ideal of universal human equality.

It Will Be Very Difficult to Genetically Engineer Human Beings

Still, we should not panic. This isn't a scientific dissertation on genetics, but it is worth noting briefly in our contemplation of the moral questions involving human genetic engineering that significantly altering our heritable components would not only be quite difficult technically, but might never be successfully accomplished in human beings. Why is that? With very few exceptions, our genes do not perform a single function. Rather, they usually serve multiple purposes, from coding for proteins to just turning other genes on and off. To put it another way, genes perform a great many and intricate tasks in the cell, often in communication with other genes. Fully understanding that genetic Gordian Knot in the human makeup—and without much chance of discovering a simple, bold solution to the problem, as Alexander the Great did in the telling of the proverb— may well be impossible.

Stanford physician and bioethicist William Hurlbut—appointed by George W. Bush to serve on the President's Council on Bioethics—explains: "It is unrealistic and dangerous if we think we can manipulate genes and obtain a product by design. Not only will we be sorely disappointed" by our inability to achieve that goal, but "it would be very likely the engineering would cause a 'break in the link of the chain,' for example, causing a protein to be missed. This, could, in turn, affect many traits than those the gene editor is targeting. In other words, biotechnologists will rarely be able to just tweak a gene and get one thing they are seeking." More challengingly, gene editing in humans "could create more problems than you are trying to solve."[8]

Hurlbut's warning should give us all great pause. If we permit such experimentation to continue and bring babies into this world whose genomes have been redesigned, we may face the duty to provide these research subjects—because that is what they are—extensive social and medical support for their entire lives. Consider the burdens they would carry through no choice of their own. Engineered children could become the subjects of bone-crunching notoriety that has the power to distort personalities and destroy lives. Moreover, they could be in danger of contracting congenital illnesses or disabling syndromes caused inadvertently by the editing process. For example, the Chinese twin girls referenced earlier may have a lower chance of contracting HIV, but concomitantly, may be more susceptible to premature death.[9] Alarmingly, three separate studies have discovered that genetically altering human embryos using the CRISPR technique created significant unwanted and unplanned genomic changes.[10] That alone should have us tapping the brakes.

Are Objections to Genetic Engineering Antiscience?

Whenever there is an intense public controversy over the moral propriety of the newest cutting-edge biotechnology, advocates for open experimentation or bounteous public funding of such projects label those who are skeptical as being antiscience. The point of the canard is to stifle discourse and protect established orthodoxies.

For example, during the embryonic stem-cell debate over modest federal funding restrictions imposed by President George W. Bush in 2001, advocates for more liberal NIH grant rules charged Bush both with being antiscience and seeking to impose his religious beliefs on science. Neither was true. Bush's objections were ethical, not religious.[11]

Given the significant beneficent scientific potential of CRISPR, such as ameliorating genetic disease—not to mention the billions that could be made by scientists and biotechnology companies from its applications—we can expect the antiscience trope to be wielded against those who seek to regulate human genetic engineering to corral it within acceptable ethical boundaries. So, let's briefly analyze the antiscience slur before discussing the moral questions involved with editing the human genome.

Describing opponents as antiscience is an ad hominem attack, a means to short circuit actual moral debate by branding the opponent as irrational, theocratic, or reactionary—toward the end that the adversary's opinions are dismissed as anti-progress. In this context, the charge is inaccurate on several levels. Science, properly understood, is a powerful *method* for gaining and applying knowledge about the workings of the physical universe. Its tools are observation, careful measurement, testing, falsification, and the like. No reasonable person really opposes those tools of learning nor most of the societal benefits that science provides.

Those who casually wield the antiscience epithet in debate are usually conflating opposition to science with disagreements about the proper *ethical* parameters of experiments. That is a different thing altogether. Returning to the embryonic stem-cell research example, the dispute wasn't about whether embryonic stem cells might produce scientifically valuable knowledge and medically useful outcomes. Most opponents of the experiments admitted that it could. Rather, their opposition was focused on the morality of destroying human embryos—nascent human beings—and whether similar benefits could be derived through other more ethical means, such as adult stem cells that do not require the destruction of embryos. Reasonable people were found on both sides of that controversy. But it was no more antiscience to oppose embryonic stem-cell studies based on ethical concerns than it was anti-ethics to support the experiments. Indeed, the induced pluripotent stem-cell breakthrough that created stem cells without destroying embryos is a vivid demonstration that the best science is accompanied by stellar ethics.[12]

Human Exceptionalism Is the Key to Analyzing the Morality of Biotechnology

So how should we analyze the ethics of CRISPR when applied in human beings? Well, that depends. How's that for taking an unequivocal stand?

There is a good reason for my wishy-washiness. Biotechnology is a classic double-edged sword, having both the potential to increase biological knowledge exponentially and alleviate tremendous human suffering—but also to trigger great social, metaphysical,

and corporeal harm. Moreover, developing the ability to genetically engineer people could be accomplished in ways entirely respectful of human dignity or through techniques that demean both the experimenter and the subject of the experiment. Thus, the morality of the enterprise itself depends on how we learn to do it, whether it can be done safely, and the ways in which we will apply the knowledge so derived.

There isn't adequate space here to fully explore those questions with regard to every proposed application of CRISPR and other biotechnologies. But we can establish the principles to apply in making such judgments.

It seems to me that analyzing such questions *through the prism of human exceptionalism* offers the simplest and most effective approach to discerning the answers to these often-difficult questions. This promises an analytical approach that would permit ethical experimentation, the development of beneficent biotechnological applications, and, if we followed our ideals, would inhibit us from crossing into unacceptable ethical terrain.

What do I mean by human exceptionalism? The term describes the dual moral aspects of our nature as human beings. First, it assumes that *we have unique and elevated value* among all known life forms, and concomitantly, that each of us has inherent equal moral worth. In this sense, human exceptionalism is the necessary predicate to universal human rights. Second, as the only truly moral species known to exist, *only we have duties*—including to never treat our fellow human beings as mere things to be manipulated, oppressed, and controlled. Hence, human experimentation is unethical when it treats the human research subjects as mere objects to be manipulated or exploited.

This approach would require us to ponder the following questions:

- Does the experiment or application support or detract from the intrinsic equal dignity of the affected human being or experimenter?

- Does the experiment or biotechnological application fulfill our human duties to treat each other with compassion and as moral equals?

- What is the nature and seriousness of the problem that the researchers hope to ameliorate or the information to be obtained?

- Are there any reasonable alternatives?

- Finally, what are the risks? What are the potential harms to the edited person, other people, the environment, or planetary life, etc.?

If all that sounds like a nuanced approach, it is. Absent clear ethical horrors such as the Tuskegee experiments—in which African-Americans with syphilis were told they were being treated but were left to suffer the consequences of the disease so researchers could see "the natural history of untreated syphilis"[13]—it is often difficult to make blanket ethical judgments about most areas of scientific inquiry.

Would Gene Editing Human Beings Be Ethical?

Now we are finally getting to the nub of the issue. Is it ethical to conduct gene editing on human beings? The answer, it seems to me, depends on the kind of gene editing we are analyzing.

Somatic Gene Editing: Somatic gene editing would take place in a fetus or born person.

If successful, it would treat a genetic malady by correcting the disease-causing defect. For example, Huntington's disease is an inherited fatal condition, one of the rare afflictions caused by a single gene that afflicts people in young adulthood through middle age. If the defective gene could be erased and a healthy replacement inserted, under a human exceptionalism analysis, would it be ethical?

Please excuse the nutshell approach required by space considerations, but this is how it seems to me:

- The treatment would no more detract from the intrinsic dignity of the patient than any other ethical intervention using medicinal drugs or surgery.

- Curing illness fulfills our obligations to treat the afflicted as we would want to be treated.

- Huntington's disease is always lethal and causes tremendous suffering before death. Therefore, curing it is a worthy cause for gene editing.

- There is no cure or prevention if one has the defective gene. Thus, genetic treatments may provide the only possible remedy.

- There would be no harm to human life or the environment. It is likely that unexpected side effects from the alteration would not pass down through the generations or affect anyone other than the treated patient.

Thus, under a human exceptionalism analysis, somatic gene editing to treat Huntington's would absolutely be ethical.

Of course, most analyses would not be that elementary. What about using CRISPR for an *enhancing* purpose—such as increasing intelligence in a normally healthy and able-bodied person? Generally, the analysis would reach a different conclusion. Again, please excuse the nutshell approach:

- Genetically enhancing people, as opposed to treating illness, has tremendous eugenics implications and could undermine universal human equality by communicating the message that a smarter human being is a better human being.

- We have no duty to increase intelligence.

- Being a normally healthy and able-bodied person is not a malady that would justify the resources required to learn how to enhance or the potential risk to the person involved with gene editing.

- No alternatives are needed because there is no malady or pathology being corrected.

- There would probably be no harm to other people or the environment from the enhancement. But the person engineered could suffer extreme injury if the editing goes wrong.

Under the above analysis and the traditional "do no harm" ethic of the Hippocratic Oath, I would consider an enhancing application of CRISPR to be unethical—although I recognize that those with a more libertarian bent might reach a different conclusion. Either way, the intrinsic equal and inherent dignity of the enhanced person would not be affected—a question we will discuss in more detail below.

Germline Gene Editing: Germline genetic engineering presents a different ethical picture altogether. Germline engineering is accomplished either by editing the gametes (eggs, sperm) or early embryos before cells differentiate into different types of tissues. This means that the genome so created will permeate the entire organism, including the gametes, through which the changes will enter the being's progeny.

Germline engineering doesn't only affect the organism being modified, but those alterations flow down the generations to the redesigned person's descendants. In other words, a mistake or unintended consequence resulting from somatic editing only involves the person so engineered. But the consequences of germline engineering—known, unknown, and unintended—could go on to affect the human genome generally.

Keeping the above in mind—and remembering that few gene alterations would be limited to the expression of a single gene or discreet organismal trait—let's do our nutshell human exceptionalism analysis:

- With very few exceptions, such as fixing a Huntington's disease gene, germline genetic engineering would undermine the equal dignity of the human being so manipulated because it would impose a future upon that person designed to fulfill the *desires of the designers*. This is particularly true if the modifications were not to prevent illness—always the initial justification for breaking ethical boundaries—but to create a child enhanced to have predetermined talents or characteristics.

- Engineering progeny would reduce procreation to a matter of manufacturing, complete with blueprints, engineering design according to the desires of the editor's customers, as well as quality-control approaches that would victimize children—born or unborn—not deemed to "measure up."

- Early germline editing would probably involve attempts to prevent disease. But as many books advocating eugenic engineering make abundantly clear, the ultimate primary purpose of germline genetic editing would be human enhancement.[14]

- Preventing the birth of babies with single gene-caused genetic disease can already be accomplished with existing approaches such as sperm sorting or IVF followed by preimplantation genetic diagnosis.[15] (Would-be parents worried about passing on genetic diseases can also adopt.) But most genetic maladies (syndromes)—even those that are caused by only one gene—create a constellation of problems because, as we have noted previously, single genes do many things. (This is why many genetically caused syndromes are untreatable.) Thus, with few exceptions, a gene editor would be unable to tweak the embryo to obtain the one desired discreet result.

- The harm to the genetically engineered human being could be profound. Having been treated as so much potter's clay and designed according to the parents' desires, the child would be bound by the naked power of genes to fulfill the desires of his or her manufacturer.

In the end, what would be achieved from enhancing human beings? Yes, people might,

one distant day, be redesigned to have greater athleticism, musical or artistic talent, or better looks than we possess today. And the children of such children might inherit those traits. But consider the great moral cost. Our children would, in essence, become slaves predisposed by the sheer power of biology to pursue a life that *we designed for them.*

Yes, it is true that children today can be compelled to pursue endeavors not of their choosing. But eventually, they can rebel—piano lessons may be abandoned, the baseball glove put permanently in the closet, the premed major dropped in exchange for studies in artistic design or automobile mechanics. In contrast, how would children mutiny against gene enhancements that would inexorably push them with the power of sheer biology in parentally predetermined directions? That could mark an end to true freedom for humans. Gene editing of progeny would open the door to a new eugenics with very sharp teeth.

In summary, if we can learn how to safely and effectively engage in somatic gene editing to treat genetic diseases and disabilities via means of ethical experiments, I think we should. Changing discreet genetic expression in a child or adult would be little different than giving the patient a strong medicine or surgical intervention. But germline genetic engineering and editing for purposes of enhancements or for eugenics purposes should be entirely beyond the pale.

How Should We Consider the Products of Gene Editing Experiments?

That doesn't conclude our contemplations. Let us assume that Wesley's wisdom is ignored and that, in the coming decades, scientists engineer progeny or adults into novel-genetic beings—as some transhumanists want. What if we have among us beings who have been genetically endowed with the eyesight of a hawk, the strength of a horse, or the lifespan of a redwood tree? Indeed, what if these individuals did not even appear human any longer? How should we judge and treat them?

This raises the question of what it means to be human: We are far more than the sum of our genetic makeup. For example, people with Down syndrome have an extra chromosome from the normal human complement, resulting in developmental disability and health problems. But other than perhaps the most outrageous bigot or eugenicist, no one believes that these precious people are not human.

This would also be true of someone who has been genetically engineered, whether for health or enhancement. In other words, we should treat and perceive such a human being just as we would any other person. Moreover, such seemingly radical alterations would be superficial because they would not alter the *moral nature of the individual* so contrived. The engineered person's genes would still be wholly (or mostly) human. The DNA and RNA would be expressed in distinctly human ways. More importantly, the moral natures of these individuals would remain "us."

This raises the question of what distinguishes human beings from animals.[16] For Christians and others of theistic faith, the answer might be that only we are made in the likeness and image of God. As a Christian, I accept that distinction, but whatever it means to be made in God's image, our creation in that regard is not about genes. God is incorporeal (let's not get into the nature of Jesus here) and moreover, "God is love,"[17] which is not a biological characteristic. Thus, a gene-edited human being would remain as much a child of God as every one of us.

But the purpose here is not to dwell on spiritual matters.[18] Human exceptionalism does not depend on religious belief. Indeed, whether our unique and morally relevant characteristics flow from the processes of blind evolution, resulted out of the mind of God, or via other mechanisms, the unique importance of being human can and should be robustly supported by rationally examining the cogent differences between humans and all other known life forms.

The acceptance of human beings as standing at the pinnacle of the moral hierarchy of life should be—and once was—uncontroversial. After all, what other species in the known history of life has attained the wondrous capacities of human beings? What other species has transcended the tooth-and-claw world of naked natural selection to the point that, at least to some degree, we now control nature instead of nature controlling us? What other species builds civilizations, records history, creates art, makes music, thinks abstractly, communicates in complex languages, envisions and fabricates machinery, improves life through science and engineering, or explores the deeper truths found in philosophy and religion? What other species has true freedom?

Perhaps the most important distinction between us and the beasts of the field—as animals were once called—is our moral agency. *Only we have duties.* As the philosopher Hans Jonas put it so well, "something like an 'ought to' can issue only from man and is alien to everything outside him."[19]

University of Berkeley physiologist Charles S. Nicoll made the same point in his article "A Physiologist's Views on the Animal Rights/Liberation Movement":

The belief that there are no morally relevant differences between us and animals ignores…significant and uniquely human qualities. We are the only species that has developed moral codes to judge our behavior, especially our behavior toward each other and toward animals. We are the only species that can make moral judgments and enter into moral contracts with other reasoning beings who understand the concept of morality and rights. We are moral agents: animals are not. The difference between us and animals is clearly a morally relevant one.[20]

Indeed, this uniquely human capacity to empathize with and appreciate "the other" is one of the best things about us.

None of that would change in the gene-edited individual. Because we *are* unquestionably the unique species—the only species capable of even contemplating ethical issues and assuming duties to animals and the environment—*we uniquely are capable of apprehending the difference between right and wrong, good and evil, proper and improper conduct.* Making biological shifts in the human being would not eradicate those traits from their inherent natures. Thus, excluding enhanced or edited people from the moral community of humanity would be a grave moral wrong.

The Question of "Upgraded" Animals

Gene editing can be—and is being—performed on animals. Animal/human chimeras—that is animals containing human genes—are used ubiquitously in medical research and other scientific studies. Some of these experiments are also using human brain cells, raising the fear that animals may be "upgraded" to have human intelligence and other characteristics.

There are enemies of human exceptionalism who yearn for such experiments. For example, Richard Dawkins, the biologist and atheism proselytizer, has yearned for the creation of a half-human, half-chimpanzee being that would shatter human exceptionalism and "change everything." He wrote,

> What if we were to fashion a chimera of 50% human and 50% chimpanzee cells and grow it into adulthood? That would change everything. That would change everything. Maybe it will?
>
> The human genome and the chimpanzee genome are now known in full. Intermediate genomes of varying proportions can be interpolated on paper. Moving from paper to flesh and blood would require embryological technologies that will probably come on stream during the lifetime of some of my readers. I think it will be done, and an approximate reconstruction of the common ancestor of ourselves and chimpanzees will be brought to life. The intermediate genome between the reconstituted "ancestor" and modern humans would, if implanted in an embryo, grow into something like a reborn *Australopithecus*: Lucy the Second. And that would (dare I say will?) change everything.[21]

Creating such a creature—and I truly doubt one could ever be manufactured—would profoundly violate human exceptionalism. Not that it would erase our belief in the unique nature of man as people like Dawkins hope, but it would be an unspeakably cruel violation of our duty to treat animals humanely. Chimps are social beings. Creating such an individual with presumably the intelligence to know it was so profoundly "different" would result in a life of unbearable loneliness and emotional isolation.

But what if one did come into existence? The creature would not be a human being biologically—its gene expression would be different, for example—or morally. But if such a monstrous and unethical "Island of Dr. Moreau" experiment were ever successfully conducted, because *we are human beings*, we would have the sacred obligation to treat these unfortunate beings humanely and with great respect as defined by their capacities, capabilities, emotions, and ability to suffer. But for our sakes and the sakes of those hypothetical hybrid creations, may such creatures never exist!

We Need Humility and Control

Biotechnology researchers are creating *the most powerful technologies invented* since the splitting of the atom, including CRISPR. And yet there are few concerted national or international discussions outside voluntary guidelines published by the research community to manage these experiments or to draw firm boundary lines. Indeed, other than some government funding restrictions, scientists are generally ethically bound only by their own consciences. That is unacceptable.

Time and scientists wait for no one! Yet the world dithers on creating legally enforceable regulations to govern biotechnology that would permit us to obtain the many benefits it potentially offers without unleashing the whirlwind. That is an abdication of leadership and our duty to leave a safe and thriving world to our posterity.

We need strong doses of humility and prudence. Having an "anything goes" attitude or leaving our biotechnological future "up to the experts" are not wise or sustainable approaches. They are not only potentially dangerous but morally fraught. We are, after all, the species that built the "unsinkable" *Titanic* and allowed the eugenics movement to thrive. *It is time for the world to focus* and control the application of CRISPR and other biotechnologies before the science controls us.

How Should Christians Think About Origins?

Richard G. Howe

When the subject of origins is mentioned, several debates come to mind. There is the perennial debate that rages between those who hold that the universe and the Earth are young (6,000 to 10,000 years) and those who hold that they are relatively old (4.5 billion years for the Earth; 13.8 billion years for the universe).[1] There is also the debate between creationism (young or old), theistic evolution, and naturalistic evolution (Darwinism).[2]

Creationism is the view that the universe and everything in it was caused by the creative will of God as recorded in the Bible. Theistic evolution holds that while God caused the universe to come into existence, the processes that have governed the development of the universe from its primordial state to complex life forms are entirely the laws of nature. Theistic evolution maintains that there is no need to posit any act of God subsequent to creation to explain the current complexity of the universe. This would include the origin of biological life itself. Miraculous intervention plays no part in the history of the universe— at least those aspects of the universe with which science is concerned.

Naturalistic evolution (Darwinism) typically embraces scientism, which maintains that everything that can be known, at least in principle, about the universe can be known entirely along the contours of the natural sciences, and that such questions about the universe are entirely indifferent to the question of God's existence or involvement. Indeed, for some, the arguments for naturalistic evolution serve to counter the traditional arguments for God's existence by seeking to show that theistic arguments explain nothing that is not better explained by the natural sciences.[3]

In many situations, these debates have generated more heat than light. However, the debates are important, serving to unpack the various dimensions of origins. Many vital philosophical and theological truths are at stake, and the Christian should understand the practical and spiritual purposes of these debates.

As evidenced throughout this volume, many evangelicals have a high regard for the natural sciences and the role scientific knowledge can play in advancing our understanding of and appreciation for God's creation. Admittedly, there have been elements within conservative or fundamentalist Christianity that have rejected many of the tools and methods of the natural sciences, considering them to be enemies of the authority of the Bible. Historically, however, the Christian church has been the friend of science.[4] Indeed, science within Western culture has flourished largely because of Christianity. While in some quarters one may find the view that religion and science have been at war, no reputable historian of science—Christian or not—thinks this.[5] As such, evangelicals will happily accede to many, if not most, of the findings of modern science.

But if and when the claims of science seem to conflict with sound reason or the teachings of the Bible, it is important for the Christian to stake his claim for truth. Other articles in this volume deal with specific questions surrounding the science of these debates in light of the Christian faith. This chapter focuses on two things: certain principles that govern the nature of these debates about origins, and why the issue of origins matters.

The Nature of the Debates Surrounding Origins

God has revealed himself through both his creation (known as general revelation) and his prophets, apostles, and the Lord Jesus Christ (known as special revelation)—what we now know as the Bible.[6] In addition to the truths specifically about God that are knowable through creation and the Bible, there are many other truths about God's creation that we can discover.[7] As Christians, we desire to be conscientious about mining those truths about God's creation as skillfully as possible. As these debates about origins attest, people sometimes read the evidence differently and come to varied conclusions.

Evangelicals generally agree that science has erred when it seeks to establish certain conclusions that deny the historicity of Genesis 1–11, including when science contends for an eternal universe, or denies the historical Adam and Eve (affirming a common ancestor for all biological life on Earth), or denies the great flood (whether local or global), or denies the historical Abraham, and more. The same holds for subsequent elements of the Old Testament and for specific assaults on the veracity of the New Testament. Such challenges can arise from philosophy, astronomy, archaeology, and other fields. While not every point that evangelicals would defend against attacks from science have a direct bearing on the issue of origins, many do.

An important point of contention is where and how a certain viewpoint is a matter of biblical authority or of biblical hermeneutics. What is meant by "biblical authority" is not whether the Bible has authority in all that it teaches. No evangelical should deny that it does. Rather, there are disagreements within evangelicalism about what the Bible means in certain instances. The most conspicuous example of this "debate about the debate" is whether the Bible unequivocally teaches that the Earth is young and that the days of the creation narrative are literal 24-hour days.

Here is another example: No Christian should ever contend that the Bible does not teach a historical Jesus. Persons who deny that Jesus was a historical person would have to say that the Bible was wrong. They could hardly defend the premise that the Gospels were never meant to teach anything but some kind of allegory as if the figure of Jesus is merely a literary device.[8] In this case, if someone denied the historical Jesus, they would be denying biblical authority and inerrancy.

Here is a final illustration before tying this back to origins: Suppose two dispensational evangelicals were disputing exactly when the rapture will occur in relation to the Great Tribulation. One might earnestly contend for a pre-tribulation rapture while the other might be just as earnest for his mid-tribulation position. What would not happen in their debate is that one would accuse the other of denying the authority of the Bible altogether. Instead, each would accuse the other of misinterpreting the Bible (biblical hermeneutics). It is possible for evangelicals to agree on biblical authority and inerrancy, acknowledging the other's commitment to them, and yet disagree on the meaning of certain biblical passages.

It is disputed within evangelicalism whether the Bible teaches a young Earth and literal 24-hour days of creation. Some on the young-Earth side accuse those on the old-Earth side of denying the authority of the Bible. They treat the issue of a young Earth and literal 24-hour days of creation as being on par with the issue of the historicity of Jesus (to use the prior example). For them, to say that the Earth and universe are old and that the days of the creation narrative need not be taken as literal 24-hour days is tantamount to denying the authority of the Bible. They refuse to allow it to be a matter of biblical interpretation.[9]

This is unfortunate. As a young-Earth creationist myself, I strive to disabuse some of my fellow young-Earth creationists of this position.[10] While holding to a young Earth and literal 24-hour days of the creation narrative, I acknowledge that those evangelicals who are old-Earth creationists can fully embrace a strong view of biblical authority and inerrancy.[11] There clearly are things that cannot be denied in the Bible while maintaining biblical authority. Further, there clearly are things that can be disputed by evangelicals while acknowledging each other's fealty to biblical authority and inerrancy. The debate over young Earth versus old Earth is a matter of biblical interpretation and not of biblical authority. There is a line between creationism (young or old) and theistic evolution.[12] (For details on "theistic evolution," see chapter 39 by Jay Richards, "Is Theistic Evolution a Viable Option for Christians?") To deny the special creation of humans is to reject what the Bible clearly teaches both historically and theologically.

Why Origins Matters

As human beings, there are many things we think we know about our origins. Our views have been shaped and informed by various sources ranging from mere observation of the world around us, to deep philosophical speculation, to heightened insights from God's revelation of himself. Voices range from the ancients (both East and West), from the natural philosophers (later called *scientists*), to the prophets and apostles, to the Lord Jesus Christ himself. As Christians, we are one of those voices. We believe not only that scientists can weigh in on the subject of origins, but also that God himself has revealed certain truths to us about origins. There may not always be the unanimity among the voices that we would like there to be.

While strategically avoiding any attempt to adjudicate the more central disagreements in defending the faith (e.g., whether the universe is eternal or temporal,[13] whether a finitely temporal universe requires a beginning,[14] whether the universe requires a cause,[15] whether there are multiverses[16]), let us look at a few thoughts about origins. Keep in mind that while evangelical Christians take the Bible to be definitive regarding much of what we believe about origins, this does not mean that even within such narrow boundaries of evangelicalism one will always find comfortable unanimity.

For evangelicals, the subject of origins is, for the most part, coextensive with the Christian doctrine of creation. From the very beginning of the Bible, one is confronted with the truth that the heavens and the Earth have not always existed. Christians hold this truth to have been revealed to us by God through the Hebrew prophets and echoed by the New Testament writers. Only God is from everlasting to everlasting.[17] For much of Christian history, we have held this view in opposition to many of the voices surrounding us who held to an eternal universe.[18] It has only been in the modern era that the natural philosophers—that is, the scientists—have corroborated our views that indeed the universe came into existence a finite time ago.[19]

While the scientists debate whether there is a definitive answer to the question about our ultimate origin (even if there is a general consensus among them, right or wrong, about human origins), let us look at what we can conclude about many things around us from the few details of origins that the Bible and sound reason give us. Many of the doctrines distinctive to the Christian worldview are grounded in a doctrine of creation. Further, many of the doctrines of non-Christian worldviews are repudiated by the same doctrine.

Certainly there are many important issues and debates that may not be directly settled by this appeal to creation.[20] Despite what we admit we cannot know by an appeal to creation as such, the doctrine of creation is important because it grounds many vital beliefs that impact the practical issues of life. Some of these beliefs are held in common with many of our fellow human beings who are not believers. Because of this, we can sometimes offer an apologetic for the Christian worldview by leveraging arguments

based on these common beliefs and by showing how these truths are accounted for only by the Christian worldview. What is more, the truths entailed by the biblical doctrine of creation vividly set Christianity apart from many other belief systems vying for attention. Of course, pointing out how creation entails this wide array of positions is not itself a demonstration of the truthfulness of these positions. Giving specific arguments for the truth of each of these points is not the goal here. Rather, this chapter seeks to explore the wide range of truths that Christianity affirms by its initial claim of creation.

The Doctrine of Creation Gives Us Our Fundamental Understanding of the Nature of God

In each of the subheadings, the term *fundamental* is used deliberately. This is not to say that everything one would need to say about a given point is easily extrapolated from a bare doctrine of creation. It is certainly the case (more for some points, less for others) that sound reason and other portions of Scripture must be brought to bear on a given point if we seek to understand it as fully as we can. The aim here is merely to show that these points, which grow more robust when reflected upon more widely, are nested in the doctrine of creation, even if in a nascent manner. In addition, the use of the term *repudiate* is deliberate. By saying a certain point is repudiated by the doctrine of creation is not to suggest that the point has been wholly falsified, but rather, that it is opposed by the Judeo-Christian understanding of creation. To accomplish falsification, one would need more argument and evidence.

God's Existence

The Christian doctrine of creation tells us that God exists. Genesis opens with these words: "In the beginning, God created the heavens and the earth."[21] With the event of creation, God is already there. The New Testament repeats the same truth, connecting it vividly with Jesus Christ in John 1:1: "In the beginning was the Word, and the Word was with God, and the Word was God." The parallel is deliberate. The God of creation who was there before the beginning is the Word who was there with God and who is God. This Word is the Lord Jesus Christ.[22] Denying the event of creation allows the atheist to avoid coming to terms with the arguments for God's existence: most notably the cosmological argument (God is the cause of the cosmos), the design argument (God is the cause of the complexity of the cosmos), and the teleological argument (God directs creation toward its proper end).[23] The first line of defense of the Christian faith is addressing the truths attested from the very first verses in the Bible. With this, we see that the doctrine of creation repudiates atheism and affirms theism.[24]

God's Attributes

Not only is God's existence entailed by the doctrine of the creation, but also God's attributes are seen and understood through creation. Romans 1:20 tells us, "His invisible attributes, namely, his eternal power and divine nature, have been clearly perceived, ever since the creation of the world, in the things that have been made."[25] Given that God is the creator of all other existing things, it follows that God himself is not only uncreated but transcends the finite universe. This is a repudiation of finite godism and an affirmation that God is the transcendent God of classical theism.[26] Further, it follows that God is timeless because he is the maker of both time and space.[27]

Isaiah 40:13-14 tells us that God is independent and sovereign—he needs no other

to assist, advise, or counsel him. "Who has measured the Spirit of the LORD, or what man shows him his counsel? Whom did he consult, and who made him understand? Who taught him the path of justice, and taught him knowledge, and showed him the way of understanding?" As the sovereign, God is provident over all creation.[28] He shows his loving care for the human race.[29] At the same time, his treatment of the nations is according to his righteousness.[30] From these few passages, we see that God's creation testifies to his existence, eternality, power, deity, goodness, righteousness, glory, and providence.[31]

The Doctrine of Creation Gives Us Our Fundamental Understanding of the Nature of the Universe

It is undeniable that the physical universe exists external to our minds. This repudiates certain New Thought and New Age doctrines that deny the reality of the physical world.[32] The arguments for God's existence, together with the testimony of Scripture, point not only to God as the creator, but also show that the universe was not made from any preexisting material. Hebrews 11:3 is quite clear: "By faith we understand that the universe was created by the word of God, so that what is seen was not made out of things that are visible."

God is not a grand fashioner or artisan who worked on preexisting materials—what Plato called the *demiurge*.[33] Neither did God somehow create the universe out of himself as *creation ex deo* would have it. Instead, he alone is the maker of all things in creation. The truth of the matter is *creation ex nihilo*—creation out of nothing.

Because the creation includes the material universe, this repudiates eternal dualism (the idea that matter and spirit are eternal) and philosophical materialism/scientism (the idea that the fundamental elements that comprise the physical universe are all that exist). Both of these ideas are common within atheism.[34] The doctrine of creation also repudiates Mormonism's understanding of reality when it asserts an eternal "element."[35] Because God did not make creation out of himself, this repudiates pantheism (the idea that everything is god) and various pantheistic New Thought/New Age philosophies as well as certain forms of witchcraft.[36] The fact that the universe is created repudiates those philosophies that deify the universe, including, again, pantheism, as well as Gaiaism (Mother Earth as a goddess), witchcraft, and paganism.[37]

The creation of the universe also repudiates any notions that the universe is eternal.[38] Further, having been created by an infinitely wise and rational God, various forms of the "new" physics (falsely so called) as well as chaos models are repudiated. This includes the New Age forms of physics.[39] Finally, the fact that the universe is created by an all-good God entails that the universe itself is good. Its goodness is attested to numerous times in the creation narrative.[40] This repudiates Gnosticism and the obligations of that form of asceticism that illicitly seeks to reject all of the amenities of the physical world as being in opposition to the godly Christian life.[41]

The Doctrine of Creation Gives Us Our Fundamental Understanding of the Nature of Mankind

As obvious as it is to some, and as hard as it is to grasp for others, the doctrine of creation tells us that we as humans are not God, nor are we "gods" in the relevant sense of the term.[42] This repudiates many Eastern religions that regard humans as divine, as well as various New Age groups and new religious movements.[43] The doctrine of creation rules out both human preexistence and reincarnation.[44]

We also see from the doctrine of creation that humans were created separate from other creatures and that we are both physical and spiritual.[45] This repudiates theistic evolution's view of human origins. It also repudiates the reductivism that is common in contemporary views of human nature and exhibited (in its worst form) in the psychological theory of behaviorism.[46] In affirming our dual nature, while carefully guarding the relative importance of each, the Scriptures carefully delineate those aspects of our lives that track along the spiritual and those that track along the material/physical. Second Corinthians 4:16 tells us, "We do not lose heart. Though our outer self is wasting away, our inner self is being renewed day by day." With this, we see that humans, while certainly animals in the biological sense of the term, are more than merely animals given our spiritual (immaterial) aspect and given our being formed in the image of God. Endowed as we are with rationality and free will, human beings are unique among all life on Earth.

Another truth about humans that we learn from the doctrine of creation is that God has created people distinctively male and female. The realization of this truth has never been more needed than it is today. The male and female distinction is integral to our very natures as human beings. A proper understanding of this repudiates androgyny (the blending of male and female), unisexuality (the denial of the male/female distinction,), transgenderism, the alleged fluidity of sexuality or gender, and, in certain respects, radical feminism.[47]

Two types of distinctions we see throughout the Scriptures are real distinctions versus functional distinctions. The biological distinctions of sexuality are real distinctions. A person is either really male or really female. In contrast, some evangelicals maintain that there is a functional distinction in the respective roles that men and women occupy within the nuclear family and the local church. This is an ongoing dispute within some areas of evangelicalism.[48] Both sides, however, agree that neither real distinctions nor functional distinctions compromise the equal worth of men and women as human beings before each other and before God.

In spite of those particulars over which we can disagree, we who are evangelicals are in solidarity in our defense of the institution of marriage and the role the nuclear family is designed to play in human society. Indeed, evangelical Christians share this commitment to the nuclear family with certain political elements and even with some non-Christian elements of our society.[49] But it is not enough to acknowledge that there is a distinction between male and female. We see further that men and women are made for each other exclusively. The Bible knows nothing of a marriage that is anything but between a man and a woman. Genesis 2:24 cannot be clearer: "Therefore a man shall leave his father and his mother and hold fast to his wife, and they shall become one flesh." This sacred truth was reiterated by the Lord Jesus himself in Matthew 19:5-6. This repudiates same-sex marriage. And because the sexual relationship is intended by God to be confined to the marriage relationship, this repudiates homosexuality.[50]

Last, in light of the important distinctions we can identify and the roles men and women and fathers and mothers are designed to occupy in God's economy, evangelicals are also in solidarity that both males and females are created in God's image and are of equal dignity and worth in the eyes of God.

Genesis 1:27 makes this clear: "God created man in his own image, in the image of God he created him; male and female he created them." Even in our fallen state as humans, we nevertheless retain that image of God.[51] It behooves all of us as Christians to uphold that dignity and worth. Here, all sexism is repudiated.

The Doctrine of Creation Gives Us Our Fundamental Understanding of Mankind's Relationship to the Creator

Given what we have already discovered from the doctrine of creation, it is evident that as humans we owe our everything to God. God not only has created us, but he sustains us as well.[52] Our well-being and flourishing occur only as we align ourselves with what God has intended for us. This is especially true for us eternally. That alignment in this life is best understood along the lines of the natural moral law and the commands that God has given us as Christians in the Scriptures.[53] That alignment in the next life is possible only by the imputed righteousness of God by faith.[54] We see here that secular humanism and human autonomy are repudiated. We are not here for our own interests. Instead, as the Westminster Shorter Catechism expresses it, "Man's chief end is to glorify God, and enjoy him forever."[55]

The Doctrine of Creation Gives Us Our Fundamental Understanding of Mankind's Relationship to the Universe

Several things follow from the fact that the universe is not divine but is God's creation. First, nature is not something to be worshipped. This further repudiates various New Age, Wiccan, and pagan ideas.[56] Instead, we are to have dominion over nature. In Genesis 1:28, God told Adam and Eve, "Be fruitful and multiply and fill the earth and subdue it, and have dominion over the fish of the sea and over the birds of the heavens and over every living thing that moves on the earth." This repudiates radical forms of environmentalism, where the interests of humans are subjugated to the interests of animals or elements. Our dominion over nature, however, does come with certain responsibilities. Being created in God's image means, among other things, that we are responsible stewards over what God has entrusted to us. Our managing of environmental resources should always be in service to the cause of Christ and to the glory of God.[57] This repudiates a wanton environmental exploitation.

Second, we see the mandate for humans to "be fruitful and multiply." Such a mandate flies in the face of sources that urge married couples to reconsider having a large number of children (or even any children at all), insisting that large human families are detrimental to the health and stability of the global environment.[58] The mindset that insists humans are intrinsically harmful to the health of the planet makes it easier for some to warm up to the ideas of abortion and euthanasia. We see, then that the Christian doctrine of creation repudiates such radical population control.

The Doctrine of Creation Gives Us Our Fundamental Understanding of Mankind's Relationship to Mankind

We see that the creation is not something to be worshipped in place of the Creator. Despite the fact that humans are unique within God's creation in that we are created in the image of God, it nevertheless follows that humans are not to be worshipped either. As the old wall poster said (adjusting for grammar): "Two Foundational Facts of Human Enlightenment. 1. There is a God. 2. You are not He." Any worship directed toward another human or to a created thing

is idolatry. Lest it seem incredible that any human would worship any person other than the God/Man Jesus Christ, consider the somber warning Paul gives us in 2 Thessalonians 2:3-4 of what is to come:

> Let no one deceive you in any way. For that day [the day of the Lord] will not come, unless the rebellion comes first, and the man of lawlessness is revealed, the son of destruction, who opposes and exalts himself against every so-called god or object of worship, so that he takes his seat in the temple of God, proclaiming himself to be God.

From the very beginning of creation, the evil of this coming one has already been repudiated.

But in disavowing human divinity, we need to be careful not to go to the other extreme and fail to come to terms with the significance of us having been created in God's image. Our having descended from Adam and Eve means that all humans are equally valuable and are imbued with dignity. This repudiates undue egoism (an inflated view of one's self in relation to others[59]) and all forms of racism (the idea that one race is superior to other races). Human life is sacred. Murder in all its forms, including abortion and euthanasia, are repudiated.

So How Should We Think About Origins?

From the doctrine of creation we can see the affirmation and foundation of profound truths about God and his creation and an ineluctable repudiation of toxic ideas and actions. The Christian doctrine of creation repudiates atheism, finite godism, New Thought/New Age, pantheism, paganism,

Gaiaism, philosophical materialism, an eternal universe, the "new" physics/chaos, asceticism, Gnosticism, preexistence, reincarnation, reductivism, behaviorism, androgyny, unisexuality/nonbinary, transgenderism, gender fluidity, radical feminism, same-sex marriage, homosexuality, sexism, human autonomy, secular humanism, radical environmentalism, environmental exploitation, radical population control, idolatry, egoism, human divinity, racism, murder, abortion, and euthanasia.

How Should We Think About the Age of the Earth?

David Haines

T he creation account contained in the early chapters of Genesis has become the focus of many Christians (especially with North American evangelicals) who seek to discover the age of the Earth. Because of this quest, there have emerged a number of different opinions proposed by believers, theologians, scientists, and philosophers in relation to origin science and the proper interpretation of the biblical creation narrative. While this chapter favors an old-Earth perspective, it will review some of the arguments for and against these various views, highlighting points of agreement and disagreement and offering concepts to consider when thinking about the age of the Earth.

Points of Agreement

In spite of the variety of interpretations and the ensuing vigorous debate, we must not lose sight of the great and widespread unity shared by believers regarding the core doctrines of the Christian faith for the last 2,000 years. These include the Trinity, deity of Christ, virgin birth, and Christ's atoning death on the cross, physical resurrection, ascension, bodily return, and the miraculous nature of his life. What is more, whether one believes in a young Earth or old Earth, there are several points of agreement between evangelical Christians who believe in the inspiration, authority, and inerrancy of Bible:

1. There exists an intelligent powerful supreme being (God).

2. God is the first cause (primary/efficient cause) of the heavens and the Earth.

3. God, through a series of speech-acts (divine fiat), spoke creation into existence out of nothing (*ex nihilo*).

4. God has revealed himself through both general revelation (nature) and special revelation (the Bible).

5. The Bible is inerrant, although man's interpretation of it may at times be fallible.

6. Likewise, man's interpretation of the "book of nature" through science is imperfect and often subject to error.

7. Genesis represents a historic, accurate account of how God created the universe, Earth, and life.

8. Each "day" (Hebrew: *yom*) of creation in Genesis 1 was historical, not mythical.

9. Each "day" contains a span of time (i.e., either 12 hours, 24 hours, or an indefinite length of time, etc.).

10. The Bible doesn't explicitly or directly state the ages of the universe or the Earth.

11. Creation was not completed all at once; rather, it was accomplished through an intentional sequence of pronouncements over six "days."

12. The inorganic aspects (time, space, matter, rocks, water, land, light) of creation were created prior to the organic parts (fish, animals, man).

13. Unguided naturalistic evolution contradicts the Genesis creation narrative.

14. According to Romans 1:20, God's existence and power is "clearly seen" in "what has been made"—i.e., we can see evidence of design in nature.

15. Creation manifests design (i.e., final causality in nonrational beings, specified complexity, beauty, etc.).

16. Design is evidence of an intelligent designer.

17. We can reject naturalism as an explanation of origins.

18. God specially created a historical Adam and Eve who served as the sole progenitors of all humankind, and

who rebelled against God at the fall, plunging humanity into sin.

19. One's view of the age of the Earth should not hinder Christian fellowship or be used as a test for salvation or orthodoxy (Romans 14).

Of course, there are additional points of agreement. We would be remiss in our failure to recognize that there is passion on both sides of the issue, which makes sharing ideas on how to think about the age of the Earth extremely difficult. But from the outset, it's worth emphasizing points 4, 5, 6, and 7 above: *Among both young-Earth and old-Earth proponents, we find Christians who are strongly committed to the inerrancy of Scripture and the historicity of Genesis.* Of course, some Christians—especially those who tend to hold to theistic evolution—reject inerrancy or the historicity of Genesis. But there are many Christians who reject the grand evolutionary story yet embrace an either old-Earth or young-Earth viewpoint *and still hold a high view of Scripture.* Thus, embracing an old-Earth perspective need not entail rejecting inerrancy or the historicity of Genesis.

My present thoughts are in no way to be understood as the complete and final word on the subject, for this would grossly underestimate the complexities involved in the discussion. Far be it from me to claim to have solved the issue once and for all. Rather, hopefully the thoughts presented here will serve as a springboard for future discussions on the topic.

Variety of Perspectives

Despite the differing viewpoints, most Christian theologians and scientists fall within one of two categories: young-Earth creationism or old-Earth creationism.[1]

Traditionally, an articulation of the young-Earth perspective finds its inspiration in the views of seventeenth-century Irish Bishop James Ussher (1581–1656). His attempt to construct a chronology, used in the early *Scofield Reference Bible*, was based on the patriarchal genealogies and lifespans in Genesis that placed creation at about 4004 BC, Noah's flood at 2500 BC, and Abraham's birth at 2166 BC. Jonathan Sarfati, in his book *Refuting Evolution*, states that compared to the mainstream scientific viewpoint wherein the Earth was formed about 4–5 billion years ago, "basing one's ideas on the Bible gives a very different picture. The Bible states that man was made six days after creation, about 6,000 years ago."[2] He goes on to provide reasons for thinking that the universe is in fact quite young: "Creationists ultimately date the earth using the chronology of the Bible. This is because they believe that this is an accurate eyewitness account of world history, which can be shown to be consistent with much data."[3] In his book *Refuting Compromise*, which is aimed at old-Earth creationists such as Hugh Ross, Sarfati provides evidence intended to show that the young-Earth position was held by many theologians, and even important early modern scientists, up to the nineteenth century.[4] More recently, scientists from groups like the Institute for Creation Research and Answers in Genesis have offered geological, genetic, and other various data sets believed to support the young-Earth view.

Many young-Earth creationists argue that although it is possible to be a Christian without adhering to the young-Earth position, it is impossible to respect the authority, inspiration, and inerrancy of Scripture and hold any position other than young-Earth creationism. Jason Lisle, for example, states that "six days of creation is a corollary of the inerrancy and perspicuity of Scripture."[5] He also implies that a literal interpretation of the six days of creation requires a young universe of roughly 6,000 years old.[6] This would mean, for Lisle, that to reject the young-Earth creationist perspective would be equivalent to also rejecting the authority, inspiration, inerrancy, and perspicuity of Christian Scripture.

On the other hand, the old-Earth perspective points to the viewpoint held by a majority of scientists that the universe is nearly 14 billion years old,[7] the Earth is 4.5–4.6 billion years old,[8] and man does not show up until near the very end of that time period. A significant number of Christian theologians and scientists argue not only that the Bible allows for the possibility that the universe is as old as some scientists suggest, but also that the natural sciences should be used in helping to interpret the Scriptures on these important subjects. Hugh Ross, for example, notes this concerning the beginning of the universe:

> In 1991, the best available measurements produced a date of 16 ± 3 billion years. In 2001, scientists updated the measurement to a date of 14 ± 1 billion years. Today, they measure it to be 13.79 ± 0.06 billion years old. The increasingly precise dates determined by astronomers trouble atheists intent on explaining life by natural means alone. The dates are too recent. But at the same time, they're far too ancient to help creationists intent on defending a six-consecutive-24-hour-creation-days interpretation of Genesis 1.[9]

For Hugh Ross, the claim that the universe is this old is entirely consistent with a clear and proper interpretation of Scripture. Norman Geisler and Ronald Brooks note,

"Really, only a few young earth advocates care to fix a date like that [around 4000 B.C. for the creation of the cosmos]…Of course, there are many Creationists who argue for an old earth."[10] Leading Christian apologists who do not hold to a young-Earth perspective, such as Ross, K. Scott Oliphint, John Lennox, and others, argue that it is possible to both accept scientific claims concerning the antiquity of the universe and theological claims that the Christian Scriptures are inspired, inerrant, and therefore authoritative. Oliphint, for example, in response to Lisle, says, "The issue between us has its focus not on what Scripture is, but on what Scripture says. We both affirm that Scripture is the inerrant Word of God."[11] Norman Geisler notes that "a faithful commitment to the authority of Scripture" does not necessarily "lead one to a young earth interpretation" because "they are different issues. One may believe in the authority (and inerrancy) of Scripture and yet hold to different interpretations of it. What Scripture is and how it should be interpreted are two different issues."[12]

The differences seen within the main views of the age of the Earth in Genesis 1 center on the Hebrew word *yom*, which is typically translated as "day." The main question is, What is the length of time meant by this term? Some suggest the presence of a numerical sequence (i.e., first day, second day, third day, etc.) is a clue that points to an interpretation of a 12-hour period of daylight or 24-hour day. Does it refer to a 12-hour period as signified by the time span between the "evening and morning," or a 24-hour period representing a full day that includes *all* evening hours and *all* morning hours, or some indefinite length of time, perhaps even millions of years? Answers to these questions have produced several viewpoints on the proper length of a *yom* in the original Hebrew text of the Genesis creation narrative and the age of the Earth. These perspectives, all of which hold to a literal creation, claim to be supported by science, Scripture, and sound hermeneutical principles. They represent the main old- and young-Earth positions (and derivations from them). These include but are not limited to:

- *24-Hour Young-Earth Theory:* The consecutive days of creation describe a 24-hour period of time, thus placing the six days of creation within a 144-hour time span. There are no long time spans in the creation event. Creation occurred less than 10,000 years ago.

- *Old-Earth Theory:* The days of Genesis 1 refer to indefinite periods of time, perhaps millions of years, because the context of Scripture and the biblical usage of the term "day" allows a flexible interpretation relating to its length.

- *Gap Theory:* There exists an unknown gap of time between Genesis 1:1 and 1:2, prior to the start of the successive creation days (which could be 24-hour days with no indefinite time span in between the days or longer periods of time).

- *Ideal Time Theory:* Creation only *appears* old but is actually very young. For example, Adam was created as a fully formed man, but was as young as a newborn baby. Trees only appear to be old because they were fashioned fully mature. These age-dating factors could be telling observers that the Earth only appears to be billions of years old but is actually only thousands of years old.

- *Day-Age Theory:* The days of creation are unspecified periods of time placed in chronological order, which could be literal 24-hour time periods but allow for long periods of time between each day.

- *Framework Theory:* Views the creation time descriptions as an ancient literary device framing the creative acts in logical order with terms like "day" and "evening and morning" into a compact literary form (e.g., like a sentence, paragraph, or chapter). This is to convey to later readers that there was a beginning and ending to successive periods of time in logical order.

- *Revelation-Day Theory:* The days of creation refer not to a literal time schedule when God created, but to the time it took God to *reveal* the past literal creation event to Moses—namely six days of revelation.

Many proponents of these views argue that a normal grammatical-historical interpretive approach to the Genesis narrative can at the same time hold to the historic six days of creation and account for long periods of time. The only view(s) that would be inconsistent with Scripture would be one that advocates a mythical (nonhistorical) view of creation or, arguably, one that involves strictly unguided natural causes to explain origins.

For this debate, it is of no little importance that the author of one of the most important defenses of the authority and inspiration of Christian Scripture, B.B. Warfield, not only thinks that: (1) the age of the universe is of absolutely no interest to the Christian theologian,[13] and it is properly a question that is to be determined by the natural sciences,[14] but also, (2) that there is no true conflict between the Bible and the natural sciences on this matter.

So what is a person to think? Is it necessary, in order to hold to a high view of Scripture, to also adhere to a young- or old-Earth viewpoint, or are there other ways of reading the biblical texts that do not compromise the inspiration, inerrancy, and authority of Scripture? How should we approach the question of the age of the Earth?

Before going further, it is worth noting that this question is an excellent example of how science and Christian theology can *appear* to be in conflict but need not be articulated in such a way as to be in *real* conflict. This issue is also an interesting test case for how we understand the relationship between the natural sciences and theology. An appropriate understanding of this issue will require a proper understanding of the nature of both scientific research and biblical interpretation, as well as interaction with both hermeneutical and scientific data.

Preliminary Issues

Our discussion will take for granted that the Bible is the inspired, inerrant, and authoritative Word of God. All branches of Christianity firmly believe that the Holy Spirit guides the church universal in the proper interpretation of Scripture. However, most Christian theologians also believe that, as human interpreters of divinely inspired Scripture, they are prone to error. Augustine famously said that he wished to believe only in those things he said which turned out to be true. Aquinas echoed this sentiment in his *Summa Theologiae*, where he noted that though the divinely inspired Scriptures are absolutely without error and fully authoritative for the establishment of doctrine, all those Christian theologians who write about the Scriptures are to be used only as probable authorities.[15] These theologians, in which Aquinas would include not only himself but also all those who preceded and came after him, including the Pope, are only probable authorities because they are prone to error. As such, it must be remembered that, per traditional Christian theology, there is an important distinction between the words of Scripture and our interpretation of those words. The Scriptures cannot err and are always authoritative; all human attempts to interpret the Scriptures are prone to error and are authoritative only as they rightly explain the meaning of Scripture. This should instill humility in the interpreter.

Some complications that are inherent in interpretation may include: (1) the fact that the original author may hold different assumptions about natural phenomena; (2) a single word may have many different usages, and we cannot just assume that because we use a word in a certain way that an ancient author used it in the same way; and (3) it is not always easy to determine the genre of literature being used by an author, and even when we have determined it, we have yet to determine how the genre affects the interpretation (i.e., though poetic texts often use metaphorical language, it is entirely possible for aspects of a particular poem to be interpreted literally).

This implies that although some of the teachings of Scripture can be determined such that to disbelieve them is to cease to be Christian,[16] others need to be held with only relative certitude, as it is entirely possible that future discoveries may require changing how we interpret certain texts.[17] It is also important to note that throughout the history of biblical interpretation, natural knowledge has always played an important role. Indeed, it is impossible to rightly interpret Christian Scripture without natural knowledge (provided by accurate observations of the cosmos).[18] As such, it is necessary to take the discoveries of the natural sciences into consideration when interpreting the Scriptures. In sum, human interpretation of the Scriptures is not infallible, inerrant, or inspired. Human interpreters may err, and thus must maintain a humble posture in relation to their interpretations of Scripture.

What about science? It is also important to remember that the sensible universe, as created by God, is intelligible. We are able to study the cosmos and gain knowledge of it and of ourselves. However, as with the interpretation of Christian Scripture, our observations and the conclusions that we arrive at through our observations are not necessarily inerrant. For this reason, most scientists hold that the discoveries of the natural sciences are only tentative, as future research and discoveries may require them to be modified or rejected. Indeed, one of the reasons philosophical skepticism is such a powerful position is that we are easily deceived not only by appearances, but also by our own desires, presuppositions, assumptions, and values.

Some philosophers of science have gone so far as to deny the possibility of discovering any nonrelativistic truth claims through scientific research—as all scientific discoveries are made by humans who are socially, culturally, linguistically, and religiously situated and often import their values into their science (frequently without realizing it), which causes them to rule out possible theories or to lean towards certain conclusions a priori (i.e., prior to experience).[19] Most scientists, however, would say that there are a certain number of scientific and philosophical truths that, if rejected, may hinder one's ability to seriously engage in the scientific enterprise (e.g., that nature operates according to regular laws, that the Earth is a sphere, that gravity operates everywhere, etc.).

However, as the history of the natural sciences has amply demonstrated, future discoveries may require major revisions or even wholesale rejection of ideas that were once widely accepted and considered settled theories (e.g., the Ptolemaic model of the solar system, Newtonian physics, Galen's medical theories, etc.). In light of the tentative nature of scientific theories (even those which are, at any given time, supposedly settled) and the human tendency toward both biased interpretation of observational data and error, it is best to maintain a humble posture in relation to the discoveries and theories of the natural sciences. With these notions in mind, we turn our attention to how to avoid some common mistakes when we think about the age of the Earth.

Mistakes to Avoid

Those who engage in discussions about the age of the Earth will recognize that there are many mistakes that can be made while navigating the scientific and hermeneutical principles and data. There are several basic concepts to consider while highlighting certain mistakes to avoid.

Don't assume that winning the debate is more important than loving the person. Passions can flare when engaging in debate, and this is when we must be reminded that the manner and attitude with which we hold our views can be nearly as important as the content of the views themselves. The world is watching, and hot tempers do not reflect what Christ is like. The greatest commandments are to love the Lord and love your neighbor as yourself. Romans 14 provides much good guidance here.

Don't assume your data and interpretation are the final word on the matter. There is no doubt that a proper interpretation of Scripture is informed by the best scientific data available. A bit of humility is always advisable because, as scientific research and theorization is, by nature, inductive, scientific data will be periodically updated and informed by further study and experimentation. I'm sure that those who believed the sun revolved around the Earth felt confident in their view with the best science of their day. However, further scientific data offered by Galileo's observations and equations eventually changed many minds to recognize that the Earth revolves around the sun (heliocentricity). This humility will serve us well in eternity as we share heavenly bliss with those with whom we currently disagree, but whose theories may later be supported by better data.

Don't assume that genealogies and patriarchal lifespans are an exact measure for the age of the Earth. As many Christian theologians and archaeologists have pointed out, the biblical genealogies do not allow us to put a time stamp on the cosmos, the Earth, or the human species. Because of the genealogical gaps present and linguistic idiosyncrasies, a reliable calculation of timespans is at best very difficult. In English, there are many words that describe familial relationships with exactness, such as father, grandfather, cousin, uncle, and other precise words. However, the ancient Hebrew language has far fewer terms to describe these relationships. For instance, the same Hebrew word used translated "son" (*ben*) can refer to a son, grandson, great-grandson, or descendant (cf. Genesis 29:5; Daniel 5:22). What is more, the Hebrew word translated "father" (*ab*) can mean father, grandfather, great-grandfather, forefather, clan, tribe, or ancestor (cf. Genesis 28:13; Joshua 19:47; Jeremiah 35:6).

Similarly, the Hebrew word translated "begat" (*yalad*) in the Genesis (and Mosaic) genealogies can be used for ancestral descendants (cf. Exodus 6:20; Numbers 26:59; 1 Chronicles 24:40; Numbers 3:27-28 with Numbers 4:36), making it extremely difficult to measure intervening gaps. Even the most detailed attempt is a perilous process because the genealogies do not *require* us to understand them without gaps. The process of omitting some names in the genealogies is called *telescoping*, which was done for the sake of brevity, while at the same time including names that are more significant.

The calculations used to arrive at the age of the Earth are based upon the genealogies of Genesis 5, 10, and 11, and assume that (1) there are no gaps in the genealogies, (2) the purpose of the genealogies is to provide a precise, person-by-person chronological line all the way back to Adam and Eve, and (3) our translations of the relevant terms in the genealogies are accurate. (With regard to the age of the universe and Earth, one might also add that such interpretations require that prior to the creation of Adam and Eve, the universe and Earth had existed for only a handful of 24-hour days.[20]) It is important to note that if any one of these assumptions is found to be untenable, then attempting to calculate the age of the Earth based upon the genealogies is a hopeless endeavor.

It turns out that all three of these assumptions have been challenged by conservative Christian theologians and biblical exegetes who adhere to and defend the inspiration, inerrancy, and authority of Scripture—such as B.B. Warfield. Warfield, in his article on the antiquity of humanity, clearly demonstrates that there are gaps in the genealogies and that we simply cannot know how many generations fit into these gaps.[21] Warfield notes, in relation to the genealogy given in Matthew 1:2-17,

> A comparison with the parallel records in the Old Testament will quickly reveal the fact that the three kings, Ahazia, Joash, and Amaziah are passed over and Joram is said to have begotten Uzziah, his great-great-grandson. The other genealogies of Scripture present similar phenomena; and as they are carefully scrutinized, it becomes even clearer that as they do not pretend to give complete lists of generations, they cannot be intended to supply a basis for chronological calculation, and it is illegitimate and misleading to attempt to use them for that purpose.[22]

Geisler points out that biblical scholarship has demonstrated that there are gaps in all of the biblical genealogies, saying, "There is a least one generation missing in the Genesis 5 and 11 genealogy which appears to be closed."[23] In other words, even in Genesis 5 and 11, which give the appearance of being closed genealogies, there is at least one gap.

What is more, John D. Currid, commenting on the Genesis 5 genealogy, notes, "The problem is that genealogies are not exhaustive or comprehensive lists in the Bible. They are highly selective...It is likely, then, that each name in Genesis 5 marks a separate landmark of a distinguished person in the line of Seth. It is simply a general chronology."[24] In order to demonstrate the error of the first assumption, it is only necessary to show that there is at least one gap. Currid's comment implies that there may be many gaps in the biblical genealogies. This warrants at least some caution when attempting to use genealogies to arrive at the age of the Earth or the age or the antiquity of the human race.

Regarding the second assumption, other scholars have addressed this issue, including John H. Sailhamer. For example, he demonstrates that the purpose of the author of Genesis, in relation to the genealogies in Genesis 5, 10, and 11, is not to give an exhaustive genealogy, but a *representative* genealogy.[25]

In relation to the third assumption, Norman Geisler notes that "there are ways to understand the text of Genesis 11 that do allow for gaps. The formula phrase 'and X lived so many years and begat Y' can mean 'and X lived so many years and became the ancestor of Y.' This is not speculation, for in Matthew 1:8 ('Jehoram begat Uzziah') it means precisely this."[26] The fact that within the genealogies there are alternative translations of the key terms implying chronology demonstrates that the third assumption is false. There are alternative translations of the key terms in the genealogies that do not imply precise chronology, but rather, lineage. This indicates there may be more gaps in the chronologies than we originally suspected.

Don't assume that the age of the Earth is an essential doctrine of Christianity. The Earth may, indeed, be young—a whole lot younger than the natural sciences suggest. But it may also be as old as the natural sciences propose, and there are legitimate ways of reading the biblical creation narrative (without compromising the inspiration, inerrancy, or authority of scripture) that allow for an extremely old universe and Earth—perhaps, as Geisler has said, even billions of years.[27] Indeed, we must separate and clearly distinguish between the essential doctrine and a secondary issue present in the debate: (1) *that* God created the cosmos and all that is in it is

an essential doctrine; (2) all claims to know the time *when* God created the universe and *how long* it took him to create it is a secondary issue. The Bible and natural revelation both clearly testify to the first (Romans 1:19-21), but neither the Bible nor natural revelation offer any irrefutable or final guidance as to the second.

Don't assume the time of creation is more important than the fact of creation. That God created the heavens and the Earth is the crucial aspect of the creation narrative. This affirms the source and the (intrinsic and extrinsic) causal connection between God and the material finite universe that stands in contradiction to the mythical creation accounts. Without this causal connect, Christians would be stuck with the unenviable (if not impossible) task of demonstrating how something could come to be from nothing. Genesis 1:1 offers a narrative consistent with the scientific law of causality, which states that "everything that has a beginning (an effect) had a cause (that which produces an effect)."

The fact of creation also stands in opposition to unguided Darwinian evolution because each creative day required God's intelligent intervention (i.e., the spoken word). This intervention is reaffirmed when God speaks twice on the third day as he transitioned from inorganic, nonliving creation (water and land) and organic, living creation (grass, herbs, trees) (Genesis 1:9-13). And again, on the sixth day there is outside intervention by God to create animals and another intervention for the creation of man.

This narrative presents the reader with intelligent divine guidance in the creation event, something totally foreign to purely naturalistic evolutionary models. Without this intervention one cannot progress through the sequence of creative acts, nature being incapable of accomplishing these transitions. Especially important is the *intrinsic* (attributes God shares) causal connection between God and material creation, and between God and animals, and uniquely between God and man. It is from this causal relationship that the image of God is retained by human beings, giving life sanctity and value because of the analogy (similarity) of being, as well as supporting the basis for Western morality and the doctrine of marriage, among other things.

These important connections between God and his creation make clear that though the *time* of creation in Scripture is uncertain, the *fact* of creation is without question present and necessary.

Don't assume that science and the Bible are in conflict. Because of the vigorous debate between scientists and theologians, it has been long asserted that science and Scripture are mutually exclusive domains. "Never the twain shall meet" is the old saying, according to some. Separating scientific inquiry from our understanding of the Bible is flawed thinking for several reasons.

First, when discussing the age of the Earth from passages in Genesis, science can inform our understanding of difficult texts much like it did to solve questions surrounding the Earth's position in the solar system five centuries ago (geocentrism versus heliocentrism). That is to say, prior to the seventeenth century, most believed in a fixed Earth (geocentric solar system) based on certain biblical texts (1 Samuel 2:8; Psalm 104:5; Ecclesiastes 1:5) and the best science of their day. However, today's best science has informed us to the extent that we now interpret those same passages with a view to a moving Earth (heliocentric solar system). Perhaps, in time, more scientific information may help us to finally solve the age of the Earth question.

Second, the domain of science (which is the material world of nature) and the domain of theology (which is the Bible) were both authored by God. Because God doesn't make mistakes, neither of the two domains can contradict the other. But there is certainly disagreement between scientists and theologians, so where does the conflict exist? The contradiction lies between the interpretations of nature by fallible scientists and the interpretations of the Bible by fallible theologians. Thus, the contradiction exists not at the domain level, but at the interpretive level.

Don't assume that a word's normal meaning must be its exclusive meaning whenever it is used. While the smallest unit of *meaning* is a sentence, individual words have *usage*, and depending upon how a word is used in its context, it can have a range of meanings. For example, the meaning of the word *run* is unknown unless you provide a context in which the word is used. *Run* could mean a drip of paint on a wall, a tear in hosiery, a score at a baseball game, jogging down the street, flowing water, or even to operate machinery, all of which have different meanings.

Similarly, it seems clear from Scripture that the Hebrew word translated "day" (*yom*) can have many usages depending on the context. In Genesis 1:5, "day" is often taken to refer to only the period of daylight within a single 24-hour day, but soon thereafter, in Genesis 2:2, *yom* seemingly means an indefinite period of time, where we read that on the seventh "day" God rested from his work of creation. Most believe this seventh day has continued up to the present moment. Obviously, as you read this chapter, God is still at work *in* his creation but has ceased from his work *of* creation (i.e., bringing new things into being). This is an example of how a "day" can mean a long period of time, as in other phrases such as "in that day…," "the day of the Lord," "and on the third day he will raise us up" (Hosea 6:2).

For example, Genesis 2:4 refers back to "the day" (*yom*) when creation occurred despite earlier passages revealing there were six days (plural) of creation, thus "the day" refers to a period of time spanning six days. What is more, *yom* is not used with the definite article "the" until days six and seven, meaning that days one through five can be translated as "a day" or "day one" and "day two" and so forth, but not "*the* nth day" until days six and seven. (Indeed, since many young-Earth interpretations of Genesis 1 hold that the sun did not exist until day four, one wonders how the yom's of days one through three could signify a 12-hour period of daylight.) Simply put, the text doesn't say how long the days were nor how long God took between days to speak. This is not to say that the word *yom* cannot refer to a 24-hour period—it can! In fact, the normal and most frequent usage of *yom* is exactly that. This typical usage, however, doesn't mean the days of creation *must* be interpreted this way in Genesis 1.

Some suggest that the attending numerical series describing the *first* day, *second* day, *third* day, etc., means that each consecutive day was a 24-hour period. However, caution is warranted because in Scripture, sometimes when numerical series are present they refer to long periods of time (e.g., Hosea 6:1-2). In addition, the phrase "evening and morning" poses a similar problem because the seventh day omits the phrase. Why is this omitted if all the days are to be understood as 24-hour periods? Further, if taken in a strict literal sense, the phrase may only refer to the end of one day and the morning early hours of the next.

In the final analysis, it appears the words and phrases in the Genesis creation narrative

may certainly be interpreted as 24-hour days, but we are not necessarily required to interpret them as such.

Don't assume that an old-Earth view is friendly to unguided Darwinian evolution. From the nineteenth century to today, a plethora of different time periods ranging from millions to billions of years have been proposed by scientists for the age of the Earth and the length of time it would take for various natural events to occur.[28] We could add to these suggestions more recent estimates, including astronomer Carl Sagan's suggestion that approximately 15 billion years have elapsed since the big bang that brought our universe into existence,[29] that the Earth formed 4.6 billion years ago,[30] and that humans evolved into existence about 3.6 million years ago;[31] theoretical physicist and cosmologist Stephen Hawking's estimate of between 10 and 20 billion years for the formation of the universe;[32] and paleontologist G.G. Simpson's estimate that modern humans arrived on the scene relatively recently—perhaps a few tens of thousands of years ago.[33] Australian philosopher of science Peter Godfrey-Smith recently described *the current picture* as saying that life originated on Earth some 3.5 billion years ago, that multicellular organisms appeared some 800 million years ago, "vertebrates (over 500 million years ago), flowering plants (over 120 million years ago), and [*Homo sapiens*] (about 300,000 years ago). That much is pure history. It is based not just on biological data, but on geology, physics, and other sciences."[34] The present standard accepted view within geology for the age of the Earth is 4.5–4.6 billion years,[35] and the standard accepted view within cosmology for the age of the universe is just shy of 14 billion years.[36]

One might think that such a great age of the Earth provides sufficient time for life to evolve via Darwinian processes. In fact, the evidence shows it is insufficient. In a recent book on the Cambrian explosion, Stephen C. Meyer notes that intelligent design (ID) proponents Michael Behe and David Snoke published an article in 2004[37] in which they used accepted theories and methods in population genetics to estimate that "based upon relevant mutation rates, known human population sizes, and generation times, the time required for two coordinated mutations to occur in the hominid line" would be hundreds of millions of years, "Yet humans are thought to have diverged from a common ancestor [with apes] only 6 million years ago."[38] Such a calculation would pose major problems for the standard picture of Darwinian biological evolution. Therefore, as Meyer notes, two neo-Darwinist mathematical biologists at Cornell University, Rick Durrett and Deena Schmidt, attempted to refute Behe's thesis.[39] Meyer points out,

Although they calculated a shorter waiting time than Behe did, their result nevertheless underscored the implausibility of relying on the neo-Darwinian mechanism to generate coordinated mutations during the relevant evolutionary timescale. Their calculation suggested that it would take not several hundred million years, but "only" 216 million years to generate and fix two coordinated mutations in the hominid line—more than thirty times the amount of time available to produce humans and chimps and all their distinctive complex adaptations and differences from their inferred common ancestor.[40]

Don't assume that supporting an old-Earth view is equivalent to being an evolutionist. The main implication of Behe's calculations is that the mechanisms of neo-Darwinian evolution do not allow for the unguided Darwinian evolution of the human species in the standard time span that is allowed by the fossil record and modern evolutionary theory. Because the standard picture and standard timeframe are never fully settled within the domain of the natural sciences, it is not necessary, in order to be rational, to uncritically accept any one timeframe that may be proposed by the natural sciences at any one time. One may maintain a healthy skepticism on this subject.

That being said, it should be noted that most ID proponents, such as Meyer and Behe, do agree that the universe is significantly older than young-Earth creationists suggest.[41] This does not make them naturalists, nor is it fair to lump them in with standard evolutionists. What separates ID proponents from neo-Darwinian materialists is that ID proponents present arguments based on observations from the natural sciences to show that some type of intelligent designer is necessary for various aspects of nature, such as the very existence of the universe, the fine-tuning of physics, and many complex features of biological life.

Don't assume that an ancient age of the universe is bad for Christian apologetics or arguments for a designer. If the universe had a beginning, who or what brought it into existence?[42] Robert Jastrow, prominent astronomer and founder of the Goddard Institute at NASA, in his book *God and the Astronomers*, notes a number of observations and calculations that imply that the universe had a beginning, including Einstein's theory of general relativity coupled with Slipher's calculations,[43] Hubble and Humason's discovery based on the red shift of light that "*the farther away a galaxy is, the faster it* moves,"[44] and the lifetime (birth and death) of stars.[45]

Further, astrophysical measurements over the past few decades have shown that the cosmic microwave background radiation in the universe matches very closely to what is predicted by the big bang.[46] The big bang model of the universe suggests it had a beginning, and if so, then according to the Kalam cosmological argument, the universe was caused. Numerous philosophers have recognized that if the universe had a beginning, then it is relatively simple to demonstrate the existence of some form of divine being. In light of this evidence, Jastrow noted, "For the scientist who has lived by his faith in the power of reason, the story ends like a bad dream. He has scaled the mountains of ignorance; he is about to conquer the highest peak; as he pulls himself over the final rock, he is greeted by a band of theologians who have been sitting there for centuries."[47] This evidence is closely tied to evidence for fine-tuning, a powerful argument for cosmic design wherein the laws and constants of the universe appear carefully crafted for advanced life to exist. All of these arguments, of course, entail the view that the universe is billions of years old.

In other words, if we are willing to consider the possibility that the universe is billions of years old, this opens us up to scientific discoveries—theories that are well accepted within mainstream science—that constitute some of the most compelling arguments for a divine designer ever uncovered by science. Of course, the fact that an argument makes for a compelling apologetic does not therefore make it true, but this also ought to alleviate fears that accepting an old earth and universe is somehow hostile to a Christian worldview.

Don't assume that "creation with the appearance of age" easily accommodates all the evidence for an old universe. The modern-day findings of mainstream science, in general, point to an ancient universe (some 14 billion years old), an ancient Earth (some 4.5 billion years old), and to the antiquity of humanity (figures ranging between 10,000–20,000 years to hundreds of thousands of years to a few million years). All of these proposed time periods are based upon research performed by a wide variety of scientists—some with atheistic assumptions, but others with clear Christian convictions.[48] Each argument needs to be analyzed for itself, but it is worth noting that none of them are entirely foolproof.

For example, arguments for the antiquity of the cosmos based upon our ability to view stars that are millions or billions of light-years away assume (1) that the speed of light in a vacuum is fixed and constant and that it has never sped up or slowed down; and (2) that the light rays coming from the stars were not created at the same time as their sources. Both of these assumptions may be true, and, if so, then the natural sciences (as we currently understand them) have demonstrated both that the universe is ancient and that it had a beginning. However, these assumptions must both be true in order to prove the great age of the universe on the basis of the speed of light.[49] There are also assumptions related to the accuracy of geological radiometric dating methods, which are used to estimate the age of the Earth (e.g., radioactive decay rates over Earth's history must remain constant; daughter products were not present at the time of creation to give the appearance of age).[50]

"Creation with the appearance of age" has been invoked occasionally by young-Earth creationists to accommodate evidence for an old Earth and universe. "Couldn't God create Adam with a navel?," or "Couldn't God create light from very distant stars 'in transit' so they could be seen on Earth?" may seem like compelling arguments wherein God could create a mature creation that appears old, even though it isn't. In some cases, these arguments may hold up without making it appear as though God was deceptive.

However, consider one example and what it means for the truthfulness of God if the universe is in fact young: supernovae. In his article "Star Light & the Age of the Universe," Christian apologist Greg Koukl points out that young-Earth creationists sometimes argue that "[l]ight from stars millions of light years away from the earth was created in transit. Observers on earth (Adam and Eve) could see the star instantly, in spite of the great distances, in spite of the fact that the universe was only days old." But what about supernovae? Supernovae are events observed throughout human history wherein a star exhausts its fuel and explodes, producing a brilliant explosion of light and other forms of deadly radiation in the sky. Dozens of these events have been recorded throughout history—often from stars millions of light-years away, whose light should have otherwise taken millions of years to arrive on Earth. (Everyone agrees that these stars are millions of light-years away from simple geometrical calculations based upon astronomical measurements.) Koukl claims these supernova events pose a difficult question for the young-Earth creationist viewpoint:

Here is my question: Have supernovas ever happened? If the young-earth explanation for the apparent age of the universe is true, then the answer is no. There have never been any supernovas. Such an explosion close enough to be actually witnessed by us during human history (within, say, 10,000 light years)

would have obliterated life on our planet. All those that appear further away merely represent light created in transit and not the supernova itself.

According to Koukl, this leads to the troubling implication that "God has fabricated images of events that never happened, but passes them off as if they did."[51] It seems most consistent with God's truthful character,[52] he argues, to accept that supernovae are real events that occurred millions of years ago, rather than phantoms that God created to give the universe a false appearance of age. Koukl concludes: "On the other hand, if we see real stars when we look heavenward, and those stars are a billion light years away, then they must have existed a billion years ago. If our eyes and our instruments can be trusted, then the universe is ancient. When the Bible tells us to behold the stars, it speaks the truth."[53]

Don't assume you have the final answer even if the evidence appears to weigh heavily in your favor. Finally, as mentioned earlier, due to the nature of scientific speculation and theorization, it is entirely possible (as incredible as it may seem to many today) that future discoveries may turn evolutionary science and the timespans it suggests on its head. Thus, the call for humility in how we hold our viewpoints. Remember that each generation of scientists thought that their theories and arguments were essentially foolproof, speaking with conviction, until later generations made discoveries that overturned earlier theories. As such, it is entirely acceptable to hold out in the possibility that future scientific discoveries will conclusively vindicate either a young-Earth or old-Earth theory with finality as it did the debate between the "fixed" Earthers and the "moving" Earthers during the time of Galileo.

The Fact of Creation

How, then, should we think about the age of the Earth? In light of what we have seen in this chapter, there are a number of conclusions that should be addressed.

First, regardless of whether we hold to a young- or old-Earth viewpoint, it is absolutely necessary to distinguish, in our discussions, between the age of the universe, the age of the Earth, and the antiquity of the human race. Even the young-Earth creationist must admit that the cosmos existed at least five days before humans.

Second, we must be willing to humbly consider the offerings of both biblical interpretation and the observations of the natural sciences, recognizing the fallibility and tentative character of both with regard to theories of origins. We must be willing to change our opinions, when necessary, without compromising our belief in the inspiration, inerrancy, and authority of Scripture. To differ on the age of the Earth based on what is revealed in Genesis 1 is not a matter of inerrancy, but rather, a question of textual interpretation. To categorize some Christians as compromising inerrancy because of holding to both inerrancy and a young or old Earth is to fail to recognize the distinction between inerrancy and hermeneutics, which marks the difference between what Scripture *is* and how we interpret what Scripture *says*. Of course the Bible is infallible, but our interpretations are not! Christians must always remember that God is truth, and to be human brings the possibility of error.

Third, we can and must defend the creation, by divine fiat, of the entire cosmos and all that is within it, all while humbly noting that our own personal views and convictions on how and when (including the age of the cosmos, the age of the Earth, and the antiquity of the human race) are not questions of orthodoxy, nor necessary biblical truths, but questions of personal conviction based upon our limited knowledge of the relevant facts.

Fourth, the matters relating to the age of the cosmos and Earth and the antiquity of the human species are not essential doctrines, nor is it necessary to have the "right answers" in order to properly interpret Christian Scripture and defend all the essential doctrines. There are a number of different approaches to this question that do not jeopardize the historic Christian faith. This is one of those areas in which Christians need to be reminded to "major on the majors, and minor on the minors." As we mentioned earlier, the *fact* of creation (versus evolution) is more important than the *time* of creation. When we make that our emphasis, then no matter whether God created everything in six seconds, six days, or six billion years, he still receives the glory for creating. An overly dogmatic adherence to any one theory of age is unhelpful both for relationships with those who are in the church and for relationships with those who are outside.[54]

Finally, the age question should not be used as a criterion for Christian fellowship nor as a test for biblical orthodoxy or inerrancy, as no one is denying the *fact* and *historical* nature of creation. Both camps have far more in common than the differences that distinguish them, including all the essentials of the faith and the promise that both will spend eternity together in the bliss of our Lord Jesus Christ.

How Have Christians Helped to Advance Science?

Henry "Fritz" Schaefer III

Many educated people are of the opinion that there has been terrible warfare between science and Christianity. Let us attempt to put this question of the relationship between science and Christianity in the broadest and most reasonable perspective possible. We begin by noting that the rapprochement between science and other intellectual pursuits has not always been easy. For example, the book *Literature* by Susan Gallagher and Roger Lundin states, "Because in recent history, literature has often found itself in opposition to science, to understand modern views about literature, we must recognize the dominance of science in our culture. For several centuries, scientists have set the standards of truth for Western culture. And their undeniable usefulness in helping us organize, analyze, and manipulate facts has given them an unprecedented importance in modern society."[1]

For example, John Keats, the great English romantic poet, did not like Isaac Newton's view of reality. He said it threatened to destroy all the beauty in the universe. He feared that a world in which myths and poetic visions had vanished would become a barren and uninviting place. In his poem "Lamia" he talks about this destructive power. In this poem, he called science "philosophy," so I will replace the word *philosophy* with *science* so as not to confuse the twenty-first-century reader:

> Do not all charms fly
> At the mere touch of cold science?
> There was an awful rainbow once
>   in heaven:
> We know her woof, her texture;
>   she is given
> In the dull catalog of common things.
> Science will clip an Angel's wings,
> Conquer all mysteries by rule
>   and line,
> Empty the haunted air and
>   gnomed mine—
> Unweave a rainbow...

My point is that there has been friction between science and virtually every other intellectual endeavor since the appearance of modern science as a newcomer on the scene around AD 1500. So it would be surprising if there were not some heated exchanges between science and Christianity. What I am describing is "the new kid on the block" syndrome in colloquial North American English.

Has Science Disproved God?

Nevertheless, the position is commonly stated that "science has disproved God." C.S. Lewis, in the autobiography of his early life, *Surprised by Joy*, said that he believed the above statement. He talked about the atheism of his early years on the faculty at Oxford University and credits it to science. Lewis writes, "You will understand that my [atheism] was inevitably based on what I believed to be the findings of the sciences, and those findings, not being a scientist, I had to take on trust—in fact, on authority."[2] What Lewis was saying is that somebody told him that science had disproved God; and he believed it, even though he knew nothing about science.

A more balanced view of this question was given by one of my scientific heroes, Erwin Schrödinger (1887–1961). He was perhaps the most important of the founders of wave mechanics and the originator of what is now the most important equation in science, Schrödinger's equation. Schrödinger declared,

> I am very astonished that the scientific picture of the real world is very deficient. It gives a lot of factual information, puts all our experience in a magnificently consistent order, but it is ghastly silent about all and sundry that is really near to our heart, that really matters to us. It cannot tell us a word about red and blue, bitter and sweet, physical pain and physical delight, knows nothing of beautiful and ugly, good or bad, God and eternity.

Science sometimes pretends to answer questions in these domains, but the answers are very often so silly that we are not inclined to take them seriously.[3]

Scientists do tell some interesting stories about religion. A good one is from *Chemistry in Britain*, which is something like the *Time* magazine of the chemical profession in England. Talking about the release of a new book on science policy in July 1989, *Chemistry in Britain* explores an interesting idea:

If God applied to the government for a research grant for the development of a heaven and an Earth, he would be turned down on the following grounds:

- His project is too ambitious;

- He has no previous track record;

- His only publication is a book, not a paper in a refereed journal;

- He refuses to collaborate with his biggest competitor;

- His proposal for a heaven and an Earth is all up in the air.[4]

Some Alternatives to Belief in the Sovereign God of the Universe

I present here two examples of notable atheists. The first is Lev Landau, the most brilliant Soviet physicist of the twentieth century. Landau received the 1962 Nobel Prize in Physics for his research on liquid helium. Moreover, Landau was named a Hero of Socialist Labor by the Soviet government. He was also the author of many famous physics textbooks with his coworker E.M. Lifshitz. I used some of these books as an undergraduate at MIT. A story about Landau by his good friend and biographer I.M. Khalatnikov appeared in the May 1989 issue of *Physics Today*. Khalatnikov writes: "The last time I saw Landau was in 1968 after he had undergone an operation. His health had greatly deteriorated. Lifshitz and I were summoned by the hospital. We were informed that there was practically no chance he could be saved. When I entered his ward, Landau was lying on his side, his face turned to the wall. He heard my steps, turned his head, and said, 'Khalat, please save me.' Those were the last words I heard from Landau. He died that night."[5]

The second example is Subrahmanyan Chandrasekhar, the famous astrophysicist who won the Nobel Prize in Physics in 1983. He was a faculty member at the University of Chicago for most of his life. At the back of his biography is an unusual interview. Chandrasekhar began the dialogue, saying, "In fact, I consider myself an atheist...But I have a feeling of disappointment because the hope for contentment and a peaceful outlook on life as the result of pursuing a goal has remained largely unfulfilled."

His biographer, K.C. Wali, responded with astonishment: "What?! I don't understand. You mean, single-minded pursuit of science, understanding parts of nature and comprehending nature with such enormous success still leaves you with a feeling of discontentment?"

Chandrasekhar continued in a serious way, saying, "I don't really have a sense of fulfillment. All I have done seems to not be very much."

The biographer sought to lighten up the discussion a little, saying that everybody has the same sort of feelings. But Chandrasekhar would not let him escape, saying, "Well it may be. But the fact that other people experience it doesn't change the fact that one is experiencing it. It doesn't become less personal on that account." And Chandrasekhar's final statement, which I urge every potential young scientist to ponder, reads, "What is true from my own personal case is that I simply don't have that sense of harmony which I had hoped for when I was young. And I have persevered in science for over fifty years. The time I have devoted to other things is minuscule."[6]

Is It Possible to Be a Scientist and a Christian?

The question I want to explore is the one that I was asked by a young man after my first freshman chemistry class at Berkeley: Is it possible to be a scientist and a Christian? The student and his high school chemistry teacher obviously thought it was not possible.

Let me begin from what some might call neutral ground by quoting two people with no particular theistic inclinations. The first individual is C.P. Snow (1905–1980). Snow remains well known in intellectual circles as the author of an essay titled "The Two Cultures and the Scientific Revolution." Snow was a physical chemist—actually, a spectroscopist—at Cambridge University. About halfway through his career he discovered that he also was a gifted writer, and he began writing novels. One is a story about university life at Cambridge or Oxford, called *The Masters*, and I recommend reading it. Snow became quite comfortable as a result of the royalties he received from his novels and was able to sit in a unique position between the world of the sciences and the world of literature. From this perspective, Snow wrote, "Statistically, I suppose, slightly more scientists are in religious terms unbelievers, compared with the rest of the intellectual world, though there are plenty who are religious, and that seems to be increasingly so among the young."[7] So is it possible to be a scientist and a Christian? Snow answered in the affirmative.

Richard Feynman (1918–1988), awarded the Nobel Prize in Physics in 1965, was a most remarkable person. Perhaps some of my readers have seen his book of anecdotes *Surely You're Joking, Mr. Feynman*. Some nine years before receiving the Nobel Prize, he said that "many scientists *do* believe in both science and God, in a perfectly consistent way."[8] So is it possible to be a scientist and a Christian? Yes, according to Richard Feynman, an outspoken atheist.

A good summary statement in this regard is by Alan Lightman, who has written a very well-received book titled *Origins: The Lives and Worlds of Modern Cosmologists*. Dr. Lightman is an MIT professor who published this seminal work with the Harvard University Press. A critical paragraph in the book states: "References to God or divine purpose continued in the scientific literature until the middle to late 1800s. It seems likely that the studied lack of religious references after this time resulted more from a change in social and professional convention among scientists rather than from any change in underlying thought. Indeed, contrary to popular myth, scientists appear to have the same range of attitudes about religious matters as does the general public."[9]

Now someone could regard the above statement as strictly anecdotal. Many people like statistics better than anecdotes. So let me present the results of a poll of the members of the scientific professional society Sigma Xi. A total of 3,300 scientists responded to the survey, so the conclusions are certainly beyond statistical uncertainty. The description of the survey in the November 7, 1988 issue of *Chemical & Engineering News* reads. "Scientists are anchored in the U. S. mainstream."[10]

The article states that half of the scientists polled participate in religious activities regularly. A less exhaustive survey (*The Scientist*, May 19, 2003) finds 52 percent of biologists identifying themselves as Christians. More recently, Professor Elaine Ecklund of Rice University has written an excellent book *Secularity and Science: What Scientists Around the World Really Think About Religion*, published by Oxford University Press in 2019. Therein she examines the religious views of 10,000 scientists. Ecklund found that 65 percent of rank-and-file scientists in the US identify themselves as Christians. So it seems that whatever it is that causes people to adopt or reject religious inclinations is unrelated to having an advanced degree in science.

Let us go a little deeper with a statement from Michael Polanyi (1891–1976), professor of chemistry and later of philosophy at the University of Manchester. His son, John Polanyi, won the Nobel Prize in 1986. I think that it may be true that when John Polanyi's scientific accomplishments, which have been truly magnificent, have been mostly forgotten, the impact of his father's work will continue.

Michael Polanyi was a great physical chemist at the University of Manchester. About halfway through his career, he switched over to philosophy, and particularly the philosophy of science. He was equally distinguished there. His books, the most influential of which is called *Personal Knowledge*, are not easy to read, but are very worthwhile. He was of Jewish descent, raised in Budapest, Hungary. About the same time that he began the switch from chemistry to philosophy, he joined the Roman Catholic Church. A typical Michael Polanyi statement reads,

I shall re-examine here the suppositions underlying our belief in science and propose to show that they are more extensive than is usually thought. They will appear to coextend with the entire spiritual foundations of man, and to go to the very root of his social existence. Hence I will urge our belief in science should be regarded as a token of much wider convictions.[11]

If you read further, you will probably come to the same conclusion that I draw. Polanyi points out that the observer is always there in the laboratory. He or she always makes conclusions. He or she is never neutral. Every scientist brings presuppositions to his or her work. A scientist, for example, never questions the basic soundness of the scientific method. This faith of the scientist arose historically from the Christian belief that God the Father created a perfectly orderly universe. Next I will provide some concrete evidence for this conclusion.

Why Might a Scientist Become a Christian?

I will ask this question several times in the course of the present chapter. Physics Nobelist Eugene Wigner (1902–1995) once noted "the unreasonable effectiveness of mathematics" and remarked, "The miracle of the appropriateness of the language of mathematics for the formulation of the laws of physics is a wonderful gift that we neither understand nor deserve."[12] Interestingly, Wigner, like Polanyi, was a man of Jewish origin who found his way into nominal Christendom—in his case, Protestantism. Indirectly, Wigner hinted that the intelligibility of the universe points to a sovereign creator God. Thus, mathematical physics can be an answer to the question we pose in this section. The laws of nature look just as

if they have been selected as the most simple and elegant principles of intelligible change by a wise creator. Belief in the intelligibility of nature strongly suggests the existence of a cosmic mind who can construct nature in accordance with rational laws. Dr. Keith Ward, Regius Professor at Oxford University, well stated (see for example, his 2009 book *The God Conclusion*):

> Thus appeal to the general intelligibility of nature, its structuring in accordance with mathematical principles which can be understood by the human mind, suggests the existence of a creative mind, a mind of vast wisdom and power. Science is not likely to get started if one thinks that the universe is just a chaos of arbitrary events, or if one thinks there are many competing gods, or perhaps a god who is not concerned with elegance or rational structure. If one believes those things, one will not expect to find general rational laws, and so one will probably not look for them. It is perhaps no accident that modern science really began with the clear realization that the Christian God was a rational creator, not an arbitrary personal agent...[13]

I need to be clear that it is not only persons with Christian sympathies who acknowledge the remarkable intelligibility of the universe. For example, Sheldon Glashow (Nobel Prize in Physics, 1979) stated in 1990, "Many scientists are deeply religious in one way or another, but all of them have a certain rather peculiar faith—they have a faith in the underlying simplicity of nature; a belief that nature is, after all, comprehensible and that one should strive to understand it as much as we can." However, without a belief in the sovereign God of the universe, one may project such general observations in questionable directions. For example, Glashow continues, "Now this faith in simplicity, that there are simple rules—a few elementary particles, a few quantum rules to explain the structure of the world—is completely irrational and completely unjustifiable."[14]

Science Developed in a Christian Environment

I like to begin with an outrageous statement that always causes reaction. This statement is from a British scientist, Robert Clark (1906–1984), and it appears in his book *Christian Belief and Science*:

> However we may interpret the fact, scientific development has only occurred in a Christian culture. The ancients had brains as good as ours. In all civilizations—Babylonia, Egypt, Greece, India, Rome, Persia, China and so on—science developed to a certain point and then stopped. It is easy to argue speculatively that, perhaps, science might have been able to develop in the absence of Christianity, but in fact, it never did. And no wonder. For the non-Christian world believed that there was something ethically wrong about science. In Greece, this conviction was enshrined in the legend of Prometheus, the fire-bearer and prototype scientist who stole fire from heaven, thus incurring the wrath of the gods.[15]

I would prefer if Dr. Clark had said that Christian culture had "sustained" scientific development in the first sentence quoted above. I think he went a little too far here, but his words certainly give people something to think about.

A frequent objection to Clark's statement is that science made significant progress in the Middle East under Islam during the Middle Ages. This is of course true, but why did these early scientific contributions fail to be "sustained"? In his important 2002 book appropriately titled *What Went Wrong? The Clash Between Islam and Modernity in the Middle East*, Professor Bernard Lewis of Princeton University addressed this critical question. The inability of science to continue under Islam is perhaps best illustrated by the fate of the great observatory built in Galata, Istanbul, in 1577. This observatory gave every promise of being comparable to that of the Danish scientist Tycho Brahe (1546–1601), who revolutionized astronomy. In *What Went Wrong?*, Lewis relates that the observatory at Galata was razed to the ground by an elite corps of Turkish troops, by order of the sultan, on the recommendation of the Chief Mufti (Islamic leader) of Istanbul. For the next 300 years there was no modern observatory in the Islamic world.

Let us explore the idea presented by Polanyi, Ward, and Clark in their statements that modern science grew up in a Christian environment. I was taught in my childhood that Francis Bacon (1561–1626) discovered the scientific method. The higher critics have now gotten into the history of science, and some claim that Bacon stole the scientific method from a multitude of others and merely popularized it. We must leave that dilemma to the science historians to settle. One of Francis Bacon's most frequently quoted statements is known as the "Two Books" manifesto. These words from Bacon have been highly influential and the subject of a magnificent recent essay by Professor Thomas Lessl. Bacon said, "Let no man...think or maintain that a man can search too far or be too well studied in the book of God's word or in the book of God's works."[16] Bacon is talking about the Bible as the book of God's word and nature as the book of God's works. He encouraged people to learn as much as possible about both. So, right here in the earliest days of the scientific method, we have a statement of the compatibility of science with the 66 books of the Hebrew Bible and New Testament. I have taken Bacon's advice personally, having read through the Bible more than 30 times since I became a Christian in 1973.

Johannes Kepler (1571–1630) was a brilliant mathematician, physicist, and astronomer. Kepler posited the idea of elliptical orbits for planets and is considered the discoverer of the laws of planetary motion. He was a devout Lutheran Christian. When he was asked, "Why do you engage in science?," Kepler answered that he desired, in his scientific research, "to obtain a sample test of the delight of the Divine Creator in His work and to partake of His joy."[17] This has been restated in many ways by other people, to think God's thoughts after him, to know the mind of God, and so on. Kepler might be mistakenly considered a follower of deism. But he elsewhere clarified, "Only and alone in the service of Jesus Christ. In him is all refuge and solace."[18]

Blaise Pascal (1623–1662) was a magnificent scientist. He is the father of the mathematical theory of probability and combinatorial analysis. He provided the essential link between the mechanics of fluids and the mechanics of rigid bodies. And he is, in my opinion, the only physical scientist to have made profound contributions to Christian thinking. Many of these thoughts are found in his little book known as the *Pensées*, which I was required to read as a sophomore at MIT.

The school was trying to civilize us geeks, but a few years later, MIT decided that it was not working; so students are no longer required to take as many humanities courses as they used to.

Pascal's theology was centered on the person of Jesus Christ as Savior and based on personal experience. He stated that God makes people "conscious of their inward wretchedness, and His infinite mercy, who unites Himself to their inmost soul, who fills it with humility and joy, with confidence and love, who renders them incapable of any other end than Himself."[19]

Robert Boyle (1627–1691) was perhaps the first chemist. He gave the first operational definition of an element, demonstrating enormous ingenuity as he constructed experiments in support of the atomistic hypothesis. Many of my freshman chemistry students remember Boyle's law. I sometimes return to Berkeley for a week during the summer, and every once in a while I will meet one of my former chemistry students on the campus. They typically ask "Didn't you used to be Professor Schaefer?" And I ask them in return, "What do you remember from my freshman chemistry course?" Occasionally they will say: "pV = nRT." In such cases, I know that my teaching was fabulously successful. This, of course, is the ideal gas law, of which Boyle's law is a critical part.

Robert Boyle was a busy person. He wrote many books, including *The Wisdom of God Manifested in the Works of Creation*. He personally endowed an annual lectureship promoted to the defense of Christianity against indifferentism and atheism. He was a good friend of Richard Baxter, one of the great Puritan theologians. He was also the governor of the Corporation for the Spread of the Gospel of Jesus Christ in New England.

Although I disagree with the finding, a

recent academic poll concerning the most important person in history gave that honor to Sir Isaac Newton (1642–1727). Newton was a mathematician, physicist, codiscoverer with Liebnitz of calculus, and the founder of classical physics. He was the first of the three great theoretical physicists to date. He also investigated many other subjects. Newton tried very hard to do chemistry but was less than successful. He wrote more words about theology than science. Still in print is his book about the return of Jesus Christ, entitled *Observations on the Prophecy of Daniel and the Revelation of St. John.*

One of Newton's most frequently quoted statements is, "This most beautiful system of the sun, planets and comets could only proceed from the counsel and dominion of an intelligent and powerful Being."[20] Based on these words, one might assume that Newton was a deist; however, other statements by Newton like the following show this not to be true: "I find more sure marks of authenticity in the Bible than in any profane history whatsoever."[21]

In fact, one may more reasonably conclude that Newton was a biblical literalist rather than a deist. Edward B. Davis writes that for Newton it was not enough that an article of faith could be deduced from Scripture. He quotes Newton as stating, "'It must be exprest in the very form of sound words in which it was delivered by the Apostles,' for men were apt to 'run into partings about deductions. All the old Heresies lay in deductions; the true faith was in the [biblical] text.'"[22]

George Trevelyan (1906–1996), the distinguished secular historian who wrote *English Social History*, summarized the contributions I have been discussing as follows:

Robert Boyle, Isaac Newton and the early members of the Royal Society

were religious men, who repudiated the skeptical doctrines of Thomas Hobbes. But they familiarized the minds of their countrymen with the idea of law in the universe and with scientific methods of enquiry to discover truth. It was believed that these methods would never lead to any conclusions inconsistent with Biblical history and miraculous religion. Newton lived and died in that faith.[23]

Beyond the Eighteenth Century

My very favorite among these legendary figures and probably the greatest experimental scientist of all time is Michael Faraday (1791–1867). The 200th anniversary of Faraday's birth was celebrated in 1991 at the Royal Institution (the multidisciplinary scientific research laboratory in London, of which Faraday was the director). There was an interesting article published in this context by my friend Sir John Meurig Thomas, who said that if Michael Faraday had lived into the era of the Nobel Prize, he would have been worthy of *eight* Nobel Prizes. Faraday discovered benzene and electromagnetic induction, invented the generator, and was the main architect of the classical field theory of electromagnetism.

Let me contrast the end of Faraday's life with the end of Lev Landau's life, which we read about earlier. As Faraday lay on his deathbed, a friend and well-wisher came by and asked, "Sir Michael, what speculations have you now?" This friend was trying to introduce some cheer into the situation. Of course the passion of Faraday's career had consisted of making "speculations" about science and then dashing into the laboratory to either prove or disprove them. It was a reasonable thing for a friend to ask in a difficult situation. Faraday took the question seriously. He replied, "Speculations, man, I have none! I have certainties. I thank God that I don't rest my dying head upon speculations, for I know whom I have believed and am persuaded that he is able to keep that which I have committed unto him against that day."[24]

The first time I used this statement in a public setting more than 30 years ago, a bright eyed and bushy-tailed young person in the front row of the audience burst out, "I've heard that before, and I am delighted to know that it was Michael Faraday who first spoke those words." As gently as possible, I informed him that the words were first penned by St. Paul some 1,900 years earlier to express his confidence in Jesus Christ. Michael Faraday had a firm grasp of the New Testament.

The second of the three great theoretical physicists of all time would certainly have to be James Clerk Maxwell (1831–1879). In *A Biographical Dictionary of Scientists* (1982), Trevor Williams summarized Maxwell's career this way:

> Maxwell possessed all the gifts necessary for revolutionary advances in theoretical physics: a profound grasp of physical reality, great mathematical ability, total absence of preconceived notions, a creative imagination of the highest order. He possessed in addition the gift to recognize the right task to which to apply his genius—the mathematical interpretation of Faraday's concept of electromagnetic field. His successful completion of this task, resulting in the field equations bearing his name, constituted one of the great creative achievements of the human intellect...[25]

Those who have thought deeply about the history and philosophy of science (e.g., Michael Polanyi and Thomas Kuhn) would disagree with one statement made above. If Maxwell indeed had a "total absence of preconceived notions," he would have accomplished a total absence of science.

Although Maxwell's equations are indeed one of the great achievements of the human intellect, as a member of the MIT sophomore physics class during the 1963–1964 academic year, I probably would have described them in different language at the time. However, just before our first examination in electromagnetism, one of the members of our class had a brilliant idea. This entrepreneurial wag had 900 T-shirts printed with Maxwell's equations embossed in large script. The entire class showed up to the first exam dressed in this unusual garb. Maxwell's equations were plainly visible from every seat in the auditorium. Our class averaged 95 percent on the first electromagnetism exam! Regrettably, the professor was distinctly unhappy. The average score on the second exam, despite our awesome T-shirts, was 15 percent. Never mess with a professor.

On June 23, 1864, James Clerk Maxwell wrote,

> Think what God has determined to do to all those who submit themselves to his righteousness and are willing to receive his gift [the gift of eternal life in Jesus Christ]. They are to be conformed to the image of His Son, and when that is fulfilled, and God sees they are conformed to the image of Christ, there can be no more condemnation.[26]

Maxwell and Charles Darwin were contemporaries. Many wondered what a committed Christian such as Maxwell thought of Darwin's ideas. In fact, Maxwell once was invited to attend a meeting on the Italian Riviera in February to discuss new developments in science and the Bible. If you have ever spent time in Cambridge, England, you know it is very gloomy there in the wintertime. If I had been a member of the Cambridge faculty, I would have taken every opportunity to go to the Italian Riviera at this time of the year. However, Maxwell turned down the invitation, explaining in his letter of declination, "The rate of change of scientific hypotheses is naturally much more rapid than that of Biblical interpretation. So that if an interpretation is founded on such a hypothesis, it may help to keep the hypothesis above ground long after it ought to be buried and forgotten."[27]

This is sage advice. One example of this is the steady-state theory, which was popularized by Fred Hoyle and others. It was for decades one of the two competing theories of the origin of the universe. The steady-state hypothesis basically said that what you see is what was always there. It became less tenable in 1965 with the observation of microwave background radiation by Arnold Penzias and Robert Wilson. There are almost no cosmologists left who believe in the steady-state hypothesis. But it is amusing to go back to about 1960, find biblical commentaries on the book of Genesis, and see the ways in which an unfortunate few explained how the steady-state hypothesis can be reconciled with the first chapter of Genesis. Any reasonable person can see that the Genesis account is describing a creation from nothing (ex nihilo), so it takes a vivid imagination to reconcile a beginning in space, time, and history with the now-discredited steady-state hypothesis.

By the second half of the twenty-first century, should planet Earth still be here, the steady-state hypothesis will be dead and nearly forgotten. And those commentaries will probably still be available in libraries, but few people will be able to understand them. This is an excellent example of the important point made by James Clerk Maxwell well more than a century ago about ideas that "ought to be buried and forgotten."

One of my favorite cartoons was published by Sidney Harris a few years ago in *The American Scientist*. Two distinguished elderly scientists are staring unhappily at an obscure equation on a blackboard. One of them delivers the punchline: "What is most depressing is the realization that everything we believe will be disproved in a few years." I hope that is not true of my students' research in quantum chemistry. I don't think it will be the case, but there is an important element of reality to this—science is inherently a tentative activity. As scientists, we come to understandings that are always subject, at the very least, to further refinements.

Of course, not all biographers of these pioneers of modern physical science spoke positively of their Christian convictions. For example, James Crowther states in his biography of Faraday and Maxwell, "The religious decisions of Faraday and Maxwell were inelegant, but effective evasions of social problems that distracted and destroyed the qualities of the works of many of their ablest contemporaries."[28] In context, what Crowther is saying is that because they were Christians, Maxwell and Faraday did not become alcoholics, womanizers, or social climbers, to enumerate the disabling sins of a number of gifted scientists of the same era.

I need to insert a little organic chemistry here so that my colleagues on the organic side will know that I am paying some attention to them as well. William Henry Perkin

(1838–1907) was perhaps the first great synthetic organic chemist. He was the discoverer of the first synthetic dye, known as Perkin's mauve or aniline purple. Prior to Perkin's discovery, the use of the color purple had been extremely expensive and often limited to persons of royal descent. He is the person for whom an important journal, the *Perkin Transactions of the Royal Society of Chemistry (London)*, was named. In the year 1873, at the age of 35, Perkin sold his highly profitable business and retired to private research and church missionary ventures. One of the more humorous responses to the contents of this chapter, shared in a lecture a few years back, was a suggestion from the audience that I follow William Henry Perkin's example in this respect.

Perkin was carrying out research on unsaturated acids three days prior to his death, brought on by the sudden onset of appendicitis and double pneumonia. The following account is given in Simon Garfield's 2002 Perkin biography titled *Mauve*. On his deathbed, William Henry Perkin stated, "The children are in Sunday School. Give them my love, and tell them always to trust Jesus."[29] He then sang the first verse of the magisterial Isaac Watts hymn "When I Survey the Wondrous Cross." When he reached the last line, which reads "And pour contempt on all my pride," Perkin declared "Proud? Who could be proud?"[30]

One can find the name George Stokes (1819–1903) in most any issue of the *Journal of Chemical Physics*, the most prestigious journal in my field. In recent years, Coherent Anti-Stokes Raman Spectroscopy (CARS) has been a subject of much scholarly investigation. Stokes was one of the great pioneers of spectroscopy, fluorescence, and the study of fluids. He held one of the most distinguished chairs in the academic world for more than

50 years, the Lucasian Professorship of Mathematics at Cambridge University. This position was held earlier by Sir Isaac Newton and was the chair occupied until recently by Stephen Hawking. Stokes was also the president of the Royal Society of London.

Stokes wrote on a range of matters beyond chemistry and physics. Concerning the question of miracles, Stokes wrote in his book *Natural Theology*, published in 1891, "Admit the existence of God, of a personal God, and the possibility of miracles follows at once. If the laws of nature are carried out in accordance with His will, He who willed them may will their suspension. And if any difficulty should be felt as to their suspension, we are not even obliged to suppose that they have been suspended."[31]

William Thomson (1824–1907) was later known as Lord Kelvin. He was recognized as the leading physical scientist and the greatest science teacher of his time. His early papers on electromagnetism and his papers on heat provide enduring proof of his scientific genius. He was a Christian with a strong faith in God and the Bible. In a speech at University College in 1903, Kelvin stated, "Do not be afraid of being free thinkers. If you think strongly enough, you will be forced by science to the belief in God."[32]

In 1897, J.J. Thomson (1856–1940) identified and characterized the electron, one of the most profound discoveries in the history of science. Thomson was for many years the Cavendish Professor of Physics at Cambridge University. The old Cavendish Laboratory still sits in the middle of the beautiful Cambridge campus. So many remarkable discoveries were made in the old Cavendish that it has essentially become a museum. A total of something like a dozen Nobel Prizes resulted from research done in that laboratory.

When the old Cavendish was opened by James Clerk Maxwell in 1874, he had a Latin phrase from Psalm 111 carved over the front door. Perhaps 30 years ago I had my daughter Charlotte, who subsequently graduated from Stanford University in classics, translate this phrase for me. Then we walked out into the Cambridge countryside where the shiny new Cavendish Laboratory was dedicated in 1973. Placed over the front door is the very same phrase, but this time in English: "The works of the LORD are great, sought out of all them that have pleasure therein." Thomson made the following statement in *Nature* (a journal in which I have published): "In the distance tower still higher [scientific] peaks which will yield to those who ascend them still wider prospects, and deepen the feeling, the truth of which is emphasised by every advance in science, that 'Great are the works of the Lord.'"[33] For the brilliant Thomson, the bottom line in science was that the works of the Lord are magnificent.

Those who know my research will not be surprised that this chapter must include at least one theoretical chemist. Let's make it three in the paragraphs that follow.

Charles Coulson (1910–1974) was one of the principal architects of the molecular orbital theory. I had the privilege of meeting Coulson just once, at the Canadian Theoretical Chemistry Conference in Vancouver in 1971. He probably would have received the Nobel Prize in Chemistry, but he did not pass the usual first test. This typical first hurdle to getting the Nobel Prize is to live to be 65 years old—a wonderful excuse for people comfortably below that threshold who have not made the trip to Stockholm. The second test, far more challenging, is to have done something very important when you were 30–40 years old. Coulson indeed did profoundly significant work when he was in his

thirties, but he died at 64, which precluded his chances of getting the Nobel Prize.

Coulson, a professor of mathematics and theoretical chemistry at Oxford University for many years, was also a Methodist lay minister. Norman March (a good friend), successor to the renamed Coulson Chair of Theoretical Chemistry, was also a Methodist lay minister. Alas, upon the retirement of Professor March a few years ago, a suitable Methodist lay minister could not be found for the Coulson Chair at Oxford. So the university settled for an Anglican Christian, Professor Mark Child (also a friend). Charles Coulson was a spokesman for Christians in academic science and the originator of the term "God of the gaps," now widely used in philosophical circles.

From the biographical memoirs of the Royal Society of London (1974), following Charles Coulson's death, we read Coulson's own description of his conversion to faith in Jesus Christ in 1930 as a 20-year-old student at Cambridge University:

> There were some ten of us and together we sought for God and together we found Him. I learnt for the first time in my life that God was my friend… God became real to me—utterly real—I knew Him and I could talk with Him as I never imagined it possible before. And these prayers…were the most glorious moment of the day. Life had a purpose and that purpose coloured everything.[34]

Coulson's experience was fairly similar to my own, 43 years later, as a young professor at Berkeley. It would be arresting if I could say that there was a thunderclap from heaven, God spoke to me in audible terms, and hence I became a Christian. However, it did not happen that way. The apostle Paul's

Damascus Road encounter with Jesus was the exception rather than the rule. But I did (and still do, some 40 years later) experience the same perception that Coulson described. My life has a purpose in Jesus Christ, and that purpose colors everything.

The Why Question

Before we move on exclusively to contemporary scientists, let us explore some of the reasons for the pattern I have described thus far. Namely, why did sustained scientific development occur first in a Christian environment? The best answers I have seen to this question were formulated by my University of Georgia chemistry colleague Professor Wesley Allen. His (slightly modified) five answers to this question are as follows:

1. If Christianity is true, the universe is real, not illusory. The universe is thus the product of a God whose character is immutable, at variance with pantheistic notions that place inherent distrust in sensory experience in a mercurial world.

2. If Christianity is true, the universe, being divinely created, is of inherent value and thus worthy of study. This conclusion supplants any zeitgeist (spirit of the times) that would view science as a mere intellectual pastime.

3. If Christianity is true, nature itself is not divine, and thus humanity may probe it free of fear. This was an important realization in early eras dominated by superstitions about the natural environment. Worship and ultimate reverence are reserved for the Creator, not the creation, nor humans as creatures therein.

4. If Christianity is true, mankind, formed in the image of God, can discover order in the universe by rational interpretation. That is, the codes of nature can be unveiled and read. Without such faith, science might never have developed because it might have appeared impossible in principle.

5. If Christianity is true, the form of nature is not inherent within nature, but rather a divine command imposed from outside nature. Thus, the details of the world must be uncovered by observation rather than by mere rational musing because God is free to create according to his own purposes. In this way science was liberated from Aristotelian rationalism, whereby the creator was subjected to the dictates of reason constructed by humans. Such gnosticism, which transformed speculation into dogma, undermined the open-endedness of science. To be sure, Christianity holds that God is a perfectly rational being who cannot act inconsistently with his character. But this principle only places partial constraints on his creative activity, which science must be free to discover in all its diversity.

Contemporary Scientists

Robert Griffiths (1937–), a member of the US National Academy of Sciences, is the Otto Stern Professor of Physics at Carnegie Mellon University. He received one of the most coveted awards of the American Physical Society in 1984 for his work on statistical mechanics and thermodynamics.

The magazine *Physics Today* reported that Griffiths is an evangelical Christian who is an amateur theologian and who helps teach a course at Carnegie Mellon on Christianity and science. I find this to be particularly intriguing because for the last 20 years, I have taught a freshman seminar at the University of Georgia on the same subject. In the April 3, 1987 issue of *Christianity Today*, Professor Griffiths made the interesting statement: "If we need an atheist for a debate, I go to the philosophy department. The physics department isn't much use."[35]

At the University of California at Berkeley, where I was a professor for 18 years, we had 50 chemistry professors. But for many years, there was only one who was willing to publicly identify himself as an atheist, my good friend Robert Harris, with whom I still have occasional discussions about spiritual things—usually during my annual summer week back on the Berkeley campus. After one such discussion perhaps 35 years ago, Bob told me he might have to rethink his position and become an agnostic. I thought to myself, *Okay, Bob, one step at a time.* But Bob came back to me a week later, firmly reinstalled in the atheist camp. A more recent addition to the Berkeley chemistry faculty is a second open atheist, Richard Saykally. Rich is also a close friend, soundly disproving the notion that disagreements about ultimate questions necessarily lead to personal rancor.

For many years, Richard Bube (1927–2018) was the chairman of the Department of Materials Science at Stanford University. No less than 56 Stanford graduate students received their PhD degrees under Professor Bube's direction. Bube carried out foundational research in solid-state physics concerning semiconductors, the photoelectronic properties of materials, photovoltaic devices (solar cells), and amorphous materials. He seconds Robert Griffiths's aforementioned statement, noting, "There are probably as many atheistic truck drivers as atheistic scientists."[36] Bube was long a spokesperson for evangelical Christians in academic life, serving for many years as editor of the journal *Perspectives on Science and Christian Faith*, published by the American Scientific Affiliation, of which I am a fellow. Bube formerly taught a second-year undergraduate course at Stanford called "Issues in Science and Christianity."

Another member of the US National Academy of Sciences is John Suppe, a noted professor of geology at Princeton University. John is an outstanding scholar in the area of plate tectonics and structural geology, the deformation of the Earth's crust. Vaguely aware of his own spiritual needs, he began attending services in the Princeton chapel, then reading the Bible and other Christian books. He eventually committed himself to Christ and, remarkably, had his first real experience of Christian fellowship in Taiwan, where he went as a recipient of a grant from the prestigious Guggenheim Fellowship. I have spoken before the Christian faculty forum at the National Taiwan University in Taipei, so I know personally that they are a good group.

> Suppe makes some interesting comments concerning the evolution controversies: Some non-scientist Christians, when they meet a scientist, feel called on to debate evolution. That is definitely the wrong thing to do. If you know scientists and the kinds of problems they have in their lives: pride, selfish ambition, jealousy; that's exactly the kind of thing Jesus talked about, and which he came to resolve (by His death on the cross).

Science is full of people with very strong egos who get into conflicts with each other... The gospel is the same for scientists as for anyone. Evolution is basically a red herring. If scientists are looking for meaning in their lives, it won't be found in evolution.[37]

Suppe certainly provides a good practical starting point about evolution.

My candidate for the outstanding experimental scientist of the twentieth century is Charles Townes (1915–2015), who received the 1964 Nobel Prize in Physics for his discovery of the laser. However, I must confess to some possible bias, since Professor Townes is the only plausible candidate for scientist of the century that I know personally. But the laser is a discovery that has significantly impacted the life of every person who reads these words. Dr. Townes almost received a second Nobel Prize for his observation of the first interstellar molecule. The study of interstellar molecules has subsequently become a major part of astrophysics, affecting even my own research. Townes was the provost at MIT when I was an undergraduate, and later a colleague (but in the physics department) during my 18 years on the faculty at Berkeley.

At Berkeley, every PhD oral examination requires four faculty members from the candidate's own department and a fifth committee member from an outside department. In chemical physics, which is actually a part of the chemistry department at Berkeley, the "outside" committee member is almost inevitably a physics faculty member. This puts a significant strain on some of the physics faculty because only a few of them are sufficiently knowledgeable about chemistry to serve on such committees, of which about 30 must be constituted each year. So it was not unusual for this particular subset of the physics faculty to come up with highly original reasons why they were unavailable for such two-hour ordeals. But Charlie Townes was never such a one. He always served the chemistry department cheerfully, although his duties in Washington and elsewhere were legion. And his demeanor on these committees was always gentlemanly to a tee. His questions to the chemistry PhD students were almost inevitably thoughtful and designed to bring out the very best in a quaking student.

Dr. Townes wrote an autobiography entitled *Making Waves*, a pun referring to the wavelike phenomena that scientifically describe lasers. The book was published in 1995 by the American Institute of Physics, and I recommend it. Charlie mentioned his church involvement and then said,

> You may well ask, "Just where does God come into this?" Perhaps my account may give you some answer, but to me that's almost a pointless question. If you believe in God at all, there is no particular "where." He's always there—everywhere. He's involved in all of these things. To me, God is personal yet omnipresent—a great source of strength who has made an enormous difference to me.[38]

Arthur Schawlow (1921–2000) won the Nobel Prize in Physics in 1981 for his work in laser spectroscopy. Schawlow served until his death as a professor at Stanford and was a truly beloved figure in the physics community. He did not hesitate to identify himself as a Protestant Christian. And he made this unusual statement, which I suspect might only be made by a scientist: "We are fortunate to have the Bible, and especially the New Testament, which tells so much about God in widely accessible human terms."[39] I know that Schawlow believed that his experimental studies of molecular spectroscopy were also telling him something about God's creative powers. The contrast with the New Testament accounts of the life of Jesus was that Schawlow did not think that his scientific research was providing information about God in "widely accessible human terms."

John Polkinghorne (1930–) was the chaired Professor of Mathematical Physics at Cambridge University from 1968 to 1979. This is the "other" chair of theoretical physics at Cambridge, in addition to the chair held by Stephen Hawking. In 1979, Polkinghorne made an abrupt career switch, enrolling in theological studies before becoming an Anglican priest. Then in 1986, Polkinghorne returned to Cambridge, first as dean of Trinity Hall and later becoming president of Queen's College. Queen's College sits next to St. Catherine's College, where I have stayed in Cambridge during visits to my longtime scientific collaborator and close friend theoretical chemistry professor Nicholas Handy, also an Anglican. Perhaps needless to say, John Polkinghorne has been outspoken about spiritual matters: "I take God very seriously indeed. I am a Christian believer, and I believe that God exists and has made Himself known in Jesus Christ."[40]

I would like to make a specific point here as it relates to biology and to a more general question. For decades, the world's greatest living observational cosmologist was Allan Sandage (1926–2010), an astronomer at the Carnegie Institution in Pasadena, California. Sandage was called the "grand old man of cosmology" by The New York Times when he won the highly lucrative 1991 Crafoord Prize given by the Royal Swedish Academy of Sciences.[41] This prize is given to a cosmologist every sixth year and is viewed by the Swedish

Academy as equivalent to the Nobel Prize. Sandage committed his life to Jesus Christ at the age of 50. In the Alan Lightman book noted earlier, Dr. Sandage was asked the old question "Can a person be a scientist and also be a Christian?" Sandage's affirmative response was expected, but he provided a surprising focus: "The world is too complicated in all its parts and interconnections to be due to chance alone. I am convinced that the existence of life with all its order in each of its organisms is simply too well put together."[42]

Sandage was the person responsible for the best current scientific estimate of the age of the universe, perhaps 13.8 billion years. Yet when this brilliant astrophysicist was asked to explain how one can be a scientist and a Christian, he turned not to cosmology but biology. Which brings me full circle to the question I addressed earlier from the perspective of mathematical physics and the intelligibility of the universe: Why might a scientist become a Christian? The answer from biology is that the extraordinary complexity and high information content of even the simplest living thing (the simplest self-replicating biochemical system) points to a sovereign creator God.

As mentioned earlier, a typically important ingredient to receiving the Nobel Prize in Chemistry is the attainment of age 65. For example, my good friend John Pople (1925–2004), a serious Methodist Christian, received the Nobel Prize in quantum chemistry at the age of 73 in 1998. He shared it with Walter Kohn, who was 75 at the time. However, the Nobel Prize in Physics is often given to much younger individuals. William Phillips (1949–) received this at the age of 48 for the development of methods to cool and trap atoms with laser light. On the announcement date, October 15, 1997, Phillips was participating in a conference on high-powered telescopes in Long Beach, California. At the mandatory press conference, Phillips spoke these words: "God has given us an incredibly fascinating world to live in and explore."[43]

Phillips formed and sings in the gospel choir of the Fairhaven United Methodist Church, a multiracial congregation of about 300 in Gaithersburg, Maryland. He also teaches Sunday school and leads Bible studies. If you took the time to delve further into the October 1997 media reports, you would find out that on Saturday afternoons, Phillips and his wife often drive into central Washington, DC, to pick up a blind, elderly African-American woman to take her grocery shopping and then to dinner.

Allow me just once more to ask the critical question, Why might a scientist become a Christian? My third answer is the remarkable fine-tuning of the universe. Let me draw a picture by citing three persons with no obvious theistic inclinations. Paul Davies, an excellent popularizer of science, states, "The present arrangement of matter indicates a very special choice of initial conditions."[44] Now, if language means anything, "a very special choice" implies that someone or something is doing the choosing. Stephen Hawking elaborates: "In fact, if one considers the possible constants and laws that could have emerged, the odds against a universe that has produced life like ours are immense."[45]

The always-quotable Fred Hoyle (1915–2001) added, "A common sense interpretation of the facts suggests that a super intellect has monkeyed with physics, as well as with chemistry and biology, and that there are no blind forces worth speaking about in nature."[46] My own view is that all three of these skeptics, in their own unique ways, unintentionally supported the position put forth by the apostle Paul nearly two millennia earlier: "His invisible attributes, namely, his eternal power and divine nature, have been clearly perceived, ever since the creation of the world, in the things that have been made" (Romans 1:20).

Two Common Questions

Prior to my concluding remarks, I would like to respond to two questions that are frequently raised. The first is this: Given the evidence you present, why do so many persist in the belief that it is not possible to be a scientist and a Christian? Although I am about as far from being a conspiracy theorist as possible, I conclude that part of the problem is indeed misrepresentation.

The respected British science historian Colin Russell described T.H. Huxley's important role in these developments in his scholarly account in the April 1989 issue of *Science and Christian Belief*,[47] the quarterly publication of the Victoria Institute, of which I have been a member for the past 30 years. Here, I would like to focus on a famous book published in 1896, entitled *The History of the Warfare of Science with Theology*. Of course, given the origins of modern science, such a title sounds rather silly. The author of this polemic was Andrew Dickson White, the first president of Cornell University, the first North American university founded on purely secular terms. The most famous passage in White's book reads, "[John] Calvin took the lead in his *Commentary on Genesis*, by condemning all who asserted that the earth is not the centre of the universe. He clinched the matter by the usual reference to the first verse of the ninety-third Psalm and asked, 'Who will venture to place the authority of Copernicus above that of the Holy Spirit?'"[48]

Perhaps needless to say, that statement does not make John Calvin look good. Nor was that Andrew Dickson White's intention. However, the truth of this matter has been brought forth by Dr. Alister McGrath, a distinguished professor at the University of Oxford. In his definitive 1990 biography of Calvin, McGrath wrote,

> This assertion [by White] is slavishly repeated by virtually every writer on the theme "religion and science," such as Bertrand Russell in his *History of Western Philosophy*. Yet it may be stated categorically that Calvin wrote no such words [in his Genesis commentary] and expressed no such sentiments in any of his known writings. The assertion that he did is first to be found, characteristically unsubstantiated, in the writings of the nineteenth century Anglican dean of Canterbury, Frederick William Farrar (1831–1903).[49]

It would be fair to ask what Calvin really thought of Copernicus's heliocentric theory of the solar system. The honest answer is that we do not know. Calvin probably wasn't familiar with the work of Copernicus, who was hardly a household name in France or Switzerland during the former's lifetime. But in his preface to his friend Pierre Olivetan's translation (1534) of the New Testament into French, Calvin wrote, "The whole point of Scripture is to bring us to a knowledge of Jesus Christ. And having come to know Him (with all that this implies), we should come to a halt and not expect to learn more."

A second frequently asked question is, Okay, but how does being a Christian change your science? Having first heard this question at Stanford University 30 years ago, I have had plenty of time to develop a good response. But I must confess that I cannot qualitatively improve on the answer given by the brilliant Notre Dame historian George Marsden in his 1997 Oxford University Press book *The Outrageous Idea of Christian Scholarship*. My answer modifies Marsden's only slightly. In science, a person's Christian faith can have a significant bearing on scholarship in at least four ways:

1. One's Christian faith may be a factor in motivating a scientist to do his or her work well. This is not to deny that some atheist or agnostic scholars may be just as motivated to work with just as much integrity. For any particular scholar, however, Christianity may be an important motivator.

2. One's Christian faith may help determine the applications one sees for his or her scholarship. One may carry out research in anything from materials science to molecular biology with the hope that it may contribute to the well-being of others. Again, the fact that some atheists are also altruistic does not negate the Christian contribution to altruism.

3. Such motives may help shape a subfield, specialty, or the questions a person asks about his or her research. For example, I readily confess that my scientific interest in interstellar molecules was initially inspired by the oft-repeated claims that this field will ultimately explain the origin of life. Ultimate questions tend to be of special interest to Christians.

4. When on occasion the scientist is asked to reflect on the wider implications of his or her scholarship,

faith may have an important bearing on how that person sees the field, or its assumptions, fitting into a larger framework of meaning.[50]

Concluding Remarks

My collection of scientists with Christian commitments is far from exhaustive. The publication of this chapter will likely bring me correspondence from near and far with excellent new examples of the genre. Of course I will be happy to attempt to incorporate new material. Please do send cards and letters. But I should mention now a few others not included above. Among chemists, professors Andrew Bocarsly (Princeton University) and James Tour (Rice University) have given Christian testimonies that have touched my heart and mind in a special way. Both Andy and Jim were born into Jewish families and gave their lives to Jesus Christ during their undergraduate years. Bocarsly is an inorganic photochemist turned catalysis pioneer and Tour a synthetic organic chemist turned materials scientist. Jim Tour's work on fullerenes, bucky tubes, and more generally nanochemistry is definitely on track for a trip to Stockholm some December, perhaps a decade from now. In 2019, professor Troy Van Voorhis (1976–) became the head of the MIT Department of Chemistry. Van Voorhis is a deeply committed Christian.

A few other Christians brilliantly pursuing science can be mentioned as well. The list is so long I restrict myself to professors from the two great British universities, Oxford and Cambridge. From Oxford I note statistical mechanician Ard Louis, laser physicist Paul Ewart, nanomaterials scientist Andrew Briggs, and mathematician John Lennox. The list from Cambridge includes, among others, molecular biologist Denis Alexander,

paleontologist Simon Conway Morris, theoretical physicist John Barrow, and geophysicist Robert White.

The present discussion has focused primarily on physics and chemistry for the obvious reason that I am a chemical physicist. This is my professional life. I suspect that it would be possible to present a similar discussion for the biological sciences. For example, biochemistry professor David Cole was the steadfast leader of the Christian faculty organization during my years on the Berkeley campus. Francis Collins is one of the most outstanding research biologists of our generation. While a professor at the University of Michigan, Collins discovered the cystic fibrosis gene. For the period 1993–2008, Collins was the director of the successful Human Genome Project. Collins currently has an even more demanding job, director of the entire National Institutes of Health (NIH).

Collins's paper "The Human Genome Project: Tool of Atheistic Reductionism or Embodiment of the Christian Mandate to Heal?" appeared in a 1999 issue of the journal *Science and Christian Belief*. Collins introduced his paper in this way:

> Let me begin by saying a brief word about my own spiritual path. I did not come from a strongly Christian home. I was raised in a home where faith was not considered particularly relevant, sent to church to learn music, but instructed that it would be best to avoid the theology. I followed those instructions well and went off to college with only the dimmest idea of what saving faith in Jesus Christ was all about. What little glimmers of faith I might have possessed were quickly destroyed by the penetrating questions of my freshman dorm colleagues who, as one

will do at that phase in life, took great delight in destroying any remnants of superstition, which is what they considered faith to be. I went on to study chemistry and physics, and then to do a Ph.D. in chemical physics, and became quite an obnoxious atheist with whom you would not have enjoyed having lunch, because I too felt it was part of my mission to point out that all that really mattered could be discerned by science, and everything else was irrelevant…Fortunately through the guidance of some very patient people, who tolerated a lot of insolent questions, I was led to read C.S. Lewis and then the Bible, and so was led to understand many of the concepts that had completely eluded me before, and I gave my life to Christ 20 years ago.[51]

I hope this chapter has given you a flavor of the history of scientists who were and are Christians. Those of you who have taken a freshman chemistry or physics course will surely recognize many of the names of the great scientists mentioned in this chapter. In fact, the reason this chapter took shape as it did was to present mini-sketches of the spiritual lives of scientists with whom my Berkeley freshman chemistry students would be familiar. There is a tremendous tradition, past and present, of distinguished scientist Christians. It gives me great joy to be a small part of that continuing tradition. And perhaps I have given you sufficient evidence that you will never again believe that it is difficult to be a scientist and a Christian.

In closing, following the example of Oxford professor Charles Coulson in his public lectures on science and the Christian faith, I encourage you to consider the advice of Psalm 34:8 "Taste and see that the LORD is good!"

How Can We Make
Sense of Natural Evil?

William A. Dembski

Satan's fall, according to orthodox Christian theology, preceded humanity's fall—after all, Satan tempted humanity into its fall, so the sin of Satan must in some sense be causally prior to the sin of humanity. Nonetheless, regardless of the ultimate origin of evil (whether it be Satan's rebellion or something still prior to it, such as in Calvinist theology the sovereign decrees of God), Christian theology has also taught that all evil in this physical world traces back to humanity and its sin. If you will, humanity is the keeper of the gate through which evil has access to the world. In consequence, the fall of humanity becomes the failure of the gatekeeper to maintain proper control of the gate. This view of humanity's role in relation to evil holds true regardless of the ultimate source of evil.

The view that all evil in the world ultimately traces back to human sin used to be part and parcel of a Christian worldview. As the *Catholic Encyclopedia* notes in its entry on evil:

Christian philosophy has, like the Hebrew, uniformly attributed moral

and physical evil to the action of created free will. Man has himself brought about the evil from which he suffers by transgressing the law of God, on obedience to which his happiness depended…[T]he errors of mankind, mistaking the true conditions of its own well-being, have been the cause of moral and physical evil.[1]

Moral evil here is the evil that we as humans, by misusing our will, commit against each other, against creation, and ultimately against God. What the *Catholic Encyclopedia* here calls *physical evil* is typically also called *natural evil* and refers to the disordering of nature whereby sickness, death, predation, parasitism, famine, etc., occur in nature apart from humans willing it or otherwise directly bringing it about. Classic Christian orthodoxy teaches that natural evil is as much a consequence of human sin as moral evil.

This view of natural evil has become hard to maintain in our day. Because of scientific discoveries over the last 200 years, many converging lines of evidence point to a long history of the universe and life on Earth that predates the arrival of humans and yet in which natural evil was rampant. For the young-Earth creationist, tracing natural evil to humanity's sin and the fall is unproblematic because the creation occurs 6,000 years ago in six literal 24-hour days: everything is perfect during that time and for a few subsequent moments until humanity sins and God curses humanity and nature, thereby making natural evil a consequence of human moral evil.

But in a world that is much older and in which natural evil reigns prior to the arrival of humans, how can natural evil be ascribed to human moral evil, as classic Christian orthodoxy has traditionally taught? Many theologians and scientists these days simply give up on trying to explain natural evil as a consequence of moral evil and the fall. Instead, they simply see natural evil as a cost God incurs in creating the natural world. The natural world, some of these theologians teach, has a certain freedom of activity that makes natural evil unavoidable, or at least frees God from responsibility for it. But such a view is, on its face, implausible. God, Christianity teaches, will resurrect the saved in new bodies and place them into paradise. But paradise is clearly without natural evil, so why can't God simply create a world that is a paradise, or at least something close to it?

Along with most of the scientific community, I accept standard astrophysical and geological dating (12 billion years for the universe, 4.5 billion years for the Earth). Thus, given my old-Earth creationist view, the problem of natural evil confronts me in a way that it doesn't the young-Earth creationist. Young-Earth creationism presents a chronology that aligns the order of creation with a traditional conception of the fall: God creates a perfect world, God places humans in that world, they sin, and the world goes

haywire. In this chronology, theology and history move in sync with the first human sin predating and being causally responsible for natural as well as moral evil. Yet if the bulk of natural history predates humans by billions of years and if over the last 600 million years multi-celled animals have been emerging, competing, fighting, preying, parasitizing, exterminating, and going extinct, then young-Earth creationism's harmony of theology and history becomes insupportable. In that case, natural history, as described by modern science, appears irreconcilable with the order of creation as described by Genesis.

Many young-Earth creationists therefore reject current scientific chronologies for the age of the Earth and universe, with some even arguing that a young-Earth position makes for better science. I personally have found such arguments for a young Earth unconvincing, slicing and dicing the scientific evidence to support a young-Earth position even though any reasonable interpretation of the evidence would, in my view, suggest a far older Earth and universe. Yet if science is against a young Earth, the history of biblical interpretation is for it. Indeed, young-Earth creationism was overwhelmingly the position of the church from the church fathers through to the Reformers. Yet as Christians we have an obligation, as the apostle Paul put it in 2 Timothy 2:15, to "rightly divide" (i.e., interpret) the Scriptures. But what ought to guide our interpretation of the Scriptures? Clearly, our knowledge of the world plays some role. Our knowledge of physics from the seventeenth century onwards, for instance, has rendered geocentrism untenable, and yet there are plenty of passages in the Bible that might be taken to suggest geocentrism.

In trying to balance the science of the day with the interpretation of Scripture, I therefore often return to an observation of the nineteenth-century Princeton theologian Charles Hodge. Early in his *Systematic Theology*, he noted that even though Scripture is true, our interpretations of it can be in error; as a consequence, it can be a trial for the church when long-held interpretations are thrown into question. As he put it,

> Christians have commonly believed that the earth has existed only a few thousands of years. If geologists finally prove that it has existed for myriads of ages, it will be found that the first chapter of Genesis is in full accord with the facts, and that the last results of science are embodied on the first page of the Bible. *It may cost the church a severe struggle to give up one interpretation and adopt another*, as it did in the seventeenth century [when the Copernican system displaced the Ptolemaic system of the universe], but no real evil need be apprehended. The Bible has stood, and still stands in the presence of the whole scientific world with its claims unshaken.[2]

Double Creation

My challenge in this chapter is to trace natural evil back to moral evil and the fall, yet in a world where natural evil has reigned for eons before the arrival of humans, thereby rejecting not only young-Earth creationism but also contemporary theological moves that attempt to dismiss natural evil as something independent of human sin. How can natural evil precede the first human sin and yet result from it? Contemporary science firmly holds that the Earth and universe are not thousands but billions of years old, that humans have been around for only a

minuscule portion of that time, and that before their arrival natural evils abounded. Let's therefore turn to how natural evil could chronologically precede the first human sin and yet be a consequence of it.

To that end, I'll offer a reading of Genesis 1–3 that attempts to reconcile a traditional understanding of the fall (which traces all evil in the world, including natural evil, to human sin) with a mainstream understanding of geology and cosmology (which regards the Earth and universe as billions of years old, and therefore makes natural evil predate humanity). The key to this reading is to interpret the days of creation in Genesis as conceptual divisions in the divine logic of creation—i.e., the key episodes by which God conceives and plots out creation. Genesis 1 is therefore not to be interpreted as ordinary chronological time, but rather as time from the vantage of God's purposes. Accordingly, the days of creation are neither exact 24-hour days (as in young-Earth creationism) nor epochs in natural history (as in typical old-Earth creationism) nor even a literary device (as in the literary framework theory).

Essentially what I'm proposing here is to view creation as a double creation. The idea of creation as double creation goes back to antiquity. We see it implicit in the Lord's Prayer, in which we pray that God's will be done on Earth (second creation) as it is done in heaven (first creation). Plato expressed this concept in the *Timaeus* when he had the demiurge (Plato's world architect) organize the physical world (second creation) so that it conformed to patterns residing in the abstract world of ideas (first creation). Aquinas developed this thought with his notion of exemplary causation. An exemplary cause is a pattern or model employed by an intelligence for producing a patterned effect (the exemplary cause being the first creation, its implementation being the second creation).

The key example of exemplary causation within Christian theology is the creation of the world according to a divine plan, with that plan consisting of ideas eternally present in the mind of God. Even in our own human experience of creation, creation is always double: first there is the formation of ideas in the mind, and then there is the implementation of those ideas—that is, the transition from thought to thing. My argument hinges on treating Genesis 1 as dealing with the first creation (i.e., what happens in the mind of God) rather than the second creation (i.e., its implementation in space and time).

On the approach I'm espousing, the days of creation in Genesis thus become actual (literal!) episodes in the divine creative activity. They represent key divisions in the divine order of creation, with one episode building logically on its predecessor. As a consequence, their description as chronological days needs to be viewed as an instance of the common scriptural practice of employing physical realities to illuminate deeper spiritual realities (cf. John 3:12). John Calvin referred to this practice as God condescending to our limited understanding. The justification for this practice is that the physical world, as a divine creative act, provides a window into the life and mind of God, the one who created it. The general principle here is that the things a creator makes and does invariably reflect the thought and character of the creator.

If, as I'm suggesting, the Genesis days represent key episodic divisions in the divine logic of creation, a widely cited reason for treating the days of creation as strict 24-hour periods dissolves. Young-Earth creationists sometimes insist that the author of Exodus,

in listing the Ten Commandments, could only be justified in connecting sabbath observance to the days of creation if the days of creation were successive 24-hour chronological days (see Exodus 20:11, where sabbath observance is justified in terms of God's creation of the world in six days and then resting on the seventh). But if the days of creation refer to basic divisions in the divine order of creation, then sabbath observance reflects a fundamental truth about the creation of the world. Specifically, because days form a basic division in the way humans experience time, sabbath observance becomes a way of getting us, who are made in the image of God, to recognize the significance of human work and rest in light of God's work and rest in creation. Without this sabbatarian perspective, we cannot understand the proper place of work or rest in human life.

Yet from a purely chronological perspective, there is nothing particularly fitting or distinctive about God creating the world in six 24-hour days. God could presumably have created the same world using very different chronologies (in his *Literal Commentary on Genesis*, Augustine entertained the possibility of God creating everything in one chronological instant). By contrast, if the Genesis days follow God's logic of creation, this gives greater force to sabbath observance, requiring humans to observe the Sabbath because it reflects fundamental divisions in the divine order of creation and not because it underscores purely contingent facts about the chronology of creation (a chronology that God could have altered in any number of ways to effect the same purposes in creation).

In God's logic of creation, the six days of creation become unashamedly anthropocentric. Genesis clearly teaches that humans are the end of creation. For instance, Genesis describes the creation as merely "good" before humans are created, but describes it as "very good" only after they are created. God's activity in creation is therefore principally concerned with forming a universe that will serve as a home for humans. Although this anthropocentrism sits uneasily with many in our culture, it is not utterly foreign to it. Indeed, the intelligibility of the physical world by means of our intellects and, in particular, by means of such intellectual feats as mathematics, suggests that we live in a meaningful world whose meaning was placed there for our benefit.

Interpreting the creation days as episodes or key divisions in the divine logic of creation now leads to the following reading of Genesis: On the first day, the most basic form of energy is created: light. With all matter and energy ultimately convertible to and from light, day one describes the beginning of physical reality. With the backdrop of physical reality in place, God devotes days two and three to ordering the Earth so that it will provide a suitable home for humanity. On these days, God confines the Earth's water to appropriate locations and forms the plants on which humans and other animals will depend for their sustenance. On day four, God situates the Earth in a wider cosmic context. On day five, animals that inhabit the sea and sky are created. And finally, on day six, animals that inhabit dry land are created, most notably human beings. Finally, on day seven, God rests from his activity in creation. Genesis 1 omits and abbreviates many details of (the first) creation. Also, in this view of the creation days as logical or conceptual divisions in the mind of God, we are not given detailed insight into how the divine purposes of creation were implemented chronologically (in the second creation).

The Fall

The key question that now needs to be addressed is how to position the fall of humanity (i.e., the sin of Adam and Eve) within this reading of Genesis that sees the days as logically dependent steps in the divine creative plan. In answering this question, we need to bear in mind that Genesis 1 describes God's original design plan for creation (the first creation). The fall and its consequences, in constituting a subversion of that design plan through human rebellion, elicits no novel creative activity from God. The fall represents the entrance of evil into the world, and evil is always parasitic, never creative. Indeed, all our words for evil presuppose a good that has been subverted. Impurity presupposes purity, unrighteousness presupposes righteousness, deviation presupposes a way (i.e., a *via*) from which we've departed, sin (the Greek *hamartia*) presupposes a target that was missed, etc. This is not to deny or trivialize evil. Rather, it is to put evil in proper perspective.

God's immediate response to the fall is therefore not to create anew but damage control. In the fall, humans rebelled against God and thereby invited evil into the world. The challenge God faces in controlling the damage resulting from this original sin is how to make humans realize the full extent of their sin so that, in the fullness of time, they can fully embrace the redemption in Christ and thus experience full release from sin. For this reason, God does not merely allow personal or moral evil (the disordering of our souls and the sins we commit as a consequence) to run their course *subsequent to* the fall. In addition, God also brings about natural evils (e.g., death, predation, parasitism, disease, drought, famines, earthquakes, hurricanes, etc.), letting them run their course *prior to* the first human sin (i.e., the fall). Thus, God himself disorders the creation, *making it defective on purpose*. God disorders the world not merely as a matter of justice (to bring judgment against human sin as required by God's holiness), but even more significantly as a matter of redemption (to bring humanity to its senses by making us realize the gravity of sin).

This conceptual reading of the Genesis days as describing the first rather than second creation preserves the young-Earth creationist emphasis on tracing all evil in the world to human sin: God creates a perfect world, God places humans in that world, they sin, and the world goes haywire. But this reading does raise the question of how to make sense of the fall chronologically. Humans do not merely exist conceptually in the divine mind; they exist chronologically in space and time, and the fall occurred in space and time. To understand how the fall occurred chronologically and how God could act preemptively to anticipate the fall by allowing natural evils to rage prior to it, we need to take seriously that the drama of the fall takes place in a *segregated area*. Genesis 2:8 refers to this area as a *garden* planted by God (i.e., the garden of Eden). Now, ask yourself why God would need to plant a garden in a perfect world untouched by natural evil. In a perfect world, wouldn't the whole world be a garden? And why, once humans sin, do they have to be expelled from this garden and live outside it, where natural evil is present (thorns, etc.)?

Proponents of the Documentary Hypothesis for the Pentateuch (JDEP) describe the juxtaposition of Genesis 1:1–2:3 and Genesis 2:4–3:24 as a kludge of two disparate and irreconcilable creation stories (the days of creation versus humanity's creation and fall in the garden). But in fact, the second creation account, situated in the garden, is just what's needed for *conceptual*

days and *chronological days* to converge in the fall. If we accept that God acts preemptively to anticipate the fall, then in the chronology leading up to the fall, the world has already experienced, in the form of natural evil, the consequences of human sin. This seems to raise a difficulty, however, because for humans who have yet to sin to come into a world in which natural evil rages seems to put them at a disadvantage, tempting and opposing them with evils for which they are not (yet) responsible. The garden of Eden, as a segregated area in which the effects of natural evil are not evident (hence its designation as paradise), provides the way out of this difficulty.

In the garden of Eden, Adam and Eve simultaneously inhabit two worlds—two worlds intersect in the garden. In the one world, the world God originally intended, the garden is part of a larger world that is perfect and includes no natural evils. In the other world, the world that became corrupt through natural evils that God brought about by acting preemptively to anticipate the fall, the garden is a safe haven that, in the conscious experience of Adam and Eve, matches up exactly with their conscious experience in the perfect world, the one God originally intended. In the originally intended world, there are no pathogenic microbes and, correspondingly, there is no need for Adam and Eve to have an immune system that wards off these microbes. In the imperfect world, whose imperfection results from God acting preemptively to anticipate the fall, both pathogenic microbes and human immune systems exist. Yet, in their garden experience, Adam and Eve never become conscious of that difference. Only after they sin and are ejected from the garden do they become conscious of the difference. Only then do they glimpse the world they might have inhabited but lost, a world symbolized by the tree of life. Only then do they realize the tragedy they now face by being cast into a world full of natural evil devoid of a tree that could grant them immortality.

Why doesn't God grant Adam and Eve immortality despite the fall? The ancient myth of Tithonus and Eos captures the problem facing God. Eos (Latin Aurora), the goddess of dawn, is married to Tithonus, who is human and mortal. She asks Zeus to make Tithonus immortal but forgets to ask that Zeus also grant him eternal youth. As a consequence, Tithonus grows older and older, ultimately becoming completely decrepit. The lesson here is that immortality must not be used to immortalize corruption—instead of attenuating corruption, immortality intensifies it. In enforcing mortality on humans by ejecting them from a garden that has a source of immortality (the tree of life) at its center, God limits human corruption and, in the protevangelium (Genesis 3:15), promises a way out of that corruption. Thus, given our corruption through sin, mortality becomes a grace and a benefit.

Entering the Garden

A final question now remains regarding this reading of Genesis: How did the first humans gain entry into the garden? There are two basic options here: special creation and evolving creation. In the first, God specially creates the first humans inside the garden. In the second, the first humans evolve from primate ancestors outside the garden and then are brought into it. Both views require direct divine action. In the former, God specially creates the first humans from scratch. In the latter, God introduces existing human-like beings from outside the garden but then transforms their consciousness so that they become rational moral agents made in God's

image. With an evolving creation, this transformation of consciousness by God upon entry into the garden is essential to this logic-of-creation reading of Genesis. For if the first humans bore the full image and likeness of God outside the garden prior to the fall, they would have been exposed to the natural evils present there—evils for which they were *not yet* responsible. This would be problematic because humanity's responsibility and culpability in the fall depends on the fall occurring without undue temptations or pressures. These temptations and pressures are absent inside the garden of Eden but not outside.

The approach to natural evil described in this chapter is part of a larger project developed in my book *The End of Christianity: Finding a Good God in an Evil World*. There, I develop a full-blown theodicy that attempts to make sense of God's goodness in the face of evil from the vantage of classic Christian orthodoxy, yet without requiring a young-Earth interpretation of Genesis. The focus in this chapter is more circumscribed, simply trying to understand and explain natural evil from an old-Earth perspective. Theologically, natural evil is not a problem from a young-Earth perspective. Yet once the young-Earth perspective comes under scientific challenge, the problem of natural evil resurfaces.

Christians who feel secure in their young-Earth perspective may think there's little of use in this chapter. But such a feeling of security is, in my view, ill-founded. I taught for years in Southern Baptist seminaries where the young-Earth position was the majority position. The Southern Baptist world is large and people can make their living by residing entirely in it. It's a secure environment provided you can stay in it. But if you must venture outside of it, you can expect secular scientists and liberal theologians to try to shake your worldview.

Some young-Earth creationists successfully maintain their young-Earth perspective in the face of such challengers, often by adopting a fideist approach that embraces a literalist reading of Genesis regardless of what science may say about the age of the Earth and universe. But others who venture beyond their "safe environment" may lose not only their young-Earth perspective but also their Christian faith entirely. Thinking that their faith depends on the truth of a young Earth, they give up their faith when they feel it's scientifically insupportable.

The point of this chapter is to show that you don't need to be a young-Earth creationist to maintain a traditional view of natural evil, the traditional view being one that sees natural evil as a consequence of moral evil and the fall. Yet once one takes seriously that the Earth is old and that natural evil predates the arrival of humans, the problem of natural evil takes on new force. That's why the young-Earth creationists' first objection to an old Earth is not the science arguing for an old Earth, but how to explain the existence of natural evil prior to the sin of Adam and Eve, especially in the light of Romans 5:12, which explicitly traces all evil to human sin. This chapter's approach to understanding natural evil takes this objection seriously and attempts to answer it.

Yet whether the approach outlined in this chapter seems convincing will depend on how committed one is to a young-Earth position as well as how deeply one feels the challenge of natural evil and the need to trace it back to human sin (moral evil). I've had Christian colleagues who hold to an old Earth but simply don't see in natural evil anything that requires serious explanation. This seems to me naïve. For all the pain and suffering

caused by moral evil, natural evil has caused at least as much. The coronavirus pandemic has sensitized us to the problem of natural evil, but the actual mortality and suffering it has produced is small compared to past plagues. The Great Influenza of 1918–1920 is estimated by some to have killed 100 million worldwide, more than the two World Wars, Stalin's purges, and Mao's murders. The Black Death from 1347–1351 is thought to have killed a third of the population of Europe (to say nothing of Asia and Africa).

Resolving Natural Evil

In the absence of God and creation, natural evil becomes simply a brute fact about the universe to which we assign the moral term *evil* because we experience the universe in guises that appear to us cruel and destructive. But in the face of God and creation, natural evil is not just the source of much human and animal suffering, but also a moral fact that we cannot simply leave without explanation. We need to know how moral and natural evil relate.

According to Christian theology, history has two key events: the fall of humanity in the rebellion of Adam and Eve, and the redemption of humanity by Jesus in the cross and resurrection. Redemption redresses the fall, but we still live in a world where the fall is not fully redressed. We still confront moral and natural evil, and as Christians we need to understand how natural evil is a consequence of moral evil. In light of modern science, which sees young-Earth creationism as implausible, an approach to natural evil as outlined in this chapter is therefore necessary to the believer who wishes to maintain classic Christian orthodoxy.

48

Should Christians Embrace Human Enhancement Science and Technology That Extends Mental and Physical Limitations?

Miguel Angel Endara

Imagine a future in which worn-out human body parts are replaced in a similar way that worn-out car parts are replaced.[1]

Imagine a future in which human beings are seamlessly connected with computers through an ultra-thin mesh of electrodes implanted in the brain.[2]

Let's go a step further: Imagine a future in which humans achieve "immortality" by having their "brains" uploaded to the cloud.[3]

Even more, imagine a future in which human beings are no longer biological but may take on different types of bodies, such as a swarm of miniscule networking robots in the shape of a body.[4]

Does all this sound like science fiction? It *is* science fiction—presently at least, though it is doubtful these could become science fact. These ideas are already the actual goals of private companies and futuristic visionaries who have spent much time and capital to make scientific and technological advances in these directions.

If we ever arrive at this imagined technologically driven futuristic world where these advances exist or, even a less fantastic one, will they be a boon or bane for human existence? If we arrive at such a world, will we then experience all we want: fulfillment, joy, and peace in perpetuity? Will we enter a golden age of humanity, our utopia, or our Shangri-La? Or would we compromise human nature and human flourishing? Even worse, would we enter a humanly degrading dystopic age? Further, how should Christians evaluate these kinds of advances in human enhancement science and technology— particularly those that extend mental and physical limitations?

In this chapter, I offer a critique of the technologically driven futuristic vision of those who would like to see human beings "evolve" past their natural mental and physical limitations. I will base my critique on virtue ethics, a theory of ethics in which character development is the chief goal. And I will end with a brief theological critique.

Freedom to Be Old, Ugly, and Impotent

To begin to answer the questions posed above, let us consider a scene from Aldous Huxley's *Brave New World*.

This novel envisions a futuristic technologically driven utopia hundreds of years from now, in which sickness and old age are no more and everyone is in a constant state of "happiness."

In one scene, John, nicknamed the Savage, challenges the workers coming off a shift to rebel against the technologically driven society because it enslaves them. As the Savage speaks to them, he throws out the workers' ration of soma (a drug used to enhance the sense of well-being) that is about to be distributed to them. This act raises the ire of the workers, nearly causing a riot. Police officers then employ emergency means of soma distribution so that the workers are once again pacified, reentering their stupor of well-being as they cry, kiss, and hug one another.[5]

The social engineer and Resident Controller, Mustapha Mond, takes the Savage along with two other "dissidents" into his office and explains that individual freedom and personal happiness are incompatible. Further, social happiness necessitates sacrificing some valuable things, such as personal freedom, Shakespeare, the search for truth, and the Bible.[6]

In what I take to be the philosophical apex of the novel, the Savage declares, "I'm claiming the right to be unhappy. Not to mention the right to grow old and ugly and impotent…"[7]

What? Who, clothed and in their right mind, would not want to be happy or do away with these universal human afflictions? Who wants to have a *right* to be unhappy and grow old, ugly, and impotent?

What, exactly, is the point that Huxley makes through the Savage? Huxley tells us that doing away with universal afflictions sometimes comes with a hefty price tag. In this case, the price is the loss of art, religion, and even science. Further, the price is also strict population control through an artificial process that manufactures individual human beings to belong to one of five castes. Through these means, the Resident Controllers set up the well-structured society of the *Brave New World*, a society in which citizens take predetermined social roles that revolve around the orderly production and consumption of material goods. This purely materialistic lifestyle results in a dehumanized state of enslavement to the state and mindless hedonism (the pursuit of physical pleasure as the highest good).

C.S. Lewis seems to agree with Huxley's idea that notable scientific advances sometimes foster the loss of freedom, though in a different way. In his book *The Abolition of Man*, Lewis claims that if humans, through "Man's conquest of Nature," attain the power to modify its descendants, this power will be used to manipulate people, thereby enslaving them. "Man's conquest of Nature…means the rule of a few hundreds of men over billions upon billions of men," he writes. "Each new power won by man is a power over man as well."[8]

Is it true that, just as there is a hefty price to pay for utopian social harmony and bliss as portrayed in *Brave New World*, so also there may be a hefty price to pay in our world if great advances are made in mental and physical human enhancement technology? Also, is it true that because "each new power won by man is a power over man," human enhancement technology will enslave us?

To answer these questions, let us take a more in-depth look into the future that human enhancement visionaries offer us.

Transhumanism, Enhancement, and Therapy

Human enhancement futurists are sometimes called *transhumanists*. In this chapter, I use the term to refer to persons who believe in and promote the idea that science and technology ought to be used to "improve" the human species. In other words, transhumanists would like to see human beings move beyond their natural mental and physical limitations.[9] For example, Humanity+, an international nonprofit educational organization, asserts in its mission statement that it "advocates the ethical use of technology to expand human capacities. In other words, we want people to be better than well."[10]

Also, I use the terms *therapy* and *enhancement* the way academic bioethicists use them. *Therapy* refers to the use of science and technology to restore the *normal* state of health and fitness to the mind and body whose capacities might be disabled or compromised through disease, disability, or impairment.[11] On the other hand, *enhancement* refers to using science and technology to intentionally augment, by direct intervention, the natural powers of the body or brain.[12]

Science and technology give us many marvelous futuristic-sounding examples of therapy. First, ABC News reported about a quadriplegic patient who can move her robotic limb with her thoughts using tiny sensors, each one the size of a baby aspirin, implanted in her brain.[13] Second, CBS News reported on a blind patient who was able to see using a retinal prosthesis device, or bionic eye.[14] An electronic stimulator was implanted in one of the patient's eyes and connected to a camera affixed to special glasses. The eyesight was not perfect, but it allowed the patient to see flashes of light that he would have to learn to interpret. Third, *The Washington Post* reported about research scientists who have

turned green spinach leaves into heart tissue that beats. However, the tissue beats for only up to three weeks. The hope is to eventually create human hearts for heart transplant patients.[15] While some may consider these technological advancements morally questionable, the aim or purpose of these is to restore human function, and thus we may label these as therapeutic.

We began this chapter with examples of enhancement envisioned by transhumanist companies and individuals. Let us revisit these.

First, the Methuselah Foundation, a company whose strategic goal is "making 90 the new 50 by 2030," has as one of its "pathfinding strategies" the replacement of worn-out body parts, such as organs, cartilage, bones, and vasculature, in a similar way that we replace worn-out car parts.[16]

Second, Elon Musk purchased Neuralink in February 2017 for "developing ultra-high bandwidth brain-machine interfaces to connect humans and computers."[17] He calls these electrode interfaces *neural lace*. The goal is to allow humans to upload information directly from a computer.[18] Neural lace, then, will "make us smarter, improve our memory, help with decision-making and eventually provide an extension of the human mind." Musk believed, in 2017, that we were only 10 to 15 years away from these goals.[19]

Third, the startup company Nectome, founded in 2016, is currently working on a commercial application for preserving brains—or, to be more precise, preserving the structure of the brain—and uploading them to the cloud.[20] How does Nectome intend to do this? The "[b]lood flow to the brain will be replaced with embalming chemicals that preserve the neuronal structure..."[21] This process of *vitrification* turns the brain into glass. The main problem with this method is that it kills the patient. Another problem is that, as of March 2018, the company has not figured out how they may upload or revive brains.

Fourth, Ray Kurzweil, who may be the best-known transhumanist, envisions that by 2045, it will be possible to create a happy marriage between the human brain and artificial intelligence that he calls *the singularity*.[22] One way to reach the singularity, according to Kurzweil, is through *smart dust*. This neologism refers to swarms of intelligent networked *nanobots*.[23] A nanobot is a type of robot that uses nanotechnology, an emerging technology that, through the direction of the brain, manipulates matter at the atomic level to simulate anything, including a human body.

Are Transhumanists "Playing God"?

One of the first thoughts that may come to our minds when we hear of transhumanist goals is that they are playing God. Transhumanists have taken on the role of God in that they want to refashion or recreate the human being. In using advances in science and technology, they believe that they know best regarding what it would take for us to become more advanced beings. In other words, transhumanists, it seems, want to take matters into their own hands and come up with version 2.0 of the human being. Their hubris, detractors claim, makes them think they can be as God, which was the sin of Satan and Eve and Adam.[24]

However, it may not be the case that they want to play God. Instead, many transhumanists would claim that their goal is merely to help out humanity. As they contemplate our human condition and notice

the many scourges that visit upon us, such as disease, decay, and death, they want to provide a remedy to our plight, a solution to our condition. Now, for the first time in the history of humanity, we may have opportunities to resolve our perennial dissatisfaction with our state, thereby improving the prospects for our flourishing—or at least that is the claim.[25] Transhumanists believe, therefore, that instead of playing God, they "must redesign humanity so that our ruinous flaws can be eliminated."[26]

Even if transhumanists are not seeking to play God, are they doing it anyway? In a sense, they believe that they know best how to recreate human beings or how to improve humanity. And, yes, having such ideas does imply at least a certain amount of hubris. However, our main concern should not be so much with hubris but with freedom and flourishing. Will a transhumanist future lead to flourishing or will it compromise people by shackling them?

Consequentialist Concerns with Human Enhancements

In the report *Beyond Enhancement*, the President's Council on Bioethics explains that there are four obvious and common concerns with scientific and technological advances in human enhancement.

The first concern is with *health*.[27] Will such enhancements place our health at risk? We know that taking pharmacological products for the brain and body, such as mood brighteners and steroids, for example, will result in a compromise of mental and physical health. What will happen if we discover more remarkable types of enhancements? Will these affect our health to a greater degree? As the President's Council states, "no biological agent used for purposes of self-perfection or self-satisfaction is likely to be entirely safe."[28]

The second concern is with *unfairness*.[29] Enhancements give persons, especially those participating in competition, unfair advantage over others. For example, athletes who engage in blood doping and students who use stimulants while taking SAT exams have unfair advantages over their competitors.

The third concern is with *equality of access* or distributive justice.[30] Certain persons will have the economic means and other resources to access new and novel enhancements while others will not be able to do so. This lack of distributive justice will create disparities amongst the social classes, including economic ones.

The fourth concern is with *liberty*.[31] We may imagine despotism within families, where parents are more concerned with reaching their goals than with jeopardizing the well-being of their children. Or, we may consider tyrannical regimes that coerce their people in some way or other—Lewis's concern. Or, we may imagine more subtle means of coercion where society places great pressure on having a high IQ level or a degree of athletic prowess or looking a certain way to get a job and be accepted as a normal member of society.

These types of concerns are what ethicists might call consequential or utilitarian ones. They deal with what may (or may not) occur as a result of implementing the science and technology of human enhancement. These concerns have to do with political and societal states of affairs. In other words, it is not of necessity that the problems in question will come to pass. The concerns, then, aren't about the enhancements themselves but about the consequences of implementing them in a particular social and political environment. These are real concerns that we should take seriously, for they describe many possible detrimental states of affairs for individuals and society.

Human Freedom Through Virtue

My main concern, though, is not a consequentialist one but an in-principle one that relates to human freedom. In *Brave New World*, the Resident Controllers enslave the citizens by not allowing them to live as individuals but, instead, forcing them to live lives with purely materialistic attachments. Specifically, I'm concerned with the freedom to be true to our humanity and our understanding of what it means to be human. Genuine human freedom is not purely subjective, constrained only by the will. Instead, it is an objective freedom that consists of acquiring the ability to develop and achieve excellence in our human powers or capacities so that we may meet our human needs or exigencies, thereby flourishing.

Freedom, in this sense, results from the acquisition of patterns or habits of behavior—as well as the accompanying habits of thought and feeling—that support the development of our nature. These are settled dispositions or character traits that result in a *second nature*. We call these habits *virtues*. These are not easy to acquire, for they require much wrestling with our appetites, desires, and previous well-established character traits. Like the building of a bicep, we gain the virtues through determination and consistent practice.

Virtues such as patience, consistency, and beneficence, amongst others, develop our humanity. Classically speaking, the main or fundamental virtues consist of three *moral virtues* and an *intellectual virtue*. Temperance, courage, and justice are moral virtues that result from physical activity and practice. Wisdom is an intellectual virtue that results from learning. Once we learn to habitually think, feel, and act in accord with, for example, the ideals of temperance, courage, justice, and wisdom, we become temperate, courageous, just, and wise. That is, we acquire new habits that express themselves as predispositions. As we develop or perfect these habits, we experience satisfaction and fulfillment in thinking, feeling, and acting in accord with these, for they fulfill our human needs. As we do this, we gain genuine human freedoms.

An example is in order. We are aesthetic beings; we seek to create and experience beauty. Insofar as we succeed at creating beauty we experience inspiration, delight, and fulfillment, for the creation of beauty meets a human need. One way to experience aesthetic fulfillment is to excel as a piano player. To do this, we ought not to merely seek the freedom from external impediments that would prevent us from exercising the freedom to strike any key we want at any time and in any sequence, for this will only result in auditory anarchy. Instead, to excel as a piano player, we are to train ourselves to play every piano key in a suitable sequence and with correct technique, articulation, and musicality, thereby creating meaningful, harmonious, and expressive music. Along the way, we acquire the habits to be patient in our learning, consistent in our practice, and competent in using our musical memory. In acquiring these virtues, we've gained a genuine human freedom—the freedom to express ourselves aesthetically—thereby promoting the realization of the aesthetic feature of our being, from which fulfillment derives.

If, on the other hand, we acquire habits that compromise our humanity, we gain *vice*. Vices are obstacles to our well-being; they damage our ability to think, feel, and act in

ways that forward our flourishing. The classical vices or seven deadly sins are pride, envy, anger, sloth, greed, gluttony, and lust.

If it is true that the virtuous expression of our humanity leads to flourishing, then we need to ask: Will advancements in scientific and technological human enhancements accompany or aid the virtues in our path of flourishing? Or would we pay the hefty price of losing the possibility of gaining genuine human freedom for such advancements?

Enhancements and Flourishing

It is easy to imagine how the restoration of our health through *therapeutic* advancements can be in line with a life in which we pursue the virtues, for we see how pharmaceutical drugs and surgery, for example, restore the proper functioning of our bodies. However, we cannot say the same of enhancements. Why not? There are at least three ways that enhancements may compromise virtue and the kind of human freedom that leads to flourishing.

First, the enhancements might not correspond to human virtues. Insofar as human virtues are excellences of our human nature that we attain through determination and practice, they line up with our natural human limits. Flourishing comes not despite our limitations but because of them. In other words, it is within the context of our physical and mental limitations that we develop virtues that realize our capacities and, in turn, experience fulfillment and flourishing. On the other hand, if enhancements break through our natural physical and mental limitations, our physical and mental capacities will no longer line up with virtue and fulfillment. In this case, the script of human nature and human flourishing would have been rewritten.

For example, if we replace worn-out body parts in a similar way that we replace worn-out car parts, we may live much longer than what we currently live. But will this be good for us? Our lives have different seasons, and each season brings with it certain challenges and joys, allowing us to learn and develop in new ways. Whereas in the early stages of our life we may develop a love for learning, self-control or temperance, patience, and other virtues, as we mature, we tend to focus more on developing wisdom, justice, and benevolence. As the psalmist says, "Teach us to number our days that we may gain a heart of wisdom."[32] Would we grow weary or lethargic about life if we increased our time here on Earth by 50 or 100 or 200 years? Would we become like the person who is excited about purchasing new clothes or a new car, but with time grows weary of these, no longer finding satisfaction in them?

Consider another example. If we develop *neural lace*, electrodes that act as brain-machine interfaces, we would have the vast amounts of information on the Internet available to us. No longer would we have to work hard at developing intellectual virtues such as wisdom and diligence. All we would need to do is something akin to a Google search to access information and apply it. Because human fulfillment or satisfaction comes from developing our capacities in virtuous ways through determination and practice, the use of neural lace would rob us of the fulfillment that comes from pursuing wisdom and being diligent, and we may even grow intellectually lazy. To use an analogy, think about persons who are born into economically well-off families. Sometimes these persons are given everything they want as they grow up. As a result, some become lazy as they gain a sense of entitlement. Attitudes or habits of emotion such as entitlement and laziness are not virtues and do not lead to fulfillment, but instead detract from it.

One more example: Though it seems doubtful that it will ever happen, if we were able to develop *smart dust*—swarms of intelligently networked robots that function as our bodies—we would not have to acquire the virtues of patience, consistency, or competent use of our musical memory. Instead, we could merely program or command that our body play the piano or some other musical instrument. After all, there is a vast difference between playing the piano and merely programming one's body to do so. The difference is learning to play the piano through determination and practice versus learning to play a piano tune on an iPod. With the latter activity, we would neither gain virtue nor develop a new human freedom for aesthetic expression. We would not find nearly as much fulfillment and would not flourish.

Second, enhancements may compromise another avenue of growth, development, and satisfaction: that of developing our character through adversity. Philosophers and theologians have long recognized that we humans can grow and even thrive after suffering. Consider John Hick's soul-making theodicy (defense against the problem of evil).[33] According to this theodicy, God allows suffering to help human beings mature, develop virtue, and follow him more closely. Psychology seems to confirm Hick's idea. According to research on adversarial growth, human beings can use suffering and traumatic events as springboards for growth and development that lead to greater satisfaction and fulfillment in life.[34] If science and technology were to recreate our bodies and brains by using nanobots or electronic messages in the cloud, would we have the opportunity to experience adversarial growth? Not likely.

Third, enhancements not only do not

lead to virtue, rather, they may lead to vice. Consider a thought experiment that involves an analogy regarding the gods of ancient Greek mythology. These gods are more like superhuman men and woman than they are like the God of the Bible. Though they possess immortality, they are prone to the human frailties of pride, envy, lust, anger, and the like. Morally speaking, they are no better than mere mortals and, at times, worse due to their vast powers. For example, when Zeus (Jupiter) became filled with anger, he cruelly punished Prometheus by chaining him to a rock. There, an eagle devoured his liver, only to devour it again each day after it regrew.[35] Consider also the story of the goddess Hera who, in a jealous rage, cut up Dionysus, the son of Zeus and Persephone, into pieces, then boiled the pieces and ate them.[36] Then there was the goddess Artemis, whose great hubris demanded human sacrifices whenever there was any slight on her honor.[37]

Now, imagine that advances in science and technology made it possible for humans to attain powers like those of the gods of mythology. Specifically, let's imagine that humans were able to possess great intellectual prowess and gain superhuman physical characteristics such as strength and endurance and live for hundreds of years. How will this help humans to become more virtuous? Maybe what Homer and Hesiod, the Greek poets who wrote mythology, implied in their writings is correct: having superhuman powers and immortality does not help in overcoming vice. Indeed, humans would likely be just as vicious—or even more so—after acquiring enhancements; they would possess greater ability to harm each other. Thus scientific human enhancements would diminish virtue while increasing vice. And the consequences for society could be serious.

Human Enhancement and Human Transcendence

Transhumanists seek what all of us want: peace, joy, and fulfillment in perpetuity. They seek these things by attempting to gain a sort of omniscience and immortality. In doing so, they capture a deep yearning within all of us, the yearning for transcendence. As Leon Kass, a former chair of the President's Council on Bioethics, puts it, "The human soul yearns for, longs for, aspires to some condition, some state, some goal toward which our earthly activities are directed but which cannot be attained in earthly life."[38]

C.S. Lewis helps us to understand this longing for transcendence as he explains that beauty and nostalgia are but images of what we want. But we sometimes turn these into idols by mistaking them for the *thing itself*. So also do the transhumanists, in thinking that the attainment of greater mental and physical powers will give us what we yearn for, turn their visions into idols. At their best, beauty, nostalgia, and futuristic visions "are only the scent of a flower we have not found, the echo of a tune we have not heard, news from a country we have never yet visited."[39]

As in the case of the Resident Controllers of *Brave New World* whose only concerns were materialistic ones, transhumanists may not recognize that we are not just physical but also spiritual beings. They do not realize that transcendence will not come from science and technology. It will not come from living 100 nor even 1,000 more years, nor from having all the information and knowledge that is available on the worldwide web. Nothing in this world can truly and permanently fulfill us. We yearn for more, much more. The only way we can achieve ultimate and perpetual satisfaction is through God.

That said, I now re-ask this question: How should Christians evaluate possible technological advancements that result in human enhancement—particularly those that extend our mental and physical limitations?

A Brief Theological Critique

God created human beings for unity with him. Whether we recognize that the chief end of man is to glorify God and enjoy him forever (as stated in the Westminster Catechism) or we understand that the goal of humans is to experience the beatific vision, we will satisfy our yearning for fulfillment and transcendence only by entering into a union of love with God.[40]

Since God is omniscient, omnipotent, and omnibenevolent, he created the universe, Earth, humanity, and human nature in the best possible way to accomplish this end. He created us human beings and the Earth, in other words, in a manner that is perfectly suited to maximize the possibility of humans to recognize their need and enter into unity with him. For us to recreate humanity or help humanity "evolve" through scientific and technological enhancements, then, would be to go against God's perfect plan for us. How so? As previously stated, such enhancements would enable us to break through our mental and physical limitations. The result? Our human nature would no longer be perfectly suited to maximize the possibility of establishing unity with God.

Instead, the best way to work toward attaining transcendence is to develop our human nature not only through moral and intellectual virtues but also through the *theological* virtues of faith, hope, and love. Because the most significant human need we have is to unite with God, the most significant genuine human freedom is to attain the theological virtues. These virtues develop our spiritual capacities, thereby allowing us

to gain the freedom to enter into and deepen our unity with God.[41]

Transhumanist Dystopia

Will the technologically driven futuristic world envisioned by Elon Musk, Ray Kurzweil, and other transhumanists be a boon or a bane for human existence?

Upon entering such a world, we would most likely enter a dystopic age where, given our society, we are likely to compromise human health and the possibility of fair competition, distributive justice, and liberty. We would likely experience a loss of liberty as a result of a few coercing and enslaving the many. Moreover, we would pay a hefty price for losing the possibility of gaining genuine human freedom through moral and intellectual virtue. By overcoming universal human afflictions such as growing old, ugly, and impotent, we would compromise human flourishing by eliminating the growth of virtues that result from human limitations. We would end up preventing adversarial growth and promoting vice.

As Christians, we may agree with concerns about these consequences and go a step beyond in our critique. The futuristic visions of the transhumanists are idols that, if they ever come to pass, would compromise the possibility of experiencing union with God. Breaking through the God-ordained mental and physical limits of human nature would disrupt not only the possibility of gaining moral and intellectual virtues, but also theological virtues.

In these ways, entering an age of enhancements would be a bane for humanity. We are but sojourners, resident aliens in this lost and fallen world. As such, this world cannot provide us with the golden age of humanity or the transcendence for which we all yearn. Only through the beatific vision can we experience perpetual fulfillment, joy, and peace.[42]

Appendix

List of Important Thinkers in the History of the Natural Sciences Who Were Religious* (AD 300–1750)

David Haines

The list below documents various important thinkers in the early history of natural science who were devoutly religious. This list is not exhaustive, and the purpose here is not to provide an in-depth consideration of each thinker, but to provide a starting point for further discussion and study. The entries appear in chronological order, and each one briefly notes a scientist's standing in relation to religion or philosophy, and some of what the individual contributed to the development of the natural sciences (e.g., their involvement in the preservation of, transmission of, or contribution to knowledge).

1. **Themistius** (c. 317–390), a Byzantine Christian who preserved and transmitted the works of Aristotle on the natural sciences.

2. **Boethius** (c. 480–524), an important Christian philosopher and theologian who contributed to the natural sciences by preserving and transmitting to the Middle Ages the logical works of Aristotle, *The Elements* of Euclid, and others. He translated them into Latin, ensuring that they could be accessed and used by the Latin West.

3. **John Philophonus** (c. 490–570), a Christian Neoplatonist from Alexandria who wrote commentaries on a number of Aristotle's works (such as *Physics, Meteorology, On Generation and Corruption,* and *De Anima*). He also attempted to refute Aristotle's theory of movement in an original manner.

* I have consulted numerous sources, both primary and secondary, for the information provided in this list. Some of the more important works include David C. Lindberg, *The Beginnings of Western Science: The European Scientific Tradition in Philosophical, Religious, and Institutional Context, Prehistory to A.D. 1450* (Chicago, IL: University of Chicago Press, 2nd ed., 2007); James Hannam, *God's Philosophers: How the Medieval World Laid the Foundations of Modern Science* (London, UK: Icon Books, 2010); Rodney Stark, *For the Glory of God: How Monotheism Led to Reformations, Science, Witch-hunts, and the End of Slavery* (Princeton, NJ: Princeton University Press, 2003); John Hedley Brooke, *Science and Religion: Some Historical Perspectives,* The Cambridge History of Science Series (Cambridge, UK: Cambridge University Press, 1991); Andrew Ede and Lesley B. Cormack, *A History of Science in Society: From Philosophy to Utility* (Peterborough, ON: Broadview Press, 2004).

4. **Isidore of Seville** (c. 560–636), a Christian archbishop who wrote two books on the natural sciences: *On the Nature of Things* and *Etymologies*. He taught, among other things, that the Earth was a sphere, and wrote about the movement of the planets, the changing of seasons, the sun, and the moon.

5. **Bede** (c. 672–735), a Christian monk from England who wrote on the natural sciences in his book *On the Nature of Things*, which was based on the writings of Isidore and Pliny.

6. **Al-Khwarizmi** (c. 780–850), a Muslim mathematician who wrote a number of books, two of which are of specific interest for the natural sciences: *On the Hindu Numbers*, which introduced a new approach to calculation, and *Algebra*, which, although its contents do not resemble contemporary algebra, is the source of what we today know as algebra.

7. **Yahya ibn Abi Mansur** (c. 832), a Muslim astronomer who was in charge of important astronomical research in a number of observatories. He also produced one of the first Arabic astronomical charts.

8. **Yuhanna ibn Masawayh** (c. 777–857), a Nestorian Christian scientist who prolifically wrote treatises on medicine, discussing subjects such as sicknesses related to the eyes, fever, and leprosy.

9. **Abu Ishaq al-Kindi** (c. 866), a Muslim scientist who wrote influential works on optics.

10. **John Scottus Eriugena** (c. 810–877), an important Christian philosopher and theologian during the Carolingian renaissance. He wrote a number of important volumes, one of which was *On Nature*, in which he developed a well-articulated philosophy of nature based on Neoplatonic philosophy.

11. **Al-Battani** (c. 858–929), a Muslim astronomer and mathematician who studied the movement of the sun and the moon. He corrected the already existing star charts and gave instructions on how to build astronomical equipment, such as the sundial. He is quoted by both Kepler and Copernicus.

12. **Gerbert of Aurillac** (c. 945–1003), elected Pope Sylvester II in 999, was a French Christian theologian and mathematician who contributed to the natural sciences by recovering, translating, preserving, and transmitting to the Latin West Greek mathematics and Arabic developments in mathematics. He also studied astronomy.

13. **Abu Said al-Ala ibn Sahl** (c. 940–1000) a Muslim scientist who did experimental studies on the refraction of light.

14. **Thierry of Chartres** (c. 1150), a Christian theologian and philosopher who proposed innovative interpretations of Genesis in relation to Neoplatonist cosmological reflections.

15. **Robert Grosseteste** (1168–1253), a Christian theologian and philosopher

who commented on Aristotle's works related to the natural sciences. He also wrote numerous innovative treatises on the natural sciences and on subjects such as rainbows and light.

16. **Nasir al-Din al-Tusi** (1201–1274), a Muslim astronomer who was in charge of the important observatory at Maragha.

17. **Albert the Great** (1205–1280), a Christian theologian and philosopher who was one of the earliest members of the Dominican Order of Preachers. He is rightly known as a father of modern botany and zoology. He is also arguably the first medieval thinker to explicitly articulate the importance of personal experience and observation in scientific research.

18. **Roger Bacon** (1220–1292), a Christian theologian and scientist who joined the Franciscan order. He is often seen as competing with Albert the Great as the first medieval thinker to have articulated the scientific method (by theory and experimentation). He did a great deal of work on optics, mathematics, and other natural sciences.

19. **Kamal al-Din al-Farisi** (c. 1267–1319), a Muslim scientist who, inspired by Avicenna (Ibn Sina), contributed to the study of rainbows with a number of important observations.

20. **John Peckham** (c. 1230–1292), a Christian theologian and philosopher who became the Archbishop of Canterbury. He wrote, among other things, on the theory of optics.

21. **Thomas Bradwardine** (c. 1290–1349), a Christian theologian and eventual Archbishop of Canterbury who did work in physics and mathematics. He was probably one of the first thinkers (if not the first) to use logarithms in his research. He was one of the Merton Calculators, a group of early mathematicians of renown.

22. **William of Ockham** (1290–1349/50), a Christian Franciscan theologian and philosopher, known for, among others things, nominalism and the principle known as Ockham's razor.

23. **Richard Swineshead** (c. 1340–1355), a Christian theologian and scientist who applied mathematics to the physical world, performing important analyses of heat and speed. He was one of the Merton Calculators.

24. **William Heytesbury** (c. 1313–1373), a Christian theologian and scientist who established a general law for the calculation of movement. He was one of the Merton Calculators.

25. **John Buridan** (c. 1300–1361), a Christian priest who studied the generation of movement (impulsion, resistance, and force). He also suggested that the Earth may be moving—in rotation.

26. **Nicole Oresme** (1338–1380), a Christian theologian who contributed to discussions about the rotation of the Earth (partly by refuting most of the arguments that were proposed to show that the Earth

was immobile). He also worked on and refined the theory of speed that was proposed by the Merton Calculators.

27. **Nicholas of Cusa** (1400–1465), an important theologian and cardinal of the church who studied the movement of the planets (and the Earth) and suggested the possibility that there may be extraterrestrial life. He also gave his own articulation of the scientific method.

28. **Pierre d'Ailly** (1350–1420), a Christian theologian and cardinal of the church who wrote on geography and astronomy, and that it would be possible to cross the ocean to go to India.

29. **Johann Gutenberg** (c. 1398–1468), the Christian inventor who created the first printing press with movable letters and numbers, making it possible to print books at a lower price, thus making them easier to obtain.

30. **Nicolaus Copernicus** (1473–1543), a Catholic mathematician and astronomer who served as canon of the church and is known for proposing a new cosmological theory (heliocentrism) based on observations and discoveries from the preceding centuries. He also studied medicine and was a practicing physician.

31. **Paracelsus** (1493–1541), an unorthodox Christian who practiced alchemy, medicine, and mysticism. His real name was Théophrastus Bombastus von Hohenheim. He attempted to invent an approach to medicine that was distinctly Christian (rather than being based on the "pagan" theories of Galen).

32. **Andreas Vesalius** (1514–1564), a devout Catholic who did innovative work on human anatomy.

33. **Tycho Brahe** (1515–1546), a Protestant scientist who is known primarily for his work in astronomy, having invented or perfected a number of astronomical tools and star charts, as well as accurately noting the movements of the planets.

34. **John Napier** (1550–1617), a devout Protestant who was a physicist, mathematician, and astronomer. He is known as the person who, among other things, invented logarithms, a portable calculation tool, and popularized the use of the decimal point.

35. **Henry Briggs** (1561–1630), a devout English Protestant (a Puritan) and important mathematician who refined John Napier's work on logarithms.

36. **Gabriel Fallopius** (1523–1562), a devout Catholic who was one of the most important Italian anatomists. He studied human reproductive anatomy (the Fallopian tube is named after him) and made important contributions to our knowledge about the ear.

37. **Christoph Clavius** (1538–1612), a Catholic mathematician and astronomer and Jesuit theologian who was instrumental in the introduction of the Gregorian calendar.

38. **William Gilbert** (1540–1603), a Protestant scientist and medical doctor who studied physics and made important contributions related to magnetism (he was a pioneer in the field, publishing a six-volume work on the subject).

39. **Simon Stevin** (1548–1620), an important Protestant mathematician and physicist who made major contributions to mechanical engineering, such as theories about immobile things and force. He also contributed to the standardization of decimal fractions.

40. **François Viète** (1540–1603), an important Catholic mathematician who contributed to equation theory and introduced the first system of algebraic notation.

41. **David Fabricius** (1564–1617), a devout Protestant and Lutheran pastor who was also an important astronomer. He was the first to discover the variable star and sunspots. He corresponded frequently with Kepler.

42. **Francis Bacon** (1561–1626), a devout Anglican who, though not an active scientist, is often credited with the invention of the scientific method via induction.

43. **Galileo Galilei** (1564–1642), a devout Catholic astronomer who was a personal friend of the Pope. He refined the telescope, discovered the moons of Jupiter, argued for heliocentrism, made important contributions to the laws of motion, and innovated the application of mathematics to experimental physics.

44. **Johannes Kepler** (1571–1630), a devout Protestant astronomer and mathematician. He is primarily known for his cosmological system, improving on Copernicus's heliocentric model by proposing that the planets move in elliptical orbits around the sun.

45. **Johann Bayer** (1572–1625), a devout Protestant and an important astronomer who made important contributions to star charts.

46. **Jan Baptista van Helmont** (1577–1644), a devout Catholic who made important contributions to chemistry, physiology, and physics. He is said to have discovered carbon dioxide, demonstrated that metals are not destroyed when they are diluted in acid, and to have coined the word *gas*.

47. **William Harvey** (1578–1657), a devout Protestant who is credited with the discovery of the circulatory system that permits blood to circulate through the body, among other things.

48. **Christoph Scheiner** (1575–1650), a devout Catholic and Jesuit priest who was also an important astronomer. He is known for his study of, among other things, sunspots.

49. **William Oughtred** (1575–1660), a devout protestant and Anglican priest who was also an important mathematician. He invented the slide calculator and published a textbook on mathematics that became the primary text for teaching

mathematics in the seventeenth century.

50. **Marin Mersenne** (1588–1660), a devout Catholic and Franciscan friar who was also an important mathematician, music theorist, and philosopher. He played a major role in making scientific literature accessible. He was also a personal friend of René Descartes.

51. **Willebrord Snell** (1591–1626), a Catholic astronomer and mathematician who refined the methods for calculating the circumference of the Earth. He proposed an important method of triangulation as well as the law of refraction known as Snell's law.

52. **Pierre Gassendi** (1592–1655), a devout Catholic priest and astronomer who was the first to discover, with the help of Joseph Gaultier (who was employed by Gassendi), the Orion Nebula. He also made a number of discoveries concerning the movement of the planets. He was in constant communication with Galileo and was a major critic of Descartes.

53. **René Descartes** (1596–1650), a devout Catholic educated by the Jesuits, was one of the most important and controversial of all modern philosophers. In addition to his work in philosophy (interactions with skepticism, arguments intended to demonstrate the existence of God, etc.), he made important contributions to geometry (basically inventing analytical geometry, and showing how algebra could be used

in geometry), astronomy, optics, and mathematics.

54. **Henry Gellibrand** (1597–1663), a devout Protestant mathematician and physicist who is known, among other things, for his work on the Earth's magnetic field.

55. **Giovanni Riccioli** (1598–1671), a devout Catholic and Jesuit priest who was also an influential astronomer.

56. **Pierre Fermat** (1601–1665), an important Catholic mathematician who made contributions to probability theory, optics, and geometry. There are two mathematical theorems named for Fermat.

57. **Otto von Guericke** (1602–1686), a Protestant scientist who studied electricity and physics. He performed important experiments and demonstrations concerning atmospheric pressures.

58. **Athanasius Kircher** (1601–1680), a devout Catholic and Jesuit priest who worked on magnetism, light, and mathematics.

59. **Evangelista Torricelli** (1606–1647), a Catholic physicist and mathematician who studied under Galileo. He invented the barometer and made important discoveries in relation to vacuums and geometry.

60. **Giovanni Borelli** (1608–1679), a Catholic mathematician and physicist.

61. **Johannes Hevelius** (1611–1687), an important Catholic astronomer.

62. **John Wallis** (1616–1703), a devout Protestant pastor and mathematician who invented and used the symbol for infinity (∞) in mathematics.

63. **Jeremiah Horrocks** (1619–1641), a devout Protestant pastor who made important discoveries in astronomy.

64. **Francesco Grimaldi** (1618–1663), a devout Catholic and Jesuit priest who made contributions to mathematics and physics (he worked, among other things, on the pendulum).

65. **Edme Mariotte** (1620–1684), a devout Catholic priest, physicist, and botanist who was one of the foundational members of the Académie des Sciences in Paris. He is known for having discovered the blind spot in human vision (known as the Mariotte spot) and also for important discoveries in botany, optics and color perception, hydraulics, and physics.

66. **Blaise Pascal** (1623–1662), a devout Catholic and a member of the Jansenists. Pascal is not only known for his contributions to philosophy (he is studied as one of the most important modern philosophers) and his defense of Christianity (he is known for Pascal's wager, among other things), but also for his major contributions to mathematics and the natural sciences. He was raised and educated by his father. At the age of 16 he published a revolutionary treatise on the geometry of cones (in which he proposed what became known as Pascal's theorem). He also laid the foundations for probability theory, proved the existence of vacuums, worked on atmospheric pressures, and invented one of the first working mechanical calculators.

67. **Jean Picard** (1620–1682), a devout Catholic priest and astronomer who made important advances in his work on telescopes and astronomical measurement (such as instruments that are used for measuring the diameters of the heavenly bodies, including the Earth).

68. **Giovanni Cassini** (1625–1712), a Catholic mathematician and astronomer who is known, in part, for his study of curves. He also contributed to the study of hydraulics and mechanical engineering. He was the first to observe four of Saturn's moons as well as the gap in the ring system of Saturn (now known as the Cassini Division). He also made a number of other important astronomical discoveries.

69. **Robert Boyle** (1627–1691), a devout Protestant chemist and physicist. He is known, in part, for Boyle's Law (concerning ideal gas) and discovered, among other things, that sound does not pass through vacuums, that fire needs oxygen, and the elastic qualities of air.

70. **Marcello Malpighi** (1628–1691), a Catholic biologist and medical doctor who is known as the founder of microscopical anatomy as well as the father of embryology and physiology.

71. **John Ray** (1628–1705), a devout Protestant (Anglican) pastor who is

known as the father of natural history. He made important contributions to embryology and the physiology of plants.

72. **Christiaan Huygens** (1629–1695), a devout Protestant who was influenced by René Descartes, who visited the Huygens household often. Huygens was an important mathematician who invented the pendulum clock and discovered a better way to make lenses (allowing him to build a better telescope that enabled him to discover the first of Saturn's moons and the shape of the rings around Saturn). He published on probability theory, worked on pendulums, and proposed multiple laws in relation to these subjects.

73. **Anton Leeuwenhoek** (1632–1723), a Protestant who used microscopy to become the father of microbiology, having discovered protozoa (single-celled organisms). He was the first scientist to study muscle fibers and bacteria.

74. **Robert Hooke** (1635–1703), a devout Protestant who made important contributions to mathematics, optics, mechanics, architecture, and astronomy.

75. **Nicolas Steno** (1638–1686), a devout Catholic priest who was a pioneer in anatomy and geology. He described, among other things, the structure of crystal quartz, fossils, etc. He also discovered the parotid salivary canal, which is today known as the Stensen canal.

76. **Regnier de Graaf** (1641–1673), a Catholic anatomist who made important discoveries in embryology. He discovered the mature ovarian follicles (known as the Graafian follicles) and made important contributions to our knowledge of the reproductive organs of female mammals.

77. **Nehemiah Grew** (1641–1712), a devout Protestant who made important discoveries in plant anatomy and is sometimes known as the father of plant anatomy.

78. **Isaac Newton** (1642–1727), a devout Protestant who made important discoveries in optics (proposing influential theories of light and color) and physics (universal theory of gravitation). He also laid the foundations for integral and differential calculus.

79. **Olaus Roemer** (1644–1710), a Protestant astronomer who was the first to measure the speed of light.

80. **John Flamsteed** (1646–1719), a Protestant astronomer who is known as one of the most important practical astronomers. He was also the first Astronomer Royal of England (receiving a commission from the king of England to build the Royal Observatory at Greenwich). He also invented the notion of conical projection, which is essential to cartography.

81. **Gottfried Leibniz** (1646–1716), a devout Protestant who is one of the most important philosophers of the modern period. Aside from

his importance to philosophy, he was also a noted mathematician; he invented the contemporary system of integral and differential calculus. He also worked on dynamics, kinetic energy, and momentum, and invented a mechanical calculator.

82. **Denis Papin** (1647–1712), a devout Protestant (a Calvinist from a Huguenot family) who was an important physicist, mathematician, and inventor. He was Christiaan Huygens's assistant for a period of time, and also worked with Robert Hooke and Robert Boyle. He was also in contact with Gottfried Wilhelm Leibniz. He is especially known for his inventions, including the steam-driven piston engine, pressure cooker, submarine, air gun, and grenade launcher. He also invented the paddleboat and made major improvements to blast furnaces.

83. **Edmund Halley** (1656–1742), a somewhat liberal Protestant who is known for his discovery of the comet that bears his name (Halley's Comet). He was John Flamsteed's successor at the Royal Observatory.

Bibliography and Resources

Books and Articles

Anderson, Ryan T. *When Harry Became Sally: Responding to the Transgender Moment.* New York: Encounter Books, 2019.

Aquinas, Thomas. *St. Thomas Aquinas Summa Theologica: Complete English Edition in Five Volumes,* translated by Fathers of the English Dominican Province. Westminster, MD: Christian Classics, 1981.

Axe, Douglas. *Undeniable: How Biology Confirms Our Intuition that Life Is Designed.* New York: HarperOne, 2016.

Barbour, Ian. G. *Religion in an Age of Science.* San Francisco, CA: Harper Publishing, 1990.

Beauregard, Mario, and Denyse O'Leary. *The Spiritual Brain: A Neuroscientist's Case for the Existence of the Soul.* New York: HarperOne, 2007.

Behe, Michael J. *Darwin's Black Box: The Biochemical Challenge to Evolution.* New York: The Free Press, 1996.

Behe, Michael J. *The Edge of Evolution: The Search for the Limits of Darwinism.* New York: The Free Press, 2007.

Behe, Michael J. *Darwin Devolves: The New Science About DNA That Challenges Evolution.* New York: HarperOne, 2019.

Bethell, Tom. *Darwin's House of Cards: A Journalist's Odyssey Through the Darwin Debates.* Seattle, WA: Discovery Institute Press, 2016.

Bowler, Peter J. *Monkey Trials and Gorilla Sermons: Evolution and Christianity from Darwin to Intelligent Design.* Cambridge, MA: Harvard University Press, 2007.

Bloom, John. *The Natural Sciences: A Student's Guide.* Wheaton, IL: Crossway, 2015.

Broom, Neil. *How Blind Is the Watchmaker? Nature's Design & the Limits of Naturalistic Science.* Downers Grove, IL: InterVarsity, 1998.

Budziszewski, J. *Written on the Heart: The Case for Natural Law.* Downers Grove, IL: InterVarsity, 1997.

Campbell, John Angus, and Stephen C. Meyer, eds. *Darwinism, Design, and Public Education.* East Lansing, MI: Michigan State University Press, 2003.

Collins, C. John. *Genesis 1–4: A Linguistic, Literary, and Theological Commentary.* Phillipsburg, NJ: P&R Publishing, 2006.

Collins, C. John. *Science & Faith: Friends or Foes?* Wheaton, IL: Crossway Books, 2003.

Collins, C. John. *Did Adam and Eve Really Exist? Who They Were and Why You Should Care.* Wheaton, IL: Crossway, 2011.

Collins, C. John. *Reading Genesis Well: Navigating History, Poetry, Science, and Truth in Genesis 1–11.* Grand Rapids, MI: Zondervan, 2018.

Copan, Paul, Tremper Longman III, Christopher L. Reese, and Michael G. Strauss. *Dictionary of Christianity and Science.* Grand Rapids, MI: Zondervan, 2017.

Davies, Paul. *The Goldilocks Enigma: Why Is the Universe Just Right for Life?* Boston, MA: Houghton Mifflin Company, 2006.

Dembski, William A. *The Design Inference: Eliminating Chance Through Small Probabilities.* Cambridge, UK: Cambridge University Press, 1998.

Dembski, William A., ed. *Mere Creation: Science, Faith & Intelligent Design.* Downers Grove, IL: InterVarsity, 1998.

Dembski, William A. *Intelligent Design: The Bridge Between Science & Theology*. Downers Grove, IL: InterVarsity, 1999.

Dembski, William A. *No Free Lunch: Why Specified Complexity Cannot Be Purchased Without Intelligence*. Lanham, MD: Rowman & Littlefield, 2002.

Dembski, William A., and Michael Ruse, eds. *Debating Design: From Darwin to DNA*. Cambridge, UK: Cambridge University Press, 2004.

Dembski, William A., ed. *Uncommon Dissent: Intellectuals Who Find Darwinism Unconvincing*. Wilmington, DE: ISI Books, 2004.

Dembski, William A. *The Design Revolution: Answering the Toughest Questions About Intelligent Design*. Downers Grove, IL: InterVarsity, 2004.

Dembski, William A., and Sean McDowell. *Understanding Intelligent Design: Everything You Need to Know in Plain Language*. Eugene, OR: Harvest House, 2008.

Dembski, William A., and Jonathan Wells. *The Design of Life: Discovering Signs of Intelligence in Biological Systems*. Dallas, TX: The Foundation for Thought and Ethics, 2008.

Dembski, William A., and Jonathan Witt. *Intelligent Design Uncensored: An Easy-to-Understand Guide to the Controversy*. Downers Grove, IL: InterVarsity, 2010.

Denton, Michael. *Evolution: A Theory in Crisis*. Chevy Chase, MD: Adler & Adler, 1986.

Denton, Michael. *Nature's Destiny: How the Laws of Biology Reveal Purpose in the Universe*. New York: Free Press, 1998.

Denton, Michael. *Evolution: Still a Theory in Crisis*. Seattle, WA: Discovery Institute Press, 2016.

Denton, Michael. *Fire-Maker: How Humans Were Designed to Harness Fire and Transform Our Planet*. Seattle, WA: Discovery Institute Press, 2016.

Denton, Michael. *The Wonder of Water*. Seattle, WA: Discovery Institute Press, 2017.

Denton, Michael. *Children of Light: The Astonishing Properties of Sunlight that Make Us Possible*. Seattle, WA: Discovery Institute Press, 2018.

Denton, Michael. *The Miracle of the Cell*. Seattle, WA: Discovery Institute Press, 2020.

DeWolf, David K., John G. West, Casey Luskin, and Jonathan Witt. *Traipsing into Evolution: Intelligent Design and the Kitzmiller v. Dover Decision*. Seattle, WA: Discovery Institute Press, 2006.

Dolezal, James E. *All that Is in God: Evangelical Theology and the Challenge of Classical Christian Theism*. Grand Rapids, MI: Reformation Heritage, 2017.

Eberlin, Marcos. *Foresight: How the Chemistry of Life Reveals Planning and Purpose*. Seattle, WA: Discovery Institute Press, 2019.

Ferngren, Gary B. *Science and Religion: A Historical Introduction*, Baltimore, MD: Johns Hopkins University, 2d edition, 2017.

Feser, Edward. *The Last Superstition: A Refutation of the New Atheism*. South Bend, IN: St. Augustine's Press, 2008.

Flannery, Michael, and William A. Dembski. *Intelligent Evolution: How Alfred Russel Wallace's World of Life Challenged Darwinism*. Nashville, TN: Erasmus Press, 2020.

Fradd, Matt, and Robert Delfino. *Does God Exist? A Socratic Dialogue on the Five Ways of Thomas Aquinas*. St. Louis, MO: Enroute, 2018.

Gauger, Ann, Douglas Axe, and Casey Luskin. *Science and Human Origins*. Seattle, WA: Discovery Institute Press, 2012.

Geisler, Norman L. *Is Man the Measure? An Evaluation of Contemporary Humanism*. Grand Rapids, MI: Baker, 1983.

Geisler, Norman L., and Kerby Anderson. *Origin Science: A Proposal for the Creation-Evolution*. Grand Rapids, MI: Baker, 1987.

Geisler, Norman L., and William D. Watkins. *Worlds Apart: A Handbook on World Views*. Grand Rapids, MI: Baker, 1989.

Geisler, Norman L. *Miracles and Modern Thoughts*. Grand Rapids: Zondervan, 1982. Revised *Miracles and the Modern Mind: A Biblical Defense of Biblical Miracles*. Grand Rapids, MI: Zondervan, 1992.

Geisler, Norman L., and Ron Rhodes. *When Cultists Ask: A Popular Handbook on Cultic Misinterpretations*. Grand Rapids, MI: Baker, 1997.

Geisler Norman L., H. Wayne House, and Max Herrera. *The Battle for God: Responding to the Challenge of Neotheism*. Grand Rapids, MI: Kregel, 2001.

Geisler, Norman L. *If God, Why Evil? A New Way to Think About the Question*. Minneapolis, MN: Bethany House, 2011.

Geivett, R. Douglas, and Gary R. Habermas. *In Defense of Miracles: A Comprehensive Case for God's Action in History*. Downers Grove, IL: InterVarsity, 1997.

Gonzalez, Guillermo, and Jay W. Richards. *The Privileged Planet: How Our Place in the Cosmos Is Designed for Discovery*. Washington, DC: Regnery, 2004.

Gordon, Bruce L., and William A. Dembski, eds. *The Nature of Nature: Examining the Role of Naturalism in Science*. Wilmington, DE: ISI Books, 2011.

Grant, Edward. *God and Reason in the Middle Ages*. Cambridge, UK: Cambridge University Press, 2001.

Hannam, James. *The Genesis of Science: How the Christian Middle Ages Launched the Scientific Revolution*. Washington, DC: Regnery Publishing, 2011.

Hereen, Fred. *Show Me God: What the Message from Space Is Telling Us About God*. Olathe, KS: Day Star Productions, 2d revised edition, 2012.

Holden, Joseph M. *The Comprehensive Guide to Apologetics*. Eugene, OR: Harvest House, 2019.

House, H. Wayne, ed. *Intelligent Design 101: Leading Experts Explain the Key Issues*. Grand Rapids, MI: Kregel, 2008.

Howe Richard G. "Modern Witchcraft: It May Not Be What You Think," *Christian Research Journal* 25, no. 1 (2005), 12-21, available at http://richardghowe.com/index_htm_files/ModernWitchcraft.pdf (accessed 07/06/20).

Howe Richard G. *Intro to God's Revelation*, a workbook and DVD curriculum available from https://afastore.net/intro-to-gods-revelation-6-week-curriculum-by-dr-richard-howe.

Howe, Thomas A. "Does Genre Determine Meaning?" *Christian Apologetics Journal* 6, no. 1 (Spring 2007): 1-19.

Hunter, Cornelius, G. *Darwin's God: Evolution and the Problem of Evil*. Grand Rapids, MI: Brazos Press, 2001.

Hunter, Cornelius, G. *Darwin's Proof: The Triumph of Religion over Science*. Grand Rapids, MI: Brazos Press, 2003.

Hunter, Cornelius, G. *Science's Blind Spot: The Unseen Religion of Scientific Naturalism*. Grand Rapids, MI: Brazos Press, 2007.

Jaki, Stanley. *Science and Creation: From Eternal Cycles to an Oscillating Universe*. Fort Collins, CO: Gondolin Press, 2d edition, 2017.

Jastrow, Robert. *God and the Astronomers*. New York: Norton, 1978.

Johnson, Donald E. *Probability's Nature and Nature's Probability: A Call to Scientific Integrity*. Charleston, SC: Booksurge Publishing, 2009.

Johnson, Donald E., *Programming of Life*. Sylacauga, AL: Big Mac Publishers, 2010.

Johnson, Phillip E. *Darwin on Trial*. Downers Grove, IL: InterVarsity, 1993.

Johnson, Phillip E. *Reason in the Balance: The Case Against Naturalism in Science, Law and Education*. Downers Grove, IL: InterVarsity, 1995.

Johnson, Phillip E. *Defeating Darwinism by Opening Minds*. Downers Grove, IL: InterVarsity, 1997.

Keas, Michael. *Unbelievable: 7 Myths About the History and Future of Science and Religion*. Wilmington, DE: ISI Books, 2019.

Keathley, Kenneth, J.B. Stump, and Joe Aguirre, eds. *Old-Earth or Evolutionary Creation? Discussing Origins with Reasons to Believe and Biologos*. Downers Grove, IL: IVP Academic, 2017.

Keener, Craig S. *Miracles: The Credibility of the New Testament Accounts*, 2 vols. Grand Rapids, MI: Baker Academic, 2011.

Kemper, Gary, Hallie Kemper, and Casey Luskin. *Discovering Intelligent Design: A Journey into the Scientific Evidence*. Seattle, WA: Discovery Institute Press, 2013.

Klinghoffer, David, ed. *Signature of Controversy: Responses to Critics of Signature in the Cell*. Seattle, WA: Discovery Institute Press, 2011.

Klinghoffer, David, ed. *The Unofficial Guide to Cosmos: Fact and Fiction in Neil deGrasse Tyson's Landmark Science Series*. Seattle, WA: Discovery Institute Press, 2014.

Klinghoffer, David, ed. *Debating Darwin's Doubt: A Scientific Controversy That Can No Longer Be Denied*. Seattle, WA: Discovery Institute Press, 2015.

Leisola, Mattie, and Jonathan Witt. *Heretic: One Scientist's Journey from Darwin to Design*. Seattle, WA: Discovery Institute Press, 2018.

Lennox, John C. *God and Stephen Hawking. Whose Design Is It Anyway?* Oxford, UK: Lion Books, 2001.

Lennox, John C. *Seven Days that Divide the World: The Beginning Accordance to Genesis and Science*. Grand Rapids, MI: Zondervan, 2011.

Lewis, C.S. *Miracles: How God Intervenes in Nature and Human Affairs*. New York: Macmillan, 1947.

Lewis, Geraint F., and Luke A. Barnes. *A Fortunate Universe: Life in a Finely Tuned Cosmos*. Cambridge, MA: Cambridge University Press, 2016.

Lindberg, David C. *The Beginnings of Western Science: The European Scientific Tradition in Philosophical, Religious, and Institutional Context, Prehistory to A.D. 1450*. Chicago, IL: University of Chicago Press, 2d edition, 2007.

Lindberg, David C., and Ronald L. Number, eds. *When Science & Christianity Meet*. Chicago, IL: The University of Chicago Press, 2003.

Lo, Thomas Y., Paul Chien, Eric H. Anderson, Robert A. Alston, and Robert P. Waltzer, *Evolution and Intelligent Design in a Nutshell*. Seattle, WA: Discovery Institute Press, 2020.

Luskin, Casey. "The New Theistic Evolutionists: BioLogos and the Rush to Embrace the 'Consensus,'" *Christian Research Journal* 37 (2014): 32-41, available at https://www.equip.org/article/new-theistic-evolutionists-biologos-rush-embrace-consensus/.

Marks II, Robert J., Michael J. Behe, William A. Dembski, Bruce L. Gordon, and John C. Sanford, eds. *Biological Information: New Perspectives*. Singapore: World Scientific, 2013.

Marks II, Robert J., William A. Dembski, and Winston Ewert. *Introduction to Evolutionary Informatics*. Singapore: World Scientific, 2017.

McGrath, Alister E. *Science & Religion: A New Introduction*. 2d edition. Malden, MA and Oxford, UK: Wiley-Blackwell, 2010.

Meyer, Stephen C., Scott Minnich, Jonathan Moneymaker, Paul Nelson, and Ralph Seelke. *Explore Evolution: The Arguments for and Against Neo-Darwinism*. Melbourne, Australia: Hill House, 2007.

Meyer, Stephen C. *Signature in the Cell: DNA and the Evidence for Intelligent Design*. New York: HarperOne, 2009.

Meyer, Stephen C. *Darwin's Doubt: The Explosive Origin of Animal Life and the Case for Intelligent Design*. New York: HarperOne, 2013.

Meyer, Stephen C. *Return of the God Hypothesis: Three Scientific Discoveries Revealing the Mind Behind the Universe*. New York: HarperOne, 2021.

Moreland, J.P. *Christianity and the Nature of Science: A Philosophical Investigation*. Grand Rapids, MI: Baker, 1989.

Moreland, J.P., Stephen C. Meyer, Christopher Shaw, Ann K. Gauger, and Wayne Grudem, eds. *Theistic Evolution: A Scientific, Philosophical, and Theological Critique*. Wheaton, IL: Crossway, 2007.

Mortenson Terry, and Thane H. Ury, eds. *Coming to Grips with Genesis: Biblical Authority and the Age of the Earth*. Green Forest, AR: Master Books, 2008.

Newman, Robert C., and John L. Wiester, with Janet and Jonathan Moneymaker. *What's Darwin Got to Do with It? A Friendly Conversation About Evolution*. Downers Grove, IL: InterVarsity, 2000.

Numbers, Ronald L. *Science and Christianity in Pulpit and Pew*. Oxford, UK: Oxford University Press, 2007.

Numbers, Ronald L., ed. *Galileo Goes to Jail and Other Myths About Science and Religion*. Cambridge, MA: Harvard University Press, 2009.

O'Leary, Denyse. *By Design or by Chance? The Growing Controversy on the Origins of Life in the Universe*. Minneapolis, MN: Augsburg Books, 2004.

Osler, Margaret. *Rethinking the Scientific Revolution*. Cambridge, MA: Cambridge University Press, 2000.

Plantinga, Alvin. *Where the Conflict Really Lies: Science, Religion, & Naturalism*. Oxford, UK: Oxford University Press, 2011.

Polkinghorne, John. *Science & Theology: An Introduction*. Minneapolis, MN: SPCK/Fortress Press, 1998.

Poythress, Vern S. *Interpreting Eden: A Guide to Faithfully Reading and Understanding Genesis 1–3*. Wheaton, IL: Crossway, 2019.

Rana, Fazale and Hugh Ross. *Origins of Life: Biblical and Evolutionary Models Face Off*. Colorado Springs, CO: NavPress, 2004.

Rana, Fazale. *The Cell's Design: How Chemistry Reveals the Creator's Artistry*. Grand Rapids, MI: Baker, 2008.

Rana, Fazale, with Hugh Ross. *Who Was Adam? A Creation Model Approach to the Origin of Humanity*. Covina, CA: RTB Press, 2015.

Richards, Jay W., ed. *God and Evolution: Protestants, Catholics, and Jews Explore Darwin's Challenge to Faith*. Seattle, WA: Discovery Institute Press, 2010.

Ross, Hugh. *The Fingerprint of God: Recent Scientific Discoveries Reveal the Unmistakable Identity of the Creator*. Orange, CA: Promise Publishing, 1989.

Ross, Hugh. *The Creator and the Cosmos: How the Latest Scientific Discoveries Reveal God*. Colorado Springs, CO: NavPress, 1993.

Ross, Hugh. *Beyond the Cosmos: What Recent Discoveries in Astrophysics Reveal About the Glory and Love of God*. Colorado Springs, CO: NavPress, 1996.

Ross, Hugh. *The Genesis Question: Scientific Advances and the Accuracy of Genesis.* Colorado Springs, CO: NavPress, 2001.

Ross, Hugh. *Improbable Planet: How Earth Became Humanity's Home.* Grand Rapids, MI: Baker, 2016.

Schaefer, Henry, F. *Science and Christianity: Conflict or Coherence?* Athens, GA: The Apollos Trust, 2003.

Simmons, Geoffrey. *What Darwin Didn't Know.* Eugene, OR: Harvest House, 2004.

Simmons, Geoffrey. *Billions of Missing Links.* Eugene, OR: Harvest House, 2007.

Smith, Wesley J. *Forced Exit: The Slippery Slope from Assisted Suicide to Legalized Murder.* New York: Times Books/Random House, 1997.

Smith, Wesley J. *Culture of Death: The Assault on Medical Ethics in America.* San Francisco, CA: Encounter Books, 2000.

Smith, Wesley J. *Consumer's Guide to a Brave New World.* San Francisco, CA: Encounter Books, 2004.

Smith, Wesley J. *A Rat Is a Pig Is a Dog Is a Boy: The Human Cost of the Animal Rights Movement.* San Francisco, CA: Encounter Books, 2010.

Smith, Wesley J. *The War on Humans.* Seattle, WA: Discovery Institute Press, 2014.

Strobel, Lee. *The Case for a Creator.* Grand Rapids, MI: Zondervan, 2004.

Stump, J.B., and Alan G. Padgett, eds. *The Blackwell Companion to Science and Christianity.* Malden, MA: Wiley-Blackwell, 2012.

Thaxton, Charles B., Walter L. Bradley, and Roger L. Olsen. *The Mystery of Life's Origin: Reassessing Current Theories.* Dallas, TX: Lewis & Stanley, 1984.

Thaxton, Charles B., Walter L. Bradley, Roger L. Olsen, James Tour, Stephen C. Meyer, Jonathan Wells, Guillermo Gonzalez, Brian Miller, and David Klinghoffer. *The Mystery of Life's Origin: The Continuing Controversy.* Seattle, WA: Discovery Institute Press, 2020.

The Lutheran Church—Missouri Synod. *In Christ All Things Hold Together: The Intersection of Science and Christian Theology* (2015). A Report of the Commission on the Theology and Church Relations.

Torrance, Thomas F. *Reality and Scientific Theology.* Edinburgh, UK: Scottish Academic Press, 1985.

Ward, Peter D., and Donald Brownlee. *Rare Earth: Why Complex Life Is Uncommon in the Universe.* New York: Corpernicus, 2000.

Warfield, B. B. *Counterfeit Miracles.* Carlisle, PA: Banner of Truth Trust, 1972.

Weikart, Richard. *From Darwin to Hitler: Evolutionary Ethics, Eugenics and Racism in Germany.* New York: Palgrave Macmillan, 2004.

Weikart, Richard. *Hitler's Ethic: The Nazi Pursuit of Evolutionary Progress.* New York: Palgrave Macmillan, 2009.

Weikart, Richard. *Hitler's Religion: The Twisted Beliefs that Drove the Third Reich.* Washington, DC: Regnery, 2016.

Wells, Jonathan. *Icons of Evolution: Why Much of What We Teach About Evolution Is Wrong.* Washington, DC: Regnery, 2000.

Wells, Jonathan. *The Politically Incorrect Guide to Darwinism and Intelligent Design.* Washington, DC: Regnery, 2006.

Wells, Jonathan. *The Myth of Junk DNA.* Seattle, WA: Discovery Institute Press, 2011.

Wells, Jonathan. *Zombie Science More Icons of Evolution.* Seattle, WA: Discovery Institute Press, 2017.

West, John G. *Darwin's Conservatives: The Misguided Quest*. Seattle, WA: Discovery Institute Press, 2006.

West, John G. *Darwin Day in America: How Our Politics and Culture Have Been Dehumanized in the Name of Science*. Wilmington, DE: ISI Books, 2007.

Wiker, Benjamin, and Jonathan Witt. *A Meaningful World: How Arts and Science Reveal the Genius of Nature*. Downers Grove, IL: IVP Academic, 2006.

Witham, Larry. *By Design: Science and the Search for God*. New York: Encounter Books, 2003.

Woodward, Thomas. *Doubts About Darwin: A History of Intelligent Design*. Grand Rapids, MI: Baker, 2003.

Woodward, Thomas. *Darwin Strikes Back: Defending the Science of Intelligent Design*. Grand Rapids, MI: Baker, 2006.

Web Sources

A Parents' Guide to Intelligent Design (discovery.org/m/2018/12/Parents-Guide-to-Intelligent-Design.pdf)

A Scientific Dissent from Darwinism (dissentfromdarwin.org)

Access Research Network (arn.org)

BIO-Complexity Journal (bio-complexity.org)

Biologic Institute (biologicinstitute.org)

Center for Intelligence (centerforintelligence.org)

Darwin's Corrosive Idea: The Impact of Evolution on Attitudes About Faith, Ethics, and Human Uniqueness (discovery.org/m/2019/01/Darwins-Corrosive-Idea.pdf)

Discovery Institute (discovery.org)

Discovery Institute YouTube Science Channel (youtube.com/user/DiscoveryScienceNews)

Evolution News & Science Today (evolutionnews.org)

Evolutionary Informatics Lab (evoinfo.org)

Faith and Evolution (faithandevolution.org)

ID the Future Podcast (idthefuture.com)

IDEA (Intelligent Design and Evolution Awareness) Student Clubs (ideacenter.org)

Intelligent Design Gateway Portal (intelligentdesign.org)

Mind Matters (mindmatters.ai)

Online Intelligent Design Courses (discoveryu.org)

Science and Faith (scienceandgod.org)

The College Student's Back to School Guide to Intelligent Design (discovery.org/m/2018/12/College-Student-Back-to-School-Guide.pdf)

Uncommon Descent (Blog) (uncommondescent.com)

Walter Bradley Center (centerforintelligence.org)

Documentaries

Darwin's Dilemma: The Mystery of the Cambrian Fossil Record (Illustra Media) (darwinsdilemma.org)

Expelled: No Intelligence Allowed (Premise Media) (ncseexposed.org)

Flight: The Genius of Birds (Illustra Media) (flightthegeniusofbirds.com)

Human Zoos: America's Forgotten History of Scientific Racism (Discovery Institute) (humanzoos.org)

Icons of Evolution Documentary (Coldwater Media) (discovery.org/a/2125)

Living Waters: Intelligent Design in the Oceans of the Earth (Illustra Media) (livingwatersthefilm.com)

Metamorphosis: The Beauty and Design of Butterflies (Illustra Media) (metamorphosisthefilm.com)

Programming of Life (LaBarge Media) (programmingoflife.com)

The Case for a Creator (Illustra Media) (thecaseforacreator.com)

The Privileged Planet (Illustra Media) (theprivilegedplanet.com)

TrueU: The Toughest Test in College, featuring Stephen C. Meyer (Focus on the Family) (scienceandgod.org/trueu/)

Unlocking the Mystery of Life (Illustra Media) (unlockingthemysteryoflife.com)

Where Does the Evidence Lead? (Illustra Media) (wheredoestheevidencelead.com)

Notes

Foreword by Stephen C. Meyer

1. Richard Dawkins, *The Blind Watchmaker: Why the Evidence of Evolution Reveals a Universe Without Design* (New York: W.W. Norton, 1986), 1.

2. Bill Nye, *Undeniable: Evolution and the Science of Creation* (New York: St. Martin's Press, 2014), 46.

3. Dawkins, *The Blind Watchmaker*, 6.

4. Lawrence M. Krauss, *A Universe from Nothing: Why There Is Something Rather than Nothing* (New York: Free Press, 2012).

5. Stephen Hawking, *Brief Answers to Big Questions* (New York: Bantam, 2018), 29.

6. Hawking, *Brief Answers to Big Questions*, 38.

7. John G. West, *Darwin's Corrosive Idea: The Impact of Evolution on Attitudes About Faith, Ethics, and Human Uniqueness* (Seattle, WA: Discovery Institute, 2016), 2-3.

8. Stephen Jay Gould, "Nonoverlapping Magisteria," *Natural History* 106 (March 1997), 16-22; Stephen Jay Gould, *Rocks of Ages: Science and Religion in the Fullness of Life* (New York: Ballantine Books, 1999), 5.

9. Galileo and Finocchiaro, *The Essential Galileo* (Indianapolis, IN: Hackett, 2008), 119. This is in Galileo's letter to the Grand Duchess Christina (1615). Galileo used this aphorism but claimed it originated with Cardinal Cesare Baronio.

10. Donald MacKay, *The Clockwork Image: A Christian Perspective on Science* (Downers Grove, IL: InterVarsity, 1978), 51-55; Howard J. Van Till, *The Fourth Day: What the Bible and the Heavens Are Telling Us About the Creation* (Grand Rapids, MI: Eerdmans, 1985), 208-215; Howard J. Van Till, Davis A. Young, and Clarence Menninga, *Science Held Hostage: What's Wrong with Creation Science and Evolutionism* (Downers Grove, IL: InterVarsity, 1988), 39-43, 127-168. For a different interpretation of complementarity that affirms methodological autonomy of science and religion but conjoins their findings, see also Oskar Gruenwald, "Science and Religion: The Missing Link," *Journal of Interdisciplinary Studies* VI (1995): 1-23.

11. Michael Peterson provides a helpful threefold typology of perceived relationships between science and religion in *Reason and Religious Belief: An Introduction to the Philosophy of Religion,* 5th ed. (Oxford, UK: Oxford University Press, 2012), 196-216. Peterson discusses the conflict, compartmentalism, and complementarity models of science and religion interaction. He does not, however, consider the possibility that scientific evidence might support theistic belief, though that remains a logical possibility. I have, therefore, proposed (and am defending here) a fourth model called *qualified agreement* or *epistemic support*. See Stephen C. Meyer, "The Demarcation of Science and Religion," *The History of Science and Religion in the Western Tradition: An Encyclopedia*, ed. G.B. Ferngren (New York: Garland, 2000), 17-23; William A. Dembski and Stephen C. Meyer, "Fruitful Interchange or Polite Chit-Chat?: The Dialogue Between Science and Theology," *Science and Evidence for Design in the Universe, Proceedings of the Wethersfield Institute* (San Francisco, CA: St. Ignatius Press, 2002), 9, 213-234.

Chapter 2—How Do We Understand the Relationship Between Faith and Reason?

1. These five positions are adapted from chapter 17, "The Relationship Between Faith and Reason," Norman L. Geisler and Paul D. Feinberg, *Introduction to Philosophy: A Christian Perspective* (Grand Rapids, MI: Baker, 1980), 255-270. The authors offer Kant and Spinoza as examples of the "reason only" category, Kierkegaard and Barth as examples of "revelation only," liberal Christian thinkers among the ancient Alexandrian school, modern Higher Critical movement as typifying "reason over revelation," Tertullian and Cornelius Van Til as an example of "revelation over reason," and Augustine and Aquinas as examples of "revelation and reason." While they side with the "revelation and reason" position themselves, they also give conditional credence to the other positions. Compare with Mark Hanna, "Faith and Reason," *The Comprehensive Guide to Apologetics*, ed. Joseph M. Holden (Eugene, OR: Harvest House, 2018), 51.

2. To explore some of the cognitive differences between apes and humans, see John W. Oller Jr. and John L. Omdahl, "Origin of the Human Language Capacity: In Whose Image?" *The Creation Hypothesis: Scientific Evidence for an Intelligent Designer*, ed. J.P. Moreland (Downers Grove, IL: InterVarsity, 1994), 257-262; Fazale Rana, *Who Was Adam? A Creation Model Approach to the Origin of Humanity* (Covina, CA: RTB Press, 2015).

3. Although we are using the words *brain* and *mind* interchangeably here, we do not consider the two to be identical. There is a tremendous overlap and interplay between them, but mind is more than mere brain activity. The two regions of the brain known as the *rostrolateral prefrontal cortex* (RLPFC) and the *inferior parietal lobule* (IPL) are very energetic while we are performing higher reasoning. But that activity is still a mystery to neuroscientists at this time. We cannot be sure that reasoning is nothing more than brain activity in and around those brain regions. Even if it is—which we are open to—the prospect for creating a persuasive theory based purely on electro-chemical explanations for even more mysterious things like perception, consciousness, qualia, intention, and personality seems like a fool's errand. The authors of this chapter believe that some immaterial part of us outlives the body that dies (Ecclesiastes 3:18-22; 12:6-7) and our bodies will someday be replaced with a physical, resurrected body. There is a very strong unity between our material bodies and our immaterial selves. The authors of this chapter lean toward a hylomorphic (or *Thomistic substance dualism*) view of the relationship between our bodies and souls, our brains and our minds. See Norman Geisler, *Systematic Theology, Volume III: Sin and Salvation* (Minneapolis, MN: Bethany House, 2004) 52, 66-69; Edward Feser, *Philosophy of Mind: A Beginner's Guide* (London, UK: Oneworld, 2005); Edward Feser, *Aristotle's Revenge: The Metaphysical Foundations of Physical and Biological Science* (Neunkirchen-Seelscheid, Germany: Editiones Scholasticae, 2019) 20-27; J.P. Moreland, *The Soul: How We Know It's Real and Why It Matters* (Chicago, IL: Moody, 2014); Walter J. Freeman, "Nonlinear Brain Dynamics and Intention According to Aquinas," *Mind and Matter*, 6 (2008), 207-234. Putting the reasonableness of hylomorphism aside, every author in this book would be quick to agree with Behe's summation:

> … in the absence of any good answer and in the light of the success of physical science, [Cartesian] dualism was discredited, hylomorphism was forgotten…the mind was assumed to be just another physical phenomenon, no different in kind than digestion. Frankly, that's crazy. I have no answer to the problem of how the mind affects the body or the reverse, but denying your mind because you can't solve a problem is like cutting off your head to cure a headache. Whatever difficulties dualism, hylomorphism, or some other proposed explanation may have, they pale in comparison to denying mind. When you make that move, no more arguments are left, because—to the extent you are consistent—there is no more mind to reason about them.

Michael Behe, *Darwin Devolves: The New Science About DNA That Challenges Evolution* (New York: HarperOne, 2019), 276-277.

4. The game known in English as Go is also known as Igo in Japanese, Wei-chi in Chinese, and Baduk in Korean. It is played by two opponents on a 19x19 grid with 181 black pebbles and 180 white pebbles. In 2016 and 2017, Google's artificial intelligence supercomputer, aptly named AlphaGo, defeated several of the world's top Go players. Go contains far more possible move permutations than chess, making the victory for AI more impressive than the 1987 victories of IBM's supercomputer over the world's greatest chess champions. This helps confirm the conclusion that cognitive computing can be "smarter" than human minds in applications involving math and strategy.

5. Norman L. Geisler and Ronald M. Brooks, *Come, Let Us Reason: An Introduction to Logical Thinking* (Grand Rapids, MI: Baker, 1990). This book is an excellent introductory text for deductive and inductive logic.

6. Charles Peirce, a brilliant logician and philosopher of language, popularized the idea of *abductive inference*, which he also called *hypothetic inference* and *retroduction*. He describes it as an insight that should be framed as a question (hypothesis) for further testing rather than as a final conclusion. C.S. Peirce, *Philosophical Writings of Peirce* (New York: Dover Publications, 1955), 150.

7. Albert Einstein, a man of such brilliance that very few on Earth today can understand his theories on a deep level, had a reputation for using his "theta wave" state of mind to find solutions. He supposedly kept paper and pencil near his bed because his mind was working on mathematical problems in his sleep and breakthrough insights occasionally surfaced into consciousness upon waking. Some of his scientific peers had disdain for this method of reasoning. But rigorous evaluation in subsequent decades would prove that they were profound and could withstand testing. Perhaps this sheds light on two of the quotes Einstein is famous for: "We cannot solve our problems with the same level of thinking that created them," and "Imagination is more important than knowledge." Inventor Thomas Edison, painter Salvador Dali, fiction author Mary Shelly, chemist August Kekule (of benzene-ring fame), and film director James Cameron also exploited the creative state of their brains while they were in the state of hypnagogia, the state between waking (alpha wave) and sleeping (theta wave). While reasoning at its purest happens on the highest ranges of neuro-oscillation (gamma and beta), it seems that some

significant reasoning (perhaps that which fits under the umbrella of abductive) can also occur in the alpha, and perhaps even theta ranges as well.

8. Motz and Weaver set the limit for scientific inquiry to the time when the universe was 10^{-35} second old: "As one goes back to these very early epochs [of the beginning of the universe], to almost 1 trillionth of a trillionth of a trillionth of a second after the initial moment (after the universe's radius R was 0), the cosmological equations tell us that the temperature and density of the universe will keep on growing without limit as we approach the zero moment, finally becoming infinite. This state is known as the initial singularity, which has no physical meaning; the equations break down, and so we have no way, with the theory as it is, of understanding the birth of the universe. Owing to this breakdown, theoreticians begin their cosmological studies when the universe was 10^{-35} second old and its temperature was of the order of 10 thousand trillion trillion degrees Kelvin [and] smaller than its present size by a factor of 10^{28}." Lloyd Motz and Jefferson Weaver, *The Story of Physics* (New York: Avon Books, 1989), 381-382. In 1983, Stephen Hawking and James Hartle argued that because scientists cannot determine conditions of the universe before 10^{-43} seconds after its origin and therefore, in that moment of uncertainty, perhaps some unknown phenomenon interfered with general relativity in a way that allowed the universe to create itself. James B. Hartle and Stephen W. Hawking, "Wave Function of the Universe," *Physical Review D* 28 (1983), 2960-2975. Astrophysicist Hugh Ross responds, "Amazingly, astrophysicists have a reasonably good understanding of the universe's development back to when it was only 10^{-35} second old. We may see some very limited probing back to 10^{-43} second, but that represents the practical limit of research...Richard Gott has taken advantage of its infinitesimal period about which we know nothing. He has proposed an infinite loss of information about events before 10^{-43} second...With this total loss of information, he says, anything becomes possible, including the ability to make an infinite number of universes...If the universe had zero information before 10^{-43} second, how did it acquire its subsequent high information state without the input of an intelligent, personal Creator?...What we see here is another case of the 'no-God of the gaps.' It seems that the many nontheistic scientists (and others) are relying on gaps, and in this case a very minute one, to provide a way around the theistic implications of scientifically established facts." Hugh Ross, *The Creator and the Cosmos: How the Greatest Scientific Discoveries of the Century Reveal God* (Colorado Springs, CO: NavPress, 1993), 92-93.

9. Philosopher of science J.P. Moreland points out that science presupposes several philosophical things: the existence of the external world, the orderly nature of the external world and its knowability, the uniformity of nature and induction, the laws of logic, the correspondence theory of truth, the reliability of our senses and our minds, the adequacy of language to describe the world, the existence of numbers and the usefulness of mathematics, the concept of formal ontology, the existence of values, singularities, ultimate boundary conditions, and brute givens. J.P. Moreland, *Scientism and Secularism* (Wheaton, IL: Crossway, 2018), 55-69; J.P. Moreland, *Christianity and the Nature of Science: A Philosophical Investigation* (Grand Rapids, MI: Baker, 1989), 108-133. Agreeing with Moreland that scientific reasoning is predicated upon philosophical reasoning, philosopher Edward Feser begins his case against scientism positing, "Aristotelian metaphysics is not only compatible with modern science, but is implicitly presupposed by modern science," and concludes it saying, "Thus does Aristotle have his revenge against those who claim to have overthrown him in the name of modern science. But he is a magnanimous victor, providing as he does the true metaphysical foundations for the very possibility of that science." Edward Feser, *Aristotle's Revenge: The Metaphysical Foundations of Physical and Biological Science* (Neunkirchen-Seelscheid, Germany: Editiones Scholasticae, 2019), 1, 546. Biochemist Michael Behe adds that the divorce of holistic reasoning from strictly scientific reasoning provides one of the failures among scientists who fail to be reasonable: "The Enlightenment separation of science and purpose seemed like a good idea at the time, but it wasn't. Reason is a unity, and arbitrary divisions of reason can lead to cognitive disaster...If it weren't for mathematical reasoning, modern science wouldn't be possible. The same can be said for even more basic modes of thinking, such as simple logic. Deduction, induction, syllogism, the principle of sufficient reason, and more—none of those were independently demonstrated by experiment. All of them are more basic than science, and science depends on them in order to do its work." Michael Behe, *Darwin Devolves: The New Science About DNA That Challenges Evolution* (New York: HarperOne, 2019), 267-270.

10. For good introductions to the arguments for the existence of God, see chapters 12 and 13 *The Comprehensive Guide to Apologetics*, ed. Joseph M. Holden (Eugene, OR: Harvest House, 2018) and chapter 3 of Norman Geisler, *The Twelve Points that Show Christianity Is True* (Charlotte, NC: NGIM, 2016). For advanced readings consider Edward Feser, *Five Proofs for the Existence of God* (San Francisco, CA: Ignatius Press, 2017) and Norman Geisler, *God: A Philosophical Argument from Being* (Matthews, NC: Bastion Books, 2015).

11. Their logic may be delineated as follows: (1) A God who is all-powerful must have the power to end suffering. (2) A God who is all-good must desire to end suffering. (3) Suffering and evil exist today. (4) Therefore, no all-powerful, all-good God exists today. This is too shortsighted as it ignores the future. Just because God has not fully resolved the problem of evil yet doesn't mean that he is not already in the process of resolving it or that he will not completely resolve it in the future. If it is possible that God has any purpose for temporarily allowing evil and suffering to continue in the present, and has any plan to solve it anytime in the future, this argument against God from evil loses its power. The corrected argument could be stated as follows: (1) Because God is all-powerful he can end suffering. (2) Because God is all-good he will end suffering. (3) Suffering and evil exist today. (4) Therefore, God can and will solve the problem of evil sometime after today.

12. There are many possible answers to this objection to faith in supernatural explanations. First, because inductive logic offers only probability and not certainty, their conclusion is only probable at best. Second, they fail to explain why we shouldn't adopt the most reasonable answer possible (such as supernatural agency) while waiting for science to come up with a supposedly better answer to questions that seem to defy natural explanations. Third, they neglect to mention the fact that science has limits to what it can explore and can never provide explanations for some things in nature—such as the moments before, during, or immediately after the creation of the time, space, matter, and energy of our universe. As the galaxies continue to move away from one another, science will become more and more unable to observe other galaxies and answer intergalactic questions. Perhaps then science is not far from reaching more of its limits. Theoretical physicists seem to be able to come up with brilliant theories (such as string theory or multiple universe theory) that may not be verifiable by science. Fourth, in limiting themselves to naturalistic presuppositions and naturalistic answers, they are being unreasonable by taking away some of the possible reasons and trying to forbid reasoning about those possible reasons. Fifth, they refuse to be reasonable about the naturalistic explanations, like neo-Darwinian theories of evolution, that have failed miserably in explaining some mysteries—such as the origin of life and the evolution of many proteins, much less of a new organism. More than 1,000 scientists from all over the world have so far signed the Scientific Dissent from Darwinism Statement (https://dissentfromdarwin.org) in the attempt to raise awareness that Darwinian evolution cannot account for the development of life.

13. The argument for the reasonableness of a messenger of God being contingent upon his being authenticated by acts of God is elaborated best in chapters 4, 5, 8, and 9 of Norman Geisler, *The Twelve Points that Show Christianity Is True: A Handbook on Defending the Christian Faith* (Indian Trail, NC: Norm Geisler International Ministries, 2016). Also see "Jesus's Apologetic Use of Miracles" (chapter 2), Norman Geisler and Patrick Zukeran, *The Apologetics of Jesus* (Grand Rapids, MI: Baker, 2009), 27-46.

14. All but one of these points are elaborated upon in chapters 8 and 9 of Geisler, *The Twelve Points that Show Christianity Is True*, 115-146.

15. Matthew 16:13-20; Mark 8:27-30; Luke 9:18-21. Also Matthew 21:46: "They feared the crowds, because they held him to be a prophet."

16. John 7:25-27 records the faulty reasoning of some who rejected Jesus. "… this is the Christ? But we know where this man comes from, and when the Christ appears, no one will know where he comes from." Expressed in syllogistic form, the logic of their argument flowed as follows: (1) No one will know the hometown of the Messiah. (2) We know Jesus's hometown. (3) Therefore, we know Jesus cannot be the Messiah. Their form of logic was valid but the major premise was wrong and, as a result, the conclusion was wrong. The prophet Micah had said that the Messiah would come from Bethlehem Ephrathah (Micah 5:2) and this is in fact where Jesus was born. Perhaps they assumed wrongly, but somewhat understandably, that he was born in Nazareth.

17. Matthew 12:22-37; Mark 3:20-30. They were using reason not to find reasons for true propositions but to find reasons for propositions they wanted to continue believing. Their logic went something like this: (1) Satan (the leader of the demons) has the authority to cast out demons. (2) Jesus exhibited the authority to cast out demons. (3) Jesus was casting out demons by the power of Satan. Their major premise was wrong. God, who created angels and demons, has the ultimate authority over demons. Jesus gave a counterargument to them that reasoned the demonic kingdom would not prosper by a policy of demons casting out demons.

   Another way that some of these leaders were reasoning in a faulty way about Jesus (per John 7:52) was: (1) No prophets ever come from the region of the Galilee. (2) Jesus came from the region of the Galilee. (3) Therefore, Jesus is not a prophet. Jesus spent his teenage years and twenties in Galilee, but wasn't born there. They didn't bother to investigate the matter thoroughly. Their logical form was technically valid, but their incorrect premise resulted in an incorrect conclusion—and an incorrect faith.

18. Matthew 21:23–22:14; Mark 11:27–12:12; Luke 20:1-19. For insight into the cultural, religious, and logical facets of these encounters, see Kenneth E. Bailey, *Jesus Through Middle Eastern Eyes: Cultural Studies in the Gospels* (Downers Grove, IL: InterVarsity, 2008), 410-426. Also J. Dwight Pentecost *The Words and Works of Jesus Christ: A Study of the Life of Christ* (Grand Rapids, MI: Zondervan, 1981), 382-392.

19. Matthew 22:15-22; see also Mark 12:13-17; Luke 20:20-26.

20. See Geisler and Zukeran, *The Apologetics of Jesus.* Their fourth chapter catalogs some of the times Jesus used the laws of logic (the law of identity, the law of noncontradiction, and the law of excluded middle), deductive syllogisms, hypothetical syllogisms, disjunctive syllogisms, categorical syllogisms, *reduction ad absurdum* and *a fortiori* arguments. Also, according to Brad H. Young, *The Parables: Jewish Tradition and Christian Interpretation* (Grand Rapids, MI: Baker Academic, 1998), Jesus interacted heavily with the rabbinic scholarship that was prominent in the Second Temple era and the rules of logic and patterns for parables used by rabbis. His masterful use of parables to communicate truths about God from the known to the unknown for audiences of farmers and scholars alike was a highly logical (and analogical) art form.

21. Matthew 10:12-20: "Do not be anxious how you are to speak or what you are to say, for what you are to say will be given to you in that hour. For it is not you who speak, but the Spirit of your Father speaking through you" (verses 19-20). Luke 12:11-12 (see also Mark 13:11-12): "Do not be anxious about how you should defend yourself or what you should say, for the Holy Spirit will teach you in that very hour what you ought to say." Luke 21:14-15: "Settle it therefore in your minds not to meditate beforehand how to answer, for I will give you a mouth and wisdom, which none of your adversaries will be able to withstand or contradict." Compare Isaiah 54:17: "You shall refute every tongue that rises against you in judgment. This is the heritage of the servants of the LORD and their vindication from me, declares the LORD."

22. According to Acts 22:3, Paul was trained under the famous rabbi Gamaliel in Jerusalem. We get a glimpse of Gamaliel's sage reasoning in the Sanhedrin in Acts 5:34-39. Gamaliel was the grandson of the famous Jewish rabbi Hillel and did the most to establish the preeminence of the Hillel school in rabbinic Judaism. One of the things Hillel was famous for was the *middoth*, his seven rules (and multiple subrules) for logical interpretation of the Scriptures. These rules seem harmonious with many of the laws of logic and principles for interpreting texts that Aristotle is famous for. In good Pharisaic fashion, the *middoth* were expanded to 32 rules over time. Paul would have almost surely been familiar with at least the original seven. Whether he kept all of them, kept some and abandoned others, or added a few of his own lays far beyond the scope of this study. It does seem that both Paul and Jesus seemed favorable to many of these laws of logical reasoning in their persuasive argumentation. It also is important to note that there is no real qualitative division between Hebrew logic, Western logic, and Eastern logic. Much like the laws of physics are universal across continents, so are the laws of logic.

23. Paul might be echoing any of many Psalms (65:9-13; 68:9; 104; 111:1-6; 135:5-7; 136:25; 145:15-16; 147:4-18) while he spoke to the pagan Greeks in Lystra (Acts 14:17). He could also be echoing his Lord (Matthew 5:45). Paul, in turn, seems to be echoed by an anonymous twentieth-century commentator: "Human vanity can best be served by a reminder that, whatever his accomplishments, his sophistication, his artistic pretension, man owes his very existence to a six-inch layer of topsoil—and the fact that it rains." For a scientific-apologetic expansion of Paul's agrarian-society-friendly argument for design, see Hugh Ross, *Improbable Planet: How Earth Became Humanity's Home* (Grand Rapids, MI: Baker, 2016).

24. Everyone has faith. Even those who have reputations for championing science and reason and for destroying religious faith prove to have religious-like faith in something. Sam Harris, for example, argues in his book *The End of Faith: Religion, Terror, and the Future of Reason* (New York: W.W. Norton, 2004) that religions based on holy books and revealed religion (with a particular focus on the factions of Islam but with plenty of disdain to spare for factions of Christianity and Judaism) are the greatest threat to the survival of the human race. He then dismisses the secular, anti-theistic, anti-supernatural, unpardonably bloody humanism of "Stalin and Mao" as "little more than a political religion" (79). Next, he proceeds quite ironically to argue that the plan of salvation needing to be adopted in the world today is a secular, anti-theistic, anti-supernatural, humanistic, authoritarian, global political solution: "We can say it even more simply: we need a world government…It would require a degree of economic, cultural, and moral integration…World government does seem a long way off—so long that we may not survive the trip" (151). Regardless of whether his Marx-inspired political religion will somehow manage to be less murderous and inhumane than the Marx-inspired political religion of Stalin and Mao, he fails to realize that the gospel he preaches is also "little more than a political religion."

Chapter 3—Has Science Refuted Miracles and the Supernatural?

1. For a look at the role Christianity played in the rise of science, see James Hannam, *The Genesis of Science: How the Christian Middle Ages Launched the Scientific Revolution* (Washington, DC: Regnery, 2011).

2. Atheist Sam Harris put it this way: "Religious faith represents so uncompromising a misuse of the power of our minds that it forms a kind of perverse, cultural singularity—a vanishing point beyond which rational discourse proves impossible…Faith is the mortar that fills the cracks in the evidence and the gaps in the logic, and thus it is faith that keeps the whole terrible edifice of religious certainty still looming dangerously over our world." Sam Harris, *The End of Faith: Religion, Terror, and the Future of Reason* (New York: W.W. Norton, 2004), 25, 233.

3. Carol E. Cleland, "Historical science, experimental science, and the scientific method," *Geology* 29 (2001), 987-990. The earliest treatment in print of this distinction (though the expressions vary) of which I am aware is Norman L. Geisler, *Is Man the Measure? An Evaluation of Contemporary Humanism* (Grand Rapids, MI: Baker, 1983), 134-135. The distinction is also found in Charles B. Thaxton, Walter L. Bradley, and Roger Olsen, *The Mystery of Life's Origin: Reassessing Current Theories* (New York: Philosophical Library, 1984), 8, 202-214. A book-length treatment soon followed with Norman L. Geisler and Kerby Anderson, *Origin Science: A Proposal for the Creation-Evolution* (Grand Rapids, MI: Baker, 1987).

4. Geisler and Anderson, *Origin Science*, 16.

5. One may think that the question of the origin of life is repeatable in the laboratory inasmuch as the scientist could cull together certain elements, apply certain conditions, and see if the resultant is alive. The problem here is that the element of life is already a part of the experiment itself. It tacitly, if not explicitly, proves the very point for which theists are arguing—i.e., that intelligent design is required for the origin of biological life. This is best illustrated with the joke that has the scientist saying, "If I can create life here in the laboratory, then I will have proven that it did not take an intelligence to create life in the beginning!"

6. "So long as the universe had a beginning, we could suppose it had a creator. But if the universe is really completely self-contained, having no boundary or edge, it would have neither beginning nor end; it would simply be. What place, then, for a creator?" Stephen Hawking, *A Brief History of Time: From the Big Bang to Black Holes* (New York: Bantam, 1988), 140-141. Consider Hawking's later assessment with Leonard Mlodinow: "Spontaneous creation is the reason there is something rather than nothing, why the universe exists, why we exist. It is not necessary to involve God to light the blue touch paper and set the universe going." Stephen Hawking and Leonard Mlodinow, *The Grand Design* (New York: Bantam, 2010), 180. The sentiment is also defended by some philosophers: "Philosophical naturalism undertakes the responsibility for elaborating a comprehensive and coherent worldview based on experience, reason, and science, and for defending science's exclusive right to explore and theorize about *all of reality*." John Shook, "The Need for Naturalism in a Scientific Age," *Center for Inquiry* (May 18, 2020), https://centerforinquiry.org/blog/the_need_for_naturalism_in_a_scientific_age/ (accessed October 31, 2020, emphasis added).

7. "The question of the origin of the universe is one of the most exciting topics for a scientist to deal with. It reaches far beyond its purely scientific significance, since it is related to human existence, to mythology, and to religion… It hits us in the heart, as it were. The origin of the universe can be talked about not only in scientific terms, but also in poetic and spiritual language, an approach that is complementary to the scientific one. Indeed, the Judeo-Christian tradition describes the beginning of the world in a way that is surprisingly similar to the scientific model." Victor F. Weisskopf, "The Origin of the Universe," *American Scientist* 71 (Sept-Oct, 1983), 473-480, reprinted in *The World of Physics: A Small Library of the Literature of Physics from Antiquity to the Present* (New York: Simon & Schuster, 1987) 3, 300, 317. Agnostic astronomer Robert Jastrow concurs: "Recent developments in astronomy have implications that may go beyond their contribution to science itself." Robert Jastrow, "Message from Professor Robert Jastrow," http://www.leaderu.com/truth/1truth18b.html (accessed October 31, 2020).

8. For an analysis of the occult, see Richard G. Howe, "Modern Witchcraft: It May Not Be What You Think," *Christian Research Journal* 25 (2005), 12-21, http://richardghowe.com/index_htm_files/ModernWitchcraft.pdf (accessed October 31, 2020).

9. Thomas Aquinas defines a miracle this way: "Those effects are properly called miracles which are produced by God's power alone on things which have a natural tendency to the opposite effect or to a contrary mode of operation; whereas effects produced by nature, the cause of which is unknown to us or to some of us, as also those effects, produced by God, that are of a nature to be produced by none but God, cannot be called miraculous but only marvelous or wonderful." Thomas Aquinas, *On the Power of God*, II, vi, 2, trans. English Dominican

Fathers (Eugene, OR: Wipf & Stock, 2004), 164-165. C.S. Lewis concurs: "I use the word *Miracle* to mean an interference with Nature by supernatural power." C.S. Lewis, *Miracles: How God Intervenes in Nature and Human Affairs* (New York: Macmillan, 1947), 5. It should be noted that creation is not itself a miracle since, in creating the world, God did not suspend any natural laws. For a treatment of the uniqueness of creation as an event, see Herbert McCabe, *God and Evil in the Theology of St. Thomas Aquinas* (New York: Continuum, 2010), 91-102; Gaven Kerr, *Aquinas and the Metaphysics of Creation* (Oxford, UK: Oxford University Press, 2019).

10. Norman L. Geisler, *Miracles and Modern Thoughts* (Grand Rapids, MI: Zondervan, 1982) revised as *Miracles and the Modern Mind: A Biblical Defense of Biblical Miracles* (Grand Rapids, MI: Zondervan, 1992), 14.

11. Richard L. Purtill, "Defining Miracles," *In Defense of Miracles: A Comprehensive Case for God's Action in History*, eds. R. Douglas Geivett and Gary R. Habermas (Downers Grove, IL: InterVarsity, 1997), 62-63, emphasis added.

12. For an in-depth analysis of the supernatural and miracles, see Richard G. Howe, "In Defense of the Supernatural," *The Jesus Quest: The Danger from Within*, eds. Norman L. Geisler and F. David Farnell (Maitland, FL: Xulon, 2014), 621-672.

13. The apologetic system (or method) known as Classical Apologetics maintains that the existence of God must be established first (at least as a matter of principle) before the specific evidences for the Christian faith can take on their full significance. Because the evidence for Christianity in particular (among the monotheistic religions) involves an appeal to miracles such as, for example, the resurrection of Jesus, these claims take on their meaning, as a matter of principle, only within the context of theism. As a practical matter, it is certainly true that an unbeliever could go from atheism to Christianity by means of a cogent argument for the resurrection of Jesus. With God, all things are possible (Matthew 19:26).

14. This point was vividly displayed by the atheist Kai Nielsen in his debate on the existence of God with J.P. Moreland, which I had the pleasure of helping organize as a graduate student in philosophy at the University of Mississippi, Oxford, on March 24, 1988. Moreland had included evidence for the miracle of the resurrection of Jesus as part of his argument for God's existence. In his rebuttal, Nielsen countered, "Let's just suppose it were the case that Jesus was raised from the dead. Suppose you collected the bones, and they came together in some way [and] reconstituted the living Jesus…This wouldn't show that there was an infinite intelligible being. It wouldn't give you any way of being able to detect if there is a God. It would be just that a very strange happening happened…It would just be a very peculiar fact we hadn't explained and indeed lacked the scientific resources to explain." J.P. Moreland and Kai Nielsen, *Does God Exist? The Great Debate* (Nashville, TN: Thomas Nelson, 1990), 64, republished, *Does God Exist? The Debate between Theists and Atheists* (Amherst, MA: Prometheus Books, 1993), 64.

15. This was the challenge of the skeptical philosopher David Hume (1711–1776), who wrote: "In matters of religion, whatever is different is contrary; and that it is impossible the religions of Ancient Rome, of Turkey, of Siam, and of China should, all of them, be established on any solid foundation. Every miracle, therefore, pretended to have been wrought in any of these religions (and all of them abound in miracles), as its direct scope is to establish the particular system to which it is attributed; so has it the same force, though more indirectly, to overthrow every other system." David Hume, *An Enquiry Concerning Human Understanding and Concerning the Principles of Morals*, ed. L.A. Selby-Bigge, rev. by P.H. Nidditch (Oxford: Clarendon Press, 3d ed., 1975), §X, Pt. 2, 95, 121.

This objection is not so much denying the possibility or actuality of miracles as it is a challenge to how relevant miracles are in Christian apologetics. It has two dimensions to it. One is the historical dimension as to whether it is true that the world's religions "abound in miracles" as Hume claims. The other dimension is whether such supposed miracles are even coherent claims within those religions that have no doctrine of a transcendent God. I maintain that they are not coherent. Because Theravada Buddhism, as an example, has no deity, then supposedly miraculous events within Theravada Buddhism cannot really be miraculous, no matter how extraordinary they otherwise seem to be. What is more, no religion that is pantheistic (God is the universe) or polytheistic (many finite gods) can have any coherent notion of transcendence that is essential to the definition of miracle. Given my definition of miracles, then the only religions in the world that could in principle claim to have miracles are the monotheistic religions: primarily Judaism, Christianity, and Islam. As Christians, we accept the miracles of Judaism. Islam makes no appeals to miracles for its vindication. Hume's contention is wrong and his objection fails.

16. Stephen Hawking and Leonard Mlodinow, *The Grand Design* (New York: Bantam Books, 2010), 5, emphasis added.

17. For an exploration of philosophy's place in answering these and other questions that are relevant to Christian theology, see Richard G. Howe, "Defending the Handmaid: How Theology Needs Philosophy," *I Am Put Here*

for the Defense of the Gospel: Dr. Norman L. Geisler: A Festschrift in His Honor, ed. Terry L. Miethe (Eugene, OR: Pickwick, Wipf and Stock, 2016), 233-256.

18. The challenge is that the technical definition of *supernatural* leaves us without a convenient word to capture the distinction between natural (created) things like trees and natural (created) things like angels. I generally prefer the term *paranormal*, except that this term is almost always associated with malevolent spiritual (i.e., demonic) activity. Because of this, I would not use the term *paranormal* to refer to the angelic activities in the Bible. Perhaps it could suffice to use *demonic* or *paranormal* for the malevolent spiritual objects or activities and *angelic* for angels and their actions.

19. Most contemporary scholars maintain that John 5:4 was not originally part of John's Gospel but was a gloss added later. It is excluded from many English translations, including the ESV, NASB, NET, NIV, and RSV. For a textual and theological defense of the authenticity of the account in John 5 that includes v. 4, see Zane C. Hodges, "The Angel at Bethesda—John 5:4," *Bibliotheca Sacra* 136 (January-March 1979), 25-39.

20. Obviously, people would have to understand what truth is before there could be a discussion about whether Christianity is true. This is especially the case with the increasing influence of postmodernism and other philosophies and religions. Postmodernism gives rise to the nonsense that some facts of reality can be "true" for one person and not "true" for another. In this, the postmodernist is not necessarily saying that the fact itself is different for different people. It can be much more subtle and insidious by saying that our ability to know the fact is compromised such that we cannot know it objectively. Consider one historian's take on our knowledge of reality: "First, scientists tell us that our perceptions are caused by things in the world stimulating our sense receptors...This being so, our perceptions are best described as providing us with information about reality, but not necessarily mirroring it precisely...Second, our perceptions are influenced by our culture...So, our perceptions of the world are not pure sense impressions of it...Finally, our perceptions are influenced by our needs, interests, and desires... For these three reasons, at least, it is wrong to say that our perceptions simply correspond to the world." C. Behan McCullagh, *The Truth of History* (London, UK: Routledge, 1997), 71. McCullagh's position is self-refuting. One must ask whether McCullagh's own statement is failing to "mirror" reality precisely and is "influenced" by his own "culture" or by his "needs, interests, and desires" such that "it is wrong to say" that McCullagh's view "simply corresponds to the world." If what McCullagh is saying is true, then there is every reason to think that it is inaccurate given its own criteria. Again, it is self-refuting. For a handy treatment of postmodernism and other failing views of knowledge, see Paul Copan, *True for You, But Not for Me: Deflating the Slogans that Leave Christians Speechless* (Minneapolis, MN: Bethany House, 1998), republished as *True for You but Not for Me: Overcoming Objections to the Christian Faith* (Minneapolis, MN: Bethany House, 2009). Christians increasingly must be prepared to address such fundamental philosophical principles.

21. For an examination of the arguments for God's existence, see Richard G. Howe, "What Are the Classical Proofs of God's Existence?" and Thomas W. Baker, "What Are Some Other Arguments for God's Existence?," *The Harvest Handbook of Apologetics*, ed. Joseph M. Holden (Eugene, OR: Harvest House, 2018), 83-87 and 89-94, respectively. For more extensive treatments, see Edward Feser, *Five Proofs of the Existence of God: Aristotle, Plotinus, Augustine, Aquinas, Leibniz* (San Francisco, CA: St. Ignatius, 2017); Matt Fradd and Robert Delfino, *Does God Exist? A Socratic Dialogue on the Five Ways of Thomas Aquinas* (St. Louis, MO: Enroute, 2018); Richard G. Howe, "In Defense of the Supernatural," *The Jesus Quest*, 621-672; Gaven Kerr, *Aquinas's Way to God: The Proof in De Ente et Essentia* (Oxford, UK: Oxford University Press, 2015).

22. Examples include William A. Dembski, *Intelligent Design: The Bridge Between Science and Theology* (Downers Grove, IL: InterVarsity, 1999); *Mere Creation: Science, Faith, and Intelligent Design*, ed. William A. Dembski (Downers Grove, IL: InterVarsity, 1998); Norman L. Geisler and Frank Turek, *I Don't Have Enough Faith to Be an Atheist* (Wheaton, IL: Crossway Books, 2004); Ron Londen, *The God Abduction: How Scientific Discovery Strengthens the Case for a Creator* (Charlottesville, VA: Journey Group, 2014); *The Creation Hypothesis: Scientific Evidence for an Intelligent Designer*, ed. J.P. Moreland (Downers Grove, IL: InterVarsity, 1994); Nancy R. Pearcey and Charles B. Thaxton, *The Soul of Science: Christian Faith and Natural Philosophy* (Wheaton, IL: Crossway Books, 1994); Hugh Ross, *The Creator and the Cosmos: How the Greatest Scientific Discoveries of the Century Reveal God* (Colorado Springs, CO: NavPress, 1993); Hugh Ross, *Why the Universe Is the Way It Is* (Grand Rapids, MI: Baker, 2008); Robert J. Spitzer, *New Proofs for the Existence of God: Contributions of Contemporary Physics and Philosophy* (Grand Rapids, MI: Eerdmans, 2010); Lee Strobel, *The Case for a Creator: A Journalist Investigates Scientific Evidence that Points Toward God* (Grand Rapids, MI: Zondervan, 2004); J. Warner Wallace, *God's Crime*

Scene: A Cold-Case Detective Examines the Evidence for a Divinely Created Universe (Colorado Springs, CO: David C. Cook, 2015).

23. Physicist Paul Davies explains, "These days most cosmologists and astronomers back the theory that there was indeed a creation…when the physical universe burst into existence in an awesome explosion popularly known as the 'big bang.' Whether one accepts all the details or not, the essential hypothesis—that there was some sort of creation—seems, from the scientific point of view, compelling." Paul Davies, *God and the New Physics* (New York: Simon & Schuster, 1983), 10. Agnostic Astronomer Robert Jastrow observes, "Recent developments in astronomy have implications that may go beyond their contribution to science itself. In a nutshell, astronomers, studying the universe through their telescopes, have been forced to the conclusion that the world began suddenly, in a moment of creation, as the product of unknown forces." Jastrow, "Message from Professor Robert Jastrow."

24. "The laws of nature form a system that is extremely fine-tuned, and very little in physical law can be altered without destroying the possibility of the development of life as we know it. Were it not for *a series of startling coincidences* in the precise details of physical law, it seems, humans and similar life-forms would never have come into being." Stephen Hawking and Leonard Mlodinow, *The Grand Design* (New York: Bantam Books, 2010), 161, emphasis added. Instead of seeing the evidence as pointing to God, note that Hawking and Mlodinow attribute the occurrence of these initial conditions to "a series of startling coincidences."

25. "Without a doubt, the atoms and molecules that comprise living cells individually obey the laws of chemistry and physics. The enigma is the origin of so unlikely an organization of these atoms and molecules…It is apparent that 'chance' should be abandoned as an acceptable model for coding of the macromolecules essential in living systems." Charles B. Thaxton, Walter L. Bradley, and Roger Olsen, *The Mystery of Life's Origin: Reassessing Current Theories* (New York: Philosophical Library, 1984), 128, 146.

26. "At nearly the same time that computer scientists were beginning to develop machine languages, molecular biologists were discovering that living cells had been using something akin to machine code or software all along." Stephen C. Meyer, *Signature in the Cell: DNA and the Evidence for Intelligent Design* (New York: HarperOne, 2009), 110.

27. Atheist Richard Dawkins raises this kind of objection against the infinite regress argument: "Even if we allow the dubious luxury of arbitrarily conjuring up a terminator to an infinite regress and giving it a name, simply because we need one, there is absolutely no reason to endow that terminator with any of the properties normally ascribed to God." Richard Dawkins, *The God Delusion* (Boston, MA: Houghton Mifflin, 2006), 77.

28. See William A. Dembski, *Intelligent Design: The Bridge Between Science and Theology* (Downers Grove, IL: Inter-Varsity, 1999).

29. For an examination of how this objection philosophically fails, see Michael Augros, *Who Designed the Designer? A Rediscovered Path to God's Existence* (San Francisco, CA: Ignatius, 2015).

30. Matthew 8:28; Luke 1:19; 1 Timothy 4:1; Revelation 19:9-10.

31. Already we see the two sources of God's revelation—through creation (called *general revelation*) and through his prophets (called *special revelation*). For an introductory-level church study of general and special revelation with a workbook and DVD curriculum, see Richard G. Howe, *Intro to God's Revelation*, https://afastore.net/intro-to-gods-revelation-6-week-curriculum-by-dr-richard-howe (accessed October 31, 2020).

32. Richard Dawkins, *The Blind Watchmaker: Why the Evidence of Evolution Reveals a Universe Without Design* (New York: W.W. Norton, 1987), 37-38. Hugh Montefiore was at the time the Bishop of Birmingham.

33. Dawkins, *The God Delusion*, 58-59.

34. Elsewhere, Dawkins repeats the same sentiment, again without offering any justification for his dogmatism: "There is an answer to every such question [about miracles], whether or not we can discover it in practice, and it is a strictly scientific answer. The methods we should use to settle the matter, in the unlikely event that relevant evidence ever became available, would be purely and entirely scientific methods." Dawkins, *The God Delusion*, 59. Other scientists and philosophers make the same mistake. "Perhaps some cancer cures are miracles. If so, the only hope of ever demonstrating this to a doubting world would be by adopting the scientific method, with its assumption of no miracles, and showing that science was utterly unable to account for the phenomena." Daniel Dennett, *Breaking the Spell: Religion as a Natural Phenomenon* (New York: Penguin, 2006), 26. "Science is a method for deciding whether what we choose to believe has a basis in the laws of nature or not." Marcia McNutt, "The Age of Disbelief," *National Geographic* (March 2015), 40.

35. For a deft exposé of how inadequate science is in addressing the question of God's existence, see Edward Feser,

The Last Superstition: A Refutation of the New Atheism (South Bend, IN: St. Augustine's Press, 2008). The reason science is inadequate is because certain categories and data must be in place antecedent to the very existence of science as a discipline. Examples would include logic, causality, universals, goodness, free will, essences, and existence. These (and more that could be listed) are all are subjects of philosophy.

36. Rudolf Clausius, "The Second Law of Thermodynamics," *The World of Physics: A Small Library of the Literature of Physics from Antiquity to the Present* (New York: Simon & Schuster, 1987), 1, 734.

37. Physicist Max Planck describes it this way: "If, for instance, an exchange of heat by conduction takes place between two bodies of different temperature, the first law, or the principle of the conservation of energy, merely demands that the quantity of heat given out by the one body shall be equal to that taken up by the other." Max Planck, "The Second Law of Thermodynamics," *Treatise on Thermodynamics*, trans. Alexander Ogg (London, UK: Longmans, Green, and Co., 1903), 79-88 reprinted in *The World of Physics* 1, 775. Physicist Paul Davies describes the second law as thus: "The second law of thermodynamics...says, roughly speaking, that in any change, the Universe becomes a slightly more disorderly place...This natural tendency towards disintegration and chaos is evident all around us: people grow old...stars burn out, clocks run down." Paul Davies, "Chance or Choice: Is the Universe an Accident?," *New Scientist* 80 (1978), 506, as cited in W.R. Bird, *The Origin of Species Revisited: Theories of Evolution and of Abrupt Appearance* (Nashville, TN: Regency, 1991), I, 397.

38. For an examination of the nature of miracles, other objections against miracles, and the historicity of miracles, see R. Douglas Geivett and Gary R. Habermas, *In Defense of Miracles: A Comprehensive Case for God's Action in History* (Downers Grove, IL: InterVarsity, 1997); Norman L. Geisler, *Miracles and Modern Thoughts* (Grand Rapids, MI: Zondervan, 1982) revised as *Miracles and the Modern Mind: A Biblical Defense of Biblical Miracles* (Grand Rapids, MI: Zondervan, 1992); Norman L. Geisler, *Signs and Wonders* (Wheaton, IL: Tyndale House, 1988); Craig S. Keener, *Miracles: The Credibility of the New Testament Accounts* (Grand Rapids, MI: Baker Academic, 2011); B.B. Warfield, *Counterfeit Miracles* (Carlisle, PA: Banner of Truth Trust, 1972).

Chapter 4—Is Christianity at War with Science?

1. Jerry Coyne, "Did Christianity (and other religions) promote the rise of science?," *Why Evolution Is True* (October 18, 2013), https://whyevolutionistrue.wordpress.com/2013/10/18/did-christianity-and-other-religions-promote-the-rise-of-science (accessed August 28, 2020).

2. David C. Lindberg, "The Medieval Church Encounters the Classical Tradition: Saint Augustine, Roger Bacon, and the Handmaiden Metaphor," *When Science and Christianity Meet*, eds. David C. Lindberg and Ronald L. Numbers (Chicago, IL: University of Chicago Press, 2003), 17; Bruce S. Eastwood, "Early-Medieval Cosmology, Astronomy, and Mathematics," *Cambridge History of Science: Volume 2*, eds. David C. Lindberg, Michael H. Shank (Cambridge, UK: Cambridge University Press, 2013), 302-322.

3. St. Augustine, *Contra Faustum Manichaeum* 32.20, as cited in Peter Harrison, "The Bible and the Emergence of Modern Science," *Science and Christian Belief* 18 (2006), 118.

4. Eastwood, "Early-Medieval Cosmology, Astronomy, and Mathematics," 305.

5. Eastwood, "Early-Medieval Cosmology, Astronomy, and Mathematics," 307.

6. Charles Burnett, "The Twelfth-Century Renaissance," *Cambridge History of Science: Volume 2*, eds. David C. Lindberg, Michael H. Shank (Cambridge, UK: Cambridge University Press, 2013), 365-384.

7. Walter Roy Laird, "Change and Motion," *Cambridge History of Science: Volume 2*, eds. David C. Lindberg, Michael H. Shank (Cambridge, UK: Cambridge University Press, 2013), 404-435.

8. David C. Lindberg and Katherine H. Tachau, "The Science of Light and Color, Seeing and Knowing," *Cambridge History of Science: Volume 2*, eds. David C. Lindberg, Michael H. Shank (Cambridge, UK: Cambridge University Press, 2013), 485-511.

9. A. Mark Smith, *From Sight to Light: The Passage from Ancient to Modern Optics* (Chicago, IL: University of Chicago Press, 2014), inside jacket synopsis. See Johannes Kepler and William H. Donahue, *Optics: Paralipomena to Witelo and Optical Part of Astronomy* (Santa Fe, NM: Green Lion Press, 2000).

10. Edward Grant, "Science and the Medieval University," *Rebirth, Reform, and Resilience: Universities in Transition 1300–1700*, eds. James M. Kittelson and Pamela J. Transue (Columbus, OH: Ohio State University Press, 1984).

11. Josiah Hesse, "Flat Earthers keep the faith at Denver conference," *The Guardian* (November 18, 2018), https://www.theguardian.com/us-news/2018/nov/18/flat-earthers-keep-the-faith-at-denver-conference (accessed August 28, 2020).

12. Neil deGrasse Tyson (January 28, 2016), https://twitter.com/neiltyson/status/692939759593865216 (accessed August 28, 2020).

13. John William Draper, *History of the Conflict Between Religion and Science* (New York: D. Appleton, 1874), 157-159.

14. "Bill Nye Speaks at the 2010 AHA Conference: Part 3/3" (June 6, 2010), https://www.youtube.com/watch?v=S4dZWbFs8T0 (accessed August 28, 2020).

15. C.S. Lewis, "Dogma and the Universe," *God in the Dock: Essays on Theology and Ethics*, ed. Walter Hooper (Grand Rapids, MI: Eerdmans, 1970), 39-42.

16. "Cosmos: Possible Worlds," *National Geographic*, https://www.nationalgeographic.com/tv/shows/cosmos-possible-worlds (accessed August 28, 2020).

17. "Does God Exist? William Lane Craig vs. Christopher Hitchens—Full Debate [HD]," Biola University, September 24, 2014, https://www.youtube.com/watch?v=0tYm41hb48o (accessed August 28, 2020).

18. Dennis Danielson, *The Book of the Cosmos* (New York: Basic Books, 2001), 106.

19. Danielson, *The Book of the Cosmos*, 150.

20. See Guillermo Gonzalez and Jay W. Richards, *The Privileged Planet: How Our Place in the Cosmos Is Designed for Discovery* (Washington, DC: Regnery, 2004); David Klinghoffer, "Gonzalez: 'Worlds Like This Are Hard to Come By,'" *Evolution News* (April 30, 2019), https://evolutionnews.org/2019/04/gonzalez-worlds-like-this-are-hard-to-come-by (accessed August 28, 2020).

21. Stephen E. Schneider and Thomas T. Arny, *Pathways to Astronomy*, 5th ed. (New York: McGraw-Hill, 2018), 629.

22. Neil deGrasse Tyson, *Astrophysics for People in a Hurry* (Waterville, ME: Thorndike Press, 2017), chapter 1. See also Neil deGrasse Tyson, foreword to Jeffrey O. Bennett Megan O. Donahue, Nicholas Schneider, and Mark Voit, *The Cosmic Perspective*, 8th ed. (Boston, MA: Pearson, 2017), xxviii.

23. Gregory Allen Schrempp, *The Ancient Mythology of Modern Science: A Mythologist Looks (Seriously) at Popular Science Writing* (Montreal, Canada: McGill-Queen's University Press, 2012).

24. Steve Nadler, "Who Tried to Kill Spinoza?," *Jewish Review of Books* (Winter 2019), https://jewishreviewofbooks.com/articles/4991/who-tried-to-kill-spinoza (accessed August 28, 2020).

25. B. de Jong, "Spinoza museum buitenlands bezoek," Vereniging Het Spinozahuis (September 28, 2018), https://www.spinozahuis.nl/en/spinoza-museum (accessed August 28, 2020).

26. Nancy Ellen Abrams and Joel R. Primack, "Einstein's View of God," http://physics.ucsc.edu/cosmo/primack_abrams/htmlformat/Einstein4.html (accessed August 28, 2020).

27. "Baruch Spinoza," *Stanford Encyclopedia of Philosophy*, https://plato.stanford.edu/entries/spinoza (accessed August 28, 2020).

28. "Baruch Spinoza," *Stanford Encyclopedia of Philosophy*.

29. Michael Keys, "Three Big Ways Christianity Supported the Rise of Modern Science Explained by Historian Michael Keas," *Discovery Science* (March 13, 2020), https://www.youtube.com/watch?v=HHcF-ffKkeg (accessed August 28, 2020).

30. Eric Schliesser, "Spinoza and the Philosophy of Science: Mathematics, Motion, and Being," *The Oxford Handbook of Spinoza*, ed. Michael Della Rocca (Oxford: Oxford University Press, 2017), https://www.oxfordhandbooks.com/view/10.1093/oxfordhb/9780195335828.001.0001/oxfordhb-9780195335828-e-020 (accessed August 28, 2020).

31. Alison Peterman, "Spinoza on Physical Science," *Philosophy Compass* 9 (2014), 214-223.

32. Peterman, "Spinoza on Physical Science."

33. Peterman, "Spinoza on Physical Science."

34. Galileo Galilei and Maurice A. Finocchiaro, *The Essential Galileo* (Indianapolis, IN: Hackett Publishing, 2008), 183.

35. Michael Newton Keas, *Unbelievable: 7 Myths About the History and Future of Science and Religion* (Wilmington, DE: ISI Books, 2019), 159.

36. Michael J. Crowe, *The Extraterrestrial Life Debate, Antiquity to 1915: A Source Book* (Notre Dame, IN: University of Notre Dame Press, 2008), 67.

Chapter 5—Does Science Conflict with Biblical Faith?

1. Julian Baggini, "Science Is at Odds with Christianity," *Debating Christian Theism*, eds. J.P. Moreland, Chad Meister, and Khaldoun A. Sweis (Oxford, UK: Oxford University Press, 2013), 320-321.

2. Baggini, "Science Is at Odds with Christianity," 319-320.

3. John Hedley Brooke, *Science and Religion: Some Historical Perspectives* (Cambridge, UK: Cambridge University Press, 1991, reprinted 1999), 2-5.

4. Brooke, *Science and Religion*, 2.

5. John Polkinghorne, *One World: The Interaction of Science and Theology* (Philadelphia, PA & London, UK: Templeton Foundation Press, 1986, reprinted 2007), xv.

6. James Hannam, *God's Philosophers: How the Medieval World Laid the Foundations of Modern Science* (London, UK: Icon Books, 2009), 309.

7. Alfred North Whitehead says, "Giordano Bruno was the martyr: though the cause for which he suffered was not that of science, but that of free imaginative speculation." Alfred North Whitehead, *Science and the Modern World* (New York: The Free Press, 1925, reprinted 1967), 1.

8. Brooke, *Science and Religion*, 39.

9. Hannam, *God's Philosophers*, 306.

10. Hannam, *God's Philosophers*, 326.

11. Hannam, *God's Philosophers*, 318-319.

12. Andrew Ede and Lesley B. Cormack, *A History of Science in Society: From Philosophy to Utility* (Peterborough, ON: Broadview Press, 2004), 134.

13. Hannam, *God's Philosophers*, 312-314.

14. Ede and Cormack, *A History of Science in Society*, 134; Hannam, *God's Philosophers*, 314.

15. Hannam, *God's Philosophers*, 325.

16. Edward Grant, *The Foundations of Modern Science in the Middle Ages: Their Religious Institutional, and Intellectual Contexts* (Cambridge, UK: Cambridge University Press, 1996), 203-204.

17. Hannam, *God's Philosophers*, 316.

18. Ede and Cormack note that in *The Dialogue*, "he used his theory of the tides as a proof of the motion of the Earth and, hence, of the Copernican doctrine. Although the theory was quite wrong-headed and convinced no one, it demonstrated to those reading the book that Galileo was indeed defending Copernicanism and so breaking the Interdict of 1616, which prohibited holding the view that the Copernican system was a proven fact." Ede and Cormack, *A History of Science in Society*, 135; Hannam, *God's Philosophers*, 325. It is worth noting here that Galileo was not seeking to prove Copernicus's theory against a church that held exclusively to the Ptolemaic system. On the contrary, during his own lifetime, and partially because of his discoveries, many scientists had started to accept that the Ptolemaic system was inadequate. This meant that another system needed to be found, and many astronomers had turned to the model of Tycho Brahe rather than that of Copernicus. Hannam, *God's Philosophers*, 312-313.

19. Brooke, *Science and Religion*, 37. Ede and Cormack agree on this point, noting, "He believed that his real enemies were not the Church authorities, but 'philosophers'—the Aristotelians who argued that only they had the right to make truth claims about the world." Ede and Cormack, *A History of Science in Society*, 134.

20. Hannam, *God's Philosophers*, 327-328; Ede and Cormack, *A History of Science in Society*, 135.

21. Ede and Cormack, *A History of Science in Society*, 135-136.

22. Whitehead, *Science and the Modern World*, 2.

23. Brooke, *Science and Religion*, 37. Ede and Cormack think that in the case of Galileo at least, it was not a matter of science versus religion, but of the obedience of a fervent Catholic to the church. Ede and Cormack, *A History of Science in Society*, 135.

24. Brooke, *Science and Religion*, 40.

25. Hannam, *God's Philosophers*, 313.

26. David C. Lindberg, *The Beginning of Western Science: The European Scientific Tradition in Philosophical, Religious, and Institutional Context, Prehistory to A.D. 1450* (Chicago, IL: University of Chicago Press, 2d ed., 2007), 224.

27. Carl Sagan, for example, proposes that evolution brought humans into existence some 3.6 million years ago. Carl Sagan, *Cosmos* (New York: Ballantine Books, 1980), 286.

28. Carl Sagan states the mainstream view that the universe burst into existence some 15 billion years ago and the Earth formed about 4.6 billion years ago. Sagan, *Cosmos*, 286.

29. Sagan estimates some "fifteen billion years of cosmic evolution." Sagan, *Cosmos*, 240. See also Carl Sagan, *Broca's Brain: Reflections on the Romance of Science* (New York: Random House, 1974, reprinted 1979), 295.

30. The main exceptions to this claim are the *Westminster Confession of Faith*, ch. 4, a. 1, and the *London Baptist Confession of Faith*, 1689, ch. 4, a. 1, which clearly state that it is necessary to believe that God created "the world, and all things therein whether visible or invisible, in the space of six days."

31. C.S. Lewis, *Miracles* (London, UK: Collins-Fontana books, 1960), 106.

32. Lewis, *Miracles*, 106.

33. In some cases, we may appeal to different natural events that may have made the action more understandable; however, because rational agents are endowed with free will, no appeal to chemicals, the movement of the planets, or other natural phenomena can fully explain the event. At some point we will need to fall back upon the will of the agent.

34. David Hume, "An Enquiry Concerning Human Understanding," *Enquiries Concerning Human Understanding and Concerning the Principles of Morals*, ed. P.H. Nidditch (Oxford, UK: Clarendon Press, 3d ed., 1975; reprinted 1979), 110.

35. Hume, "An Enquiry Concerning Human Understanding," 111.

36. Hume, "An Enquiry Concerning Human Understanding," 115.

37. Richard Whateley, *Historic Doubts Relative to Napoleon Bonaparte* (London, UK: John W. Parker, 1849). Whateley uses the very principles by which Hume purports to demonstrate the impossibility of miracles to demonstrate the impossibility of the life of Napoleon Bonaparte. He, in this way, shows that Hume's "proof against miracles" is false because it "demonstrates" too much.

38. Plato, *Phaedo*, trans. G.M.A. Grube (Indianapolis, IN: Hackett, 1977). One of the main arguments of this book intends to show that the soul is immortal and persists after death. Later Platonists accepted and developed Plato's arguments further. Plotinus, *Enneads* 1.1 (53), 1.9 (16), 4.1-4.9; Proclus, *The Elements of Theology*, trans. E.R. Dodds (Oxford, UK: Clarendon Press, 2nd ed., 2004), 161-185. Though Aristotle is sometimes said to have denied personal immortality, some of his comments led medieval Muslim, Jewish, and Christian theologians to believe that Aristotle had argued, and provided sufficient evidence to prove, that the human intellect was both independent of the body, and immortal. Aristotle, *On the Soul*, bk. 3, ch. 5, 430a20-26. Moses Maimonides was a medieval Jewish Aristotelian who understood the soul to be immortal. Moses Maimonides, *The Guide for the Perplexed*, pt. 1, ch. XLI.

39. Lindberg, *The Beginning of Western Science*, 236.

40. Baggini, "Science Is at Odds with Christianity," 314.

41. Baggini, "Science Is at Odds with Christianity," 314.

42. Baggini, "Science Is at Odds with Christianity," 315.

43. Baggini, "Science Is at Odds with Christianity," 315.

44. Baggini, "Science Is at Odds with Christianity," 315. This seems an enormous unproven assumption to make, and upon which one would base their "hopes" for future explanation.

45. Baggini, "Science Is at Odds with Christianity," 315, emphasis added.

46. Though, in my humble opinion, the old tired arguments of those who adhere to strict materialism are beginning to fall by the wayside.

47. Baggini, "Science Is at Odds with Christianity," 317.

48. Lindberg, *The Beginning of Western Science*, 210.

49. Lindberg, *The Beginning of Western Science*, 213.

50. Ronald L. Numbers, "Science Without God: Natural Laws and Christian Beliefs," *The Nature of Nature: Examining the Role of Naturalism in Science*, eds. Bruce L. Gordon and William A. Dembski (Wilmington, DE: ISI Books, 2011), 63.

51. Baggini, "Science Is at Odds with Christianity," 317-319.

52. Psalm 19:1-4; Romans 1:19-20; Acts 14:15-17; Acts 17; etc.

53. Thomas Aquinas, *Faith, Reason and Theology: Questions I-IV of His Commentary on the* De Trinitate *of Boethius*, trans. Armand Maurer (Toronto, Canada: PIMS, 1987), 48.

54. Thomas Aquinas, *Summa Theologiae I*, q. 1, a. 2, *respondeo*. Unless otherwise stated, I use the translation done by the Fathers of the English Dominican Province (Notre Dame, IN: Ave Maria Press, 1948, reprinted 1981).

55. Aquinas, *Summa Theologiae*, q. 1, a. 5, *respondeo*.

56. Aquinas, *Summa Theologiae*, q. 1, a. 1.

57. Many thanks to Benoît Côté and Veronique Cloutier, who both read through this article and provided many helpful suggestions and critiques. This article was improved because of their comments.

Chapter 6—Did Christianity Help Give Rise to Science?

1. For those seeking to go deeper into this topic, Gary B. Ferngren's *Science and Religion: A Historical Introduction* (Baltimore, MD: The Johns Hopkins University Press, 2002), is an excellent place to start. David C. Lindberg's *The Beginnings of Western Science* (Chicago, IL: University of Chicago Press, 1992), is a classic, although reviewers note that some sections of the second edition have given in to political correctness.

2. As quoted in Louis E. Van Norman, *Poland: The Knight Among Nations* (New York: Fleming H. Revell Company, 1907), 290.

3. Stephen Gaukroger, "Francis Bacon," *Encyclopedia of Philosophy* (2005), https://www.encyclopedia.com/people/philosophy-and-religion/philosophy-biographies/francis-bacon (accessed June 29, 2020).

4. Jeff Hecht, *Fiber Optic History* (2016), https://www.jeffhecht.com/history.html (accessed July 2, 2020).

5. Eugene Wigner, "The Unreasonable Effectiveness of Mathematics in the Natural Sciences," *Communications in Pure and Applied Mathematics* 13 (I) (February 1960), 1-14.

6. See Stanley Jaki, *Science and Creation: From Eternal Cycles to an Oscillating Universe*, 2d ed. (Fort Collins, CO: Gondolin Press, 2017).

7. William B. Jensen, "Fritz Haber," *Encyclopaedia Britannica* (2020), https://www.britannica.com/biography/Fritz-Haber (accessed July 3, 2020).

Chapter 7—Can a Christian Be a Scientist (and Vice Versa)?

1. Richard Carrier, "Christianity Was Not Responsible for Modern Science," *The Christian Delusion: Why Faith Fails*, ed. John W. Loftus (Amherst, NY: Prometheus Books, 2010), 413, emphasis in original.

2. Carrier, "Christianity Was Not Responsible for Modern Science," 414.

3. Martin Heidegger, *Introduction to Metaphysics*, trans. Gregory Fried and Richard Polt (New Haven, CT: Yale University Press, 2000), 8.

4. Richard Dawkins, *The God Delusion* (Boston, MA: Houghton Mifflin, 2006), 323.

5. Dawkins, *The God Delusion*, 320-321.

6. Dawkins, *The God Delusion*, 321.

7. Dawkins, *The God Delusion*, 319.

8. Heidegger, *Introduction to Metaphysics*, 7-8.

9. Augustine, "The Trinity," *Augustine: Later Works*, trans. and ed. John Burnaby (Philadelphia, PA: The Westminster Press, 1955), 45.

10. Augustine, "The Spirit and the Letter," *Augustine: Later Works*, 238.

11. Thomas Aquinas, *On Being and Essence*, trans. Armand Maurer, 2d ed. (Toronto, ON: PIMS, 1968), 28.

12. James Hannam, *God's Philosophers: How the Medieval World Laid the Foundations of Modern Science* (London, UK: Icon Books, 2009), 288.

13. Hannam, *God's Philosophers*, 288.

14. Hannam, *God's Philosophers*, 291.

15. See David Haines and Andrew Fulford, "The Metaphysics of Scripture," *Philosophy and the Christian: The Quest for Wisdom in the Light of Christ*, ed. Joseph Minich (Lincoln, NE: The Davenant Press, 2018), 9-49.

16. Mario De Caro, "Varieties of Naturalism," *The Waning of Materialism*, eds. Robert C. Koons and George Bealer (Oxford, UK: Oxford University Press, 2010), 366.

17. J.P. Moreland, *Scientism and Secularism: Learning to Respond to a Dangerous Ideology* (Wheaton, IL: Crossway, 2018), 23.

18. Angus J.L. Menuge, "Against Methodical Naturalism," *The Waning of Materialism*, eds. Robert C. Koons and George Bealer (Oxford, UK: Oxford University Press, 2010), 375. See also Ernan McMullin, "Varieties of Methodical Naturalism," *The Nature of Nature: Examining the Role of Naturalism in Science*, eds. Bruce L. Gordon and William A. Dembski (Wilmington, DE: ISI Books, 2011), 82-94.

19. David C. Lindberg, *The Beginnings of Western Science: The European Scientific Tradition in Philosophical, Religious, and Institutional Context, Prehistory to A.D. 1450*, 2d ed. (Chicago, IL: University of Chicago Press, 2007), 210.

20. Lindberg, *The Beginnings of Western Science*, 210-211.

21. Lindberg, *The Beginnings of Western Science*, 211-212.

22. Lindberg, *The Beginnings of Western Science*, 212.

23. Ronald L. Numbers, "Science Without God: Natural Laws and Christian Belief," *The Nature of Nature*, eds. Gordon and Dembski, 63.

24. Numbers, "Science Without God," 63.

25. Numbers, "Science Without God," 64.

26. Albert the Great, *On the Causes of the Properties of the Elements*, trans. Irven M. Resnick (Milwaukee, WI: Marquette University Press, 2010), 20.

27. Hannam, *God's Philosophers*, 318.

28. John Hedley Brooke, *Science and Religion: Some Historical Perspectives* (Cambridge, UK: Cambridge University Press, 1991), 37.

29. Willem Jacob's Gravesande, *Mathematical Elements of Natural Philosophy*, trans. J.T. Desaguliers (London, UK: W. Innys, T. Longman, and T. Shewell in Pater-noster-row, 1747), 1:2.

30. For the stories of a number of contemporary scientists who are both devout Christians and excellent scientists, see *Scientists Who Believe: 21 Tell Their Own Stories*, eds. Eric C. Barrette and David Fisher (Chicago, IL: Moody, 1984).

31. John Lennox, *Can Science Explain Everything?* (Epsom, UK: The Good Book Company, 2019), 27.

32. Ian Sample and Stephen Hawking, "Interview: Stephen Hawking: 'There is no heaven; it's a fairy story,'" *The Guardian* (May 15, 2011), https://www.theguardian.com/science/2011/may/15/stephen-hawking-interview -there-is-no-heaven (accessed November 3, 2020).

33. Lennox, *Can Science Explain Everything?*, 30.

34. Many thanks to Benoît Côté and Veronique Cloutier, who both read through this article and provided many helpful suggestions and critiques.

Chapter 8—What Is the Biblical and Scientific Case for a Historical Adam and Eve?

1. Wayne Grudem, "Theistic Evolution Undermines Twelve Creation Events and Several Crucial Christian Doctrines," *Theistic Evolution: A Scientific, Philosophical, and Theological Critique*, eds. J.P. Moreland, Stephen C. Meyer, Christopher Shaw, Ann K. Gauger, and Wayne Grudem (Wheaton, IL: Crossway, 2017), 783-838.

2. R. Laird Harris, Gleason L. Archer Jr., and Bruce K. Waltke, *Theological Wordbook of the Old Testament, Vol. 1* (Chicago, IL: Moody, 1980), 396.

3. Harris, Archer, and Waltke, *Theological Wordbook of the Old Testament, Vol. 1*, 116-117.

4. R. Laird Harris, Gleason L. Archer Jr., and Bruce K. Waltke, *Theological Wordbook of the Old Testament, Vol. 2* (Chicago, IL: Moody, 1980), 701-702.

5. Harris, Archer, and Waltke, *Theological Wordbook of the Old Testament, Vol. 1*, 127-128.

6. For example, see Dennis R. Venema and Scot McKnight, *Adam and the Genome: Reading Scripture After Genetic Science* (Grand Rapids, MI: Brazos Press, 2017).

7. A helpful example of the range of views on the historical Adam and Eve in light of human evolution can be found in Jeff Hardin, "Biology and Theological Anthropology: Friends or Foes," *BioLogos* (December 11, 2019), https://wp.biologos.org/wp-content/uploads/2019/12/Biology-and-Theological-Anthropology-Friends-or-Foes .pdf (accessed November 6, 2020).

8. S. Joshua Swamidass, *The Genealogical Adam and Eve: The Surprising Science of Universal Ancestry* (Downers Grove, IL: IVP Academic, 2019).

9. John D. Currid, "Theistic Evolution Is Incompatible with the Teachings of the Old Testament," *Theistic Evolution: A Scientific, Philosophical, and Theological Critique*, eds. J. P. Moreland, Stephen C. Meyer, Christopher Shaw, Ann K. Gauger, and Wayne Grudem (Wheaton, IL: Crossway, 2017), 853.

10. Kenneth R. Samples, "The Original Couple: What Is the Range of Viable Positions Concerning Adam and Eve?," *Old-Earth or Evolutionary Creation? Discussing Origins with Reasons to Believe and BioLogos*, eds. Kenneth Keathley, J.B. Stump, and Joe Aguirre (Downers Grove, IL: IVP Academic, 2017), 58.

11. Samples, "The Original Couple," 58-59.

12. William Henry Green, "Primeval Chronology," *Bibliotheca Sacra* (April 1890), 285-303, http://genevaninstitute .org/syllabus/unit-two-theology-proper/lesson-5-the-decree-of-creation/primeval-chronology-by-dr-william -henry-green/ (accessed November 6, 2020).

13. Daniel J. Dyke and Hugh Henry, "Q&A: What Is the Purpose of the Numbers in the Genesis Genealogies?," *Today's New Reason to Believe* (March 27, 2014), https://reasons.org/explore/blogs/todays-new-reason-to-believe/read/ tnrtb/2014/03/27/q-a-what-is-the-purpose-of-the-numbers-in-the-genesis-genealogies (accessed November 6, 2020).

14. Dyke and Henry, "Genesis Genealogies."

15. Kenneth D. Keathley and Mark F. Rooker, *40 Questions About Creation and Evolution*, series ed. Benjamin L. Merkle (Grand Rapids, MI: Kregel Academic, 2014), 169-178.

16. Harris, Archer, and Waltke, *Theological Wordbook, Vol. 1*, 5-6.

17. Harris, Archer, and Waltke, *Theological Wordbook, Vol. 1*, 378-379.

18. K.A. Kitchen, *On the Reliability of the Old Testament* (Grand Rapids, MI: Eerdmans, 2003), 440-441.

19. Rebecca L. Cann, "Y Weigh In Again on Modern Humans," *Science* 341 (August 2, 2013), 465-467; G. David Poznik et al., "Sequencing Y Chromosomes Resolves Discrepancy in Time to Common Ancestor of Males Versus Females," *Science* 341 (August 2, 2013), 562-565; Paolo Francalacci et al., "Low-Pass DNA Sequencing of 1200 Sardinians Reconstructs European Y-Chromosome Phylogeny," *Science* 341 (August 2, 2013), 565-569.

20. Simon Neubauer, Jean-Jacques Hublin, and Philipp Gunz, "The Evolution of Modern Human Brain Shape," *Science Advances* 4 (January 24, 2018), eaao5961.

21. Fazale Rana with Hugh Ross, *Who Was Adam? A Creation Model Approach to the Origin of Humanity* (Covina, CA: RTB Press, 2015), 271-276.

22. Renaud Kaeuffer et al., "Unexpected Heterozygosity in an Island Mouflon Population Founded by a Single Pair of Individuals," *Proceedings of the Royal Society B* 274 (December 2006), 527-533; Hiroki Goto et al., "A Massively Parallel Sequencing Approach Uncovers Ancient Origins and High Genetic Variability of Endangered Przewalski's Horses," *Genome Biology and Evolution* 3 (2011), 1096-1106; Catherine Lippé, Pierre Dumont, and Louis Bernatchez, "High Genetic Diversity and No Inbreeding in the Endangered Copper Redhorse, *Moxostoma hubbsi* (Catostomidae, Pisces): The Positive Sides of a Long Generation Time," *Molecular Ecology* 15 (June 2006), 1769-1780; Frank Hailer et al., "Bottlenecked But Long-Lived: High Genetic Diversity Retained in White-Tailed Eagles upon Recovery from Population Decline," *Biology Letters* 2 (June 2006), 316-319; Jaana Kekkonen, Mikael Wikström, and Jon E. Brommer, "Heterozygosity in an Isolated Population of a Large Mammal Founded by Four Individuals Is Predicted by an Individual-Based Genetic Model," *Plos One* 7 (September 2012), e43482.

23. Ola Hössjer and Ann Gauger, "A Single-Couple Human Origin Is Possible," *BIO-Complexity* 2019 (1).

24. See Thomas Suddendorf, *The Gap: The Science of What Separates Us from Other Animals* (New York: Basic Books, 2013).

25. Fazale Rana with Hugh Ross, *Who Was Adam? A Creation Model Approach to the Origin of Humanity* (Covina, CA: RTB Press, 2015), 313-333.

26. Johan J. Bolhuis et al., "How Could Language Have Evolved?," *Plos Biology* 12 (August 26, 2014), e1001934, doi:10.1371/journal.pbio.1001934.

Chapter 9—On Science and Scientism: What Insights Does C.S. Lewis Offer?

1. C.S. Lewis, *Miracles* (New York: Macmillan, 1947), 135.

2. Quoted in R.L. Green and Walter Hooper, *C.S. Lewis: A Biography* (London, UK: Collins Fount, 1979), 173.

3. Quoted in Jean Mayer, "Science Without Conscience," *American Scholar* 41 (Spring 1972), 265.

4. E.F. Schumacher, *A Guide for the Perplexed* (New York: Harper & Row, 1977), 4-5.

5. C.S. Lewis, *The Weight of Glory* (New York: Macmillan, 1962), 72.

6. C.S. Lewis, *The Voyage of the Dawn Treader* (New York: Macmillan, 1952), 180.

7. Lewis, *The Weight of Glory*, 71.

8. C.S. Lewis, *The Abolition of Man* (New York: Macmillan, 1965), 3.

Chapter 10—How Has Evil Been Done in the Name of Science?

1. For an interesting documentary exploring this theme, see *Human Zoos*, directed by John West, https://www .youtube.com/watch?v=nY6Zrol5QEk (accessed June 13, 2020).

2. John H. Evans, *What Is a Human? What the Answers Mean for Human Rights* (Oxford, UK: Oxford University Press, 2016).

3. I discuss the impact of both biological and environmental determinism on modern thought in *The Death of Humanity: And the Case for Life* (Washington, DC: Regnery Faith, 2016), chapters 3-4. On scientific racism, see Nancy Stepan, *The Idea of Race in Science: Great Britain, 1800-1960* (Hamden, CN: Archon Books, 1982).

4. Charles Darwin, *The Descent of Man*, 2 vols. (Princeton, NJ: Princeton University Press, 1981), 1:3.

5. Darwin, *The Descent of Man*, 1:201.

6. Adrian Desmond and James Moore, *Darwin's Sacred Cause: How a Hatred of Slavery Shaped Darwin's Views on Human Evolution* (Boston, MA: Houghton Mifflin Harcourt, 2009), 318.

7. Ernst Haeckel, *Natürliche Schöpfungsgeschichte* (Berlin, Germany: Georg Reimer, 1868), frontispiece, 555.

8. Ernst Haeckel, *Die Lebenswunder: Gemeinverständliche Studien über Biologische Philosophie* (Stuttgart, Germany: Alfred Kröner, 1904), 451-452.

9. Ernst Haeckel, *Ewigkeit. Weltkriegsgedanken über Leben und Tod, Religion und Entwicklungslehre* (Berlin, Germany: Georg Reimer, 1917), 35-36, 110-111, 120-123; quotes at 85-86, 35.

10. Quoted in Horst Drechsler, *Let Us Die Fighting: The Struggle of the Herero and Nama Against German Imperialism (1884-1915)*, trans. Bernd Zöllner (London, UK: Zed Press, 1980), 167-168, n. 6.

11. Quoted in Peter Schmitt-Egner, *Kolonialismus und Faschismus. Eine Studie zur historischen und begrifflichen Genesis faschistischer Beweusstseinsformen am deutschen Beispiel* (Giessen, Germany: 1975), 125.

12. I discuss this and other examples of Darwinist militarism and racial extermination in *From Darwin to Hitler: Evolutionary Ethics, Eugenics, and Racism in Germany* (New York: Palgrave Macmillan, 2004), chapters 9-10.

13. Quoted in Wolfgang Bialas, *Moralische Ordnungen des Nationalsozialismus* (Göttingen, Germany: Vandenhoeck & Ruprecht, 2014), 43.

14. Benoit Massin, "The 'Science of Race,'" *Deadly Medicine: Creating the Master Race* (Washington, DC: United States Holocaust Memorial Museum, 2004).

15. Richard Weikart, "Hitler's Struggle for Existence Against Slavs: Anthropology, Racial Theory and Vacillations in Nazi Policy toward Czechs and Poles," *Eradicating Differences: The Treatment of Minorities in Nazi-Dominated Europe*, ed. Anton Weiss-Wendt (Cambridge, UK: Cambridge Scholars Publishing, 2010), 61-83.

16. Sheila Faith Weiss, *The Nazi Symbiosis: Human Genetics and Politics in the Third Reich* (Chicago, IL: University of Chicago Press, 2010), 104.

17. Charles B. Davenport, *Eugenics: The Science of Human Improvement by Better Breeding* (New York: Henry Holt, 1910), 33-34.

18. Ian Dowbiggin, *A Merciful End: The Euthanasia Movement in Modern America* (New York: Oxford University Press, 2003), 8; N.D.A. Kemp, *'Merciful Release': The History of the British Euthanasia Movement* (Manchester, UK: Manchester University Press, 2002), 19; Hans-Walter Schmuhl, *Rassenhygiene, Nationalsozialismus, Euthanasie. Von der Verhütung zur Vernichtung ,lebensunwerten Lebens' 1890–1945* (Göttingen, Germany: Vandenhoek und Ruprecht, 1987), 18-19.

19. Hans-Joachim Lang, *Die Frauen von Block 10: Medizinische Versuche in Auschwitz* (Hamburg, Germany: Hoffmann und Campe Verlag, 2011).

20. Ernst Klee, *Auschwitz, die NS-Medizin und ihre Opfer* (Frankfurt, Germany: S. Fischer, 1997).

21. Robert Jütte with Wolfgang Eckart, Hans-Walter Schmuhl, and Winfried Süss, *Medizin und Nationalsozialismus: Bilanz und Perspektiven der Forschung* (Göttingen, Germany: Wallstein Verlag, 2011), 124.

22. James H. Jones, "The Tuskegee Syphilis Experiment," *Man, Medicine, and the State: The Human Body as an Object of Government Sponsored Medical Research in the 20th Century*, ed. Wolfgang U. Eckart (Stuttgart, Germany: Franz Steiner Verlag, 2006), 252-260.

23. Ulf Schmidt, "Medical Ethics and Human Experiments at Porton Down: Informed Consent in Britain's Biological and Chemical Warfare Experiments," *History and Theory of Human Experimentation: The Declaration of Helsinki and Modern Medical Ethics*, eds. Ulf Schmidt and Andreas Frewer (Stuttgart, Germany: Franz Steiner Verlag, 2007), 283-305.

Chapter 11—How Can We Use Science in Apologetics?

1. Carl Sagan, *Cosmos* (New York: Ballantine, 1980), 1.

2. Alex Rosenberg, *The Atheist's Guide to Reality: Enjoying Life Without Illusions* (New York: W.W. Norton, 2012), 190-191.

3. Del Ratzsch, quoted in Jay Wesley Richards, "How Phil Johnson Changed my Mind," *Darwin's Nemesis: Phillip Johnson and the Intelligent Design Movement*, ed. William A. Dembski (Downers Grove, IL: InterVarsity Press, 2006), 53.

4. See Stephen C. Meyer, *Signature in the Cell: DNA and the Evidence for Intelligent Design* (San Francisco, CA: HarperOne, 2010), 397-415.

5. C.S. Lewis, "Is Theology Poetry" (Quebec, Canada: Samizdat University Press, 1944, reprinted 2014), http://augustinecollective.org/wp-content/uploads/2016/06/1.2-Is-Theology-Poetry-Reading.pdf (accessed November 8, 2020).

6. Christian philosopher William Lane Craig has done the most work in developing this modern form of the Kalam cosmological argument. See William Lane Craig and Quentin Smith, *Theism, Atheism, and Big Bang Cosmology* (Oxford, UK: Clarendon Books, 1996).

7. For an excellent recent treatment, see Luke A. Barnes and Geraint F. Lewis, *The Cosmic Revolutionary's Handbook* (Cambridge, UK: Cambridge University Press, 2020).

8. Stephen Hawking, *A Brief History of Time: From the Big Bang to Black Holes* (Toronto, Canada: Bantam, 1988), 132.

9. For brief explanations of the examples listed here, see Jay W. Richards, "List of Fine-Tuning Parameters," Discovery Institute, https://www.discovery.org/m/securepdfs/2018/12/List-of-Fine-Tuning-Parameters-Jay-Richards.pdf (accessed November 9, 2020); Geraint A. Lewis and Luke Barnes, *A Fortunate Universe: Life in a Fine-Tuned Cosmos* (Cambridge, UK: Cambridge University Press, 2016); Robin Collins, "The Teleological Argument: An Exploration of the Fine-tuning of the Cosmos," *Blackwell Companion to Natural Theology*, eds. William Lane Craig and J.P. Moreland, (Oxford, UK: Blackwell Publications, 2009); John Barrow and Frank Tipler, *The Anthropic Cosmological Principle* (Oxford, UK: Oxford University Press, 1986); Roger Penrose, *The Road to Reality: A Complete Guide to the Laws of the Universe* (New York: Vintage, 2007).

10. Guillermo Gonzalez and Jay W. Richards, *The Privileged Planet: How Our Place in the Cosmos Is Designed for Discovery* (Washington, DC: Regnery, 2004; rev. ed. 2019).

11. Henry Quastler, *The Emergence of Biological Organization* (New Haven, CT: Yale University Press, 1964), 16.

12. Stephen C. Meyer, *Signature in the Cell: DNA and the Evidence for Intelligent Design* (San Francisco, CA: HarperOne, 2009).

13. Douglas Axe, *Undeniable: How Biology Confirms Our Intuition That Life Is Designed* (San Francisco, CA: HarperOne, 2016).

14. Michael J. Behe, *Darwin's Black Box: The Biochemical Challenge to Evolution,* 2d ed. (New York: The Free Press, 2006).

15. Michael J. Behe, *The Edge of Evolution: The Search for the Limits of Darwinism* (New York: The Free Press, 2007).

16. Michael J. Behe, *Darwin Devolves: The New Science About DNA That Challenges Evolution* (New York: The Free Press, 2019).

17. Stephen C. Meyer, *Darwin's Doubt: The Explosive Origin of Animal Life and the Case for Intelligent Design* (San Francisco, CA: HarperOne, 2013).

18. John Lennox, "John Lennox Busts a Myth About Religion, Faith and Science," *Solas* (March 18, 2019), https://www.solas-cpc.org/john-lennox-busts-a-myth-about-religion-faith-and-science/ (accessed November 9, 2020).

19. Rodney Stark, *For the Glory of God: How Monotheism Led to Reformations, Science, Witch-Hunts, and the End of Slavery* (Princeton, NJ: Princeton University Press, 2003).

Chapter 12—What About the Historical Relationship Between Christianity and Science?

1. This chapter is based on Dr. House's article "Christianity and Science," *Encyclopedia of Christianity in the United States, Vol. 4*, eds. George Kurian and Mark Lamport (Lanham, MD: Rowman & Littlefield, 2016), 2037-2042, used by permission.

2. Stephen C. Meyer, "Qualified Agreement Modern Science & the Return of the 'God Hypothesis,'" *Science and Christianity: Four Views*, ed. Richard F. Carlson (Downers Grove, IL: InterVarsity, 2000), 127-174.

3. Matthew C. Harrison, *In Christ All Things Hold Together: The Intersection of Science & Christian Theology* (St. Louis, MO: Concordia, 2015), 49.

4. James Hannam, *The Genesis of Science: How the Christian Middle Ages Launched the Scientific Revolution* (Washington, DC: Regency, 2011).

5. Thomas F. Torrance, *Reality and Scientific Theology* (Edinburgh, UK: Scottish Academic Press, 1985), 4.

6. Torrance, *Reality and Scientific Theology*, 5.

7. In this context by *creationist* I simply mean someone who believes in a creator.

8. J.P. Moreland, *Christianity and the Nature of Science: A Philosophical Investigation* (Grand Rapids, MI: Baker, 1989), 3.

9. Neal Gillespie, *Charles Darwin and the Problem of Creation* (Chicago, IL: University of Chicago Press, 1979), 19.

10. Robert Jastrow, *God and the Astronomers* (New York: Norton, 1978), 115-116.

Chapter 13—What Is the Evidence for Intelligent Design and What Are Its Theological Implications?

1. Richard Dawkins, *The Blind Watchmaker: Why the Evidence of Evolution Reveals a Universe Without Design* (New York: Norton, 1986), 1.

2. Charles Darwin, *The Life and Letters of Charles Darwin, Vol. 1*, ed. Francis Darwin (New York: Appleton, 1887), 278-279.

3. Francisco J. Ayala, "Darwin's Greatest Discovery: Design without Designer," *Proceedings of the National Academy of Sciences USA* 104 (May 15, 2007), 8567-8573.

4. Richard Dawkins, *River Out of Eden: A Darwinian View of Life* (New York: Basic, 1995), 17.

5. Bill Gates, *The Road Ahead* (New York: Viking, 1995), 188.

6. Leroy Hood and David Galas, "The digital code of DNA," *Nature* 421 (2003), 444-448.

7. Stephen Meyer, *Signature in the Cell: DNA and the Evidence for Intelligent Design* (San Francisco, CA: HarperOne, 2009), 173-323.

8. Henry Quastler, *The Emergence of Biological Organization* (New Haven, CT: Yale University Press, 1964), 16.

9. William A. Dembski, *The Design Inference: Eliminating Chance Through Small Probabilities* (Cambridge, MA: Cambridge University Press, 1998), 36-66.

10. Paul Davies, *The Cosmic Blueprint* (New York: Simon & Schuster, 1988), 203.

11. Fred Hoyle, "The Universe: Past and Present Reflections," *Annual Review of Astronomy and Astrophysics* 20 (1982), 16.

12. A few physicists have proposed that if our bubble universe bumped into another bubble universe, it would leave detectable patterns in the cosmic microwave background radiation (CMBR). See Joshua Sokol, "A Brush with a Universe Next Door," *New Scientist* 228 (October 31, 2015), 8-9. Roger Penrose has made a similar claim for his conformal cyclic cosmology (CCC) model, in which the universe goes through infinitely many cycles, with the future time-like infinity of each earlier iteration being identified with the big bang singularity of the next. For a popular account, see Roger Penrose, *Cycles of Time: An Extraordinary New View of the Universe* (New York: Alfred A. Knopf, 2011). He argues that observed "hot spots" in the CMBR represent evidence of interaction between the different modes of the universe in its collapsing and expanding phases. Specifically, he sees hot spots in the CMBR as evidence of the collapse of black holes prior to the beginning of our universe in its present expansion phase. See Roger Penrose, "On the Gravitization of Quantum Mechanics 2: Conformal Cyclic Cosmology," *Foundations of Physics* 44 (2014), 873-890. Even so, his model does not, strictly speaking, represent a multiverse model, since the universes exist in succession, not in parallel.

13. Clifford Longley, "Focusing on Theism," *London Times* (January 21, 1989), 10.

14. Stephen C. Meyer, *Return of the God Hypothesis: Compelling Scientific Evidence for the Existence of God* (San Francisco, CA: HarperOne, 2021).

15. Guillermo Gonzalez and Jay Richards, *The Privileged Planet: How Our Place in the Cosmos Is Designed for Discovery* (Washington, DC: Regnery, 2004), 293-311.

16. Francis Crick, *Life Itself: Its Origin and Nature* (New York: Simon & Schuster), 88, 95-166. See also F.H.C. Crick and L.E. Orgel, "Directed Panspermia," *Icarus* 19 (1973), 341-346.

17. Sir Fred Hoyle and N.C. Wickramasinghe, *Evolution from Space: A Theory of Cosmic Creationism* (New York: Touchstone), 35-50.

18. See Richard Dawkins quoted in *Expelled: No Intelligence Allowed* (Premise Media, 2008).

19. Crick, *Life Itself*, 88.

20. See Elliott Sober, "Intelligent design theory and the supernatural—The 'god or extraterrestrials' reply," *Faith and Philosophy* 24 (2007), 72-82. Sober, a philosophical naturalist who rejects the case for intelligent design, argues that *if* one does accept the argument for intelligent design in biology (from irreducible complexity), it makes more sense to affirm a supernatural designer than an extraterrestrial one. He argues that the "minimalist case" for intelligent design, when supplemented with a few additional and plausible premises (such as, for example, "the universe is finite"), leads logically to the conclusion that a transcendent intelligent designer must exist.

21. In *Return of the God Hypothesis*, I also argue that theism provides a better explanation than deism, pantheism, panenthesim, and pansychism for the key facts that we have about biological and cosmological origins.

Chapter 14—Is Our Intuition of Design in Nature Correct?

1. Russell Powell and Steve Clarke, "Religion as an evolutionary byproduct: A critique of the standard model," *The British Journal for the Philosophy of Science* 63 (2012), 457-486.

2. Job 42:3.

3. Plato, *Timaeus* (trans. Benjamin Jowett, Project Gutenberg e-Book), https://www.gutenberg.org/files/1572/1572 -h/1572-h.htm (accessed August 24, 2020).

4. Plutarch, "Fortune" (translated by Frank Cole Babbitt), *Moralia*, vol. 2, Loeb Classical Library (Cambridge, MA: Harvard University Press, 1928), 87.

5. William Paley, *Natural Theology or Evidences of the Existence and Attributes of the Deity* (Philadelphia, PA: John Morgan, 1802).

6. Michael J. Behe, *Darwin's Black Box: The Biochemical Challenge to Evolution* (New York: Free Press, 1996).

7. Stephen C. Meyer, *Signature in the Cell: DNA and the Evidence for Intelligent Design* (New York: HarperOne, 2009).

8. Romans 1:19-20.

9. Romans 1:20.

10. Acts 2:25-36.

11. Alison Gopnik, "See Jane Evolve: Picture Books Explain Darwin," *Wall Street Journal* (April 18, 2014), http:// online.wsj.com/news/articles/SB10001424052702304311204579505574046805070 (accessed August 24, 2020).

12. Douglas Axe, *Undeniable: How Biology Confirms Our Intuition That Life Is Designed* (New York: HarperOne, 2016).

13. Richard Dawkins, *The Blind Watchmaker: Why the Evidence of Evolution Reveals a Universe Without Design* (New York: Norton, 1986).

14. Psalm 14:1.

15. Romans 1:20.

16. Charles Darwin, *On the Origin of Species by Means of Natural Selection, or the Preservation of Favoured Races in the Struggle for Life* (London, UK: John Murray, 1859), chapter IV, 81.

17. Axe, *Undeniable*, 97.

18. Hugo De Vries, *Species and Varieties–Their Origin by Mutation* (Chicago, IL: Open Court Publishing, 1904).

19. George Wald, "The Origin of Life," *Scientific American* 191 (August 1954), 44-53.

20. Dawkins, *The Blind Watchmaker*, 9, emphasis in original.

21. A growing number of researchers who affiliate under the heading *The Third Way* have "expressed their concerns regarding natural selection's scope and…believe that other mechanisms are essential for a comprehensive understanding of evolutionary processes." "List of scientists who think that a fresh look at evolution is needed." The Third Way of Evolution, https://www.thethirdwayofevolution.com/people (accessed August 24, 2020).

22. Michael J. Behe, *Darwin Devolves: The New Science about DNA that Challenges Evolution* (New York: HarperOne, 2019). See also Michael J. Behe, "Experimental Evolution, Loss-of-Function Mutations, and 'The First Rule of Adaptive Evolution,'" *The Quarterly Review of Biology* 85 (December 2020), 419-445.

23. "Fully Automatic Biscuit and Cookie Production Line," Taizy.com, https://www.taizyfoodmachine.com/full -automatic-biscuit-production-line-oem.html (accessed August 24, 2020).

Chapter 15—What Is Intelligent Design and How Should We Defend It?

1. Douglas D. Axe, "Extreme Functional Sensitivity to Conservative Amino Acid Changes on Enzyme Exteriors,"

Journal of Molecular Biology, 301 (2000), 585-595; Douglas D. Axe, "Estimating the Prevalence of Protein Sequences Adopting Functional Enzyme Folds," *Journal of Molecular Biology* 341 (2004), 1295-1315.

2. Transcript of testimony of Scott Minnich, *Kitzmiller v. Dover* (M.D. Pa., PM Testimony, November 3, 2005), 103-112; Robert M. Macnab, "Flagella," in *Escherichia Coli and Salmonella Typhimurium: Cellular and Molecular Biology, Vol. 1*, eds. Neidhardt et al. (Washington, DC: American Society for Microbiology, 1987), 73-74.

3. Roger Penrose and M. Gardner, *The Emperor's New Mind: Concerning Computers, Minds, and the Laws of Physics* (Oxford, UK: Oxford University Press, 2002).

4. Bonnie Azab Powell, "'Explore as much as we can': Nobel Prize winner Charles Townes on evolution, intelligent design, and the meaning of life," *UC Berkeley NewsCenter* (June 17, 2005), https://www.berkeley.edu/news/media/releases/2005/06/17_townes.shtml (accessed October 26, 2020).

5. For example see Mark J. Pallen and Nicholas J. Matzke, "From The Origin of Species to the origin of bacterial flagella," *Nature Reviews Microbiology* 4 (2006), 784-790.

6. Philip Kitcher, *Living with Darwin: Evolution, Design, and the Future of Faith* (Oxford, UK: Oxford University Press, 2007), 57, 58, 62, 129.

7. Richard Dawkins, *A Devil's Chaplain: Reflections on Hope, Lies, Science, and Love* (Boston, MA: Houghton Mifflin, 2004), 99.

8. The ENCODE Project Consortium, "An integrated encyclopedia of DNA elements in the human genome," *Nature* 489 (September 6, 2012), 57-74.

9. For a review of some of this research, see Jonathan Wells, *The Myth of Junk DNA* (Seattle, WA: Discovery Institute Press, 2011).

10. Moleirinho et al., "Evolutionary Constraints in the ß-Globin Cluster: The Signature of Purifying Selection at the d-Globin (*HBD*) Locus and Its Role in Developmental Gene Regulation," *Genome Biology and Evolution* 5 (2013), 559-571.

11. Richard Dawkins, in "Jonathan Sacks and Richard Dawkins at BBC RE:Think festival 12 September 2012," http://www.youtube.com/watch?v=roFdPHdhgKQ [12:57-13:11] (accessed December 2, 2020).

12. Charles R. Marshall, "When Prior Belief Trumps Scholarship," *Science* 341 (September 20, 2013), 1344.

13. For a review and summary of this debate see Casey Luskin, "A Listener's Guide to the Meyer-Marshall Radio Debate: Focus on the Origin of Information Question," *Evolution News* (December 4, 2013), https://evolution news.org/2013/12/a_listeners_gui/ (accessed October 26, 2020).

14. Peter Heger et al., "The genetic factors of bilaterian evolution," *eLife* 9 (2020), e45530; Jordi Paps and Peter W.H. Holland, "Reconstruction of the ancestral metazoan genome reveals an increase in genomic novelty," *Nature Communications* 9 (2018), 1730.

15. For a discussion see David Klinghoffer, "Leading Theistic Evolutionist Darrel Falk, Past BioLogos President, Praises Darwin's Doubt," *Evolution News* (September 12, 2014), https://evolutionnews.org/2014/09/break through_le/ (accessed October 26, 2020).

16. See "Bibliographic and Annotated List of Peer-Reviewed Publications Supporting Intelligent Design," https://www.discovery.org/id/peer-review/ (accessed October 26, 2020).

17. Evolutionary Informatics Lab, https://evoinfo.org/ (accessed March 16, 2021).

18. For a good example of this, see Casey Luskin, "Unintended Consequences: How Hostile Responses to Darwin's Doubt Turned a Thoughtful Reader Against Darwinian Evolution," *Evolution News* (November 26, 2013), https://evolutionnews.org/2013/11/unintended_cons/ (accessed November 30, 2020).

19. See Inna Kouper, "Science blogs and public engagement with science: practices, challenges, and opportunities," *Journal of Science Communication* 9 (March 2010); Dale L. Sullivan, "Keeping the Rhetoric Orthodox: Forum Control in Science," *Technical Communication Quarterly* 9 (Spring 2000), 125-146. See also Virginia Heffernan, "Unnatural Science," *New York Times* (July 30, 2010), https://www.nytimes.com/2010/08/01/magazine/01FOB -medium-t.html (accessed October 26, 2020)

20. For further rebuttals to the *Kitzmiller v. Dover* ruling see "The Truth About the Dover Intelligent Design Trial," www.traipsingintoevolution.com (accessed October 26, 2020) or "Ten Dover Myths," *Evolution News*, https://evolutionnews.org/tag/ten-dover-myths/ (accessed October 27, 2020).

21. For example, the judge made many inaccurate (or at least easily challenged) scientific claims about the evolution of the bacterial flagellum, the blood clotting cascade, and the origin of new genetic information. For details,

see Casey Luskin, "Do Car Engines Run on Lugnuts? A Response to Ken Miller & Judge Jones's Straw Tests of Irreducible Complexity for the Bacterial Flagellum," *Discovery Institute* (April 19, 2006), https://www.discovery .org/a/3718/ (accessed October 27, 2020); Casey Luskin, "Kenneth Miller, Michael Behe, and the Irreducible Complexity of the Blood Clotting Cascade Saga," *Discovery Institute* (January 1, 2010), https://www.discovery .org/a/14081/ (accessed October 27, 2020); Casey Luskin, "The NCSE, Judge Jones, and Citation Bluffs About the Origin of New Functional Genetic Information" *IDEA Center*, http://www.ideacenter.org/contentmgr/show details.php/id/1501 (accessed October 27, 2020).

22. John West and David DeWolf, "A Comparison of Judge Jones' Opinion in *Kitzmiller v. Dover* with Plaintiffs' Proposed 'Findings of Fact and Conclusions of Law'," Discovery Institute (December 12, 2006), https://www .discovery.org/m/2019/07/Comparing_Jones_and_ACLU.pdf (accessed October 27, 2020).

23. As anti-ID legal scholar Jay Wexler writes, "The part of *Kitzmiller* that finds ID not to be science is unnecessary, unconvincing, not particularly suited to the judicial role, and even perhaps dangerous both to science and to freedom of religion." Jay D. Wexler, "*Kitzmiller* and the 'Is It Science?' Question," *First Amendment Law Review* 5 (2006), 90-111.

24. Michael Shermer, *How We Believe: The Search for God in an Age of Science* (New York: W.H. Freeman, 2000), 242-245.

25. See The Elie Wiesel Foundation for Humanity: Nobel Laureates Initiative (September 9, 2005), https://web .archive.org/web/20060207230425/http://media.ljworld.com/pdf/2005/09/15/nobel_letter.pdf (accessed Octo- ber 26, 2020); Francisco J. Ayala, "Darwin's greatest discovery: Design without designer," *Proceedings of the National Academy of Sciences* 104 (May 15, 2007), 8567-8573.

26. BioLogos has framed the differences between ID and theistic evolution as follows:

    1. We are skeptical about the ability of biological science to prove the existence of an Intelligent Designer (whom we take to be the God of the Bible), while ID advocates are confident.

    2. We find unconvincing those attempts by ID theorists to scientifically confirm God's activity in natural history, while ID theorists believe they have sufficiently demonstrated it.

    3. We see no biblical reason to view natural processes (including natural selection) as having removed God from the process of creation. It is all God's and it is all intelligently designed. Those in the ID movement for the most part reject some or all of the major conclusions of evolutionary theory.

    "How Is BioLogos different from Evolutionism, Intelligent Design, and Creationism?," https://web.archive.org/ web/20120114022530/http://biologos.org/questions/biologos-id-creationism (accessed October 26, 2020).

27. Francis Collins, *The Language of God: A Scientist Presents Evidence for Belief* (New York: Free Press, 2006), 205.

28. "How Is BioLogos different from Evolutionism, Intelligent Design, and Creationism?"

29. Robert Bishop and Robert O'Connor, "Doubting the Signature: Stephen Meyer's case for intelligent design," *Books & Culture* (October 17, 2014), http://www.booksandculture.com/articles/2014/novdec/doubting-signature .html?paging=off (accessed October 26, 2020).

30. William Provine, "No free will," *Catching up with the Vision*, ed. Margaret W. Rossiter (Chicago, IL: University of Chicago Press, 1999), S123.

31. Rachel Held Evans, "13 Things I Learned at the BioLogos Conference," June 16, 2010, https://web.archive .org/web/20140830134350/http://biologos.org/blog/13-things-i-learned-at-the-biologos-conference/ (accessed October 26, 2020).

32. See Casey Luskin, "The New Theistic Evolutionists: BioLogos and the Rush to Embrace the 'Consensus,'" *Christian Research Journal* 37 (2014), 32-41, https://www.equip.org/article/new-theistic-evolutionists-biologos -rush-embrace-consensus/ (accessed October 27, 2020).

33. Erik Strandness, "From Theistic Evolution to Intelligent Design: Why I changed my mind," *Unbelievable Blog at Patheos* (December 1, 2020), https://www.patheos.com/blogs/unbelievable/2020/12/from-theistic-evolution -to-intelligent-design-why-i-changed-my-mind/ (accessed December 2, 2020).

Chapter 16—What Is the Positive Case for Design?

1. Stephen Jay Gould, "Evolution and the triumph of homology: Or, why history matters," *American Scientist* 74 (1986), 61.

2. Stephen C. Meyer, *Signature in the Cell: DNA and the Evidence for Intelligent Design* (New York: HarperOne, 2009), 154.

3. Charles Lyell, *Principles of Geology: Being an Inquiry How Far the Former Changes of the Earth's Surface Are Referable to Causes Now in Operation* (London, UK: John Murray, 1835).

4. William A. Dembski, *The Design Inference: Eliminating Chance Through Small Probabilities* (Cambridge, UK: Cambridge University Press 1998), 62.

5. William A. Dembski, "Intelligent Design as a Theory of Information," *Intelligent Design Creationism and Its Critics: Philosophical, Theological, and Scientific Perspectives*, ed. Robert T. Pennock (Cambridge, MA: MIT Press), 553-573.

6. Gary Kemper, Hallie Kemper, and Casey Luskin, *Discovering Intelligent Design: A Journey into the Scientific Evidence* (Seattle, WA: Discovery Institute Press, 2013).

7. Stephen C. Meyer, "The origin of biological information and the higher taxonomic categories," *Proceedings of the Biological Society of Washington* 117 (2004), 213-239.

8. Scott A. Minnich and Stephen C. Meyer, "Genetic Analysis of Coordinate Flagellar and Type III Regulatory Circuits in Pathogenic Bacteria," *Proceedings of the Second International Conference on Design & Nature, Rhodes Greece*, eds. M.W. Collins and C.A. Brebbia (Southampton, UK: WIT Press, 2004).

9. William A. Dembski, *No Free Lunch: Why Specified Complexity Cannot Be Purchased Without Intelligence* (Lanham, MD: Rowman & Littlefield, 2002), 239-310; Michael Behe, *Darwin's Black Box: The Biochemical Challenge to Evolution* (New York: Free Press, 1996), 51-73; Minnich and Meyer, "Genetic analysis of coordinate flagellar and type III regulatory circuits in pathogenic bacteria"; A.C. McIntosh, "Information and Entropy—Top-Down or Bottom-Up Development in Living Systems?," *International Journal of Design & Nature and Ecodynamics* 4 (2009), 351-385; A.C. McIntosh, "Evidence of Design in Bird Feathers and Avian Respiration," *International Journal of Design & Nature and Ecodynamics* 4 (2009), 154-169.

10. Douglas D. Axe, "Extreme Functional Sensitivity to Conservative Amino Acid Changes on Enzyme Exteriors," *Journal of Molecular Biology* 301 (2000), 585-595; Douglas D. Axe, "Estimating the Prevalence of Protein Sequences Adopting Functional Enzyme Folds," *Journal of Molecular Biology* 341 (2004), 1295-1315; Ann K. Gauger et al., "Reductive Evolution Can Prevent Populations from Taking Simple Adaptive Paths to High Fitness," *BIO-Complexity* 2010 (2); Kirk K. Durston et al., "Measuring the functional sequence complexity of proteins," *Theoretical Biology and Medical Modelling* 4 (2007), 47; Ann K. Gauger and Douglas D. Axe, "The Evolutionary Accessibility of New Enzyme Functions: A Case Study from the Biotin Pathway," *BIO-Complexity* 2011 (1); M.A. Reeves, A.K. Gauger, and D.D. Axe, "Enzyme Families-Shared Evolutionary History or Shared Design? A Study of the GABA-Aminotransferase Family," *BIO-Complexity* 2014 (4).

11. Stephen C. Meyer, Marcus Ross, Paul Nelson, and Paul Chien, "The Cambrian Explosion: Biology's Big Bang," *Darwinism, Design, and Public Education*, eds. John A. Campbell and Stephen C. Meyer (East Lansing, MI: Michigan State University Press, 2003), 367, 386.

12. Stephen C. Meyer, *Darwin's Doubt: The Explosive Origin of Animal Life and the Case for Intelligent Design* (New York: HarperOne, 2013), 373, 375.

13. Stephen C. Meyer, "The Cambrian Information Explosion," *Debating Design: From Darwin to DNA*, eds. M. Ruse and W. Dembski (Cambridge, MA: Cambridge University Press, 2004), 371-391; W.E. Lonnig, "Dynamic genomes, morphological stasis, and the origin of irreducible complexity," *Dynamical Genetics*, eds. V. Parisi, V. De Fonzo, and F. Aluffi-Pentini (2004), 101-119; McIntosh, "Evidence of Design in Bird Feathers and Avian Respiration."

14. C.P. Hickman, L.S. Roberts, and F.M. Hickman, *Integrated Principles of Zoology* 8th ed. (St. Louis, MO: Times Mirror/Moseby College Publishing, 1988), 866.

15. R.S.K. Barnes, P. Calow, and P.J.W. Olive, *The Invertebrates: A New Synthesis*, 3d ed. (London, UK: Blackwell Scientific, 2001), 9-10. See also Douglas H. Erwin and James W. Valentine, *The Cambrian Explosion: The Construction of Animal Biodiversity* (Greenwood Village, CO: Roberts and Company Publishers, 2013); Meyer, *Darwin's Doubt*; Samuel A. Bowring et al., "Calibrating Rates of Early Cambrian Evolution," *Science* 261 (September 3, 1993), 1293-1298; Kevin J. Peterson et al., "MicroRNAs and metazoan macroevolution: insights into canalization, complexity, and the Cambrian explosion," *BioEssays* 31 (2009), 736. For documentation on the sudden appearance of fossils in the Cambrian explosion, see Casey Luskin, "How 'Sudden' Was the Cambrian Explosion? Nick Matzke Misreads Stephen Meyer and the Paleontological Literature; *New Yorker* Recycles Misrepresentation," *Evolution News* (July 16, 2013), https://evolutionnews.org/2013/07/how_sudden_was_/ (accessed October 27, 2020).

16. A. Cooper and R. Fortey, "Evolutionary Explosions and the Phylogenetic Fuse," *Trends in Ecology and Evolution* 13 (1998), 151-156.

17. S. De Bodt et al., "Genome duplication and the origin of angiosperms," *Trends in Ecology and Evolution* 20 (November 2005), 591-597; P.R. Crane et al., "The origin and early diversification of angiosperms," *Nature* 374: (March 2, 1995), 27-33.

18. See Niles Eldredge, *The Monkey Business: A Scientist Looks at Creationism* (New York: Washington Square Press, 1982).

19. For details, see chapter 32 in this volume, "Do Fossils Demonstrate Human Evolution?"; Casey Luskin, "Human Origins and the Fossil Record"; Ann Gauger, Douglas Axe, and Casey Luskin, *Science and Human Origins* (Seattle, WA: Discovery Institute Press, 2012), 45-83; Casey Luskin, "Missing Transitions: Human Origins and the Fossil Record," *Theistic Evolution: A Scientific, Philosophical, and Theological Critique*, eds. J. P. Moreland, Stephen C. Meyer, Christopher Shaw, Ann K. Gauger, and Wayne Grudem (Wheaton, IL: Crossway, 2017), 437-473.

20. Paul Nelson and Jonathan Wells, "Homology in Biology," *Darwinism, Design, and Public Education*, 303-322.

21. William A. Dembski and Jonathan Witt, *Intelligent Design Uncensored: An Easy-to-Understand Guide to the Controversy* (Downers Grove, IL: InterVarsity, 2010), 85.

22. Günter Bechly and Stephen C. Meyer, "The Fossil Record and Universal Common Ancestry," *Theistic Evolution: A Scientific, Philosophical, and Theological Critique*, eds. J.P. Moreland, Stephen C. Meyer, Christopher Shaw, Ann K. Gauger, and Wayne Grudem (Wheaton, IL: Crossway, 2017), 331-361.

23. R. Quiring et al., "Homology of the eyeless gene of *Drosophila* to the *Small eye* in mice and *Aniridia* in humans," *Science* 265 (August 5, 1994), 785-789; D.B. Wake et al., "Homoplasy: from detecting pattern to determining process and mechanism of evolution," *Science* 331 (February 25, 2011), 1032-1035.

24. P.-A. Christin et al., "Causes and evolutionary significance of genetic convergence," *Trends in Genetics* 26 (2010), 400-405; Y. Li et al., "The hearing gene Prestin unites echolocating bats and whales," *Current Biology* 20 (January 2010), R55-R56; G. Jones, "Molecular Evolution: Gene Convergence in Echolocating Mammals," *Current Biology* 20 (January 2010), R62-R64.

25. For example, see W. Ford Doolittle, "Phylogenetic Classification and the Universal Tree," *Science* 284 (June 25, 1999), 2124-2128; W. Ford Doolittle, "Uprooting the Tree of Life," *Scientific American* (February 2000), 90-95; Arcady R. Mushegian et al., "Large-Scale Taxonomic Profiling of Eukaryotic Model Organisms: A Comparison of Orthologous Proteins Encoded by the Human, Fly, Nematode, and Yeast Genomes," *Genome Research* 8 (1998), 590-598; James H. Degnan and Noah A. Rosenberg, "Gene Tree Discordance, Phylogenetic Inference and the Multispecies Coalescent," *Trends in Ecology and Evolution* 24 (2009), 332-340; Eric Bapteste et al., "Networks: Expanding Evolutionary Thinking," *Trends in Genetics* 29 (August 2013), 439-441; Graham Lawton, "Why Darwin Was Wrong about the Tree of Life," *New Scientist* (January 21, 2009), 34-39; Trisha Gura, "Bones, Molecules, or Both?" *Nature* 406 (July 20, 2000), 230-233; Anna Marie A. Aguinaldo et al., "Evidence for a Clade of Nematodes, Arthropods, and Other Moulting Animals," *Nature* 387 (May 29, 1997), 489-493; Erich D. Jarvis et al., "Whole-Genome Analyses Resolve Early Branches in the Tree of Life of Modern Birds," *Science* 346 (December 12, 2014), 1320-1331; Ewen Callaway, "Flock of Geneticists Redraws Bird Family Tree," *Nature* 516 (December 11, 2014), 297; Emma C. Teeling and S. Blair Hedges, "Making the Impossible Possible: Rooting the Tree of Placental Mammals," *Molecular Biology and Evolution* (2013); James E. Tarver et al., "The Interrelationships of Placental Mammals and the Limits of Phylogenetic Inference," *Genome Biology and Evolution* 8 (2006), 330-344; Jeffrey H. Schwartz and Bruno Maresca, "Do Molecular Clocks Run at All? A Critique of Molecular Systematics," *Biological Theory* 1 (December 2006), 357-371; Maximilian J. Telford, "Fighting Over a Comb," *Nature* 529 (January 21, 2016), 286; Antonis Rokas, "My Oldest Sister Is a Sea Walnut?" *Science* 342 (December 13, 2013), 1327-1329; Benjamin J. Liebeskind et al., "Complex Homology and the Evolution of Nervous Systems," *Trends in Ecology and Evolution* 31 (February 2016), 127-135; Amy Maxmen, "Evolution: You're Drunk: DNA Studies Topple the Ladder of Complexity," *Nautilus* 9 (January 30, 2014), http://nautil.us/issue/9/time/evolution-youre-drunk (accessed October 28, 2020); David A. Legg et al., "Cambrian Bivalved Arthropod Reveals Origin of Arthrodization," *Proceedings of the Royal Society B* 279 (2012), 4699-4704; Mark S. Springer et al., "Endemic African Mammals Shake the Phylogenetic Tree," *Nature* 388 (July 3, 1997), 61-64; William J. Murphy et al., "Molecular Phylogenetics and the Origins of Placental Mammals," *Nature* 409 (February 1, 2001), 614-618; F. Keith Barker et al., "Phylogeny and Diversification of the Largest Avian Radiation," *Proceedings of the National Academy of Sciences USA* 101 (July 27, 2004), 11040-11045; Rodolphe Tabuce et al., "Early Tertiary

Mammals from North Africa Reinforce the Molecular Afrotheria Clade," *Proceedings of the Royal Society B* 274 (2007), 1159-1166.

26. Liliana M. Dávalos et al., "Understanding phylogenetic incongruence: lessons from phyllostomid bats," *Biological Reviews of the Cambridge Philosophical Society* 87 (2012), 991-1024.

27. Jonathan Wells, "Using Intelligent Design Theory to Guide Scientific Research," *Progress in Complexity, Information, and Design* 3.1 (November 2004).

28. William Dembski, "Intelligent Science and Design," *First Things* 86 (October 1998), 21-27.

29. See reviews in Jonathan Wells, *The Myth of Junk DNA* (Seattle, WA: Discovery Institute Press, 2011).

30. The ENCODE Project Consortium, "An integrated encyclopedia of DNA elements in the human genome," *Nature* 489 (September 6, 2012), 57-74.

31. Ed Yong, "ENCODE: the rough guide to the human genome," *Discover Magazine* (September. 5, 2012), https://www.discovermagazine.com/the-sciences/encode-the-rough-guide-to-the-human-genome (accessed October 28, 2020).

32. See for example Richard Sternberg, "On the Roles of Repetitive DNA Elements in the Context of a Unified Genomic-Epigenetic System," *Annals of the NY Academy of Science*, 981 (2002), 154-188; James A. Shapiro and Richard Sternberg, "Why repetitive DNA is essential to genome function," *Biological Reviews of the Cambridge Philosophical Society*, 80 (2005), 227-250; McIntosh, "Information and Entropy—Top-Down or Bottom-Up Development in Living Systems?"; S. Hirotsune et al., "An expressed pseudogene regulates the messenger-RNA stability of its homologous coding gene," *Nature* 423 (2003), 91-96; G. Lev-Maor et al., "The birth of an alternatively spliced exon: 3' splice-site selection in Alu exons," *Science* 300 (May 23, 2003), 1288-1291; W.T. Gibbs, "The Unseen Genome: Gems among the Junk," *Scientific American* (November 2003); M.S. Hakimi et al., "A chromatin remodelling complex that loads cohesin onto human chromosomes," *Nature* 418 (2002), 994-998; T.A. Morrish et al., "DNA repair mediated by endonuclease-independent LINE-1 retrotransposition," *Nature Genetics* 31 (June 2002), 159-165; E.S. Balakirev and F.J. Ayala, "Pseudogenes, Are They 'Junk' or Functional DNA?," *Annual Review of Genetics* 37 (2003), 123-151; E. Pennisi, "Shining a Light on the Genome's 'Dark Matter,'" *Science*, 330 (December 2010), 1614; A.B. Conley et al., "Retroviral promoters in the human genome," *Bioinformatics* 24 (2008), 1563-1567; G.J. Faulkner et al., "The regulated retrotransposon transcriptome of mammalian cells," *Nature Genetics* 41 (April 19, 2009), 563-571.

33. W.T. Gibbs, "The Unseen Genome, Gems Among the Junk," *Scientific American* (November 2003); W. Makalowski, "Not Junk After All," *Science* 300 (May 23, 2003), 1246-1247; Seth W. Cheetham et al., "Overcoming challenges and dogmas to understand the functions of pseudogenes," *Nature Reviews Genetics* 21 (2020), 191-201.

34. Meyer, "The Cambrian Information Explosion."

35. Meyer, *Darwin's Doubt*, 362-363.

36. Geraint Lewis and Luke Barnes, *A Fortunate Universe: Life in a Finely Tuned Cosmos* (Cambridge, UK: Cambridge University Press, 2016), 109.

37. John Leslie, *Universes* (London, UK: Routledge, 1989), 37, 51.

38. Alan Guth, "Inflationary Universe: a possible solution to the horizon and flatness problems," *Physical Review D* 23 (1981), 347-356; Leslie, *Universes*, 3, 29.

39. Paul Davies, *The Accidental Universe* (Cambridge, UK: Cambridge University Press, 1982), 89; Leslie, *Universes*, 29.

40. Leslie, *Universes*, 5, 31.

41. Roger Penrose and Martin Gardner, *The Emperor's New Mind: Concerning Computers, Minds, and the Laws of Physics* (Oxford, UK: Oxford University Press, 2002), 444-445; Leslie, *Universes*, 28.

42. Charles Townes as quoted in Bonnie Azab Powell, "'Explore as much as we can': Nobel Prize winner Charles Townes on evolution, intelligent design, and the meaning of life," *UC Berkeley NewsCenter* (June 17, 2005), https://www.berkeley.edu/news/media/releases/2005/06/17_townes.shtml (accessed October 26, 2020).

43. Kenneth R. Miller, *Kitzmiller v. Dover*, Day 1 PM Testimony (September 26, 2005).

44. *Kitzmiller v. Dover*, 400 F. Supp. 2d 707, 738 (M.D. Pa. 2005).

45. Charles R. Marshall, "When Prior Belief Trumps Scholarship," *Science* 341 (September 20, 2013), 1344.

46. Denis O. Lamoureux, "Intelligent Design Theory: The God-of-the-Gaps Rooted in Concordism," *Perspectives on Science and Christian Faith* 70 (June 2018), 113-132, emphasis in original.

47. Andreas Østergaard Jacobsen, "Uncovering Creation as the Marvelous Symphony," March 10, 2021, https://biologos.org/personal-stories/uncovering-creation-as-the-marvelous-symphony/ (accessed March 24, 2021).

48. Michael Behe, *Darwin's Black Box: The Biochemical Challenge to Evolution* (New York: Free Press, 2006 reprint), 263-264.

49. See Meyer, *Darwin's Doubt*, 169-254; Axe, "Extreme Functional Sensitivity to Conservative Amino Acid Changes on Enzyme Exteriors"; Axe, "Estimating the Prevalence of Protein Sequences Adopting Functional Enzyme Folds"; Steinar Thorvaldsen and Ola Hössjer, "Using statistical methods to model the fine-tuning of molecular machines and systems," *Journal of Theoretical Biology* 501 (2020), 110352.

50. Winston Ewert, "The Dependency Graph of Life," *BIO-Complexity* 2018 (3).

51. Ewert, "The Dependency Graph of Life."

52. Richard Dawkins, *The Blind Watchmaker* (New York: Norton, 1996), 94.

53. Susumu Ohno, "So much 'junk' DNA in our genome," *Evolution of Genetic Systems*, ed. H.H. Smith (New York: Gordon and Breach), 366-370.

54. Richard Dawkins, *The Selfish Gene* (Oxford, UK: Oxford University Press, 1976), 44-45.

55. W.F. Doolittle and Carmen Sapienza, "Selfish genes, the phenotype paradigm and genome evolution," *Nature*, 284 (April 17, 1980), 601-603.

56. Leslie Orgel and Francis Crick, "Selfish DNA: the ultimate parasite," *Nature*, 284 (April 17, 1980), 604-706.

57. Kenneth Miller, "Life's Grand Design," *Technology Review* 97 (February/March 1994), 24-32.

58. Quoted in W.T. Gibbs, "Unseen Genome: Gems Among the Junk," *Scientific American* (November 2003).

59. Makalowski, "Not Junk After All."

60. Cheetham et al., "Overcoming challenges and dogmas to understand the functions of pseudogenes."

61. Forrest Mims, "Rejected Letter to the Editor to *Science*" (December 1, 1994), http://forrestmims.org/publications.html (accessed October 29, 2020).

62. Dembski, "Intelligent Science and Design."

63. Alexander J. Gates et al., "A wealth of discovery built on the Human Genome Project—by the numbers," *Nature* 590 (February 11, 2021), 212-215.

64. For example, see Axe, "Extreme Functional Sensitivity to Conservative Amino Acid Changes on Enzyme Exteriors"; Axe, "Estimating the Prevalence of Protein Sequences Adopting Functional Enzyme Folds"; Michael J. Behe and David W. Snoke, "Simulating Evolution by Gene Duplication of Protein Features That Require Multiple Amino Acid Residues," *Protein Science* 13 (2004), 2651-2664; Douglas D. Axe, "The Case Against a Darwinian Origin of Protein Folds," *Bio-Complexity* 2010 (1); Gauger and Axe, "The Evolutionary Accessibility of New Enzyme Functions: A Case Study from the Biotin Pathway"; Reeves et al., "Enzyme Families–Shared Evolutionary History or Shared Design? A Study of the GABA-Aminotransferase Family."

65. Stephen Jay Gould, "Evolution's erratic pace," *Natural History* 86 (May 1977), 12-16.

66. See Niles Eldredge and Stephen Jay Gould, "Punctuated Equilibria: An Alternative to Phyletic Gradualism," *Models in Paleobiology*, ed. Thomas J.M. Schopf (San Francisco, CA: Freeman, Cooper & Company, 1972), 82-115.

67. For further elaborations of problems with punctuated equilibrium, see Casey Luskin, "Punctuated Equilibrium and Patterns from the Fossil Record," *IDEA Center* (September 18, 2004), http://www.ideacenter.org/contentmgr/showdetails.php/id/1232 (accessed October 27, 2020); Casey Luskin, "Finding Intelligent Design in Nature," *Intelligent Design 101: Leading Experts Explain the Key Issues*, ed. H. Wayne House (Grand Rapids, MI: Kregel, 2008), 67-112; Meyer, *Darwin's Doubt*; Casey Luskin, "Pseudogenes," *Dictionary of Christianity and Science*, eds. Paul Copan, Tremper Longman III, Christopher L. Reese, and Michael G. Strauss (Grand Rapids, MI: Zondervan, 2017), 549-550.

68. George F.R. Ellis, "Does the Multiverse Really Exist?," *Scientific American* (August 2011).

69. Bruce Gordon, "Balloons on a String: A Critique of Multiverse Cosmology," *The Nature of Nature*, eds. Bruce Gordon and William A. Dembski (Wilmington, DE: ISI Books, 2011), 558-585.

70. Kenneth R. Miller, *Kitzmiller v. Dover*, Day 2 AM Testimony (September 27, 2005).

71. Kenneth R. Miller, *Only a Theory: Evolution and the Battle for America's Soul* (New York: Viking Penguin, 2008), 87.

72. Axe, "Extreme Functional Sensitivity to Conservative Amino Acid Changes on Enzyme Exteriors"; Axe, "Estimating the Prevalence of Protein Sequences Adopting Functional Enzyme Folds"; Behe and Snoke, "Simulating Evolution by Gene Duplication of Protein Features That Require Multiple Amino Acid Residues"; Axe, "The Case Against a Darwinian Origin of Protein Folds"; Gauger and Axe, "The Evolutionary Accessibility of New Enzyme Functions: A Case Study from the Biotin Pathway"; Reeves et al., "Enzyme Families-Shared Evolutionary History or Shared Design? A Study of the GABA-Aminotransferase Family"; Thorvaldsen and Hössjer, "Using statistical methods to model the fine-tuning of molecular machines and systems."

73. Guillermo Gonzalez and Donald Brownlee, "The Galactic Habitable Zone: Galactic Chemical Evolution," *Icarus* 152 (2001), 185-200; Guillermo Gonzalez, Donald Brownlee, and Peter D. Ward, "Refuges for Life in a Hostile Universe," *Scientific American* (2001), 62-67; Guillermo Gonzalez and Jay Wesley Richards, *The Privileged Planet: How Our Place in the Cosmos Is Designed for Discovery* (Washington, DC, Regnery, 2004); Guillermo Gonzalez, "Setting the Stage for Habitable Planets," *Life* 4 (2014), 34-65; D. Halsmer, J. Asper, N. Roman, and T. Todd, "The Coherence of an Engineered World," *International Journal of Design & Nature and Ecodynamics* 4 (2009), 47-65.

74. William A. Dembski, *The Design Inference*; William A. Dembski and Robert J. Marks II, "Bernoulli's Principle of Insufficient Reason and Conservation of Information in Computer Search," *Proceedings of the 2009 IEEE International Conference on Systems, Man, and Cybernetics* (October 2009), 2647-2652; William A. Dembski and Robert J. Marks II, "The Search for a Search: Measuring the Information Cost of Higher Level Search," *Journal of Advanced Computational Intelligence and Intelligent Informatics* 14 (2010), 475-486; Øyvind Albert Voie, "Biological function and the genetic code are interdependent," *Chaos, Solitons and Fractals* 28 (2006), 1000-1004; McIntosh, "Information and Entropy —Top-Down or Bottom-Up Development in Living Systems?"

75. Behe and Snoke, "Simulating evolution by gene duplication of protein features that require multiple amino acid residues"; Ann K. Gauger, Stephanie Ebnet, Pamela F. Fahey, and Ralph Seelke, "Reductive Evolution Can Prevent Populations from Taking Simple Adaptive Paths to High Fitness," *BIO-Complexity* 2010 (2).

76. William A. Dembski and Robert J. Marks II, "Conservation of Information in Search: Measuring the Cost of Success," *IEEE Transactions on Systems, Man, and Cybernetics-Part A: Systems and Humans* 39 (September 2009), 1051-1061; Winston Ewert, William A. Dembski, and Robert J. Marks II, "Evolutionary Synthesis of Nand Logic: Dissecting a Digital Organism," *Proceedings of the 2009 IEEE International Conference on Systems, Man, and Cybernetics* (October 2009); Dembski and Marks, "Bernoulli's Principle of Insufficient Reason and Conservation of Information in Computer Search"; Winston Ewert, George Montanez, William Dembski and Robert J. Marks II, "Efficient Per Query Information Extraction from a Hamming Oracle," *42nd South Eastern Symposium on System Theory* (March 2010), 290-297; Douglas D. Axe, Brendan W. Dixon, and Philip Lu, "*Stylus*: A System for Evolutionary Experimentation Based on a Protein/Proteome Model with Non-Arbitrary Functional Constraints," *Plos One* 3 (June 2008), e2246.

77. Jonathan Wells, "Using Intelligent Design Theory to Guide Scientific Research"; William Dembski and Jonathan Wells, *The Design of Life: Discovering Signs of Intelligence in Living Systems* (Dallas, TX: Foundation for Thought and Ethics, 2008).

78. Meyer, "The origin of biological information and the higher taxonomic categories"; Kirk K. Durston, David K.Y. Chiu, David L. Abel, Jack T. Trevors, "Measuring the functional sequence complexity of proteins," *Theoretical Biology and Medical Modelling* 4 (2007), 47; David K.Y. Chiu and Thomas W.H. Lui, "Integrated Use of Multiple Interdependent Patterns for Biomolecular Sequence Analysis," *International Journal of Fuzzy Systems* 4 (September 2002), 766-775.

79. Minnich and Meyer. "Genetic Analysis of Coordinate Flagellar and Type III Regulatory Circuits in Pathogenic Bacteria"; McIntosh, "Information and Entropy—Top-Down or Bottom-Up Development in Living Systems?"

80. Jonathan Wells, "Do Centrioles Generate a Polar Ejection Force?," *Rivista di Biologia / Biology Forum*, 98 (2005), 71-96; Scott A. Minnich and Stephen C. Meyer, "Genetic analysis of coordinate flagellar and type III regulatory circuits in pathogenic bacteria," *Proceedings of the Second International Conference on Design & Nature Rhodes Greece* (2004); Behe, *Darwin's Black Box*; Lönnig, "Dynamic genomes, morphological stasis, and the origin of irreducible complexity."

81. Lönnig, "Dynamic genomes, morphological stasis, and the origin of irreducible complexity"; Nelson and Jonathan Wells, "Homology in Biology"; Ewert, "The Dependency Graph of Life"; John A. Davison, "A Prescribed Evolutionary Hypothesis," *Rivista di Biologia/Biology Forum* 98 (2005), 155-166; Ewert, "The Dependency Graph of Life."

82. Sherman, "Universal Genome in the Origin of Metazoa: Thoughts About Evolution"; Albert D.G. de Roos, "Origins of introns based on the definition of exon modules and their conserved interfaces," *Bioinformatics* 21 (2005), 2-9; Albert D.G. de Roos, "Conserved intron positions in ancient protein modules," *Biology Direct* 2 (2007), 7; Albert D.G. de Roos, "The Origin of the Eukaryotic Cell Based on Conservation of Existing Interfaces," *Artificial Life* 12 (2006), 513-523.

83. Meyer et al., "The Cambrian Explosion: Biology's Big Bang"; Meyer, "The Cambrian Information Explosion"; Meyer, "The origin of biological information and the higher taxonomic categories"; Lönnig, "Dynamic genomes, morphological stasis, and the origin of irreducible complexity."

84. Richard Sternberg, "DNA Codes and Information: Formal Structures and Relational Causes," *Acta Biotheoretica* 56 (September 2008), 205-232; Voie, "Biological function and the genetic code are interdependent"; David L. Abel and Jack T. Trevors, "Self-organization vs. self-ordering events in life-origin models," *Physics of Life Reviews* 3 (2006), 211-228.

85. Richard Sternberg, "On the Roles of Repetitive DNA Elements in the Context of a Unified Genomic– Epigenetic System"; Jonathan Wells, "Using Intelligent Design Theory to Guide Scientific Research"; Josiah D. Seaman and John C. Sanford, "Skittle: A 2-Dimensional Genome Visualization Tool," *BMC Informatics* 10 (2009), 451.

Chapter 18—Have Science and Philosophy Refuted Free Will?

1. See Alfred R. Mele, *Free: Why Science Hasn't Disproved Free Will* (Oxford, UK: Oxford University Press, 2014).

2. Eleonore Stump, *Aquinas* (New York: Routledge, 2005)

3. See Mele, *Free*; Stump, *Aquinas*; Edward Feser, *The Philosophy of Mind: A Short Introduction* (Oxford, UK: Oneworld Publications, 2006).

4. For a discussion see Brian Greene, *The Fabric of the Cosmos: Space, Time and the Texture of Reality* (New York: Random House, 2004).

5. For a discussion see Greene, *The Fabric of the Cosmos*.

6. Sam Harris, *Free Will* (New York: Free Press, 2012).

7. Wilder Penfield, *Mystery of the Mind: A Critical Study of Consciousness and the Human Brain* (Princeton, NJ: Princeton Legacy Library, 1975).

8. Benjamin Libet, *Mind Time: The Temporal Factor in Consciousness* (Cambridge, MA: Harvard University Press, 2004), 123-156.

9. C.S. Soon, M. Brass, H.J. Heinze, and J.D. Haynes, "Unconscious determinants of free decisions in the human brain," *Nature Neuroscience* 11 (2008), 543-545.

10. I. Fried, R. Mukamel, and G. Kreiman, "Internally generated preactivation of single neurons in human medial frontal cortex predicts volition," *Neuron* 69 (2011), 548-562.

11. Penfield, *Mystery of the Mind*.

12. See M.R. Bennett and P.M.S. Hacker, *History of Cognitive Neuroscience* (West Sussex, UK: Wiley-Blackwell, 2013), 11-14.

13. Adrian M. Owen, Martin R. Coleman, Melanie Boly, Matthew H. Davis, Steven Laureys, and John D. Pickard, "Detecting awareness in the vegetative state," *Science* 313 (September 8, 2006), 1402.

14. See Pim van Lommel, *Consciousness Beyond Life: The Science of Near-Death Experience* (New York: HarperOne, 2007).

15. Daniel W. Wegner, *The Illusion of Conscious Will* (Cambridge, MA: MIT Press, 2003).

16. Hannah Arendt, *The Origins of Totalitarianism* (New York: Harcourt, 1973).

Chapter 19—Can Materialism Explain Human Consciousness?

1. Edward Feser, *Aquinas: A Beginner's Guide* (Oxford, UK: Oneworld Publications Oxford, 2009), chapter 5.

2. Edward Feser, *The Philosophy of Mind: A Short Introduction* (Oxford, UK: Oneworld Publications, 2006).

3. Noam Chomsky, *What Kind of Creatures Are We?* (New York: Columbia University Press, 2011).

4. John Searle, *Mind: A Brief Introduction* (Oxford, UK: Oxford University Press, 2004).

5. Daniel C. Dennett, *The Intentional Stance* (Cambridge, MA: MIT Press, 1980).

6. Wilder Penfield, *Mystery of the Mind: A Critical Study of Consciousness and the Human Brain* (Princeton, NJ: Princeton Legacy Library, 1975).

7. See M.R. Bennett and P.M.S. Hacker, *History of Cognitive Neuroscience* (West Sussex, UK: Wiley-Blackwell, 2013), 11-14.

8. Adrian M. Owen, Martin R. Coleman, Melanie Boly, Matthew H. Davis, Steven Laureys, and John D. Pickard, "Detecting awareness in the vegetative state," *Science* 313 (September 8, 2006), 1402.

9. See Pim van Lommel, *Consciousness Beyond Life: The Science of Near-Death Experience* (New York: HarperOne, 2007).

10. Barbara Bradley Hagerty, "Decoding the Mystery of Near-Death Experiences," *NPR* (May 22, 2009), https://www.npr.org/templates/story/story.php?storyId=104397005 (accessed November 13, 2020); Charlotte Begg, "In 1991, Pam Reynolds went into brain surgery. She says she could hear and see the whole thing," *MamaMia* (September 5, 2020), https://www.mamamia.com.au/pam-reynolds/ (accessed November 13, 2020).

11. Elizabeth Kubler-Ross, *On Life After Death* (Berkeley/Toronto: Celestial Arts, 1991).

12. For details see Harriet Dempsey-Jones, "Neuroscientists put the dubious theory of 'phrenology' through rigorous testing for the first time," *The Conversation* (January 22, 2018), https://theconversation.com/neuroscientists-put-the-dubious-theory-of-phrenology-through-rigorous-testing-for-the-first-time-88291 (accessed November 13, 2020); O. Parker Jones, F. Alfaro-Almagro, S. Jbabdi, "An empirical, 21st century evaluation of phrenology," *Cortex* 106 (September 2018), 26-35.

13. Benjamin Libet, *Mind Time: The Temporal Factor in Consciousness* (Cambridge, MA: Harvard University Press, 2004), 123-156.

14. Chomsky, *What Kind of Creatures Are We?*

15. Quoted in Michael Studdert-Kennedy, "The particulate origins of language generativity: from syllable to gesture," *Approaches to the Evolution of Language*, eds. J. Hurford, M. Studdert-Kennedy, and C. Knight (Cambridge, UK: Cambridge University Press, 1998), 202-221.

16. See, for example, Noam Chomsky quoted in "Things No Amount of Learning Can Teach: Noam Chomsky interviewed by John Gliedman," *Omni* 6 (November 1983), 11, https://chomsky.info/198311__/ (accessed November 13, 2020).

Chapter 20—Does the Big Bang Support Cosmic Design?

1. Simon Singh, *Big Bang: The Origin of the Universe* (New York: Harper Perennial, 2005), 79. Singh provides an excellent history of the origin and development of the big bang theory.

2. For a more detailed discussion, see Stephen C. Meyer, *Return of the God Hypothesis: Three Scientific Discoveries Revealing the Mind Behind the Universe* (San Francisco, CA: HarperOne, 2021).

3. Christine Sutton, Review of *Einstein's Universe* (BBC2, March 14, 1979), *New Scientist* 81/1145 (March 8, 1979), 784.

4. Jean-Pierre Luminet, "Lemaître's Big Bang," lecture given at Frontiers of Fundamental Physics 14, Aix Marseille University, Marseille, France (July 15-18, 2014), 10, https://arxiv.org/ftp/arxiv/papers/1503/1503.08304.pdf (accessed November 14, 2020).

5. Quotes are from Stephen C. Meyer's transcription of a private film of Sandage's remarks at "Christianity Challenges the University: An International Conference of Theists and Atheists," Dallas, Texas, February 7-10, 1985.

6. Robert Jastrow, *God and the Astronomers* (New York: Norton, 1978), 116.

7. Alexander Vilenkin, *Many Worlds in One: The Search for Other Universes* (New York: Hill and Wang, 2006), 176.

8. Stephen Hawking, *A Brief History of Time: From the Big Bang to Black Holes* (London, UK: Bantam, 1988), 140-141.

9. I am speaking mathematically because time is imaginary and not real, and I am speaking literally because the equations do not correspond to actual reality.

10. Hawking, *A Brief History of Time*, 136.

11. Stephen Hawking and Leonard Mlodinow, *The Grand Design* (New York: Bantam Books, 2010), 180.

12. Lawrence M. Krauss, *A Universe from Nothing: Why There Is Something Rather Than Nothing* (New York: Atria Books, 2012), 169.

13. J.P. Moreland, *Scaling the Secular City: A Defense of Christianity* (Grand Rapids, MI: Baker Academic, 1987), 42.

14. Hawking, *A Brief History of Time*, 174.

15. Vilenkin, *Many Worlds in One*, 205.

16. For an extended argument, see Meyer, *Return of the God Hypothesis*, chapters 12-14.

Chapter 21—How Does Fine-Tuning Make the Case for Nature's Designer?

1. John D. Barrow and Frank J. Tipler, *The Anthropic Cosmological Principle* (New York: Oxford University Press, 1986), 400.

2. James S. Trefil, *The Moment of Creation* (New York: Collier Books, 1983), 127-134.

3. John D. Barrow, *The Constants of Nature* (New York: Pantheon Books, 2002), 165-167.

4. Frank Wilczek, "Hard-Core Revelations," *Nature* 445 (January 11, 2007), 156-157.

5. Martin John Rees and Willian Hunter McCrea, "Large Numbers and Ratios in Astrophysics and Cosmology (and Discussion)," *Philosophical Transactions of the Royal Society London A* 310 (December 20, 1983), 311-322.

6. Fred Hoyle, *Galaxies, Nuclei, and Quasars* (New York: Harper & Row, 1965), 147-150; Fred Hoyle, "The Universe: Past and Present Reflection," *Annual Review of Astronomy and Astrophysics* 20 (1982), 1-36.

7. Fred Hoyle, *The Nature of the Universe*, 2d ed. (Oxford, UK: Basil Blackwell, 1952), 109; Fred Hoyle, *Astronomy and Cosmology: A Modern Course* (San Francisco, CA: W.H. Freeman, 1975), 522, 684-685; Hoyle, "The Universe: Past and Present Reflection," 3.

8. Hoyle, *The Nature of the Universe*, 111.

9. Hoyle, "The Universe: Past and Present Reflection," 16.

10. H. Oberhummer, A. Csótó, and H. Schlattl, "Stellar Production Rates of Carbon and Its Abundance in the Universe," *Science* 289 (July 7, 2000), 88-90.

11. Oberhummer, Csótó, and Schlattl, "Stellar Production Rates," 90.

12. Hugh Ross, "RTB Design Compendium (2009)," *Today's New Reason to Believe* (November 16, 2010), https://reasons.org/explore/publications/tnrtb/read/tnrtb/2010/11/16/rtb-design-compendium-2009 (accessed November 6, 2020).

13. Lawrence M. Krauss, "The End of the Age Problem, and the Case for a Cosmological Constant Revisited," *Astrophysical Journal* 501 (July 10, 1998), 461-466.

14. For the reasons why, the background technical details, and the observational verification of the cosmic inflation event, see Hugh Ross, *The Creator and the Cosmos: How the Latest Scientific Discoveries Reveal God*, 4th ed. (Covina, CA: RTB Press, 2018), 68-69.

15. Invented and designed by physicists at the Californian Institute of Technology and the Massachusetts Institute of Technology, LIGO is able to identify and measure a disturbance equivalent to one-tenth the diameter of a proton over a four-kilometer baseline. LIGO was the first instrument to detect gravity waves. LIGO ranks as the largest project undertaken to date by the National Science Foundation.

16. Stephen Hawking, *A Brief History of Time: From the Big Bang to Black Holes* (New York: Bantam Books, 1988), 126.

17. Victor J. Stenger, *God: The Failed Hypothesis* (Amherst, NY: Prometheus, 2007), 156-157.

18. In my book *Improbable Planet: How Earth Became Humanity's Home* (Grand Rapids, MI: Baker, 2016), I describe and document the basis for my claim that a carefully designed 3.8-billion-year history of plate tectonics and increasingly complex life are essential to the buildup of biodeposits on which global human civilization depends.

19. I explain the uniqueness of our location for astronomical observations in *Why the Universe Is the Way It Is* (Grand Rapids, MI: Baker, 2008), 79-93.

20. M. Mittag et al., "Chromospheric Activity and Evolutionary Age of the Sun and Four Solar Twins," *Astronomy & Astrophysics* 591 (June 2016), A89; Jorge Meléndez et al., "HIP 114328: A New Refractory-Poor and Li-Poor Solar Twin," *Astronomy & Astrophysics* 567 (July 2014), L3; Marília Carlos, Poul E. Nissen, and Jorge Meléndez, "Correlation Between Lithium Abundances and Ages of Solar Twin Stars," *Astronomy & Astrophysics* 587 (March 2016), A100; G. F. Porto de Mello et al., "A Photometric and Spectroscopic Survey of Solar Twin Stars Within 50 Parsecs of the Sun. I. Atmospheric Parameters and Color Similarity to the Sun," *Astronomy & Astrophysics* 563 (March 2014), A52; Jorge Meléndez et al., "18 Sco: A Solar Twin Rich in Refractory and Neutron-Capture Elements. Implications for Chemical Tagging," *Astrophysical Journal* 791 (July 21, 2014), 14; Hugh Ross, "Our Sun Is Still the One and Only," *Today's New Reason to Believe* blog (April 17, 2017), https://reasons.org/explore/blogs/todays-new-reason-to-believe/read/todays-new-reason-to-believe/2017/04/17/our-sun-is-still-the-one-and-only (accessed November 6, 2020).

21. T.A. Michtchenko and S. Ferraz-Mello, "Resonant Structure of the Outer Solar System in the Neighborhood of the Planets," *Astronomical Journal* 122 (July 1, 2001), 474-481; Kimmo Innanen, Seppo Mikkola, and Paul

Wiegert, "The Earth-Moon System and the Dynamical Stability of the Inner Solar System," *Astronomical Journal* 116 (October 1, 1998), 2055-2057; Ravit Helled and Peter Bodenheimer, "The Formation of Uranus and Neptune: Challenges and Implications for Intermediate-Mass Exoplanets," *Astrophysical Journal* 789 (June 17, 2014), 69; Konstantin Batygin, Michael E. Brown, and Hayden Betts, "Instability-Driven Dynamical Evolution Model of a Primordially Five-Planet Outer Solar System," *Astrophysical Journal Letters* 744 (January 1, 2012), L3; Ross, *Improbable Planet*, 43-93.

22. Ross, "RTB Design Compendium (2009)."

Chapter 22—Do We Live on a Privileged Planet?

1. Carl Sagan, *Pale Blue Dot: A Vision of the Human Future in Space* (New York: Ballantine Books, 1997), 12-13.

2. Richard Dawkins, *The Blind Watchmaker* (New York: Norton, 1986), 1.

3. Michael N. Keas, *Unbelievable: 7 Myths About the History and Future of Science and Religion* (Wilmington, DE: ISI Books, 2019); Alvin Plantinga, *Where the Conflict Really Lies: Science, Religion, and Naturalism* (New York: Oxford University Press, 2011).

4. Guillermo Gonzalez and Jay W. Richards, *The Privileged Planet: How Our Place in the Cosmos Is Designed for Discovery* (Washington, DC: Regnery, 2004).

5. Raymond L. Lee Jr. and Alistair B. Fraser, *The Rainbow Bridge: Rainbows in Art, Myth, and Science* (University Park, PA: Penn State University Press, 2001).

6. Many exoplanets are being discovered around red dwarfs, but they are in tight orbits. A famous example is the planetary system around Trappist-1, which has seven planets. The one farthest from the host star has an orbital period less than 20 days. See "TRAPPIST-1" at https://en.wikipedia.org/wiki/TRAPPIST-1 (accessed June 14, 2020).

7. Ernan McMullin, "Varieties of Methodological Naturalism," *The Nature of Nature: Examining the Role of Naturalism in Science*, eds. Bruce L. Gordon and William A. Dembski (Wilmington, DE: ISI Books, 2011).

8. Geraint F. Lewis and Luke A. Barnes, *A Fortunate Universe: Life in a Finely Tuned Universe* (Cambridge, UK: Cambridge University Press, 2016).

9. Lewis and Barnes, *A Fortunate Universe*; Bruce Gordon, "Balloons on a String: A Critique of Multiverse Cosmology," *The Nature of Nature: Examining the Role of Naturalism in Science*, eds. Bruce L. Gordon and William A. Dembski (Wilmington, DE: ISI Books, 2011).

10. Guillermo Gonzalez, "The Solar System: Favored for Space Travel," *BIO-Complexity* 2020 (1).

11. Eugene Wigner, "The Unreasonable Effectiveness of Mathematics in the Natural Sciences," reprinted in *The World Treasury of Physics, Astronomy, and Mathematics*, ed. Timothy Ferris (Boston, MA: Little, Brown, 1991), 527.

12. Melissa Cain Travis, *Science and the Mind of the Maker* (Eugene, OR: Harvest House, 2018), 208.

13. Quoted in John Hearnshaw, "Auguste Comte's Blunder: An Account of the First Century of Stellar Spectroscopy and How It Took One Hundred Years to Prove That Comte Was Wrong," *Journal of Astronomical History and Heritage*, 13(2) (2010), 90-104. Actually, some aspects of his claims were proven wrong within only a few years.

14. James Hannam, *The Genesis of Science: How the Christian Middle Ages Launched the Scientific Revolution* (Washington, DC: Regnery, 2011), 349.

Chapter 23—How Do Solar Eclipses Point to Intelligent Design?

1. Guillermo Gonzalez, "Wonderful Eclipses," *Astronomy & Geophysics* 40(3) (1999), 18-20. I presented an expanded discussion of solar eclipses in chapter 1 of Guillermo Gonzalez and Jay W. Richards, *The Privileged Planet: How Our Place in the Cosmos Is Designed for Discovery* (Washington, DC: Regnery, 2004).

2. Guillermo Gonzalez, "Mutual Eclipses in the Solar System," *Astronomy & Geophysics* 50(2) (2009), 17-19.

3. Gonzalez, "Mutual Eclipses." See Figure 4.

4. John Gribbin, *Alone in the Universe: Why Our Planet Is Unique* (New York: John Wiley & Sons, 2011).

5. Gonzalez, "Wonderful Eclipses."

6. Caleb A. Scharf, "The Solar Eclipse Coincidence," *Scientific American* blog (May 18, 2012), https://blogs.scientificamerican.com/life-unbounded/the-solar-eclipse-coincidence/ (accessed June 13, 2020). Note: His estimate for the size of the window that allows us to observe total solar eclipses is a little narrow; see previous endnote.

7. I say "approximately" because the CHZ has a width. In other words, there is some small range of distances from its host star a planet can have and still maintain liquid water on its surface.

8. Carl Wunsch, "Moon, tides, and climate," *Nature* 405 (June 15, 2000), 743-744.

9. Melaine Saillenfest, Jacques Laskar, and Gwenaël Boué, "Secular Spin-axis Dynamics of Exoplanets," *Astronomy & Astrophysics* 623 (2019), A4.

10. If the Earth were moved to a star less than 23 percent the mass of our sun, then the tidal torque produced by the star alone would be equal to the tidal torques produced by our sun and moon (see the appendix in reference to endnote 12). One could argue that tides and stabilization of the planet's rotation axis would therefore not require a large moon. However, planets around stars in this mass range are less likely to be habitable to complex life for a variety of reasons. See David Waltham, "Star Masses and Star-Planet Distances for Earth-like Habitability," *Astrobiology* 17(1) (2017), 61-77.

11. Robin Canup and Erik Asphaug, "Origin of the Moon in a Giant Impact Near the End of the Earth's Formation," *Nature* (2001), 412: 708-712. For a summary of the various versions of the giant impact theory, see "Giant-impact hypothesis," https://en.wikipedia.org/wiki/Giant-impact_hypothesis (accessed June 13, 2020).

12. David Waltham, "Is Earth Special?," *Earth-Science Reviews* 192 (2019), 445-470.

13. Waltham, "Is Earth Special?" 450-451.

14. O. Néron De Surgy and J. Laskar, "On the long-term Evolution of the Spin of the Earth," *Astronomy & Astrophysics* 318 (1997), 975-989.

15. Fernando de Sousa Mello and Amâncio César Santos Friaça, "The End of Life on Earth Is Not the End of the World: Converging to an Estimate of Lifespan of Biosphere?," *International Journal of Astrobiology* 19 (2020), 25-42. The carbon dioxide level in the atmosphere is expected to continue to decline as the sun brightens. It will eventually decline below the level that plants can continue to function, depending on the species. After about 200 million years, the level of carbon dioxide will be too low for C3 plants to function. About 85 percent of plant species are C3. This includes trees. Both the biological productivity and diversity will steadily decline over the next few hundred million years. The authors of this study assume that Earth is not strongly perturbed in their modelling. However, it is very likely Earth will be hit by asteroids and comets and have major volcanic eruptions in this time period. These perturbations could be enough to push Earth's biosphere over the edge in its anemic state. For these reasons, I estimate the future lifetime of the biosphere to be somewhat less than 500 million years.

16. Daniel Kennefick, *No Shadow of a Doubt: The 1919 Eclipse That Confirmed Einstein's Theory of Relativity* (Princeton, NJ: Princeton University Press, 2019).

Chapter 24—How Does the Intelligibility of Nature Point to Design?

1. Albert Einstein, "Physics and Reality," *Ideas and Opinions* (New York: Crown, 1954), 292. Originally published in *The Journal of the Franklin Institute* 221 (3) (1936).

2. Eugene Wigner, "The Unreasonable Effectiveness of Mathematics in the Natural Sciences," *Communications on Pure and Applied Mathematics* 13 (1) (1960), 1-14.

3. Bruce Gordon, "The Rise of Naturalism and Its Problematic Role in Science and Culture," *The Nature of Nature: Examining the Role of Naturalism in Science*, eds. Bruce L. Gordon and William A. Dembski (Wilmington, DE: ISI Books, 2011), 3-61; Edward Grant, *A History of Natural Philosophy: from the Ancient World to the Nineteenth Century* (New York: Cambridge University Press, 2007); James Hannam, *The Genesis of Science: How the Christian Middle Ages Launched the Scientific Revolution* (Washington, DC: Regnery, 2011); David C. Lindberg, *The Beginnings of Western Science: The European Scientific Tradition in Philosophical, Religious, and Institutional Context, Prehistory to A.D. 1450,* 2d ed. (Chicago, IL: University of Chicago Press, 2007); Nancy Pearcey and Charles Thaxton, *The Soul of Science: Christian Faith and Natural Philosophy* (Wheaton, IL: Crossway, 1994); Alfred North Whitehead, *Science and the Modern World* (New York: Macmillan, 1925).

4. William B. Ashworth Jr., "Christianity and the Mechanistic Universe," *When Science and Christianity Meet*, eds. David C. Lindberg and Ronald L. Numbers (Chicago, IL: University of Chicago Press, 2003), 61-84. See also Lindberg, *The Beginnings of Science.*

5. See Gordon, "The Rise of Naturalism and Its Problematic Role in Science and Culture" for an extensive discussion of these points.

6. William Lane Craig and J.P. Moreland, *Naturalism: A Critical Analysis* (New York: Routledge, 2000); Michael Rea, *World Without Design: The Ontological Consequences of Naturalism* (New York: Oxford University Press, 2002); Alvin Plantinga, "Against Materialism," *Faith and Philosophy* 23 (1) (2006), 3-32; Stewart Goetz and

Charles Taliaferro, *Naturalism* (Grand Rapids, MI: Eerdmans, 2008); Alvin Plantinga, *Where the Conflict Really Lies: Science, Religion, and Naturalism* (New York: Oxford University Press, 2011).

7. See Richard Gale and Alexander Pruss, "A New Cosmological Argument," *Religious Studies* 35 (1999), 461-476; Alexander Pruss, *The Principle of Sufficient Reason: A Reassessment* (New York: Cambridge University Press, 2006), 234-235.

8. Stephen W. Hawking, *A Brief History of Time: From the Big Bang to Black Holes* (New York: Bantam Books, (1988), 174.

9. See my other chapter in this volume, "Does the Multiverse Refute Cosmic Design?" (chapter 42).

10. See Alexander Pruss and Joshua Rasmussen, *Necessary Existence* (Oxford, UK: Oxford University Press, 2018) for an extended examination of these themes. The logical consequences of intuitions (positive or negative) regarding necessary being are interactively explorable at www.necessarybeing.com.

11. For consideration of teleology in nature, see the other chapters in Part II of this book. In respect to cosmological fine-tuning, see especially Hugh Ross's chapter in this volume, and, more rigorously and comprehensively, the work of Robin Collins and Luke Barnes, notably: Robin Collins, "Evidence for Fine-Tuning," *God and Design: The Teleological Argument and Modern Science*, ed. Neil A. Manson (New York: Routledge, 2003), 178-199; Robin Collins, "The Teleological Argument: An Exploration of the Fine-Tuning of the Universe," *The Blackwell Companion to Natural Theology* eds. William L. Craig and J.P. Moreland (Oxford, UK: Blackwell, 2009), 202-281; Robin Collins, "The Fine-Tuning Evidence Is Convincing," *Debating Christian Theism*, eds. J.P. Moreland, Chad Meister, and Khaldoun A. Sweis (New York: Oxford University Press, 2013), 35-46; Luke A. Barnes, "The Fine-Tuning of the Universe for Intelligent Life," *Publications of the Astronomical Society of Australia* 29 (2012), 529-564; Luke A. Barnes, "A Reasonable Little Question: A Formulation of the Fine-Tuning Argument," *Ergo* 6 (42) (2020), 1220-1257.

12. Examples of these necessitarian approaches are: (1) laws as broad logical necessities: A. Bird, "The Dispositional Conception of Law," *Foundations of Science* 10 (4) (2005), 353-370; (2) laws as relationships among universals: D. Armstrong, *What Is a Law of Nature?* (Cambridge, UK: Cambridge University Press, 1983); (3) laws as causal powers: R. Harré and E. Madden, *Causal Powers: A Theory of Natural Necessity* (Oxford, UK: Blackwell, 1975), and J. Bigelow and R. Pargetter, *Science and Necessity* (Cambridge, UK: Cambridge University Press, 1990).

13. See David Lewis, "New Work for a Theory of Universals," *Australasian Journal of Philosophy* 61 (1983), 343-377; David Lewis, "Humean Supervenience Debugged," *Mind* 103 (1994), 473-490.

14. I've addressed this issue in a number of places. For example, see Bruce L. Gordon, "The Necessity of Sufficiency: The Argument from the Incompleteness of Nature," *Two Dozen (or So) Arguments for God: The Plantinga Project*, eds. Jerry L. Walls and Trent Dougherty (New York: Oxford University Press, 2018), 417-445; Bruce L. Gordon, "The Incompatibility of Physicalism with Physics," *Christian Physicalism: Philosophical Theological Criticisms*, eds. Joshua Farris and Keith Loftin (New York: Lexington Books, 2018), 371-402.

15. M. Scully and K. Drühl, "Quantum eraser: A proposed photon correlation experiment concerning observation and 'delayed choice' in quantum mechanics," *Physical Review A* 25 (1982), 2208-2213; Y.-H. Kim, R. Yu, S.P. Kulik, Y.H. Shih, and M.O. Scully, "Delayed Choice Quantum Eraser," *Physical Review Letters* 84 (1) (2000), 1-5.

16. X.-S Ma *et al.*, "Quantum erasure with causally disconnected choice," *Proceedings of the National Academy of Sciences (USA)* 110 (4) (2013), 1221-1226.

17. This conclusion also follows from the nonlocality demonstrated by violations of Bell inequalities. See Bruce L. Gordon, "A Quantum-Theoretic Argument Against Naturalism," *The Nature of Nature: Examining the Role of Naturalism in Science*, eds. Bruce L. Gordon and William A. Dembski (Wilmington, DE: ISI Books, 2011), 179-214, especially 183-189.

18. See David Malament, "In Defense of Dogma: Why there cannot be a relativistic quantum mechanics of (localizable) particles," *Perspectives on Quantum Reality: Non-Relativistic, Relativistic, and Field-Theoretic,* ed. R. Clifton (Dordrecht, Netherlands: Kluwer Academic Publishers, 1996), 1-9.

19. Hans Halvorson and Robert Clifton, "No place for particles in relativistic quantum theories?," *Philosophy of Science* 69 (2002), 1-28.

20. M. Fuwa et al., "Experimental Proof of Nonlocal Wavefunction Collapse for a Single Particle Using Homodyne Measurement," *Nature Communications* 6 (2015), 6665.

21. Y. Aharonov et al., "Quantum Cheshire Cats," *New Journal of Physics* 15 (2013), 113018.

22. T. Denkmayr et al., "Observation of a quantum Cheshire Cat in a matter-wave interferometer experiment," *Nature Communications* 5 (2014), 4492.

23. R. Lapkiewicz et al., "Experimental non-classicality of an indivisible quantum system," *Nature* 474 (2011), 490-493.

24. S. Eibenberger et al., "Matter-wave interference of particles selected from a molecular library with masses exceeding 10,000 amu," *Physical Chemistry and Chemical Physics* 15 (2013), 14696-14700.

25. J. Friedman et al., "Quantum superposition of distinct macroscopic states," *Letters to Nature* 406 (2000), 43-46.

26. This is the so-called "many-worlds" interpretation of quantum physics. It had its genesis in the work of Hugh Everett but has since been elaborated to grapple with its technical problems. David Wallace, *The Emergent Multiverse: Quantum Theory According to the Everett Interpretation* (Oxford, UK: Oxford University Press, 2012), ably represents the state of the art. An excellent compendium of critical discussions is Simon Saunders, Jonathan Barrett, Adrian Kent, and David Wallace eds., *Many Worlds? Everett, Quantum Theory, & Reality* (New York: Oxford University Press, 2010). For a discussion and evaluation of this and other multiverse hypotheses, see my chapter, "Does the Multiverse Refute Cosmic Design?" in this volume (chapter 42).

27. If time permitted, we could also examine how ongoing research in quantum gravity furthers the case for theistic quantum idealism. In particular, just as we have discussed how matter is rendered insubstantial and merely phenomenological in regular quantum physics, so too the very fabric of spacetime is rendered merely phenomenological in the context of quantum gravity. More specifically, the general quantum-gravitational picture forming sees spacetime as emerging from immaterial information that cannot, in principle, reside within space-time itself. See B. Swingle and M. van Raamsdonk, "Universality of Gravity from Entanglement" (2014), https://arxiv.org/pdf/1405.2933.pdf (accessed July 8, 2020); C. Cao, S. Carroll, and S. Michalakis, "Space from Hilbert Space: Recovering Geometry from Bulk Entanglement," *Physical Review D* 95 (2017), 024031.

28. For more extensive and more technical treatments of these and related issues, see Bruce L. Gordon, "Idealism and Science: The Quantum-Theoretic and Neuroscientific Foundations of Reality," *The Routledge Handbook of Idealism and Immaterialism*, eds. Joshua R. Farris and Benedikt-Paul Göcke (London, UK: Routledge, 2021); Bruce L. Gordon, "Consciousness and Quantum Information," *Minds, Brains, and Consciousness*, eds. Brian Krouse and Cristi Cooper (Seattle, WA: Discovery Institute Press, forthcoming).

29. Dallas Willard, *Knowing Christ Today: Why We Can Trust Spiritual Knowledge* (New York: HarperOne, 2009), 15, emphasis in original.

30. See C.S. Lewis, *Miracles: A Preliminary Study* (New York: Macmillan; 1947); Ronald Nash, "Miracles and Conceptual Systems," *In Defense of Miracles: A Comprehensive Case for God's Action in History*, in eds. R. Douglas Geivett and Gary Habermas (Downers Grove, IL: IVP Academic, 1997), 115-131; Alvin Plantinga, *Warrant and Proper Function* (New York: Oxford University Press, 1993); Alvin Plantinga, "Against Materialism," *Faith and Philosophy* 23 (1) (2006), 3-32; Alvin Plantinga, "Evolution Versus Naturalism," *The Nature of Nature: Examining the Role of Naturalism in Science*, eds. Bruce L. Gordon and William A. Dembski (Wilmington, DE: ISI Books, 2011), 137-151; Alvin Plantinga, *Where the Conflict Really Lies: Science, Religion, and Naturalism* (New York: Oxford University Press, 2011); Richard Taylor, *Metaphysics* (Englewood Cliffs, NJ: Prentice-Hall, 1963).

31. The *locus classicus* for the modern version of the argument is chapter 12 of Alvin Plantinga's *Warrant and Proper Function* (New York: Oxford University Press, 1993). Critical analyses, defenses, and then responses from Plantinga himself can be found in James Beilby, ed., *Naturalism Defeated? Essays on Plantinga's Evolutionary Argument Against Naturalism* (Ithaca, NY: Cornell University Press, 2002).

32. See J. Mark, B. Marion, and D. Hoffman, "Natural Selection and Veridical Perceptions," *Journal of Theoretical Biology* 266 (2010), 504-515. Note that a "fitness function" is a mathematical expression that characterizes, in terms of the performance of something relative to its alternatives, how close that thing is to achieving a certain goal, for example, survival. What Marion and Hoffman showed was that organisms functioning on the basis of accurate (veridical) representations of an objective environment can be outcompeted and driven to extinction by organisms with arbitrary functions tuned to environmental utility (usefulness) rather than veridicality. This means that useful perceptions readily diverge from and replace true perceptions as organisms struggle to survive.

33. Abduction infers the truth of the *best* explanation—i.e., a range of possible explanations for something is considered and the one offering the most likely explanation is accepted. See "Abduction," *Stanford Encyclopedai of Philosophy*, https://plato.stanford.edu/entries/abduction/ (accessed March 31, 2021).

34. Seth Lloyd, "Computational Capacity of the Universe," *Physical Review Letters* 88 (23) (2002), 237901.

35. William Dembski, "Specification: The Pattern that Signifies Intelligence," *Philosophia Christi* 7 (2005), 299-343.

36. Winston Ewert, William Dembski, and Robert Marks, "On the Improbability of Algorithmic Specified Complexity," *Proceedings of the 2013 45th Southeastern Symposium on System Theory (SSST)*, IEEE (2013), 68-70.

37. Robert M. Hazen, Patrick L. Griffin, James M. Carothers, and Jack W. Szostak, "Functional Information and the Emergence of Biocomplexity," *Proceedings of the National Academy of Sciences (USA)*, 104 (2007), 8574-8581.

38. Michael Behe, *Darwin's Black Box: The Biochemical Challenge to Evolution* (New York: Free Press, 1996).

39. George Montañez, "A Unified Model of Complex Specified Information," *BIO-Complexity* 2018 (4).

40. For a brief description of statistical hypothesis testing, see "Hypothesis Testing," *Statistics Solutions*, https://www.statisticssolutions.com/hypothesis-testing/ (accessed March 31, 2021).

41. A likelihood ratio test compares the goodness of fit of two competing statistical models based on the ratio of their likelihoods (individual goodness of fit to the data). In the present context, one of the hypotheses is the null hypothesis that nothing statistically significant is happening and the other is the alternative hypothesis that the observed structure is most likely the product of intelligent design. See Montañez, "A Unified Model of Complex Specified Information," for a full technical account.

42. Those familiar with the literature on statistical hypothesis testing and, more particularly, disputes over the mathematics of design inferences, will be aware that advocates of decision-making and hypothesis testing deriving from Bayes' theorem are critical of hypothesis-testing approaches deriving from the work of Fisher, Neyman, and Pearson. In this regard, it is worth noting that design inferences from Fisherian likelihood ratios can be recast in a Bayesian framework and they lead to the same conclusions.

Chapter 25—Did Life First Arise by Purely Natural Means (Abiogenesis)?

1. See for example Ronak Gupta, "The 7 biggest unsolved mysteries in science," *Digit* (May 26, 2015), https://www.digit.in/features/general/7-greatest-unsolved-problems-in-science-26132.html (accessed November 18, 2020).

2. See for example Philip Ball, "10 Unsolved Mysteries in Chemistry," *Scientific American* (October 2011), https://www.scientificamerican.com/article/10-unsolved-mysteries/ (accessed November 18, 2020).

3. Technically the official line from neo-Darwinian evolutionists is that evolution knows nothing of "progress" and does not necessarily move from "simple to more complex." Nonetheless, it is also true that the grand arc of the evolutionary story moves from simpler organisms toward more complex ones. In this evolutionary story, biological and organic systems began with a single self-replicating molecule and ended up at us. Evolutionary theorists sometimes try to trivialize this clear progression by calling it "bouncing off the lower wall of complexity," but it cannot be denied that their story entails a march towards greater complexity. See for example Stephen Jay Gould, *Full House: The Spread of Excellence from Plato to Darwin* (New York: Three Rivers, 1996).

4. R.L. Devonshire, *The Life of Pasteur*, translated R. Vallery-Radot (New York: Doubleday, 1920), 109.

5. Alexander I. Oparin, *Proiskhozhdenie Zhizni* (Moscow, Russia: Izd. Moskovski Rabochii, 1924), translated as *Origin of Life* by S. Morgulis (New York: Macmillan, 1938).

6. J.B.S. Haldane, "Origin of Life," *Rationalist Annual* 148 (1929), 3-10. For a discussion of Haldane's views, see Stéphane Tirard, "J.B.S. Haldane and the origin of life," *Journal of Genetics* 96 (November 2017), 735-739.

7. Y.D. Bernal, "The Physical Basis of Life," paper presented before British Physical Society in 1949, found in *The Physical Basis of Life* (London, UK: Routledge, 1951).

8. Erwin Schrödinger, *What Is Life? The Physical Aspect of the Living Cell* (Cambridge, UK: Cambridge University Press, 1944).

9. Harold C. Urey, *The Planets: Their Origin and Development* (New Haven, CT: Yale University Press, 1952).

10. Stanley L. Miller, "A Production of Amino Acids Under Possible Primitive Earth Conditions," *Science* 117 (May 15, 1953), 528-529.

11. Richard A. Kerr, "Origin of Life: New Ingredients Suggested," *Science* 210 (October 3, 1980), 42-43.

12. Jon Cohen, "Novel Center Seeks to Add Spark to Origins of Life," *Science* 270: 1925-1926 (December 22, 1995).

13. Adam P. Johnson, "The Miller Volcanic Spark Discharge Experiment," *Science* 322 (October 17, 2008), 404.

14. Kevin Zahnle, Laura Schaefer, and Bruce Fegley, "Earth's Earliest Atmospheres," *Cold Spring Harbor Perspectives in Biology* 2(10), a004895 (October 2010) ("Geochemical evidence in Earth's oldest igneous rocks indicates that the redox state of the Earth's mantle has not changed over the past 3.8 Gyr"); Dante Canil, "Vanadian in peridotites, mantle redox and tectonic environments: Archean to present," *Earth and Planetary Science Letters* 195:75-90 (2002).

15. David W. Deamer, "The First Living Systems: a Bioenergetic Perspective," *Microbiology & Molecular Biology Reviews* 61:239 (1997).

16. National Research Council Space Studies Board, *The Search for Life's Origins* (Washington, DC: National Academy Press, 1990).

17. Deborah Kelley, "Is It Time to Throw Out 'Primordial Soup' Theory?," NPR (February 7, 2010).

18. Nick Lane, John F. Allen, and William Martin, "How did LUCA make a living? Chemiosmosis in the origin of life," *BioEssays* 2 (2010), 271-280.

19. Committee on the Limits of Organic Life in Planetary Systems, Committee on the Origins and Evolution of Life, National Research Council, *The Limits of Organic Life in Planetary Systems* (Washington, DC: National Academy Press, 2007), 60.

20. Stanley Miller and Jeffrey Bada, "Submarine hot springs and the origin of life," *Nature* 334 (August 18, 1988), 609-611.

21. John Horgan, "In the Beginning," *Scientific American* 264 (February 1991), 116-125. Horgan is discussing the research of Miller and Bada in Miller and Bada, "Submarine hot springs and the origin of life."

22. Jeffrey L. Bada, "New insights into prebiotic chemistry from Stanley Miller's spark discharge experiments," *Chemical Society Review* 42 (2013), 2186-2196.

23. Koichiro Matsuno and Eiichi Imai, "Hydrothermal Vent Origin of Life Models," *Encyclopedia of Astrobiology*, eds. Gargaud M. et al. (Berlin, Germany: Springer, 2015), 1162-1166.

24. Horgan, "In the Beginning."

25. Miller and Bada, "Submarine hot springs and the origin of life." See also Stanley L. Miller and Antonio Lazcano, "The Origin of Life—Did It Occur at High Temperatures?," *Journal of Molecular Evolution* 41 (1995), 689-692.

26. Matsuno and Imai, "Hydrothermal Vent Origin of Life Models"; Deborah S. Kelley et al., "An off-axis hydrothermal vent field near the Mid-Atlantic Ridge at 30°N," *Nature* 412 (July 12, 2001), 145-149; Deborah S. Kelley et al., "A Serpentinite-Hosted Ecosystem: The Lost City Hydrothermal Field," *Science* 307 (March 4, 2005), 1428-1434.

27. Norio Kitadai and Shigenori Maruyama, "Origins of building blocks of life: A review," *Geoscience Frontiers* 9 (2018), 1117-1153.

28. Harold S. Bernhardt and Warren P. Tate, "Primordial soup or vinaigrette: did the RNA world evolve at acidic pH?," *Biology Direct* 7 (2012), 4.

29. Bernhardt and Tate, "Primordial soup or vinaigrette?"

30. Sidney W. Fox, John R. Jungck, and Tadayoshi Nakashima, "From Protenoic Microsphere to Contemporary Cell: Formation of Internucleotide and Peptide Bonds by Protenoid Particles," *Origins of Life* 5 (1974), 227-237.

31. Emanuele Astoricchio, Caterina Alfano, Lawrence Rajendran, Piero Andrea Temussi, and Annalisa Pastore, "The Wide World of Coacervates: From the Sea to Neurodegeneration," *Trends in Biochemical Sciences* 45 (August 2020), 706-717.

32. Zhu Hua, "On the Origin of Life: A Possible Way from Fox's Microspheres into Primitive Life," *Symbiosis* 4 (2018), 1-7.

33. Jane B. Reece, Lisa A. Urry, Michael L. Cain, Steven A. Wasserman, Peter V. Minorsky, and Robert B. Jackson, *Cambell's Biology*, 9th ed. (Boston, MA: Pearson, 2011), 125.

34. James Tour, "An Open Letter to My Colleagues," *Inference Review: International Review of Science* 3 (2017), 2.

35. Tour, "An Open Letter to My Colleagues."

36. Statements made by Stanley Miller at a talk given by him for a UCSD Origins of Life seminar class on January 19, 1999 (the talk was attended and notated by the author of this article).

37. Steven A. Benner, "Paradoxes in the Origin of Life," *Origins of Life and Evolution of Biospheres* 44 (2014), 339-343.

38. Benner, "Paradoxes in the Origin of Life."

39. Benner, "Paradoxes in the Origin of Life."

40. Robert Shapiro, quoted in Richard Van Noorden, "RNA world easier to make," *Nature News* (May 13, 2009), http://www.nature.com/news/2009/090513/full/news.2009.471.html (accessed November 18, 2020).

41. James Tour, "Are Present Proposals on Chemical Evolutionary Mechanisms Accurately Pointing Toward First Life?," *Theistic Evolution: A Scientific, Philosophical, and Theological Critique*, eds. Edited by J.P. Moreland, Stephen C. Meyer, Christopher Shaw, Ann K. Gauger, and Wayne Grudem (Wheaton, IL: Crossway, 2017), 165-191.

42. Michael P. Robertson and Gerald F. Joyce, "The Origins of the RNA World," *Cold Spring Harbor Perspectives in Biology* 4 (May 2012), a003608.

43. See Stephen C. Meyer, *Signature in the Cell: DNA and the Evidence for Intelligent Design* (New York: HarperOne, 2009), 304.

44. Harold S Bernhardt, "The RNA world hypothesis: the worst theory of the early evolution of life (except for all the others)," *Biology Direct* 7 (2012), 23.

45. Jack W. Szostak, David P. Bartel, and P. Luigi Luisi, "Synthesizing Life," *Nature*, 409 (January 18, 2001), 387-390; Tomonori Totani, "Emergence of life in an inflationary universe," *Scientific Reports* 10 (2020), 1671.

46. Robert Shapiro, "A Replicator Was Not Involved in the Origin of Life," *IUBMB Life* 49 (2000), 173-176.

47. Robert Shapiro, "A Simpler Origin for Life," *Scientific American* (June 2007), 46-53.

48. Totani, "Emergence of life in an inflationary universe."

49. Benner, "Paradoxes in the Origin of Life."

50. Benner, "Paradoxes in the Origin of Life."

51. Jeremy England quoted in Natalie Wolchover, "A New Physics Theory of Life," *Quanta Magazine* (January 22, 2014), https://www.quantamagazine.org/a-new-thermodynamics-theory-of-the-origin-of-life-20140122/ (accessed November 18, 2020).

52. Brian Miller, "Hot Wired," *Inference Review: International Review of Science* 5 (May 2020), https://inference-review.com/article/hot-wired (accessed November 18, 2020).

53. Frank B. Salisbury, "Doubts About the Modern Synthetic Theory of Evolution," *American Biology Teacher* (September 1971), 33: 335-338.

54. J.T. Trevors and D.L. Abel, "Chance and necessity do not explain the origin of life," *Cell Biology International*, (2004), 28: 729-739.

55. George Wald, "The Origin of Life," *Scientific American* (August 1954), 44-53.

56. Steering Committee on Science and Creationism, National Academy of Sciences, *Science and Creationism: A View from the National Academy of Sciences* (Washington, DC: National Academy Press, 1999), 5.

57. Ronny Schoenberg, Balz S. Kamber, Kenneth D. Collerson, and Stephen Moorbath, "Tungsten isotope evidence from, 3.8-Gyr metamorphosed sediments for early meteorite bombardment of the Earth," *Nature* 418 (July 25, 2002), 403-405.

58. Norman H. Sleep, Kevin J. Zahnlet, James F. Kasting, and Harold J. Morowitz, "Annihilation of ecosystems by large asteroid impacts on the early Earth," *Nature* 342 (November 9, 1989), 139-442; Kevin A. Maher and David J. Stevenson, "Impact frustration of the origin of life," *Nature* 331 (February 18, 1988), 612-614; Norman H. Sleep and Kevin Zahnle, "Refugia from asteroid impacts on early Mars and the early Earth," *Journal of Geophysical Research* 103 (November 25, 1998), 28, 529-528, 544; E.G. Nisbet and N.H. Sleep, "The habitat and nature of early life," *Nature* 409 (February 22, 2001), 1083-1091.

59. Matthew S. Dodd, Dominic Papineau, Tor Grenne, John F. Slack, Martin Rittner, Franco Pirajno, Jonathan O'Neil, and Crispin T.S. Little, "Evidence for early life in Earth's oldest hydrothermal vent precipitates," *Nature* 543 (March 2, 2017), 60-64.

60. Stephen Jay Gould, "An Early Start," *Natural History* 87 (February 1978), 10.

61. Cyril Ponnamperuma quoted in Fred Hoyle and Chandra Wickramasinghe, *Evolution from Space* (New York: Simon & Schuster, 1981), 76

62. Francis Crick, *Life Itself: Its Origin and Nature* (New York: Touchstone, 1981), 88,

63. Alonso Ricardo and Jack W. Szostak, "Life on Earth," *Scientific American* (September 2009), 54-61.

64. George M. Whitesides, "Revolutions in Chemistry: Priestley Medalist George M. Whitesides' Address," *Chemical and Engineering News* 85 (March 26, 2007), 12-17.

65. Conor Myhrvold, "Three Questions for George Whitesides," *MIT Technology Review* (September 3, 2012), https://www.technologyreview.com/2012/09/03/184017/three-questions-for-george-whitesides/ (accessed November 18, 2020).

66. Antonio Lazcano, "Origin of Life," *Encyclopedia of Astrobiology*, eds. M. Gargaud et al. (Berlin, Germany: Springer, 2011), 1184.

67. Richard Dawkins, *The Ancestors Tale: A Pilgrimage to the Dawn of Evolution* (New York: Houghton Mifflin, 2004), 613.

68. Eugene V. Koonin, *The Logic of Chance: The Nature and Origin of Biological Evolution* (Upper Saddle River, NJ: FT Press, 2011), 391.

69. Charles B. Thaxton, Walter L. Bradley, and Roger L. Olsen, *The Mystery of Life's Origin: Reassessing Current Theories* (Dallas, TX: Lewis and Stanley, 1984), 193, 197.

70. Stephen C. Meyer, "Evidence of Intelligent Design in the Origin of Life," *The Mystery of Life's Origin: The Continuing Controversy* (Seattle, WA: Discovery Institute Press, 2020), 455-456.

Chapter 26—What Are the Top Scientific Problems with Evolution?

1. Theodosius Dobzhansky, "Nothing in Biology Makes Sense Except in the Light of Evolution," *The American Biology Teacher* 35 (1973), 125-129.

2. Richard Dawkins, "Put Your Money on Evolution," *The New York Times* (April 9, 1989), section VII, 35, https://www.nytimes.com/1989/04/09/books/in-short-nonfiction.html (accessed August 23, 2020).

3. Charles Darwin, *On the Origin of Species by Means of Natural Selection*, 1st ed. (London, UK: John Murray, 1859), 6, http://darwin-online.org.uk/content/frameset?pageseq=508&itemID=F373&viewtype=side (accessed August 23, 2020).

4. Francis Darwin, ed., *The Life and Letters of Charles Darwin, Including an Autobiographical Chapter* (London, UK: John Murray, 1887), volume I, 309, http://darwin-online.org.uk/content/frameset?pageseq=327&itemID=F1452.1&viewtype=side (accessed August 23, 2020).

5. Charles Darwin, *Origin of Species,* 1st ed. (1859), 434, http://darwin-online.org.uk/content/frameset?pageseq=453&itemID=F373&viewtype=side (accessed August 23, 2020).

6. Darwin, *Origin of Species,* 1st ed., 435.

7. Charles Darwin, *Origin of Species,* 4th ed. (1866), 513, http://darwin-online.org.uk/content/frameset?pageseq=545&itemID=F385&viewtype=side (accessed August 23, 2020).

8. Darwin, *Origin of Species,* 1st ed., 435.

9. Jonathan Wells, *Zombie Science: More Icons of Evolution* (Seattle, WA: Discovery Institute Press, 2017), 44-47.

10. Thomas J. Givnish, "New evidence on the origin of carnivorous plants," *Proceedings of the National Academy of Sciences USA* 112 (2015), 10-11.

11. Leonardo O. Alvarado-Cárdenas, Enrique Martínez-Meyer, Teresa P. Feria, Luis E. Eguiarte, Héctor M. Hernández, Guy Midgley, and Mark E. Olson, "To converge or not to converge in environmental space: Testing for similar environments between analogous succulent plants of North America and Africa," *Annals of Botany* 111 (2013), 1125-1138.

12. David B. Wake, "Homoplasy, homology and the problem of 'sameness' in biology," *Novartis Symposium 222—Homology*, eds. G.K. Bock and G. Cardew (Chichester, UK: John Wiley & Sons, 1999), 45.

13. Ronald H. Brady, "On the independence of systematics," *Cladistics* 1 (1985), 113-126.

14. Wells, *Zombie Science*, 42.

15. Simon Conway Morris, *Life's Solution: Inevitable Humans in a Lonely Universe* (Cambridge, NY: Cambridge University Press, 2003), 283, 327.

16. *Merriam-Webster's* definition of "fossil," https://www.merriam-webster.com/dictionary/fossil (accessed August 23, 2020).

17. Charles Darwin, *Origin of Species,* 1st ed., 281-282, http://darwin-online.org.uk/content/frameset?pageseq=299&itemID=F373&viewtype=side (accessed August 23, 2020).

18. Darwin, *Origin of Species,* 1st ed., 308, http://darwin-online.org.uk/content/frameset?pageseq=326&itemID=F373&viewtype=side (accessed August 23, 2020).

19. James W. Valentine, Stanley M. Awramik, Philip W. Signor, and Peter M. Sadler, "The biological explosion at the Precambrian-Cambrian boundary," *Evolutionary Biology* 25 (1991), 279-356.

20. Niles Eldredge and Stephen Jay Gould, "Punctuated equilibria: An alternative to phyletic gradualism," *Models in Paleobiology*, ed. Thomas J. M. Schopf (San Francisco, CA: Freeman Cooper, 1972), 82-115.

21. Stephen Jay Gould, *The Structure of Evolutionary Theory* (Cambridge, MA: Harvard University Press, 2002), 759.

22. Gareth Nelson, "Presentation to the American Museum of Natural History" (1969), in David M. Williams and Malte C. Ebach, "The reform of palaeontology and the rise of biogeography," *Journal of Biogeography* 31 (2004), 685-712.

23. Henry Gee, *In Search of Deep Time: Beyond the Fossil Record to a New History of Life* (New York: The Free Press, 1999), 113, 116-117.

24. *Merriam-Webster*'s definition of "phylogeny," https://www.merriam-webster.com/dictionary/phylogeny (accessed August 23, 2020).

25. Wells, *Zombie Science*, 32-33.

26. James A. Lake, "The order of sequence alignment can bias the selection of tree topology," *Molecular Biology and Evolution* 8 (1991), 378–385; Wells, *Zombie Science*, 35-36.

27. Charles Darwin, *Origin of Species*, 1st ed., 130, http://darwin-online.org.uk/content/frameset?pageseq=148&ite mID=F373&viewtype=side (accessed August 23, 2020).

28. Antonis Rokas, Dirk Krüger, and Sean B. Carroll, "Animal evolution and the molecular signature of radiations compressed in time," *Science* 310 (2005), 1933-1938.

29. Liliana Dávalos, Andrea Cirranello, Jonathan Geisler, and Nancy Simmons, "Understanding phylogenetic incongruence: Lessons from phyllostomid bats," *Biological Reviews of the Cambridge Philosophical Society* 87 (2012), 991-1024.

30. Charles Darwin, *Origin of Species*, 1st ed., 6, http://darwin-online.org.uk/content/frameset?pageseq=21&itemI D=F373&viewtype=side (accessed August 23, 2020).

31. Darwin, *Origin of Species*, 1st ed., 90, http://darwin-online.org.uk/content/frameset?pageseq=105&itemID=F3 73&viewtype=side (accessed August 23, 2020).

32. Darwin, *Origin of Species*, 1st ed., 32-34, http://darwin-online.org.uk/content/frameset?pageseq=47&itemID= F373&viewtype=side (accessed August 23, 2020).

33. Theodosius Dobzhansky, *Genetics and the Origin of Species* (New York: Columbia University Press, 1937), 12.

34. H.B.D. Kettlewell, "Darwin's missing evidence," *Scientific American* 200 (1959), 48–53.

35. Jonathan Wells, "Second Thoughts About Peppered Moths: This classical story of evolution by natural selection needs revising," *The Scientist* 13 (May 24, 1999), https://www.discovery.org/a/590/ (accessed August 23, 2020); Jonathan Wells, *Icons of Evolution* (Washington, DC: Regnery, 2000), 137-157.

36. Judith Hooper, *Of Moths and Men: Intrigue, Tragedy and the Peppered Moth* (London, UK: Fourth Estate, 2002); Wells, *Zombie Science*, 63-66.

37. Peter T. Boag and Peter R. Grant, "Intense natural selection in a population of Darwin's finches (Geospizinae) in the Galápagos," *Science* 214 (1981), 82-85.

38. H. Lisle Gibbs and Peter R. Grant, "Oscillating selection on Darwin's finches," *Nature* 327 (1987), 511-513.

39. Wells, *Icons of Evolution*, 159-175.

40. Hugo de Vries, *Species and Varieties, Their Origin by Mutation*, 2d ed. (Chicago, IL: Open Court Press, 1906), 825-826, https://www.gutenberg.org/files/7234/7234-h/7234-h.htm (accessed August 23, 2020).

41. Francis H.C. Crick, "On protein synthesis," *Symposia of the Society for Experimental Biology* 12 (1958), 138-163.

42. François Jacob, *The Logic of Life*, trans. Betty E. Spillmann (Princeton, NJ: Princeton University Press, 1973), 3.

43. Jacques Monod, quoted in Horace Freeland Judson, *The Eighth Day of Creation* (New York: Simon & Schuster, 1979), 217.

44. Michael Behe, *Darwin Devolves: The New Science About DNA That Challenges Evolution* (New York: Free Press, 2019).

45. Jonathan Wells, "Membrane Patterns Carry Ontogenetic Information That Is Specified Independently of DNA," *BIO-Complexity* 2014 (2); Jonathan Wells, "Why DNA Mutations Cannot Accomplish What Neo-Darwinism Requires," *Theistic Evolution: A Scientific, Philosophical, and Theological Critique*, eds. J.P. Moreland, Stephen C. Meyer, Christopher Shaw, Ann K. Gauger, and Wayne Grudem (Wheaton, IL: Crossway, 2017), 237-256.

46. Thomas Cavalier-Smith, "The membranome and membrane heredity in development and evolution," *Organelles, Genomes and Eukaryote Phylogeny*, eds. Robert P. Hirt and David S. Horner (Boca Raton, FL: CRC Press, 2004), 335–351.

47. Ernst Mayr, *The Growth of Biological Thought* (Cambridge, MA: Harvard University Press, 1982), 403.

48. Douglas J. Futuyma, *Evolution* (Sunderland, MA: Sinauer Associates, 2005), 401.

49. Jerry A. Coyne and H. Allen Orr, *Speciation* (Sunderland, MA: Sinauer Associates, 2004), 25.

50. Mayr, *The Growth of Biological Thought*, 273; Coyne and Orr, *Speciation*, 26-35.

51. Arne Müntzing, "Cytogenetic Investigations on Synthetic *Galeopsis tetrahit*," *Hereditas* 16 (1932), 105-154.

52. Futuyma, *Evolution*, 398.

53. Richard Goldschmidt, *The Material Basis of Evolution* (New Haven, CT: Yale University Press, 1940), 8, 396.

54. Charles Darwin, *Origin of Species,* 1st ed., 52, http://darwin-online.org.uk/content/frameset?pageseq=67&item ID=F373&viewtype=side (accessed August 23, 2020).

55. Jonathan Wells, *The Politically Incorrect Guide to Darwinism and Intelligent Design* (Washington, DC: Regnery, 2006), 52-55.

56. Keith Stewart Thomson, "Natural Selection and Evolution's Smoking Gun," *American Scientist* 85 (1997), 516-518.

57. Scott F. Gilbert, John M. Opitz, and Rudolf A. Raff, "Resynthesizing Evolutionary and Developmental Biology," *Developmental Biology* 173 (1996), 357-372.

58. Alan H. Linton, "Scant Search for the Maker," *The Times Higher Education Supplement* (April 20, 2001), Book Section, 29.

59. Lynn Margulis and Dorion Sagan, *Acquiring Genomes: A Theory of the Origins of Species* (New York: Basic Books, 2002), 32.

60. Darwin, *Origin of Species,* 1st ed., 459.

61. Richard Dawkins, *The Greatest Show on Earth: The Evidence for Evolution* (New York: Free Press, 2009), vii.

Chapter 27—How Does Irreducible Complexity Challenge Darwinism?

1. M.J. Schiefsky, "Galen's teleology and functional explanation," *Oxford Studies in Ancient Philosophy* 33 (2007), 369-400.

2. A. Schierbeek, *Measuring the Invisible World: The Life and Works of Antoni van Leeuwenhoek* (London, UK: Abelard-Schuman Publishing, 1959), 171.

3. William Paley, *Natural Theology, or, Evidences of the Existence and Attributes of the Deity, Collected from the Appearances of Nature* (Philadelphia, PA: J. Faulder, 1809), 433, http://darwin-online.org.uk/content/frameset?itemID =A142&pageseq=1&viewtype=text (accessed November 20, 2020).

4. Ernst Mayr, *What Evolution Is* (New York: BasicBooks, 2001), 86.

5. St. George Mivart *On the Genesis of Species* (London, UK: Macmillan, 1871).

6. R.A. Fisher, *The Genetical Theory of Natural Selection* (London, UK: Oxford University Press, 1930).

7. J. Farley, *The Spontaneous Generation Controversy from Descartes to Oparin* (Baltimore, MD: Johns Hopkins University Press, 1977).

8. N.M. Taylor et al., "Structure of the T4 baseplate and its function in triggering sheath contraction," *Nature* 533 (2016), 346-352. See the animation therein, "Structural transformation of the baseplate upon binding to the host membrane."

9. Michael J. Behe, Darwin's Black Box: *The Biochemical Challenge to Evolution* (New York: The Free Press, 1996).

10. "Machine," *The Free Dictionary* by Farlex, https://www.thefreedictionary.com/machine (accessed November 20, 2020).

11. Michael J. Behe, *Darwin Devolves: The New Science about DNA That Challenges Evolution* (New York: HarperOne, 2019).

12. Behe, *Darwin Devolves.*

13. "Design," *The Free Dictionary* by Farlex, https://www.thefreedictionary.com/design (accessed November 20, 2020).

14. Paley, *Natural Theology,* 1.

15. Michael J. Behe, The Biochemical Argument for Design, in *Contemporary Arguments in Natural Theology.* ed. C. Ruloff (London, UK: Bloomsbury Press, in press).

Chapter 28—Can New Proteins Evolve?

1. This is true for bacterial genes, anyway. Higher life forms use more complex processes to make their proteins, making it much harder, in some cases, to predict the proteins from the genome sequences.

2. Robert K. Nakamoto, Joanne A. Baylis Scanlon, and Marwan K. Al-Shawi, "The Rotary Mechanism of the ATP Synthase," *Archives of Biochemistry and Biophysics* 476 (2008), 43-50.

3. For example, see Sean V. Taylor, Kai U. Walter, Peter Kast, and Donald Hilvert, "Searching sequence space for

protein catalysts," *Proceedings of the National Academy of Sciences USA* 98 (2001), 10596-10601; John F. Reidhaar-Olson, and Robert T. Sauer, "Functionally acceptable substitutions in two a-helical regions of l repressor," *Proteins: Structure Function, and Genetics* 7 (1990), 306-316.

4. Douglas D. Axe, "Estimating the prevalence of protein sequences adopting functional enzyme folds," *Journal of Molecular Biology* 314 (2004), 1295-1315.

5. Thomas Nagel, *Mind and Cosmos: Why the Materialist Neo-Darwinian Conception of Nature Is Almost Certainly False* (New York: Oxford University Press, 2012), 5.

6. Nagel, *Mind and Cosmos*, 5.

7. Nagel, *Mind and Cosmos,* 9.

8. Richard Dawkins, *The Blind Watchmaker: Why the Evidence of Evolution Reveals a Universe Without Design* (New York: Norton, 1986), 9, emphasis in original.

9. Nagel, *Mind and Cosmos*, 9.

10. Nagel, *Mind and Cosmos*, 32.

11. Nagel, *Mind and Cosmos*, 33.

12. Nagel, *Mind and Cosmos*, 92, quoting Roger White, "Does Origins of life research rest on a mistake?" *Noûs* 41 (2003), 453-477.

13. Nagel, *Mind and Cosmos*, 25.

14. This would explain Dawkins's sponsorship of ads on London buses that read, "There's probably no God. Now stop worrying and enjoy your life." *Daily Mail Reporter*, "'There's probably no God…now stop worrying and enjoy your life': Atheist group launches billboard campaign," *Daily Mail* (January 7, 2009), https://www.dailymail.co.uk/news/article-1106924/Theres-probably-God--stop-worrying-enjoy-life-Atheist-group-launches-billboard-campaign.html (accessed August 24, 2020).

15. Thomas Nagel, *The Last Word* (New York: Oxford, 1997), 130-131.

16. See Douglas Axe, "Is Our Intuition of Design in Nature Correct?" (this volume); Douglas Axe, *Undeniable: How Biology Confirms Our Intuition That Life Is Designed* (New York: HarperOne, 2016).

17. Axe, "Estimating the prevalence of protein sequences adopting functional enzyme folds."

18. See Axe, "Is Our Intuition of Design in Nature Correct?"; Axe, *Undeniable*.

Chapter 29—Does the Evidence Support Universal Common Ancestry?

1. David A. Baum, Stacey DeWitt Smith, and Samuel S. S. Donovan, "The Tree-Thinking Challenge," *Science* 310 (November 11, 2005), 979-980, emphasis added.

2. Baum et al., "The Tree-Thinking Challenge."

3. For a discussion of this, see Casey Luskin, "Confusing Evidence for Common Ancestry with Evidence for Random Mutation and Natural Selection," *Evolution News & Views* (September 29, 2011), https://evolutionnews.org/2011/09/confusing_evidence_for_common_/ (accessed November 30, 2020).

4. As Michael Behe explains, "modern Darwinists point to evidence of common descent and erroneously assume it to be evidence of the power of random mutation." Michael J. Behe, *The Edge of Evolution: The Search for the Limits of Darwinism* (New York: Free Press, 2007), 95.

5. Karl Giberson and Francis Collins, *The Language of Science and Faith* (Downers Grove, IL: InterVarsity, 2011), 49.

6. Karl W. Giberson, *Saving Darwin: How to be a Christian and Believe in Evolution* (New York: HarperOne, 2008), 53.

7. This is not to say that affirming universal common ancestry requires belief in a full-throated Darwinian history of life. Rather, because neo-Darwinism requires common ancestry to be true, if common ancestry is false, then so is the dominant evolutionary paradigm of biological origins.

8. For example, see Douglas L. Theobald, "A formal test of the theory of universal common ancestry," *Nature* 465 (May 13, 2010), 219-222, stating: "UCA is now supported by a wealth of evidence from many independent sources, including: (1) the agreement between phylogeny and biogeography; (2) the correspondence between phylogeny and the palaeontological record; (3) the existence of numerous predicted transitional fossils; (4) the hierarchical classification of morphological characteristics; (5) the marked similarities of biological structures with different functions (that is, homologies); and (6) the congruence of morphological and molecular phylogenies."

9. National Center for Science Education, "Evolution on Islands," *NCSE*, https://ncse.ngo/marsupials (accessed

October 31, 2020). For a rebuttal, see Casey Luskin, "The NCSE's Biogeographic Conundrums: A Defense of Explore Evolutions Treatment of Biogeography," *ExploreEvolution.com* (January 19, 2010), http://www.exploreevolution.com/exploreEvolutionFurtherDebate/2010/01/the_ncses_biogeographic_conund.php (accessed October 31, 2020).

10. Alfred L Rosenberger and Walter Carl Hartwig, "New World Monkeys," *Encyclopedia of Life Sciences* (Nature Publishing Group, 2001).

11. Carlos G. Schrago and Claudia A.M. Russo, "Timing the origin of New World monkeys," *Molecular Biology and Evolution* 20 (2003), 1620-1625; John J. Flynn and A.R. Wyss, "Recent advances in South American mammalian paleontology," *Trends in Ecology and Evolution* 13 (November 1998), 449-454; C. Barry Cox and Peter D. Moore, *Biogeography: An Ecological and Evolutionary Approach* (Oxford, UK: Blackwell Science, 1993), 185.

12. Adrienne L. Zihlman, *The Human Evolution Coloring Book* (Napa, CA: Harper Collins, 2000), 4-11.

13. John G. Fleagle and Christopher C. Gilbert, "The Biogeography of Primate Evolution: The Role of Plate Tectonics, Climate and Chance," *Primate Biogeography: Progress and Prospects*, eds. Shawn M. Lehman and John G. Fleagle (New York: Springer, 2006), 393-394.

14. Zihlman, *The Human Evolution Coloring Book*, 4-11.

15. Fleagle and Gilbert, "The Biogeography of Primate Evolution," 394.

16. Layal Liverpool, "Monkeys made their way from Africa to South America at least twice," *New Scientist* (April 9, 2020), https://www.newscientist.com/article/2240325-monkeys-made-their-way-from-africa-to-south-america-at-least-twice/ (accessed March 25, 2020).

17. Fleagle and Gilbert, "The Biogeography of Primate Evolution," 394-395, emphasis added.

18. Fleagle and Gilbert, "The Biogeography of Primate Evolution," 404.

19. Fleagle and Gilbert, "The Biogeography of Primate Evolution," 403-404.

20. Walter Carl Hartwig, "Patterns, Puzzles and Perspectives on Platyrrhine Origins," *Integrative Paths to the Past: Paleoanthropological Advances in Honor of F. Clark Howell*, eds. Robert S. Corruccini and Russell L. Ciochon (Englewood Cliffs, NJ: Prentice Hall, 1994), 76, 84.

21. Fleagle and Gilbert, "The Biogeography of Primate Evolution," 395.

22. John C. Briggs, *Global Biogeography* (Amsterdam, Netherlands: Elsevier Science, 1995), 93.

23. Susan Fuller, Michael Schwarz, and Simon Tierney, "Phylogenetics of the allodapine bee genus *Braunsapis*: historical biogeography and long-range dispersal over water," *Journal of Biogeography* 32 (2005), 2135-2144; Anne D. Yoder et al., "Ancient single origin of Malagasy primates," *Proceedings of the National Academy of Sciences* 93 (May 1996), 5122-5126; Peter M. Kappeler, "Lemur Origins: Rafting by Groups of Hibernators?," *Folia Primatol* 71 (2000), 422-425; Christian Roos, Jürgen Schmitz, and Hans Zischler, "Primate jumping genes elucidate strepsirrhine phylogeny," *Proceedings of the National Academy of Sciences* 101 (July 20, 2004), 10650-10654; Philip D. Rabinowitz and Stephen Woods, "The Africa-Madagascar connection and mammalian migrations," *Journal of African Earth Sciences* 44 (2006), 270-276; Anne D. Yoder et al., "Single origin of Malagasy Carnivora from an African ancestor," *Nature* 421 (February 13, 2003), 734-737.

24. Richard John Huggett, *Fundamentals of Biogeography*, 2d ed. (London, UK: Routledge, 1998), 39.

25. G. John Measey et al. "Freshwater paths across the ocean: molecular phylogeny of the frog *Ptychadena newtoni* gives insights into amphibian colonization of oceanic islands," *Journal of Biogeography* 34 (2007), 7-20.

26. Alan de Queiroz, "The resurrection of oceanic dispersal in historical biogeography," *Trends in Ecology and Evolution* 20 (February 2005), 68-73.

27. Giancarlo Scalera, "Fossils, frogs, floating islands and expanding Earth in changing-radius cartography—A comment to a discussion on *Journal of Biogeography*," *Annals of Geophysics* 50 (December 2007), 789-798.

28. de Queiroz, "The resurrection of oceanic dispersal in historical biogeography."

29. de Queiroz, "The resurrection of oceanic dispersal in historical biogeography."

30. Donald Prothero, *Bringing Fossils to Life* (Boston, MA: McGraw Hill, 1998), vii.

31. Donald Prothero, *Evolution: What the Fossils Say and Why It Matters* (New York: Columbia University Press, 2007), xx.

32. Prothero, *Evolution*, 44.

33. See Casey Luskin, "How 'Sudden' Was the Cambrian Explosion?" *Debating Darwin's Doubt: A Scientific Controversy that Can No Longer Be Denied*, ed. David Klinghoffer (Seattle, WA: Discovery Institute Press, 2015), 75-88.

34. R.S.K. Barnes, P. Calow, and P.J.W. Olive, *The Invertebrates: A New Synthesis* (London, UK: Blackwell Scientific, 3d ed., 2001), 9-10.

35. Martin Scheffer, *Critical Transitions in Nature and Society* (Princeton, NJ: Princeton University Press, 2009), 169-170.

36. Walter Etter, "Patterns of Diversification and Extinction," *Handbook of Paleoanthropology: Principles, Methods, and Approaches*, eds. Winfried Henke and Ian Tattersall, 2d ed. (Heidelberg, Germany: Springer-Verlag, 2015), 351-415.

37. Arthur Strahler, *Science and Earth History: The Evolution/Creation Controversy* (Buffalo, NY: Prometheus Books, 1987), 408-409.

38. Steven M. Stanley, *Macroevolution: Pattern and Process* (Baltimore, MD: Johns Hopkins University Press, 1998), 251.

39. Niles Eldredge, *Life Pulse: Episodes from the Story of the Fossil Record* (New York: Facts on File, 1987), 69, 81.

40. Michael J. Everhart, "Rapid evolution, diversification and distribution of mosasaurs (Reptilia; Squamata) prior to the K-T boundary," *Oceans of Kansas Paleontology* (June 11, 2010), http://oceansofkansas.com/rapidmosa.html (accessed October 31, 2020).

41. Richard M. Bateman et al., "Early Evolution of Land Plants: Phylogeny, Physiology, and Ecology of the Primary Terrestrial Radiation," *Annual Review of Ecology and Systematics* 29 (1998), 263-292.

42. David Grimaldi and Michael S. Engel, *Evolution of the Insects* (Cambridge, UK: Cambridge University Press, 2005), 302; Michael Krauss, "The Big Bloom—How Flowering Plants Changed the World," *National Geographic* (July 2002), 102-121, https://www.nationalgeographic.com/science/prehistoric-world/big-bloom/ (accessed October 31, 2020); Stanley A. Rice, *Encyclopedia of Evolution* (New York: Checkmark Books, 2007), 70.

43. Stefanie De Bodt, Steven Maere, and Yves Van de Peer, "Genome duplication and the origin of angiosperms," *Trends in Ecology and Evolution* 20 (2005), 591-597.

44. Jennifer A. Clack, *Gaining Ground: The Origin and Evolution of Tetrapods* (Bloomington, IN: Indiana University Press), 278, 327; Thomas S. Kemp, *The Origin of Higher Taxa: Palaeobiological, Developmental, and Ecological Perspectives* (Chicago, IL: Oxford University Press and University of Chicago Press, 2016), 157.

45. Michael Balter, "Pint-Sized Predator Rattles the Dinosaur Family Tree," *Science* 331 (January 14, 2011), 134.

46. Frank Gill, *Ornithology*, 3d ed. (New York: W.H. Freeman, 2007), 42.

47. Peter D. Ward, *Out of Thin Air: Dinosaurs, Birds, and Earth's Ancient Atmosphere* (Washington, DC: Joseph Henry Press, 2006), 224; Grimaldi and Engel, *Evolution of the Insects*, 37; Rice, *Encyclopedia of Evolution*, 6; Edwin H. Colbert, *Evolution of the Vertebrates: A History of the Backboned Animals Through Time* (New York: John Wiley & Sons, 1969), 123; Niles Eldredge, *Macroevolutionary Dynamics: Species, Niches, & Adaptive Peaks* (New York: McGraw Hill, 1989), 44; Robert A. Martin, *Missing Links: Evolutionary Concepts & Transitions Through Time* (Boston, MA: Jones and Bartlett, 2004), 135, 139, 179; Marc Godinot, "Fossil Record of the Primates from the Paleocene to the Oligocene," in *Handbook of Paleoanthropology*, 1137-1259; Maureen A. O'Leary et al., "The Placental Mammal Ancestor and the Post-K-Pg Radiation of Placentals," *Science* 339 (February 8, 2013), 662-667.

48. Niles Eldredge, *The Monkey Business: A Scientist Looks at Creationism* (New York: Washington Square Press, 1982), 65.

49. See chapter 16, "What Is the Positive Case for Design?" of this volume, or see chapter 7 in Stephen C. Meyer, *Darwin's Doubt: The Explosive Origin of Animal Life and the Case for Intelligent Design* (New York: HarperOne, 2013), 136-152.

50. Jeffrey H. Schwartz, *Sudden Origins: Fossils, Genes, and the Emergence of Species* (New York: John Wiley & Sons, 1999), 3.

51. For example, see Jane B. Reece, Lisa A. Urry, Michael L. Cain, Steven A. Wasserman, Peter V. Minorsky, Robert B. Jackson, *Campbell Biology*, 9th ed. (San Francisco, CA: Pearson, 2011), 14; Mark Ridley, *The Problems of Evolution* (Oxford, UK: Oxford University Press, 1985), 10-11; Benjamin Lewin, *Genes VII* (Oxford, UK: Oxford University Press, 2000), 169; Nicholas H. Barton et al., *Evolution* (Cold Spring Harbor, NY: Cold Spring Harbor Laboratory Press, 2007), 66; Scott Freeman and Jon C. Herron, *Evolutionary Analysis* (Upper Saddle River, NJ: Prentice Hall, 1998), 59; Mark Ridley, *Evolution*, 3d ed. (Malden, MA: Blackwell Publishing, 2004), 66.

52. For a list of known variants to the standard genetic code, see Andrzej (Anjay) Elzanowski and Jim Ostell, "The Genetic Codes," *Taxonomy Browser*, National Center for Biotechnology Information, https://www.ncbi.nlm.nih.gov/Taxonomy/taxonomyhome.html/index.cgi?chapter=cgencodes (accessed October 31, 2020). See also Robin

D. Knight, Stephen J. Freeland, and Laura F. Landweber, "Rewiring the Keyboard: Evolvability of the Genetic Code," *Nature Reviews Genetics* 2 (January 2001), 49-58.

53. Theobald, "A formal test of the theory of universal common ancestry."

54. Comments by William Martin in review of Eugene V. Koonin and Yuri I. Wolf, "The common ancestry of life," *Biology Direct* 5 (2010), 64.

55. Paul Nelson and Jonathan Wells, "Homology in Biology," *Darwinism, Design, and Public Education*, eds. John Angus Campbell and Stephen C. Meyer (East Lansing, MI: Michigan State University Press, 2003), 316.

56. As one paper found, "several core components of the bacterial [DNA] replication machinery are unrelated or only distantly related to the functionally equivalent components of the archaeal/eukaryotic replication apparatus," leading them to suggest "DNA replication likely evolved independently in the bacterial and archaeal/eukaryotic lineages." Detlef D. Leipe, L. Aravind, and Eugene V. Koonin, "Did DNA replication evolve twice independently?" *Nucleic Acids Research* 27 (1999), 3389-3401. Even more striking, another paper compared the genomes of 1,000 different prokaryotic organisms and found that "of the 1000 genomes available, *not a single protein is conserved across all genomes.*" Karin Lagesen, Dave W. Ussery, and Trudy M. Wassenaar, "Genome update: the 1000th genome—a cautionary tale," *Microbiology* 156 (March 2010), 603-608, emphasis added.

57. For example, see Carl Zimmer, *Evolution: The Triumph of an Idea* (New York: HarperCollins/WGBH, 2001), 102-103.

58. W. Ford Doolittle, "Phylogenetic Classification and the Universal Tree," *Science* 284 (June 25, 1999), 2124-2128.

59. W. Ford Doolittle, "Uprooting the Tree of Life," *Scientific American* (February 2000), 90-95.

60. Arcady R. Mushegian et al., "Large-Scale Taxonomic Profiling of Eukaryotic Model Organisms: A Comparison of Orthologous Proteins Encoded by the Human, Fly, Nematode, and Yeast Genomes," *Genome Research* 8 (1998), 590-598.

61. James H. Degnan and Noah A. Rosenberg, "Gene tree discordance, phylogenetic inference and the multispecies coalescent," *Trends in Ecology and Evolution* 24 (2009), 332-340.

62. Eric Bapteste et al., "Networks: expanding evolutionary thinking," *Trends in Genetics* 29 (August 2013), 439-441.

63. Liliana M. Dávalos et al., "Understanding phylogenetic incongruence: lessons from phyllostomid bats," *Biological Reviews of the Cambridge Philosophical Society* 87 (2012), 991-1024.

64. Graham Lawton, "Why Darwin was wrong about the tree of life," *New Scientist* (January 21, 2009), 34-39.

65. Lawton, "Why Darwin was wrong about the tree of life."

66. Lawton, "Why Darwin was wrong about the tree of life."

67. Lawton, "Why Darwin was wrong about the tree of life."

68. Lawton, "Why Darwin was wrong about the tree of life."

69. Carl Woese, "The Universal Ancestor," *Proceedings of the National Academy of Sciences* 95 (June 1998), 6854-9859.

70. Trisha Gura, "Bones, molecules, or both?," *Nature* 406 (July 20, 2000), 230-233.

71. Anna Marie A. Aguinaldo et al., "Evidence for a clade of nematodes, arthropods and other moulting animals," *Nature* 387 (May 29, 1997), 489-493.

72. Erich D. Jarvis et al., "Whole-genome analyses resolve early branches in the tree of life of modern birds," *Science* 346 (December 12, 2014), 1320-1331.

73. Ewen Callaway, "Flock of geneticists redraws bird family tree," *Nature* 516 (December 11, 2014), 297.

74. Marketa Zvelebil and Jeremy O. Baum, *Understanding Bioinformatics* (New York: Garland Science, 2008), 239, emphasis added. See also Elliott Sober and Michael Steele, "Testing the Hypothesis of Common Ancestry," *Journal of Theoretical Biology* 218 (2002), 395-408 ("Whether one uses cladistic parsimony, distance measures, or maximum likelihood methods, the typical question is which tree is the best one, not whether there is a tree in the first place."). UC Berkeley Museum of Paleontology's introductory page on cladistics similarly states:
   - What assumptions do cladists make?
   - There are three basic assumptions in cladistics:
   - Any group of organisms are related by descent from a common ancestor.
   "An Introduction to Cladistics," http://www.ucmp.berkeley.edu/clad/clad1.html (accessed October 31, 2020).

75. Jeffrey H. Schwartz and Bruno Maresca, "Do Molecular Clocks Run at All? A Critique of Molecular Systematics," *Biological Theory* 1 (December 2006), 357-371, emphasis added.

76. Michael Syvanen, "Evolutionary Implications of Horizontal Gene Transfer," *Annual Review of Genetics* 46 (2012), 339-356, emphases added.

77. See Nicholas Matzke, "Meyer's Hopeless Monster, Part II," *Panda's Thumb* (June 19, 2013), http://pandasthumb .org/archives/2013/06/meyers-hopeless-2.html (accessed October 31, 2020).

78. David A. Legg, Mark D. Sutton, Gregory D. Edgecombe, Jean-Bernard Caron, "Cambrian bivalved arthropod reveals origin of arthrodization," *Proceedings of the Royal Society B* 279 (2012), 4699-4704; Derek Briggs and Richard Fortey, "The Early Radiation and Relationships of the Major Arthropod Groups," *Science* 246 (October 13, 1989), 241-243.

79. Mark S. Springer et al., "Endemic African mammals shake the phylogenetic tree," *Nature* 388 (July 3, 1997), 61-64.

80. F. Keith Barker et al., "Phylogeny and diversification of the largest avian radiation," *Proceedings of the National Academy of Sciences* 101 (July 27, 2004), 11040-11045.

81. Rodolphe Tabuce et al., "Early Tertiary mammals from North Africa reinforce the molecular Afrotheria clade," *Proceedings of the Royal Society B* 274 (2007), 1159-1166.

82. Shanan S. Tobe, Andrew C. Kitchener, Adrian M.T. Linacre, "Reconstructing Mammalian Phylogenies: A Detailed Comparison of the Cytochrome b and Cytochrome Oxidase Subunit I Mitochondrial Genes," *Plos One* (November 2010), e14156.

83. Richard Dawkins, *The Blind Watchmaker* (New York: Norton, 1996), 94.

84. Winston Ewert, "The Dependency Graph of Life," *BIO-Complexity* 2018 (3).

85. Ewert, "The Dependency Graph of Life."

86. Charles Darwin, "To Asa Gray," *Darwin Correspondence Project* (September 10, 1860), http://www.darwinproject .ac.uk/entry-2910 (accessed October 31, 2020).

87. Holt Science & Technology, *Life Science* (Austin, TX: Holt, Rinehart, and Winston, 2001), 183.

88. See "Early Evolution and Development: Ernst Haeckel," *Understanding Evolution*, https://evolution.berkeley .edu/evolibrary/article/0_0_0/history_15 (accessed October 31, 2020).

89. Elizabeth Pennisi, "Haeckel's Embryos: Fraud Rediscovered," *Science* 277 (September 5, 1997), 1435.

90. See Casey Luskin, "Haeckel's Fraudulent Embryo Drawings Are Still Present in Biology Textbooks—Here's a List," *Evolution News & Views* (April 3, 2015), http://www.evolutionnews.org/2015/04/haeckels_fraudu094971 .html (accessed October 31, 2020).

91. Stephen Jay Gould, "Abscheulich! (Atrocious!)," *Natural History* (March 2000), 42-49.

92. Pennisi, "Haeckel's Embryos: Fraud Rediscovered."

93. Andres Collazo, "Developmental Variation, Homology, and the Pharyngula Stage," *Systematic Biology* 49 (2000), 3-18.

94. Alex T. Kalinka et al., "Gene expression divergence recapitulates the developmental hourglass model," *Nature* 468 (December 9, 2010), 811-816 (internal citations removed).

95. Brian Hall, "Phylotypic stage or phantom: is there a highly conserved embryonic stage in vertebrates?" *Trends in Ecology and Evolution* 12 (December 1997), 461-463.

96. PZ Myers, "Casey Luskin, smirking liar," *Pharyngula* (May 8, 2009), http://scienceblogs.com/pharyn gula/2009/05/08/casey-luskin-smirking-liar/ (accessed October 31, 2020).

97. Michael K. Richardson et al., "There is no highly conserved embryonic stage in the vertebrates: implications for current theories of evolution and development," *Anatomy and Embryology* 196 (1997), 91-106.

98. Richardson et al., "There is no highly conserved embryonic stage in the vertebrates."

99. Olaf Bininda-Emonds, Jonathan Jeffery, and Michael Richardson, "Inverting the hourglass: quantitative evidence against the phylotypic stage in vertebrate development," *Proceedings of the Royal Society of London B* 270 (2003), 341-346.

100. PZ Myers, "Jonathan MacLatchie collides with reality again," *Pharyngula* (June 17, 2011), http://scienceblogs .com/pharyngula/2011/06/17/jonathan-maclatchie-collides-w/ (accessed October 31, 2020).

101. PZ Myers, "Jonathan MacLatchie really is completely ineducable," *Pharyngula* (June 25, 2011), http://science blogs.com/pharyngula/2011/06/25/jonathan-maclatchie-really-is/ (accessed October 31, 2020).

102. Myers, "Jonathan MacLatchie collides with reality again."

103. See the discussion in Stephen C. Meyer, Paul A. Nelson, Jonathan Moneymaker, Ralph Seelke, and Scott

Minnich, *Explore Evolution: The Arguments for and Against Neo-Darwinism* (Melbourne, Australia: Hill House, 2007). See also Syvanen, "Evolutionary Implications of Horizontal Gene Transfer"; W. Ford Doolittle, "The practice of classification and the theory of evolution, and what the demise of Charles Darwin's tree of life hypothesis means for both of them," *Philosophical Transactions of The Royal Society B* 364 (2009), 2221-2228; Malcolm S. Gordon, "The Concept of Monophyly: A Speculative Essay," *Biology and Philosophy* 14 (1999), 331-348; Eugene V. Koonin, "The Biological Big Bang model for the major transitions in evolution," *Biology Direct* 2 (2007), 21; Vicky Merhej and Didier Raoult, "Rhizome of life, catastrophes, sequence exchanges, gene creations, and giant viruses: how microbial genomics challenges Darwin," *Frontiers in Cellular and Infection Microbiology* 2 (August 28, 2012), 113; Didier Raoult, "The post-Darwinist rhizome of life," *The Lancet* 375 (January 9, 2010), 104-105; Carl R. Woese, "On the evolution of cells," *Proceedings of the National Academy of Sciences* 99 (June 25, 2002), 8742-8747; Lawton, "Why Darwin was wrong about the tree of life."

104. Eugene Koonin, "The Origin at 150: Is a New Evolutionary Synthesis in Sight?," *Trends in Genetics* 25 (2009), 473-475.

105. Didier Raoult, "There is no such thing as a tree of life (and of course viruses are out!)," *Nature Reviews Microbiology* 7 (2009), 615.

106. Syvanen, "Evolutionary Implications of Horizontal Gene Transfer."

107. Koonin, "The Biological Big Bang model for the major transitions in evolution."

Chapter 30—Can Universal Common Descent Be Tested?

1. Charles Darwin letter to Asa Gray of (May 11, 1863), https://www.darwinproject.ac.uk/letter/DCP-LETT-4153 .xml (November 14, 2020), emphases in original.

2. Darwin used the phrase "Tree of Life" in his original edition of *Origin of Species*. Charles Darwin, *On the Origin of Species* (Cambridge MA: Harvard University Press, 1964, facsimile of 1st ed., 1859), 130.

3. Darwin, *On the Origin of Species*, 484.

4. Immediately after arguing in his classic *On Growth and Form* (1942) that "'a principle of discontinuity,' then, is inherent in all our classifications, whether mathematical, physical, or biological…and to seek for stepping-stones across the gaps between is to seek in vain, for ever," D'Arcy Thompson pledges his fealty to UCD. "This is no argument," he writes in the very next sentence, "against the theory of evolutionary descent" D'Arcy Thompson, *On Growth and Form* (Cambridge, UK: Cambridge University Press, 1945), 1094.

5. "What Is Evolution?," BioLogos, https://biologos.org/common-questions/what-is-evolution (accessed 14 August 2020).

6. Niles Eldredge and Joel Cracraft, *Phylogenetic Patterns and the Evolutionary Process* (New York: Columbia University Press, 1980), 2.

7. Thomas S. Kemp, "Models of diversity and phylogenetic reconstruction," *Oxford Surveys in Evolutionary Biology Volume 2*, eds. Richard Dawkins and Matt Ridley (Oxford, UK: Oxford University Press, 1986), 135-157.

8. Mark Ridley, *The Problems of Evolution* (Oxford, UK: Oxford University Press, 1985), 1.

9. Peter Ax, *The Phylogenetic System* (New York: John Wiley, 1987), 1.

10. Russell Doolittle, "New Perspectives on Evolution Provided by Protein Sequences," *New Perspectives on Evolution*, eds. Leonard Warren and Hilary Koprowski Wistar Symposium Series Volume 4, (New York: Wiley-Liss, 1991), 165-173.

11. Bernard Davis, "Molecular Genetics and the Foundations of Evolution," *Perspectives in Biology & Medicine* 28 (1985), 251-268.

12. Ridley, *The Problems of Evolution*, 11.

13. James Watson et al., *Molecular Biology of the Gene*, 4th ed. (Menlo Park, CA: Benjamin/Cummings, 1987), 453.

14. Ralph Hinegardner and Joseph Engelberg, "Rationale for a Universal Genetic Code," *Science* 142 (1963), 1083-1085.

15. Hinegardner and Engelberg, "Rationale for a Universal Genetic Code."

16. Syozo Osawa, Akira Muto, Thomas Jukes, and Takeshi Ohama, "Evolutionary changes in the genetic code," *Proceedings of the Royal Society of London B* 241 (1990), 19-28.

17. Thomas Fox, "Diverged genetic codes in protozoans and a bacterium," *Nature* 314 (1985), 132-133.

18. Fox, "Diverged genetic codes in protozoans and a bacterium."

19. Thomas Fox, "Natural Variation in the Genetic Code," *Annual Review of Genetics* 21 (1987), 67-91.

20. Fox, "Natural Variation in the Genetic Code."

21. F. Caron, "Eucaryotic codes," *Experientia* 46 (1990), 1106-1117.

22. List derived from Andrzej (Anjay) Elzanowski and Jim Ostell, The Genetic Codes, NCBI, https://www.ncbi.nlm .nih.gov/Taxonomy/Utils/wprintgc.cgi (accessed November 16, 2020).

23. Fox, "Natural Variations in the Genetic Code," emphasis added.

24. Since the discovery of variant codes, an array of alternative proposals for code evolution from a common ancestral code have been proposed. None, however, has achieved canonical standing: "The evolutionary mechanisms that lead to codon reassignment and emergence of deviant codes," observe Koonin and Novozhilov, "are not thoroughly understood, but clearly they must involve changes in tRNA specificity or to the evolution of new specificities in the case of stop codon recruitment." See E. Koonin and A. Novozhilov, "Origin and Evolution of the Universal Genetic Code," *Annual Review of Genetics* 51 (2017), 41-62, 49. Experimental modifications of the code in bacteria necessitate meticulous bio-engineering by investigators, to carry the cells over the gulf of inviability from the universal code to a variant code. No experimental evidence exists of random mutations to the code causing novel codon assignments to evolve viably.

25. Osawa et al., "Evolutionary changes."

26. Osawa et al., "Evolutionary changes."

27. Caron, "Eucaryotic codes."

28. Patrick Keeling, "Genomics: Evolution of the Genetic Code," *Current Biology* 26 (2016), R851-853.

29. Lars Vogt, "The unfalsifiability of cladograms and its consequences," *Cladistics* 24 (February 2008), 62-73, emphases in original, internal citations omitted.

30. Harold I. Brown, *Perception, Theory and Commitment* (Chicago, IL: University of Chicago Press, 1979), 106.

31. Kevin de Queiroz and Michael Donoghue, "Phylogenetic Systematics or Nelson's Version of Cladistics," *Cladistics* 6 (1990), 61-75.

32. de Queiroz and Donoghue, "Phylogenetic Systematics or Nelson's Version of Cladistics."

33. Eugene Koonin, "The Biological Big Bang Model for the major transitions in evolution," *Biology Direct* 2 (2007), 21.

34. Francisco J. Ayala, "The Theory of Evolution: Recent Successes and Challenges," *Evolution and Creation*, ed. Ernan McMullin (Notre Dame, IN: Notre Dame University Press, 2007), 89.

35. See Eugene V. Koonin, *The Logic of Chance: The Nature and Origin of Biological Evolution* (Upper Saddle River, NJ: FT Press, 2012), Box 3-2, 70-71, citing data primarily from Marina V. Omelchenko, Michael Y. Galperin, Yuri I. Wolf, and Eugene V. Koonin, "Non-homologous isofunctional enzymes: A systematic analysis of alternative solutions in enzyme evolution," *Biology Direct* 5 (2010), 31.

36. Mario R. Capecchi, "Polypeptide chain termination in vitro: Isolation of a release factor," *Proceedings of the National Academy of Sciences USA* 58 (1967), 1144-1151. Except sometimes they recognize stop codons. Biology is unruly. If the species in question, such as the ciliated protozoan *Tetrahymena*, carries a variant genetic code, its translation machinery will break the usual rules. *Tetrahymena* possesses a single stop (UGA) and assigns UAA and UAG, stops in the universal code, to the amino acid glutamine. In 1967, however, these discoveries lay more than 20 years in the future.

37. Pavel Baranov et al., "Diverse bacterial genomes encode an operon of two genes, one of which is an unusual class-I release factor that potentially recognizes atypical mRNA signals other than normal stop codons," *Biology Direct* 1 (2006), 28.

38. Eugene Koonin, "Evolution of the Genomic Universe," *Genetics, Evolution and Radiation*, eds. Victoria L. Korogodina, Carmel E Mothersill, Sergey G. Inge-Vechtomov, and Colin B. Seymour (Basel, Switzerland: Springer International, 2006), 413-440.

Chapter 31—Does the Fossil Record Demonstrate Darwinian Evolution?

1. See J.S. Weiner, *The Piltdown Forgery,* 50th anniversary edition (Oxford, UK: Oxford University Press, 2003).

2. Christopher P. Sloan, "Feathers for *T. rex*?," *National Geographic* 196 (November 1999), 98-107; Xu Xing, "Feathers for *T. rex*?," *National Geographic* 197 (March 2000).

3. Richard Dawkins, *The Greatest Show on Earth: The Evidence for Evolution* (New York: Free Press, 2009), 155, emphasis in original.

4. See Stephen C. Meyer, *Darwin's Doubt: The Explosive Origin of Animal Life and the Case for Intelligent Design* (New York: HarperOne, 2013). Meyer explains in that book that Darwin had doubts about his own theory precisely because of the lack of fossil evidence for the earliest animal fossils, today known as the Cambrian explosion.

5. For a more detailed discussion of the explosions and discontinuities in the history of life, see Günter Bechly and Stephen C. Meyer, "The Fossil Record and Universal Common Ancestry," *Theistic Evolution: A Scientific, Philosophical, and Theological Critique*, eds. J.P. Moreland, Stephen C. Meyer, Christopher Shaw, Ann K. Gauger, and Wayne Grudem (Wheaton, IL: Crossway, 2017), 331-361.

6. Ronny Schoenberg, Balz S. Kamber, Kenneth D. Collerson, and Stephen Moorbath, "Tungsten isotope evidence from 3.8-Gyr metamorphosed sediments for early meteorite bombardment of the Earth," *Nature* 418 (July 25, 2002), 403-405.

7. Matthew S. Dodd, Dominic Papineau, Tor Grenne, John F. Slack, Martin Rittner, Franco Pirajno, Jonathan O'Neil, and Crispin T. S. Little, "Evidence for early life in Earth's oldest hydrothermal vent precipitates," *Nature* 543 (March 2, 2017), 60-64.

8. Bing Shen, Lin Dong, Shuhai Xiao, and Michał Kowalewski, "The Avalon Explosion: Evolution of Ediacara Morphospace," *Science* 319 (January 4, 2008), 81-84.

9. See Mark A.S. McMenamin, *The Garden of Ediacara: Discovering the First Complex Life* (New York: Columbia University Press, 1998).

10. J. Madeleine Nash, "Evolution's Big Bang: When Life Exploded," *Time* (December 4, 1995), 66-74.

11. For more detailed reviews of the Cambrian explosion, see Meyer, *Darwin's Doubt*; Stephen C. Meyer, "The origin of biological information and the higher taxonomic categories," *Proceedings of the Biological Society of Washington* 117 (2004), 213-239; *Debating Darwin's Doubt: A Scientific Controversy That Can No Longer Be Denied*, ed. David Klinghoffer (Seattle, WA: Discovery Institute Press, 2015).

12. James O'Donoghue, "The Ordovician: Life's second big bang," *New Scientist* 198 (June 11, 2008), https://www.newscientist.com/article/mg19826601-700-the-ordovician-lifes-second-big-bang/ (accessed November 20, 2020).

13. Richard M. Bateman, Peter R. Crane, William A. DiMichele, Paul R. Kenrick, Nick P. Rowe, Thomas Speck, and William E. Stein, "Early evolution of land plants: Phylogeny, Physiology, and Ecology of the Primary Terrestrial Radiation," *Annual Review of Ecology and Systematics* 29 (1998), 263-292.

14. Grzegorz Niedźwiedzki, Piotr Szrek, Katarzyna Narkiewicz, Marek Narkiewicz, and Per E. Ahlberg, "Tetrapod trackways from the early Middle Devonian period of Poland," *Nature* 463 (January 7, 2010), 43-48; Philippe Janvier and Gaël Clément, "Muddy tetrapod origins," *Nature* 463 (January 7, 2010), 40-41.

15. See Maria Sibylla Merian, *The Wondrous Transformation of Caterpillars* (London, UK: Scholar Press, reprinted 1978).

16. Massimo Bernardi, Piero Gianolla, Fabio Massimo Petti, Paolo Mietto, and Michael J. Benton, "Dinosaur diversification linked with the Carnian Pluvial Episode," *Nature Communications* 9 (2018), 1499.

17. See Massimo Bernardi quoted in an official press release about the study at "University of Bristol, Dinosaurs ended—and originated—with a bang!," *University of Bristol News and Features* (April 16, 2018), http://www.bristol.ac.uk/news/2018/april/dinosaurs-ended-and-originated-with-a-bang-.html (accessed November 20, 2020).

18. See William Friedman, "The Meaning of Darwin's Abominable Mystery," *American Journal of Botany* 96 (2009), 5-21.

19. Alan Feduccia, "'Big bang' for tertiary birds?," *Trends in Ecology and Evolution* 18 (April 2003), 172-176.

20. Nancy B. Simmons, "An Eocene Big Bang for Bats," *Science* 307 (January 28, 2005), 527-528.

21. John Hawks, Keith Hunley, Sang-Hee Lee, and Milford Wolpoff, "Population Bottlenecks and Pleistocene Human Evolution," *Molecular Biology and Evolution* 17 (2000), 2-22.

22. "New Study Suggests Big Bang Theory of Human Evolution," *University of Michigan News Service* (January 10, 2000), http://www.umich.edu/~newsinfo/Releases/2000/Jan00/r011000b.html (accessed November 20, 2020).

23. See for example John J. Shea, "The Middle Stone Age archaeology of the Lower Omo Valley Kibish Formation: Excavations, lithic assemblages, and inferred patterns of early *Homo sapiens* behavior," *Journal of Human Evolution* 55 (2008), 448-485.

24. Ola Hössjer and Ann Gauger, "A Single-Couple Human Origin is Possible," *BIO-Complexity* 2019 (1). See also Ola Hössjer, Ann Gauger, and Colin Reeves, "Genetic Modeling of Human History Part 1: Comparison of Common

Descent and Unique Origin Approaches," *BIO-Complexity* 2016 (3); Ola Hössjer, Ann Gauger, and Colin Reeves, "Genetic Modeling of Human History Part 2: A Unique Origin Algorithm," *BIO-Complexity* 2016 (4).

25. For example see Jonathon R. Stone and Gregory A. Wray, "Rapid Evolution of *cis*-Regulatory Sequences via Local Point Mutations," *Molecular Biology and Evolution* 18 (2001), 1764-1770; Michael J. Behe and David W. Snoke, "Simulating evolution by gene duplication of protein features that require multiple amino acid residues," *Protein Science* 13 (2004), 2651-2664; Michael Lynch, "Simple evolutionary pathways to complex proteins," *Protein Science* 14 (2005), 2217-2225; Rick Durrett and Deena Schmidt, "Waiting for Two Mutations: With Applications to Regulatory Sequence Evolution and the Limits of Darwinian Evolution," *Genetics* 180 (November 2008), 1501-1509; Sarah Behrens and Martin Vingron, "Studying the Evolution of Promoter Sequences: A Waiting Time Problem," *Journal of Computational Biology* 17 (2010), 1591-1606; John Sanford, Wesley Brewer, Franzine Smith, and John Baumgardner, "The waiting time problem in a model hominin population," *Theoretical Biology and Medical Modelling* 12 (2015), 18.

26. Durrett and Schmidt, "Waiting for Two Mutations."

27. See Rui-Dong Yang, Jia-Ren Mao, Wei-Hua Zhang, Li-Jun Jiang, and Hui Gao, "Bryophyte-like Fossil (Parafunaria sinensis) from Early-Middle Cambrian Kaili Formation in Guizhou Province, China," *Acta Botanica Sinica* 46 (2006), 180-185; Michael S. Engel and David A. Grimaldi, "New light shed on the oldest insect," *Nature* 427 (February 12, 2004), 627-630.

28. For a more detailed discussion of biogeography and the origin of New World monkeys, see Casey Luskin's chapter in this volume, "Does the Evidence Support Universal Common Ancestry?"

29. Sophie E. Harrison, Mark S. Harvey, Steve J.B. Cooper, Andrew D. Austin, and Michael G. Rix, "Across the Indian Ocean: A remarkable example of transoceanic dispersal in an austral mygalomorph spider," *Plos One* 12 (2017), e0180139.

30. For a discussion of debates over the origin of insect wings, see David E. Alexander, *On the Wing: Insects, Pterosaurs, Birds, Bats and the Evolution of Animal Flight* (Oxford, UK: Oxford University Press, 2015).

31. For a review of this debate, see Xing Xu and Susan Mackem, "Tracing the Evolution of Avian Wing Digits," *Current Biology* 23 (June 17, 2013), R538-R544.

32. Catharina Clewing, Frank Riedel, Thomas Wilke, and Christian Albrecht, "Ecophenotypic plasticity leads to extraordinary gastropod shells found on the 'Roof of the World'," *Ecology and Evolution* 5 (2015), 2966-2979.

33. Pincelli M. Hull and Richard D. Norris, "Evidence for abrupt speciation in a classic case of gradual evolution," *Proceedings of the National Academy of Sciences* 106 (December 15, 2009), 21224-21229.

34. Famous paleontologist Simon Conway Morris wrote two books about this conundrum. See Simon Conway Morris, *Life's Solution: Inevitable Humans in a Lonely Universe* (Cambridge, UK: Cambridge University Press, 2005); Simon Conway Morris, *The Runes of Evolution: How the Universe became Self-Aware* (West Conshohocken, PA: Templeton Press, 2015).

Chapter 32—Do Fossils Demonstrate Human Evolution?

1. Ronald Wetherington, testimony before Texas State Board of Education (January 21, 2009). Original recording on file with author, SBOECommtFullJan2109B5.mp3, time index 1:52:00-1:52:44.

2. Richard Lewontin, *Human Diversity* (New York: Scientific American Library, 1995), 163.

3. Stephen Jay Gould, *The Panda's Thumb: More Reflections in Natural History* (New York: Norton, 1980), 126.

4. See Alton Biggs et al., National Geographic Society, *Biology: The Dynamics of Life* (New York: Glencoe/McGraw Hill, 2000), 442-443.

5. Biggs et al., *Biology: The Dynamics of Life*; Esteban E. Sarmiento, Gary J. Sawyer, and Richard Milner, *The Last Human: A Guide to Twenty-Two Species of Extinct Humans* (New Haven, CT: Yale University Press, 2007); Richard Potts and Christopher Sloan, *What Does It Mean to Be Human?* (Washington, DC: National Geographic, 2010); Carl Zimmer, *Smithsonian Intimate Guide to Human Origins* (Toronto, Canada: Madison Press, 2005).

6. Earnest Albert Hooton, *Up from the Ape*, rev. ed. (New York: Macmillan, 1946), 329.

7. Paige Williams, "Digging for Glory," *The New Yorker* (June 27, 2016), http://www.newyorker.com/magazine/2016/06/27/lee-berger-digs-for-bones-and-glory (accessed October 26, 2020); Donald Johanson and Blake Edgar, *From Lucy to Language* (New York: Simon & Schuster, 1996).

8. Mark Davis, "Into the Fray: The Producer's Story," *PBS NOVA Online* (February 2002), http://www.pbs.org/wgbh/nova/neanderthals/producer.html (accessed October 26, 2020).

9. Henry Gee, "Return to the Planet of the Apes," *Nature* 412 (July 12, 2001), 131-132.

10. Bernard Wood and Mark Grabowski, "Macroevolution in and around the Hominin Clade," *Macroevolution: Explanation, Interpretation, and Evidence*, eds. Serrelli Emanuele and Nathalie Gontier (Heidelberg, Germany: Springer, 2015), 347-376.

11. Michel Brunet et al., "*Sahelanthropus* or '*Sahelpithecus*'?," *Nature* 419 (October 10, 2002), 582.

12. Michel Brunet et al., "A new hominid from the Upper Miocene of Chad, Central Africa," *Nature* 418 (July 11, 2002), 145-151. See also Michel Brunet et al., "New material of the earliest hominid from the Upper Miocene of Chad," *Nature* 434 (April 7, 2005), 752-755.

13. Smithsonian Natural Museum of Natural History, "*Sahelanthropus tchadensis*," https://humanorigins.si.edu/evidence/human-fossils/species/sahelanthropus-tchadensis (accessed November 30, 2020).

14. "Skull Find Sparks Controversy," *BBC News* (July 12, 2002), http://news.bbc.co.uk/2/hi/science/nature/2125244.stm (accessed October 26, 2020).

15. Milford Wolpoff et al., "*Sahelanthropus* or '*Sahelpithecus*'?" *Nature* 419 (October 10, 2002), 581-582.

16. Roberto Macchiarelli et al., "Nature and relationships of *Sahelanthropus tchadensis*," *Journal of Human Evolution* 149 (2020), 102898.

17. Macchiarelli et al., "Nature and relationships of *Sahelanthropus tchadensis*."

18. Madelaine Böhme, quoted in Michael Marshall, "Our supposed earliest human relative may have walked on four legs," *New Scientist* (November 18, 2020), https://www.newscientist.com/article/mg24833093-600-our-supposed-earliest-human-relative-may-have-walked-on-four-legs/ (accessed November 30, 2020).

19. Bob Yirka, "Study of partial left femur suggests *Sahelanthropus tchadensis* was not a hominin after all," *Phys.org* (November 24, 2020), https://phys.org/news/2020-11-partial-left-femur-sahelanthropus-tchadensis.html (accessed November 30, 2020).

20. Potts and Sloan, *What Does It Mean to Be Human?*, 38.

21. John Noble Wilford, "Fossils May Be Earliest Human Link," *New York Times* (July 12, 2001), http://www.nytimes.com/2001/07/12/world/fossils-may-be-earliest-human-link.html (accessed October 26, 2020).

22. John Noble Wilford, "On the Trail of a Few More Ancestors," *New York Times* (April 8, 2001), http://www.nytimes.com/2001/04/08/world/on-the-trail-of-a-few-more-ancestors.html (accessed October 26, 2020).

23. Leslie Aiello and Mark Collard, "Our Newest Oldest Ancestor?" *Nature* 410 (March 29, 2001), 526-527.

24. K. Galik et al., "External and Internal Morphology of the BAR 1002'00 *Orrorin tugenensis* Femur," *Science* 305 (September 3, 2004), 1450-1453.

25. Sarmiento, Sawyer, and Milner, *The Last Human*, 35.

26. Tim White, quoted in Ann Gibbons, "In Search of the First Hominids," *Science* 295 (February 15, 2002), 1214-1219.

27. Jennifer Viegas, "'Ardi,' Oldest Human Ancestor, Unveiled," *Discovery News* (October 1, 2009), https://web.archive.org/web/20110613073934/http://news.discovery.com/history/ardi-human-ancestor.html (accessed October 26, 2020).

28. Randolph Schmid, "World's Oldest Human-Linked Skeleton Found," *NBC News* (October 1, 2009), https://www.nbcnews.com/id/wbna33110809 (accessed October 26, 2020).

29. Ann Gibbons, "Breakthrough of the Year: *Ardipithecus ramidus*," *Science* 326 (December 18, 2009), 1598-1599.

30. Gibbons, "New Kind of Ancestor," 36-40.

31. White, quoted in Gibbons, "In Search of the First Hominids," 1214-1219, 1215-1216.

32. Michael Lemonick and Andrea Dorfman, "Ardi Is a New Piece for the Evolution Puzzle," *Time* (October 1, 2009), http://content.time.com/time/magazine/article/0,9171,1927289,00.html (accessed October 26, 2020).

33. Gibbons, "New Kind of Ancestor," 36-40, 39.

34. Esteban Sarmiento, "Comment on the Paleobiology and Classification of *Ardipithecus ramidus*," *Science* 328 (May 28, 2010), 1105b.

35. Gibbons, "New Kind of Ancestor," 36-40.

36. Bernard Wood and Terry Harrison, "The Evolutionary Context of the First Hominins," *Nature* 470 (February 17, 2011), 347-352.

37. Thomas C. Prang, Kristen Ramirez, Mark Grabowski, and Scott A. Williams, "*Ardipithecus* hand provides

evidence that humans and chimpanzees evolved from an ancestor with suspensory adaptations," *Science Advances* 7 (February 24, 2021), eabf2474.

38. New York University, "Fossils may look like human bones: Biological anthropologists question claims for human ancestry," *ScienceDaily* (February 16, 2011), https://www.sciencedaily.com/releases/2011/02/110216132034.htm (accessed October 26, 2020).

39. See Eben Harrell, "Ardi: The Human Ancestor Who Wasn't?," *Time* (May 27, 2010), http://content.time.com/time/health/article/0,8599,1992115,00.html (accessed October 26, 2020).

40. Harrell, "Ardi: The Human Ancestor Who Wasn't?"

41. Henry McHenry and Katherine Coffing, "*Australopithecus* to *Homo*: Transformations in Body and Mind," *Annual Review of Anthropology* 29 (2000), 125-146.

42. John Roach, "Fossil Find Is Missing Link in Human Evolution, Scientists Say," *National Geographic News* (April 13, 2006), https://web.archive.org/web/20060423155712/http://news.nationalgeographic.com/news/2006/04/0413_060413_evolution.html (accessed October 26, 2020).

43. Seth Borenstein, "Fossil Discovery Fills Gap in Human Evolution," *NBC News* (April 12, 2006), https://www.nbcnews.com/id/wbna12286206 (accessed October 26, 2020).

44. See Tim White et al., "Asa Issie, Aramis, and the Origin of *Australopithecus*," *Nature* 440 (April 13, 2006), 883-889.

45. White et al., "Asa Issie, Aramis, and the Origin of *Australopithecus*."

46. Bernard Wood, "Evolution of the Australopithecines," *The Cambridge Encyclopedia of Human Evolution*, eds. Steve Jones, Robert Martin, and David Pilbeam (Cambridge, UK: Cambridge University Press, 1992), 231-240.

47. Wood, "Evolution of the Australopithecines."

48. Mark Collard and Leslie Aiello, "From Forelimbs to Two Legs," *Nature* 404 (March 23, 2000), 339-340.

49. Collard and Aiello, "From Forelimbs to Two Legs." See also Brian Richmond and David Strait, "Evidence That Humans Evolved from a Knuckle-Walking Ancestor," *Nature* 404 (March 23, 2000), 382-385.

50. Jeremy Cherfas, "Trees Have Made Man Upright," *New Scientist* 97 (January 20, 1983), 172-177.

51. Richard Leakey and Roger Lewin, *Origins Reconsidered: In Search of What Makes Us Human* (New York: Anchor, 1993), 195.

52. Leakey and Lewin, *Origins Reconsidered*, 193-194.

53. Donald Johanson et al., "Morphology of the Pliocene Partial Hominid Skeleton (A.L. 288-1) From the Hadar Formation, Ethiopia," *American Journal of Physical Anthropology* 57 (1982), 403-451.

54. François Marchal, "A New Morphometric Analysis of the Hominid Pelvic Bone," *Journal of Human Evolution* 38 (March 2000), 347-365.

55. M.M. Abitbol, "Lateral View of *Australopithecus afarensis*: Primitive Aspects of Bipedal Positional Behavior in the Earliest Hominids," *Journal of Human Evolution* 28 (March 1995), 211-229 (internal citations removed).

56. Fred Spoor et al., "Implications of Early Hominid Labyrinthine Morphology for Evolution of Human Bipedal Locomotion," *Nature* 369 (June 23, 1994), 645-648.

57. Timothy Bromage and M. Christopher Dean, "Re-Evaluation of the Age at Death of Immature Fossil Hominids," *Nature* 317 (October 10, 1985), 525-527.

58. Ronald Clarke and Phillip Tobias, "Sterkfontein Member 2 Foot Bones of the Oldest South African Hominid," *Science* 269 (July 28, 1995), 521-524.

59. Peter Andrews, "Ecological Apes and Ancestors," *Nature* 376 (August 17, 1995), 555-556.

60. C.E. Oxnard, "The Place of the Australopithecines in Human Evolution: Grounds for Doubt?" *Nature* 258 (December 4, 1975), 389-395.

61. Yoel Rak, Avishag Ginzburg, and Eli Geffen, "Gorilla-Like Anatomy on *Australopithecus afarensis* Mandibles Suggests *Au. afarensis* Link to Robust Australopiths," *Proceedings of the National Academy of Sciences* 104 (April 17, 2007), 6568-6572.

62. Leslie Aiello, quoted in Leakey and Lewin, *Origins Reconsidered: In Search of What Makes Us Human*, 196. See also Bernard Wood and Mark Collard, "The Human Genus," *Science* 284 (April 2, 1999), 65-71.

63. See Alan Walker and Pat Shipman, *Wisdom of the Bones: In Search of Human Origins* (New York: Alfred Knopf, 1996), 133.

64. Jeffrey Schwartz and Ian Tattersall, "Defining the Genus *Homo*," *Science* 349 (August 28, 2015), 931-932.

65. Ian Tattersall, "The Many Faces of *Homo habilis*," *Evolutionary Anthropology* 1 (1992), 33-37.

66. See F. Spoor et al., "Implications of New Early *Homo* Fossils from Ileret, East of Lake Turkana, Kenya," *Nature* 448 (August 9, 2007), 688-691; Seth Borenstein, "Fossils Paint Messy Picture of Human Origins," *NBC News* (August 8, 2007), https://www.nbcnews.com/id/wbna20178936 (accessed October 26, 2020).

67. Wood and Collard, "The Human Genus"; see also Mark Collard and Bernard Wood, "Defining the Genus *Homo*," in *Handbook of Paleoanthropology*, 2107-2144.

68. Sigrid Hartwig-Scherer and Robert Martin, "Was 'Lucy' More Human than Her 'Child'? Observations on Early Hominid Postcranial Skeletons," *Journal of Human Evolution* 21 (1991), 439-449.

69. Hartwig-Scherer and Martin, "Was 'Lucy' More Human than Her 'Child'?"

70. Walker and Shipman, *Wisdom of the Bones*, 132, 130.

71. Sigrid Hartwig-Scherer, "Apes or Ancestors?" *Mere Creation: Science, Faith and Intelligent Design*, ed. William A. Dembski (Downers Grove, IL: InterVarsity, 1998), 226.

72. David McKenzie and Hamilton Wende, "*Homo naledi*: New Species of Human Ancestor Discovered in South Africa," *CNN* (September 10, 2015), http://www.cnn.com/2015/09/10/africa/homo-naledi-human-relative -species/ (accessed October 26, 2020).

73. Rachel Reilly, "Is This the First Human? Extraordinary Find in a South African Cave Suggests Man May Be Up to 2.8 Million Years Old," *Daily Mail* (September 10, 2015), http://www.dailymail.co.uk/sciencetech/ article-3228991/New-species-ancient-human-discovered-Fossilised-remains-15-bodies-unearthed-South -African-cave.html (accessed October 26, 2020).

74. "Trove of Fossils from a Long-Lost Human Ancestor Is Greatest Find in Decades," *PBS Newshour* (September 10, 2015), https://www.pbs.org/newshour/show/new-family (accessed October 26, 2020).

75. University of the Witwatersrand, "The Hand and Foot of *Homo naledi*," *ScienceDaily* (October 6, 2015), http:// www.sciencedaily.com/releases/2015/10/151006123631.htm (accessed October 26, 2020).

76. Berger et al., "*Homo naledi*, a New Species of the Genus *Homo* from the Dinaledi Chamber, South Africa," *eLife* 4 (2015), e09560.

77. Tracy Kivell et al., "The Hand of *Homo naledi*," *Nature Communications* 6 (October 6, 2015), 8431.

78. W.E.H. Harcourt-Smith et al., "The Foot of *Homo naledi*," *Nature Communications* 6 (October 6, 2015), 8432.

79. American Museum of Natural History, "Foot Fossils of Human Relative Illustrate Evolutionary 'Messiness' of Bipedal Walking," *ScienceDaily* (October 6, 2015), http://www.sciencedaily.com/releases/2015/10/151006131938 .htm (accessed October 26, 2020).

80. Berger et al., "*Homo naledi*, a New Species of the Genus *Homo*."

81. Berger et al., "*Homo naledi*, a New Species of the Genus *Homo*."

82. Harcourt-Smith et al., "The Foot of *Homo naledi*."

83. Kate Wong, "First of Our Kind," *Scientific American* (November 1, 2012), http://www.scientificamerican.com/ article/first-of-our-kind-2012-12-07/ (accessed October 26, 2020). See also Brandon Bryn, "*Australopithecus sediba* May Have Paved the Way for *Homo*," *AAAS News* (September 8, 2011), http://www.aaas.org/news/science -australopithecus-sediba-may-have-paved-way-homo (accessed October 26, 2020).

84. Ann Gibbons, "A Human Smile and Funny Walk for *Australopithecus sediba*," *Science* 340 (April 12, 2013), 132-133. See also Nadia Ramlagan, "Human Evolution Takes a Twist with *Australopithecus sediba*," *AAAS News* (April 11, 2013), http://www.aaas.org/news/science-human-evolution-takes-twist-australopithecus-sediba (accessed October 26, 2020).

85. Peter Schmid et al., "Mosaic Morphology in the Thorax of *Australopithecus sediba*," *Science* 340 (April 12, 2013), 1234598; Charles Choi, "Humanity's Closest Ancestor Was Pigeon-Toed, Research Reveals," *LiveScience* (April 11, 2013), https://www.livescience.com/28656-closest-human-ancestor-was-pigeon-toed.html (accessed October 26, 2020).

86. Caroline Vansickle et al., "Primitive Pelvic Features in a New Species of *Homo*," The 85th Annual Meeting of the American Association of Physical Anthropologists, 2016, http://meeting.physanth.org/program/2016/session39/ vansickle-2016-primitive-pelvic-features-in-a-new-species-of-homo.html (accessed October 26, 2020).

87. Berger et al., "*Homo naledi*, a New Species of the Genus *Homo*."

88. Ed Yong, "6 Tiny Cavers, 15 Odd Skeletons, and 1 Amazing New Species of Ancient Human," *The Atlantic* (September 10, 2015), http://www.theatlantic.com/science/archive/2015/09/homo-naledi-rising-star-cave-hominin /404362/ (accessed October 26, 2020).

89. Andrew Du and Zeresenay Alemseged, "Temporal evidence shows Australopithecus sediba is unlikely to be the ancestor of Homo," *Science Advances* 5 (May 8, 2019), eaav9038; Tim White, "Five's a Crowd in Our Family Tree," *Current Biology* 23 (February 4, 2013), R112-R115; William Kimbel, "Hesitation on Hominin History," *Nature* 497 (May 30, 2013), 573-574; Gibbons, "Human Smile and Funny Walk for *Australopithecus sediba*"; Gibbons, "Who Was *Homo habilis*?" *Science* 332 (June 17, 2011), 1370-1371; Nicholas Wade, "New Fossils May Redraw Human Ancestry," *New York Times* (September 8, 2011), http://www.nytimes.com/2011/09/09/science/09fossils .html (accessed October 26, 2020); John Noble Wilford, "Some Prehumans Feasted on Bark instead of Grasses," *New York Times* (June 27, 2012), http://www.nytimes.com/2012/06/28/science/australopithecus-sediba -preferred-forest-foods-fossil-teeth-suggest.html (accessed October 26, 2020).

90. Du and Alemseged, "Temporal evidence shows Australopithecus sediba is unlikely to be the ancestor of Homo."

91. Carl Zimmer, "Yet Another 'Missing Link,'" *Slate* (April 8, 2010), http://www.slate.com/articles/health_and _science/science/2010/04/yet_another_missing_link.single.html (accessed October 26, 2020).

92. Michael Balter, "Candidate Human Ancestor from South Africa Sparks Praise and Debate," *Science* 328 (April 9, 2010), 154-155.

93. See Kate Wong, "Debate Erupts over Strange New Human Species," *Scientific American* (April 8, 2016), http:// www.scientificamerican.com/article/debate-erupts-over-strange-new-human-species/ (accessed October 26, 2020); Tanya Farber, "Professor's Claims Rattle *Naledi*'s bones," *Sunday Times* (April 24, 2016), http://www .timeslive.co.za/sundaytimes/stnews/2016/04/24/Professors-claims-rattle-Naledis-bones (accessed October 26, 2020); Aurore Val, "Deliberate Body Disposal by Hominins in the Dinaledi Chamber, Cradle of Humankind, South Africa?," *Journal of Human Evolution* 96 (2016), 145-148.

94. Kate Wong, "Mysterious New Human Species Emerges from Heap of Fossils," *Scientific American* (September 10, 2015), http://www.scientificamerican.com/article/mysterious-new-human-species-emerges-from-heap-of -fossils/ (accessed October 26, 2020).

95. Quoted in Yong, "6 Tiny Cavers."

96. University of Colorado Anschutz Medical Campus, "Ancient ancestor of humans with tiny brain discovered," *ScienceDaily* (September 10, 2015), https://www.sciencedaily.com/releases/2015/09/150910084610.htm (accessed October 26, 2020).

97. University of the Witwatersrand, "*Homo naledi*'s surprisingly young age opens up more questions on where we come from," *ScienceDaily* (May 9, 2017), https://www.sciencedaily.com/releases/2017/05/170509083554.htm (accessed October 26, 2020). See also Dirks et al., "The age of *Homo naledi* and associated sediments in the Rising Star Cave, South Africa," *eLife* 6 (2017), e24231.

98. James Kidder, "What Homo Naledi Means for the Study of Human Evolution," *BioLogos* (May 30, 2017), http:// biologos.org/blogs/guest/what-homo-naledi-means-for-the-study-of-human-evolution (accessed October 26, 2020).

99. See Chris Stringer, "Human Evolution: The Many Mysteries of *Homo naledi*," *eLife* 4 (2015), e10627; Daniel Curnoe, "What About *Homo naledi*'s Geologic Age?," *Phys.org* (September 15, 2015), http://phys.org/news/2015 -09-opinion-homo-naledi-geologic-age.html (accessed October 26, 2020).

100. Walker and Shipman, *Wisdom of the Bones*, 134.

101. Berhane Asfaw et al., "Remains of *Homo erectus* from Bouri, Middle Awash, Ethiopia," *Nature* 416 (March 21, 2002), 317-320.

102. William Kimbel and Brian Villmoare, "From *Australopithecus* to *Homo*: The Transition that Wasn't," *Philosophical Transactions of the Royal Society B* 371 (2016), 20150248.

103. Kimbel and Villmoare, "From *Australopithecus* to *Homo*: The Transition that Wasn't."

104. Stanley A. Rice, *Encyclopedia of Evolution* (New York: Checkmark, 2007), 241.

105. Franz Wuketits, "Charles Darwin, Paleoanthropology, and the Modern Synthesis," *Handbook of Paleoanthropology*, 97-125, 116.

106. Dean Falk, "Hominid Brain Evolution: Looks Can Be Deceiving," *Science* 280 (June 12, 1998), 1714.

107. Specifically, *Homo erectus* is said to have intermediate brain size, and *Homo ergaster* is said to have a *Homo*-like postcranial skeleton with a smaller, more australopith-like brain size.

108. Terrance Deacon, "Problems of Ontogeny and Phylogeny in Brain-Size Evolution," *International Journal of Primatology* 11 (1990), 237-282. See also Terrence Deacon, "What Makes the Human Brain Different?," *Annual*

Review of Anthropology 26 (1997), 337-357; Stephen Molnar, *Human Variation: Races, Types, and Ethnic Groups*, 5th ed. (Upper Saddle River, NJ: Prentice Hall, 2002), 189.

109. Christof Koch, "Does Brain Size Matter?," *Scientific American Mind* (January/February, 2016), 22-25.

110. See Wood and Collard, "The Human Genus."

111. Marchal, "New Morphometric Analysis of the Hominid Pelvic Bone."

112. Robin Dennell and Wil Roebroeks, "An Asian Perspective on Early Human Dispersal from Africa," *Nature* 438 (Dec. 22/29, 2005), 1099-1104.

113. John Hawks, Keith Hunley, Sang-Hee Lee, and Milford Wolpoff, "Population Bottlenecks and Pleistocene Human Evolution," *Molecular Biology and Evolution* 17 (2000), 2-22.

114. Hawks et al., "Population Bottlenecks and Pleistocene Human Evolution."

115. Hawks et al., "Population Bottlenecks and Pleistocene Human Evolution."

116. Daniel E. Lieberman, David R. Pilbeam, and Richard W. Wrangham, "The Transition from *Australopithecus* to *Homo*," *Transitions in Prehistory: Essays in Honor of Ofer Bar-Yosef*, eds. John J. Shea and Daniel E. Lieberman (Cambridge, MA: Oxbow, 2009), 1.

117. Lieberman et al., "The Transition from *Australopithecus* to *Homo*."

118. Alan Turner and Hannah O'Regan, "Zoogeography: Primate and Early Hominin Distribution and Migration Patterns," in *Handbook of Paleoanthropology*, 623-642.

119. Kimbel, "Hesitation on Hominin History."

120. Ernst Mayr, *What Makes Biology Unique?: Considerations on the Autonomy of a Scientific Discipline* (Cambridge, UK: Cambridge University Press, 2004), 198.

121. "New Study Suggests Big Bang Theory of Human Evolution," *University of Michigan News Service* (January 10, 2000), http://www.umich.edu/~newsinfo/Releases/2000/Jan00/r011000b.html (accessed October 26, 2020).

122. Eric Delson, "One Skull Does Not a Species Make," *Nature* 389 (October 2, 1997), 445-446; Hawks et al., "Population Bottlenecks and Pleistocene Human Evolution"; Emilio Aguirre, "*Homo erectus* and *Homo sapiens*: One or More Species?," *100 Years of Pithecanthropus: The Homo erectus Problem 171 Courier Forschungsinstitut Senckenberg*, ed. Jens Lorenz (Frankfurt, Germany: Courier Forschungsinstitut Senckenberg, 1994), 333-339; Milford H. Wolpoff, et al., "The Case for Sinking *Homo erectus*: 100 Years of *Pithecanthropus* Is Enough!," *100 Years of Pithecanthropus*, 341-361.

123. See Hartwig-Scherer and Martin, "Was 'Lucy' More Human than Her 'Child'?"

124. William R. Leonard, Marcia L. Robertson, and J. Josh Snodgrass, "Energetic Models of Human Nutritional Evolution," *Evolution of the Human Diet: The Known, the Unknown, and the Unknowable*, ed. Peter S. Ungar (Oxford, UK: Oxford University Press, 2007), 344-359.

125. Kevin G. Hatala et al., "Footprints Reveal Direct Evidence of Group Behavior and Locomotion in *Homo erectus*," *Scientific Reports* 6 (2016), 28766.

126. Max-Planck-Gesellschaft, "*Homo erectus* walked as we do," *Science Daily* (July 12, 2016), https://www.sciencedaily.com/releases/2016/07/160712110444.htm (accessed October 26, 2020).

127. Spoor et al., "Implications of Early Hominid Labyrinthine Morphology for Evolution of Human Bipedal Locomotion."

128. William Leonard and Marcia Robertson, "Comparative Primate Energetics and Hominid Evolution," *American Journal of Physical Anthropology* 102 (February 1997), 265-281. See also Leslie C. Aiello and Jonathan C.K. Wells, "Energetics and the Evolution of the Genus *Homo*," *Annual Review of Anthropology* 31 (2002), 323-338.

129. Moreover, "Although the relative brain size of *Homo erectus* is smaller than the average for modern humans, it is outside of the range seen among other living primate species." William R. Leonard, Marcia L. Robertson, and J. Josh Snodgrass, "Energetics and the Evolution of Brain Size in Early Homo," *Guts and Brains: An Integrative Approach to the Hominin Record*, ed. Wil Roebroeks (Leiden, Germany: Leiden University Press, 2007), 29-46.

130. Daniel Everett, "Did *Homo erectus* speak?," *Aeon* (February 28, 2018), https://aeon.co/essays/tools-and-voyages-suggest-that-homo-erectus-invented-language (accessed October 26, 2020).

131. Donald Johanson and Maitland Edey, *Lucy: The Beginnings of Humankind* (New York: Simon & Schuster, 1981), 144.

132. References for cranial capacities cited in Table 1 are as follows: Gorilla and chimpanzee: Stephen Molnar, *Human Variation: Races, Types, and Ethnic Groups*, 4th ed. (Upper Saddle River, NJ: Prentice Hall, 1998), 203.

Australopithecus: Glenn Conroy et al., "Endocranial Capacity in an Early Hominid Cranium from Sterkfontein, South Africa," *Science* 280 (June 12, 1998), 1730-1731; Wood and Collard, "Human Genus." *Homo habilis*: Wood and Collard, "Human Genus." *Homo erectus*: Molnar, *Human Variation*, 203; Wood and Collard, "The Human Genus." Neanderthals: Molnar, *Human Variation: Races, Types, and Ethnic Groups*, 4th ed., 203; Molnar, *Human Variation: Races, Types, and Ethnic Groups*, 5th ed., 189. *Homo sapiens* (modern man): Molnar, *Human Variation*, 203; E.I. Odokuma et al., "Craniometric Patterns of Three Nigerian Ethnic Groups," *International Journal of Medicine and Medical Sciences* 2 (February 2010), 34-37; Molnar, *Human Variation*, 5th ed., 189.

133. See Wood and Collard, "The Human Genus."

134. Marc Kaufman, "Modern Man, Neanderthals Seen as Kindred Spirits," *The Washington Post* (April 30, 2007), http://www.washingtonpost.com/wp-dyn/content/article/2007/04/29/AR2007042901101_pf.html (accessed October 26, 2020).

135. Michael Lemonick, "A Bit of Neanderthal in Us All?" *Time* (April 25, 1999), http://content.time.com/time/magazine/article/0,9171,23543,00.html (accessed October 26, 2020).

136. Joe Alper, "Rethinking Neanderthals," *Smithsonian* (June 2003).

137. Molnar, *Human Variation: Races, Types, and Ethnic Groups*, 5th ed., 189.

138. B. Arensburg et al., "A Middle Palaeolithic Human Hyoid Bone," *Nature* 338 (April 27, 1989), 758-760.

139. Alper, "Rethinking Neanderthals"; Kate Wong, "Who Were the Neanderthals?" *Scientific American* (August 2003), 28-37; Erik Trinkaus and Pat Shipman, "Neanderthals: Images of Ourselves," *Evolutionary Anthropology* 1 (1993), 194-201; Philip Chase and April Nowell, "Taphonomy of a Suggested Middle Paleolithic Bone Flute from Slovenia," *Current Anthropology* 39 (August/October 1998), 549-553; Tim Folger and Shanti Menon, "…Or Much Like Us?" *Discover* (January 1997), http://discovermagazine.com/1997/jan/ormuchlikeus1026 (accessed October 26, 2020); C.B. Stringer, "Evolution of Early Humans," *The Cambridge Encyclopedia of Human Evolution*, 248.

140. Chase and Nowell, "Taphonomy of a Suggested Middle Paleolithic Bone Flute from Slovenia"; Folger and Menon, "…Or Much Like Us?"

141. Notes in *Nature* 77 (April 23, 1908), 587.

142. Jessica Ruvinsky, "Cavemen: They're Just Like Us," *Discover* (January 2009), http://discovermagazine.com/2009/jan/008 (accessed October 26, 2020).

143. Erik Trinkaus and Cidália Duarte, "The Hybrid Child from Portugal," *Scientific American* (August 2003), 32. It is worth noting that some paleoanthropologists disagree about the existence of human-Neanderthal hybrids.

144. Rex Dalton, "Neanderthals May Have Interbred with Humans," *Nature News* (April 20, 2010), http://www.nature.com/news/2010/100420/full/news.2010.194.html (accessed October 26, 2020).

145. Delson, "One Skull Does Not a Species Make."

146. Kaufman, "Modern Man, Neanderthals Seen as Kindred Spirits."

147. Fazale Rana and Hugh Ross, *Who Was Adam?: A Creation Model Approach to the Origin of Man* (Colorado Springs, CO: NavPress, 2005).

148. Hartwig-Scherer, "Apes or Ancestors," 220.

149. Wood and Grabowski, "Macroevolution in and around the Hominin Clade."

150. Paul Mellars, "Neanderthals and the Modern Human Colonization of Europe," *Nature* 432 (November 25, 2004), 461-465; April Nowell, "From a Paleolithic Art to Pleistocene Visual Cultures (Introduction to Two Special Issues on 'Advances in the Study of Pleistocene Imagery and Symbol Use')," *Journal of Archaeological Method and Theory* 13 (2006), 239-249. Others call this abrupt appearance a "revolution." See Ofer Bar-Yosef, "The Upper Paleolithic Revolution," *Annual Review of Anthropology* 31 (2002), 363-393.

151. Randall White, *Prehistoric Art: The Symbolic Journey of Humankind* (New York: Abrams, 2003), 11, 231.

152. Rice, *Encyclopedia of Evolution*, 104, 187, 194.

153. Robert Kelly and David Thomas, *Archaeology*, 5th ed. (Belmont, CA: Wadsworth Cengage Learning, 2010), 303.

154. Bar-Yosef, "Upper Paleolithic Revolution."

155. Nicholas Toth and Kathy Schick, "Overview of Paleolithic Archaeology," in *Handbook of Paleoanthropology*, 2441-2464.

156. Marc Hauser et al., "The Mystery of Language Evolution," *Frontiers in Psychology* 5 (May 7, 2014), 401.

Chapter 33—Is Evolutionary Psychology a Legitimate Way to Understand Our Humanity?

1. "Evolutionary psychology," *ScienceDaily*, https://www.sciencedaily.com/terms/evolutionary_psychology.htm (accessed November 22, 2020).

2. *See* "Sexual selection," *Understanding Evolution*, https://evolution.berkeley.edu/evolibrary/article/evo_28 (accessed November 22, 2020).

3. "Social Darwinism," *History.com*, https://www.history.com/topics/early-20th-century-us/social-darwinism (accessed November 22, 2020).

4. "Sociobiology," *Stanford Encyclopedia of Philosophy*, https://plato.stanford.edu/entries/sociobiology/ (accessed November 22, 2020).

5. Miriam D. Rosenthal, "Sociobiology: Laying the Foundation for a Racist Synthesis," *The Harvard Crimson* (February 8, 1977), https://www.thecrimson.com/article/1977/2/8/sociobiology-laying-the-foundation-for-a/ (accessed November 22, 2020).

6. Ben Leach, "Shopping is 'throwback to days of cavewomen,'" *Telegraph* (February 25, 2009), https://www.telegraph.co.uk/news/newstopics/howaboutthat/4803286/Shopping-is-throwback-to-days-of-cavewomen.html (accessed November 22, 2020).

7. Steve Johnson, "Have you heard? Gossip can be good," *Chicago Tribune* (January 5, 2009).

8. Nick Collins, "Trivial anger a 'result of modern life,'" *Telegraph* (June 21, 2013), https://www.telegraph.co.uk/news/science/science-news/10132346/Trivial-anger-a-result-of-modern-life.html (accessed November 22, 2020).

9. Denyse O'Leary, "Dissecting the caveman theory of psychology," *MercatorNet* (August 9, 2009), https://mercatornet.com/dissecting_the_caveman_theory_of_psychology/8552/ (accessed November 22, 2020).

10. Kate Douglas, "The evolutionary mysteries of religion and orgasms," *New Scientist* (July 4, 2012), https://www.newscientist.com/article/mg21528722-400-the-evolutionary-mysteries-of-religion-and-orgasms/ (accessed November 22, 2020).

11. Sharon Begley, "Can We Blame Our Bad Behavior on Stone-Age Genes?," *Newsweek* (June 19, 2009), https://www.newsweek.com/can-we-blame-our-bad-behavior-stone-age-genes-80349 (accessed November 22, 2020).

12. Christopher Boehme, "Banks gone bad: Our evolved morality has failed us," *New Scientist* (March 20, 2013), https://www.newscientist.com/article/mg21729090-200-banks-gone-bad-our-evolved-morality-has-failed-us/ (accessed November 22, 2020).

13. Patrick McNamara, "How sex rules our dreams," *Aeon* (April 25, 2014), https://aeon.co/essays/was-freud-right-about-dreams-all-along (accessed November 22, 2020).

14. Melissa Hogenboom, "Why does the human brain create false memories?," *BBC* (September 29, 2013), https://www.bbc.com/news/science-environment-24286258 (accessed November 22, 2020).

15. Rheana Murray, "Men lusting after younger women caused menopause: study," *New York Daily News* (June 14, 2013), https://www.nydailynews.com/life-style/health/men-caused-menopause-study-article-1.1372814 (accessed November 22, 2020).

16. Denyse O'Leary, "Scientists clash over origin of monogamy," *MercatorNet* (August 14, 2013), https://mercatornet.com/scientists_clash_over_origin_of_monogamy/15267/ (accessed November 22, 2020).

17. "The Evolution of PMS: It May Exist to Break up Infertile Relationships," *Macquarie University* (August 12, 2014), https://www.mq.edu.au/newsroom/2014/08/12/the-evolution-of-pms-it-may-exist-to-break-up-infertile-relationships/ (accessed November 22, 2020).

18. Helen Fisher, *Why We Love: The Nature and Chemistry of Romantic Love* (New York: Henry Holt, 2005).

19. Rick Hanson, *Hardwiring Happiness: The New Brain Science of Contentment, Calm, and Confidence* (New York: Harmony, 2013).

20. Michael Graziano, "The first smile," *Aeon* (August 13, 2014), https://aeon.co/essays/the-original-meaning-of-laughter-smiles-and-tears (accessed November 22, 2020).

21. Jesse Prinz, "How wonder works," *Aeon* (June 21, 2013), https://aeon.co/essays/why-wonder-is-the-most-human-of-all-emotions (accessed November 22, 2020).

22. Alan S. Miller and Satoshi Kanazawa, "Ten Politically Incorrect Truths About Human Nature," *Psychology Today* (July 1, 2007), https://www.psychologytoday.com/us/articles/200707/ten-politically-incorrect-truths-about-human-nature (accessed November 22, 2020).

23. Cordelia Fine, *Delusions of Gender: How Our Minds, Society, and Neurosexism Create Difference* (New York: Norton, 2010), 208.

24. V.S. Ramachandran, "Why Do Gentlemen Prefer Blondes?" *Medical Hypotheses* 48 (January 1997), 19-20.

25. Sissela Bok, "Human Kind: Is selflessness in our nature?," *The American Scholar* 79 (Autumn 2010), 104-108, reviewing Oren Harman, *The Price of Altruism: George Price and the Search for the Origins of Kindness* (New York: Norton, 2010).

26. Denyse O'Leary, "An Evolutionary Challenge: Explaining Away Compassion, Philanthropy, and Self-Sacrifice," *Evolution News and Science Today* (January 9, 2015), https://evolutionnews.org/2015/01/an_evolutionary_1/ (accessed November 22, 2020).

27. Emily Boring, "The Problem with Altruism," *The Scope: The Blog of the Yale Scientific Magazine* (November 13, 2016), https://yalescientific.org/thescope/2016/11/the-problem-with-altruism/ (accessed November 22, 2020).

28. Boring, "The Problem with Altruism."

29. Natasha Gilbert, "Altruism can be explained by natural selection," *Nature News* (August 25, 2010), https://www.nature.com/news/2010/100825/full/news.2010.427.html (accessed November 22, 2020). The paper being discussed is Martin A. Nowak, Corina E. Tarnita, and Edward O. Wilson, "The evolution of eusociality," *Nature* 466 (August 26, 2010), 1057-1062.

30. Michael Marshall, "Sparks fly over origin of altruism," *New Scientist* (September 29, 2010), https://www.newscientist.com/article/mg20827804-100-sparks-fly-over-origin-of-altruism/ (accessed November 22, 2020).

31. Jerry Coyne, "A misguided attack on kin selection," *Why Evolution Is True* blog (August 30, 2010), https://whyevolutionistrue.com/2010/08/30/a-misguided-attack-on-kin-selection/ (accessed November 22, 2020).

32. Michael Gazzaniga, "Evolution Revolution: Group think helped produce ant colonies and beehives—and the heights of human culture," *Wall Street Journal* (April 6, 2012), https://www.wsj.com/articles/SB10001424052702303404704577311553569846904 (accessed November 22, 2020).

33. John Gray, "The Knowns and the Unknowns," *The New Republic* (April 19, 2012), https://newrepublic.com/article/102760/righteous-mind-haidt-morality-politics-scientism (accessed November 22, 2020).

34. University of Pennsylvania, "Generosity leads to evolutionary success, biologists show," *ScienceDaily* (September 2, 2013), https://www.sciencedaily.com/releases/2013/09/130902162716.htm (accessed November 22, 2020).

35. National Institute for Mathematical and Biological Synthesis (NIMBioS), "Altruism or manipulated helping? Altruism may have origins in manipulation," *ScienceDaily* (August 19, 2013), https://www.sciencedaily.com/releases/2013/08/130819090218.htm (accessed November 22, 2020).

36. Kenneth Taylor, "Psychological vs. Biological Altruism," *PhilosophyTalk* (June 11, 2010), https://www.philosophytalk.org/blog/psychological-vs-biological-altruism (accessed November 22, 2020).

37. Manpal Singh Bhogal, Niall Galbraith, and Ken Manktelow, "Sexual selection and the evolution of altruism: males are more altruistic and cooperative towards attractive females," *Letters on Evolutionary Behavioral Science* 7 (2016), 10-13.

38. Steven Arnocky and Pat Barclay, "Altruistic People Have More Sexual Partners," *Scientific American* (November 15, 2016), https://www.scientificamerican.com/article/altruistic-people-have-more-sexual-partners/ (accessed November 22, 2020).

39. University of Zurich, "The evolutionary roots of human altruism," *ScienceDaily* (August 27, 2014), https://www.sciencedaily.com/releases/2014/08/140827092002.htm (accessed November 22, 2020).

40. Jill Suttie, "Why Does Altruism Exist?," *Greater Good Magazine* (April 15, 2015), https://greatergood.berkeley.edu/article/item/does_altruism_exist (accessed November 22, 2020), reviewing Sloan Wilson, *Does Altruism Exist?* (New Haven, CT: Yale University Press, 2015).

41. Denis Noble, "Neo-Darwinism, the Modern Synthesis and selfish genes: are they of use in physiology?," *Journal of Physiology* 589 (2011), 1007-1015.

42. David Dobbs, "Die, selfish gene, die," *Aeon* (December 3, 2013), https://aeon.co/essays/the-selfish-gene-is-a-great-meme-too-bad-it-s-so-wrong (accessed November 22, 2020).

43. "The nature of altruism makes it all too easy to drift from a scientific to a political, philosophical, and even a religious approach to this subject." Lee Alan Dugatkin, "Inclusive Fitness Theory from Darwin to Hamilton," *Genetics* 176 (July 2007), 1375-1380.

44. Denyse O'Leary, "Evolutionary Conundrum: Is Religion a Useful, Useless, or Harmful Adaptation?," *Evolution*

News and Science Today (April 3, 2015), https://evolutionnews.org/2015/04/evolutionary_co/ (accessed November 22, 2020).

45. Jonathan Haidt, "Have We Evolved to Be Religious?," *Time* (March 27, 2012), https://ideas.time.com/2012/03/27/have-we-evolved-to-be-religious/ (accessed November 22, 2020).

46. Ilkka Pyysiäinen and Marc Hauser, "The origins of religion: evolved adaptation or by-product?," *Trends in Cognitive Sciences* 14 (March 2010), 104-109.

47. Jerry A. Coyne, "Science, Religion, and Society: The Problem of Evolution in America," *Evolution* 66 (August 2012), 2654-2663.

48. Andy Coghlan, "Dear God, please confirm what I already believe," *New Scientist* (November 30, 2009), https://www.newscientist.com/article/dn18216-dear-god-please-confirm-what-i-already-believe/ (accessed November 22, 2020).

49. David Stove, *Darwinian Fairytales: Selfish Genes, Errors of Heredity, and Other Fables of Evolution* (New York: Encounter Books, 1995, reprinted 2007).

50. Jerry Fodor and Massimo Piatelli-Palmarini, *What Darwin Got Wrong* (New York: Farrar, Straus and Giroux, 2010).

51. *Alas, Poor Darwin: Arguments Against Evolutionary Psychology*, eds. Hilary Rose and Steven Rose (New York: Vintage, 2001).

52. Sharon Begley, "Why Scientists Need to Change Their Minds," *Newsweek* (January 2, 2009), https://www.newsweek.com/begley-why-scientists-need-change-their-minds-78309 (accessed November 22, 2020).

53. Begley, "Can We Blame Our Bad Behavior on Stone-Age Genes?"

54. Andrew Scull, "Psychiatry's Legitimacy Crisis," *Los Angeles Review of Books* (August 8, 2012), https://lareviewofbooks.org/article/psychiatrys-legitimacy-crisis (accessed November 22, 2020), reviewing Jerome C. Wakefield and Allan V. Horwitz, *All We Have to Fear: Psychiatry's Transformation of Natural Anxieties into Mental Disorders* (Oxford, UK: Oxford University Press, 2012).

55. Hank Campbell, "Evolutionary Psychology Can't Be Wrong, Says Evolutionary Psychologist," *Science 2.0* (December 14, 2011), https://www.science20.com/science_20/blog/evolutionary_psychology_cant_be_wrong_says_evolutionary_psychologist-85539 (accessed November 22, 2020).

56. Subrena E. Smith, "Is Evolutionary Psychology Possible?," *Biological Theory* 15 (2020), 39-49.

57. Smith, "Is Evolutionary Psychology Possible?"

58. Smith, "Is Evolutionary Psychology Possible?"

59. Aaron Goetz and Kayla Causey, "Sex Differences in Perceptions of Infidelity: Men Often Assume the Worst," *Evolutionary Psychology* 7 (2009), 253-263.

60. Smith, "Is Evolutionary Psychology Possible?"

61. Several examples of unexpected support are discussed at "Philosopher flattens evolutionary psychology," *Mind Matters News* (May 22, 2020), https://mindmatters.ai/2020/05/philosopher-flattens-evolutionary-psychology/ (accessed November 22, 2020).

62. Smith, "Is Evolutionary Psychology Possible?"

63. Elizabeth Fernandez, "Why 'Fatherhood' Is Unique to Humans Among the Primates," *Forbes* (May 17, 2020), https://www.forbes.com/sites/fernandezelizabeth/2020/05/17/why-fatherhood-is-unique-to-humans-among-the-primates/ (accessed November 20, 2020).

64. Boston College, "Reexamining the origins of human fatherhood," *ScienceDaily* (May 13, 2020), https://www.sciencedaily.com/releases/2020/04/200428093510.htm (accessed November 22, 2020); Ingela Alger, Paul L. Hooper, Donald Cox, Jonathan Stieglitz, and Hillard S. Kaplan, "Paternal provisioning results from ecological change," *Proceedings of the National Academy of Sciences* 117 (May 2020), 10746-10754.

65. Arianne Shahvisi, "Pregnant Women 'Nest.' But There's Nothing Biological About It," *Psyche* (July 22, 2020), https://psyche.co/ideas/pregnant-women-nest-but-theres-nothing-biological-about-it (accessed November 22, 2020).

66. Shahvisi, "Pregnant Women 'Nest.' But There's Nothing Biological About It."

67. Martin Graff, "Will the Coronavirus Threat Lead to Female Infidelity?," *Psychology Today* (May28, 2020), https://www.psychologytoday.com/us/blog/love-digitally/202005/will-the-coronavirus-threat-lead-female-infidelity (accessed November 22, 2020).

68. Denyse O'Leary, "How naturalism rots science from the head down," *Evolution News and Science Today* (May 12, 2017), https://evolutionnews.org/2017/05/how-naturalism-rots-science-from-the-head-down/ (accessed November 22, 2020).

Chapter 34—Does Darwinism Make Theological Assumptions?

1. Charles Darwin, *On the Origin of Species*, 6th ed. (London, UK: John Murray, 1872), 47.

2. Chris Cosans, "Was Darwin a Creationist?," *Perspectives in Biology and Medicine* 48 (2005), 362-371.

3. This number could be higher because multiple arguments are sometimes bundled together. On the other hand, some arguments are repeated. Cornelius Hunter, "Darwin's Principle: The Use of Contrastive Reasoning in the Confirmation of Evolution," *History of the Philosophy of Science* 4 (2014), 106-149.

4. Hunter, "Darwin's Principle."

5. Darwin, *Origin of Species*, 66.

6. Darwin, *Origin of Species*, 122.

7. Robert Chambers, *Vestiges of the Natural History of Creation* (London, UK: John Churchill, 1844).

8. Alfred Russel Wallace, *The Wonderful Century: Its Successes and Its Failures* (New York: Dodd, Mead, 1898), 138.

9. Charles Darwin, reprinted in *The Foundations of the Origin of Species: Two essays written in 1842 and 1844*, ed. Francis Darwin (Cambridge, UK: Cambridge University Press, 1909), 172-173.

10. Charles Darwin, "B notebook" (1838) reprinted in *Metaphysics, Materialism, and the Evolution of Mind: Early Writings of Charles Darwin*, ed. P.H. Barrett (Chicago, IL: University of Chicago Press, 1980), 216; Charles Darwin, "Notebook B: Transmutation of species" (1837–1838), transcribed by Kees Rookmaaker (2007), based on P.H. Barrett. "A transcription of Darwin's first notebook [B] on 'Transmutation of species,'" *Bulletin of the Museum of Comparative Zoology, Harvard* 122 (1960), 245-296, corrected against the microfilm. Deletions, punctuation, and page numbers were added. Corrections by John van Wyhe.

11. Cornelius Hunter, *Darwin's God: Evolution and the Problem of Evil* (Grand Rapids, MI: Brazos, 2001); Cosans, "Was Darwin a Creationist?"; Elliott Sober, *Evidence and Evolution: The Logic Behind the Science* (Cambridge, UK: Cambridge University Press, 2008); Kenneth C. Waters, "The arguments in the *Origin of Species*," *The Cambridge Companion to Darwin*, 2d ed., eds. Jonathan Hodge and Gregory Radick (Cambridge, UK: Cambridge University Press, 2009), 116-139.

12. For a discussion see Hunter, "Darwin's Principle."

13. Cosans, "Was Darwin a Creationist?," 364.

14. Ernst Mayr, *Toward a New Philosophy of Biology: Observations of an Evolutionist* (Cambridge, MA: Harvard University Press, 1988), 192.

15. Stephen Jay Gould, "The Panda's Thumb," *The Panda's Thumb: More Reflections in Natural History* (New York: Norton, 1980), 19-26.

Chapter 35—How Has Darwinism Negatively Impacted Society?

1. Daniel Dennett, *Darwin's Dangerous Idea: Evolution and the Meanings of Life* (New York: Touchstone, 1995), 18.

2. Paul Barrett et al., *Charles Darwin's Notebooks, 1836–1844* (New York: Cornell University Press, 1987), "Notebook B," #74, 189.

3. Barrett et al., *Charles Darwin's Notebooks*, #207, 222-223.

4. Quoted in Robert Lee Hotz, "Full Sequence of Fly's Genes Deciphered," *Los Angeles Times* (March 24, 2000), https://www.latimes.com/archives/la-xpm-2000-mar-24-mn-12253-story.html (accessed November 24, 2020).

5. Quoted in Maggie Fox, "Fly Gene Map May Have Many Uses, Scientists Say," *Reuters* (March 23, 2000), http://dailynews.yahoo.com/h/nm/20000323/sc/fly_uses_2.html (accessed April 3, 2000).

6. Patricia Reaney, "Are You Man or Mouse? Check Your Genes…," *Reuters* (December 4, 2002), http://story.news.yahoo.com/news?tmpl=story2&cid=570&u=/nm/20021204/ (accessed December 4, 2003).

7. Derek E. Wilman, Monica Uddin, Guozhen Liu, Lawrence Grossman, and Morris Goodman, "Implications of natural selection in shaping 99.4% nonsynonmnous DNA identity between humans and chimpanzees: Enlarging genus *Homo*," *Proceedings of the National Academy of Sciences* 100 (June 10, 2003), 7181-7188. Goodman is identified as the contributor of this article to *Proceedings*.

8. John Derbyshire, "What's So Scary About Evolution?—for Both Right and Left, a Lot," *Taki's Magazine* (May

19, 2008), http://www.johnderbyshire.com/Opinions/HumanSciences/darwin.html (accessed November 24, 2020), emphasis in original.

9. Quoted in Johann Hari, "Peter Singer—an Interview," originally run in *The Independent* (January 7, 2004), https://web.archive.org/web/20060317041348/http://www.johannhari.com/archive/article.php?id=410 (accessed November 24, 2020).

10. Nora Barlow, ed., *The Autobiography of Charles Darwin, 1809–1882, with Original Omissions Restored* (New York: Norton, 1969), 87.

11. Letter from Nobel Laureates to Kansas State Board of Education (September 9, 2005), https://web.archive .org/web/20051103170647/http://media.ljworld.com/pdf/2005/09/15/nobel_letter.pdf (accessed November 24, 2020).

12. George Gaylord Simpson, *The Meaning of Evolution: A Study of the History of Life and of Its Significance for Man*, rev. ed. (New Haven, CT: Yale University Press, 1967), 345.

13. Charles Darwin, *On the Origin of Species by Means of Natural Selection*, 1st ed. (London, UK: John Murray, 1859), 490.

14. Stephen Jay Gould, *Ontogeny and Phylogeny* (Cambridge, MA: Belknap Press / Harvard, 1977), 127.

15. Charles Darwin, *The Descent of Man, and Selection in Relation to Sex, Vol. I* (Princeton, NJ: Princeton University Press, 1981), 109-110.

16. Darwin, *The Descent of Man*, 201. For a further discussion of Darwinian racism, see John G. West, *Darwin Day in America: How Our Politics and Culture Have Been Dehumanized in the Name of Science*, expanded paperback edition (Wilmington, DE: ISI Books, 2015), 144-150 and accompanying notes.

17. Richard Weikart, *From Darwin to Hitler: Evolutionary Ethics, Eugenics, and Racism in Germany* (New York: Palgrave Macmillan, 2004), 106-107.

18. Charles Davenport, "Scientific Cooperation with Nature: Eugenics," typescript in Charles Davenport Papers, Collection B D27 (Philadelphia, PA: American Philosophical Society), 2.

19. Quoted in Weikart, *From Darwin to Hitler*, 206.

20. See Nancy J. Parezo and Don D. Fowler, *Anthropology Goes to the Fair: The 1904 Louisiana Purchase Exposition* (Lincoln, NE: Nebraska University Press, 2007); *Human Zoos: America's Forgotten History of Scientific Racism* (Seattle, WA: Discovery Institute Press, 2018), https://www.youtube.com/watch?v=nY6Zrol5QEk (accessed November 24, 2020). More generally, see Pascal Blanchard, Nicolas Bancel et al., *Human Zoos: Science and Spectacle in the Age of Colonial Empires*, trans. Teresa Bridgeman (Liverpool, UK: Liverpool University Press, 2008).

21. See *Human Zoos: America's Forgotten History of Scientific Racism*; Pamela Newkirk, *Spectacle: The Astonishing Life of Ota Benga* (New York: Amistad, 2015); Phillips Verner Bradford and Harvey Blume, *Ota Benga: The Pygmy in the Zoo* (New York: Delta, 1992).

22. West, *Darwin Day in America*, 134-135.

23. See Charlotte Hunt-Grubbe, "The Elementary DNA of Dr. Watson," *The Sunday Times* (October 14, 2007), https://www.thetimes.co.uk/article/the-elementary-dna-of-dr-watson-gllb6w2vpdr (accessed November 24, 2020); James D. Watson, *Avoid Boring People: Lessons from a Life in Science* (New York: Alfred Knopf, 2007), 326.

24. *Human Zoos: America's Forgotten History of Scientific Racism*; David Klinghoffer, "Evolution and the Alt-Right," *Evolution News and Science Today* (April 14, 2016), https://evolutionnews.org/2016/04/evolution_and_t_1/ (accessed November 24, 2020); David Klinghoffer, "Evolution and the Alt-Right, Continued," *Evolution News and Science Today* (September 1, 2017), https://evolutionnews.org/2017/09/evolution-and-the-alt-right-continued/ (accessed November 24, 2020).

25. Darwin, *Descent of Man*, 168.

26. Daniel J. Kevles, *In the Name of Eugenics: Genetics and the Uses of Human Heredity* (Cambridge, MA: Harvard University Press, 1995), xiii.

27. West, *Darwin Day in America*, 161.

28. Edward M. East, *Heredity and Human Affairs* (New York: Charles Scribner's Sons, 1927), 237.

29. Edwin Conklin, "Value of Negative Eugenics," *Journal of Heredity* 6 (December 1915), 539-540.

30. Mark H. Haller, *Eugenics: Hereditarian Attitudes in American Thought* (New Brunswick, NJ: Rutgers University Press, 1963), 141.

31. "Remembering the 'forgotten victims' of Nazi 'euthanasia' murders," *Deutsche Welle* (January 26, 2017), https://

www.dw.com/en/remembering-the-forgotten-victims-of-nazi-euthanasia-murders/a-37286088 (accessed November 24, 2020); Leo Alexander, "Medical Science Under Dictatorship," *New England Journal of Medicine* 241 (July 14, 1949), 39-47; Michael Burleigh, *Death and Deliverance: "Euthanasia" in Germany 1900–1945* (Cambridge, UK: Cambridge University Press, 1994).

32. See West, *Darwin Day in America,* 325-333.

33. Testimony of Dr. James Neel, May 20, 1981, in *The Human Life Bill: Hearings Before the Subcommittee on Separation of Powers of the Committee on the Judiciary, United States Senate, Ninety-Seventh Congress, First Session, on S. 158, a Bill to Provide that Human Life Shall be Deemed to Exist from Conception, April 23, 24; May 20, 21; June 1, 10, 12 and 18.* Serial No. J-97-16 (Washington, DC: U. Government Printing Office, 1982), 77.

34. Carl Sagan and Ann Druyan, "Is It Possible to Be Pro-Life And Pro-Choice?," *Parade* (April 22, 1990), 6.

35. Christopher Hitchens, *God Is Not Great: How Religion Poisons Everything* (New York: Twelve, 2007), 221.

36. Alexander Sanger, *Beyond Choice: Reproductive Freedom and the 21st Century* (New York: Public Affairs, 2004), 292.

37. Jerry Coyne, "Should one be allowed to euthanize severely deformed or doomed newborns?," *Why Evolution Is True* (July 13, 2017), https://whyevolutionistrue.com/2017/07/13/should-one-be-allowed-to-euthanize-severely-deformed-or-doomed-newborns/ (accessed November 24, 2020).

38. Coyne, "Should one be allowed to euthanize severely deformed or doomed newborns?"

39. Wesley J. Smith, *The War on Humans* (Seattle, WA: Discovery Institute Press, 2014).

40. Eric R. Pianka, "The Vanishing Book of Life on Earth," 21, http://www.zo.utexas.edu/courses/bio373/Vanishing.Book.pdf (accessed November 24, 2020).

41. Pianka, "The Vanishing Book of Life on Earth, 10.

42. Pianka, "The Vanishing Book of Life on Earth, 19.

43. Pianka, "The Vanishing Book of Life on Earth, 17; Jamie Mobley, "Doomsday: UT prof says death is imminent," *Seguin Gazette-Enterprise* (February 27, 2010), http://seguingazette.com/news/article_11e24a27-e97e-53f0-9d3b-29875b3d72e7.html (accessed November 24, 2020).

44. Christopher Manes, *Green Rage: Radical Environmentalism and the Unmaking of Civilization* (Boston, MA: Little, Brown, 1990), 142.

45. James Lee, "The Discovery Channel MUST broadcast to the world their commitment to save the planet and to do the following IMMEDIATELY," archived at https://web.archive.org/web/20101220024835/http://savetheplanetprotest.com/ (accessed November 24, 2020).

46. Michael Ruse and E.O. Wilson, "The Evolution of Ethics," *Religion and the Natural Sciences: The Range of Engagement,* ed. James E. Huchingson (New York: Harcourt Brace Jovanovich, 1993), 310.

47. West, *Darwin Day in America,* 51-55.

48. Cesare Lombroso, *Crime: Its Causes and Remedies,* trans. Henry Horton (Montclair, NJ: Patterson Smith, 1968), 376.

49. Enrico Ferri, "The Positive School of Criminology," *Criminology: A Book of Readings,* eds. Clyde Vedder, Samuel Koenig, and Robert Clark (New York: The Dryden Press, 1953), 137-138.

50. Robert Wright, *The Moral Animal: Evolutionary Psychology and Everyday Life* (New York: Vintage Books, 1995), 350.

51. Wright, *The Moral Animal,* 37.

52. Wright, *The Moral Animal,* 88.

53. Darwin, *Descent of Man, Vol. II,* 362. Also see discussion in John G. West, *Darwin's Conservatives: The Misguided Quest* (Seattle, WA: Discovery Institute Press, 2006), 19-32; West, *Darwin Day in America,* 23-42.

54. John G. West, *Darwin's Corrosive Idea: The Impact of Evolution on Attitudes About Faith, Ethics, and Human Uniqueness* (Seattle, WA: Discovery Institute Press, 2016), 10.

55. See West, *Darwin Day in America,* 271-290.

56. Christopher Ryan, quoted in Thomas Rogers, "'Sex at Dawn': Why monogamy goes against our nature," *Salon* (June 27, 2010), https://www.salon.com/2010/06/27/sex_at_dawn_interview/ (accessed November 24, 2020). See also Christopher Ryan and Cacilda Jetha, *Sex at Dawn: How We Mate, Why We Stray, and What It Means for Modern Relationships* (New York: Harper Perennial, 2011); Christopher Ryan and Calcida Jetha, "Open Marriage:

We Don't Believe in Monogamy," *The Times* (July 24, 2010), https://www.thetimes.co.uk/article/open-marriage -we-dont-believe-in-monogamy-vz6lm77bfzc (accessed November 24, 2020).

57. Richard Dawkins, *The Blind Watchmaker: Why the Evidence of Evolution Reveals a Universe Without Design* (New York: Norton, 1996), 6.

58. Edward O. Wilson, *Consilience: The Unity of Knowledge* (New York: Alfred Knopf, 1998), 241.

59. David Barash, "God, Darwin, and My Biology Class," *The New York Times* (September 27, 2014), https://www .nytimes.com/2014/09/28/opinion/sunday/god-darwin-and-my-college-biology-class.html (accessed November 24, 2020).

60. West, *Darwin's Corrosive Idea*, 7, 10. The Dawkins quote is from Richard Dawkins, *River Out of Eden: A Darwinian View of Life* (New York: Basic Books, 1995), 133.

61. John Polkinghorne, *Quarks, Chaos, and Christianity*, rev. ed. (Chestnut Ridge, NY: Crossroad Publishing, 2006), 113.

62. Kenneth R. Miller, *Finding Darwin's God: A Scientist's Search for Common Ground Between God and Evolution* (New York: HarperCollins, 1999), 272; see also 244.

63. George V. Coyne, S.J., "The Dance of the Fertile Universe," 7, https://web.archive.org/web/20051104052227/ http://www.aei.org/docLib/20051027_HandoutCoyne.pdf (accessed November 24, 2020).

64. Karl Giberson, *Saving Darwin: How to Be a Christian and Believe in Evolution* (New York: HarperOne, 2008), 12.

65. Francis S. Collins, *The Language of God: A Scientist Presents Evidence for Belief* (New York: Free Press, 2006), 205.

66. See John G. West, "Nothing New Under the Sun: Theistic Evolution, the Early Church, and the Return of Gnosticism, Part 1," *God and Evolution: Protestants, Catholics, and Jews Explore Darwin's Challenge to Faith*, ed. Jay Richards (Seattle, WA: Discovery Institute Press, 2010), 33-52; *The Patristic Understanding of Creation: An Anthology of Writings from the Church Fathers on Creation and Design*, eds. William A. Dembski, Wayne J. Downs, and Father Justin B.A. Frederick (Riesel, TX: Erasmus Press, 2008).

67. Dennett, *Darwin's Dangerous Idea*, 63.

68. For recent examples, see Stephen C. Meyer, *Darwin's Doubt: The Explosive Origin of Animal Life and the Case for Intelligent Design* (New York: HarperOne, 2013); Michael Behe, *Darwin Devolves: The New Science About DNA That Challenges Evolution* (New York: HarperOne, 2019); Douglas Axe, *Undeniable: How Biology Confirms Our Intuition That Life Is Designed* (New York: HarperOne, 2017); Michael Denton, *Evolution: Still a Theory in Crisis* (Seattle, WA: Discovery Institute Press, 2016).

69. Robert B. Laughlin, *A Different Universe: Reinventing Physics from the Bottom Down* (New York: Basic Books, 2005), 168-169.

Chapter 36—Do Scientists Have the Intellectual Freedom to Challenge Darwinism?

1. Quoted in Lincoln Barnett, "J. Robert Oppenheimer," *Life* (October 10, 1949), 136.

2. Giuseppe Sermonti, "Darwin Is a Prime Number," *Rivista di Biologia* 95 (2002), 10.

3. Letter from Nobel Laureates to Kansas State Board of Education (September 9, 2005), https://web.archive .org/web/20060104191526/http://media.ljworld.com/pdf/2005/09/15/nobel_letter.pdf (accessed November 24, 2020). The letter was sent out under the auspices of the Elie Wiesel Foundation.

4. See Michael Flannery, *Nature's Prophet: Alfred Russel Wallace and His Evolution from Natural Selection to Natural Theology* (Tuscaloosa, AL: University of Alabama Press, 2018); *Darwin's Heretic: Did the Co-Founder of Evolution Embrace Intelligent Design?* (Seattle, WA: Discovery Institute Press, 2011), https://www.youtube.com/ watch?v=hxvAVln6HLI (accessed November 24, 2020).

5. ColdWater Media Interview with Stephen C. Meyer, Spokane, WA (September 7, 2001).

6. For an annotated listing of relevant articles, see "Bibliography of Supplementary Resources for Ohio Science Instruction," *Discovery Institute* (Seattle, WA: Discovery Institute Press, March 11, 2002), https://www.discovery .org/a/1127/ (accessed November 24, 2020); "Bibliographic and Annotated List of Peer-Reviewed Publications Supporting Intelligent Design," *Discovery Institute* (Seattle, WA: Discovery Institute Press, July 2017), https:// www.discovery.org/m/2018/12/ID-Peer-Review-July-2017.pdf (accessed November 24, 2020).

7. W. Daniel Hillis quoted in John Brockman, *The Third Culture: Beyond the Scientific Revolution* (New York: Simon & Schuster, 1995), 26.

8. Stephen Meyer, "Danger: Indoctrination, a Scopes Trial for the 90s," *The Wall Street Journal* (December 6, 1993), https://www.discovery.org/a/93/ (accessed November 24, 2020).

9. Geoff Brumfiel, "Intelligent design: Who has designs on your students' minds?," *Nature* 434 (April 28, 2005), 1062-1065.

10. See "Intelligent Design and Academic Freedom," *All Things Considered, National Public Radio* (November 10, 2005).

11. See Gordon Gregory, "Biology instructor's doctrine draws fire," *OregonLive.com* (February 18, 2000); Gordon Gregory, "Creationist instructor likely will lose his job," *OregonLive.com* (March 28, 2000); Julie Foster, "Biology professor forced out; Pointed to flaws in theory of evolution, encouraged critical thinking," *WorldNetDaily.com* (April 14, 2000), https://web.archive.org/web/20010427122836/http://www.wnd.com/news/article.asp?ARTICLE_ID=17856 (accessed November 25, 2020).

12. Haley's former department chair Bruce McClelland, quoted in Gregory, "Biology instructor's doctrine draws fire."

13. Fred Heeren, "The Lynching of Bill Dembski: Scientists say the jury is out—so let the hanging begin," *The American Spectator* 33 (November 2000), 44-51.

14. Testimony of Nancy Bryson before the Texas State Board of Education, *Transcript of the Public Hearing Before the Texas State Board of Education, September 10, 2003, Austin, Texas* (Austin, TX: Chapman Court Reporting Service, 2003), 504-505.

15. Email from University of Kentucky physicist Thomas Troland, quoted in Casey Luskin, "E-mails in Gaskell Case Show That Darwin Skeptics Need Not Apply to the University of Kentucky," *Evolution News and Views* (February 10, 2011), https://evolutionnews.org/2011/02/e-mails_in_gaskell_case_show_t/ (accessed November 24, 2020).

16. Casey Luskin, "Evidence of Discrimination Against Martin Gaskell Due to His Views on Evolution," *Evolution News and Views* (December 15, 2010), https://evolutionnews.org/2010/12/evidence_of_discrimination_aga/ (accessed November 24, 2020).

17. "Refereed Publications [of Eric Hedin]," Ball State University, https://web.archive.org/web/20130526183917/http://cms.bsu.edu/-/media/WWW/DepartmentalContent/Physics/PDFs/Hedin/PublicationsHedin%20(3).pdf (accessed November 24, 2020).

18. John G. West, "Misrepresenting the Facts about Eric Hedin's 'Reading List'," *Evolution News and Views* (July 11, 2013), https://evolutionnews.org/2013/07/misrepresenting/ (accessed November 24, 2020).

19. Joshua Youngkin, "What Does Eric Hedin Really Teach? Self-Professed Agnostic Speaks Out About 'Boundaries of Science' Seminar," *Evolution News and Views* (August 2, 2013), https://evolutionnews.org/2013/08/what_does_eric/ (accessed November 24, 2020); Joshua Youngkin, "Hedin Witness #3: 'This Course Made Me a Better Learner,'" *Evolution News and Views* (August 9, 2013), https://evolutionnews.org/2013/08/hedin_witness_3/ (accessed November 24, 2020); Joshua Youngkin, "Dr. Hedin's Student Could Teach Ball State University a Thing or Two," *Evolution News and Views* (July 16, 2013), https://evolutionnews.org/2013/07/what_happened_i/ (accessed November 24, 2020).

20. David Klinghoffer, "At Ball State University, Intimidation Campaign Against Physicist Gets Troubling Results," *Evolution News and Views* (May 22, 2013), https://evolutionnews.org/2013/05/at_ball_state_u/ (accessed November 24, 2020).

21. John G. West, "Questions Raised About Impartiality of Panel Reviewing Ball State University Professor's Course," *Evolution News and Views* (June 25, 2013), https://evolutionnews.org/2013/06/review_panel_or/ (accessed November 24, 2020); Joshua Youngkin, "Indiana Professors Question Ball State University's Disregard For Rules on Academic Freedom," *Evolution News and Views* (August 25, 2013), https://evolutionnews.org/2013/08/indiana_profess/ (accessed November 24, 2020); John G. West, "Clarifying the Issues At Ball State: Some Questions and Answers," *Evolution News and Views* (September 13, 2013), https://evolutionnews.org/2013/09/clarifying_the_/ (accessed November 24, 2020).

22. "Atheist Rift!!," *BSU Freethought Alliance: The Official Blog of Ball State University Freethought Alliance* (October 23, 2009), http://freethoughtbsu.blogspot.com/2009/10/atheist-rift.html (accessed November 24, 2020).

23. John G. West, "Ball State President's Orwellian Attack on Academic Freedom," *Evolution News and Views* (August 1, 2013), https://evolutionnews.org/2013/08/ball_state_pres/ (accessed November 24, 2020).

24. John West, "Scandal Brewing at Baylor University? Denial of Tenure to Francis Beckwith Raises Serious Questions About Fairness and Academic Freedom," *Evolution News and Views* (March 28, 2006), https://evolutionnews.org/2006/03/scandal_at_baylor_university_d/ (accessed November 24, 2020).

25. See, for example, Francis J. Beckwith, *Law, Darwinism, and Public Education: The Establishment Clause and the Challenge of Intelligent Design* (Lanham, MD: Rowman & Littlefield, 2003); Francis J. Beckwith, "Science and

Religion Twenty Years After *McLean v. Arkansas*: Evolution, Public Education, and the New Challenge of Intelligent Design," *Harvard Journal of Law & Public Policy* 26 (Spring 2003), 455-499; Francis J. Beckwith, "Public Education, Religious Establishment, and the Challenge of Intelligent Design," *Notre Dame Journal of Law, Ethics, & Public Policy* 17 (2003), 461-519; Francis J. Beckwith, "A Liberty Not Fully Evolved?: The Case of Rodney LeVake and the Right of Public School Teachers to Criticize Darwinism," *San Diego Law Review* 39 (November/ December 2002), 1311-1325.

26. Robert Crowther, "Welcome News as Scholar Francis Beckwith Is Granted Tenure at Baylor," *Evolution News & Views* (September 27, 2006), https://evolutionnews.org/2006/09/welcome_news_as_scholar_franci/ (accessed November 24, 2020).

27. For information about the Bryan Leonard case, see Catherine Candinsky, "Evolution debate re-emerges: Doctoral student's work was possibly unethical, OSU professors argue," *The Columbus Dispatch* (June 9, 2005); "Attack on OSU Graduate Student Endangers Academic Freedom," *Discovery Institute* (April 18, 2005), https://www.discovery.org/a/2661/ (accessed November 24, 2020); "Professors Defend Ohio Grad Student Under Attack by Darwinists," *Discovery Institute* (July 11, 2005), https://www.discovery.org/a/2715/ (accessed November 24, 2020).

28. For information and documentation about the Coppedge case, see Robert Crowther, "Trial to Begin in Intelligent Design Discrimination Lawsuit against NASA's Jet Propulsion Lab," *Evolution News and Views* (March 5, 2012), https://evolutionnews.org/2012/03/trial_to_begin_/ (accessed November 24, 2020); "Facts of the Coppedge Lawsuit Contradict the Spin from Jet Propulsion Lab and National Center for Science Education," *Evolution News and Views*, March 12, 2012, https://evolutionnews.org/2012/03/facts_of_the_co/ (accessed November 24, 2020); Joshua Youngkin, "Why Did NASA's JPL Discriminate Against David Coppedge and Why Does It Matter?" *Evolution News and Views* (November 22, 2011), https://evolutionnews.org/2011/11/what_happened_t/ (accessed November 24, 2020).

29. See David Klinghoffer, "The Branding of a Heretic," *The Wall Street Journal*, January 28, 2005, https://www.wsj.com/articles/SB110687499948738917 (accessed November 24, 2020). For more information about the controversy surrounding the publication of the journal article supportive of intelligent design, see "Sternberg, Smithsonian, Meyer, and the Paper That Started It All," https://www.discovery.org/a/2399/ (accessed November 24, 2020); Richard Sternberg, "Smithsonian Controversy," http://www.richardsternberg.com/smithsonian.php (accessed November 24, 2020).

30. Letter to Richard Sternberg from the US Office of Special Counsel, August 5, 2005, available at http://www.richardsternberg.com/smithsonian.php?page=letter (accessed November 24, 2020). Also see Klinghoffer, "The Branding of a Heretic."

31. *Intolerance and the Politicization of Science at the Smithsonian: Smithsonian's Top Officials Permit the Demotion and Harassment of Scientist Skeptical of Darwinian Evolution,* Staff Report Prepared for the Hon. Mark Souder, Chairman, Subcommittee on Criminal Justice, Drug Policy and Human Resources (Washington, DC: US House of Representatives, Committee on Government Reform, December 11, 2006), 3, 20-21, https://www.discovery.org/m/securepdfs/2020/11/IntoleranceandthePoliticizationofScienceattheSmithsonian.pdf (accessed November 26, 2020).

32. *Intolerance and the Politicization of Science at the Smithsonian*, 4.

33. *Intolerance and the Politicization of Science at the Smithsonian*, 5-6.

34. *Intolerance and the Politicization of Science at the Smithsonian*, 22, emphasis in original. The congressional report further explained, "Dr. Sues hoped that the NCSE could unearth evidence that Dr. Sternberg had misrepresented himself as a Smithsonian employee, which would have been grounds for his dismissal as a Research Associate: 'As a Research Associate, Sternberg is not allowed to represent himself as an employee of the Smithsonian Institution, and, if he were to do so, he would forfeit his appointment.'"

35. *Intolerance and the Politicization of Science at the Smithsonian*, 23, emphasis in original.

36. Quoted in Michael Powell, "Editor Explains Reasons for 'Intelligent Design' Article," *The Washington Post* (August 19, 2005), A19.

37. Quoted in *Rodney LeVake vs. Independent School District #656,* State of Minnesota Court of Appeals, C8-00-1613 (May 8, 2001); https://web.archive.org/web/20130314100547/http://www.lawlibrary.state.mn.us/archive/ctappub/0105/c8001613.htm (accessed November 24, 2020). Additional information on the LeVake case can be found in James Kilpatrick, "Case of Scientific Heresy is Doomed," *Augusta Chronicle* (December 23, 2001), A4.

The Minnesota Court of Appeals found that the school district's interest in maintaining its curriculum overrode LeVake's First Amendment interest in teaching material critical of Darwinian evolution.

38. John G. West, *Darwin Day in America: How Our Politics and Culture Have Been Dehumanized in the Name of Science* (Wilmington, DE: ISI Books, 2007), 231-232, 234-238.

39. Peter Machamer, Marcello Pera, and Aristedes Baltas, *Scientific Controversies: Philosophical and Historical Perspectives* (New York: Oxford University Press, 2000), 6. It should be noted that the authors here are summarizing what they call the "historical-social view of science" that has developed among scholars since the late 1950s. Although they think this view "has done more justice to science as actually practiced than the logical positivists' rational reconstructions," they also believe that it has shortcomings of its own. Machamer et al., *Scientific Controversies*, 6-7.

40. See the helpful discussion of the suppression of dissenting views in Gordon Moran, *Silencing Scientists and Scholars in Other Fields: Power, Paradigm Controls, Peer Review, and Scholarly Communication* (Greenwich, CT: Ablex Publishing, 1998).

41. Jonathan Wells, "Critics Rave Over *Icons of Evolution*: A Response to Published Reviews," *Discovery Institute* (June 12, 2002), https://www.discovery.org/a/1180/ (accessed November 24, 2020).

42. Michael Ruse, "How Evolution Became a Religion," *National Post* (May 13, 2000), B1. See also Michael Ruse, *Darwinism as Religion* (New York: Oxford University Press, 2016).

43. For examples, see Brian Leiter, "Biology Textbooks under Attack," *The Leiter Reports* (August 11, 2003), https://web.archive.org/web/20060622172656/http://webapp.utexas.edu/blogs/archives/bleiter/000146.html (accessed November 24, 2020); "Statement of the Reverend Mark Belletini," Ohio Citizens for Science, http://ecology.cwru.edu/ohioscience/state-belletini.asp (accessed July 6, 2002).

44. P.Z. Myers, Comment #35130, *Panda's Thumb* (June 14, 2005), https://web.archive.org/web/20060219014629/http://www.pandasthumb.org/archives/2005/06/a_new_recruit.html (accessed November 24, 2020).

45. P.Z. Myers, "Perspective," *Pharyngula* (August 4, 2005), https://web.archive.org/web/20051212081222/http://pharyngula.org/index/weblog/comments/perspective/ (accessed November 25, 2020).

46. Comment posted by bobxxxx, *The Nevada Sagebrush* (October 20, 2009), https://web.archive.org/web/20091101165531/http://nevadasagebrush.com/blog/2009/10/19/intelligent-design-theory-insults-science-religion/ (accessed November 24, 2020).

47. Comment posted by Jumbalaya, *The Nevada Sagebrush* (October 21, 2009), https://web.archive.org/web/20091101165531/http://nevadasagebrush.com/blog/2009/10/19/intelligent-design-theory-insults-science-religion/ (accessed November 24, 2020).

48. Casey Luskin, "Bullies-R-Us: How 'Freethought Oasis' Threatened 'Disruption' and Pressured a College into Cancelling Intelligent Design Course," *Evolution News and Views* (December 10, 2013), https://evolutionnews.org/2013/12/freethought-oasis-bullies/ (accessed November 24, 2020).

49. "California Science Center Pays $110,000 to Settle Intelligent Design Discrimination Lawsuit," *Evolution News and Views* (August 29, 2011), https://evolutionnews.org/2011/08/california_science_center_pays/ (accessed November 24, 2020); Robert Crowther, "Los Angeles Times Reporting on Lawsuit Against California Science Center for Cancelling Intelligent Design Film," *Evolution News and Views* (December 29, 2009), https://evolution news.org/2009/12/los_angeles_times_reporting_on/ (accessed November 24, 2020); John G. West, "Why the California Science Center's Censorship of Pro-Intelligent Design Film Is a Big Deal," *Evolution News and Views* (December 30, 2009), https://evolutionnews.org/2009/12/why_the_california_science_cen/ (accessed November 24, 2020); John G. West, "California Science Center Engaged in Illegal Cover-Up to Hide the Truth About Its Censorship of Pro-Intelligent Design Film," *Evolution News and Views* (January 4, 2010), https://evolutionnews.org/2010/01/california_science_center_enga/ (accessed November 24, 2020).

50. Anika Smith, "Darwinists Launch Cyber Attack Against Intelligent Design Website," *Evolution News and Views* (October 28, 2009), https://evolutionnews.org/2009/10/darwinists_launch_cyber_attack/ (accessed November 24, 2020). For a later cyberattack, see John G. West, "Cyber Attacks Attempt to Shut Down Discovery Institute's Websites on Day of Event Challenging Darwinism," *Evolution News and Views* (September 23, 2013), https://evolutionnews.org/2010/09/cyber_attacks_attempt_to_shut_/ (accessed November 26, 2020).

51. David Gelernter, "Giving Up Darwin," *The Claremont Review of Books* (Spring 2019), https://claremontreviewofbooks.com/giving-up-darwin/ (accessed November 24, 2020).

52. For a sampling of reactions, see Jerry Coyne, "David Gelernter Is Wrong About Ditching Darwin," *Quillette* (September 9, 2019), https://quillette.com/2019/09/09/david-gelernter-is-wrong-about-ditching-darwin/

(accessed November 24, 2020); George Weigel, "Getting Beyond Darwin," *First Things* (August 21, 2019), https://www.firstthings.com/web-exclusives/2019/08/getting-beyond-darwin (accessed November 24, 2020); Melanie Phillips, "Darwinism, Judaism and the Clash Between Science and Religion" (September 6, 2019), https://www.melaniephillips.com/darwinism-judaism-clash-science-religion/ (accessed November 24, 2020).

53. See "A Scientific Dissent from Darwinism," https://dissentfromdarwin.org/ (accessed November 24, 2020).

54. Quoted in "100 Scientists, National Poll Challenge Darwinism" (Seattle, WA: Discovery Institute, September 24, 2001), http://www.reviewevolution.com/press/pressRelease_100Scientists.php (accessed November 24, 2020).

55. Quoted in "80 Years After Scopes Trial New Scientific Evidence Convinces Over 400 Scientists That Darwinian Evolution is Deficient," *Discovery Institute* (July 18, 2005), https://www.discovery.org/m/2005/07/scopes-evolution-update.pdf (accessed November 24, 2020).

56. Quoted in *Icons of Evolution* documentary (DVD) (Palmer Lake, CO: ColdWater Media, 2002).

57. Philip S. Skell, "Why Do We Invoke Darwin? Evolutionary theory contributes little to experimental biology," *The Scientist* 16 (August 28, 2005), 10. To make this point, Skell quotes A.S. Wilkins, editor of the journal *BioEssays*, who has written, "Evolution would appear to be the indispensable unifying idea and, at the same time, a highly superfluous one."

58. For examples of materials submitted to educational policymakers citing peer-reviewed research on these issues, see "Bibliography of Supplementary Resources"; "The Scientific Controversy over Whether Microevolution Can Account for Macroevolution," *Discovery Institute*, https://www.discovery.org/m/2004/08/micromacrosum.pdf (accessed November 24, 2020); "The Scientific Controversy over the Cambrian Explosion," *Discovery Institute*, https://www.discovery.org/f/119/ (accessed November 24, 2020); "An Analysis of the Treatment of Homology in Biology Textbooks," *Discovery Institute*, https://www.discovery.org/m/2016/03/homologyrpt.pdf (accessed November 24, 2020); "A Preliminary Analysis of the Treatment of Evolution in Biology Textbooks," *Discovery Institute*, https://www.discovery.org/a/1522/ (accessed November 24, 2020).

59. "2016 Poll: Public Opinion on Scientific Dissent," https://freescience.today/2016/07/01/public-opinion-scientific-dissent/ (accessed November 24, 2020).

60. "2016 Poll: Public Opinion on Scientific Dissent."

61. Charles Darwin, *On the Origin of Species*, 1st ed. (London, UK: John Murray, 1859), Introduction, https://www.gutenberg.org/files/1228/1228-h/1228-h.htm (accessed November 24, 2020).

Chapter 37—Is Darwinism a Theory in Crisis?

1. Michael Denton, *Evolution: A Theory in Crisis* (Bethesda, MD: Adler & Adler, 1986).

2. Michael Denton, *Evolution: Still a Theory in Crisis* (Seattle, WA: Discovery Institute Press, 2016).

3. Stephen Dilley, "Charles Darwin's use of theology in the *Origin of Species*," *British Journal for the History of Science* 45 (2012), 29-56.

4. Thomas S. Kuhn, *The Structure of Scientific Revolutions* (Chicago, IL: University of Chicago Press, 1962).

5. Thomas S. Kuhn, *The Structure of Scientific Revolutions*, 2d ed. (Chicago, IL: University of Chicago Press, 1970), 10.

6. Kuhn, *The Structure of Scientific Revolutions*, 2d ed., 19, 93.

7. Kuhn, *The Structure of Scientific Revolutions*, 2d ed., 164-166.

8. Kuhn, *The Structure of Scientific Revolutions*, 2d ed., 24.

9. Kuhn, *The Structure of Scientific Revolutions*, 2d ed., 77-79.

10. Kuhn, *The Structure of Scientific Revolutions*, 2d ed., 153.

11. Kuhn, *The Structure of Scientific Revolutions*, 2d ed., 157-158.

12. Kuhn, *The Structure of Scientific Revolutions*, 2d ed., 144.

13. Kuhn, *The Structure of Scientific Revolutions*, 2d ed., 144.

14. Kuhn, *The Structure of Scientific Revolutions*, 2d ed., 103.

15. Kuhn, *The Structure of Scientific Revolutions*, 2d ed., 103-105, 163.

16. Carl F. von Weizsäcker, *The Relevance of Science* (New York: Harper & Row, 1964), 151-153.

17. Letter from Harvey F. Lodish to Ohio Governor Bob Taft (February 24, 2004). https://www.newswise.com/articles/ascb-president-says-creationism-does-not-belong-in-ohios-classrooms (accessed August 22, 2020).

18. Statement on the Teaching of Evolution, *American Astronomical Society* (September 20, 2005). https://aas.org/press/aas-supports-teaching-evolution (accessed August 22, 2020).

19. Statement on Teaching Alternatives to Evolution, *Biophysical Society* (November 2005). https://www.biophysics.org/policy-advocacy/stay-informed/policy-issues/evolution-1 (accessed August 22, 2020).

20. Niall Shanks, *God, the Devil, and Darwin* (New York: Oxford University Press, 2006), xi–xii.

21. Kenneth R. Miller, *Only a Theory: Evolution and the Battle for America's Soul* (New York: Viking Press, 2008), 16, 190-191.

22. Scott Todd, "A view from Kansas on that evolution debate," *Nature* 401 (1999), 423.

23. Kuhn, *The Structure of Scientific Revolutions*, 2d ed., 91.

24. Massimo Pigliucci, "Do we need an extended evolutionary synthesis?," *Evolution* 61 (2007), 2743-2749.

25. Suzan Mazur, *The Altenberg 16: An Exposé of the Evolution Industry* (Wellington, New Zealand: Scoop Media, 2009).

26. Massimo Pigliucci and Gerd B. Müller, *Evolution: The Extended Synthesis* (Cambridge, MA: MIT Press, 2010).

27. James A. Shapiro, *Evolution: A View from the 21st Century* (Upper Saddle River, NJ: FT Press Science, 2011), 134-137.

28. Kevin Laland, Tobias Uller, Marc Feldman, Kim Sterelny, Gerd B. Müller, Armin Moczek, Eva Jablonka, John Odling-Smee, Gregory A. Wray, Hopi E. Hoekstra, Douglas J. Futuyma, Richard E. Lenski, Trudy F.C. Mackay, Dolph Schluter, and Joan E. Strassmann, "Does evolutionary theory need a rethink?" *Nature* 514 (2014), 161-164.

29. Kevin N. Laland, Tobias Uller, Marcus W. Feldman, Kim Sterelny, Gerd B. Müller, Armin Moczek, Eva Jablonka, and John Odling-Smee, "The extended evolutionary synthesis: its structure, assumptions and predictions," *Proceedings of the Royal Society of London B* 282 (2015), 20151019.

30. Paul A. Nelson, "Specter of intelligent design emerges at the Royal Society meeting," *Evolution News & Views* (November 8, 2016), https://evolutionnews.org/2016/11/specter_of_inte/ (accessed August 22, 2020).

31. Paul A. Nelson and David Klinghoffer, "Scientists confirm: Darwinism is broken," *CNS News* (December 13, 2016), https://www.cnsnews.com/commentary/david-klinghoffer/scientists-confirm-darwinism-broken (accessed August 22, 2020).

32. Alejandro Fábregas-Tejeda and Francisco Vergara-Silva, "Hierarchy Theory of Evolution and the Extended Evolutionary Synthesis: Some Epistemic Bridges, Some Conceptual Rifts," *Evolutionary Biology* 45 (2018), 127-139.

33. Kuhn, *The Structure of Scientific Revolutions*, 2d ed., 93.

34. Jonathan Wells, *The Politically Incorrect Guide to Darwinism and Intelligent Design* (Washington, DC: Regnery, 2006), 143-144.

35. Jim Brown and Ed Vitagliano, "Professor Dumped over Evolution Beliefs," *Agape Press* (March 11, 2003). http://www.arn.org/docs2/news/professordumped031203.htm (accessed August 22, 2020).

36. Wells, *The Politically Incorrect Guide to Darwinism*, 190-191.

37. "Outside Professors Derail Dissertation," *Free Science*, https://freescience.today/story/bryan-leonard/ (accessed August 22, 2020).

38. "Demoted, Terminated," *Free Science*, https://freescience.today/story/david-coppedge/ (accessed August 22, 2020).

39. "Marginalized, Shown the Door," *Free Science*, https://freescience.today/story/gunter-bechly/ (accessed August 22, 2020).

40. Michael Egnor, "What Scientists Know," *Evolution News and Science Today* (May 28, 2020), https://evolutionnews.org/2020/05/what-scientists-know/ (accessed August 22, 2020).

41. David Speijer, "Bad Faith Reasoning, Predictable Chaos, and the Truth," *BioEssays* 42 (June 2020), 2000040.

42. Kuhn, *The Structure of Scientific Revolutions*, 2d ed., 146.

43. Kuhn, *The Structure of Scientific Revolutions*, 2d ed., 172-173.

44. Kuhn, *The Structure of Scientific Revolutions*, 2d ed., 205-206.

Chapter 38—Is Science the Only Means for Acquiring Truth?

1. Gary Wolf was the first to coin the phrase *new atheism* in Gary Wolf, "The Church of the Non-Believers," *Wired* (November 1, 2006), https://www.wired.com/2006/11/atheism/ (accessed November 29, 2020).

2. Richard Dawkins, "Religion's misguided missiles," *The Guardian* (September 15, 2001), https://www.theguardian .com/world/2001/sep/15/september11.politicsphilosophyandsociety1 (accessed November 29, 2020).

3. Richard Dawkins, "Time to Stand Up," *Freedom from Religion Foundation* (September 22, 2001), https://ffrf.org/ news/timely-topics/item/14035-time-to-stand-up (accessed November 29, 2020).

4. Victor J. Stenger, *The New Atheism: Taking a Stand for Science and Reason* (New York: Prometheus Books, 2009), 59.

5. Among many other examples, see Sam Harris, *The End of Faith: Religion, Terror, and the Future of Reason* (New York: Norton, 2004); Daniel Dennett, *Breaking the Spell: Religion as a Natural Phenomenon* (New York: Viking, 2006); Sam Harris, *Letter to a Christian Nation: A Challenge to the Faith of America* (New York: Alfred Knopf, 2006); Victor J. Stenger, *God—the Failed Hypothesis: How Science Shows That God Does Not Exist* (New York: Prometheus, 2007); Christopher Hitchens, *God Is Not Great: How Religion Poisons Everything* (New York: Twelve, 2007); Alex Rosenberg, *The Atheist's Guide to Reality: Enjoying Life Without Illusions* (New York: Norton, 2011); Michael Shermer, *The Believing Brain: From Ghosts and Gods to Politics and Conspiracies* (New York: Times Books, 2011). Christian authors did not remain silent responding with publications such as John C. Lennox, *God's Undertaker: Has Science Buried God?* (Oxford, UK: Lion Hudson, 2007); Antony Flew and Roy Abraham Varghese, *There Is a God: How the World's Most Notorious Atheist Changed His Mind* (New York: HarperOne, 2007); Alister E. McGrath and Joanna Collicutt McGrath, *The Dawkins Delusion? Atheist Fundamentalism and the Denial of the Divine* (Downers Grove, IL: InterVarsity, 2007); John C. Lennox, *Gunning for God: Why the New Atheists Are Missing the Target* (Oxford, UK: Lion Hudson, 2011); Alvin Plantinga, *Where the Conflict Really Lies: Science, Religion, and Naturalism* (Oxford, UK: Oxford University Press, 2011), to name a few.

6. Roger Highfield, "Future of science: 'We will have the power of the gods,'" *The Telegraph* (October 23, 2007), https://www.telegraph.co.uk/news/science/science-news/3311478/Future-of-science-We-will-have-the-power -of-the-gods.html (accessed November 29, 2020).

7. Melanie Gosling, "Forget faith, only science can save us now" (June 6, 2002), *Independent Online (IOL)*, https:// www.iol.co.za/technology/forget-faith-only-science-can-save-us-now-31067 (accessed November 29, 2020).

8. Michael Shermer, "Science Is My Savior," *Science and Spirit* 16 (July-August 2005), https://www.questia.com/ magazine/1G1-171212564/science-is-my-savior (accessed November 29, 2020).

9. Richard Dawkins, "Thoughts for the Millennium," *Microsoft Encarta Encyclopaedia 2000* (Microsoft Corporation, 1993–1999).

10. Peter Atkins, "The Limitless Power of Science," *Nature's Imagination: The Frontiers of Scientific Vision*, ed. John Cornwell (Oxford, UK: Oxford University Press, 1995), 125.

11. Peter Atkins, "Science as truth," *History of the Human Sciences* 8 (May 1995), 97-102.

12. Jerry A. Coyne, *Faith vs. Fact: Why Science and Religion Are Incompatible* (New York: Viking, 2015), 186. Similarly, Dawkins has written that scientists are "specialists in what is true about the world." Richard Dawkins, *A Devil's Chaplain: Reflections on Hope, Lies, Science, and Love* (New York: Houghton Mifflin, 2003), 284.

13. "Scientism," entry in the *Oxford English Dictionary*, http://www.oed.com/. J.P. Moreland and William Lane Craig distinguish between strong and weak versions of scientism. The former "is the view that some proposition or theory is true and/or rational to believe if and only if it is a scientific proposition or theory," while the latter weaker version will "allow for the existence of truths apart from science and are even willing to grant that they can have some minimal, positive rationality status without the support of science" yet "still hold that science is the most valuable, most serious and most authoritative sector of human learning." J.P. Moreland and William Lane Craig, *Philosophical Foundations for a Christian Worldview* (Downers Grove, IL: InterVarsity, 2003), 347. The three criticisms of this chapter will be levelled against strong scientism, with clear implications for weaker scientism.

14. This is no doubt a consequence of the fact that philosophers by and large recognize that the positivist spirit is dead. Nonetheless, the occasional philosopher has voiced their allegiance to scientism in various forms (either strong or weak). Alexander Rosenberg, for example, writes that "the methods of science are the only reliable ways to secure knowledge of anything." Alex Rosenberg, *The Atheist's Guide to Reality: Enjoying Life Without Illusions* (New York: Norton, 2011), 6.

15. Carl Sagan, *Cosmos* (New York: Random House, 1980), 4.

16. See, for example, Thomas Nagel, "The Facts Fetish," *The New Republic* (October 20, 2010), https://newrepublic .com/article/78546/the-facts-fetish-morality-science (accessed November 29, 2020); John Gray, "The Knowns

and the Unknowns," *The New Republic* (April 20, 2012), https://newrepublic.com/article/102760/righteous-mind -haidt-morality-politics-scientism (accessed November 29, 2020); David Berlinski, *The Devil's Delusion: Atheism and Its Scientific Pretensions* (New York: Basic Books, 2009).

17. "Science," in the *Oxford English Dictionary*, http://www.oed.com/.

18. Prior to the modern era, there were distinct anticipations of positivism in the writings of ancient philosophers such as the atomic theory of Democritus (460–370 BC), the Greek materialism of Epicurus (ca. 341–270 BC), and the skepticism of Sextus Empiricus (ca. AD 160–210). In the medieval era, there are also intimations of positivist philosophy with thinkers such as William of Ockham (1285–1347). A contemporary of Comte, not to be overlooked, was the well-known positivist-logician John Stuart Mill (1806–1873).

19. See Harriet Martineau, *The Positive Philosophy of Auguste Comte, Vol. I* (London, UK: George Bell & Sons, 1853).

20. August Comte, quoted in Mary Pickering, *Auguste Comte: An Intellectual Biography, Vol. I* (Cambridge, UK: Cambridge University Press, 1993), 566.

21. Empirically verifiable quantities constitute the key distinction here, as philosophy and theology do indeed consider quantities (e.g., the Christian teaching of the Trinity). For further consideration of this, see Stanley L. Jaki, "The Limits of a Limitless Science," *The Asbury Theological Journal* 54 (Spring 1999), 23-39.

22. The modern emphasis on *process* (means) as opposed to *purposes* (ends) has deep roots in the philosophical writings of David Hume and Immanuel Kant, as well as the early modern scientists instrumental in the Scientific Revolution who abandoned the Aristotelian concern for teleology in science.

23. In logic, the work of Gottlob Frege paved the way for later developments in the foundations of mathematics, such as Alfred North Whitehead and Bertrand Russell's monumental *Principia Mathematica* (Cambridge, UK: Cambridge University Press, 1910–1913).

24. In the nineteenth century, Comte's progress motif found support in Charles Darwin's theory of evolution by natural selection, itself largely influenced by the dialectical method of G.W.F. Hegel. Although Darwin limited the application of his theory to living organisms, philosophers such as Herbert Spencer sought to merge Comtean positivism with Darwinian evolution to give a complete scientific description of natural phenomena. Austrian physicist Ernst Mach went even further, suggesting that positive knowledge is not of physical reality per se but rather our sense-experience of reality. Hence, the positive or scientific stage does not describe "the phenomena of the physical universe and their laws," as Isaac Newton had suggested; rather, it describes the conceptual organizations and correlations of physical reality given in sense-experience. Albert Einstein's relativity theory as well as Erwin Schrödinger and Werner Heisenberg's revision of quantum theory seemed to confirm Mach's criticism of Newton.

25. Logical positivism was largely developed out of Germany and Austria. The Berlin Circle included Hans Reichenbach and Carl Gustav Hempel, and others, while the well-known Vienna Circle included Moritz Schlick, Rudolph Carnap, Otto Neurath, Hans Hahn, and Friedrich Waismann. Despite Ludwig Wittgenstein and Karl Raimund Popper's loose attendance, they were not official members.

26. David Hume, *An Enquiry Concerning Human Understanding* (Oxford, UK: Oxford University Press, 2000), 123.

27. See, for example, Oswald Hanfling, "Logical positivism," *Philosophy of Science, Logic and Mathematics in the Twentieth Century*, ed. Stuart G. Shanker (London, UK: Routledge, 1996), 193-194.

28. As early as the 1930s, this issue was beginning to emerge with critiques against logical positivism levelled by Wolfgang Köhler, Karl Popper and others, as well as the defection of John Dewey. By the latter half of the twentieth century, philosophers such as Noam Chomsky and Willard Van Orman Quine saw logical positivism put to rest. Of interest to Christians, the well-known book by C.S. Lewis, *The Abolition of Man* (1943), was written in response to the logical positivism of Ayer and others.

29. Étienne Gilson, *The Unity of Philosophical Experience* (San Francisco, CA: Ignatius Press, 1999), 246.

30. Stephen Hawking and Leonard Mlodinow, *The Grand Design* (New York: Bantam Books, 2011), 1.

31. Jaki, "The Limits of a Limitless Science."

32. In addition to Jaki's article previously cited, see an excellent essay against scientism in Edward Feser, *Scholastic Metaphysics: A Contemporary Introduction* (Neunkirchen-Seelscheid, Germany: Editiones Scholasticae, 2014), 10-33.

33. As Immanuel Kant famously wrote, "Thoughts without content are empty, intuitions without concepts are blind." Immanuel Kant, *Critique of Pure Reason*, trans. P. Guyer and A.W. Wood (Cambridge, UK: Cambridge University Press, 1998), 193-194.

34. Perhaps no one has articulated it as succinctly as L.M. Régis when he wrote that "it is not because a house is made up of feet and yards that it is measurable, but because it is built of materials that can be measured in feet and yards." L.M. Régis, *Epistemology*, trans. Imelda Choquette Byrne (New York: Macmillan, 1959), 333. The "feet and yards" give a scientific description of the house and its materials, but "feet and yards" do not provide a complete explanation of what the house is, nor of its materials. In a word, something has "to be" before it "can be" what science says, and knowledge of this "to be" is not scientific (more on this in *Criticism 3*).

35. Feser, *Scholastic Metaphysics*, 24.

36. Aristotle put it this way: "To say of a thing that is that it is not, or to say of a thing that is not that it is, is false; but to say of a thing that is that it is, or to say of a thing that is not that it is not, is true." Aristotle, *Metaphysics*, IV(Γ).7.1011b23-1012a17.

37. John Lennox, "Science and God," *C.S. Lewis Institute* (May 19, 2016), https://www.youtube.com/watch?v=DoLTcv-RPdM&t=4962s (accessed November 29, 2020).

38. The position I am putting forward here would disagree with the distinction drawn by Stephen Jay Gould's Non-overlapping magisterial (NOMA) model, which suggests that science and religion represent distinct spheres of inquiry. See Stephen Jay Gould, *Rock of Ages: Science and Religion in the Fullness of Life* (New York: Ballantine Books, 2002). The argument of this chapter, which will become clear by the conclusion, is that science depends upon a conceptual framework that only a specifically Judeo-Christian worldview can furnish.

39. For a fascinating study on this distinction as it relates to the sciences and humanities within the academy, see Roger Scruton, "Scientism and the Humanities," *Scientism: The New Orthodoxy*, eds., Daniel N. Robinson and Richard N. Williams (New York: Bloomsbury Academic, 2015), 131-146.

40. See Frederick Wilhelmsen, *The Paradoxical Structure of Existence* (New Brunswick, NJ: Transaction Publishers, 2015), 104.

41. For a measured study on how naturalistic evolutionary theory cannot account for the truth of things, see Alvin Plantinga, *Where the Conflict Really Lies: Science, Religion, and Naturalism* (Oxford, UK: Oxford University Press, 2011).

42. Note that while metaphysics will invariably affect one's view of science and, consequently, how they go about the "activity" of science, it should not affect science *qua* science insofar as science is an epistemological undertaking committed to observation, measurement, and interpretation of empirically verifiable things. Science does indeed depend upon metaphysical preconditions; however, its "activity" does not require either naturalistic or non-naturalistic metaphysics to be consciously maintained (it is possible to be consciously inconsistent after all). This explains why we have scientists with all sorts of different metaphysical presuppositions.

43. Lawrence M. Krauss, *A Universe from Nothing: Why There Is Something Rather Than Nothing* (New York: Free Press, 2012), xiv.

44. Hawking and Mlodinow, *The Grand Design*, 227. Writing against similar strains of logical positivist thought in his own day, Ludwig Wittgenstein well observed, "The whole modern conception of the world is founded on the illusion that the so-called laws of nature are the explanations of natural phenomena. Thus, people today stop at the laws of nature, treating them as something inviolable just as God and Fate were treated in past ages." Ludwig Wittgenstein, *Tractatus Logico-Philosophicus*, trans. D.F. Pears and B.F. McGuinness (London, UK: Routledge Classics, 2001), 6.371.

45. Philosophically speaking, there can be only two possibilities for why a thing (*x*) exists: either by necessity of its own essence which is to exist (*x* just *is*), or by necessity of something other than itself which causes it to exist (*y* creates *x*). As the former (*x* just *is*) implies the absence of a beginning for the universe (which contravenes contemporary science), it makes no sense to speak about the universe creating itself because it already is. Similarly, as the latter (*y* creates *x*) assumes the existence of something else, it makes no sense to speak of the universe creating itself because the act of creation presupposes the existence of something else (*y*) that causes the existence of the universe (*x*).

46. Would-be theories of everything (TOEs) abound under all sorts of names, from M-theory, to string theory, to the multiverse. In each case, they seek to present a self-contained mathematical model to describe fundamental reality. Interestingly, Kurt Gödel's incompleteness theorems were postulated to demonstrate the inherent logical limitations of mathematics to form a complete axiomatic system of thought, so even within mathematical logic there are demonstrations of its own incompleteness.

47. Louis de Broglie, *Physics and Microphysics: Expositions on Physics and the History and Philosophy of Science*, trans. Martin Davidson (New York: Harper & Brothers, 1955), 209.

48. Immanuel Kant, *Critique of Pure Reason*, trans. P. Guyer and A.W. Wood (Cambridge, UK: Cambridge University Press, 1998), 627.

49. See Feser, *Scholastic Metaphysics*, 17.

50. Pythagoreanism and Platonism came to influence Judaism under Philo of Alexandria, who sought to allegorically introduce number symbolism in his interpretation of Jewish texts (e.g., Philo's *On Creation of the World*).

51. The demise of logical positivism was, in part, because of some of the shocking science-only answers that were given by Ayer's principle of verificationism. For example, Ayer wrote, "Our contention is simply that, in our language, sentences which contain normative ethical symbols are not equivalent to sentences which express psychological propositions, or indeed empirical propositions of any kind…The reason why they are unanalyzable is that they are mere pseudo-concepts. The presence of an ethical symbol in a proposition adds nothing to its factorial content…" A.J. Ayer, *Language, Truth and Logic* (London, UK: Camelot Press, 1953), 105, 107-108. In other words, because fundamental ethical concepts cannot be empirically expressed, they are meaningless. So, too, with theological claims: "There cannot be any transcendent truths of religion. For the sentences which the theist uses to express such truths are not literally significant." Ayer, *Language, Truth and Logic*, 117-118. What this goes to show, in the words of Richard Lewontin, is the "willingness to accept scientific claims that are against common sense," which is "the real struggle between science and the supernatural. We take the side of science in spite of the patent absurdity of some of its constructs…because we have a prior commitment, a commitment to materialism…that materialism is absolute, for we cannot allow a Divine Foot in the door." Richard Lewontin, "Billions and Billions of Demons," *The New York Review of Books* (January 9, 1997), 31.

52. Richard Dawkins, *River Out of Eden: A Darwinian View of Life* (New York: Basic Books, 1995), 133.

53. Richard Dawkins, *The Blind Watchmaker: Why the Evidence of Evolution Reveals a Universe Without Design* (New York: Norton, 1986), 3. Of course, the difference between human beings and machines is that machines are neither alive nor dead, neither conscious nor unconscious, and therefore neither think nor feel. The danger, then, is when human beings are thought of as machines they will be treated as such (as they were under the sociopolitical Marxist regimes of the twentieth century). Dystopic visions such as Aldous Huxley's *Brave New World* (1931), George Orwell's *1984* (1949), and even contemporary cult-movie hits such as Ridley Scott's *Blade Runner* (1982), adapted from Philip K Dick's *Do Androids Dream of Electric Sheep?* (1968), and the Wachowski brothers' *The Matrix* trilogy (1999–2003) speak to these fears.

54. When touring Australia in April 2012, Richard Dawkins stated on national TV, "'Why?' is a silly question. You can ask, '*What* are the factors that led to something coming into existence?' That's a sensible question. But '*What* is the purpose of the universe?' is a silly question. It has no meaning." "Religion and Atheism," https://www.abc.net.au/qanda/religion-and-atheism/10661470 (accessed November 29, 2020). See also Richard Dawkins, *The God Delusion* (London, UK: Bantam Press, 2006), 56.

Chapter 39—Is Theistic Evolution an Option for Christians?

1. This chapter is adapted from Jay W. Richards, "Introduction: Squaring the Circle," *God and Evolution: Protestants, Catholics, and Jews Explore Darwin's Challenge to Faith*, ed. Jay W. Richards (Seattle, WA: Discovery Institute Press, 2010).

2. This is often a popular way of speaking. See, for instance, George Murphy quoted in Ted Davis, "Theistic Evolution: History and Beliefs," The BioLogos Foundation (October 15, 2018), https://biologos.org/articles/series/science-and-the-bible/theistic-evolution-history-and-beliefs (accessed November 10, 2020). In the last several years, there have been a number of books that have defended some version of this thesis. See, for example, Denis Alexander, *Creation or Evolution: Do We Have to Choose?* (Oxford, UK: Monarch Books, 2008); Ted Peters and Martinez Hewlett, *Can You Believe in God and Evolution? A Guide for the Perplexed* (Nashville, TN: Abingdon Press, 2006); *Perspectives on an Evolving Creation*, ed. Keith Miller (Grand Rapids, MI: Eerdmans, 2003); Denis O. Lamoureux, *Evolutionary Creation: An Evangelical Approach to Evolution* (Eugene, OR: Wipf & Stock, 2008); Karl W. Giberson, *Saving Darwin: How to Be a Christian and Believe in Evolution* (New York: HarperOne, 2008). Jeremy Manier reports on a number of the recent thinkers who try to reconcile God and evolution:

> The religious response is simple, some believers say: God used evolution to create us. But invoking God's direct guidance raises daunting scientific hurdles. Ever since the publication in 1859 of Darwin's "Origin of Species," religious writers have tried to cram the idea of design back into evolution, often without success.

Jeremy Manier, "The New Theology," *Chicago Tribune* (January 20, 2008), https://www.chicagotribune.com/news/ct-xpm-2008-01-20-0801120310-story.html (accessed November 10, 2020).

3. Aristotle distinguished between four "causes," that is, factors that explain or are responsible for something else. To put it simply, the material cause explains what something is made of, the formal cause explains what something is, the final cause explains the ultimate purpose toward which something tends, and the efficient cause explains where something came from or what or who produced it.

4. William A. Dembski, "Intelligent Design: A Brief Introduction," *Evidence for God*, eds. William A. Dembski and Michael R. Licona (Grand Rapids, MI: Baker, 2010), 104.

5. Daniel R. Brooks and E.O. Wiley, *Evolution as Entropy* (Chicago, IL: University of Chicago Press, 1986), xi, quoted in M.A. Corey, *Back to Darwin: The Scientific Case for Deistic Evolution* (Lanham, MD: University Press of America, 1994), 3.

6. G.G. Simpson, *The Meaning of Evolution: A Study of the History of Life and of Its Significance for Man*, rev. ed. (New Haven, CT: Yale University Press, 1967), 345.

7. Francis Darwin, *Life and Letters of Charles Darwin, Vol. 1* (New York: Appleton, 1887), 280, 283-284, 278-279.

8. For a discussion of the relationship between creation and conservation, see Jonathan Kvanvig, "Creation and Conservation," *Stanford Encyclopedia of Philosophy* (2017), http://plato.stanford.edu/entries/creation-conservation/ (accessed November 10, 2020).

9. Here's how *Merriam-Webster* defines "theism": "belief in the existence of a god or gods; specifically: belief in the existence of one God viewed as the creative source of the human race and the world who transcends yet is immanent in the world." *Merriam-Webster*, https://www.merriam-webster.com/dictionary/theism (accessed November 10, 2020).

10. This seems to be the position of theologian John Haught, for instance, who follows the thought of Teilhard de Chardin. See John Haught, *Making Sense of Evolution: Darwin, God, and the Drama of Life* (Philadelphia, PA: Westminster/John Knox Press, 2010); John F. Haught, *God After Darwin: A Theology of Evolution* (Boulder, CO: Westview Press, 2007).

11. Beyond this point, the details get complicated. There is an extensive literature dating back to the medieval Catholic scholastics who discussed the various modes of divine action and how God's action relates to natural causation. The three main views are occasionalism, mere conservationism, and concurrentism. In occasionalism, God alone brings about every effect in nature. In mere conservationism, according to Alfred Freddoso, "God contributes to the ordinary course of nature solely by creating and conserving natural substances along with their active and passive causal powers or capacities." In concurrentism,

> a natural effect is produced immediately by *both* God *and* created substances, so that, contrary to occasionalism, secondary agents make a genuine causal contribution to the effect and in some sense determine its specific character by virtue of their own intrinsic properties, whereas, contrary to mere conservationism, they do so only if God cooperates with them contemporaneously as an immediate cause in a certain "general" way which goes beyond the conservation of the relevant agents, patients, and powers, and which renders the resulting effect the immediate effect of both God and the secondary causes.

Alfred J. Freddoso, "God's General Concurrence with Secondary Causes: Pitfalls and Prospects," *American Catholic Philosophical Quarterly* 67 (1994), 131-156, http://www.nd.edu/~afreddos/papers/pitfall.htm (accessed November 10, 2020). See also Alfred J. Freddoso, "God's General Concurrence with Secondary Causes: Why Conservation Is Not Enough," *Philosophical Perspectives* 5 (1991), 553-585, http://www.nd.edu/~afreddos/papers/conserv.htm (accessed November 10, 2020). Among the Abrahamic faiths, some version of concurrentism (though not necessarily designated as such) is by the far the most common position. However, many modern theistic evolutionists seem inclined toward mere conservationism. In that sense, these traditional debates are relevant to our current discussion. For our purposes, however, the crucial point is that God in the theistic view is free to act in a variety of ways both within the created order and as creator of that order.

12. Stephen C. Meyer and Michael Newton Keas, "The Meanings of Evolution," *Darwinism, Design, and Public Education*, eds. John Angus Campbell and Stephen C. Meyer (Lansing, MI: Michigan State University Press, 2004), 135-156.

13. See the explanation for the meaning of *evolution* from The BioLogos Foundation, which seeks to give a Christian interpretation and defense of evolution. The explanation begins with "change over time," then goes on to fill out

the definition with common descent and the Darwinian mechanism. But it quickly slips from defining the term to presenting the details as if they were uncontested facts.

"What is evolution?," The BioLogos Foundation, https://web.archive.org/web/20091007082550/https://biologos .org/questions/what-is-evolution/ (accessed November 10, 2020).

14. See discussion of this point in the comments of Thomas Cudworth, "Olive Branch from Karl Giberson," *Uncommon Descent* (April 15, 2010), https://uncommondescent.com/intelligent-design/olive-branch-from-karl -giberson/ (accessed November 10, 2020).

15. Niall Ferguson, *The Ascent of Money: A Financial History of the World* (New York: Penguin Books, 2008).

16. I am not referring to Darwin's personal beliefs, which seem to have varied over time. I am referring to the actual arguments Darwin made in the *Origin*, which use deistic rather than atheistic or materialistic premises. Whether this reflected Darwin's personal beliefs or merely a rhetorical strategy is a separate and difficult question. See, for instance, Neil C. Gillespie, *Charles Darwin and the Problem of Creation* (Chicago, IL: University of Chicago Press, 1979), and M.A. Corey, *Back to Darwin: The Scientific Case for Deistic Evolution* (Lanham, MD: University Press of America), 6-25.

17. Ernst Mayr, Foreword, in Michael Ruse, *Darwinism Defended: A Guide to the Evolution Controversy* (Reading, MA: Addison-Wesley, 1982), xi–xii.

18. William A. Dembski, "Converting Matter into Mind: Alchemy and the Philosopher's Stone in Cognitive Science," *Perspectives on Science and Christian Faith* 42 (1990), 202-226, https://www.asa3.org/ASA/PSCF/1990/ PSCF12-90Dembski.html (accessed November 10, 2020).

19. Richard Dawkins, *The Blind Watchmaker: Why the Evidence of Evolution Reveals a Universe Without Design* (New York: Norton, 1996), 6.

20. Timothy Keller, *The Reason for God: Belief in God in an Age of Skepticism* (New York: Dutton, 2008), 94.

21. Keller, *The Reason for God*, 87.

22. Keller is quoting from David Atkinson, *The Message of Genesis 1–11* (Downers Grove, IL: InterVarsity, 1990), 31.

23. This is in a white paper written for Tim Keller, "Christian, Evolution, and Christian Laypeople," The BioLogos Foundation, https://wp.biologos.org/wp-content/uploads/2019/02/Keller_white_paper-compressed.pdf (accessed November 10, 2020).

Chapter 40—Will Intelligent Machines Rise Up and Overtake Humanity?

1. "Is Artificial Intelligence Set to Become Art's Next Medium?" *Christie's* (December 12, 2018), https://www .christies.com/features/A-collaboration-between-two-artists-one-human-one-a-machine-9332-1.aspx (accessed August 24, 2020).

2. "Google AI defeats human Go champion," BBC News (May 25, 2017), https://www.bbc.com/news/technology -40042581 (accessed August 24, 2020); Carnegie Mellon University, "AI beats professionals in six-player poker," *ScienceDaily* (July 11, 2019), https://www.sciencedaily.com/releases/2019/07/190711141343.htm (accessed August 24, 2020).

3. Ken Dilanian and Courtney Kube "Airport informants, overhead drones: How the U.S. killed Soleimani," *NBC News* (January 10, 2020), https://www.nbcnews.com/news/mideast/airport-informants-overhead-drones-how -u-s-killed-soleimani-n1113726 (accessed August 24, 2020).

4. Charlie Gao, "The Ultimate Weapon of War No One Is Talking About," *The National Interest* (January 25, 2019), https://nationalinterest.org/blog/buzz/ultimate-weapon-war-no-one-talking-about-42497 (accessed August 24, 2020).

5. Robert J. Marks, *The Case for Killer Robots* (Seattle, WA: Discovery Institute Press, 2020).

6. Ray Kurzweil, *The Singularity Is Near: When Humans Transcend Biology* (London, UK: Penguin, 2005).

7. Ray Kurzweil, *The Age of Spiritual Machines: When Computers Exceed Human Intelligence* (New York: Penguin, 2000).

8. Yuval Noah Harari. *Homo Deus: A Brief History of Tomorrow* (New York: Random House, 2016).

9. Arjun Kharpal, "Stephen Hawking says A.I. could be 'worst event in the history of our civilization,'" *CNBC* (November 6, 2017), https://www.cnbc.com/2017/11/06/stephen-hawking-ai-could-be-worst-event-in-civiliza tion.html (accessed August 24, 2020).

10. Henry Kissinger, "How the Enlightenment Ends," *The Atlantic* (August 30. 2019), https://www.theatlantic.com/

magazine/archive/2018/06/henry-kissinger-ai-could-mean-the-end-of-human-history/559124/ (accessed August 24, 2020).

11. Samuel Gibbs, "Elon Musk: artificial intelligence is our biggest existential threat," *The Guardian* (October 27, 2014), https://www.theguardian.com/technology/2014/oct/27/elon-musk-artificial-intelligence-ai-biggest-existential -threat (accessed August 24, 2020).

12. Stephen Hawking, Stephen William, and Roger Penrose, "The singularities of gravitational collapse and cosmology," *Proceedings of the Royal Society of London A: Mathematical and Physical Sciences* 314 (1970), 529-548.

13. Roger Penrose, *The Emperor's New Mind: Concerning Computers, Minds, and the Laws of Physics* (Oxford, UK: Oxford University Press, 1989), 16.

14. "Quotations Roger Penrose," https://www-history.mcs.st-andrews.ac.uk/Quotations/Penrose.html (accessed August 24, 2020).

15. Gregory Chirikjian, "Help Wanted: For the Cognitive Era," *JHU Engineering Magazine* (May 19, 2017), https:// engineering.jhu.edu/magazine/2017/05/help-wanted-for-the-cognitive-era/ (accessed August 24, 2020).

16. Satya Nadella, Greg Shaw, and Jill Tracie Nichols, *Hit Refresh: The Quest to Rediscover Microsoft's Soul and Imagine a Better Future for Everyone* (New York: Harper Business), 207.

17. Some of the material in this chapter is borrowed from the author's works posted on MindMatters.AI, including "Advice to Physicists: 'Shut Up and Do Physics'" (May 25, 2020), https://mindmatters.ai/2020/05/advice-to -physicists-shut-up-and-do-physics/ (August 24, 2020) and the paper "Linear Complexity Can Generate an Exponential Explosion of Contingencies: Implications for General Artificial Intelligence," currently under review for publication with two coauthors.

18. Robert J. Marks II, William A. Dembski, and Winston Ewert, *Introduction to Evolutionary Informatics* (Singapore: World Scientific, 2017).

19. Thomas M. Cover and Joy A. Thomas, *Elements of Information Theory* (Hoboken, NJ: John Wiley & Sons, 1999).

20. Meilan Solly, "This Poker-Playing A.I. Knows When to Hold 'Em and When to Fold 'Em," *Smithsonian* (July 19, 2019), https://www.smithsonianmag.com/smart-news/poker-playing-ai-knows-when-hold-when-fold -em-180972643/ (accessed August 24, 2020).

21. Rob Toews, "GPT-3 Is Amazing—And Overhyped," *Forbes* (July 19, 2020), https://www.forbes.com/sites/ robtoews/2020/07/19/gpt-3-is-amazingand-overhyped/ (accessed August 24, 2020).

22. Quoted in Jorge R. Barrio, "Consensus Science and the Peer Review," *Molecular Imaging and Biology* 11 (2009), 293.

23. George Gilder, "The Materialist Superstition," *The Intercollegiate Review* (Spring 1996), 6-14.

24. Gregory J. Chaitin, "Algorithmic information theory," *IBM Journal of Research and Development* 21 (1977), 350-359.

25. Peter D. Grünwald and Paul MB Vitányi, "Algorithmic information theory," *Handbook of the Philosophy of Information* (2008), 281-320.

26. Debra Bell, "Celebrities Who Have Testified to Congress," *US News & World Report* (September 24, 2010), https://www.usnews.com/news/slideshows/celebrities-who-have-testified-to-congress (accessed August 25, 2020).

27. Laura Ingraham, *Shut Up and Sing: How Elites from Hollywood, Politics, and the Media Are Subverting America* (Washington, DC: Regnery, 2003).

28. Keiran Southern, "'Most of you spent less time in school than Greta Thunberg'—Ricky Gervais's best and most shocking jokes from the Golden Globes," *Independent.ie* (January 6, 2020), https://www.independent .ie/entertainment/movies/most-of-you-spent-less-time-in-school-than-greta-thunberg-ricky-gervaiss-best-and -most-shocking-jokes-from-the-golden-globes-38837007.html (accessed August 25, 2020).

29. Matt McFarland, "Elon Musk: 'With artificial intelligence we are summoning the demon,'" *The Washington Post* (October 24, 2014), https://www.washingtonpost.com/news/innovations/wp/2014/10/24/elon-musk-with -artificial-intelligence-we-are-summoning-the-demon/ (accessed August 25, 2020).

30. Tom Maloney, "Elon Musk Soars Past Warren Buffett on Billionaires Ranking," *Bloomberg News* (July 10, 2020), https://www.bloomberg.com/news/articles/2020-07-10/elon-musk-rockets-past-warren-buffett-on-billionaires -ranking (accessed March 16, 2021); Robert Frank, "Elon Musk is now the richest person in the world, passing Jeff Bezos," CNBC (January 7, 2021), https://www.cnbc.com/2021/01/07/elon-musk-is-now-the-richest-person-in

-the-world-passing-jeff-bezos-.html (accessed March 16, 2021); Annie Palmer, "Jeff Bezos overtakes Elon Musk to reclaim spot as world's richest person," *CNBC* (February 16, 2021), https://www.cnbc.com/2021/02/16/jeff -bezos-reclaims-spot-as-worlds-richest-person-from-elon-musk.html (accessed March 16, 2021).

31. Catherine Clifford, "Facebook CEO Mark Zuckerberg: Elon Musk's doomsday AI predictions are 'pretty irresponsible,'" *CNBC* (July 24, 2017), https://www.cnbc.com/2017/07/24/mark-zuckerberg-elon-musks -doomsday-ai-predictions-are-irresponsible.html (accessed August 25, 2020).

32. Rich Karlgaard, "Why Technology Prophet George Gilder Predicts Big Tech's Disruption" *Forbes* (February 8, 2018), https://www.forbes.com/sites/richkarlgaard/2018/02/09/why-technology-prophet-george-gilder-predicts -big-techs-disruption/ (accessed August 25, 2020).

33. "Bill Gates: I think we do need to worry about artificial intelligence," *Fox Business* YouTube channel (January 22, 2016), https://www.youtube.com/watch?v=EmfrMKLwr3k (accessed August 25, 2020).

34. Rory Cellan-Jones, "Stephen Hawking warns artificial intelligence could end mankind," *BBC News* (December 2, 2014), bbc.com/news/technology-302290540.

35. Stephen Hawking, "Gödel and the End of Physics," TAMU lecture (2017), http://yclept.ucdavis.edu/course/215c .S17/TEX/GodelAndEndOfPhysics.pdf (accessed August 27, 2020).

36. Linghe Kong, Muhammad Khurram Khan, Fan Wu, Guihai Chen, and Peng Zeng, "Millimeter-wave wireless communications for IoT-cloud supported autonomous vehicles: Overview, design, and challenges," *IEEE Communications* 55 (2017), 62-68.

37. Neil McBride, "The ethics of driverless cars," *ACM SIGCAS Computers and Society* 45 (2016), 179-184.

38. Aarian Marshall, "The Uber Crash Won't Be the Last Shocking Self-Driving Death," *Wired* (March 31, 2018), https://www.wired.com/story/uber-self-driving-crash-explanation-lidar-sensors/ (accessed August 28, 2020).

39. Brendan Dixon, "Star Self-Driving Truck Firm Shuts; AI Not Safe Enough Soon Enough," *Mind Matters News* (March 24, 2020), https://mindmatters.ai/2020/03/star-self-driving-truck-firm-shuts-ai-not-safe-enough-soon -enough/ (accessed August 28, 2020).

40. Paul Scharre, *Army of None: Autonomous Weapons and the Future of War* (New York: Norton, 2018).

41. Marks, *The Case for Killer Robots*.

42. Roger Penrose identified quantum collapse as nonalgorithmic and conjectured this might be the source of humans' nonalgorithmic abilities. There has been no research performed yet, and Penrose's highly speculative hypothesis remains untested.

43. Psalm 8:6; see also Genesis 1.

44. See Genesis 1:27 and 1 Corinthians 2:16.

45. Psalm 139:14.

Chapter 41—Can Panspermia Explain the Origin of Life?

1. Francis Crick and Leslie Orgel, "Directed Panspermia," *Icarus* 19 (1973), 341-346.

2. Francis Crick, *Life Itself: Its Origin and Nature* (New York: Simon & Schuster, 1981), 88.

3. Two other types of panspermia discussed in the literature are pseudo-panspermia and necropanspermia. Pseudo-panspermia is the transport of the building blocks of life, such as amino acids. Necropanspermia is the transport of dead organisms. I don't consider either of these as effective ways to infect another world. For this reason, I will not consider these.

4. W.T.B. Kelvin, *Popular Lectures and Addresses; Constitution of Matter, Vol. 2: Geology and General Physics* (New York: Macmillan, 1894).

5. S. Arrhenius, *Worlds in the Making: The Evolution of the Universe* (New York: Harper & Row, 1908).

6. D.C. Golden, D.W. Ming, R.V. Morris et al., "Evidence for exclusively inorganic formation of magnetite in Martian meteorite ALH84001," *American Mineralogist* 89 (2004), 681-695; J. Martel, D. Young, H.-H. Peng et al., "Biomimetic Properties of Minerals and the Search for Life in the Martian Meteorite ALH84001," *Annual Reviews of Earth and Planetary Sciences* 40 (2012), 167-193.

7. J.C. Armstrong, L.E. Wells, G. Gonzalez, "Rummaging Through Earth's Attic for Remains of Ancient Life," *Icarus* 160 (2002), 183-196; L.E. Wells, J.C. Armstrong, G. Gonzalez, "Reseeding of Early Earth by Impacts of Returning Ejecta During the Late Heavy Bombardment," *Icarus* 162 (2003), 38-46.

8. J.J. Bellucci, A.A. Nemchin, M. Grange et al., "Terrestrial-like Zircon in a Clast from an Apollo Breccia," *Earth and Planetary Science Letters* 510 (2019), 173-185.

9. G. Horneck, D.M. Klaus, R.L. Mancinelli, "Space Microbiology," *Microbiology and Molecular Biology Reviews* 74(1) (2010), 121-156.

10. N. Novikova, E. Deshevaya, M. Levinskikh, N. Polikarpov, "Study of the effects of the outer space environment on dormant forms of microorganisms, fungi and plants in the 'Expose-R' experiment." *International Journal of Astrobiology* 14(1) (2015), 137-142.

11. P.S. Wesson, "Panspermia Past and Present: Astrophysical and Biophysical Conditions for the Dissemination of Life in Space," *Space Science Reviews* 156 (2010), 239-252.

12. V. Duda, N. Suzina, Polivtseva, V; A. Boronin, "Ultramicrobacteria: Formation of the Concept and Contribution of Ultramicrobacteria to Biology," *Microbiology* 81 (4) (2012), 379-390.

13. R.J. Worth, S. Sigurdsson, C.H. House, "Seeding Life on the Moons of the Outer Planets via Lithopanspermia," *Astrobiology* 13(12) (2014), 1155-1165.

14. Worth et al., "Seeding Life."

15. Worth et al., "Seeding Life."

16. Wells et al., "Reseeding of Early Earth."

17. Wells et al., "Reseeding of Early Earth."

18. Wells et al., "Reseeding of Early Earth."

19. Wells et al., "Reseeding of Early Earth."

20. Yoko Ohtomo, Takeshi Kakegawa, Akizumi Ishida, Toshiro Nagase, and Minik T. Rosing, "Evidence for Biogenic Graphite in Early Archaean Isua Metasedimentary Rocks," *Nature Geoscience* 7 (2014), 25-28.

21. "Exoplanet Exploration: Planets Beyond Our Solar System," https://exoplanets.nasa.gov/ (accessed June 13, 2020).

22. "'Oumuamua," https://en.wikipedia.org/wiki/%CA%BBOumuamua (accessed June 13, 2020).

23. "2I/Borisov," https://en.wikipedia.org/wiki/2I/Borisov (accessed June 13, 2020).

24. Worth et al., "Seeding Life"; J. Melosh, "Exchange of Meteorites (and Life?) Between Stellar Systems," *Astrobiology* 3(1) (2003), 207-215.

25. Worth et al., "Seeding Life"; Melosh, "Exchange of Meteorites."

26. F.C. Adams and D.N. Spergel, "Lithopanspermia in Star-Forming Clusters," *Astrobiology* 5 (2005), 497-514.

27. R.G. Martin and M. Livio, "On the Formation and Evolution of Asteroid Belts and Their Potential Significance for Life," *Monthly Notices of the Royal Astronomical Society* 428(1) (2013), L11-L15. This study argues for the importance of an asteroid belt in the delivery of water to the terrestrial planets. An asteroid belt is also required as a source of impactors for panspermia. Having a gas giant in a large orbit, like Jupiter, permits an asteroid belt to exist in a planetary system.

28. Melosh, "Exchange of Meteorites."

29. C.J. Lada and E.A. Lada, "Embedded Clusters in Molecular Clouds," *Annual Review Astronomy & Astrophysics* 41 (2003), 57-115.

30. Adams and Spergel, "Lithopanspermia in Star-Forming Clusters."

31. I. Ginsburg, M. Lingam, A. Loeb, "Galactic Panspermia," *Astrophysical Journal Letters* 868 (2018), L12.

32. Melosh, "Exchange of Meteorites."

33. Ginsburg et al., "Galactic Panspermia."

34. E. Willerslev, A.J. Hansen, R. Ronn et al., "Long-term Persistence of Bacterial DNA," *Current Biology* 14(1) (2004), R9-R10.

35. A.V. Shatilovich, A.V. Tchesunov, T.V. Neretina, I.P. Grabardnik, S.V. Gubin, T.A. Vishnivetskaya, T.C. Onstott, E.M. Rivkina, "Viable Nematodes from Late Pleistocene Permafrost of the Kolyma River Lowland," *Doklady Biological Sciences* 480 (2018), 100-102.

36. T.K. Lowenstein, B.A. Schubert, M.N. Timofeeff, "Microbial Communities in Fluid Inclusions and Long-Term Survival in Halite," *GSA Today* 21(1) (January 2011), 4-9.

37. Lowenstein et al., "Microbial Communities in Fluid Inclusions," 7.

38. Lowenstein et al., "Microbial Communities in Fluid Inclusions," 7.

Chapter 42—Does the Multiverse Refute Cosmic Design?

1. The conditions that make our universe habitable—i.e., consistent with the existence of life—are very special. Multiverse explanations seek a naturalistic answer for why our universe has these special characteristics by inventing ways that there could be enough universes with randomly generated initial conditions, laws, and constants to make these special characteristics the inevitable result of an undirected process. In other words, multiverse theories require *multiple trials for viability*—the generation of innumerably many different universes—to explain away the excruciatingly fine-tuned characteristics of our universe. In this respect, multiverse theories can be regarded as the cosmological cousins of theories of undirected evolution in biology.

2. Quantum physics sets aside familiar conceptions of motion and the interaction of bodies and introduces acts of measurement and probabilities for observational outcomes in an *irreducible* way not ameliorated by appealing to our limited knowledge. The state of a quantum system is described by an abstract mathematical object called a *wavefunction*. As long as the system is unobserved, the wave function develops deterministically, but it only specifies the *probability* that various observables (like position or momentum) will have a particular value when measured. Quantum entanglement is a physical phenomenon that occurs as quanta interact or share spatial proximity in a manner that "entangles" their wavefunctions so that individual quanta cannot be described independently of the state of the others, even when they are separated by a large distance. A couple of helpful popular expositions not just for these concepts but for others later in this chapter are Frank Wilczek, "Entanglement Made Simple," *Quanta Magazine* (April 28, 2016), https://www.quantamagazine.org/entanglement-made-simple-20160428/ (accessed July 8, 2020); and K.C. Cole. "Wormholes Untangle a Black Hole Paradox," *Quanta Magazine* (April 24, 2015), https://www.quantamagazine.org/wormhole-entanglement-and-the-firewall-paradox-20150424 (accessed July 8, 2020).

3. See Carlo Rovelli, "The strange equation of quantum gravity," arXiv.org (2015), https://arxiv.org/pdf/1506.00927.pdf (accessed July 8, 2020).

4. Alexander Vilenkin, "Birth of inflationary universes," *Physical Review D* 27 (1983), 2848-2851.

5. See the discussion of the principle of sufficient reason in my companion chapter in this volume, "How Does the Intelligibility of Nature Point to Design?" The relevant formulation of the principle here is that *every contingent state of affairs has an explanation.*

6. Alexander Vilenkin, "Creation of universes from nothing," *Physics Letters B* 117 (1982), 25-28; James Hartle and Stephen Hawking, "Wave function of the universe," *Physical Review D* 28 (1983), 2960-2975.

7. Everyone is familiar with a curved 2-dimensional space—consider the surface of a sphere, for example. A curved 2-dimensional space, while appearing 2-dimensional from within the space, curves in the third dimension, however. Similarly, a curved 3-dimensional space is curved in the fourth dimension. As described in general relativity, we live in a 4-dimensional spacetime in which (as John Wheeler summarized it), spacetime tells matter how to move, and matter tells spacetime how to curve.

8. The name for this mathematical space is *superspace*. Complex numbers are numbers expressed in the form a + bi, where a and b are real numbers, and i is a solution of the equation $x^2 = -1$; that is, i is the square root of -1.

9. You can think of the different paths the universe might take as being like waves in water. If the waves have the same height/depth, then when the crest of one wave meets (is *superposed* on) the trough of the other—a situation known as *destructive interference*—the waves disappear and the surface of the water is calm. Mathematically described, different stable universal histories come out of the universal wavefunction by a process like that, yielding parallel realities. In quantum cosmology, *destructive interference* in the mathematical space describing the hypothesized primordial quantum vacuum—what quantum physicists would call *decoherence*—gives our universe a stable appearance and history from our current vantage point. This story is part of the many-worlds interpretation of quantum theory, about which we will have more to say below.

10. Imaginary time results from multiplying time, *t*, by the imaginary number *i*, which is the square root of -1, and then substituting this product for *t* in the equations. This changes the structure of spacetime in general relativity in an unrealistic way, making time act like a fourth spatial dimension. In his popular book *A Brief History of Time*, Hawking speaks of this procedure as if it were a mere computational expedient, treating it as if it were merely *instrumental*, but then inconsistently interprets its significance realistically. I discuss this point below in the text.

11. Hartle and Hawking, "Wave function of the universe."

12. Stephen Hawking, "Quantum cosmology," *300 Years of Gravitation*, eds. S.W. Hawking and W. Israel (Cambridge, UK: Cambridge University Press, 1987), 631-651.

13. Hawking, "Quantum cosmology," 646-647.

14. Stephen Hawking, *A Brief History of Time: From the Big Bang to Black Holes* (New York: Bantam Books, 1988), 174.

15. The process of decoherence through destructive interference is explained briefly in note 8 above.

16. So-called after Hugh Everett (1930–1982), who first proposed it in 1957 in his doctoral dissertation at Princeton. John Wheeler (1911-2008) was his advisor.

17. David Deutsch, "Quantum Theory of Probability and Decisions," *Proceedings of the Royal Society A* 455 (1999), 3129-3137; David Wallace, "A formal proof of the Born rule from decision-theoretic assumptions," arXiv.org (2009), https://arxiv.org/pdf/0906.2718v1.pdf (accessed July 8, 2020); Simon Saunders, "Derivation of the Born rule from operational assumptions," *Proceedings of the Royal Society A* 460 (2004), 1771-1788.

18. Murray Gell-Mann and James Hartle, "Quantum mechanics in the light of quantum cosmology," *Complexity, Entropy, and the Physics of Information*, ed. W.H. Zurek (Boston, MA: Addison-Wesley, 1990), https://arxiv.org/ftp/arxiv/papers/1803/1803.04605.pdf (accessed July 8, 2020); Simon Saunders, "Decoherence, relative states, and evolutionary adaptation," *Foundations of Physics* 23 (1993), 1553-1585; Simon Saunders, "Time, quantum mechanics, and decoherence," *Synthese* 102 (1995), 235-266; James B. Hartle, "The quasi-classical realms of this quantum universe," *Foundations of Physics* 41 (2011), 982-1006, https://arxiv.org/pdf/0806.3776.pdf (accessed July 8, 2020).

19. Hawking, *Brief History of Time*, 141.

20. Joe Feldbrugge, Jean-Luc Lehners, and Neil Turok, "Lorentzian Quantum Cosmology," *Physical Review D* 95 (2017), 103508, https://arxiv.org/pdf/1703.02076.pdf (accessed July 8, 2020); Joe Feldbrugge, Jean-Luc. Lehners, and Neil Turok, "No smooth beginning for spacetime," *Physical Review Letters* 119 (2017), 171301, https://arxiv.org/pdf/1705.00192.pdf (accessed July 8, 2020); Joe Feldbrugge, Jean-Luc. Lehners, and Neil Turok, "No rescue for the no boundary proposal: Pointers to the future of quantum cosmology," *Physical Review D* 97 (2018), 023509, https://arxiv.org/pdf/1708.05104.pdf (accessed July 8, 2020).

21. J.J. Halliwell, J.B. Hartle, and T. Hertog, "What Is the No Boundary Wavefunction of the Universe?" *Physical Review D* 99 (2018), 043526, https://arxiv.org/pdf/1812.01760.pdf (accessed July 8, 2020).

22. Juan Maldacena, quoted in Natalie Wolchover, "Physicists Debate Hawking's Idea that the Universe Had No Beginning," *Quanta Magazine* (June 6, 2019), https://www.quantamagazine.org/physicists-debate-hawkings -idea-that-the-universe-had-no-beginning-20190606/ (accessed July 8, 2020).

23. A very nice semipopular introduction to the ins and outs of loop quantum gravity and loop quantum cosmology is provided in Jim Baggott, *Quantum Space: Loop Quantum Gravity and the Search for the Structure of Space, Time, and the Universe* (Oxford, UK: Oxford University Press, 2018).

24. Note this involves a superposition of histories in the universal wavefunction, so the MWI (Everett interpretation) plays an essential role in loop quantum gravity and loop quantum cosmology too.

25. Daniel Sudarsky, "Quantum Gravity Induced Granularity of Space-Time and Lorentz Invariance Violation," *Proceedings of the EPS-13 Conference: Beyond Einstein—Physics for the 21st Century*, eds. A.M. Cruise and L. Ouwehand (Bern, Switzerland, 2005), 8.1-8.5, http://adsabs.harvard.edu/full/2006ESASP.637E...8S (accessed July 8, 2020); David Mattingly, "Modern Tests of Lorentz Invariance," *Living Reviews in Relativity* 8 (2005), 5, https://link.springer.com/article/10.12942/lrr-2005-5 (accessed July 8, 2020).

26. Rodolfo Gambini and Jorge Pullin, "Nonstandard optics from quantum space-time," Physical Review D 59 (12) (1999), 124021, https://arxiv.org/pdf/gr-qc/9809038 (accessed July 8, 2020); A.A. Abdo et al., "Testing Einstein's special relativity with Fermi's short hard gamma-ray burst GRB090510," *Nature* 462 (2009), 331-334; Floyd W. Stecker, "A new limit on Planck scale Lorentz violation from gamma-ray burst polarization," *Astroparticle Physics* 35 (2) (2011), 95-97, https://arxiv.org/pdf/1102.2784.pdf (accessed July 8, 2020); S. Liberati, "Tests of Lorentz invariance: a 2013 update," *Classical and Quantum Gravity* 30 (13) (2013), 133001, https://arxiv.org/pdf/1304.5795.pdf (accessed July 8, 2020).

27. Carlo Rovelli, "Loop Quantum Gravity," *Physics World* (November 2003), 1-5, http://igpg.gravity.psu.edu/people/Ashtekar/articles/rovelli03.pdf (accessed July 8, 2020).

28. Carlo Rovelli and Simone Speziale, "Lorentz covariance of loop quantum gravity," *Physical Review D* 83 (2010), article 104029, https://arxiv.org/pdf/1012.1739.pdf (accessed July 8, 2020).

29. Rodolfo Gambini and Jorge Pullin, "Emergence of string-like physics from Lorentz invariance in loop quantum

gravity," *International Journal of Modern Physics D* 23 (12) (2014), 1442023, https://arxiv.org/pdf/1406.2610.pdf (accessed July 8, 2020).

30. The horizon problem deals with the fact that the cosmic microwave background radiation (CMBR)—the radiative echo of the big bang—has the same temperature to one part in 100,000 from every direction in the sky. Standard big bang cosmology allows no opportunity for its temperature to reach this fine-tuned equilibrium. The flatness problem confronts the fact that our observable universe is *vastly* flatter than expected, as this also begs explanation. Inflationary cosmology was originally invented to address these peculiarities in big bang cosmology.

31. Other recent attempts to revive bounce scenarios that are drawing attention, notably those by Anna Ijjas and by Peter Graham, David Kaplan, and Surjeet Rajendran, rely on highly artificial and implausible conditions that reject inflation and thread the needle of the Hawking-Penrose singularity theorems in a highly fine-tuned manner. See Anna Ijjas, "Space-time slicing in Horndeski theories and its implications for non-singular bouncing solutions," *Journal of Cosmology and Astroparticle Physics* 007-007 (2018), https://arxiv.org/pdf/1710.05990.pdf (accessed July 8, 2020); Peter Graham, David Kaplan, and Surjeet Rajendran, "Born again universe," *Physical Review D* 97 (4) (2018), 044003, https://arxiv.org/ftp/arxiv/papers/1709/1709.01999.pdf (accessed July 8, 2020).

32. Albert Einstein and Nathan Rosen, "The Particle Problem in the General Theory of Relativity," *Physical Review* 48 (1935), 73-77.

33. Albert Einstein, Boris Podolsky, and Nathan Rosen, "Can Quantum-Mechanical Description of Physical Reality Be Considered Complete?" *Physical Review* 47 (1935), 777-780.

34. Juan Maldacena and Leonard Susskind, "Cool horizons for black holes," *Fortschritte der Physik* 61 (9) (2013), 781-811, https://arxiv.org/pdf/1306.0533.pdf (accessed July 8, 2020).

35. Leonard Susskind, "Copenhagen vs. Everett, Teleportation, and ER = EPR," *Fortschritte der Physik*, 64 (2016) (6-7), 551-564, https://arxiv.org/pdf/1604.02589.pdf (accessed July 8, 2020). See also ChunJun Cao, Sean M. Carroll, and Spyridon Michalakis, "Space from Hilbert Space: Recovering Geometry from Bulk Entanglement," *Physical Review D* 95 (2017), 024031, https://arxiv.org/pdf/1606.08444.pdf (accessed July 8, 2020); Brian Swingle and Mark van Raamsdonk, "Universality of Gravity from Entanglement" (2014), https://arxiv.org/pdf/1405.2933.pdf (accessed July 8, 2020).

36. Hrant Gharibyan and Robert F. Penna, "Are entangled particles connected by wormholes? Evidence for the ER = EPR conjecture from entropy inequalities," *Physical Review D* 89 (2014), 066001, https://arxiv.org/pdf/1308.0289.pdf (accessed July 8, 2020).

37. Don Page and William Wootters, "Evolution without evolution: Dynamics described by stationary observables," *Physical Review D* 27 (1983), 2885-2892; William Wootters, "'Time' replaced by quantum correlations," *International Journal of Theoretical Physics* 23 (8) (1984), 701-711.

38. V. Giovannetti, S. Lloyd, and L. Maccone, "Quantum Time," *Physical Review D* 92 (2015), 045033.

39. E. Moreva, G. Brida, M. Gramegna, V. Giovannetti, L. Maccone, and M. Genovese, "Time from quantum entanglement: An experimental illustration," *Physical Review A* 89 (2014), 052122; E. Moreva, G. Brida, M. Gramegna, V. Giovannetti, L. Maccone, and M. Genovese, "The time as an emergent property of quantum mechanics, a synthetic description of a first experimental approach," *Journal of Physics: Conference Series* 626 (2015), 012019.

40. Saint Augustine, *The Confessions of St. Augustine*, trans. John K. Ryan (New York: Doubleday, 1960; 397). Particularly insightful passages may be found in XI.15, 254; XI.16, 255; XI.17, 256; XI.18, 258; XI.26, 264.

41. Specifically, Alan Guth invented cosmic inflation in 1980 to explain why the temperature of the cosmic microwave background radiation was the same throughout the observable universe to one part in 100,000, and why the density of mass energy resulting from the big bang yielded a universe that was locally flat to an extraordinary degree (explanatory demands known respectively as the "horizon" and "flatness" problems). See note 30.

42. Roger Penrose, "Singularities and Time-Asymmetry," *General Relativity: An Einstein Centenary*, eds. Stephen Hawking and Werner Israel (Cambridge, UK: Cambridge University Press 1979), 581-638; Roger Penrose, "Time-asymmetry and quantum gravity," *Quantum Gravity 2*, eds. Christopher Isham, Roger Penrose, and Dennis Sciama (Oxford, UK: Clarendon Press, 1989), 245-272; Roger Penrose, *The Road to Reality: A Complete Guide to the Laws of the Universe* (New York: Alfred Knopf, 2004), 726-734.

43. Sean Carroll and Heywood Tam, "Unitary Evolution and Cosmological Fine-Tuning" (2010), https://arxiv.org/pdf/1007.1417v1.pdf (accessed July 8, 2020).

44. Anna Ijjas, Paul J. Steinhardt, and Abraham Loeb, "Inflationary paradigm in trouble after Planck2013," *Physics*

Letters B 723 (2013), 261-266, https://arxiv.org/pdf/1304.2785.pdf (accessed July 8, 2020); Anna Ijjas, Paul J. Steinhardt, and Abraham Loeb, "Inflationary schism after Planck2013," *Physics Letters B* 736 (2014), 142-146, https://arxiv.org/pdf/1402.6980.pdf (accessed July 8, 2020); and Anna Ijjas, Paul J. Steinhardt, and Abraham Loeb, "Pop goes the universe," *Scientific American*, 316 (1) (2017), 32-39.

45. Roger Penrose, *The Road to Reality* (New York: Vintage Books, 2004), 746-757.

46. Ijjas, Steinhardt, and Loeb, "Pop goes the universe," 38.

47. Stephen Hawking and Don N. Page, "How Probable Is Inflation?" *Nuclear Physics B* 298 (1987), 789-809.

48. J.A. Peacock, *Cosmological Physics* (Cambridge, UK: Cambridge University Press, 1999), 342, 503.

49. E.R. Harrison, "Fluctuations at the Threshold of Classical Cosmology," *Physical Review D1* (10) (1970), 2726-2730.

50. Y.B. Zeldovich, "A hypothesis, unifying the structure and the entropy of the Universe," *Monthly Notices of the Royal Astronomical Society* 160 (1972), 7-8.

51. Ijjas, Steinhardt, and Loeb 2013, 2014, 2017 (see note 44).

52. Note that cosmic gravitational waves have *nothing* to do with those detected by LIGO in 2015 produced by the merging of black holes in the contemporary universe.

53. Ijjas, Steinhardt, and Loeb, "Pop goes the universe."

54. Leonard Susskind, "The Anthropic Landscape of String Theory" (2003), https://arxiv.org/pdf/hep-th/0302219 .pdf (accessed July 8, 2020); Leonard Susskind, *The Cosmic Landscape: String Theory and the Illusion of Intelligent Design* (New York: Little, Brown, 2006); Steven Weinberg, "Living in the Multiverse," *The Nature of Nature: Examining the Role of Naturalism in Science*, eds. Bruce L. Gordon and William A. Dembski (Wilmington, DE: ISI Books, 2011), 547-557. See also the interesting discussion of the historical background to all these developments in Helge Kragh, *Higher Speculations: Grand Theories and Failed Revolutions in Physics and Cosmology* (Oxford, UK: Oxford University Press, 2011).

55. Ongoing research into the mathematical relationship among the five anomaly-free classes of string theories led, in 1994, to the discovery of an eleventh unifying dimension, resulting in a new theoretical construct that physicists call *M-theory* (*M* for "membrane" or "mystery," or even "mother of all theories").

56. The cosmological constant in Einstein's field equations for general relativity affects the expansion rate of spacetime. In 1998, it was discovered this expansion rate was accelerating, so the cosmological constant needs to be positive.

57. See Casey Luskin, "Just-So Stories," *Dictionary of Christianity and Science*, eds. Paul Copan et al. (Grand Rapids, MI: Zondervan, 2017), 396.

58. Arvind Borde, Alan Guth, and Alexander Vilenkin, "Inflationary spacetimes are not past-complete," *Physical Review Letters* 90 (2003), 151301, https://arxiv.org/pdf/gr-qc/0110012.pdf (accessed July 8, 2020).

59. Max Tegmark, "Infinity Is a Beautiful Concept—and It's Ruining Physics," *This Idea Must Die: Scientific Theories that Are Blocking Progress*, ed. John Brockman (New York: Harper Perennial, 2015), 48-51. It's good to see Tegmark pulling back from his advocacy of the multiverse. Not too long ago, he was advocating it in a particularly extreme form: the *mathematical universe hypothesis*. See Max Tegmark, "The Multiverse Hierarchy," *Universe or Multiverse?*, ed. Bernard Carr (Cambridge, UK: Cambridge University Press, 2007), 99-126, https://arxiv.org/pdf/0905.1283. pdf (accessed July 8, 2020); Max Tegmark, *Our Mathematical Universe: My Quest for the Ultimate Nature of Reality* (New York: Alfred Knopf, 2014). The mathematical universe hypothesis (MUH) asserts that mathematical existence is the *same thing* as physical existence and every possible consistent mathematical structure instantiates a physical reality. There is no need to winnow mathematical structures as quantum cosmologists winnowed mathematical space in constructing their models because *every* consistent mathematical reality exists. Obviously, this extreme reification of abstract mathematics lacks any justification beyond the suggestion that it explains how a universe's law-structures necessary for life happen to exist, but it also destroys scientific rationality, undermining induction even more radically than inflation. If every consistent mathematical structure is instantiated, there's no reason laws might not change form discontinuously and randomly across time and space, so long as they remain consistent within nonoverlapping spacetime patches. The reality we're experiencing now could arbitrarily and discontinuously change in the next instant. Down this path lies insanity, so it's good to see Tegmark recognizing this and searching for a better way.

60. Lisa Dyson, Matthew Kleban, and Leonard Susskind, "Disturbing Implications of a Cosmological Constant," *Journal of High Energy Physics* 0210 (2002), 011, https://arxiv.org/pdf/hep-th/0208013v3.pdf (accessed July 8, 2020); Raphael Bousso and Ben Freivogel, "A Paradox in the Global Description of the Multiverse," *Journal of*

High-Energy Physics 0706 (2007), 018, https://arxiv.org/pdf/hep-th/0610132.pdf (accessed July 8, 2020); Andrei Linde, "Sinks in the Landscape, Boltzmann Brains, and the Cosmological Constant Problem," *Journal of Cosmology and Astroparticle Physics* 0701 (2007), 022, https://arxiv.org/pdf/hep-th/0611043.pdf (accessed July 8, 2020).

61. Alan Guth, "Eternal Inflation and Its Implications," *The Nature of Nature: Examining the Role of Naturalism in Science*, eds. Bruce L. Gordon and William A. Dembski (Wilmington, DE: ISI Books, 2011), 487-505.

62. See also the critique in Bruce L. Gordon, "Inflationary Cosmology and the String Multiverse," *New Proofs for the Existence of God: Contributions of Contemporary Physics and Philosophy*, ed. Robert J. Spitzer (Grand Rapids, MI: Eerdmans, 2010), 75-103; Bruce L. Gordon, "Balloons on a String," *The Nature of Nature: Examining the Role of Naturalism in Science*, eds. Bruce L. Gordon and William A. Dembski (Wilmington, DE: ISI Books, 2011), 558-601; Penrose, *The Road to Reality*, 746-757; Paul Steinhardt, "The Inflation Debate," *Scientific American* 34 (4) (2011), 36-43.

63. Jim Baggott, *Farewell to Reality: How Modern Physics Has Betrayed the Search for Scientific Truth* (New York: Pegasus Books, 2013); Gordon, "Inflationary Cosmology and the String Multiverse"; Gordon, "Balloons on a String"; Lee Smolin, *The Trouble with Physics: The Rise of String Theory, the Fall of a Science, and What Comes Next* (New York: Houghton Mifflin, 2006); Alexander Unzicker and Sheilla Jones, *Bankrupting Physics: How Today's Top Scientists Are Gambling Away Their Credibility* (New York: Palgrave Macmillan, 2013); Peter Woit, *Not Even Wrong: The Failure of String Theory and the Search for Unity in Physical Law* (New York: Basic Books, 2006).

64. Natalie Wolchover, "Supersymmetry Fails Test, Forcing Physics to Seek New Ideas," *Scientific American* (2012), https://www.scientificamerican.com/article/supersymmetry-fails-test-forcing-physics-seek-new-idea/ (accessed July 8, 2020); Geraint F. Lewis and Luke A. Barnes, *A Fortunate Universe: Life in a Finely Tuned Cosmos* (Cambridge, UK: Cambridge University Press, 2016) 63; Ethan Siegel, "Why Supersymmetry May Be the Greatest Failed Prediction in Particle Physics History," *Forbes* (February 12, 2019), https://www.forbes.com/sites/startswithabang/2019/02/12/why-supersymmetry-may-be-the-greatest-failed-prediction-in-particle-physics-history/ (accessed July 8, 2020).

65. See Bruce L. Gordon, "Divine Action and the World of Science: What Cosmology and Quantum Physics Teach Us About the Role of Providence in Nature," *Journal of Biblical and Theological Studies* 2 (2) (2017), 247-298, especially 258-259.

66. See, most trenchantly, Robin Collins, "The Multiverse Hypothesis: A Theistic Perspective," *Universe or Multiverse?*, ed. Bernard Carr (Cambridge, UK: Cambridge University Press, 2007), 459-480; a more idiosyncratic view is offered by Don N. Page, "Does God So Love the Multiverse?," *Science and Religion in Dialogue, Vol. 1*, ed. Melville Y. Stewart (Oxford, UK: Wiley-Blackwell, 2010), 380-395, https://arxiv.org/pdf/0801.0246.pdf (accessed July 8, 2020).

67. That theism provides the best explanation for reality as we know it is the theme of a new book by my good friend Stephen C. Meyer, for which I served as a consultant. I highly recommend you check it out: Stephen C. Meyer, *Return of the God Hypothesis: Compelling Scientific Evidence for the Existence of God* (New York: HarperOne, 2021).

Chapter 43—What About Human Exceptionalism and Genetic Engineering?

1. Aldous Huxley, *Brave New World* (New York: Harper Perennial, 1998), xi, from the Foreword.

2. Huxley, *Brave New World*, xi.

3. Erik Sherman, "Chinese Researchers Claim to Have Genetically Engineered the First HIV-Immune Babies," *Fortune* (November 26, 2018), https://fortune.com/2018/11/26/chinese-researchers-genetic-engineering-babies-crispr-hiv/ (accessed November 30, 2020).

4. Gregory Stock, *Redesigning Humans: Our Inevitable Genetic Future* (New York: Houghton Mifflin, 2002), 194, 195.

5. Howard L. Kaye, "Anxiety and Genetic Manipulation: A Sociological View," *Perspectives in Biology and Medicine* 41 (Summer 1998), 483-490.

6. Leon Kass, *Life, Liberty and the Defense of Dignity: The Challenge for Bioethics* (San Francisco, CA: Encounter Books, 2002), 33.

7. Leon Kass, "Preventing a Brave New World," *New Republic* (May 21, 2001).

8. Wesley J. Smith, Interview with William Hurlbut (July 10, 2020).

9. Rob Stein, "2 Chinese Babies with Edited Genes May Face Higher Risk of Premature Death," *NPR* (June 3, 2019), https://www.npr.org/sections/health-shots/2019/06/03/727957768/2-chinese-babies-with-edited-genes-may-face-higher-risk-of-premature-death (accessed November 30, 2020).

10. Bob Yirka, "Three new studies show unwanted changes in human embryo genome after CRISPR-Cas9 editing," *Phys.org* (June 29, 2020), https://phys.org/news/2020-06-unwanted-human-embryo-genome-crispr-cas9.html (accessed November 30, 2020).

11. President George W. Bush thought it was morally wrong to destroy a human embryo for the purpose of experimenting upon its cells. To avoid the federal government from providing a financial incentive to do that, he restricted NIH funding to embryonic stem cell lines already in existence as of the date of his executive order. Bush's policy was repealed by President Barack Obama in 2009. But it is worth noting that by that time, embryonic stem cell research was fading in light of the great advances in stem cell science brought about by the induced pluripotent stem cell process that did not destroy embryos. In my view, the Bush stem cell policy was one of the great successes of the Bush presidency.

12. It is worth noting that the induced pluripotent stem cell process has been so successful in this regard that its inventor received the Nobel Prize in Medicine in 2012. For more information on this amazing scientific success, see Dana G. Smith, "Reflecting on the Discovery of the Decade: Induced Pluripotent Stem Cells," *Gladstone Institutes* (February 26, 2016), https://gladstone.org/news/reflecting-discovery-decade-induced-pluripotent-stem-cells (accessed November 30, 2020).

13. "U.S. Health Service Syphilis Study at Tuskegee," *Wikipedia*, https://en.wikipedia.org/wiki/U.S._Public_Health_Service_Syphilis_Study_at_Tuskegee (accessed November 30, 2020).

14. For example, see Stock, *Redesigning Humans*.

15. This is not to say that in vitro fertilization (IVF) followed by preimplantation genetic testing (PGD) is ethical. The process involves fertilizing embryos in a Petri dish and then ensuring that only embryos without genetic disease are implanted. I don't believe PGD is generally ethical because the process usually involves discarding unused embryos as medical waste, turning them over to researchers for experimentation, or putting them in frozen stasis where they may linger for years. There is little need to go even further down the unethical path to *Brave New World* by allowing germline genetic engineering of embryos.

16. I am well aware that human beings are animals biologically. But in this context, I distinguish between *humans* and *animals* to maintain a crucial moral distinction. For this reason, I choose to refrain from using the ubiquitous *nonhuman animals* because I believe the point of that term is to blur that crucial moral distinctions between us and fauna.

17. 1 John 4:8 and 1 John 4:16.

18. Some of the material below is adapted from Wesley J. Smith, *A Rat Is a Pig Is a Dog is a Boy: The Human Cost of the Animal Rights Movement* (New York: Encounter Books, 2010).

19. Hans Jonas, *The Phenomenon of Life: Toward a Philosophical Biology* (Evanston, IL: Northwestern University Press, 1966), 283.

20. Charles S. Nicoll, "A Physiologist's Views on the Animal Rights/Liberation Movement," *The Physiologist* 34 (1991), 303-315.

21. Richard Dawkins, "How Would You Feel About a Half-Human and Half-Chimp Hybrid?," *The Guardian* (January 2, 2009), https://www.theguardian.com/science/blog/2009/jan/02/richard-dawkins-chimpanzee-hybrid (accessed November 30, 2020).

Chapter 44—How Should Christians Think About Origins?

1. For a defense of a young Earth/universe position, see *Coming to Grips with Genesis: Biblical Authority and the Age of the Earth*, eds. Terry Mortenson and Thane H. Ury (Green Forest, AR: Master Books, 2008). For a defense of an old Earth/universe position, see Hugh Ross, *Navigating Genesis: A Scientist's Journey Through Genesis 1–11* (Covina, CA: RTB Press, 2014).

2. For a treatment of theistic evolution, see *Theistic Evolution: A Scientific, Philosophical, and Theological Critique*, eds. J.P. Moreland, Stephen C. Meyer, Christopher Shaw, Ann K. Gauger, and Wayne Grudem (Wheaton, IL: Crossway, 2017). For a treatment of naturalistic evolution, see Michael J. Behe, *Darwin's Black Box: The Biochemical Challenge to Evolution* (New York: Free Press, 1996); Michael Denton, *Evolution: A Theory in Crisis* (Bethesda, MD: Adler & Adler, 1986); Phillip Johnson, *Darwin on Trial* (Washington, DC: Regnery Gateway, 1991); Phillip Johnson, *Reason in the Balance: The Case Against Naturalism in Science, Law, and Education* (Downers Grove, IL: InterVarsity, 1995); *Mere Creation: Science, Faith and Intelligent Design*, ed. William A. Dembski (Downers Grove, IL: InterVarsity, 1998).

3. For a defense of naturalistic evolution, see Richard Dawkins, *The Blind Watchmaker: Why the Evidence of Evolution*

Reveals a Universe Without Design (New York: Norton, 1987). For Dawkins's extension of the argument to be an argument for atheism, see Richard Dawkins, *The God Delusion* (Boston, MA: Houghton Mifflin, 2006).

4. For a defense of how Christianity played an integral role in the rise of science, see James Hannam, *The Genesis of Science: How the Christian Middle Ages Launched the Scientific Revolution* (Washington, DC: Regnery, 2011).

5. The prominent historian of science Ronald L. Numbers points out how the "religion versus science" myth got its impetus: "The greatest myth in the history of science and religion holds that they have been in a state of constant conflict. No one bears more responsibility for promoting this notion than two nineteenth-century American polemicists: Andrew Dickson White (1832–1918) and John William Draper (1811–1882)." Ronald L. Numbers, *Galileo Goes to Jail and Other Myths about Science and Religion*, ed. Ronald L. Numbers (Cambridge, MA: Harvard University Press, 2009), 1-2.

6. For an introductory-level church study of general and special revelation with a workbook and DVD curriculum available, see Richard G. Howe, *Intro to God's Revelation*, https://afastore.net/intro-to-gods-revelation-6-week-curriculum-by-dr-richard-howe (accessed October 31, 2020).

7. As a classical philosophical realist (Thomist), I would argue that some of the truths that sound reason can demonstrate include the reality of natures (e.g., human nature) as opposed to evolutionary gradualism that incrementally gives rise to kinds or essences; teleology as opposed to a mechanist view of the physical world; the reality of causality, especially along the contours of act/potency; the reality of good and moral good; the reality of human rationality and free will; and more.

8. This counters Christians who would deny biblical inerrancy and seek to show its irrelevancy when they suggest that historical narratives such as the Gospels could be taken as nonhistorical allegory and still be consistent with the doctrine of inerrancy. Such a position ignores how the doctrine of inerrancy factors in the literary genre such that historical narrative (while not denying that it could employ figures of speech) is inerrant precisely because the events of the narrative are actual history. For an examination of this from the International Council of Biblical Inerrancy (ICBI), see *Hermeneutics, Inerrancy & the Bible: Papers from ICBI Summit II*, eds. Earl D. Radmacher and Robert D. Preus (Grand Rapids, MI: Academie Books/ICBI, 1984); Thomas A. Howe, "Does Genre Determine Meaning?" *Christian Apologetics Journal* 6 (Spring 2007), 1-19.

9. For example, young-Earth creationists may argue like this: "Many Christians simply will not believe the history recorded in Genesis 1, no matter how clear the text is, because they place more faith in men than in God. They will either reject Genesis outright, or worse, they will 'reinterpret' the Bible to match the secular notion of billions of years." Tim Chaffey and Jason Lisle, *Old-Earth Creationism on Trial: The Verdict Is In* (Green Forest, AR: Master Books, 2010), 110. Contrast this sentiment with the comments in note 8. To take the days of Genesis as long periods of time—in and of themselves, billions of years have nothing to do with being secular—is not the same as taking the days allegorically or as otherwise nonhistorical. While I might agree with Chaffey and Lisle in their ultimate conclusion about the days of Genesis being 24-hour days, it remains that Chaffey and Lisle fail to recognize what, on the one hand, is a legitimate matter of biblical interpretation that falls within the boundaries of biblical authority and inerrancy and what, on the other hand, would be a denial of biblical authority altogether. For a treatment of this aspect of the debate, see John C. Lennox, *Seven Days that Divide the World: The Beginning Accordance to Genesis and Science* (Grand Rapids, MI: Zondervan, 2011).

10. I had the opportunity to discuss this issue with Ken Ham at the 2017 National Conference on Christian Apologetics (NCCA) at Southern Evangelical Seminary. The video is available at "God's Word or Man's Word: From Where Must Apologetics Begin?" YouTube (June 25, 2019), https://www.youtube.com/watch?v=RkE0jfryX0w (accessed July 25, 2020). I also had the honor of participating in a panel discussion at the 2018 NCCA on whether the age of the Earth and the days of Genesis were a matter of biblical authority or biblical interpretation with Huge Ross, Randy Guliuzza, Norman Geisler, and Richard Land. The video is available at "Is the Age of the Earth a Matter of Biblical Authority?" (October 13, 2018), https://www.facebook.com/SESNationalConference/videos/10638267237785251 (accessed July 26, 2020).

11. Henry Morris, the founder and then director of the Institute of Creation Research, a young-Earth science and faith ministry, signed the "Chicago Statement" of the International Council on Biblical Inerrancy in 1978—see https://library.dts.edu/Pages/TL/Special/ICBI_1_typed.pdf (accessed October 31, 2020), 2, middle column, line 16. In 1982, the council convened on the specific matter of inerrancy and hermeneutics (including inerrancy and science) and "left open the question of the age of the earth on which there is no unanimity among evangelicals and which was beyond the purview of this conference." R.C. Sproul and Norman L. Geisler, *Explaining Biblical Inerrancy: Official Commentary on the ICBI Statements* (Matthews, NC: Bastion Books, 2013), 83, http://www

.isca-apologetics.org/sites/default/files/Explaining%20Biblical%20Inerrancy.pdf (accessed October 31, 2020). Morris did not sign the resultant statement.

12. Regarding Article XXII of ICBI's statement on hermeneutics, Geisler commented, "The use of the term 'creation' was meant to exclude the belief in macro-evolution, whether of the atheistic or theistic varieties." Sproul and Geisler, *Explaining Biblical Inerrancy*, 83.

13. Atheist scientist Victor Stenger allows for an eternal universe (or an eternal universe anterior to this universe): "The observations confirming the big bang do not rule out the possibility of a prior universe…In that model [of James Hartle and Stephen Hawking], the universe has no beginning or end in space or time. In the scenario I presented, our universe is described as having 'tunneled' through the chaos at the Planck time from a prior universe that existed for all previous time." Victor J. Stenger, *God—the Failed Hypothesis: How Science Shows that God Does Not Exist* (Amherst, MA: Prometheus, 2007), 125-126.

14. Stephen Hawking argues for such a view. "So long as the universe had a beginning, we could suppose it had a creator. But if the universe is really completely self-contained, having no boundary or edge, it would have neither beginning nor end; it would simply be. What place, then, for a creator?" Stephen Hawking, *A Brief History of Time: From the Big Bang to Black Holes* (New York: Bantam, 1988), 140-141. Revisiting the point much later, Hawking, together with Leonard Mlodinow, repeated the sentiment: "Spontaneous creation is the reason there is something rather than nothing, why the universe exists, why we exist. It is not necessary to involve God to light the blue touch paper and set the universe going." Stephen Hawking and Leonard Mlodinow, *The Grand Design* (New York: Bantam, 2010), 180. For a Christian response to Hawking's position, see William Lane Craig, "'What Place, Then, for a Creator?' Hawking on God and Creation," *British Journal for the Philosophy of Science* 41 (1990), 473-491.

15. "In his writings, [William Lane] Craig takes the first premise [whatever begins to exist has a cause] to be self-evident, with no justification other than common, everyday experience. That's the type of experience that tells us the world is flat. In fact, physical events at the atomic and subatomic level are observed to have no evident cause." Stenger, *God: The Failed Hypothesis*, 123-124. Philosopher Quentin Smith has argued that the universe could have come into existence from nothing completely uncaused. See note 18.

16. For a creationist treatment of the multiverse theory, see Jeffrey A. Zweerink, *Who's Afraid of the Multiverse?* (Covina, CA: RTB Press, 2008).

17. Psalm 90:2: "Before the mountains were brought forth, or ever you had formed the earth and the world, from everlasting to everlasting you are God."

18. Throughout history, theories of the origin of the universe have ranged from an eternal universe (Aristotle, *Physics*, Bk. VIII) to a universe that began without a cause (Quentin Smith and William Lane Craig, *Theism, Atheism and Big Bang Cosmology* [Oxford, UK: Oxford University Press, 1993]), to a universe that began self-caused (Quentin Smith, "The Reason the Universe Exists Is that It Caused Itself to Exist," *Philosophy* 74, no. 290 [October 1999], 579-586), to a universe that is finite in time but which nevertheless never had a temporal beginning (Hawking, *A Brief History of Time*).

19. For a mathematical/philosophical defense that the creation has not existed from eternity, see William Lane Craig, *The Kalam Cosmological Argument* (New York: Macmillan, 1979), republished (Eugene, OR: Wipf & Stock, 2000).

20. Questions about what should be the proper form of government or what should be the proper form of the economy are not questions we can settle by an appeal merely to the doctrine of creation. Not only are there important questions that fall outside the doctrine of creation proper, there are also other important doctrines within Christianity that arise from other portions of revealed truth. The understanding of the holy Trinity, the hope of salvation offered to us as sinners, the second coming of Christ, and other truths are truths we cherish yet could not know merely from creation per se.

21. All passages are from the English Standard Version (Wheaton, IL: Crossway, 2009).

22. Paul repeats this truth in Colossians 1:15-16: "He [Jesus] is the image of the invisible God, the firstborn of all creation. For by him all things were created, in heaven and on earth, visible and invisible, whether thrones or dominions or rulers or authorities—all things were created through him and for him."

23. For a thorough historical survey of the cosmological argument, see William Lane Craig, *The Cosmological Argument from Plato to Leibniz* (Eugene, OR: Wipf & Stock, 2001). For a contemporary treatment of the classical arguments for God's existence, see Edward Feser, *Five Proofs of the Existence of God: Aristotle, Plotinus, Augustine,*

Aquinas, Leibniz (San Francisco, CA: St. Ignatius, 2017). For a treatment of Thomas Aquinas's arguments, see *St. Thomas Aquinas on the Existence of God: The Collected Papers of Joseph Owens*, ed. John R. Catan (Albany, NY: State University of New York Press, 1980); Gaven Kerr, *Aquinas's Way to God: The Proof in* De Ente et Essentia (Oxford, UK: Oxford University Press, 2015). For a delightful informal account of Aquinas's "Five Ways," see Matt Fradd and Robert Delfino, *Does God Exist? A Socratic Dialogue on the Five Ways of Thomas Aquinas* (St. Louis, MO: Enroute, 2018).

24. For a helpful exploration of various worldviews and their contrast with Christian theism, see Norman L. Geisler and William D. Watkins, *Worlds Apart: A Handbook on World Views* (Grand Rapids, MI: Baker, 1989), originally published as *Perspectives: Understanding and Evaluating Today's World Views* (San Bernardino, CA: Here's Life, 1984).

25. The indictment on the human race is that the creation is perspicuous. It is not so much that fallen man misses the evidence, it is that fallen man rebels against the knowledge that is evident from creation. This rebellion cascades to an ultimate atheism, as Romans 1 attests. By suppressing the truth in unrighteousness (verse 18), fallen man knows God but does not glorify him as God (verse 21); changes the glory of the incorruptible God into the corruptible (verse 23); exchanges the truth of God for the lie (verse 25); and finally disapproves of even retaining the knowledge of God at all (verse 28). The presence of false religions is not so much a testament to man's search for God as it is a testament to man's flight from God. "Though man may desire and create for himself a deity who meets his needs and provides him with innumerable benefits, he will not desire a God who is holy, omniscient, and sovereign." R.C. Sproul, *If There Is a God, Why Are There Atheists?* (Minneapolis, MN: Bethany, 1978). Previously published under the title *The Psychology of Atheism*. Revised and republished as *If There's a God, Why Are There Atheists? Why Atheists Believe in Unbelief* (Fearn, Scotland: Christian Focus, 2018).

26. The expression *classical theism* is a reference to the view about God's nature that maintains that God possesses all the superlative attributes that historic Christianity has affirmed of him, including simplicity, immutability, impassibility, omnipotence, omniscience, omnipresence, all good, and more. For important treatments of classical theism in light of the drift away from it within contemporary evangelicalism, see Norman L. Geisler, H. Wayne House, and Max Herrera, *The Battle for God: Responding to the Challenge of Neotheism* (Grand Rapids, MI: Kregel, 2001) and James E. Dolezal, *All That Is in God: Evangelical Theology and the Challenge of Classical Christian Theism* (Grand Rapids, MI: Reformation Heritage, 2017). For citations showing the voices in church history affirming this classical view of God and examples of contemporary departures from that view, see the presentation titled "God Fading Away," http://richardghowe.com/index_htm_files/GodFadingAway16x9.pdf (accessed October 31, 2020).

27. First Timothy 6:15-16: "He who is the blessed and only Sovereign, the King of kings and Lord of lords, who alone has immortality, who dwells in unapproachable light, whom no one has ever seen or can see. To him be honor and eternal dominion. Amen." This point is corroborated by Colossians 1:17: "He is before all things, and in him all things hold together."

28. "The Lord has established His throne in the heavens, and His kingdom rules over all" (Psalm 103:19).

29. In Acts 14:16-17, Paul instructed his audience in Lystra, "In past generations he allowed all the nations to walk in their own ways. Yet he did not leave himself without witness, for he did good by giving you rains from heaven and fruitful seasons, satisfying your hearts with food and gladness."

30. Romans 2:14-15: "When Gentiles, who do not have the law, by nature do what the law requires, they are a law to themselves, even though they do not have the law. They show that the work of the law is written on their hearts, while their conscience also bears witness, and their conflicting thoughts accuse or even excuse them."

31. Very often it is God's goodness, more than other attributes, that is challenged by skeptics and atheists. This is occasioned largely because of the problem of evil. For a treatment of the notion of goodness as such and the goodness of God, see Jan A. Aertsen, "The Convertibility of Being and Good in St. Thomas Aquinas," *New Scholasticism* 59 (1985), 449-470; Jan A. Aertsen, *Medieval Philosophy and the Transcendentals: The Case of Thomas Aquinas* (Leiden, Netherlands: Brill, 1996), 290-334; *Being and Goodness: The Concept of the Good in Metaphysics and Philosophical Theology*, ed. Scott MacDonald (Ithaca, NY: Cornell University Press, 1991). For a treatment of God's goodness in light of the problem of evil, see Brian Davies, *Thomas Aquinas on God and Evil* (Oxford, UK: Oxford University Press, 2011); Norman L. Geisler, *If God, Why Evil? A New Way to Think About the Question* (Minneapolis, MN: Bethany House, 2011); Herbert McCabe, *God and Evil in the Theology of St. Thomas Aquinas* (New York: Continuum, 2010); John F.X. Knasas, *Aquinas and the Cry of Rachel: Thomistic Reflections on the Problem of Evil* (Washington, DC: The Catholic University of America Press, 2013).

32. The New Thought movement of the late nineteenth and early twentieth century spawned several mind science occult groups. One of the pioneers of New Thought was Mary Baker Eddy (1821–1910), who founded the new religious movement called Christian Science (not to be confused with the same expression that refers to the legitimate study of the natural sciences by Christians). She maintained, "The fading forms of matter, the mortal body and material earth, are the fleeting concepts of the human mind…Mortals must look beyond fading, finite forms, if they would gain the true sense of things." Mary Baker Eddy, *Science and Health with Key to the Scriptures* (Boston, MA: The Trustees under the Will of Mary Baker Eddy, 1934), "Creation," 263.31, 264.7. For a treatment of new religious movements and their contrast with historic, biblical Christianity, see Norman L. Geisler and Ron Rhodes, *When Cultists Ask: A Popular Handbook on Cultic Misinterpretations* (Grand Rapids, MI: Baker, 1997); Walter Martin, *The Kingdom of the Cults* (Minneapolis, MN: Bethany House, rev. ed., 1985).

33. Through the mouth of the interlocutor Timaeus, the ancient Greek philosopher Plato (c. 428 BC–348 BC) talks about the Demiurge (Greek, *dēmiourgos*) in his dialog *Timaeus*, 28a-b. Sometimes the Greek word is translated "creator," but here the notion clearly has rather to do with one who fashions or sculpts something out of already existing material. See *Plato: The Collected Dialogues*, eds. Edith Hamilton and Huntington Cairns (Princeton, NJ: Princeton University Press, 1961), 1161. Only later (no doubt because of the influence of Christianity) did the word become used to mean "creator." See Henry George Liddell and Robert Scott, *A Greek-English Lexicon* (Oxford, UK: Clarendon Press, 1976), s.v., "δημιουργ-ειον [dēmiourg-eion]," 386.

34. Atheist George H. Smith embraces such a physicalism. "The notion of an 'immaterial being' entails a contradiction and cannot be expressed in positive terms. We cannot imagine an 'immaterial being' because the concept of 'matter' is essential to our concept of 'being.'…'Immaterial' does not describe another kind of existence; it negates the concept of existence as we understand it." George H. Smith, *Atheism: The Case Against God* (Buffalo, NY: Prometheus Books, 1989), 67.

35. Mormon "apostle" Bruce R. McConkie (1915–1985) explains the Mormon view: "Spirit element has always existed; it is co-eternal with God. Portions of the self-existence spirit element were born as spirit children, or in other words the intelligence which cannot be created or made because it is self-existent, is organized into intelligences." Bruce R. McConkie, *Mormon Doctrine*, 2d ed. (Salt Lake City, UT: Bookcraft, 1966), s.v., "spirit element," 751. For a treatment of the Mormon worldview, see *The New Mormon Challenge: Responding to the Latest Defenses of a Fast-Growing Movement*, gen eds. Francis J. Beckwith, Carl Mosser, and Paul Owens (Grand Rapids, MI: Zondervan, 2002), 90-266.

36. The contemporary American witch Margot Adler (1946–2014) maintains, "The world is holy. Nature is holy. The body is holy. Sexuality is holy. The mind is holy. The imagination is holy. You are holy. A spiritual path that is not stagnant ultimately leads one to the understanding of one's own divine nature Thou art God. Divinity is imminent in all Nature. It is as much within you as without." Margot Adler, *Drawing Down the Moon: Witches, Druids, Goddess-Worshippers, and Other Pagans in America Today* (Boston, MA: Beacon Press, 1986), ix. For a Christian analysis of witchcraft, see Richard G. Howe, "Modern Witchcraft: It May Not Be What You Think," http://richardghowe.com/index_htm_files/ModernWitchcraft.pdf (accessed October 31, 2020).

37. Solitary witch Scott Cunningham observes, "The entire planet is a manifestation of Goddess energy, a tangible example of the powers of Mother Nature. Wiccans may revere Her in this aspect as Gaea, Demeter, Astarte, Kore, and by many other names." Scott Cunningham, *The Truth About Witchcraft Today* (St. Paul, MN: Llewellyn Publications, 1988), 72.

38. Some may find it interesting, if not surprising, that Thomas Aquinas's cosmological argument is indifferent as to whether the universe is eternal or not. While he certainly believed that the universe was not eternal since this truth was revealed in Scripture, his argument nevertheless demonstrates that even if the universe never came into being but had existed from all eternity past, it would still need God as its creator and sustainer. For a brief treatment of Aquinas's arguments for God's existence, see Edward Feser, *Aquinas: A Beginner's Guide* (London, UK: Oneworld Publications, 2009), 62-130.

39. Examples include Fritjof Capra, *The Tao of Physics* (New York: Bantam Books, 1975); Michael Talbot, *Mysticism and the New Physics* (New York: Bantam Books, 1981); Gary Zukav, *The Dancing Wu Li Masters: An Overview of the New Physics* (New York: HarperOne, 2001). For a short treatment of a common misunderstanding of quantum physics that can give rise to this New Age view of physics, see James Trefil, "Physics Demystified," *Science Digest* (April 1983), 94-95. In personal correspondence with Professor Trefil in June of 2020, he informed me that he still holds the views he defends in this article.

40. Genesis 1:4, 10, 12, 18, 21, 25, 31.

41. Colossians 2:18-23: "Let no one disqualify you, insisting on asceticism and worship of angels, going on in detail about visions, puffed up without reason by his sensuous mind, and not holding fast to the Head, from whom the whole body, nourished and knit together through its joints and ligaments, grows with a growth that is from God. If with Christ you died to the elemental spirits of the world, why, as if you were still alive in the world, do you submit to regulations—'Do not handle, Do not taste, Do not touch' (referring to things that all perish as they are used)—according to human precepts and teachings? These have indeed an appearance of wisdom in promoting self-made religion and asceticism and severity to the body, but they are of no value in stopping the indulgence of the flesh."

42. Some cults have argued that in John 10:34, Jesus is acknowledging that humans are gods. "Jesus answered them, 'Is it not written in your Law, "I said, you are gods"?'" Jesus is actually quoting Psalm 82:6, which says, "I said, 'You are gods, sons of the Most High, all of you.'" Verse 7 follows up with, "Nevertheless, like men you shall die, and fall like any prince." In the Old Testament, the expression of being a god is sometimes used of instances where a human is an official representative of God to people. Exodus 7:1 says, "The Lord said to Moses, 'See, I have made you like God to Pharaoh, and your brother Aaron shall be your prophet.'" Another example is how the judges in Israel were regarded, for example, in Psalm 82. Regarding Jesus's point in John 10, Norman Geisler and Thomas Howe comment, "The title of 'gods' is not addressed to everyone, but only to these judges about whom Jesus said are those to 'whom the word of God came' (v. 35). Jesus was showing that if the OT Scriptures could give some divine status to divinely appointed judges, why should they find it incredible that He should call Himself the Son of God? Thus, Jesus was giving a defense for His own deity, not for the deification of man." Norman Geisler and Thomas Howe, *When Critics Ask: A Popular Handbook on Bible Difficulties* (Wheaton, IL: Victor Books, 1992), 417.

43. That we are God (Atman is Brahman) is a key doctrine of Upanishadic Hinduism. "Having given up the false identification of the Self [Atman] with the senses and the mind, and knowing the Self to be Brahman [God], the wise, on departing this life, become immortal." Kena Upanishad, 1.1-2, as cited in International Religious Foundation, *World Scripture: A Comparative Anthology of Sacred Texts* (New York: Paragon House, 1995), 75. Mormonism also teaches that we can become gods: "Here, then, is eternal life—to know the only wise and true God; and you have to go learn how to be gods yourselves…the same as all gods have done before you…from exaltation to exaltation…until you arrive at the station of a god." *Teachings of the Prophet Joseph Smith*, 345-347, as cited in McConkie, *Mormon Doctrine*, s.v., "Godhood," 321.

44. For a critique of the claim that reincarnation is compatible with the Bible, see Richard G. Howe, "Does the Bible Teach Reincarnation? A Response to Joe Fisher's 'The Lost Cord of Christianity,'" http://richardghowe.com/index_htm_files/DoestheBibleTeachReincarnation.pdf (accessed October 31, 2020).

45. Genesis 2:7 tells us, "The Lord God formed the man of dust from the ground and breathed into his nostrils the breath of life, and the man became a living creature." This body/soul relationship is referred to in philosophy as dualism. There have been two influential views of dualism in Christianity. The most common view found today among both professional Christian philosophers and the general Christian population is substance dualism. Substance dualism maintains that humans are composed of two different substances: material (body) and immaterial (soul). It is not uncommon for substance dualists to regard the soul as the essence of the human and the body more or less like a vessel that the soul inhabits. Substance dualism arose largely from the influence of the philosophy of René Descartes (1596–1650). The other view is that of Thomas Aquinas (1224/5–1274). It is known by the rather obscure title of *hylomorphic dualism*. The title comes from the combination of two Greek words *hulē* ("matter") and *morphē* ("form") referencing the metaphysical doctrines of Matter and Form in the philosophy of Aristotle. Aquinas qualified Aristotle's view to accommodate his Christianity, particularly with regard to the doctrine that the soul survives the death of the body—a position for which Aristotle was notorious among the medieval Christians for rejecting. For a treatment of Aquinas's view, see George P. Klubertanz, *The Philosophy of Human Nature* (New York: Appleton-Century-Crofts, 1953), republished as *The Philosophy of Human Nature* (Neuknirchen-Seelscheid, Germany: Editiones Scholasticae, 2014); Christina Van Dyke, "Not Properly a Person: The Rational Soul and 'Thomistic Substance Dualism,'" *Faith and Philosophy* 26 (April 2009), 186-204.

46. Reductivism is the view that human nature and human behavior can be understood and (when necessary) managed entirely in terms of the basic physical elements that constitute the human body. Atheist Victor Stenger puts it this way: "In any case, whether reductive or not, the emergent properties of the purely physical brain and body do not survive their deaths. The nonreductive physicalist soul is not an immortal immaterial soul—not even a moral immaterial soul. Once again it appears that a God with a traditional attribute of the monotheistic God,

one who endows humans with immortal immaterial souls, does not exist." Victor J. Stenger, *God: The Failed Hypothesis*, 106.

47. For a treatment of transgenderism, see Ryan T. Anderson, *When Harry Became Sally: Responding to the Transgender Moment* (New York: Encounter Books, 2019).

48. The two sides are egalitarianism and complementarianism. For a defense of egalitarianism, see *Discovering Biblical Equality: Complementarity Without Hierarchy*, eds. Ronald W. Pierce and Rebecca Merrill Groothuis (Downers Grove, IL: IVP Academic, 2005). For a defense of complementarianism, see *Recovering Biblical Manhood and Womanhood: A Response to Evangelical Feminism*, eds. John Piper and Wayne Grudem (Wheaton, IL: Crossway Books, 2012).

49. For a defense of the sacredness of marriage and the culture war on marriage, see Ryan T. Anderson, *Truth Over-ruled: The Future of Marriage and Religious Freedom* (Washington, DC: Regnery, 2015).

50. For a brief natural law (see note 53) treatment of same-sex marriage, see TFP Committee on American Issues, *Defending a Higher Law: Why We Must Resist Same-Sex "Marriage" and the Homosexual Movement* (Spring Grove, PA: The American Society for the Defense of Tradition, Family and Property, 2004), 137-146.

51. James 3:9: "With it we bless our Lord and Father, and with it we curse people who are made in the likeness of God."

52. Colossians 1:17: "He [Christ] is before all things, and in him all things hold together."

53. The natural (or common) moral law has been known as natural law theory. It is the idea that many of God's moral requirements for us are knowable through human reason apart from Scripture. Romans 2:14-15 tells us, "When Gentiles, who do not have the law, by nature do what the law requires, they are a law to themselves, even though they do not have the law. They show that the work of the law is written on their hearts, while their conscience also bears witness, and their conflicting thoughts accuse or even excuse them." For a treatment of natural law theory, see Thomas Aquinas, *Treatise on Law* (Cambridge, MA: Hackett Publishing, 2000); J. Budziszewski, *Written on the Heart: The Case for Natural Law* (Downers Grove, IL: InterVarsity, 1997); J. Daryl Charles, *Retrieving the Natural Law: A Return to Moral First Things* (Grand Rapids, MI: Eerdmans, 2008); *Natural Law and Evangelical Political Thought*, eds. Jesse Covington, Bryan McGraw, and Micah Watson (Lanham, MD: Lexington Books, 2013); Stephen J. Grabill, *Rediscovering the Natural Law in Reformed Theological Ethics* (Grand Rapids, MI: Eerdmans, 2006); David VanDrunen, *Divine Covenants and the Moral Order: A Biblical Theology of Natural Law* (Grand Rapids, MI: Eerdmans, 2014).

54. Romans 4:4-5: "Now to the one who works, his wages are not counted as a gift but as his due. And to the one who does not work but believes in him who justifies the ungodly, his faith is counted as righteousness."

55. *The Westminster Standards* (Philadelphia, PA: Great Commission Publications, 2003), 71.

56. Margot Adler describes it this way: "Neo-pagans look at *religion* differently: they often point out that the root of the word means 'to relink' and 'to connect,' and therefore refers to any philosophy that makes deep connections between human beings and the universe." She goes on, "For many Pagans, *pantheism* implies much the same thing as animism. It is a view that divinity is inseparable from nature and that deity is immanent in nature." Adler, *Drawing Down the Moon*, 12, 25, emphasis in original.

57. First Corinthians 10:31: "Whether you eat or drink, or whatever you do, do all to the glory of God."

58. See, for example, such organization as Population Connection, https://www.populationconnection.org/ (accessed October 31, 2020).

59. Romans 12:3: "By the grace given to me I say to everyone among you not to think of himself more highly than he ought to think, but to think with sober judgment, each according to the measure of faith that God has assigned."

Chapter 45—How Should We Think About the Age of the Earth?

1. The term *creationism* here should be taken as referring only to the claim "God created the universe," not as taking a position on when, by what method or process, or how long it took.

2. Jonathan Sarfati, *Refuting Evolution* (Green Forest, AR: Master Books, 1999, reprinted 2000), 103.

3. Sarfati, *Refuting Evolution*, 115.

4. Jonathan Sarfati, *Refuting Compromise* (Green Forest, AR: Master Books, 2004), 107-139.

5. Jason Lisle, "Presuppositional Response," *Christian Apologetics Journal* 11 (Fall 2018), 154.

6. Lisle, "Presuppositional Response," 151-153.

7. Luke A. Barnes and Geraint F. Lewis, *The Cosmic Revolutionary's Handbook (Or: How to Beat the Big Bang)* (Cambridge, UK: Cambridge University Press, 2020), 2.

8. G. Brent Dalrymple, *The Age of the Earth* (Redwood City, CA: Stanford University Press, 1991), 401.

9. Hugh Ross, *A Matter of Days: Resolving a Creation Controversy*, 2d ed. (Covina, CA: RTB Press, 2015); Hugh Ross, *The Creator and the Cosmos: How the Latest Scientific Discoveries Reveal God*, 4th ed. (Covina, CA: RTB Press, 2018), 65, 86.

10. Norman L. Geisler and Ronald M. Brooks, *When Skeptics Ask: A Handbook on Christian Evidences* (Grand Rapids, MI: Baker, 1990, reprinted 2008), 230.

11. K. Scott Oliphint, "Covenantal Reply," *Christian Apologetics Journal* 11 (Fall 2018), 161.

12. Norman Geisler, "Review," *Christian Apologetics Journal* 11 (Fall 2018), 167.

13. B.B. Warfield, "On the Antiquity and the Unity of the Human Race," *Studies in Theology: The Works of Benjamin B. Warfield*, ed. John E. Meeter (Grand Rapids, MI: Baker, 1932, reprinted 2000) 9, 235-236.

14. Warfield, "On the Antiquity and the Unity of the Human Race," 245.

15. Thomas Aquinas, *Summa Theologiae I*, q. 1, a. 10.

16. For Aquinas, the 14 articles of the Apostles' Creed are infallible and must be believed.

17. This has already taken place on numerous occasions in the history of biblical interpretation, both in relation to archaeological/historical claims and scientific claims.

18. David Haines, "Biblical Interpretation and Natural Knowledge: A Key to Solving the Protestant Problem," *Reforming the Catholic Tradition: The Whole Word for the Whole Church*, ed. Joseph Minich (Leesburg, VA: The Davenant Press, 2019), 99-134.

19. For some interaction with this school of thought, see Christopher Norris, *Against Relativism: Philosophy of Science, Deconstruction and Critical Theory* (Oxford, UK: Blackwell Publishers, 1997; reprinted 1998).

20. There are more than a dozen views offered by old-Earth creationists to account for the time span (deep time) between the origin of the universe and the creation of Adam and Eve. Among these are four main views that include (1) the "days" of Genesis 1 are long periods of time; (2) a long gap of time between Genesis 1:1 and 1:2 (called "gap" theories); (3) a long period of time prior to the days of creation in the Genesis 1 narrative; and (4) allowing for long periods of time *in between* the literal 24-hour days of creation (known as alternate day-age theories).

21. Warfield, "On the Antiquity and the Unity of the Human Race," 238-244. See also Norman L. Geisler, "Genealogies, Open or Closed," *Baker Encyclopedia of Christian Apologetics* (Grand Rapids, MI: Baker Academic, 1999, reprinted 2006), 267-270; Gleason Archer, *A Survey of Old Testament Introduction*, 2d ed. (Chicago, IL: Moody, 1985), 202-205; John D. Currid, *Genesis 1:1–25:18, A Study Commentary on Genesis* (Grand Rapids, MI: Evangelical Press, 2003; reprinted 2015) 1, 168; Kenton L. Sparks, "Genesis 1–11 as Ancient Historiography," *Genesis: History, Fiction, or Neither? Three Views on The Bible's Earliest Chapters*, ed. Charles Halton (Grand Rapids, MI: Zondervan, 2015), 118-122.

22. Warfield, "On the Antiquity and the Unity of the Human Race," 238-239.

23. Geisler, "Genealogies, Open or Closed," 268.

24. Currid, *Genesis 1:1–25:18*, 168.

25. John H. Sailhamer, *The Pentateuch as Narrative: A Biblical-Theological Commentary* (Grand Rapids, MI: Zondervan, 1992), 118-119, 130-133. John D. Currid agrees; see Currid, *Genesis 1:1–25:18*, 167-168, 233-235.

26. Geisler, "Genealogies, Open or Closed," 269-270.

27. Geisler, "Genesis, Days of," 273. See also Lennox, *Seven Days That Divide the World*, 53.

28. Many of these historical viewpoints are reviewed in Warfield, "On the Antiquity and the Unity of the Human Race."

29. Carl Sagan, *Cosmos* (New York: Random House, 1980), 337.

30. Carl Sagan, *Broca's Brain: Reflections on the Romance of Science* (New York: Ballantine Books, 1974, reprinted 1979), 175-177. See also Sagan, *Cosmos*, 339. Richard Dawkins agrees with Sagan, referring to roughly 4.5 billion years as the age of the Earth. Richard Dawkins, *The Blind Watchmaker: Why the Evidence of Evolution Reveals a Universe Without Design* (New York: Penguin Books, 1986; reprinted 2006), 145.

31. Sagan, *Cosmos*, 345.

32. Stephen Hawking, *Une Brève Histoire du Temps: Du big bang aux trous noirs*, trans. Isabelle Naddeo-Souriau (Paris, France: Flammarion, 1989), 142.

33. George Gaylord Simpson, *The Meaning of Evolution* (New York: Mentor Books, 1951; reprinted 1958), 14, 24-25, 42.

34. Peter Godfrey-Smith, "Information and the Argument from Design," *Intelligent Design Creationism and Its Critics*, ed. Robert T. Pennock (Cambridge, MA: MIT University Press, 2001), 585.

35. G. Brent Dalrymple, *The Age of the Earth* (Redwood City, CA: Stanford University Press, 1991), 401.

36. Barnes and Lewis, *The Cosmic Revolutionary's Handbook (Or: How to Beat the Big Bang)*, 2.

37. Michael J. Behe and David W. Snoke, "Simulating Evolution by Gene Duplication of Protein Features That Require Multiple Amino Acid Residues," *Protein Science* 13 (2004), 2651-2664.

38. Stephen C. Meyer, *Darwin's Doubt: The Explosive Origin of Animal Life and the Case for Intelligent Design* (New York: HarperOne, 2013), 248. Meyer notes that other biologists have corroborated Behe and Snoke's findings. See Ann Gauger and Douglas Axe, "The Evolutionary Accessibility of New Enzyme Functions: A Case Study from the Biotin Pathway," *BIO-Complexity* 2011 (1); Ann Gauger et al., "Reductive Evolution Can Prevent Populations from Taking Simple Adaptive Paths to High Fitness," *BIO-Complexity* 2010 (2); Michael J. Behe, *The Edge of Evolution: The Search for the Limits of Darwinism* (New York: Free Press, 2007).

39. Rick Durrett and Deena Schmidt, "Waiting for Two Mutations: With Applications to Regulatory Sequence Evolution and the Limits of Darwinian Evolution," *Genetics* 180 (2008), 1501-1509.

40. Meyer, *Darwin's Doubt*, 249.

41. See Meyer, *Darwin's Doubt*, 71-72. See also Casey Luskin, "How 'Sudden' Was the Cambrian Explosion?," *Debating Darwin's Doubt: A Scientific Controversy That Can No Longer be Denied*, ed. David Klinghoffer (Seattle, WA: Discovery Institute Press, 2015), 77. Here, Luskin notes that Meyer accepts the traditional "10-million-year duration" of the Cambrian explosion.

42. Robert Jastrow, *God and the Astronomers* (New York: Norton, 1978), 111-116.

43. Jastrow, *God and the Astronomers*, 24-29.

44. Jastrow, *God and the Astronomers*, 47, emphasis in original. See also pages 48-49, 85-86.

45. Jastrow, *God and the Astronomers*, 105-110.

46. Neil F. Comins, *Discovering the Essential Universe*, 4th ed. (New York: W.H. Freeman, 2009), 406.

47. Jastrow, *God and the Astronomers*, 107.

48. For example, the scientist who first proposed the big bang theory was a devout Belgian Roman Catholic priest, Georges Lemaître.

49. Geisler, "Genesis, Days of," 273.

50. Geisler, "Genesis, Days of," 273.

51. Greg Koukl, "Star Light & the Age of the Universe," *Stand to Reason* (July 1, 1998), https://www.str.org/w/star-light-the-age-of-the-universe (accessed December 15, 2020).

52. Numbers 23:19; John 14:6-7; John 17:17.

53. Koukl, "Star Light & the Age of the Universe."

54. Many thanks to Benoît Côté and Veronique Cloutier, who read this article and provided many helpful suggestions and critiques.

Chapter 46—How Have Christians Helped to Advance Science?

1. Susan V. Gallagher and Roger Lundin, *Literature Through the Eyes of Faith* (Washington, DC: Christian College Coalition, 1989), xiv.

2. C.S. Lewis, *Surprised by Joy: The Shape of My Early Life* (Orlando, FL: Harcourt Brace, 1955), 174.

3. Erwin Schrödinger, *Nature and the Greeks and Science and Humanism* (Cambridge, UK: Cambridge University Press, 1954, reprinted 1996), 95.

4. *Chemistry in Britain* 25, no. 7 (July 1989), 663.

5. I.M. Khalatnikov, "Reminiscences of Landau," *Physics Today* 42 (May 1989), 34-41.

6. Kameshwar C. Wali, *Chandra: A Biography of S. Chandrasekhar* (Chicago, IL: University of Chicago Press, 1991), 304-306.

7. C.P. Snow, *The Two Cultures* (Cambridge, UK: Cambridge University Press, 1959, reprinted 2007), 9-10.

8. Richard P. Feynman, "The Relation of Science and Religion," *Engineering and Science* 19 (June 1956), 20-23, emphasis in original.

9. Alan Lightman, *Origins: The Lives and Worlds of Modern Cosmologists* (Cambridge, MA: Harvard University Press, 1992).

10. Richard Seltzer, "Poll draws portrait of U.S. scientists' views," *Chemical & Engineering News* 66 (November 7, 1988), 6.

11. Michael Polanyi, *Science, Faith, and Society* (London, UK: Oxford University Press, 1946), 7.

12. Eugene Wigner, "The Unreasonable Effectiveness of Mathematics in the Natural Sciences," *Communications on Pure and Applied Mathematics* 13 (1960), 1-14.

13. Keith Ward, "Why God Must Exist," *Science & Christian Belief* 11 (1999), 5-13.

14. Sheldon Glashow quoted in W. Mark Stuckey, "Science, Religion, Templeton Prize," *Physics Today* 54 (August 2001), 72, 74.

15. Robert Edward David Clark, *Christian Belief and Science: A Reconciliation and a Partnership* (London, UK: The English Universities Press, 1960).

16. Francis Bacon, *Of the Proficience and Advancement of Learning Divine and Human* (London, UK: Henrie Tomes, 1605), in *The Works of Francis Bacon, Lord Chancellor of England*, eds. James Spedding, Robert Leslie Ellis, and Douglas Denon Heath (London, UK: Longmans & Co., 1876), 268.

17. Quoted in Schaefer, *Science and Christianity: Conflict or Coherence?*, 16.

18. Kepler quoted in John Hudson Tiner, *Johannes Kepler: Giant of Faith and Science* (Milford, MI: Mott Media, 1977), 193.

19. Blaise Pascal, Pensées (Project Gutenberg, reprinted 2006), 154, https://www.gutenberg.org/files/18269/18269 -h/18269-h.htm (accessed December 1, 2020).

20. Isaac Newton, *General Scholium to the Principia,* 3rd ed. (1726), http://isaac-newton.org/general-scholium/ (accessed December 1, 2020).

21. Newton, as quoted in Henry Morris, *Men of Science, Men of God* (Green Forest, AR: Master Books, 1982), 32.

22. Newton, as quoted in Edward B. Davis, "Newton's Rejection of the 'Newtonian World View': The Role of Divine Will in Newton's Natural Philosophy," *Science & Christian Belief* 3 (October 1991), 103-117.

23. George M. Trevelyan, *English Social History: A Survey of Six Centuries from Chaucer to Queen Victoria,* 2d ed. (London, UK: Longmans, Green, 1946), 257.

24. Quoted in Schaefer, *Science and Christianity: Conflict or Coherence?*, 18.

25. Trevor Williams, entry for James Clerk Maxwell in *The Biographical Dictionary of Scientists* (London, UK: A & C Black, 1982), 358.

26. Maxwell, as quoted in Matthew Stanley, "By design: James Clerk Maxwell and the evangelical unification of science," *The British Journal for the History of Science* 45 (March 2012), 57-73.

27. Maxwell quoted in Theodore M. Porter, "A statistical survey of gases: Maxwell's social physics," *Historical Studies in the Physical Sciences* 12 (1981), 77-116.

28. James G. Crowther, *British Scientists of the Nineteenth Century* (New York: Routledge, 1935, reprinted 2009), 262.

29. Simon Garfield, *Mauve: How One Man Invented a Color That Changed the World* (New York: Norton, 2002), 136-137.

30. Garfield, *Mauve: How One Man Invented a Colour That Changed the World*, 137.

31. George Stokes, *Natural Theology: The Gifford Lectures* (London, UK: Adam and Charles Black, 1891), 23-24.

32. Kelvin, as quoted in "Lord Kelvin on Science and Religion," *The Indian Review* 4 (July 1903), 447.

33. J.J. Thomson, "Inaugural Address by Prof. Sir J.J. Thomson, M.A., LLD., D.Sc., F.R.S., President of the Association," *Nature* 81 (August 26, 1909), 248-257, 257.

34. Charles Coulson, as quoted in S.L. Altmann and E.J. Bowen, "Charles Alfred Coulson 1910–1974," *Biographical Memoirs of Fellows of the Royal Society* 20 (December 1974), 75-134, 76-77.

35. Robert Griffiths quoted in Tim Stafford, "Cease-fire in the Laboratory," *Christianity Today* (April 3, 1987), https://www.christianitytoday.com/ct/1987/april-3/cease-fire-in-laboratory.html (accessed December 1, 2020).

36. Richard Bube, as quoted in Stafford, "Cease-fire in the Laboratory."

37. John Suppe, as quoted in Stafford, "Cease-fire in the Laboratory."

38. Charles H. Townes, *Making Waves* (Woodbury, NY: American Institute of Physics Press, 1995), 203.

39. Arthur L. Schawlow, as quoted in Henry Margenau and Roy A. Varghese, *Cosmos, Bios, Theos: Scientists Reflect on Science, God, and the Origins of the Universe, Life, and Homo Sapiens* (Peru, IL: Open Court, 1992), 107.

40. Quoted in Schaefer, *Science and Christianity: Conflict or Coherence?*, 29.

41. See John Noble Wilford, "Sizing Up the Cosmos: An Astronomer's Quest," *The New York Times* (March 12, 1991), https://www.nytimes.com/1991/03/12/science/sizing-up-the-cosmos-an-astronomer-s-quest.html (accessed December 1, 2020).

42. The quote is also found in Allan Sandage, "A Scientist Reflects on Religious Belief," *Truth: An Interdisciplinary Journal of Christian Thought* 1 (1985), http://www.leaderu.com/truth/1truth15.html (accessed December 1, 2020).

43. William Phillips, as quoted in *Perspectives on Science and Christian Faith* 50 (September 1998), 175.

44. Paul Davies, as quoted in Bill Durbin, "How It All Began," *Christianity Today* (August 12, 1988), https://www.christianitytoday.com/ct/1988/august-12/christianity-today-institute-how-it-all-began.html (accessed December 1, 2020).

45. Stephen Hawking as quoted in Durbin, "How It All Began."

46. Fred Hoyle, "The Universe: Past and Present Reflections," *Engineering & Science* (November 1981), 8-12, 12.

47. Colin A. Russell, "The Conflict Metaphor and Its Social Origins," *Science and Christian Belief* 1 (April 1989), 3-26.

48. Andrew Dickson White, *A History of the Warfare of Science with Theology in Christendom, Vol. I* (originally published New York: Appleton, 1896; reprinted Amherst, NY: Prometheus Books, 1993), 127.

49. Alister E. McGrath, *A Life of John Calvin: A Study in the Shaping of Western Culture* (Oxford, UK: Blackwell, 1990), xiv.

50. See George M. Marsden, *The Outrageous Idea of Christian Scholarship* (Oxford, UK: Oxford University Press, 1997), 63-64.

51. Francis Collins, "The Human Genome Project: Tool of Atheistic Reductionism or Embodiment of the Christian Mandate to Heat?," *Science & Christian Belief* 11 (October 1999), 99-111, 99.

Chapter 47—How Can We Make Sense of Natural Evil?

1. "Evil," *The Catholic Encyclopedia* (New York: Robert Appleton Company, 1909), http://www.newadvent.org/cathen/05649a.htm (accessed October 30, 2020).

2. Charles Hodge, *Systematic Theology* (reprinted Grand Rapids, MI: Eerdmans, 1981) I, 171, emphasis added.

Chapter 48—Should Christians Embrace Human Enhancement Science and Technology That Extends Mental and Physical Limitations?

1. Methuselah Foundation, "New Parts for People," https://www.mfoundation.org/who-we-are#return-on-mission (accessed February 13, 2019).

2. Sarah Marsh, "Neurotechnology, Elon Musk and the goal of human enhancement," *The Guardian* (January 1, 2018), https://www.theguardian.com/technology/2018/jan/01/elon-musk-neurotechnology-human-enhancement-brain-computer-interfaces (accessed February 9, 2019). Futurist Ray Kurzweil calls the merging of artificial intelligence and human beings the singularity. See, Dom Galeon, "Ray Kurzweil: AI Will Not Displace Humans, It's Going to Enhance Us," https://futurism.com/former-general-motors-executive-approaching-automotive-era (accessed February 9, 2019).

3. Alex Hern, "Startup wants to upload your brain to the Cloud but has to kill you to do it," *The Guardian* (March 14, 2018), https://www.theguardian.com/technology/2018/mar/14/nectome-startup-upload-brain-the-cloud-kill-you (accessed February 1, 2019).

4. Charles T. Rubin, *The Eclipse of Man* (New York: New Atlantis Books, 2014), 84.

5. See Aldous Leonard Huxley, *Brave New World*, chapter 15, http://idph.com.br/conteudos/ebooks/BraveNewWorld.pdf (accessed February 6, 2019).

6. Huxley, *Brave New World*, chapter 16.

7. Huxley, *Brave New World*, chapter 16, 163.

8. C.S. Lewis, *The Abolition of Man* (Quebec, Canada: Samizdat University Press, 2014), 30.

9. Transhumanists are especially hopeful that a recent gene-editing scientific and technological breakthrough called CRISPR-Cas9 will help them advance their goals. This breakthrough may also help scientists to develop therapies that help humans regain lost bodily and mental functions due to genetic anomalies. For more information on

CRISPR-Cas9, see the NIH's US Library of Reference article, "What are genome editing and CRSPR-Cas9?" https://ghr.nlm.nih.gov/primer/genomicresearch/genomeediting (accessed May 11, 2019).

10. Humanity+, "What is the Mission of Humanity+?," https://humanityplus.org/about/mission/ (accessed March 06, 2019).

11. The President's Council on Bioethics, *Beyond Therapy*, 2003, 13, https://repository.library.georgetown.edu/bitstream/handle/10822/559341/beyond_therapy_final_webcorrected.pdf (accessed February 8, 2019). The President's Council claimed that this distinction is problematic due to the ambiguity of the words *enhancement* and *therapy* (14-16). Nonetheless, it is a "fitting beginning and useful shorthand for calling attention to the problem," and so for this chapter, I will keep the distinction.

12. The President's Council on Bioethics, *Beyond Therapy*.

13. Katie Moisse, "Paralyzed Woman Moves Robotic Arm with Her Mind," *ABC News* (May 16, 2012), https://abcnews.go.com/Health/w_MindBodyNews/paralyzed-woman-moves-robotic-arm-mind/story?id=16353993 (accessed February 8, 2019). This link comes from Michael Guillen, *The End of Life as We Know It* (Washington, DC, Salem Books, 2018), note 13.

14. Jessica Firger, "Blind man sees for first time with 'bionic eye,'" *CBS News* (October 15, 2014), https://www.cbsnews.com/news/blind-man-sees-for-first-time-with-bionic-eye/ (accessed February 8, 2019).

15. Ben Guarino, "Scientists convert spinach leaves into human heart tissue—that beats," *The Washington Post* (March 27, 2017), https://www.washingtonpost.com/news/morning-mix/wp/2017/03/27/scientists-convert-spinach-leaves-into-human-heart-tissue-that-beats/ (accessed February 8, 2019). This link was taken from Guillen, *The End of Life as We Know It*, note 105.

16. Methuselah Foundation, "New Parts for People."

17. Marsh, "Neurotechnology."

18. Marsh, "Neurotechnology."

19. Marsh, "Neurotechnology."

20. Hearn, "Startup wants to upload your brain."

21. Hearn, "Startup wants to upload your brain."

22. See Ray Kurzweil, "Predictions by Ray Kurzweil," http://www.kurzweilai.net/futurism-ray-kurzweil-claims-singularity-will-happen-by-2045 (accessed February 13, 2019).

23. Rubin, *The Eclipse of Man*.

24. See Genesis 3:5 and Isaiah 14:12-14.

25. Rubin, *The Eclipse of Man* , 6-7.

26. Rubin, *The Eclipse of Man*, 2. Although Rubin is not a transhumanist, here, he nicely summarizes one of their goals.

27. President's Council on Bioethics, *Beyond Enhancement*, 279-280.

28. President's Council on Bioethics, *Beyond Enhancement*, 280.

29. President's Council on Bioethics, *Beyond Enhancement*, 280-281.

30. President's Council on Bioethics, *Beyond Enhancement*, 281-283.

31. President's Council on Bioethics, *Beyond Enhancement*, 283-285.

32. Psalm 90:12 (NIV).

33. See, John Hick, *Evil and the God of Love* , 2d. ed. (New York: Palgrave Macmillan, 2007 (1966)).

34. P. Alex Linley and Stephen Joseph, "The Human Capacity for Growth Through Adversity," *American Psychologist* 60(3) (2005), 262-264, https://psycnet.apa.org/record/2005-03019-012 (accessed March 3, 2019).

35. *Bullfinch's Mythology* (New York: Avenel, 1978), 18.

36. Donna Rosenberg and Sorelle Baker, *Mythology and You* (Lincolnwood, IL: National Textbook Company, 1996), 84.

37. Rosenberg and Baker, *Mythology and You*, 61-62.

38. Leon R. Kass, *Life, Liberty and the Defense of Dignity* (San Francisco, CA: Encounter Books, 2002), 269.

39. "The Weight of Glory and Other Addresses," *The Essential C.S. Lewis*, ed. Lyle W. Dorsett (New York: Touchstone, 1988), 363.

40. *Westminster Shorter Catechism*, Question 1, https://reformed.org/documents/wsc/index.html?_top=https://reformed.org/documents/WSC.html (accessed March 13, 2019).

41. According to Thomas Aquinas, it is God who bestows or infuses in us the theological virtues. See the *Summa Theologiae* I-II 63.

42. On April 11, 2019, the Ethics and Religious Liberty Commission of the Southern Baptist Convention posted, on its website, a statement titled "Artificial Intelligence: An Evangelical Statement of Principles," at https://erlc.com/resource-library/statements/artificial-intelligence-an-evangelical-statement-of-principles (accessed April 12, 2019). Many of the most prominent leaders in relevant fields signed off on the Statement. The statement contains 12 biblically referenced statements, some of which are about enhancement science and technology. This statement is a good place to begin to do further investigation on human enhancement from a biblical/theological perspective.